Applications (continued from front endsheets)

An Applied Approach

BRIEF

CALCULUS

with
CalcChat® & CalcView™

10e

An Applied Approach

BRIEF

CALCULUS

with
CalcChat® & CalcView™

10e

Ron Larson

The Pennsylvania State University
The Behrend College

With the assistance of David C. Falvo

The Pennsylvania State University
The Behrend College

CENGAGE
Learning®

Australia • Brazil • Mexico • Singapore • United Kingdom • United States

CENGAGE
Learning

Brief Calculus: An Applied Approach with CalcChat & CalcView
Tenth Edition
Ron Larson

Product Director: Terry Boyle

Product Manager: Rita Lombard

Content Developer: Erin Brown, Spencer Arritt

Product Assistant: Kathryn Schrumpf

Marketing Manager: Julie Schuster

Content Project Manager: Jennifer Risden

Manufacturing Planner: Doug Bertke

Production Service: Larson Texts, Inc.

Photo Researcher: Lumina Datamatics

Text Researcher: Lumina Datamatics

Text Designer: Larson Texts, Inc.

Cover Designer: Larson Texts, Inc.

Cover Image: Rawpixel/Shutterstock.com
                molaruso/Shutterstock.com

Compositor: Larson Texts, Inc.

For product information and technology assistance, contact us at
Cengage Learning Customer & Sales Support, 1-800-354-9706.

For permission to use material from this text or product, submit all requests online at **www.cengage.com/permissions.**
Further permissions questions can be e-mailed to
permissionrequest@cengage.com.

Library of Congress Control Number: 2015944034

Student Edition:
ISBN-13: 978-1-305-86092-6

Loose Leaf Edition:
ISBN-13: 978-1-305-95326-0

Cengage Learning
20 Channel Center Street
Boston, MA 02210
USA

Cengage Learning is a leading provider of customized learning solutions with employees residing in nearly 40 different countries and sales in more than 125 countries around the world. Find your local representative at **www.cengage.com.**

Cengage Learning products are represented in Canada by Nelson Education, Ltd.

To learn more about Cengage Learning Solutions, visit **www.cengage.com.**
Purchase any of our products at your local college store or at our preferred online store **www.cengagebrain.com.**

QR Code is a registered trademark of Denso Wave Incorporated

Printed in the United States of America
Print Number: 06   Print Year: 2020

Contents

7 Functions of Several Variables   419

Appendices

*Available at the text-specific website **CengageBrain.com**

Welcome to the Tenth Edition of *Brief Calculus: An Applied Approach with CalcChat & CalcView*! I am proud to present this new edition to you. As with all editions, I have been able to incorporate many useful comments from you, our user. In this edition, I introduce several new features and revise others. You will still find what you expect— a pedagogically sound, mathematically precise, and comprehensive textbook that includes a multitude of business and life sciences applications.

I am pleased and excited to offer you two brand new websites with this edition— **CalcView.com** and **LarsonAppliedCalculus.com.** Both websites were created with the goal of providing you with the resources needed to master Calculus. **CalcView.com** contains worked-out solution videos for selected exercises in the book, and **LarsonAppliedCalculus.com** offers multiple resources to supplement your learning experience. Best of all, these websites are completely *free*.

A theme throughout the book is **"IT'S ALL ABOUT YOU."** Please pay special attention to the study aids with a red **U**. These study aids will help you learn calculus, use technology, refresh your algebra skills, and prepare for tests. For an overview of these aids, check out CALCULUS & YOU on page 0. In each exercise set, quiz, and test, be sure to notice the reference to **CalcChat.com.** At this free site, you can download a step-by-step solution to any odd-numbered exercise. You can also work with a tutor, free of charge, during the hours posted at the site. Over the years, thousands of students have visited the site for help.

New To This Edition

 ™

The website **CalcView.com** contains video solutions of selected exercises. Calculus instructors progress step-by-step through solutions, providing guidance to help you solve the exercises. You can use your smartphone's QR Code® reader to scan the code ▓ and go directly to a video solution. Or you can access the videos at **CalcView.com.**

NEW LarsonAppliedCalculus.com

This companion website offers multiple tools and resources to supplement your learning. Access to these features is *free*. Watch videos explaining concepts from the book, explore examples, take a diagnostic test, view solutions to the checkpoint problems, and much more.

QR Code is a registered trademark of Denso Wave Incorporated

NEW Data Spreadsheets

Download these editable spreadsheets from **LarsonAppliedCalculus.com** and use the data to solve exercises.

REVISED Exercise Sets

The exercise sets have been carefully and extensively examined to ensure they are rigorous, relevant, and cover all topics necessary to understand the fundamentals of Calculus. The exercises have been reorganized and titled so that you can better see the connections between examples and exercises. Multi-step, real-life exercises reinforce problem-solving skills and mastery of concepts by giving you the opportunity to apply the concepts in real-life situations.

Trusted Features

HOW DO YOU SEE IT? Exercise

The *How Do You See It?* exercise in each section presents a real-life problem that you will solve by visual inspection using the concepts learned in the lesson.

 CalcChat®

For the past several years, an independent website— **CalcChat.com**—has been maintained to provide free solutions to all odd-numbered problems in the text. Thousands of students have visited the site for practice and help with their homework from live tutors.

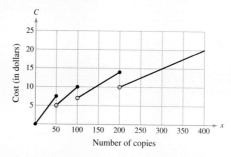

76. **HOW DO YOU SEE IT?** The graph shows the cost C (in dollars) of making x photocopies at a copy shop.

(a) Does $\lim_{x \to 50} C$ exist? Explain your reasoning.

(b) Does $\lim_{x \to 150} C$ exist? Explain your reasoning.

(c) You have to make 200 photocopies. Would it be better to make 200 or 201? Explain your reasoning.

Chapter Opener

Each *Chapter Opener* highlights a real-life problem from an example in the chapter, showing a graph related to the data and describing the math concept used to solve the problem.

Section Opener

Each *Section Opener* highlights a real-life problem in the exercises, showing a graph for the situation with a description of how you will use the math of the section to solve the problem.

Section Objectives

A bulleted list of learning objectives provides you with the opportunity to preview what will be presented in the upcoming section.

Definitions and Theorems

All definitions and theorems are highlighted for emphasis and easy recognition.

Business Capsule

Susie Wang and Ric Kostick graduated in 2002 from the University of California at Berkeley with degrees in mathematics. Together they launched a cosmetics brand called 100% Pure, which uses fruit and vegetable pigments to color cosmetics and uses only organic ingredients for the purest skin care. The company grew quickly and now has annual sales of over $40 million. Wang and Kostick attribute their success to applying what they learned from their studies. "Mathematics teaches you logic, discipline, and accuracy, which help you with all aspects of daily life," says Ric Kostick.

49. Research Project Use your school's library, the Internet, or some other reference source to research the opportunity cost of attending graduate school for 2 years to receive a Masters of Business Administration (MBA) degree rather than working for 2 years with a bachelor's degree. Write a short paper describing these costs.

Checkpoint

Paired with every example, the *Checkpoint* problems encourage immediate practice and check your understanding of the concepts presented in the example. Answers to all *Checkpoint* problems appear at the back of the text to reinforce understanding of the skill sets learned.

Business Capsule

Business Capsules appear at the end of selected sections. These capsules and their accompanying research project highlight business situations related to the mathematical concepts covered in the chapter.

SUMMARIZE

The *Summarize* feature at the end of each section helps you organize the lesson's key concepts into a concise summary, providing you with a valuable study tool.

STUDY TIP

These hints and tips can be used to reinforce or expand upon concepts, help you learn how to study mathematics, caution you about common errors, address special cases, or show alternative or additional steps to a solution of an example.

TECH TUTOR

The *Tech Tutor* gives suggestions for effectively using tools such as calculators, graphing calculators, and spreadsheet programs to help deepen your understanding of concepts, ease lengthy calculations, and provide alternate solution methods for verifying answers obtained by hand.

ALGEBRA TUTOR

The *Algebra Tutor* appears throughout each chapter and offers algebraic support at point of use. This support is revisited in a two-page algebra review at the end of the chapter, where additional details of example solutions with explanations are provided.

SKILLS WARM UP

The *Skills Warm Up* appears at the beginning of the exercise set for each section. These problems help you review previously learned skills that you will use in solving the section exercises.

Project

The projects at the end of selected sections involve in-depth applied exercises in which you will work with large, real-life data sets, often creating or analyzing models. These projects are offered online at **LarsonAppliedCalculus.com.**

47. Project: ATM Surcharge Fee For a project analyzing the average ATM surcharge fee in the United States from 2002 to 2014, visit this text's website at *LarsonAppliedCalculus.com.* *(Source: Bankrate, Inc.)*

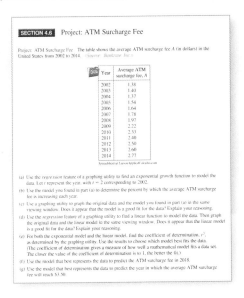

Instructor Resources

Media

Complete Solutions Manual

The *Complete Solutions Manual* provides worked-out solutions for all exercises in the text, including Checkpoints, Quiz Yourself, Test Yourself, and Tech Tutors.

Turn the Light On with MindTap for Larson's *Calculus: An Applied Approach*

Through personalized paths of dynamic assignments and applications, MindTap is a digital learning solution and representation of your course that turns cookie cutter into cutting edge, apathy into engagement, and memorizers into higher-level thinkers.

The Right Content: With MindTap's carefully curated material, you get the precise content and groundbreaking tools you need for every course you teach.

Personalization: Customize every element of your course—from rearranging the Learning Path to inserting videos and activities.

Improved Workflow: Save time when planning lessons with all of the trusted, most current content you need in one place in MindTap.

Tracking Students' Progress in Real Time: Promote positive outcomes by tracking students in real time and tailoring your course as needed based on the analytics.

Learn more at **cengage.com/mindtap.**

Instructor Companion Site

Everything you need for your course in one place! This collection of book-specific lecture and class tools is available online at **cengage.com/login.** Access and download PowerPoint presentations, images, solutions, videos, and more.

Cengage Learning Testing Powered by Cognero (ISBN: 978-1-3058-8230-0) is a flexible, online system that allows you to author, edit, and manage test bank content, create multiple test versions in an instant, and deliver tests from your LMS, your classroom, or wherever you want. This is available online at **cengage.com/login.**

Student Resources

Print

Student Solutions Manual

ISBN 13: 978-1-305-86099-5

The *Student Solutions Manual* provides complete worked-out solutions to all odd-numbered exercises in the text. In addition, the solutions of all Checkpoint, Quiz Yourself, Test Yourself, and Tech Tutor exercises are included.

Media

MindTap for Larson's *Calculus: An Applied Approach*

MindTap is a digital representation of your course that provides you with the tools you need to better manage your limited time, stay organized, and be successful. You can complete assignments whenever and wherever you are ready to learn with course material specially customized for you by your instructor and streamlined in one proven, easy-to-use interface. With an array of study tools, you will get a true understanding of course concepts, achieve better grades, and set the groundwork for your future courses.

Learn more at **cengage.com/mindtap**.

CengageBrain.com

To access additional course materials and companion resources, please visit **CengageBrain.com.** At the **CengageBrain.com** home page, search for the ISBN of your title (from the back cover of your book) using the search box at the top of the page. This will take you to the product page where free companion resources can be found.

Acknowledgments

I would like to thank my colleagues who have helped me develop this program. Their encouragement, criticisms, and suggestions have been invaluable to me.

I would particularly like to thank the following reviewers of this and previous editions:

Nasri Abdel-Aziz, *State University of New York College of Environmental Sciences and Forestry;* Carol Achs, *Mesa Community College;* Alejandro Acuna, *Central New Mexico Community College;* Lateef Adelani, *Harris-Stowe State University, Saint Louis;* Frederick Adkins, *Indiana University of Pennsylvania;* Polly Amstutz, *University of Nebraska at Kearney;* George Anastassiou, *University of Memphis;* Judy Barclay, *Cuesta College;* Sheeny Behmard, *Chemeketa Community College;* Jean Michelle Benedict, *Augusta State University;* Dona Boccio, *Queensborough Community College;* George Bradley, *Duquesne University;* David Bregenzer, *Utah State University;* Ben Brink, *Wharton County Junior College;* William Burgin, *Gaston College;* Mary Chabot, *Mt. San Antonio College;* Joseph Chance, *University of Texas—Pan American;* Jimmy Chang, *St. Petersburg College;* John Chuchel, *University of California;* Derron Coles, *Oregon State University;* Miriam E. Connellan, *Marquette University;* William Conway, *University of Arizona;* Karabi Datta, *Northern Illinois University;* Keng Deng, *University of Louisiana at Lafayette;* Liam Donohoe, *Providence College;* Roger A. Engle, *Clarion University of Pennsylvania;* David French, *Tidewater Community College;* Randy Gallaher, *Lewis & Clark Community College;* Perry Gillespie, *Fayetteville State University;* Jose Gimenez, *Temple University;* Betty Givan, *Eastern Kentucky University;* Walter J. Gleason, *Bridgewater State College;* Shane Goodwin, *Brigham Young University of Idaho;* Mark Greenhalgh, *Fullerton College;* Harvey Greenwald, *California Polytechnic State University;* Karen Hay, *Mesa Community College;* Raymond Heitmann, *University of Texas at Austin;* Larry Hoehn, *Austin Peay State University;* William C. Huffman, *Loyola University of Chicago;* Kala Iyer, *Los Angeles Valley College;* Arlene Jesky, *Rose State College;* Raja Khoury, *Collin County Community College;* Ronnie Khuri, *University of Florida;* Bernadette Kocyba, *J. Sergeant Reynolds Community College;* Duane Kouba, *University of California—Davis;* James A. Kurre, *The Pennsylvania State University;* Melvin Lax, *California State University—Long Beach;* Norbert Lerner, *State University of New York at Cortland;* Yuhlong Lio, *University of South Dakota;* Peter J. Livorsi, *Oakton Community College;* Bob Lombard, *Evergreen Valley College;* Ivan Loy, *Front Range Community College;* Peggy Luczak, *Camden County College;* Lewis D. Ludwig, *Denison University;* Samuel A. Lynch, *Southwest Missouri State University;* Augustine Maison, *Eastern Kentucky University;* Andrea Marchese, *Pace University;* Kevin McDonald, *Mt. San Antonio College;* Ronda McDonald, *Colorado Mesa University;* Earl H. McKinney, *Ball State University;* Randall McNiece, *San Jacinto College;* Philip R. Montgomery, *University of Kansas;* John Nardo, *Oglethorpe University;* Mike Nasab, *Long Beach City College;* Karla Neal, *Louisiana State University;* Benselamonyuy Ntatin, *Austin Peay State University;* James Osterburg, *University of Cincinnati;* Darla Ottman, *Elizabethtown Community & Technical College;* William Parzynski, *Montclair State University;* Scott Perkins, *Lake Sumter Community College;* Laurie Poe, *Santa Clara University;* Maijian Qian, *California State University, Fullerton;* Adelaida Quesada, *Miami Dade College—Kendall;* Brooke P. Quinlan, *Hillsborough Community College;* David Ray, *University of Tennessee at Martin;* Rita Richards, *Scottsdale Community College;* Stephen B. Rodi, *Austin Community College;* Carol Rychly, *Augusta State University;*

Yvonne Sandoval-Brown, *Pima Community College;* Richard Semmler, *Northern Virginia Community College—Annandale;* Bernard Shapiro, *University of Massachusetts—Lowell;* Mike Shirazi, *Germanna Community College;* Rick Simon, *University of La Verne;* Judy Smalling, *St. Petersburg College;* Jane Y. Smith, *University of Florida;* Billie Steinkamp, *Arkansas Northeastern College;* Marvin Stick, *University of Massachusetts—Lowell;* Eddy Stringer, *Tallahassee Community College;* DeWitt L. Sumners, *Florida State University;* Devki Talwar, *Indiana University of Pennsylvania;* Linda Taylor, *Northern Virginia Community College;* Stephen Tillman, *Wilkes University;* Jay Wiestling, *Palomar College;* Jonathan Wilkin, *Northern Virginia Community College;* Carol G. Williams, *Pepperdine University;* John Williams, *St. Petersburg College;* Ted Williamson, *Montclair State University;* Melvin R. Woodard, *Indiana University of Pennsylvania;* Carlton Woods, *Auburn University at Montgomery;* Jan E. Wynn, *Brigham Young University;* Robert A. Yawin, *Springfield Technical Community College;* Charles W. Zimmerman, *Robert Morris College*

My thanks to Robert Hostetler, The Pennsylvania State University, The Behrend College, Bruce Edwards, University of Florida, and David Heyd, The Pennsylvania State University, The Behrend College, for their significant contributions to previous editions of this text.

I would also like to thank the staff at Larson Texts, Inc. who assisted with proofreading the manuscript, preparing and proofreading the art package, and checking and typesetting the supplements.

On a personal level, I am grateful to my spouse, Deanna Gilbert Larson, for her love, patience, and support. Also, a special thanks goes to R. Scott O'Neil.

If you have suggestions for improving this text, please feel free to write to me. Over the past two decades I have received many useful comments from both instructors and students, and I value these comments very highly.

Ron Larson, Ph.D.
Professor of Mathematics
Penn State University
www.RonLarson.com

CALCULUS & YOU

Every feature in this text is designed to help you learn calculus. Whenever you see a red **U**, pay special attention to the study aid. These study aids represent years of experience in teaching students *just like you*.   Ron Larson

STUDY TIP

The notation $\partial z/\partial x$ is read as "the partial derivative of z with respect to x," and $\partial z/\partial y$ is read as "the partial derivative of z with respect to y."

The *Study Tips* occur at point of use throughout the text. They represent **common questions** that students ask me, **insights** into understanding concepts, and **alternative ways to look at concepts**. For instance, the *Study Tip* at the left provides insight on how to read mathematical notation.

TECH TUTOR

If you have access to a symbolic integration utility, try using it to find antiderivatives.

The *Tech Tutors* give suggestions on how you can use various types of technology to help understand the material. This includes **graphing calculators**, **computer graphing programs**, and **spreadsheet programs** such as Excel. For instance, the *Tech Tutor* at the left points out that some calculators and some computer programs are capable of symbolic integration.

ALGEBRA TUTOR

Finding intercepts involves solving equations. For a review of some techniques for solving equations, see page 71.

Throughout years of teaching, I have found that the greatest stumbling block to success in calculus is a weakness in algebra. Each time you see an *Algebra Tutor*, please read it carefully. Then, flip ahead to the referenced page and give yourself a chance to enjoy a brief **algebra refresher**. It will be time well spent.

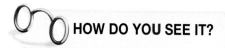 **HOW DO YOU SEE IT?**

The *How Do You See It?* question in each exercise set helps you **visually summarize concepts** without messy computations.

SUMMARIZE

The *Summarize* outline at the end of each section asks you to write each learning objective in **your own words**.

SKILLS WARM UP

The *Skills Warm Up* exercises that precede each exercise set will help you **review previously learned skills**.

SUMMARY AND STUDY STRATEGIES

The *Summary and Study Strategies*, coupled with the Review Exercises are designed to help you organize your thoughts as you **prepare for a chapter test**.

QUIZ YOURSELF

The *Quiz Yourself* occurs midway in each chapter. Take each of these quizzes as you would **take a quiz in class**.

TEST YOURSELF

The *Test Yourself* occurs at the end of each chapter. All questions are answered so you can **check your progress**.

0

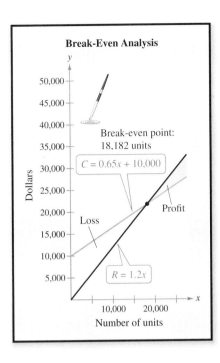

Break-Even Analysis

Break-even point: 18,182 units

$C = 0.65x + 10,000$

Profit

Loss

$R = 1.2x$

Dollars

Number of units

Example 5 on page 15 shows how the point of intersection of two graphs can be used to find the break-even point for a company manufacturing and selling a product.

1 Functions, Graphs, and Limits

iStockphoto.com/Yuri_Arcurs

1.1 The Cartesian Plane and the Distance Formula

■ Plot points in a coordinate plane and represent data graphically.
■ Find the distance between two points in a coordinate plane.
■ Find the midpoint of a line segment connecting two points.
■ Translate points in a coordinate plane.

The Cartesian Plane

Just as you can represent real numbers by points on a real number line, you can represent ordered pairs of real numbers by points in a plane called the **rectangular coordinate system,** or the **Cartesian plane,** after the French mathematician René Descartes (1596–1650).

The Cartesian plane is formed by using two real number lines intersecting at right angles, as shown in Figure 1.1. The horizontal real number line is usually called the *x*-axis, and the vertical real number line is usually called the *y*-axis. The point of intersection of these two axes is the **origin,** and the two axes divide the plane into four parts called **quadrants.**

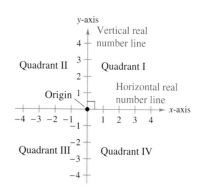

The Cartesian Plane
FIGURE 1.1

Each point in the plane corresponds to an **ordered pair** (*x*, *y*) of real numbers *x* and *y*, called **coordinates** of the point. The *x*-**coordinate** represents the directed distance from the *y*-axis to the point, and the *y*-**coordinate** represents the directed distance from the *x*-axis to the point, as shown in Figure 1.2.

The notation (*x*, *y*) denotes both a point in the plane and an open interval on the real number line. The context will tell you which meaning is intended.

EXAMPLE 1   Plotting Points in the Cartesian Plane

Plot the points

$$(-1, 2), \quad (3, 4), \quad (0, 0), \quad (3, 0), \quad \text{and} \quad (-2, -3).$$

SOLUTION   To plot the point

$$(-1, 2)$$

x-coordinate ──┘ └── *y*-coordinate

imagine a vertical line through −1 on the *x*-axis and a horizontal line through 2 on the *y*-axis. The intersection of these two lines is the point (−1, 2). The other four points can be plotted in a similar way and are shown in Figure 1.3.

✓ **Checkpoint 1**   *Worked-out solution available at LarsonAppliedCalculus.com*

Plot the points

$$(-3, 2), \quad (4, -2), \quad (3, 1), \quad (0, -2), \quad \text{and} \quad (-1, -2).$$

In Exercise 29 on page 9, you will use a line graph to estimate the Dow Jones Industrial Average.

y-axis

FIGURE 1.2

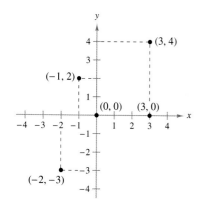

FIGURE 1.3

iStockphoto.com/Gradyreese

Using a rectangular coordinate system allows you to visualize relationships between two variables. In Example 2, data are represented graphically by points plotted in a rectangular coordinate system. This type of graph is called a **scatter plot.**

EXAMPLE 2   Sketching a Scatter Plot

The numbers E (in millions of people) of private-sector employees in the United States from 2005 through 2013 are shown in the table, where t represents the year. Sketch a scatter plot of the data.   *(Source: U.S. Bureau of Labor Statistics)*

DATA	t	2005	2006	2007	2008	2009	2010	2011	2012	2013
	E	112	114	116	115	109	108	110	112	115

Spreadsheet at LarsonAppliedCalculus.com

SOLUTION   To sketch a scatter plot of the data given in the table, represent each pair of values by an ordered pair

$$(t, E)$$

and plot the resulting points, as shown in Figure 1.4. For instance, the first pair of values is represented by the ordered pair

$$(2005, 112).$$

Note that the break in the t-axis indicates that the numbers between 0 and 2005 have been omitted, and the break in the E-axis indicates that the numbers between 0 and 104 have been omitted.

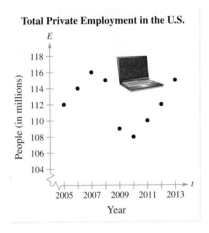

FIGURE 1.4

✓ *Checkpoint 2*   *Worked-out solution available at LarsonAppliedCalculus.com*

The numbers E (in thousands of people) of employees in the consumer lending industry in the United States from 2005 through 2013 are shown in the table, where t represents the year. Sketch a scatter plot of the data.   *(Source: U.S. Bureau of Labor Statistics)*

DATA	t	2005	2006	2007	2008	2009	2010	2011	2012	2013
	E	113	118	119	110	97	91	87	91	95

Spreadsheet at LarsonAppliedCalculus.com

In Example 2, $t = 1$ could have been used to represent the year 2005. In that case, the horizontal axis would not have been broken, and the tick marks would have been labeled 1 through 9 (instead of 2005 through 2013).

The scatter plot in Example 2 is one way to represent the given data graphically. Another technique, a *bar graph*, is shown in the figure at the right. If you have access to a graphing utility, try using it to represent the data given in Example 2 graphically.

Another way to represent data is with a *line graph* (see Exercise 29).

The Distance Formula

Recall from the Pythagorean Theorem that, for a right triangle with hypotenuse of length c and sides of lengths a and b, you have

$$a^2 + b^2 = c^2 \qquad \text{Pythagorean Theorem}$$

as shown in Figure 1.5. Note that the converse is also true. That is, if $a^2 + b^2 = c^2$, then the triangle is a right triangle.

Suppose you want to determine the distance d between two points

$$(x_1, y_1) \quad \text{and} \quad (x_2, y_2)$$

in the plane. These two points can form a right triangle, as shown in Figure 1.6. The length of the vertical side of the triangle is

$$|y_2 - y_1|$$

and the length of the horizontal side is

$$|x_2 - x_1|.$$

By the Pythagorean Theorem, you can write

$$d^2 = |x_2 - x_1|^2 + |y_2 - y_1|^2$$
$$d = \sqrt{|x_2 - x_1|^2 + |y_2 - y_1|^2}$$
$$d = \sqrt{(x_2 - x_1)^2 + (y_2 - y_1)^2}.$$

This result is the **Distance Formula.**

Pythagorean Theorem
FIGURE 1.5

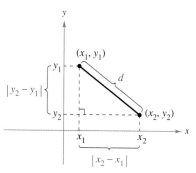

Distance Between Two Points
FIGURE 1.6

The Distance Formula

The distance d between the points (x_1, y_1) and (x_2, y_2) in the plane is

$$d = \sqrt{(x_2 - x_1)^2 + (y_2 - y_1)^2}.$$

EXAMPLE 3 Finding a Distance

Find the distance between the points $(-2, 1)$ and $(3, 4)$.

SOLUTION Let $(x_1, y_1) = (-2, 1)$ and $(x_2, y_2) = (3, 4)$. Then apply the Distance Formula as shown.

$$d = \sqrt{(x_2 - x_1)^2 + (y_2 - y_1)^2} \qquad \text{Distance Formula}$$
$$= \sqrt{[3 - (-2)]^2 + (4 - 1)^2} \qquad \text{Substitute for } x_1, y_1, x_2, \text{ and } y_2.$$
$$= \sqrt{(5)^2 + (3)^2} \qquad \text{Simplify.}$$
$$= \sqrt{34} \qquad \text{Simplify.}$$
$$\approx 5.83 \qquad \text{Use a calculator.}$$

So, the distance between the points is about 5.83 units. Note in Figure 1.7 that a distance of 5.83 looks about right.

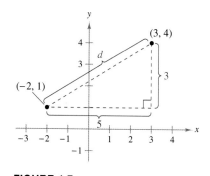

FIGURE 1.7

✓ *Checkpoint 3* *Worked-out solution available at LarsonAppliedCalculus.com*

Find the distance between the points $(-2, 1)$ and $(2, 4)$. ■

FIGURE 1.8

EXAMPLE 4   Verifying a Right Triangle

Use the Distance Formula to show that the points

$$(2, 1), \quad (4, 0), \quad \text{and} \quad (5, 7)$$

are vertices of a right triangle.

SOLUTION   The three points are plotted in Figure 1.8. Using the Distance Formula, you can find the lengths of the three sides as shown below.

$$d_1 = \sqrt{(5 - 2)^2 + (7 - 1)^2} = \sqrt{9 + 36} = \sqrt{45}$$
$$d_2 = \sqrt{(4 - 2)^2 + (0 - 1)^2} = \sqrt{4 + 1} = \sqrt{5}$$
$$d_3 = \sqrt{(5 - 4)^2 + (7 - 0)^2} = \sqrt{1 + 49} = \sqrt{50}$$

Because

$$d_1^2 + d_2^2 = 45 + 5 = 50 = d_3^2$$

you can apply the converse of the Pythagorean Theorem to conclude that the triangle must be a right triangle.

✓ *Checkpoint 4*   *Worked-out solution available at LarsonAppliedCalculus.com*

Use the Distance Formula to show that the points $(2, -1)$, $(5, 5)$, and $(6, -3)$ are vertices of a right triangle. ■

The figures provided with Examples 3 and 4 were not really essential to the solution. *Nevertheless*, it is strongly recommended that you develop the habit of including sketches with your solutions—even when they are not required.

EXAMPLE 5   Finding the Length of a Pass

In a football game, a quarterback throws a pass from the 5-yard line, 20 yards from one sideline. The pass is caught by a wide receiver on the 45-yard line, 50 yards from the same sideline, as shown in Figure 1.9. How long is the pass?

SOLUTION   You can find the length of the pass by finding the distance between the points $(20, 5)$ and $(50, 45)$.

$$d = \sqrt{(50 - 20)^2 + (45 - 5)^2} \qquad \text{Distance Formula}$$
$$= \sqrt{900 + 1600} \qquad \text{Simplify.}$$
$$= 50 \qquad \text{Simplify.}$$

So, the pass is 50 yards long.

✓ *Checkpoint 5*   *Worked-out solution available at LarsonAppliedCalculus.com*

A quarterback throws a pass from the 10-yard line, 10 yards from one sideline. The pass is caught by a wide receiver on the 30-yard line, 25 yards from the same sideline. How long is the pass? ■

FIGURE 1.9

STUDY TIP

In Example 5, the scale along the goal line showing distance from the sideline does not normally appear on a football field. However, when you use coordinate geometry to solve real-life problems, you are free to place the coordinate system in any way that is convenient for the solution of the problem.

The Midpoint Formula

To find the **midpoint** of the line segment that joins two points in a coordinate plane, find the average values of the respective coordinates of the two endpoints.

> **The Midpoint Formula**
>
> The midpoint of the line segment joining the points (x_1, y_1) and (x_2, y_2) is
>
> $$\text{Midpoint} = \left(\frac{x_1 + x_2}{2}, \frac{y_1 + y_2}{2} \right).$$

EXAMPLE 6   Finding the Midpoint of a Line Segment

Find the midpoint of the line segment joining the points

$$(-5, -3) \quad \text{and} \quad (9, 3).$$

SOLUTION   Let $(x_1, y_1) = (-5, -3)$ and $(x_2, y_2) = (9, 3)$.

$$\text{Midpoint} = \left(\frac{x_1 + x_2}{2}, \frac{y_1 + y_2}{2} \right) \quad \text{Midpoint Formula}$$

$$= \left(\frac{-5 + 9}{2}, \frac{-3 + 3}{2} \right) \quad \text{Substitute for } x_1, y_1, x_2, \text{ and } y_2.$$

$$= (2, 0) \quad \text{Simplify.}$$

The midpoint of the line segment is $(2, 0)$, as shown in Figure 1.10.

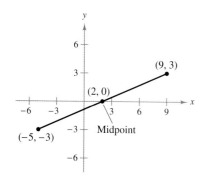

FIGURE 1.10

✔ **Checkpoint 6**   Worked-out solution available at LarsonAppliedCalculus.com

Find the midpoint of the line segment joining the points

$$(-6, 2) \quad \text{and} \quad (2, 8).$$

EXAMPLE 7   Estimating Annual Revenues

McDonald's Corporation had annual revenues of about \$27.0 billion in 2011 and about \$28.1 billion in 2013. Without knowing any additional information, estimate the 2012 annual revenues. *(Source: McDonald's Corp.)*

SOLUTION   One solution to the problem is to assume that revenues followed a linear pattern. Then you can estimate the 2012 revenues by finding the midpoint of the line segment connecting the points $(2011, 27.0)$ and $(2013, 28.1)$.

$$\text{Midpoint} = \left(\frac{x_1 + x_2}{2}, \frac{y_1 + y_2}{2} \right) \quad \text{Midpoint Formula}$$

$$= \left(\frac{2011 + 2013}{2}, \frac{27.0 + 28.1}{2} \right) \quad \text{Substitute for } x_1, y_1, x_2, \text{ and } y_2.$$

$$= (2012, 27.55) \quad \text{Simplify.}$$

So, you can estimate that the 2012 revenues were about \$27.55 billion, as shown in Figure 1.11. (The actual 2012 revenues were about \$27.6 billion.)

McDonald's Annual Revenues

FIGURE 1.11

✔ **Checkpoint 7**   Worked-out solution available at LarsonAppliedCalculus.com

Kellogg Company had annual sales of about \$13.2 billion in 2011 and about \$14.8 billion in 2013. Without knowing any additional information, estimate the 2012 annual sales. *(Source: Kellogg Co.)*

Translating Points in the Plane

Much of computer graphics consists of transformations of points in a coordinate plane. One type of transformation, a translation, is illustrated in Example 8. Other types of transformations include reflections, rotations, and stretches.

EXAMPLE 8   **Translating Points in the Plane**

Figure 1.12(a) shows the vertices of a parallelogram. Find the vertices of the parallelogram after it has been translated four units to the right and two units down.

SOLUTION   To translate each vertex four units to the right, add 4 to each x-coordinate. To translate each vertex two units down, subtract 2 from each y-coordinate.

Original Point	*Translated Point*
$(1, 0)$	$(1 + 4, 0 - 2) = (5, -2)$
$(3, 2)$	$(3 + 4, 2 - 2) = (7, 0)$
$(3, 6)$	$(3 + 4, 6 - 2) = (7, 4)$
$(1, 4)$	$(1 + 4, 4 - 2) = (5, 2)$

The translated parallelogram is shown in Figure 1.12(b).

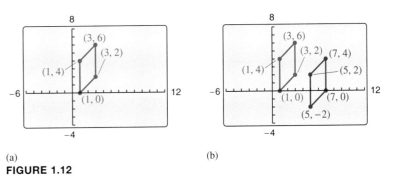

(a)   (b)

FIGURE 1.12

✓ *Checkpoint 8*   Worked-out solution available at LarsonAppliedCalculus.com

Find the vertices of the parallelogram in Example 8 after it has been translated two units to the left and four units down.   ■

SUMMARIZE   (Section 1.1)

1. Describe the Cartesian plane *(page 2)*. For an example of plotting points in the Cartesian plane, see Example 1.

2. Describe a scatter plot *(page 3)*. For an example of a scatter plot, see Example 2.

3. State the Distance Formula *(page 4)*. For examples of using the Distance Formula, see Examples 3, 4, and 5.

4. State the Midpoint Formula *(page 6)*. For an example of using the Midpoint Formula, see Example 6.

5. Describe a real-life example of how the Midpoint Formula can be used to estimate annual revenues *(page 6, Example 7)*.

6. Describe how to translate points in the Cartesian plane *(page 7)*. For an example of translating points in the Cartesian plane, see Example 8.

Many movies now use extensive computer graphics, much of which consists of transformations of points in two- and three-dimensional space. The photo above is from *The Amazing Spider-Man*. The movie's animators used computer graphics to design the scenery, characters, motion, and even the lighting throughout much of the film.

SKILLS WARM UP 1.1

The following warm-up exercises involve skills that were covered in a previous course. You will use these skills in the exercise set for this section. For additional help, review Appendix A.3.

In Exercises 1–6, simplify the expression.

1. $\dfrac{5 + (-4)}{2}$

2. $\dfrac{-3 + (-1)}{2}$

3. $\sqrt{(3 - 6)^2 + [1 - (-5)]^2}$

4. $\sqrt{(-2 - 0)^2 + [-7 - (-3)]^2}$

5. $\sqrt{27} + \sqrt{12}$

6. $\sqrt{8} - \sqrt{18}$

In Exercises 7–10, solve for x or y.

7. $\dfrac{x + (-5)}{2} = 7$

8. $\dfrac{-7 + y}{2} = -3$

9. $\sqrt{(3 - x)^2 + (7 - 4)^2} = \sqrt{45}$

10. $\sqrt{(6 - 2)^2 + (-2 - y)^2} = \sqrt{52}$

Exercises 1.1

See *CalcChat.com* for tutorial help and worked-out solutions to odd-numbered exercises.

 Plotting Points in the Cartesian Plane In Exercises 1 and 2, plot the points in the Cartesian plane. *See Example 1.*

1. $(-5, 3), (1, -1), (-2, -4), (2, 0), (1, 4)$

2. $(0, -4), (5, 1), (-3, 5), (2, -2), (-6, -1)$

 Finding a Distance and the Midpoint of a Line Segment In Exercises 3–12, (a) plot the points, (b) find the distance between the points, and (c) find the midpoint of the line segment joining the points. *See Examples 1, 3, and 6.*

3. $(3, 1), (5, 5)$

4. $(-3, 2), (3, -2)$

5. $(-3, 7), (1, -1)$

6. $(2, 2), (4, 14)$

7. $(2, -12), (8, -4)$

8. $(-5, -2), (7, 3)$

9. $\left(\frac{1}{2}, 1\right), \left(-\frac{3}{2}, -5\right)$

10. $\left(\frac{2}{3}, -\frac{1}{3}\right), \left(\frac{5}{6}, 1\right)$

11. $(0, -4.8), (0.5, 6)$

12. $(5.2, 6.4), (-2.7, 1.8)$

 Verifying a Right Triangle In Exercises 13–16, (a) find the length of each side of the right triangle and (b) show that these lengths satisfy the Pythagorean Theorem. *See Example 4.*

13.

14.

15.

16.

 Verifying a Polygon In Exercises 17–20, show that the points form the vertices of the indicated polygon. (A rhombus is a quadrilateral whose sides have the same length.)

17. Right triangle: $(0, 1), (3, 7), (4, -1)$

18. Isosceles triangle: $(1, -3), (3, 2), (-2, 4)$

19. Rhombus: $(0, 0), (1, 2), (2, 1), (3, 3)$

20. Parallelogram: $(0, 1), (3, 7), (4, 4), (1, -2)$

 Finding Values In Exercises 21 and 22, find the value(s) of x such that the distance between the points is 5.

21. $(1, 0), (x, -4)$

22. $(2, -1), (x, 2)$

Finding Values In Exercises 23 and 24, find the value(s) of y such that the distance between the points is 8.

23. $(-3, 0), (-5, y)$

24. $(4, -6), (4, y)$

25. Sports A soccer player passes the ball from a point that is 18 yards from an endline and 12 yards from a sideline. The pass is received by a teammate who is 42 yards from the same endline and 50 yards from the same sideline, as shown in the figure. How long is the pass?

26. Sports The first soccer player in Exercise 25 passes the ball to another teammate who is 37 yards from the same endline and 33 yards from the same sideline. How long is the pass?

⨏ Graphing Data In Exercises 27 and 28, use a graphing utility to graph a scatter plot, a bar graph, and a line graph to represent the data. Describe any trends that appear.

27. Consumer Trends The numbers (in billions) of individuals using the Internet in the world for 2006 through 2013 are shown in the table. *(Source: International Telecommunications Union)*

Year	2006	2007	2008	2009
Individuals	1.151	1.365	1.561	1.751

Year	2010	2011	2012	2013
Individuals	2.032	2.271	2.510	2.710

Spreadsheet at LarsonAppliedCalculus.com

28. Consumer Trends The numbers (in millions) of cellular telephone subscribers in the United States for 2006 through 2013 are shown in the table. *(Source: CTIA-The Wireless Association)*

Year	2006	2007	2008	2009
Subscribers	233.0	255.4	270.3	285.6

Year	2010	2011	2012	2013
Subscribers	296.3	316.0	326.5	335.7

Spreadsheet at LarsonAppliedCalculus.com

29. Dow Jones Industrial Average The graph shows the Dow Jones Industrial Average for common stocks. *(Source: S&P Dow Jones Indices LLC)*

(a) Estimate the Dow Jones Industrial Average for March 2013, July 2013, and July 2014.

(b) Estimate the percent increase or decrease in the Dow Jones Industrial Average from December 2013 to January 2014.

30. Home Sales The graph shows the median sales prices (in thousands of dollars) of existing one-family homes sold in the United States from 2006 through 2013. *(Source: National Association of Realtors)*

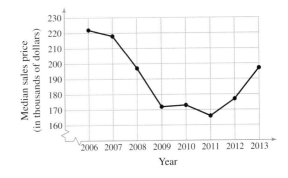

(a) Estimate the median sales prices of existing one-family homes for 2007, 2009, and 2012.

(b) Estimate the percent increase or decrease in the median value of existing one-family homes from 2011 to 2012.

The symbol ⨏ indicates an exercise in which you are instructed to use graphing technology or a symbolic computer algebra system. The solutions of other exercises may also be facilitated by use of appropriate technology.

31. Revenue and Profit The revenues and profits of Buffalo Wild Wings for 2011 and 2013 are shown in the table. (a) Use the Midpoint Formula to estimate the revenue and profit in 2012. (b) Then use your school's library, the Internet, or some other reference source to find the actual revenue and profit for 2012. (c) Did the revenue and profit increase in a linear pattern from 2011 to 2013? Explain your reasoning. (d) What were the expenses during each of the given years? (e) How would you rate the growth of Buffalo Wild Wings from 2011 to 2013? *(Source: Buffalo Wild Wings, Inc.)*

Year	2011	2012	2013
Revenue (millions of $)	784.5		1266.7
Profit (millions of $)	50.4		71.6

32. Revenue and Profit The revenues and profits of Walt Disney Company for 2011 and 2013 are shown in the table. (a) Use the Midpoint Formula to estimate the revenue and profit in 2012. (b) Then use your school's library, the Internet, or some other reference source to find the actual revenue and profit for 2012. (c) Did the revenue and profit increase in a linear pattern from 2011 to 2013? Explain your reasoning. (d) What were the expenses during each of the given years? (e) How would you rate the growth of Walt Disney Company from 2011 to 2013? *(Source: Walt Disney Company)*

Year	2011	2012	2013
Revenue (billions of $)	40.9		45.0
Profit (billions of $)	4.8		6.1

33. Economics The table shows the numbers of ear infections treated by doctors at HMO clinics of three different sizes: small, medium, and large.

Number of doctors	0	1	2	3	4
Cases per small clinic	0	20	28	35	40
Cases per medium clinic	0	30	42	53	60
Cases per large clinic	0	35	49	62	70

Spreadsheet at LarsonAppliedCalculus.com

(a) On the same coordinate plane, show the relationship between doctors and treated ear infections using *three* line graphs, where the number of doctors is on the horizontal axis and the number of ear infections treated is on the vertical axis.

(b) Compare the three relationships.

(Source: Adapted from Taylor, Economics, Fifth Edition)

34. HOW DO YOU SEE IT? The scatter plot shows the numbers of pickup trucks sold in a city from 2010 to 2015.

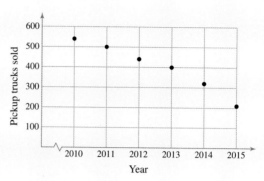

(a) In what year were 500 pickup trucks sold?

(b) About how many pickup trucks were sold in 2013?

(c) Describe the pattern shown by the data.

 Translating Points in the Plane In Exercises 35 and 36, use the translation and the graph to find the vertices of the figure after it has been translated. *See Example 8.*

35. 3 units left and 5 units down

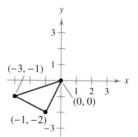

36. 2 units right and 4 units up

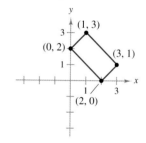

37. Using the Midpoint Formula Use the Midpoint Formula repeatedly to find the three points that divide the line segment joining (x_1, y_1) and (x_2, y_2) into four equal parts.

38. Using the Midpoint Formula Use Exercise 37 to find the points that divide the line segment joining the given points into four equal parts.

(a) $(1, -2), (4, -1)$ (b) $(-2, -3), (0, 0)$

39. Using the Midpoint Formula Show that $\left(\frac{1}{3}[2x_1 + x_2], \frac{1}{3}[2y_1 + y_2]\right)$ is one of the points of trisection of the line segment joining (x_1, y_1) and (x_2, y_2). Then, find the second point of trisection by finding the midpoint of the line segment joining

$$\left(\frac{1}{3}[2x_1 + x_2], \frac{1}{3}[2y_1 + y_2]\right) \quad \text{and} \quad (x_2, y_2).$$

40. Using the Midpoint Formula Use Exercise 39 to find the points of trisection of the line segment joining the given points.

(a) $(1, -2), (4, 1)$ (b) $(-2, -3), (0, 0)$

1.2 Graphs of Equations

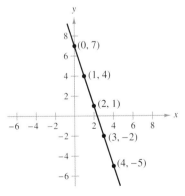

■ Sketch graphs of equations by hand.
■ Find the *x*- and *y*-intercepts of graphs of equations.
■ Write the standard forms of equations of circles.
■ Find the points of intersection of two graphs.
■ Use mathematical models to model and solve real-life problems.

The Graph of an Equation

In Section 1.1, you used a coordinate system to represent graphically the relationship between two quantities. There, the graphical picture consisted of a collection of points in a coordinate plane (see Example 2 in Section 1.1).

Frequently, a relationship between two quantities is expressed as an equation. For instance, degrees on the Fahrenheit scale are related to degrees on the Celsius scale by the equation

$$F = \frac{9}{5}C + 32.$$

In this section, you will study some basic procedures for sketching the graphs of such equations. The **graph** of an equation is the set of all points that are solutions of the equation.

EXAMPLE 1   Sketching the Graph of an Equation

Sketch the graph of $y = 7 - 3x$.

SOLUTION One way to sketch the graph of an equation is the *point-plotting method*. With this method, you construct a table of values that consists of several solution points of the equation, as shown in the table below. For instance, when $x = 0$,

$$y = 7 - 3(0) = 7$$

which implies that $(0, 7)$ is a solution point of the equation.

x	0	1	2	3	4
$y = 7 - 3x$	7	4	1	-2	-5

From the table, it follows that $(0, 7)$, $(1, 4)$, $(2, 1)$, $(3, -2)$, and $(4, -5)$ are solution points of the equation. After plotting these points, you can see that they appear to lie on a line, as shown in Figure 1.13. The graph of the equation is the line that passes through the five plotted points.

✓ *Checkpoint 1*   *Worked-out solution available at LarsonAppliedCalculus.com*

Sketch the graph of $y = 2x - 1$.   ■

Solution Points for $y = 7 - 3x$

FIGURE 1.13

In Exercise 61 on page 21, you will use a mathematical model to analyze the number of associate's degrees conferred in the United States.

STUDY TIP

Even though the sketch shown in Figure 1.13 is referred to as the graph of $y = 7 - 3x$, it actually represents only a *portion* of the graph. The entire graph is a line that would extend off the page.

TECH TUTOR

On most graphing utilities, the display screen is two-thirds as high as it is wide. On such screens, you can obtain a graph with a true geometric perspective by using a *square setting*—one in which

$$\frac{\text{Ymax} - \text{Ymin}}{\text{Xmax} - \text{Xmin}} = \frac{2}{3}.$$

One such setting is shown below. Notice that the x and y tick marks are equally spaced on a square setting, but not on a *nonsquare* setting.

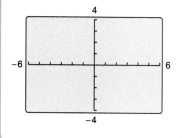

EXAMPLE 2   Sketching the Graph of an Equation

Sketch the graph of $y = x^2 - 2$.

SOLUTION   Begin by constructing a table of values, as shown below.

x	-2	-1	0	1	2	3
$y = x^2 - 2$	2	-1	-2	-1	2	7

Next, plot the points given in the table, as shown in Figure 1.14(a). Finally, connect the points with a smooth curve, as shown in Figure 1.14(b).

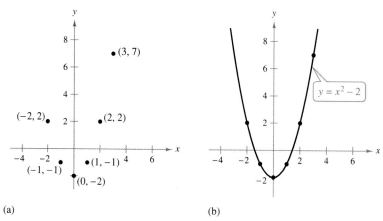

(a)   (b)

FIGURE 1.14

✔ *Checkpoint 2*   *Worked-out solution available at LarsonAppliedCalculus.com*

Sketch the graph of $y = x^2 - 4$.   ■

The graph shown in Example 2 is a **parabola.** The graph of any second-degree equation of the form

$$y = ax^2 + bx + c, \quad a \neq 0$$

has a similar shape. If $a > 0$, then the parabola opens upward, as shown in Figure 1.14(b), and if $a < 0$, then the parabola opens downward.

Note that the point-plotting technique demonstrated in Examples 1 and 2 has some shortcomings. With too few solution points, you can badly misrepresent the graph of a given equation. For instance, how would you connect the four points in Figure 1.15? Without further information, any one of the three graphs in Figure 1.16 would be reasonable.

FIGURE 1.15

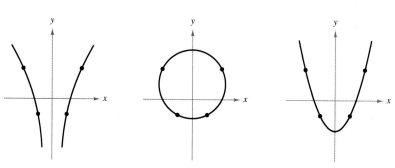

FIGURE 1.16

Intercepts of a Graph

Some solution points have zero as either the *x*-coordinate or the *y*-coordinate. These points are called **intercepts** because they are the points at which the graph intersects the *x*- or *y*-axis.

Some texts denote the *x*-intercept as simply the *x*-coordinate of the point $(a, 0)$ rather than the point itself. Likewise, some texts denote the *y*-intercept as the *y*-coordinate of the point $(0, b)$. Unless it is necessary to make a distinction, the term *intercept* will refer to either the point or the coordinate.

It is possible for a graph to have no intercepts, one intercept, or several intercepts, as shown in Figure 1.17.

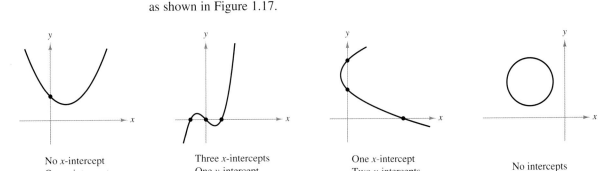

No *x*-intercept
One *y*-intercept

Three *x*-intercepts
One *y*-intercept

One *x*-intercept
Two *y*-intercepts

No intercepts

FIGURE 1.17

Finding Intercepts

1. To find **x-intercepts,** let *y* be zero and solve the equation for *x*.

2. To find **y-intercepts,** let *x* be zero and solve the equation for *y*.

EXAMPLE 3   Finding x- and y-Intercepts

Find the *x*- and *y*-intercepts of the graph of $y = x^3 - 4x$.

SOLUTION   To find the *x*-intercepts, let *y* be zero and solve for *x*.

$$x^3 - 4x = 0 \qquad \text{Let } y \text{ be zero.}$$
$$x(x^2 - 4) = 0 \qquad \text{Factor out common monomial factor.}$$
$$x(x + 2)(x - 2) = 0 \qquad \text{Factor.}$$
$$x = 0, -2, \text{ or } 2 \qquad \text{Solve for } x.$$

Because this equation has three solutions, you can conclude that the graph has three *x*-intercepts:

$$(0, 0), \quad (-2, 0), \quad \text{and} \quad (2, 0). \qquad \text{x-intercepts}$$

To find the *y*-intercepts, let *x* be zero and solve for *y*. Doing this produces

$$y = x^3 - 4x = 0^3 - 4(0) = 0.$$

This equation has only one solution, so the graph has one *y*-intercept:

$$(0, 0). \qquad \text{y-intercept}$$

(See Figure 1.18.)

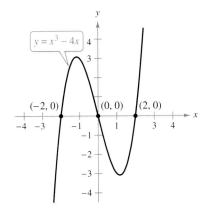

FIGURE 1.18

✓ **Checkpoint 3**   *Worked-out solution available at LarsonAppliedCalculus.com*

Find the *x*- and *y*-intercepts of the graph of $y = x^2 - 2x - 3$.

Circles

Throughout this course, you will learn to recognize several types of graphs from their equations. For instance, you should recognize that the graph of a second-degree equation of the form

$$y = ax^2 + bx + c, \quad a \neq 0$$

is a parabola (see Example 2). Another easily recognized graph is that of a **circle.**

Consider the circle shown in Figure 1.19. A point (x, y) is on the circle if and only if its distance from the center (h, k) is r. By the Distance Formula,

$$\sqrt{(x - h)^2 + (y - k)^2} = r.$$

By squaring each side of this equation, you obtain the **standard form of the equation of a circle.**

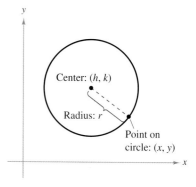

FIGURE 1.19

Standard Form of the Equation of a Circle

The **standard form of the equation of a circle** is

$$(x - h)^2 + (y - k)^2 = r^2. \qquad \text{Center at } (h, k)$$

The point (h, k) is the **center** of the circle, and the positive number r is the **radius** of the circle. The standard form of the equation of a circle whose center is the origin, $(h, k) = (0, 0)$, is

$$x^2 + y^2 = r^2. \qquad \text{Center at } (0, 0)$$

EXAMPLE 4   Finding the Equation of a Circle

The point $(3, 4)$ lies on a circle whose center is at $(-1, 2)$. Find the standard form of the equation of this circle and sketch its graph.

SOLUTION   The radius of the circle is the distance between $(-1, 2)$ and $(3, 4)$.

$$
\begin{aligned}
r &= \sqrt{[3 - (-1)]^2 + (4 - 2)^2} && \text{Distance Formula} \\
&= \sqrt{(4)^2 + (2)^2} && \text{Simplify.} \\
&= \sqrt{16 + 4} && \text{Simplify.} \\
&= \sqrt{20} && \text{Radius}
\end{aligned}
$$

Using $(h, k) = (-1, 2)$ and $r = \sqrt{20}$, the standard form of the equation of the circle is

$$
\begin{aligned}
(x - h)^2 + (y - k)^2 &= r^2 && \text{Equation of a circle} \\
[x - (-1)]^2 + (y - 2)^2 &= \left(\sqrt{20}\right)^2 && \text{Substitute for } h, k, \text{ and } r. \\
(x + 1)^2 + (y - 2)^2 &= 20. && \text{Write in standard form.}
\end{aligned}
$$

The graph of the equation of the circle is shown in Figure 1.20.

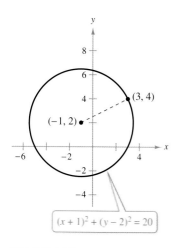

$$(x + 1)^2 + (y - 2)^2 = 20$$

FIGURE 1.20

✔ *Checkpoint 4*   *Worked-out solution available at LarsonAppliedCalculus.com*

The point $(1, 5)$ lies on a circle whose center is at $(-2, 1)$. Find the standard form of the equation of this circle and sketch its graph. ■

Points of Intersection

An ordered pair that is a solution of two different equations is called a **point of intersection** of the graphs of the two equations. For instance, Figure 1.21 shows that the graphs of

$$y = x^2 - 3 \quad \text{and} \quad y = x - 1$$

have two points of intersection: $(2, 1)$ and $(-1, -2)$. To find the points analytically, set the two y-values equal to each other and solve the equation

$$x^2 - 3 = x - 1$$

for x.

FIGURE 1.21

A common business application that involves a point of intersection is **break-even analysis.** The marketing of a new product typically requires an initial investment. When sufficient units have been sold so that the total revenue has offset the total cost, the sale of the product has reached the **break-even point.** The **total cost** of producing x units of a product is denoted by C, and the **total revenue** from the sale of x units of the product is denoted by R. So, you can find the break-even point by setting the cost C equal to the revenue R and solving for x. In other words, the break-even point corresponds to the point of intersection of the cost and revenue graphs.

EXAMPLE 5   Finding a Break-Even Point

A company manufactures a product at a cost of $0.65 per unit and sells the product for $1.20 per unit. The company's initial investment to produce the product was $10,000. Will the company break even when it sells 18,000 units? How many units must the company sell to break even?

SOLUTION   The total cost of producing x units of the product is given by

$$C = 0.65x + 10,000. \qquad \text{Cost equation}$$

The total revenue from the sale of x units is given by

$$R = 1.2x. \qquad \text{Revenue equation}$$

To find the break-even point, set the cost equal to the revenue and solve for x.

$R = C$	Set revenue equal to cost.
$1.2x = 0.65x + 10,000$	Substitute for R and C.
$0.55x = 10,000$	Subtract $0.65x$ from each side.
$x = \dfrac{10,000}{0.55}$	Divide each side by 0.55.
$x \approx 18,182$	Use a calculator.

So, the company will not break even when it sells 18,000 units. The company must sell 18,182 units before it breaks even. This result is shown graphically in Figure 1.22. Note in Figure 1.22 that sales less than 18,182 units correspond to a loss for the company $(R < C)$, whereas sales greater than 18,182 units correspond to a profit for the company $(R > C)$.

Break-Even Analysis

FIGURE 1.22

✓ *Checkpoint 5*   *Worked-out solution available at LarsonAppliedCalculus.com*

How many units must the company in Example 5 sell to break even when the selling price is $1.45 per unit?

Supply Curve
FIGURE 1.23

Demand Curve
FIGURE 1.24

Two types of equations that economists use to analyze a market are supply and demand equations. A **supply equation** shows the relationship between the unit price p of a product and the quantity supplied x. The graph of a supply equation is called a **supply curve.** (See Figure 1.23.) A typical supply curve rises because producers of a product want to sell more units when the unit price is higher.

A **demand equation** shows the relationship between the unit price p of a product and the quantity demanded x. The graph of a demand equation is called a **demand curve.** (See Figure 1.24.) A typical demand curve tends to show a decrease in the quantity demanded with each increase in price.

In an ideal situation, with no other factors present to influence the market, the production level should stabilize at the point of intersection of the graphs of the supply and demand equations. This point is called the **equilibrium point.** The x-coordinate of the equilibrium point is called the **equilibrium quantity** and the p-coordinate is called the **equilibrium price.** (See Figure 1.25.) You can find the equilibrium point by setting the demand equation equal to the supply equation and solving for x.

Equilibrium Point
FIGURE 1.25

EXAMPLE 6   Finding the Equilibrium Point

The demand and supply equations for an e-book reader are

$$p = 195 - 5.8x \qquad \text{Demand equation}$$
$$p = 150 + 3.2x \qquad \text{Supply equation}$$

where p is the price in dollars and x represents the number of units in millions. Find the equilibrium point for this market.

SOLUTION   Begin by setting the demand equation equal to the supply equation.

$$195 - 5.8x = 150 + 3.2x \qquad \text{Set equations equal to each other.}$$
$$45 - 5.8x = 3.2x \qquad \text{Subtract 150 from each side.}$$
$$45 = 9x \qquad \text{Add } 5.8x \text{ to each side.}$$
$$5 = x \qquad \text{Divide each side by 9.}$$

So, the equilibrium point occurs when the demand and supply are each five million units. (See Figure 1.26.) The price that corresponds to this x-value is obtained by substituting $x = 5$ into either of the original equations. For instance, substituting into the demand equation produces

$$p = 195 - 5.8(5) = 195 - 29 = \$166.$$

Note that when you substitute $x = 5$ into the supply equation, you obtain

$$p = 150 + 3.2(5) = 150 + 16 = \$166.$$

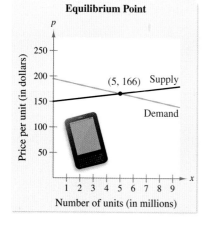

Equilibrium Point

FIGURE 1.26

✓ *Checkpoint 6*   *Worked-out solution available at LarsonAppliedCalculus.com*

The demand and supply equations for a streaming-media device are

$$p = 113 - 3.5x \qquad \text{Demand equation}$$
$$p = 89 + 2.5x \qquad \text{Supply equation}$$

where p is the price in dollars and x represents the number of units in millions. Find the equilibrium point for this market.   ■

ALGEBRA TUTOR

For help with evaluating the expressions in Example 7, see the review of order of operations on page 70.

Mathematical Models

In this text, you will see many examples of the use of equations as **mathematical models** of real-life phenomena. In developing a mathematical model to represent actual data, you should strive for two (often conflicting) goals—accuracy and simplicity.

EXAMPLE 7   Using a Mathematical Model

The table shows the annual average crude oil production (in millions of barrels per day) in the United States from 2009 through 2013.   *(Source: U.S. Energy Information Administration)*

Year	2009	2010	2011	2012	2013
Crude oil production	5.35	5.48	5.64	6.50	7.45

A mathematical model for these data is given by

$$y = 0.1671t^2 - 3.155t + 20.23$$

where y is the annual average crude oil production (in millions of barrels per day) and t is the year, with $t = 9$ corresponding to 2009. Use a graph and a table to compare the data with the model. Use the model to estimate the crude oil production in 2014.

SOLUTION   Figure 1.27 and the table below compare the data with the model. The model appears to fit the data well.

Year	2009	2010	2011	2012	2013
Crude oil production	5.35	5.48	5.64	6.50	7.45
Model	5.37	5.39	5.74	6.43	7.45

Using $t = 14$ to represent 2014, you can estimate the crude oil production in 2014 to be

$$y = 0.1671(14)^2 - 3.155(14) + 20.23 \approx 8.81 \text{ million barrels per day.}$$

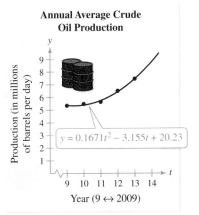

Annual Average Crude Oil Production

$y = 0.1671t^2 - 3.155t + 20.23$

Year ($9 \leftrightarrow 2009$)

FIGURE 1.27

✓ *Checkpoint 7*   *Worked-out solution available at LarsonAppliedCalculus.com*

The table shows the annual sales (in billions of dollars) for Dollar Tree stores from 2006 through 2013.   *(Source: Dollar Tree, Inc.)*

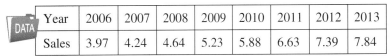

Year	2006	2007	2008	2009	2010	2011	2012	2013
Sales	3.97	4.24	4.64	5.23	5.88	6.63	7.39	7.84

Spreadsheet at LarsonAppliedCalculus.com

A mathematical model for these data is given by

$$S = 0.0294t^2 + 0.030t + 2.63$$

where S is the annual sales (in billions of dollars) and t is the year, with $t = 6$ corresponding to 2006. Use a graph and a table to compare the data with the model. Use the model to estimate the sales in 2014.

Much of your study of calculus will center around the behavior of the graphs of mathematical models. Figure 1.28 shows the graphs of six basic algebraic equations. Familiarity with these graphs will help you in the creation and use of mathematical models.

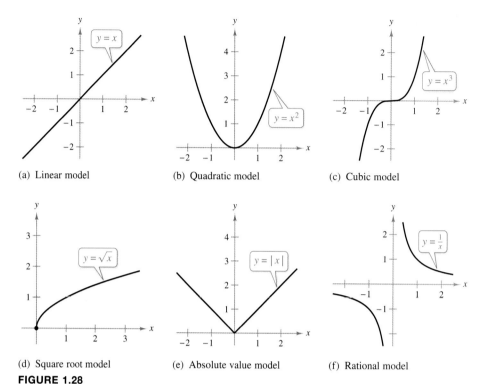

(a) Linear model   (b) Quadratic model   (c) Cubic model

(d) Square root model   (e) Absolute value model   (f) Rational model

FIGURE 1.28

SUMMARIZE (Section 1.2)

1. Describe how to sketch the graph of an equation by hand *(page 11)*. For examples of sketching graphs by hand, see Examples 1 and 2.

2. Describe how to find the *x*- and *y*-intercepts of a graph *(page 13)*. For an example of finding the *x*- and *y*-intercepts of a graph, see Example 3.

3. State the standard form of the equation of a circle *(page 14)*. For an example of finding the standard form of the equation of a circle, see Example 4.

4. Describe how to find a point of intersection of the graphs of two equations *(page 15)*. For examples of finding points of intersection, see Examples 5 and 6.

5. Describe break-even analysis *(page 15)*. For an example of break-even analysis, see Example 5.

6. Describe supply equations, demand equations, and equilibrium points *(page 16)*. For an example of a supply equation, a demand equation, and an equilibrium point, see Example 6.

7. Describe a mathematical model *(page 17)*. For an example of a mathematical model, see Example 7.

SKILLS WARM UP 1.2

The following warm-up exercises involve skills that were covered in a previous course. You will use these skills in the exercise set for this section. For additional help, review Appendices A.3 and A.4.

In Exercises 1–6, solve for y.

1. $5y - 12 = x$

2. $-y = 15 - x$

3. $x^3y + 2y = 1$

4. $x^2 + x - y^2 - 6 = 0$

5. $(x - 2)^2 + (y + 1)^2 = 9$

6. $(x + 6)^2 + (y - 5)^2 = 81$

In Exercises 7–10, evaluate the expression for the given value of x.

Expression	x-Value
7. $y = 5x$	$x = -2$
9. $y = 4x^2 - 7$	$x = 0.5$

Expression	x-Value
8. $y = 3x - 4$	$x = 3$
10. $y = 9x^2 + 9x - 5$	$x = \frac{1}{3}$

In Exercises 11–14, factor the expression.

11. $x^2 - 3x + 2$

12. $x^2 + 5x + 6$

13. $y^2 - 3y + \frac{9}{4}$

14. $y^2 - 7y + \frac{49}{4}$

Exercises 1.2

See *CalcChat.com* for tutorial help and worked-out solutions to odd-numbered exercises.

Matching In Exercises 1–6, match the equation with its graph. [The graphs are labeled (a)–(f).]

1. $y = x - 2$

2. $y = -\frac{1}{2}x + 2$

3. $y = x^2 + 2x$

4. $y = \sqrt{9 - x^2}$

5. $y = |x| - 2$

6. $y = x^3 - x$

(a)

(b)

(c)

(d)

(e)

(f)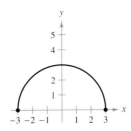

Sketching the Graph of an Equation In Exercises 7–22, sketch the graph of the equation. Use a graphing utility to verify your results. *See Examples 1 and 2.*

7. $y = 2x + 3$

8. $y = 1 - 4x$

9. $y = x^2 - 3$

10. $y = x^2 + 6$

11. $y = (x - 1)^2$

12. $y = (x + 5)^2$

13. $y = x^3 + 2$

14. $y = 1 - x^3$

15. $y = -\sqrt{x - 1}$

16. $y = \sqrt{x + 4}$

17. $y = |x + 1|$

18. $y = -|x - 2|$

19. $y = \dfrac{1}{x - 3}$

20. $y = \dfrac{1}{x + 2}$

21. $x = y^2 - 4$

22. $x = 4 - y^2$

Finding x- and y-Intercepts In Exercises 23–32, find the x- and y-intercepts of the graph of the equation. *See Example 3.*

23. $2x - y - 3 = 0$

24. $4x - 3y - 6 = 0$

25. $y = x^2 + x - 2$

26. $y = x^2 - 4x + 3$

27. $y = x^3 + 7x^2$

28. $y = x^3 - 9x$

29. $y = \dfrac{x^2 - 4}{x - 2}$

30. $y = \dfrac{x^2 + 3x}{2x}$

31. $x^2y - x^2 + 4y = 0$

32. $2x^2y + 8y - x^2 = 1$

Finding the Equation of a Circle In Exercises 33–40, find the standard form of the equation of the circle with the given characteristics and sketch its graph. *See Example 4.*

33. Center: $(0, 0)$; radius: 4

34. Center: $(0, 0)$; radius: 5

35. Center: $(2, -1)$; radius: 3

36. Center: $(-4, 3)$; radius: 2

37. Center: $(-1, 1)$; solution point: $(-1, 5)$

38. Center: $(2, -3)$; solution point: $(5, -7)$

39. Endpoints of a diameter: $(-6, -8), (6, 8)$

40. Endpoints of a diameter: $(0, -4), (6, 4)$

Finding Points of Intersection In Exercises 41–48, find the points of intersection (if any) of the graphs of the equations. Use a graphing utility to check your results.

41. $y = -x + 2, y = 2x - 1$

42. $y = -x + 7, y = \frac{3}{2}x - 8$

43. $y = -x^2 + 15, y = 3x + 11$

44. $y = x^2 - 5, y = x + 1$

45. $y = x^3, y = 2x$

46. $y = \sqrt{x}, y = x$

47. $y = x^4 - 2x^2 + 1, y = 1 - x^2$

48. $y = x^3 - 2x^2 + x - 1, y = -x^2 + 3x - 1$

Finding a Break-Even Point In Exercises 49–54, C represents the total cost (in dollars) of producing x units of a product and R represents the total revenue (in dollars) from the sale of x units. How many units must the company sell to break even? *See Example 5.*

49. $C = 0.85x + 35,000, R = 1.55x$

50. $C = 6x + 500,000, R = 35x$

51. $C = 8650x + 250,000, R = 9950x$

52. $C = 2.5x + 10,000, R = 4.9x$

53. $C = 6x + 5000, R = 10x$

54. $C = 130x + 12,600, R = 200x$

55. **Break-Even Analysis** You are setting up a part-time business with an initial investment of $21,000. The unit cost of the product is $11.50, and the selling price is $19.90.

   (a) Find equations for the total cost C (in dollars) and total revenue R (in dollars) for x units.

   (b) Find the break-even point by finding the point of intersection of the cost and revenue equations.

   (c) How many units would yield a profit of $1000?

56. **Break-Even Analysis** A 2015 Toyota Camry costs $33,500 with a gasoline engine. A 2015 Toyota Avalon costs $36,775 with a hybrid engine. The Camry gets 31 miles per gallon of gasoline and the Avalon gets 39 miles per gallon of gasoline. Assume that the price of gasoline is $2.759. *(Source: Toyota Motor Sales, U.S.A., Inc. and U.S. Energy Information Administration)*

   (a) Show that the cost C_g (in dollars) of driving the Toyota Camry x miles is

   $$C_g = 33,500 + \frac{2.759x}{31}$$

   and the cost C_h (in dollars) of driving the Toyota Avalon x miles is

   $$C_h = 36,775 + \frac{2.759x}{39}.$$

   (b) Find the break-even point. That is, find the mileage at which the hybrid-powered Toyota Avalon becomes more economical than the gasoline-powered Toyota Camry.

57. **Supply and Demand** The demand and supply equations for a fitness tracking band are given by

   $p = 205 - 4x$ Demand equation

   $p = 135 + 3x$ Supply equation

   where p is the price (in dollars) and x represents the number of units (in thousands). Find the equilibrium point for this market.

58. **Supply and Demand** The demand and supply equations for an MP3 player are given by

   $p = 190 - 15x$ Demand equation

   $p = 75 + 8x$ Supply equation

   where p is the price (in dollars) and x represents the number of units (in hundreds of thousands). Find the equilibrium point for this market.

59. **E-Book Revenue** The table shows the annual revenues (in billions of dollars) of e-books in the United States from 2009 through 2013. *(Source: Statista)*

Year	2009	2010	2011	2012	2013
Revenue	0.82	1.52	2.31	3.35	4.52

A mathematical model for the data is given by $y = 0.00333t^3 - 0.0250t^2 + 0.252t - 1.85$, where y is the annual revenue (in billions of dollars) and t is the year, with $t = 9$ corresponding to 2009.

   (a) Use a graph and a table to compare the data with the model.

   (b) Use the model to predict the revenue in 2018.

60. **HOW DO YOU SEE IT?** The graph shows the cost and revenue equations for a product.

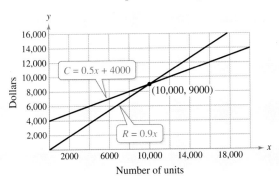

(a) For what number of units sold does the company break even?

(b) For what numbers of units sold is there a loss for the company?

(c) For what numbers of units sold is there a profit for the company?

61. Associate's Degrees A mathematical model for the numbers of associate's degrees conferred y (in thousands) from 2008 through 2012 is given by the equation $y = 7.79t^2 - 86.6t + 941$, where t represents the year, with $t = 8$ corresponding to 2008. *(Source: National Center for Education Statistics)*

(a) Use the model to complete the table.

Year	2008	2009	2010	2011	2012	2016
Degrees						

(b) This model was created using actual data from 2008 through 2012. How accurate do you think the model is in predicting the number of associate's degrees conferred in 2016? Explain your reasoning.

(c) Using this model, what is the prediction for the number of associate's degrees conferred in 2020? Do you think this prediction is valid?

62. Heart Transplants A mathematical model for the numbers of heart transplants y performed in the United States in the years 2009 through 2013 is given by $y = 19.000t^3 - 617.71t^2 + 6696.7t - 21{,}873$, where t represents the year, with $t = 9$ corresponding to 2009. *(Source: Organ Procurement and Transplantation Network)*

(a) Use a graphing utility or a spreadsheet to complete the table.

Year	2009	2010	2011	2012	2013
Transplants					

(b) Use your school's library, the Internet, or some other reference source to find the actual numbers of heart transplants in the years 2009 through 2013. Compare the actual numbers with those given by the model. How well does the model fit the data? Explain your reasoning.

(c) Using this model, what is the prediction for the number of heart transplants in 2019? Do you think this prediction is valid? What factors could affect this model's accuracy?

63. Making a Conjecture Use a graphing utility to graph the equation $y = cx + 1$ for $c = 1, 2, 3, 4$, and 5. Then make a conjecture about the x-coefficient and the graph of the equation.

64. Break-Even Point Define the break-even point for a business marketing a new product. Give examples of a linear cost equation and a linear revenue equation for which the break-even point is 10,000 units.

Finding Intercepts In Exercises 65–70, use a graphing utility to graph the equation and approximate the x- and y-intercepts of the graph.

65. $y = 0.24x^2 + 1.32x + 1.815$

66. $y = -0.56x^2 - 5.34x + 6.25$

67. $y = \sqrt{0.3x^2 - 4.3x + 5.7}$

68. $y = \sqrt{-1.21x^2 + 2.34x + 5.6}$

69. $y = \dfrac{0.2x^2 + 1}{0.1x + 2.4}$

70. $y = \dfrac{0.4x - 5.3}{0.4x^2 + 5.3}$

71. Project: Number of Stores For a project analyzing the numbers of Tiffany & Co. stores from 2004 through 2013, visit this text's website at *LarsonAppliedCalculus.com*. *(Source: Tiffany & Co.)*

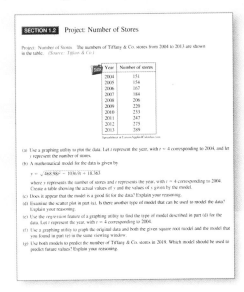

1.3 Lines in the Plane and Slope

■ Use the slope-intercept form of a linear equation to sketch graphs.
■ Find slopes of lines passing through two points.
■ Use the point-slope form to write equations of lines.
■ Find equations of parallel and perpendicular lines.
■ Use linear equations to model and solve real-life problems.

Using Slope

The simplest mathematical model for relating two variables is the **linear equation**

$$y = mx + b. \qquad \text{Linear equation}$$

This equation is called *linear* because its graph is a line. (In this text, the term *line* is used to mean *straight line*.) By letting $x = 0$, you can see that the line crosses the y-axis at

$$y = b$$

In Exercise 83 on page 32, you will use slope to analyze the average salaries of postsecondary education administrators.

as shown in Figure 1.29. In other words, the y-intercept is $(0, b)$. The steepness or slope of the line is m.

$$y = mx + b$$

Slope ⌐   ⌐ y-intercept

The **slope** of a nonvertical line is the number of units the line rises (or falls) vertically for each unit of horizontal change from left to right, as shown in Figure 1.29.

Positive slope, line rises.

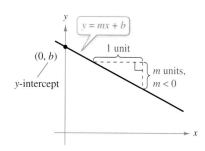

Negative slope, line falls.

FIGURE 1.29

A linear equation that is written in the form $y = mx + b$ is said to be written in **slope-intercept form.**

The Slope-Intercept Form of the Equation of a Line

The graph of the equation

$$y = mx + b$$

is a line whose slope is m and whose y-intercept is $(0, b)$.

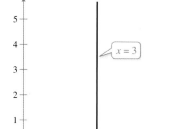

When a line is vertical, the slope is undefined.

FIGURE 1.30

A vertical line has an equation of the form

$$x = a. \qquad \text{Vertical line}$$

Because the equation of a vertical line cannot be written in the form $y = mx + b$, it follows that the slope of a vertical line is undefined, as indicated in Figure 1.30.

Once you have determined the slope and the y-intercept of a line, it is a relatively simple matter to sketch its graph.

EXAMPLE 1   Graphing Linear Equations

Sketch the graph of each linear equation.

a. $y = 2x + 1$

b. $y = 2$

c. $x + y = 2$

SOLUTION

a. This equation is written in slope-intercept form.

$$y = 2x + 1 \qquad\qquad y = mx + b$$

Because $b = 1$, the y-intercept is $(0, 1)$. Moreover, because the slope is $m = 2$, the line *rises* two units for each unit the line moves to the right, as shown in Figure 1.31(a).

b. By writing this equation in slope-intercept form

$$y = (0)x + 2 \qquad\qquad y = mx + b$$

you can see that the y-intercept is $(0, 2)$ and the slope is zero. A zero slope implies that the line is horizontal—that is, it does not rise *or* fall, as shown in Figure 1.31(b).

c. By writing this equation in slope-intercept form

$$
\begin{aligned}
x + y &= 2 &&\text{Write original equation.}\\
y &= -x + 2 &&\text{Subtract } x \text{ from each side.}\\
y &= (-1)x + 2 &&\text{Write in slope-intercept form.}
\end{aligned}
$$

you can see that the y-intercept is $(0, 2)$. Moreover, because the slope is $m = -1$, the line *falls* one unit for each unit the line moves to the right, as shown in Figure 1.31(c).

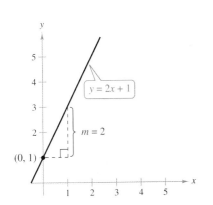

(a) When m is positive, the line rises from left to right.

FIGURE 1.31

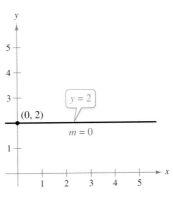

(b) When m is zero, the line is horizontal.

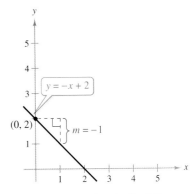

(c) When m is negative, the line falls from left to right.

✓ **Checkpoint 1**   *Worked-out solution available at LarsonAppliedCalculus.com*

Sketch the graph of each linear equation.

a. $y = 4x - 2$

b. $x = 1$

c. $2x + y = 6$

In real-life problems, the slope of a line can be interpreted as either a *ratio* or a *rate*. If the *x*- and *y*-axes have the same unit of measure, then the slope has no units and is a **ratio**. If the *x*- and *y*-axes have different units of measure, then the slope is a **rate** or **rate of change.**

EXAMPLE 2   Using Slope as a Ratio

The maximum recommended slope of a wheelchair ramp is $\frac{1}{12} \approx 0.083$. A business installs a wheelchair ramp that rises 22 inches over a horizontal length of 24 feet, as shown in the figure. Is the ramp steeper than recommended? *(Source: ADA Standards for Accessible Design)*

SOLUTION The horizontal length of the ramp is 24 feet or $12(24) = 288$ inches. So, the slope of the ramp is

$$\text{Slope} = \frac{\text{vertical change}}{\text{horizontal change}}$$

$$= \frac{22 \text{ in.}}{288 \text{ in.}}$$

$$\approx 0.076.$$

Because the slope of the ramp is less than 0.083, the ramp is not steeper than recommended. Note that the slope is a ratio and has no units.

✓ Checkpoint 2   *Worked-out solution available at LarsonAppliedCalculus.com*

The business in Example 2 installs a second ramp that rises 27 inches over a horizontal length of 26 feet. Is the ramp steeper than recommended?

EXAMPLE 3   Using Slope as a Rate of Change

A manufacturing company determines that the total cost in dollars of producing *x* units of a product is

$$C = 25x + 3500.$$

Describe the practical significance of the *y*-intercept and slope of the line given by this equation.

SOLUTION The *y*-intercept $(0, 3500)$ tells you that the cost of producing zero units is $3500. This is the **fixed cost** of production—it includes costs that must be paid regardless of the number of units produced. The slope, which is $m = 25$, tells you that the cost of producing each unit is $25, as shown in Figure 1.32. Economists call the cost per unit the **marginal cost.** If the production increases by one unit, then the "margin" or extra amount of cost is $25.

✓ Checkpoint 3   *Worked-out solution available at LarsonAppliedCalculus.com*

A small business determines that the value of a digital copier *t* years after its purchase is $V = -300t + 1500$. Describe the practical significance of the *y*-intercept and slope of the line given by this equation. ■

Production Cost

FIGURE 1.32

Finding the Slope of a Line

Given an equation of a nonvertical line, you can find its slope by writing the equation in slope-intercept form. When you are not given an equation, you can still find the slope of a line. For instance, suppose you want to find the slope of the line passing through the points (x_1, y_1) and (x_2, y_2), as shown in Figure 1.33. As you move from left to right along this line, a change of $(y_2 - y_1)$ units in the vertical direction corresponds to a change of $(x_2 - x_1)$ units in the horizontal direction. These two changes are denoted by the symbols

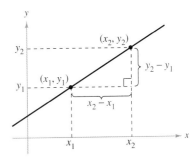

FIGURE 1.33

$$\Delta y = y_2 - y_1 = \text{the change in } y$$

and

$$\Delta x = x_2 - x_1 = \text{the change in } x.$$

(The symbol Δ is the Greek capital letter delta, and the symbols Δy and Δx are read as "delta y" and "delta x.") The ratio of Δy to Δx represents the slope of the line that passes through the points (x_1, y_1) and (x_2, y_2).

$$\text{Slope} = \frac{\Delta y}{\Delta x} = \frac{y_2 - y_1}{x_2 - x_1}$$

Be sure you see that Δx represents a single number, not the product of two numbers (Δ and x). The same is true for Δy.

The Slope of a Line Passing Through Two Points

The **slope** m of the nonvertical line passing through the points (x_1, y_1) and (x_2, y_2) is

$$m = \frac{\Delta y}{\Delta x} = \frac{y_2 - y_1}{x_2 - x_1}$$

where $x_1 \neq x_2$. Slope is not defined for vertical lines.

When this formula is used for slope, the *order of subtraction* is important. Given two points on a line, you may label either one of them as (x_1, y_1) and the other as (x_2, y_2). However, once you have done this, you must form the numerator and denominator using the same order of subtraction.

$$m = \frac{y_2 - y_1}{x_2 - x_1} \qquad m = \frac{y_1 - y_2}{x_1 - x_2} \qquad m = \frac{y_2 - y_1}{x_1 - x_2}$$

Correct · Correct · Incorrect

For instance, the slope of the line passing through the points $(3, 4)$ and $(5, 7)$ can be calculated as

$$m = \frac{7 - 4}{5 - 3} = \frac{3}{2}$$

or

$$m = \frac{4 - 7}{3 - 5} = \frac{-3}{-2} = \frac{3}{2}.$$

EXAMPLE 4   **Finding Slopes of Lines**

Find the slope of the line passing through each pair of points.

a. $(-2, 0)$ and $(3, 1)$ **b.** $(-1, 2)$ and $(2, 2)$

c. $(0, 4)$ and $(1, -1)$ **d.** $(3, 4)$ and $(3, 1)$

SOLUTION

a. Letting $(x_1, y_1) = (-2, 0)$ and $(x_2, y_2) = (3, 1)$, you obtain a slope of

$$m = \frac{y_2 - y_1}{x_2 - x_1} = \frac{1 - 0}{3 - (-2)} = \frac{1}{5}$$ ⟵ Difference in y-values
⟵ Difference in x-values

as shown in Figure 1.34(a).

b. The slope of the line passing through $(-1, 2)$ and $(2, 2)$ is

$$m = \frac{2 - 2}{2 - (-1)} = \frac{0}{3} = 0.$$ See Figure 1.34(b).

c. The slope of the line passing through $(0, 4)$ and $(1, -1)$ is

$$m = \frac{-1 - 4}{1 - 0} = \frac{-5}{1} = -5.$$ See Figure 1.34(c).

d. The slope of the vertical line passing through $(3, 4)$ and $(3, 1)$ is not defined because division by zero is undefined. [See Figure 1.34(d).]

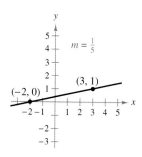

(a) Positive slope, line rises from left to right.

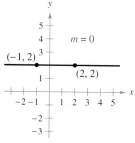

(b) Zero slope, line is horizontal.

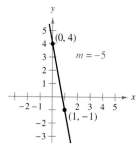

(c) Negative slope, line falls from left to right.

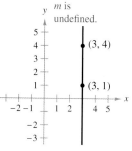

(d) Vertical line, undefined slope.

FIGURE 1.34

✓ *Checkpoint 4*   *Worked-out solution available at LarsonAppliedCalculus.com*

Find the slope of the line passing through each pair of points.

a. $(-3, 2)$ and $(5, 18)$

b. $(-2, 1)$ and $(-4, 2)$

c. $(2, -4)$ and $(-2, -4)$

Writing Linear Equations

When you know the slope of a line and the coordinates of one point on the line, you can find an equation for the line. For instance, in Figure 1.35, let (x_1, y_1) be a point on the line whose slope is m. If (x, y) is any other point on the line, then

$$\frac{y - y_1}{x - x_1} = m.$$

This equation in the variables x and y can be rewritten in the form

$$y - y_1 = m(x - x_1)$$

which is the **point-slope form** of the equation of a line.

Any two points on a nonvertical line can be used to determine the slope of the line.

FIGURE 1.35

Point-Slope Form of the Equation of a Line

The equation of the line with slope m passing through the point (x_1, y_1) is

$$y - y_1 = m(x - x_1).$$

The point-slope form is most useful for *finding* the equation of a nonvertical line. You should remember this formula—it is used throughout the text.

EXAMPLE 5   Using the Point-Slope Form

Find the slope-intercept form of the equation of the line that has a slope of 3 and passes through the point $(1, -2)$.

SOLUTION   Use the point-slope form with $m = 3$ and $(x_1, y_1) = (1, -2)$.

$$y - y_1 = m(x - x_1)$$ Point-slope form

$$y - (-2) = 3(x - 1)$$ Substitute for m, x_1, and y_1.

$$y + 2 = 3x - 3$$ Simplify.

$$y = 3x - 5$$ Write in slope-intercept form. See Figure 1.36.

So, the slope-intercept form of the equation of the line is $y = 3x - 5$.

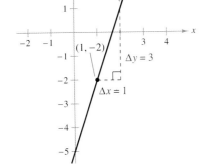

FIGURE 1.36

✓ *Checkpoint 5*   *Worked-out solution available at LarsonAppliedCalculus.com*

Find the slope-intercept form of the equation of the line that has a slope of 2 and passes through the point $(-1, 2)$.

The point-slope form can be used to find an equation of the line passing through two points (x_1, y_1) and (x_2, y_2). To do this, first find the slope of the line

$$m = \frac{y_2 - y_1}{x_2 - x_1}, \qquad x_1 \neq x_2$$

and then use the point-slope form to obtain the equation

$$y - y_1 = \frac{y_2 - y_1}{x_2 - x_1}(x - x_1).$$ Two-point form

This is sometimes called the **two-point form** of the equation of a line.

Sales Per Share

FIGURE 1.37

(a) Linear extrapolation

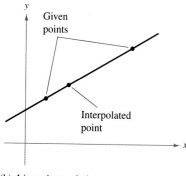

(b) Linear interpolation

FIGURE 1.38

EXAMPLE 6 **Estimating Sales Per Share**

The sales per share for Starbucks Corporation was \$15.71 in 2011 and \$17.75 in 2012. Using only this information, write a linear equation that gives the sales per share in terms of the year. Then estimate the sales per share in 2013. *(Source: Starbucks Corp.)*

SOLUTION Let $t = 11$ represent 2011. Then the two given values are represented by the data points

$$(11, 15.71) \quad \text{and} \quad (12, 17.75).$$

The slope of the line through these points is

$$m = \frac{17.75 - 15.71}{12 - 11} = 2.04.$$

Using the point-slope form, you can find the equation that relates the sales per share y and the year t to be

$$y = 2.04t - 6.73.$$

Using $t = 13$ to represent 2013, you can estimate that the 2013 sales per share was

$$y = 2.04(13) - 6.73 = 26.52 - 6.73 = 19.79.$$

According to this equation, the sales per share in 2013 was \$19.79, as shown in Figure 1.37. (In this case, the estimate is fairly good—the actual sales per share in 2013 was \$19.77.)

✓ *Checkpoint 6* *Worked-out solution available at LarsonAppliedCalculus.com*

The sales per share for Amazon.com was \$105.65 in 2011 and \$134.40 in 2012. Using only this information, write a linear equation that gives the sales per share in terms of the year. Then estimate the sales per share in 2013. *(Source: Amazon.com)* ∎

The estimation method illustrated in Example 6 is called **linear extrapolation.** Note in Figure 1.38(a) that an extrapolated point does not lie between the given points. When the estimated point lies between two given points, as shown in Figure 1.38(b), the procedure is called **linear interpolation.**

Because the slope of a vertical line is not defined, its equation cannot be written in slope-intercept form. However, every line has an equation that can be written in the **general form**

$$Ax + By + C = 0 \qquad \text{General form}$$

where A and B are not both zero. For instance, the vertical line $x = a$ can be represented by the general form

$$x - a = 0. \qquad \text{General form of vertical line}$$

The five most common forms of equations of lines are summarized below.

Equations of Lines

1. General form: $\qquad Ax + By + C = 0$

2. Vertical line: $\qquad x = a$

3. Horizontal line: $\qquad y = b$

4. Slope-intercept form: $\quad y = mx + b$

5. Point-slope form: $\qquad y - y_1 = m(x - x_1)$

Parallel and Perpendicular Lines

Slope can be used to decide whether two nonvertical lines in a plane are parallel, perpendicular, or neither.

> **Parallel and Perpendicular Lines**
>
> 1. Two distinct nonvertical lines are **parallel** if and only if their slopes are equal. That is, $m_1 = m_2$.
>
> 2. Two nonvertical lines are **perpendicular** if and only if their slopes are negative reciprocals of each other. That is, $m_1 = -1/m_2$.

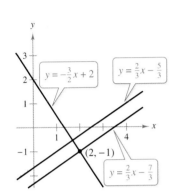

FIGURE 1.39

TECH TUTOR

On a graphing utility, lines will not appear to have the correct slopes unless you use a viewing window that has a *square setting*. For instance, try graphing the lines in Example 7 using the standard setting $-10 \le x \le 10$ and $-10 \le y \le 10$. Then reset the viewing window with the square setting $-9 \le x \le 9$ and $-6 \le y \le 6$. On which setting do the lines $y = \frac{2}{3}x - \frac{5}{3}$ and $y = -\frac{3}{2}x + 2$ appear to be perpendicular?

EXAMPLE 7   Finding Parallel and Perpendicular Lines

Find the slope-intercept form of the equations of the lines that pass through the point $(2, -1)$ and are (a) parallel to and (b) perpendicular to the line $2x - 3y = 5$.

SOLUTION   By writing the given equation in slope-intercept form

$2x - 3y = 5$	Write original equation.
$-3y = -2x + 5$	Subtract $2x$ from each side.
$y = \dfrac{2}{3}x - \dfrac{5}{3}$	Write in slope-intercept form.

you can see that it has a slope of $m = \frac{2}{3}$, as shown in Figure 1.39.

a. Any line parallel to the given line must also have a slope of $\frac{2}{3}$. So, the line through $(2, -1)$ that is parallel to the given line has the following equation.

$y - (-1) = \dfrac{2}{3}(x - 2)$	Write in point-slope form.
$y + 1 = \dfrac{2}{3}x - \dfrac{4}{3}$	Simplify.
$y = \dfrac{2}{3}x - \dfrac{4}{3} - 1$	Solve for y.
$y = \dfrac{2}{3}x - \dfrac{7}{3}$	Write in slope-intercept form.

b. Any line perpendicular to the given line must have a slope of $-\frac{3}{2}$ (because $-\frac{3}{2}$ is the negative reciprocal of $\frac{2}{3}$). So, the line through $(2, -1)$ that is perpendicular to the given line has the following equation.

$y - (-1) = -\dfrac{3}{2}(x - 2)$	Write in point-slope form.
$y + 1 = -\dfrac{3}{2}x + 3$	Simplify.
$y = -\dfrac{3}{2}x + 3 - 1$	Solve for y.
$y = -\dfrac{3}{2}x + 2$	Write in slope-intercept form.

✓ *Checkpoint 7*   *Worked-out solution available at LarsonAppliedCalculus.com*

Find the slope-intercept form of the equations of the lines that pass through the point $(2, 1)$ and are (a) parallel to and (b) perpendicular to the line $2x - 4y = 5$. ■

Extended Application: Linear Depreciation

Most business expenses can be deducted the same year they occur. One exception to this is the cost of property that has a useful life of more than 1 year, such as buildings, cars, or equipment. Such costs must be **depreciated** over the useful life of the property. When the *same amount* is depreciated each year, the procedure is called **linear depreciation** or **straight-line depreciation.** The *book value* is the difference between the original value and the total amount of depreciation accumulated to date.

EXAMPLE 8   Depreciating Equipment

Your company has purchased a $12,000 machine that has a useful life of 8 years. The salvage value at the end of 8 years is $2000. Write a linear equation that describes the book value of the machine each year.

SOLUTION   Let V represent the value of the machine at the end of year t. You can represent the initial value of the machine by the ordered pair $(0, 12,000)$ and the salvage value of the machine by the ordered pair $(8, 2000)$. The slope of the line is

$$m = \frac{2000 - 12,000}{8 - 0} = \frac{-\$1250}{1 \text{ year}} \qquad m = \frac{V_2 - V_1}{t_2 - t_1}$$

which represents the annual depreciation in *dollars per year.* Using the point-slope form, you can write the equation of the line as shown.

$$V - 12,000 = -1250(t - 0) \qquad \text{Write in point-slope form.}$$
$$V = -1250t + 12,000 \qquad \text{Write in slope-intercept form.}$$

The graph of this equation is shown in Figure 1.40.

✓ *Checkpoint 8*   Worked-out solution available at LarsonAppliedCalculus.com

Write a linear equation for the machine in Example 8 when the salvage value at the end of 8 years is $1000.   ■

Straight-Line Depreciation

$V = -1250t + 12,000$

FIGURE 1.40

SUMMARIZE   (Section 1.3)

1. State the slope-intercept form of the equation of a line *(page 22).* For an example of an equation in slope-intercept form, see Example 1.

2. Explain how to decide whether the slope of a line is a ratio or a rate of change *(page 24).* For an example of a slope that is a ratio, see Example 2. For an example of a slope that is a rate of change, see Example 3.

3. Explain how to find the slope of a line passing through two points *(page 25).* For an example of finding the slope of a line passing through two points, see Example 4.

4. State the point-slope form of the equation of a line *(page 27).* For examples of using the point-slope form, see Examples 5 and 6.

5. Explain how to decide whether two lines are parallel, perpendicular, or neither *(page 29).* For an example of finding equations of parallel and perpendicular lines, see Example 7.

6. Describe a real-life example of how a linear equation can be used to analyze the depreciation of property *(page 30, Example 8).*

SKILLS WARM UP 1.3

The following warm-up exercises involve skills that were covered in a previous course. You will use these skills in the exercise set for this section. For additional help, review Appendix A.3.

In Exercises 1 and 2, simplify the expression.

1. $\dfrac{5 - (-2)}{-3 - 4}$

2. $\dfrac{-4 - (-10)}{7 - 5}$

3. Evaluate $-\dfrac{1}{m}$ when $m = -3$.

4. Evaluate $-\dfrac{1}{m}$ when $m = \dfrac{6}{7}$.

In Exercises 5–10, solve for y in terms of x.

5. $-4x + y = 7$

6. $3x - y = 7$

7. $y - 2 = 3(x - 4)$

8. $y - (-5) = -1[x - (-2)]$

9. $y - (-3) = \dfrac{4 - (-2)}{11 - 3}(x - 12)$

10. $y - 1 = \dfrac{-3 - 1}{-7 - (-1)}[x - (-1)]$

Exercises 1.3

See *CalcChat.com* for tutorial help and worked-out solutions to odd-numbered exercises.

Estimating Slope In Exercises 1–4, estimate the slope of the line.

1.

2.

3.

4.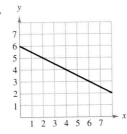

Finding the Slope and y-Intercept In Exercises 5–16, find the slope and y-intercept (if possible) of the equation of the line.

5. $y = x + 7$

6. $y = 4x + 3$

7. $5x + y = 20$

8. $2x + y = 40$

9. $7x + 6y = 30$

10. $8x + 3y = 12$

11. $3x - y = 15$

12. $2x - 3y = 24$

13. $x = 4$

14. $x + 5 = 0$

15. $y - 9 = 0$

16. $y + 1 = 0$

Graphing Linear Equations In Exercises 17–26, sketch the graph of the linear equation. Use a graphing utility to verify your result. *See Example 1.*

17. $y = -2$

18. $y = -4$

19. $y = -2x + 1$

20. $y = 3x - 2$

21. $3x + 2y = 4$

22. $4x + 5y = 20$

23. $2x - y - 3 = 0$

24. $x + 2y + 10 = 0$

25. $3x + 5y + 30 = 0$

26. $-5x + 2y - 20 = 0$

Finding Slopes of Lines In Exercises 27–40, find the slope of the line passing through the pair of points. *See Example 4.*

27. $(0, -2), (8, 0)$

28. $(-1, 0), (1, 5)$

29. $(3, -4), (5, 2)$

30. $(1, 2), (-2, 2)$

31. $(4, -1), (2, 7)$

32. $\left(\frac{11}{3}, -2\right), \left(\frac{11}{3}, -10\right)$

33. $(-8, -3), (-8, -5)$

34. $(2, -1), (-2, -5)$

35. $(-2, 6), (1, 6)$

36. $(3, -13), (-2, -3)$

37. $\left(\frac{1}{4}, -2\right), \left(-\frac{3}{8}, 1\right)$

38. $\left(-\frac{3}{2}, -5\right), \left(\frac{5}{6}, 4\right)$

39. $\left(\frac{2}{3}, \frac{5}{2}\right), \left(\frac{1}{4}, -\frac{5}{6}\right)$

40. $\left(\frac{7}{8}, \frac{3}{4}\right), \left(\frac{5}{4}, -\frac{1}{4}\right)$

Finding Points on a Line In Exercises 41–48, use the point on the line and the slope of the line to find three additional points through which the line passes. (There are many correct answers.)

	Point	Slope		Point	Slope
41.	$(2, 1)$	$m = 0$	**42.**	$(-5, -3)$	$m = 0$
43.	$(1, 7)$	$m = -3$	**44.**	$(7, -2)$	$m = 2$
45.	$(6, -4)$	$m = \frac{2}{3}$	**46.**	$(-1, -6)$	$m = -\frac{1}{2}$
47.	$(-8, 1)$	m is undefined.	**48.**	$(-3, 4)$	m is undefined.

Using Slope In Exercises 49–52, use the concept of slope to determine whether the three points are collinear.

49. $(-2, 1), (-1, 0), (2, -2)$

50. $(0, 4), (7, -6), (-5, 11)$

51. $(-2, -1), (0, 3), (2, 7)$

52. $(4, 1), (-2, -2), (8, 3)$

 Using the Point-Slope Form In Exercises 53–60, find an equation of the line that passes through the given point and has the given slope. Then sketch the line. *See Example 5.*

	Point	Slope		Point	Slope
53.	$(0, 3)$	$m = \frac{3}{4}$	**54.**	$(0, 0)$	$m = \frac{2}{3}$
55.	$(-2, 7)$	$m = 0$	**56.**	$(-2, 4)$	$m = 0$
57.	$(-1, -2)$	$m = -4$	**58.**	$(-1, -4)$	$m = -2$
59.	$\left(\frac{8}{3}, 0\right)$	$m = \frac{1}{4}$	**60.**	$\left(\frac{3}{2}, 0\right)$	$m = -\frac{1}{6}$

 Writing an Equation of a Line In Exercises 61–70, find an equation of the line that passes through the points. Then sketch the line.

61. $(4, 3), (0, -5)$ **62.** $(-2, -4), (1, 5)$

63. $(2, 3), (2, -2)$ **64.** $(6, 1), (10, 1)$

65. $(3, -1), (-2, -1)$ **66.** $(2, 5), (2, -10)$

67. $\left(-\frac{1}{2}, 4\right), \left(\frac{1}{2}, 8\right)$ **68.** $\left(-\frac{1}{4}, 1\right), \left(\frac{1}{4}, 5\right)$

69. $\left(-\frac{1}{3}, 1\right), \left(-\frac{2}{3}, \frac{5}{6}\right)$ **70.** $\left(\frac{7}{8}, \frac{3}{4}\right), \left(\frac{5}{4}, -\frac{1}{4}\right)$

Writing an Equation of a Line In Exercises 71–74, find an equation of the line with the given characteristics.

71. A vertical line through $(3, 0)$

72. A horizontal line through $(0, -5)$

73. A line with a *y*-intercept at -10 and parallel to all horizontal lines

74. A line with an *x*-intercept at -5 and parallel to all vertical lines

 Finding Parallel and Perpendicular Lines In Exercises 75–82, find equations of the lines that pass through the given point and are (a) parallel to the given line and (b) perpendicular to the given line. Then use a graphing utility to graph all three equations in the same viewing window using a square setting. *See Example 7.*

	Point	Line		Point	Line
75.	$(-3, 2)$	$x + y = 7$	**76.**	$(2, 1)$	$4x - 2y = 3$
77.	$\left(-\frac{2}{3}, \frac{7}{8}\right)$	$3x + 4y = 7$	**78.**	$\left(\frac{7}{8}, \frac{3}{4}\right)$	$5x + 3y = 0$
79.	$(-1, 0)$	$y + 3 = 0$	**80.**	$(2, 5)$	$y + 4 = 0$
81.	$(1, 1)$	$x - 2 = 0$	**82.**	$(12, -3)$	$x - 5 = 0$

83. Average Salary The graph shows the average salaries (in dollars) of postsecondary education administrators from 2008 through 2013. *(Source: U.S. Bureau of Labor Statistics)*

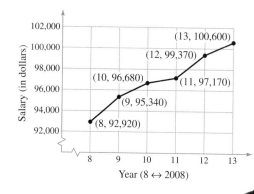

(a) Determine the time periods in which the average salary increased the greatest and the least.

(b) Find the slope of the line segment connecting the points for the years 2008 and 2013.

(c) Interpret the meaning of the slope in part (b) in the context of the problem.

84. Revenue The graph shows the revenue (in billions of dollars) for AT&T for the years 2009 through 2013. *(Source: AT&T Inc.)*

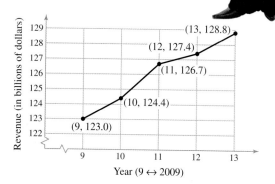

(a) Determine the years in which the revenue increased the greatest and the least.

(b) Find the slope of the line segment connecting the points for the years 2009 and 2013.

(c) Interpret the meaning of the slope in part (b) in the context of the problem.

85. Road Grade You are driving on a road that has a 6% uphill grade. This means that the slope of the road is $\frac{6}{100}$. Approximate the amount of vertical change in your position when you drive 200 feet.

86. Temperature Conversion Use the fact that water freezes at 0°C (32°F) and boils at 100°C (212°F).

(a) Write a linear equation that expresses the relationship between the temperature in degrees Celsius C and the temperature in degrees Fahrenheit F.

(b) A person has a temperature of 102.2°F. What is this temperature on the Celsius scale?

(c) The temperature in a room is 76°F. What is this temperature on the Celsius scale?

87. Population The resident population of Wisconsin (in thousands) was 5655 in 2009 and 5743 in 2013. Assume that the relationship between the population y and the year t is linear. Let $t = 0$ represent 2000. *(Source: U.S. Census Bureau)*

(a) Write a linear model for the data. What is the slope and what does it tell you about the population?

(b) Use the model to estimate the population in 2011.

(c) Use your school's library, the Internet, or some other reference source to find the actual population in 2011. How close was your estimate?

(d) Do you think your model could be used to predict the population in 2018? Explain.

88. Personal Income Personal income (in billions of dollars) in the United States was 12,430 in 2008 and 14,167 in 2013. Assume that the relationship between the personal income y and the time t (in years) is linear. Let $t = 0$ represent 2000. *(Source: U.S. Bureau of Economic Analysis)*

(a) Write a linear model for the data.

(b) Estimate the personal incomes in 2011 and 2014.

(c) Use your school's library, the Internet, or some other reference source to find the actual personal incomes in 2011 and 2014. How close were your estimates?

89. Linear Depreciation A small business purchases a piece of equipment for $1025. After 5 years, the equipment will be outdated, having no value.

(a) Write a linear equation giving the value y (in dollars) of the equipment in terms of the time t (in years), $0 \le t \le 5$.

(b) Use a graphing utility to graph the equation.

(c) Move the cursor along the graph and estimate (to two-decimal-place accuracy) the value of the equipment after 3 years.

(d) Move the cursor along the graph and estimate (to two-decimal-place accuracy) the time when the value of the equipment will be $600.

90. Linear Depreciation A hospital purchases a $500,000 magnetic resonance imaging (MRI) machine that has a useful life of 9 years. The salvage value at the end of 9 years is $77,000.

(a) Write a linear equation that describes the value y (in dollars) of the MRI machine in terms of the time t (in years), $0 \le t \le 9$.

(b) Find the value of the machine after 5 years.

(c) Find the time when the value of the equipment will be $160,000.

91. Choosing a Job As a salesperson, you receive a monthly salary of $2000, plus a commission of 7% of sales. You are offered a new job at $2300 per month, plus a commission of 5% of sales.

(a) Write linear equations for your monthly wage W (in dollars) in terms of your monthly sales S (in dollars) for your current job and for your job offer.

(b) Use a graphing utility to graph each equation and find the point of intersection. What does the point of intersection signify?

(c) You think you can sell $20,000 worth of a product per month. Should you change jobs? Explain.

92. HOW DO YOU SEE IT? Match the description of the situation with its graph. Then write the equation of the line. [The graphs are labeled (i), (ii), (iii), and (iv).]

(a) You are paying $10 per week to repay a $100 loan.

(b) An employee is paid $12.50 per hour plus $1.50 for each unit produced per hour.

(c) A sales representative receives $50 per day for food plus $0.58 for each mile traveled.

(d) A computer that was purchased for $600 depreciates $100 per year.

(i)

(ii)

(iii)

(iv)

QUIZ YOURSELF   See *CalcChat.com* for tutorial help and worked-out solutions to odd-numbered exercises.

Take this quiz as you would take a quiz in class. When you are done, check your work against the answers given in the back of the book.

In Exercises 1–3, (a) plot the points, (b) find the distance between the points, and (c) find the midpoint of the line segment joining the points.

1. $(3, -2), (-3, 1)$ **2.** $\left(\frac{1}{4}, -\frac{3}{2}\right), \left(\frac{1}{2}, 2\right)$ **3.** $(-12, 4), (6, -2)$

4. Use the Distance Formula to show that the points $(4, 0)$, $(2, 1)$, and $(-1, -5)$ are vertices of a right triangle.

5. The resident population of Georgia (in thousands) was 9810 in 2011 and 9992 in 2013. Use the Midpoint Formula to estimate the population in 2012. *(Source: U.S. Census Bureau)*

In Exercises 6–8, sketch the graph of the equation and label the intercepts.

6. $y = 5x + 2$ **7.** $y = x^2 + x - 6$ **8.** $y = |x - 3|$

In Exercises 9–11, find the standard form of the equation of the circle with the given characteristics and sketch its graph.

9. Center: $(0, 0)$; radius: 9

10. Center: $(-1, 0)$; radius: 6

11. Center: $(2, -2)$; solution point: $(-1, 2)$

12. A business manufactures a product at a cost of $4.55 per unit and sells the product for $7.19 per unit. The company's initial investment to produce the product was $12,500. How many units must the company sell to break even?

In Exercises 13–15, find an equation of the line that passes through the given point and has the given slope. Then sketch the line.

13. $(0, -3); m = 0$ **14.** $(1, 1); m = 2$ **15.** $(6, 5); m = -\frac{1}{3}$

In Exercises 16–18, find an equation of the line that passes through the points. Then sketch the line.

16. $(1, -1), (-4, 5)$ **17.** $(-2, 3), (-2, 2)$ **18.** $\left(\frac{5}{2}, 2\right), (0, 2)$

19. Find equations of the lines that pass through the point $(3, -5)$ and are

(a) parallel to the line $x + 4y = -2$.

(b) perpendicular to the line $x + 4y = -2$.

20. A company had sales of $1,330,000 in 2011 and $1,800,000 in 2015. The company's sales can be modeled by a linear equation. Predict the sales in 2019 and 2022.

21. A company reimburses its sales representatives $218 per day for lodging and meals, plus $0.56 per mile driven. Write a linear equation giving the daily cost C (in dollars) in terms of x, the number of miles driven.

22. Your annual salary was $35,700 in 2013 and $39,100 in 2015. Assume your salary can be modeled by a linear equation.

(a) Write a linear equation giving your salary S (in dollars) in terms of the time t (in years). Let $t = 13$ represent 2013.

(b) Use the linear model to predict your salary in 2020.

1.4 Functions

- Decide whether the relationship between two variables is a function.
- Find the domains and ranges of functions.
- Use function notation and evaluate functions.
- Combine functions to create other functions.
- Find inverse functions algebraically.

In Exercise 67 on page 46, you will use a function to estimate the amounts spent on prescription drugs in the United States.

Functions

In many common relationships between two variables, the value of one of the variables depends on the value of the other variable. Here are some examples.

1. The sales tax on an item depends on its selling price.

2. The distance an object moves in a given amount of time depends on its speed.

3. The area of a circle depends on its radius.

Consider the relationship between the total revenue R and the sale of x units of a product sold for $1.25 per unit. This relationship can be expressed by the equation

$$R = 1.25x.$$

In this equation, the value of R depends on the choice of x. Because of this, R is the **dependent variable** and x is the **independent variable.**

Most of the relationships that you will study in this course have the property that for a given value of the independent variable, there corresponds exactly one value of the dependent variable. Such a relationship is a **function.**

Definition of Function

A **function** is a relationship between two variables such that to each value of the independent variable there corresponds exactly one value of the dependent variable.

The **domain** of the function is the set of all values of the independent variable for which the function is defined. The **range** of the function is the set of all values taken on by the dependent variable.

As illustrated in the figure at the right, a function can be thought of as a machine that inputs values of the independent variable and outputs values of the dependent variable.

Although functions can be described by various means such as tables, graphs, and diagrams, they are most often specified by formulas or equations. For instance, the equation

$$y = 4x^2 + 3$$

describes y as a function of x. For this function, x is the independent variable and y is the dependent variable.

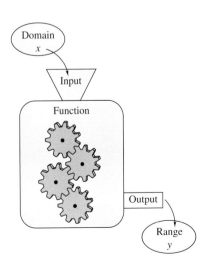

EXAMPLE 1   **Deciding Whether Equations Are Functions**

Decide whether each equation defines y as a function of x.

a. $x + y = 1$ **b.** $x^2 + y^2 = 1$

c. $x^2 + y = 1$ **d.** $x + y^2 = 1$

SOLUTION   To decide whether an equation defines a function, it is helpful to isolate the dependent variable on the left side. For instance, to decide whether the equation $x + y = 1$ defines y as a function of x, write the equation in the form

$$y = 1 - x.$$

From this form, you can see that for any value of x, there is exactly one value of y. So, y is a function of x.

Original Equation	Rewritten Equation	Test: Is y a function of x?
a. $x + y = 1$	$y = 1 - x$	Yes, each value of x determines exactly one value of y.
b. $x^2 + y^2 = 1$	$y = \pm\sqrt{1 - x^2}$	No, some values of x determine two values of y.
c. $x^2 + y = 1$	$y = 1 - x^2$	Yes, each value of x determines exactly one value of y.
d. $x + y^2 = 1$	$y = \pm\sqrt{1 - x}$	No, some values of x determine two values of y.

Note that the equations that assign two values ($\pm$) to the dependent variable for a given value of the independent variable do not define functions of x. For instance, in part (b), when $x = 0$, the equation $y = \pm\sqrt{1 - x^2}$ indicates that $y = +1$ or $y = -1$. Figure 1.41 shows the graphs of the four equations.

(a)

(b)

(c)

(d)

FIGURE 1.41

 Checkpoint 1   *Worked-out solution available at LarsonAppliedCalculus.com*

Decide whether each equation defines y as a function of x.

a. $x - y = 1$ **b.** $x^2 + y^2 = 4$ **c.** $y^2 + x = 2$ **d.** $x^2 - y = 0$ ▪

When the graph of a function is sketched, the standard convention is to let the horizontal axis represent the independent variable. When this convention is used, the test described in Example 1 has a nice graphical interpretation called the **Vertical Line Test.** This test states that if every vertical line intersects the graph of an equation at most once, then the equation defines y as a function of x. For instance, in Figure 1.41, the graphs in parts (a) and (c) pass the Vertical Line Test, but those in parts (b) and (d) do not.

The Domain and Range of a Function

The domain of a function may be described explicitly, or it may be *implied* by an equation used to define the function. For example, the function given by

$$y = \frac{1}{x^2 - 4}$$

has an implied domain that consists of all real numbers x except $x = \pm 2$. These two values are excluded from the domain because division by zero is undefined.

Another type of implied domain is that used to avoid even roots of negative numbers, as indicated in Example 2.

EXAMPLE 2   Finding the Domain and Range of a Function

Find the domain and range of each function.

a. $y = \sqrt{x - 1}$

b. $y = \begin{cases} 1 - x, & x < 1 \\ \sqrt{x - 1}, & x \geq 1 \end{cases}$

SOLUTION

a. Because $\sqrt{x - 1}$ is not defined for $x - 1 < 0$ (that is, for $x < 1$), it follows that the domain of the function is the interval

$$x \geq 1 \text{ or } [1, \infty).$$

To find the range, observe that $\sqrt{x - 1}$ is never negative. Moreover, as x takes on the various values in the domain, y takes on all nonnegative values. So, the range is the interval

$$y \geq 0 \text{ or } [0, \infty)$$

as shown in Figure 1.42(a).

b. Because this function is defined for $x < 1$ *and* for $x \geq 1$, the domain is the entire set of real numbers. This function is called a **piecewise-defined function** because it is defined by two or more equations over a specified domain. When $x \geq 1$, the function behaves as in part (a). For $x < 1$, the values of $1 - x$ are positive. So, the range of the function is

$$y \geq 0 \text{ or } [0, \infty)$$

as shown in Figure 1.42(b).

(a)

(b)

FIGURE 1.42

✓ *Checkpoint 2*   *Worked-out solution available at LarsonAppliedCalculus.com*

Find the domain and range of each function.

a. $y = \sqrt{x + 1}$ **b.** $y = \begin{cases} x^2, & x \leq 0 \\ \sqrt{x}, & x > 0 \end{cases}$ ■

A function is **one-to-one** when to each value of the dependent variable in the range there corresponds exactly one value of the independent variable in the domain. For instance, the function in Example 2(a) is one-to-one, whereas the function in Example 2(b) is not one-to-one.

Geometrically, a function is one-to-one when every horizontal line intersects the graph of the function at most once. This geometrical interpretation is the **Horizontal Line Test** for one-to-one functions. So, a graph that represents a one-to-one function must satisfy *both* the Vertical Line Test and the Horizontal Line Test.

Function Notation

When using an equation to define a function, you generally isolate the dependent variable on the left. For instance, writing the equation $x + 2y = 1$ as

$$y = \frac{1 - x}{2} \qquad \text{y is a function of x.}$$

indicates that y is the dependent variable and x is the independent variable. Using f as the name of the function, you can write this equation in **function notation** as

$$f(x) = \frac{1 - x}{2}. \qquad \text{Function notation}$$

The symbol $f(x)$ is read as "f of x," and it denotes the value of the dependent variable at the independent variable x. Be sure you understand that f is the *name* of the function, whereas $f(x)$ is the *value* of the function at x. For instance, the value of f when $x = 3$ is

$$f(3) = \frac{1 - 3}{2} = \frac{-2}{2} = -1.$$

The notation $f(3)$ is called a **function value,** and it lies in the range of f. This means that the point $(3, f(3))$ lies on the graph of f. One of the advantages of function notation is that it allows you to be less wordy. For instance, instead of asking "What is the value of y when $x = 3$?" you can ask "What is $f(3)$?"

EXAMPLE 3 Evaluating a Function

Find the values of the function

$$f(x) = 2x^2 - 4x + 1$$

when x is -1, 0, and 2. Is f one-to-one?

SOLUTION You can evaluate f at each value of x as shown.

$$x = -1: \quad f(-1) = 2(-1)^2 - 4(-1) + 1 = 2 + 4 + 1 = 7$$
$$x = 0: \quad f(0) = 2(0)^2 - 4(0) + 1 = 0 - 0 + 1 = 1$$
$$x = 2: \quad f(2) = 2(2)^2 - 4(2) + 1 = 8 - 8 + 1 = 1$$

Because two different values of x yield the same value of $f(x)$, the function is *not* one-to-one, as shown in Figure 1.43.

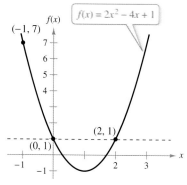

FIGURE 1.43

✓ *Checkpoint 3* Worked-out solution available at LarsonAppliedCalculus.com

Find the values of $f(x) = x^2 - 5x + 1$ when x is 0, 1, and 4. Is f one-to-one? ■

In an equation that defines a function of x, the role of the variable x is simply that of a placeholder. For instance, the function in Example 3, $f(x) = 2x^2 - 4x + 1$, could be written using rectangles instead of x.

$$f(\quad) = 2(\quad)^2 - 4(\quad) + 1$$

To evaluate $f(-2)$, replace each rectangle with -2.

$$f(-2) = 2(-2)^2 - 4(-2) + 1 = 8 + 8 + 1 = 17$$

Although f is often used as a convenient function name with x as the independent variable, you can use other symbols. For instance, these three equations all define the same function.

$$f(x) = x^2 - 4x + 7 \qquad \text{Function name is } f, \text{ independent variable is } x.$$
$$f(t) = t^2 - 4t + 7 \qquad \text{Function name is } f, \text{ independent variable is } t.$$
$$g(s) = s^2 - 4s + 7 \qquad \text{Function name is } g, \text{ independent variable is } s.$$

EXAMPLE 4   Evaluating a Function

Given $f(x) = x^2 + 7$, evaluate each expression.

a. $f(x + \Delta x)$ **b.** $\dfrac{f(x + \Delta x) - f(x)}{\Delta x}$

SOLUTION

a. To evaluate f at $x + \Delta x$, substitute $x + \Delta x$ for x in the original function.

$$f(x + \Delta x) = (x + \Delta x)^2 + 7$$
$$= x^2 + 2x\Delta x + (\Delta x)^2 + 7$$

b. Using the result of part (a), you can write

$$\frac{f(x + \Delta x) - f(x)}{\Delta x} = \frac{[(x + \Delta x)^2 + 7] - (x^2 + 7)}{\Delta x}$$
$$= \frac{x^2 + 2x\Delta x + (\Delta x)^2 + 7 - x^2 - 7}{\Delta x}$$
$$= \frac{2x\Delta x + (\Delta x)^2}{\Delta x}$$
$$= \frac{\Delta x(2x + \Delta x)}{\Delta x}$$
$$= 2x + \Delta x, \quad \Delta x \neq 0.$$

> **STUDY TIP**
>
> In Example 4(b), the expression
>
> $$\frac{f(x + \Delta x) - f(x)}{\Delta x}$$
>
> is called a *difference quotient* and has a special significance in calculus. You will learn more about this in Chapter 2.

✓ *Checkpoint 4*   *Worked-out solution available at LarsonAppliedCalculus.com*

Given $f(x) = x^2 + 3$, evaluate each expression.

a. $f(x + \Delta x)$ **b.** $\dfrac{f(x + \Delta x) - f(x)}{\Delta x}$ ■

In calculus, it is important to specify the domain of a function or expression clearly. For instance, in Example 4(b), the two expressions

$$\frac{f(x + \Delta x) - f(x)}{\Delta x} \quad \text{and} \quad 2x + \Delta x, \quad \Delta x \neq 0$$

are equivalent because $\Delta x = 0$ is excluded from the domain of each expression. Without a stated domain restriction, the two expressions would not be equivalent.

Combinations of Functions

Two functions can be combined in various ways to create new functions. For instance, given $f(x) = 2x - 3$ and $g(x) = x^2 + 1$, you can form the following functions.

$$f(x) + g(x) = (2x - 3) + (x^2 + 1) = x^2 + 2x - 2 \qquad \text{Sum}$$

$$f(x) - g(x) = (2x - 3) - (x^2 + 1) = -x^2 + 2x - 4 \qquad \text{Difference}$$

$$f(x)g(x) = (2x - 3)(x^2 + 1) = 2x^3 - 3x^2 + 2x - 3 \qquad \text{Product}$$

$$\frac{f(x)}{g(x)} = \frac{2x - 3}{x^2 + 1} \qquad \text{Quotient}$$

You can combine two functions in yet another way called a **composition.** The resulting function is called a **composite function.** For instance, given $f(x) = x^2$ and $g(x) = x + 1$, the composite of f with g is

$$f(g(x)) = f(x + 1) = (x + 1)^2.$$

This composition is denoted by $f \circ g$ and is read as "f composed with g."

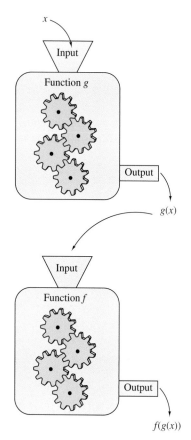

FIGURE 1.44

Definition of Composite Function

Let f and g be functions. The function given by

$$(f \circ g)(x) = f(g(x))$$

is the **composite** of f with g. The **domain** of $f \circ g$ is the set of all x in the domain of g such that $g(x)$ is in the domain of f, as indicated in Figure 1.44.

In general, the composite of f with g is not the same as the composite of g with f, as shown in the next example.

EXAMPLE 5   Forming Composite Functions

Given $f(x) = 2x - 3$ and $g(x) = x^2 + 1$, find each composite function.

a. $f(g(x))$ **b.** $g(f(x))$

SOLUTION

a. The composite of f with g is given by

$$\begin{aligned}
f(g(x)) &= 2(g(x)) - 3 & &\text{Evaluate } f \text{ at } g(x). \\
&= 2(x^2 + 1) - 3 & &\text{Substitute } x^2 + 1 \text{ for } g(x). \\
&= 2x^2 + 2 - 3 & &\text{Distributive Property} \\
&= 2x^2 - 1. & &\text{Simplify.}
\end{aligned}$$

b. The composite of g with f is given by

$$\begin{aligned}
g(f(x)) &= (f(x))^2 + 1 & &\text{Evaluate } g \text{ at } f(x). \\
&= (2x - 3)^2 + 1 & &\text{Substitute } 2x - 3 \text{ for } f(x). \\
&= 4x^2 - 12x + 9 + 1 & &\text{Expand.} \\
&= 4x^2 - 12x + 10. & &\text{Simplify.}
\end{aligned}$$

✔ **Checkpoint 5**   *Worked-out solution available at LarsonAppliedCalculus.com*

Given $f(x) = 2x + 1$ and $g(x) = x^2 + 2$, find each composite function.

a. $f(g(x))$ **b.** $g(f(x))$

Inverse Functions

Informally, the inverse function of f is another function g that "undoes" what f has done. For instance, subtraction can be used to undo addition, and division can be used to undo multiplication.

$$x \qquad\qquad f(x) \qquad\qquad g(f(x)) = x$$

Definition of Inverse Function

Let f and g be two functions such that

$$f(g(x)) = x \text{ for each } x \text{ in the domain of } g$$

and

$$g(f(x)) = x \text{ for each } x \text{ in the domain of } f.$$

Under these conditions, the function g is the **inverse function** of f. The function g is denoted by f^{-1}, which is read as "f-inverse." So,

$$f(f^{-1}(x)) = x \qquad \text{and} \qquad f^{-1}(f(x)) = x.$$

The domain of f must be equal to the range of f^{-1}, and the range of f must be equal to the domain of f^{-1}.

Be sure you understand the use of the superscript -1 to denote the inverse function f^{-1}. In this text, whenever f^{-1} is written, it *always* refers to the inverse function of f and *not* to the reciprocal of $f(x)$.

EXAMPLE 6   Finding Inverse Functions Informally

Find the inverse function of each function informally.

a. $f(x) = 2x$ **b.** $f(x) = x + 4$

SOLUTION

a. The function f *multiplies* each input by 2. To "undo" this function, you need to *divide* each input by 2. So, the inverse function of $f(x) = 2x$ is

$$f^{-1}(x) = \frac{x}{2}.$$

b. The function f *adds* 4 to each input. To "undo" this function, you need to *subtract* 4 from each input. So, the inverse function of $f(x) = x + 4$ is

$$f^{-1}(x) = x - 4.$$

Check that f and f^{-1} are inverse functions by showing that $f(f^{-1}(x)) = x$ and $f^{-1}(f(x)) = x$.

✓ **Checkpoint 6**   Worked-out solution available at LarsonAppliedCalculus.com

Find the inverse function of each function informally.

a. $f(x) = \frac{1}{5}x$ **b.** $f(x) = x - 6$ ■

The graphs of f and f^{-1} are mirror images of each other (with respect to the line $y = x$), as shown in Figure 1.45. Try using a graphing utility with a *square setting* to confirm this for each of the functions given in Example 6.

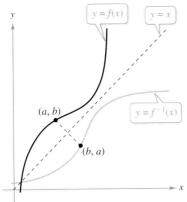

The graph of f^{-1} is a reflection of the graph of f in the line $y = x$.

FIGURE 1.45

The functions in Example 6 are simple enough so that their inverse functions can be found by inspection. The next example demonstrates a strategy for finding the inverse function of a more complicated function.

EXAMPLE 7   **Finding an Inverse Function**

Find the inverse function of

$$f(x) = \sqrt{2x - 3}.$$

SOLUTION   Begin by replacing $f(x)$ with y. Then, interchange x and y and solve for y.

$f(x) = \sqrt{2x - 3}$	Write original function.
$y = \sqrt{2x - 3}$	Replace $f(x)$ with y.
$x = \sqrt{2y - 3}$	Interchange x and y.
$x^2 = 2y - 3$	Square each side.
$x^2 + 3 = 2y$	Add 3 to each side.
$\dfrac{x^2 + 3}{2} = y$	Divide each side by 2.

So, the inverse function has the form

$$f^{-1}(\quad) = \frac{(\quad)^2 + 3}{2}.$$

Using x as the independent variable, you can write

$$f^{-1}(x) = \frac{x^2 + 3}{2}, \quad x \geq 0.$$

In the figure, note that the domain of f^{-1} is the range of f, which is $[0, \infty)$. Also, the range of f^{-1} is the domain of f, which is $\left[\frac{3}{2}, \infty\right)$.

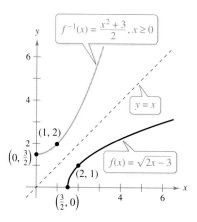

$f^{-1}(x) = \dfrac{x^2 + 3}{2}, x \geq 0$

$y = x$

$f(x) = \sqrt{2x - 3}$

$(1, 2)$ $(0, \frac{3}{2})$ $(2, 1)$ $(\frac{3}{2}, 0)$

✓ **Checkpoint 7**   *Worked-out solution available at LarsonAppliedCalculus.com*

Find the inverse function of $f(x) = x^2 + 2$ for $x \geq 0$. ■

After you have found an inverse function, you should check your results. You can check your results *graphically* by observing that the graphs of f and f^{-1} are reflections of each other in the line

$$y = x.$$

You can check your results *algebraically* by evaluating $f(f^{-1}(x))$ and $f^{-1}(f(x))$—both should be equal to x.

Check that $f(f^{-1}(x)) = x$

$$f(f^{-1}(x)) = f\left(\frac{x^2 + 3}{2}\right)$$
$$= \sqrt{2\left(\frac{x^2 + 3}{2}\right) - 3}$$
$$= \sqrt{x^2}$$
$$= x, \quad x \geq 0$$

Check that $f^{-1}(f(x)) = x$

$$f^{-1}(f(x)) = f^{-1}\left(\sqrt{2x - 3}\right)$$
$$= \frac{\left(\sqrt{2x - 3}\right)^2 + 3}{2}$$
$$= \frac{2x}{2}$$
$$= x, \quad x \geq \frac{3}{2}$$

TECH TUTOR

A graphing utility can help you check that the graphs of f and f^{-1} are reflections of each other in the line $y = x$. To do this, graph $y = f(x)$, $y = f^{-1}(x)$, and $y = x$ in the same viewing window, using a square setting.

Not every function has an inverse function. In fact, for a function to have an inverse function, it must be one-to-one.

EXAMPLE 8   **A Function That Has No Inverse Function**

Show that the function

$$f(x) = x^2 - 1$$

has no inverse function.

SOLUTION   Begin by sketching the graph of f, as shown in Figure 1.46. Note that

$$f(2) = (2)^2 - 1 = 3$$

and

$$f(-2) = (-2)^2 - 1 = 3.$$

So, f does not pass the Horizontal Line Test, which implies that it is not one-to-one and therefore has no inverse function. The same conclusion can be obtained by trying to find the inverse function of f algebraically.

$f(x) = x^2 - 1$	Write original function.
$y = x^2 - 1$	Replace $f(x)$ with y.
$x = y^2 - 1$	Interchange x and y.
$x + 1 = y^2$	Add 1 to each side.
$\pm\sqrt{x + 1} = y$	Take square root of each side.

The last equation does not define y as a function of x, and so f has no inverse function.

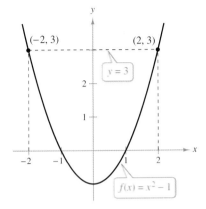

f is not one-to-one and has no inverse function.
FIGURE 1.46

✓ *Checkpoint 8*   *Worked-out solution available at LarsonAppliedCalculus.com*

Show that the function

$$f(x) = x^2 + 4$$

has no inverse function.   ■

SUMMARIZE   (Section 1.4)

1. State the definition of a function *(page 35)*. For an example of deciding whether the relationship between two variables is a function, see Example 1.

2. Explain the meanings of domain and range *(page 37)*. For an example of finding a domain and a range, see Example 2.

3. Explain the meaning of function notation *(page 38)*. For examples of using function notation and evaluating functions, see Examples 3 and 4.

4. State the definition of a composite function *(page 40)*. For an example of forming a composite function, see Example 5.

5. State the definition of an inverse function *(page 41)*. For examples of finding inverse functions, see Examples 6 and 7.

6. State when a function does not have an inverse function *(page 43)*. For an example of a function that does not have an inverse function, see Example 8.

SKILLS WARM UP 1.4

The following warm-up exercises involve skills that were covered in a previous course. You will use these skills in the exercise set for this section. For additional help, review Appendices A.3 and A.5.

In Exercises 1–6, simplify the expression.

1. $5(-1)^2 - 6(-1) + 9$

2. $(-2)^3 + 4(-2)^2 - 12$

3. $(x - 2)^2 + 5x - 10$

4. $(3 - x) + (x + 3)^3$

5. $\dfrac{1}{1 - (1 - x)}$

6. $3 + \dfrac{2x - 7}{x}$

In Exercises 7–12, solve for y in terms of x.

7. $2x + y - 6 = 11$

8. $5y - 6x^2 - 1 = 0$

9. $(y - 3)^2 = 5 + (x + 1)^2$

10. $y^2 - 4x^2 = 2$

11. $x = \dfrac{2y - 1}{4}$

12. $x = \sqrt[3]{2y - 1}$

Exercises 1.4

See *CalcChat.com* for tutorial help and worked-out solutions to odd-numbered exercises.

Deciding Whether Equations Are Functions In Exercises 1–8, decide whether the equation defines y as a function of x. *See Example 1.*

1. $x^2 + y^2 = 16$

2. $x + y^2 = 4$

3. $\frac{1}{2}x - 6y = -3$

4. $3x - 2y + 5 = 0$

5. $x^2 + y = 4$

6. $x^2 + y^2 + 2x = 0$

7. $y = |x + 2|$

8. $x^2y^2 - 3x^2 + 4y^2 = 0$

Vertical Line Test In Exercises 9–12, use the Vertical Line Test to determine whether y is a function of x.

9. $x^2 + y^2 = 9$

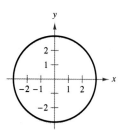

10. $x - xy + y + 1 = 0$

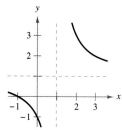

11. $x^2 = xy - 1$

12. $x = |y|$

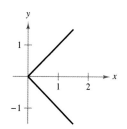

Finding the Domain and Range of a Function In Exercises 13–16, find the domain and range of the function. Use interval notation to write your result. *See Example 2.*

13. $f(x) = x^3$

14. $f(x) = \sqrt{2x - 3}$

15. $f(x) = \sqrt{4 - x^2}$

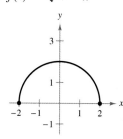

16. $f(x) = |x - 2|$

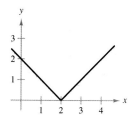

Finding the Domain and Range of a Function In Exercises 17–24, find the domain and range of the function. Use a graphing utility to verify your results. *See Example 2.*

17. $f(x) = -x^2 - 2x + 3$

18. $f(x) = 5x^3 + 6x^2 - 1$

19. $f(x) = \dfrac{|x|}{x}$

20. $f(x) = \dfrac{x}{\sqrt{x - 9}}$

21. $f(x) = \begin{cases} x - 5, & x \geq 0 \\ 3x - 5, & x < 0 \end{cases}$

22. $f(x) = \begin{cases} 3x + 2, & x < 0 \\ 2 - x, & x \geq 0 \end{cases}$

23. $f(x) = \dfrac{x - 2}{x + 4}$

24. $f(x) = \dfrac{x^2}{1 - x}$

Evaluating a Function In Exercises 25–28, evaluate the function at the specified values of the independent variable. Simplify the results. *See Example 3.*

25. $f(x) = 3x - 2$
    (a) $f(0)$ (b) $f(5)$ (c) $f(x - 1)$

26. $f(x) = x^2 - 4x + 1$
    (a) $f(-1)$ (b) $f\left(\frac{1}{2}\right)$ (c) $f(c + 2)$

27. $g(x) = 1/x$
    (a) $g\left(\frac{1}{5}\right)$ (b) $g(-0.6)$ (c) $g(x + 4)$

28. $f(x) = |x| + 4$
    (a) $f(-3)$ (b) $f(0.8)$ (c) $f(x + 2)$

 Evaluating a Function In Exercises 29–34, evaluate the difference quotient and simplify the result. *See Example 4.*

29. $f(x) = x^2 - 5x + 2$
$$\frac{f(x + \Delta x) - f(x)}{\Delta x}$$

30. $h(x) = x^2 + x + 3$
$$\frac{h(x + \Delta x) - h(x)}{\Delta x}$$

31. $g(x) = \sqrt{x + 1}$
$$\frac{g(4 + \Delta x) - g(4)}{\Delta x}$$

32. $f(x) = \frac{1}{\sqrt{x}}$
$$\frac{f(x) - f(2)}{x - 2}$$

33. $f(x) = \frac{1}{x - 2}$
$$\frac{f(x + \Delta x) - f(x)}{\Delta x}$$

34. $f(x) = \frac{1}{x + 4}$
$$\frac{f(x + \Delta x) - f(x)}{\Delta x}$$

 Combinations of Functions In Exercises 35–38, find (a) $f(x) + g(x)$, (b) $f(x) - g(x)$, (c) $f(x) \cdot g(x)$, (d) $f(x)/g(x)$, (e) $f(g(x))$, and (f) $g(f(x))$, if defined. *See Example 5.*

35. $f(x) = 2x - 5$
    $g(x) = 4 - 3x$

36. $f(x) = x^2 + 5$
    $g(x) = \sqrt{1 - x}$

37. $f(x) = x^2 + 1$
    $g(x) = x - 1$

38. $f(x) = \frac{x}{x + 1}$
    $g(x) = x^3$

39. **Composite Functions** Given $f(x) = \sqrt{x}$ and $g(x) = x^2 - 1$, evaluate or find the composite functions.
    (a) $f(g(1))$ (b) $g(f(1))$
    (c) $g\left(f\left(\frac{1}{2}\right)\right)$ (d) $f\left(g\left(-\sqrt{5}\right)\right)$
    (e) $f(g(x))$ (f) $g(f(x))$

40. **Composite Functions** Given $f(x) = 1/x$ and $g(x) = x^2 - 1$, evaluate or find the composite functions.
    (a) $f(g(2))$ (b) $g(f(2))$
    (c) $f(g(-3))$ (d) $g\left(f\left(1/\sqrt{2}\right)\right)$
    (e) $f(g(x))$ (f) $g(f(x))$

 Finding Inverse Functions Informally In Exercises 41–44, find the inverse function of f informally. Verify your results by showing that $f(f^{-1}(x)) = x$ and $f^{-1}(f(x)) = x$. *See Example 6.*

41. $f(x) = 4x$

42. $f(x) = \frac{1}{3}x$

43. $f(x) = x + 12$

44. $f(x) = x - 3$

Finding an Inverse Function In Exercises 45–56, find the inverse function of f. See *Example 7.*

45. $f(x) = 2x - 3$ 46. $f(x) = 5 - \frac{3}{4}x$

47. $f(x) = \frac{3}{2}x + 1$ 48. $f(x) = -6x - 4$

49. $f(x) = x^5$ 50. $f(x) = x^3$

51. $f(x) = \frac{1}{x}$ 52. $f(x) = -\frac{2}{x}$

53. $f(x) = \sqrt{9 - x^2}$, $0 \le x \le 3$

54. $f(x) = \sqrt{x^2 - 4}$, $x \ge 2$

55. $f(x) = x^{2/3}$, $x \ge 0$

56. $f(x) = x^{3/5}$

Determining Whether a Function Is One-to-One In Exercises 57–62, use a graphing utility to graph the function. Then use the Horizontal Line Test to determine whether the function is one-to-one. If it is, find its inverse function.

57. $f(x) = 3 - 7x$ 58. $f(x) = \sqrt{x - 2}$

59. $f(x) = x^2$ 60. $f(x) = x^4$

61. $f(x) = |x + 3|$ 62. $f(x) = -5$

63. **Graphing a Function** Use the graph of $f(x) = \sqrt{x}$ below to sketch the graph of each function.
    (a) $y = \sqrt{x} + 2$
    (b) $y = -\sqrt{x}$
    (c) $y = \sqrt{x - 2}$
    (d) $y = \sqrt{x + 3}$
    (e) $y = \sqrt{x - 4} - 1$
    (f) $y = 2\sqrt{x}$

64. **Graphing a Function** Use the graph of $f(x) = |x|$ below to sketch the graph of each function.
    (a) $y = |x| + 3$
    (b) $y = -|x|$
    (c) $y = |x - 1|$
    (d) $y = |x + 4|$
    (e) $y = |x + 2| - 1$
    (f) $y = \frac{1}{2}|x|$

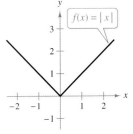

65. Writing a Function Use the graph of $f(x) = x^2$ to write an equation for each function whose graph is shown.

(a)

(b)

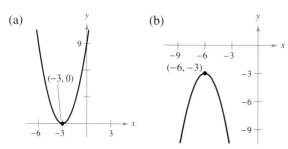

66. Writing a Function Use the graph of $f(x) = x^3$ to write an equation for each function whose graph is shown.

(a)

(b)

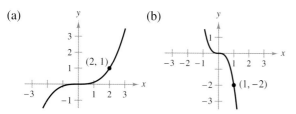

67. Prescription Drugs The amounts d (in billions of dollars) spent on prescription drugs in the United States from 2002 through 2013 (see figure) can be approximated by the model

$$d(t) = \begin{cases} 15.73t + 128.3, & 2 \le t \le 5 \\ -0.620t^2 + 18.11t + 138.8, & 6 \le t \le 13 \end{cases}$$

where t represents the year, with $t = 2$ corresponding to 2002. *(Source: U.S. Centers for Medicare & Medicaid Services)*

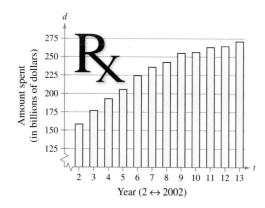

(a) Use the graph to estimate the amounts spent on prescription drugs in 2004, 2006, and 2013.

(b) Use the model to find the amounts spent on prescription drugs in 2004, 2006, and 2013. How well does the model fit the data? Explain your reasoning.

68. HOW DO YOU SEE IT? An electronically controlled thermostat in a home is programmed to lower the temperature automatically during the night. The temperature in the house T (in degrees Fahrenheit) is given in terms of t, the time in hours on a 24-hour clock (see figure).

(a) Explain why T is a function of t.

(b) Approximate $T(4)$ and $T(15)$.

(c) The thermostat is reprogrammed to produce a temperature H for which

$$H(t) = T(t - 1).$$

How does this change the temperature?

(d) The thermostat is reprogrammed to produce a temperature H for which

$$H(t) = T(t) - 1.$$

How does this change the temperature?

69. Owning a Business You own two restaurants. From 2009 through 2015, the sales R_1 (in thousands of dollars) for one restaurant can be modeled by

$$R_1 = 690 - 8t - 0.8t^2, \quad t = 9, 10, 11, 12, 13, 14, 15$$

where $t = 9$ represents 2009. During the same seven-year period, the sales R_2 (in thousands of dollars) for the second restaurant can be modeled by

$$R_2 = 458 + 0.78t, \quad t = 9, 10, 11, 12, 13, 14, 15.$$

Write a function that represents the total sales for the two restaurants. Use a graphing utility to graph the total sales function.

70. Births and Deaths From 2008 to 2012, the total numbers of births B (in thousands) and deaths D (in thousands) in the United States can be approximated by the models

$$B(t) = 4.917t^3 - 124.71t^2 + 925.9t + 2308$$

and

$$D(t) = -7.083t^3 + 222.64t^2 - 2281.8t + 10{,}104$$

where t represents the year, with $t = 8$ corresponding to 2008. Find $B(t) - D(t)$ and interpret this function. *(Source: U.S. National Center for Health Statistics)*

71. Cost The inventor of a new game believes that the variable cost for producing the game is $2.89 per unit. The fixed cost is $8000.

(a) Express the total cost C as a function of x, the number of games sold.

(b) Find a formula for the average cost per unit

$$\overline{C} = \frac{C}{x}.$$

(c) The selling price for each game is $6.89. How many units must be sold before the average cost per unit falls below the selling price?

72. Demand The demand function for a commodity is

$$p = \frac{14.75}{1 + 0.01x}, \quad x \geq 0$$

where p is the price per unit and x is the number of units sold.

(a) Find x as a function of p.

(b) Find the number of units sold when the price is $10.

73. Cost The weekly cost C of producing x units in a manufacturing process is given by $C(x) = 70x + 500$. The number of units x produced in t hours is given by $x(t) = 40t$.

(a) Find and interpret $C(x(t))$.

(b) Find the cost of 4 hours of production.

(c) After how much time does the cost of production reach $18,000?

74. Revenue For groups of 80 or more people, a charter bus company determines the rate r (in dollars per person) according to the formula

$$r = 15 - 0.05(n - 80), \quad n \geq 80$$

where n is the number of people.

(a) Express the revenue R for the bus company as a function of n.

(b) Complete the table.

n	100	125	150	175	200	225	250
R							

(c) Is the formula for the rate a good one for the company? Explain your reasoning.

75. Cost, Revenue, and Profit A company invests $98,000 for equipment to produce a new product. Each unit of the product costs $12.30 and is sold for $17.98. Let x be the number of units produced and sold.

(a) Write the total cost C as a function of x.

(b) Write the revenue R as a function of x.

(c) Write the profit P as a function of x.

76. Profit A manufacturer charges $90 per unit for units that cost $60 to produce. To encourage large orders from distributors, the manufacturer will reduce the price by $0.01 per unit for orders in excess of 100 units. (For example, an order of 101 units would have a price of $89.99 per unit, and an order of 102 units would have a price of $89.98 per unit.) This price reduction is discontinued when the price per unit drops to $75.

(a) Express the price per unit p as a function of the order size x.

(b) Express the profit P as a function of the order size x.

Restricting the Domain In Exercises 77–84, restrict the domain of the function f so that the function is one-to-one and has an inverse function. State the range of the restricted function. Then find the inverse function f^{-1}. State the domain and range of f^{-1}. (There are many correct answers.)

77. $f(x) = (x - 1)^2$ **78.** $f(x) = (x + 2)^2$

79. $f(x) = |x + 4|$ **80.** $f(x) = |x - 3|$

81. $f(x) = -2x^2 + 1$ **82.** $f(x) = \frac{1}{2}x^2 - 4$

83. $f(x) = |x + 1| - 2$ **84.** $f(x) = -|x - 2| + 3$

Business Capsule

CitiKitty, Inc. was founded in 2005 by 26-year-old Rebecca Rescate after she moved into a small apartment in New York City with no place to hide her cat's litter box. Finding no easy-to-use cat toilet training kit, she created one, and CitiKitty was born with an initial investment of $20,000. Today the company flourishes with an expanded product line. Revenues in 2014 reached $600,000.

85. Research Project Use your school's library, the Internet, or some other reference source to find information about the start-up costs of beginning a business, such as the example above. Write a short paper about the company.

1.5 Limits

■ Find limits of functions graphically and numerically.

■ Understand the definition of the limit of a function and use the properties of limits to evaluate limits of functions.

■ Use different analytic techniques to evaluate limits of functions.

■ Evaluate one-sided limits.

■ Recognize unbounded behavior of functions.

Removing Pollutants

$C = \dfrac{25p}{100 - p}$

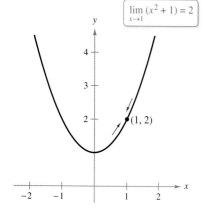

In Exercise 75 on page 59, you will use a limit to analyze the cost of removing pollutants from a small lake.

The Limit of a Function

In everyday language, people refer to a speed limit, a wrestler's weight limit, the limit of one's endurance, or stretching a spring to its limit. These phrases all suggest that a limit is a bound, which on some occasions may not be reached but on other occasions may be reached or exceeded.

Consider a spring that will break only when a weight of 10 pounds or more is attached. To determine how far the spring will stretch without breaking, you could attach increasingly heavier weights and measure the spring length s for each weight w, as shown in Figure 1.47. If the spring length approaches a value of L, then it is said that "the limit of s as w approaches 10 is L." A mathematical limit is much like the limit of a spring. The notation for a limit is

$$\lim_{x \to c} f(x) = L$$

which is read as "the limit of $f(x)$ as x approaches c is L."

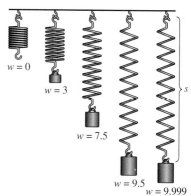

$w = 0$

$w = 3$

$w = 7.5$

$w = 9.5$ $w = 9.999$

What is the limit of s as w approaches 10 pounds?

FIGURE 1.47

EXAMPLE 1 Finding a Limit

Find the limit: $\lim_{x \to 1} (x^2 + 1)$.

SOLUTION Let $f(x) = x^2 + 1$. From the graph of f in Figure 1.48, it appears that $f(x)$ approaches 2 as x approaches 1 from either side, and you can write

$$\lim_{x \to 1} (x^2 + 1) = 2.$$

The table yields the same conclusion. Notice that as x gets closer and closer to 1, $f(x)$ gets closer and closer to 2.

$\lim_{x \to 1} (x^2 + 1) = 2$

(1, 2)

FIGURE 1.48

	x approaches 1.				*x* approaches 1.		
x	0.9	0.99	0.999	1	1.001	1.01	1.1
$f(x)$	1.810	1.980	1.998	2.000	2.002	2.020	2.210

f(x) approaches 2.   *f(x)* approaches 2.

✓ **Checkpoint 1**   *Worked-out solution available at LarsonAppliedCalculus.com*

Find the limit: $\lim_{x \to 1} (2x + 4)$.

EXAMPLE 2 Finding Limits Graphically and Numerically

Find the limit.

$$\lim_{x \to 1} f(x)$$

a. $f(x) = \dfrac{x^2 - 1}{x - 1}$ **b.** $f(x) = \dfrac{|x - 1|}{x - 1}$ **c.** $f(x) = \begin{cases} x, & x \neq 1 \\ 0, & x = 1 \end{cases}$

SOLUTION

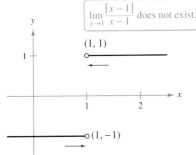

(a)

a. From the graph of f in Figure 1.49(a), it appears that $f(x)$ approaches 2 as x approaches 1 from either side. A missing point is denoted by the open dot on the graph. This conclusion is reinforced by the table. Be sure you see that *it does not matter that $f(x)$ is undefined when $x = 1$. The limit depends only on values of $f(x)$ near 1, not at 1.*

	x approaches 1.				x approaches 1.		
x	0.9	0.99	0.999	1	1.001	1.01	1.1
$f(x)$	1.900	1.990	1.999	?	2.001	2.010	2.100

$f(x)$ approaches 2. $f(x)$ approaches 2.

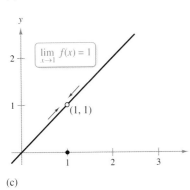

(b)

b. From the graph of f in Figure 1.49(b), you can see that $f(x) = -1$ for all values to the left of $x = 1$ and $f(x) = 1$ for all values to the right of $x = 1$. So, $f(x)$ is approaching a different value from the left of $x = 1$ than it is from the right of $x = 1$. In such situations, you can conclude that *the limit does not exist.* This conclusion is reinforced by the table.

	x approaches 1.				x approaches 1.		
x	0.9	0.99	0.999	1	1.001	1.01	1.1
$f(x)$	-1.000	-1.000	-1.000	?	1.000	1.000	1.000

$f(x)$ approaches -1. $f(x)$ approaches 1.

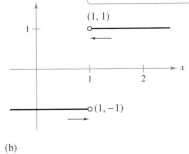

(c)

FIGURE 1.49

c. From the graph of f in Figure 1.49(c), it appears that $f(x)$ approaches 1 as x approaches 1 from either side. This conclusion is reinforced by the table. It does not matter that $f(1) = 0$. The limit depends only on values of $f(x)$ near 1, not at 1.

	x approaches 1.				x approaches 1.		
x	0.9	0.99	0.999	1	1.001	1.01	1.1
$f(x)$	0.900	0.990	0.999	?	1.001	1.010	1.100

$f(x)$ approaches 1. $f(x)$ approaches 1.

✓ *Checkpoint 2*   *Worked-out solution available at LarsonAppliedCalculus.com*

Find the limit.

$$\lim_{x \to 2} f(x)$$

a. $f(x) = \dfrac{x^2 - 4}{x - 2}$ **b.** $f(x) = \dfrac{|x - 2|}{x - 2}$ **c.** $f(x) = \begin{cases} x^2, & x \neq 2 \\ 0, & x = 2 \end{cases}$

Definition of the Limit of a Function and Properties of Limits

There are three important ideas to learn from Examples 1 and 2.

1. Saying that the limit of $f(x)$ approaches L as x approaches c means that $f(x)$ may be made *arbitrarily close* to the number L by choosing x closer and closer to c.

2. If $f(x)$ approaches the same number from *either side* of c, then the limit of $f(x)$ as x approaches c exists. However, if $f(x)$ approaches a different number from the right side of c than it does from the left side, then the limit of $f(x)$ as x approaches c *does not exist*. [See Example 2(b).]

3. The value of $f(x)$ when $x = c$ has no bearing on the existence or nonexistence of the limit of $f(x)$ as x approaches c. For instance, in Example 2(a), the limit of $f(x)$ exists as x approaches 1 even though the function f is not defined at $x = 1$.

Definition of the Limit of a Function

If $f(x)$ becomes arbitrarily close to a single number L as x approaches c from either side, then

$$\lim_{x \to c} f(x) = L$$

which is read as "the **limit** of $f(x)$ as x approaches c is L."

Many times the limit of $f(x)$ as x approaches c is simply $f(c)$, as shown in Example 1. Whenever the limit of $f(x)$ as x approaches c is

$$\lim_{x \to c} f(x) = f(c) \qquad \text{Substitute } c \text{ for } x.$$

the limit can be evaluated by **direct substitution.** (In the next section, you will learn that a function that has this property is *continuous at c*.) It is important that you learn to recognize the types of functions that have this property. Some basic ones are given in the following list.

Some Basic Limits

Let b and c be real numbers, and let n be a positive integer.

1. $\displaystyle\lim_{x \to c} b = b$ 2. $\displaystyle\lim_{x \to c} x = c$ 3. $\displaystyle\lim_{x \to c} x^n = c^n$ 4. $\displaystyle\lim_{x \to c} \sqrt[n]{x} = \sqrt[n]{c}$

In Property 4, if n is even, then c must be positive.

EXAMPLE 3   Evaluating Basic Limits

a. $\displaystyle\lim_{x \to 2} 3 = 3$

b. $\displaystyle\lim_{x \to -4} x = -4$

c. $\displaystyle\lim_{x \to 2} x^3 = 2^3 = 8$

d. $\displaystyle\lim_{x \to 9} \sqrt{x} = \sqrt{9} = 3$

✔ **Checkpoint 3**   Worked-out solution available at LarsonAppliedCalculus.com

Find the limit.

a. $\displaystyle\lim_{x \to 1} 5$ **b.** $\displaystyle\lim_{x \to 6} x$ **c.** $\displaystyle\lim_{x \to 5} x^2$ **d.** $\displaystyle\lim_{x \to -8} \sqrt[3]{x}$

By combining the basic limits from the preceding page with the properties of limits shown below, you can find limits for a wide variety of algebraic functions.

Properties of Limits

Let b and c be real numbers, let n be a positive integer, and let f and g be functions with the following limits.

$$\lim_{x \to c} f(x) = L \quad \text{and} \quad \lim_{x \to c} g(x) = K$$

1. **Scalar multiple:** $\displaystyle\lim_{x \to c} [bf(x)] = bL$

2. **Sum or difference:** $\displaystyle\lim_{x \to c} [f(x) \pm g(x)] = L \pm K$

3. **Product:** $\displaystyle\lim_{x \to c} [f(x) \cdot g(x)] = LK$

4. **Quotient:** $\displaystyle\lim_{x \to c} \frac{f(x)}{g(x)} = \frac{L}{K}, \quad K \neq 0$

5. **Power:** $\displaystyle\lim_{x \to c} [f(x)]^n = L^n$

6. **Radical:** $\displaystyle\lim_{x \to c} \sqrt[n]{f(x)} = \sqrt[n]{L}$

In Property 6, if n is even, then L must be positive.

TECH TUTOR

Symbolic computer algebra systems are capable of evaluating limits. Try using a computer algebra system to evaluate the limit given in Example 4.

EXAMPLE 4   **Finding the Limit of a Polynomial Function**

Find the limit: $\displaystyle\lim_{x \to 2} (x^2 + 2x - 3)$.

SOLUTION

$$
\begin{aligned}
\lim_{x \to 2} (x^2 + 2x - 3) &= \lim_{x \to 2} x^2 + \lim_{x \to 2} 2x - \lim_{x \to 2} 3 && \text{Apply Property 2.}\\
&= \lim_{x \to 2} x^2 + 2 \lim_{x \to 2} x - \lim_{x \to 2} 3 && \text{Apply Property 1.}\\
&= 2^2 + 2(2) - 3 && \text{Use direct substitution.}\\
&= 5
\end{aligned}
$$

✓ **Checkpoint 4**   *Worked-out solution available at LarsonAppliedCalculus.com*

Find the limit: $\displaystyle\lim_{x \to 1} (2x^2 - x + 4)$.

In Example 4, note that the limit (as $x \to 2$) of the *polynomial function*

$$p(x) = x^2 + 2x - 3$$

is simply the value of p at $x = 2$.

$$\lim_{x \to 2} p(x) = p(2) = 2^2 + 2(2) - 3 = 4 + 4 - 3 = 5$$

This is an illustration of the following important result, which states that the limit of a polynomial function can be evaluated by direct substitution.

The Limit of a Polynomial Function

If p is a polynomial function and c is any real number, then

$$\lim_{x \to c} p(x) = p(c).$$

Techniques for Evaluating Limits

Many techniques for evaluating limits are based on the following property. Let $f(x) = g(x)$ for all $x \neq c$, where c is a real number. If the limit of $g(x)$ exists as x approaches c, then the limit of $f(x)$ also exists and

$$\lim_{x \to c} f(x) = \lim_{x \to c} g(x).$$

To apply this property, recall from algebra that for a polynomial function p, $p(c) = 0$ if and only if $(x - c)$ is a factor of $p(x)$. This process is demonstrated in the next example.

EXAMPLE 5   **Finding the Limit of a Function**

Find the limit: $\displaystyle \lim_{x \to 1} \frac{x^3 - 1}{x - 1}$.

SOLUTION Note that the numerator and denominator are zero when $x = 1$. This implies that $x - 1$ is a factor of both, and you can divide out this common factor.

$$\frac{x^3 - 1}{x - 1} = \frac{(x - 1)(x^2 + x + 1)}{x - 1} \qquad \text{Factor numerator.}$$

$$= \frac{(x - 1)(x^2 + x + 1)}{x - 1} \qquad \text{Divide out common factor.}$$

$$= x^2 + x + 1, \quad x \neq 1 \qquad \text{Simplify.}$$

So, the rational function $f(x) = (x^3 - 1)/(x - 1)$ and the polynomial function $g(x) = x^2 + x + 1$ agree for all values of x other than $x = 1$, and you can find the limit as shown.

$$\lim_{x \to 1} \frac{x^3 - 1}{x - 1} = \lim_{x \to 1} (x^2 + x + 1) = 1^2 + 1 + 1 = 3$$

This result is illustrated in the two graphs below. Note that the two graphs are identical except that the graph of g contains the point $(1, 3)$, whereas this point is missing on the graph of f. (In the graph of f, the missing point is denoted by an open dot.)

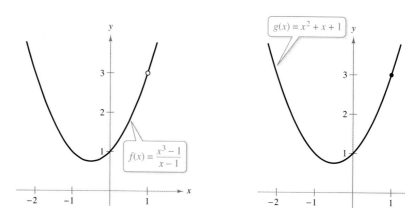

✔ **Checkpoint 5**   *Worked-out solution available at LarsonAppliedCalculus.com*

Find the limit: $\displaystyle \lim_{x \to 2} \frac{x^3 - 8}{x - 2}$.   ■

The technique used to evaluate the limit in Example 5 is called the **dividing out technique**. This technique is further demonstrated in the next example.

EXAMPLE 6 Using the Dividing Out Technique

For $\displaystyle\lim_{x \to -3} \frac{x^2 + x - 6}{x + 3}$, direct substitution fails because both the numerator and the denominator are zero when $x = -3$.

$$\lim_{x \to -3} \frac{x^2 + x - 6}{x + 3} \qquad \begin{array}{l}\longleftarrow \lim_{x \to -3}(x^2 + x - 6) = 0 \\ \longleftarrow \lim_{x \to -3}(x + 3) = 0\end{array}$$

Because the limits of both the numerator and the denominator are zero when $x = -3$, they must have a *common factor* of $x + 3$. So, for all $x \neq -3$, you can divide out this factor and find the limit as shown.

$$\lim_{x \to -3} \frac{x^2 + x - 6}{x + 3} = \lim_{x \to -3} \frac{(x - 2)(x + 3)}{x + 3} \qquad \text{Factor numerator.}$$

$$= \lim_{x \to -3} \frac{(x - 2)\cancel{(x + 3)}}{\cancel{x + 3}} \qquad \text{Divide out common factor.}$$

$$= \lim_{x \to -3} (x - 2) \qquad \text{Simplify.}$$

$$= -3 - 2 \qquad \text{Direct substitution}$$

$$= -5 \qquad \text{Simplify.}$$

This result is shown graphically in Figure 1.50. Note that the graph of f coincides with the graph of $g(x) = x - 2$, except that the graph of f has a hole at $(-3, -5)$.

✓ **Checkpoint 6**   *Worked-out solution available at LarsonAppliedCalculus.com*

Find the limit: $\displaystyle\lim_{x \to 3} \frac{x^2 + x - 12}{x - 3}$.

f is undefined when $x = -3$.

FIGURE 1.50

STUDY TIP

When you try to evaluate a limit and both the numerator and the denominator are zero, remember that you must rewrite the fraction so that the new denominator does not have 0 as its limit. One way to do this is to divide out common factors, as shown in Example 6. Another technique is to rationalize the numerator, as shown in Example 7.

EXAMPLE 7 Finding the Limit of a Function

Find the limit: $\displaystyle\lim_{x \to 0} \frac{\sqrt{x + 1} - 1}{x}$.

SOLUTION   Direct substitution fails because both the numerator and the denominator are zero when $x = 0$. Rewrite the fraction by rationalizing the numerator.

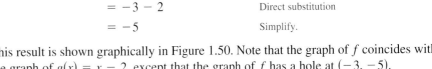

$$\frac{\sqrt{x + 1} - 1}{x} = \left(\frac{\sqrt{x + 1} - 1}{x}\right)\left(\frac{\sqrt{x + 1} + 1}{\sqrt{x + 1} + 1}\right)$$

$$= \frac{(x + 1) - 1}{x(\sqrt{x + 1} + 1)}$$

$$= \frac{\cancel{x}}{\cancel{x}(\sqrt{x + 1} + 1)}$$

$$= \frac{1}{\sqrt{x + 1} + 1}, \quad x \neq 0$$

Now evaluate the limit as shown.

$$\lim_{x \to 0} \frac{\sqrt{x + 1} - 1}{x} = \lim_{x \to 0} \frac{1}{\sqrt{x + 1} + 1} = \frac{1}{\sqrt{0 + 1} + 1} = \frac{1}{1 + 1} = \frac{1}{2}$$

✓ **Checkpoint 7**   *Worked-out solution available at LarsonAppliedCalculus.com*

Find the limit: $\displaystyle\lim_{x \to 0} \frac{\sqrt{x + 4} - 2}{x}$.

One-Sided Limits

In Example 2(b), you saw that one way in which a limit can fail to exist is when a function approaches a different value from the left of c than it approaches from the right of c. This type of behavior can be described more concisely with the concept of a **one-sided limit.**

$$\lim_{x \to c^-} f(x) = L \qquad \text{Limit from the left}$$

$$\lim_{x \to c^+} f(x) = L \qquad \text{Limit from the right}$$

The first of these two limits is read as "the limit of $f(x)$ as x approaches c from the left is L." The second is read as "the limit of $f(x)$ as x approaches c from the right is L."

EXAMPLE 8   Finding One-Sided Limits

Find the limit as $x \to 0$ from the left and the limit as $x \to 0$ from the right for the function

$$f(x) = \frac{|2x|}{x}.$$

SOLUTION   From the graph of f shown at the right, you can see that

$$f(x) = -2$$

for all $x < 0$. So, the limit from the left is

$$\lim_{x \to 0^-} \frac{|2x|}{x} = -2. \qquad \text{Limit from the left}$$

Because

$$f(x) = 2$$

for all $x > 0$, the limit from the right is

$$\lim_{x \to 0^+} \frac{|2x|}{x} = 2. \qquad \text{Limit from the right}$$

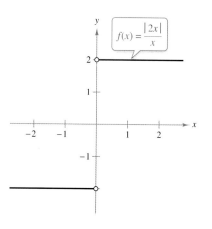

✓ **Checkpoint 8**   *Worked-out solution available at LarsonAppliedCalculus.com*

Find each limit.

a. $\displaystyle \lim_{x \to 2^-} \frac{|x - 2|}{x - 2}$

b. $\displaystyle \lim_{x \to 2^+} \frac{|x - 2|}{x - 2}$ ■

In Example 8, note that the function approaches different limits from the left and from the right. In such cases, the limit of $f(x)$ as $x \to c$ does not exist. For the limit of a function to exist as $x \to c$, *both* one-sided limits must exist and must be equal.

Existence of a Limit

If f is a function and c and L are real numbers, then

$$\lim_{x \to c} f(x) = L$$

if and only if both the left and right limits are equal to L.

TECH TUTOR

On most graphing utilities, the absolute value function is denoted by *abs*. You can verify the result in Example 8 by graphing

$$y = \frac{\text{abs}(2x)}{x}$$

in the viewing window $-3 \le x \le 3$ and $-3 \le y \le 3$.

EXAMPLE 9   Finding One-Sided Limits

Find the limit of $f(x)$ as x approaches 1.

$$f(x) = \begin{cases} 4 - x, & x < 1 \\ 4x - x^2, & x > 1 \end{cases}$$

SOLUTION Remember that you are concerned about the value of f near $x = 1$ rather than at $x = 1$. So, for $x < 1, f(x)$ is given by

$$4 - x$$

and you can use direct substitution to obtain

$$\lim_{x \to 1^-} f(x) = \lim_{x \to 1^-} (4 - x) = 4 - 1 = 3.$$

For $x > 1$, $f(x)$ is given by

$$4x - x^2$$

and you can use direct substitution to obtain

$$\lim_{x \to 1^+} f(x) = \lim_{x \to 1^+} (4x - x^2) = 4(1) - 1^2 = 4 - 1 = 3.$$

Because both one-sided limits exist and are equal to 3, it follows that

$$\lim_{x \to 1} f(x) = 3.$$

The graph in Figure 1.51 confirms this conclusion.

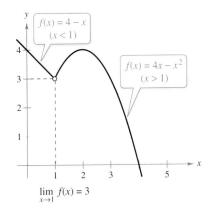

$\lim\limits_{x \to 1} f(x) = 3$

FIGURE 1.51

✓ **Checkpoint 9**   Worked-out solution available at LarsonAppliedCalculus.com

Find the limit of $f(x)$ as x approaches 0.

$$f(x) = \begin{cases} x^2 + 1, & x < 0 \\ 2x + 1, & x > 0 \end{cases}$$

EXAMPLE 10   Comparing One-Sided Limits

The daytime (9:00 A.M. to 5:00 P.M.) fees for a university's visitor parking lot are given by f, where $f(x)$ is the fee (in dollars) and x is the time (in hours). Show that the limit of $f(x)$ as x approaches 2 does not exist.

$$f(x) = \begin{cases} 4, & 0 < x \le 2 \\ 6, & 2 < x \le 3 \\ 7, & 3 < x \le 4 \\ 8, & 4 < x \le 5 \\ 9, & 5 < x \le 6 \\ 10, & 6 < x \le 8 \end{cases}$$

SOLUTION The graph of f is shown in Figure 1.52. The limit of $f(x)$ as x approaches 2 from the left is

$$\lim_{x \to 2^-} f(x) = 4 \qquad \text{Limit from the left}$$

whereas the limit of $f(x)$ as x approaches 2 from the right is

$$\lim_{x \to 2^+} f(x) = 6. \qquad \text{Limit from the right}$$

Because the one-sided limits are not equal, the limit of $f(x)$ as x approaches 2 does not exist.

Visitor Parking

FIGURE 1.52

✓ **Checkpoint 10**   Worked-out solution available at LarsonAppliedCalculus.com

Show that the limit of $f(x)$ as x approaches 3 does not exist in Example 10.

Unbounded Behavior

Example 10 shows a limit that fails to exist because the limits from the left and right differ. Another important way in which a limit can fail to exist is when $f(x)$ increases or decreases without bound as x approaches c.

EXAMPLE 11   **An Unbounded Function**

Find the limit (if possible): $\lim\limits_{x \to 2} \dfrac{3}{x - 2}$.

> **STUDY TIP**
>
> The equal sign in the statement $\lim\limits_{x \to c^+} f(x) = \infty$ does not mean that the limit exists. On the contrary, it tells you how the limit *fails to exist* by denoting the unbounded behavior of $f(x)$ as x approaches c.

SOLUTION   From the figure, you can see that $f(x)$ decreases without bound as x approaches 2 from the left and $f(x)$ increases without bound as x approaches 2 from the right. Symbolically, you can write this as

$$\lim_{x \to 2^-} \frac{3}{x - 2} = -\infty$$

and

$$\lim_{x \to 2^+} \frac{3}{x - 2} = \infty.$$

Because f is unbounded as x approaches 2, the limit does not exist. (The symbols ∞ and $-\infty$ refer to positive infinity and negative infinity, respectively. These symbols do not represent real numbers. They are convenient symbols used to describe unbounded conditions more concisely.)

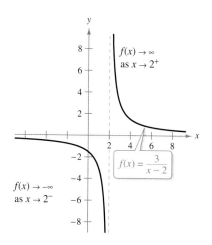

$f(x) \to \infty$ as $x \to 2^+$

$f(x) = \dfrac{3}{x - 2}$

$f(x) \to -\infty$ as $x \to 2^-$

✓ **Checkpoint 11**   *Worked-out solution available at LarsonAppliedCalculus.com*

Find the limit (if possible): $\lim\limits_{x \to -2} \dfrac{5}{x + 2}$.

SUMMARIZE   (Section 1.5)

1. State the definition of the limit of a function *(page 50)*. For examples of finding limits graphically and numerically, see Examples 1 and 2.

2. Make a list of the basic limits *(page 50)*. For examples of evaluating basic limits, see Example 3.

3. Make a list of the properties of limits *(page 51)*. For an example of using these properties to find a limit, see Example 4.

4. State the limit of a polynomial function *(page 51)*. For an example of finding the limit of a polynomial function, see the text after Example 4.

5. Describe the dividing out technique *(page 52)*. For examples of the dividing out technique, see Examples 5 and 6.

6. Describe a one-sided limit *(page 54)*. For examples of one-sided limits, see Examples 8, 9, and 10.

7. Describe the limit $\lim\limits_{x \to c} f(x)$ when $f(x)$ increases without bound as x approaches c *(page 56)*. For an example of an unbounded function, see Example 11.

SKILLS WARM UP 1.5

The following warm-up exercises involve skills that were covered in a previous course or in earlier sections. You will use these skills in the exercise set for this section. For additional help, review Appendix A.3 and Section 1.4.

In Exercises 1–4, simplify the expression by factoring.

1. $\dfrac{2x^3 + x^2}{6x}$

2. $\dfrac{x^5 + 9x^4}{x^2}$

3. $\dfrac{x^2 - 3x - 28}{x - 7}$

4. $\dfrac{x^2 + 11x + 30}{x + 5}$

In Exercises 5–8, evaluate the expression and simplify.

5. $f(x) = x^2 - 3x + 3$

    (a) $f(-1)$ (b) $f(c)$ (c) $f(x + h)$

6. $f(x) = \begin{cases} 2x - 2, & x < 1 \\ 3x + 1, & x \geq 1 \end{cases}$

    (a) $f\left(-\frac{1}{2}\right)$ (b) $f(1)$ (c) $f(t^2 + 1)$

7. $f(x) = x^2 - 2x + 2$ $\dfrac{f(1 + h) - f(1)}{h}$

8. $f(x) = 4x$ $\dfrac{f(2 + h) - f(2)}{h}$

In Exercises 9–12, find the domain and range of the function and sketch its graph.

9. $h(x) = -\dfrac{5}{x}$

10. $g(x) = \sqrt{25 - x^2}$

11. $f(x) = |x - 3|$

12. $f(x) = \dfrac{2|x|}{x}$

In Exercises 13 and 14, determine whether y is a function of x.

13. $9x^2 + 4y^2 = 49$

14. $2x^2y + 8x = 7y$

Exercises 1.5

See CalcChat.com for tutorial help and worked-out solutions to odd-numbered exercises.

Finding Limits Graphically In Exercises 1–4, use the graph to find the limit. *See Examples 1 and 2.*

1.

    (a) $\lim\limits_{x \to 2} f(x)$

    (b) $\lim\limits_{x \to -1} f(x)$

2.

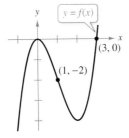

    (a) $\lim\limits_{x \to 1} f(x)$

    (b) $\lim\limits_{x \to 3} f(x)$

3.

    (a) $\lim\limits_{x \to 0} g(x)$

    (b) $\lim\limits_{x \to -1} g(x)$

4.

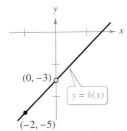

    (a) $\lim\limits_{x \to -2} h(x)$

    (b) $\lim\limits_{x \to 0} h(x)$

Finding Limits Numerically In Exercises 5–12, complete the table and use the result to estimate the limit. Use a graphing utility to graph the function to confirm your result. *See Examples 1 and 2.*

5. $\lim\limits_{x \to 6} \dfrac{2x + 3}{5}$

x	5.9	5.99	5.999	6	6.001	6.01	6.1
$f(x)$				?			

6. $\lim\limits_{x \to 1} (x^2 - 4x - 1)$

x	0.9	0.99	0.999	1	1.001	1.01	1.1
$f(x)$				?			

7. $\lim\limits_{x \to 4} \dfrac{x - 4}{x^2 - 5x + 4}$

x	3.9	3.99	3.999	4	4.001	4.01	4.1
$f(x)$				?			

8. $\lim\limits_{x \to 2} \dfrac{x - 2}{x^2 - 4}$

x	1.9	1.99	1.999	2	2.001	2.01	2.1
$f(x)$				?			

9. $\lim\limits_{x \to 0} \dfrac{\sqrt{x + 16} - 4}{x}$

x	-0.1	-0.01	-0.001	0	0.001	0.01	0.1
$f(x)$				?			

10. $\lim\limits_{x \to 0} \dfrac{\sqrt{x + 2} - \sqrt{2}}{x}$

x	-0.1	-0.01	-0.001	0	0.001	0.01	0.1
$f(x)$				?			

11. $\lim\limits_{x \to -4} \dfrac{\dfrac{1}{x + 4} - \dfrac{1}{4}}{x}$

x	-4.1	-4.01	-4.001	-4	-3.999	-3.99	-3.9
$f(x)$				?			

12. $\lim\limits_{x \to -2} \dfrac{\dfrac{1}{2} - \dfrac{1}{x + 2}}{2x}$

x	-2.1	-2.01	-2.001	-2	-1.999	-1.99	-1.9
$f(x)$				?			

 Evaluating Basic Limits In Exercises 13–20, find the limit. *See Example 3.*

13. $\lim\limits_{x \to 3} 6$

14. $\lim\limits_{x \to 5} 4$

15. $\lim\limits_{x \to -2} x$

16. $\lim\limits_{x \to 10} x$

17. $\lim\limits_{x \to 7} x^2$

18. $\lim\limits_{x \to 3} x^3$

19. $\lim\limits_{x \to 36} \sqrt{x}$

20. $\lim\limits_{x \to -1} \sqrt[3]{x}$

 Operations with Limits In Exercises 21 and 22, find the limit of (a) $f(x) + g(x)$, (b) $f(x)g(x)$, and (c) $f(x)/g(x)$, as x approaches c.

21. $\lim\limits_{x \to c} f(x) = 3$

$\lim\limits_{x \to c} g(x) = 9$

22. $\lim\limits_{x \to c} f(x) = \frac{3}{2}$

$\lim\limits_{x \to c} g(x) = \frac{1}{2}$

 Operations with Limits In Exercises 23 and 24, find the limit of (a) $\sqrt{f(x)}$, (b) $3f(x)$, and (c) $[f(x)]^2$, as x approaches c.

23. $\lim\limits_{x \to c} f(x) = 16$

24. $\lim\limits_{x \to c} f(x) = 9$

 Using Properties of Limits In Exercises 25–36, find the limit using direct substitution. *See Examples 3 and 4.*

25. $\lim\limits_{x \to -3} (2x + 5)$

26. $\lim\limits_{x \to -4} (4x + 3)$

27. $\lim\limits_{x \to 1} (1 - x^2)$

28. $\lim\limits_{x \to 2} (-x^2 + x - 2)$

29. $\lim\limits_{x \to 3} \sqrt{x + 6}$

30. $\lim\limits_{x \to 5} \sqrt[3]{x - 5}$

31. $\lim\limits_{x \to -3} \dfrac{2}{x + 2}$

32. $\lim\limits_{x \to -2} \dfrac{3x + 1}{2 - x}$

33. $\lim\limits_{x \to -2} \dfrac{x^2 - 1}{2x}$

34. $\lim\limits_{x \to -8} \dfrac{3x}{x + 2}$

35. $\lim\limits_{x \to 5} \dfrac{\sqrt{x + 11} + 6}{x}$

36. $\lim\limits_{x \to 12} \dfrac{\sqrt{x - 3} - 2}{x}$

 Finding Limits In Exercises 37–58, find the limit (if it exists). *See Examples 5, 6, 7, 9, and 11.*

37. $\lim\limits_{x \to -3} \dfrac{x^2 - 9}{x + 3}$

38. $\lim\limits_{x \to -1} \dfrac{2x^2 - x - 3}{x + 1}$

39. $\lim\limits_{x \to 2} \dfrac{x^2 + 3x - 10}{x^2 - 4}$

40. $\lim\limits_{t \to 1} \dfrac{t^2 + t - 2}{t^2 - 1}$

41. $\lim\limits_{x \to -2} \dfrac{x^3 + 8}{x + 2}$

42. $\lim\limits_{x \to -3} \dfrac{x^3 + 27}{x + 3}$

43. $\lim\limits_{\Delta x \to 0} \dfrac{2(x + \Delta x) - 2x}{\Delta x}$

44. $\lim\limits_{\Delta x \to 0} \dfrac{-3(x + \Delta x) + 3x}{\Delta x}$

45. $\lim\limits_{\Delta t \to 0} \dfrac{(t + \Delta t)^2 - 5(t + \Delta t) - (t^2 - 5t)}{\Delta t}$

46. $\lim\limits_{\Delta t \to 0} \dfrac{(t + \Delta t)^2 - 4(t + \Delta t) + 2 - (t^2 - 4t + 2)}{\Delta t}$

47. $\lim\limits_{x \to 4} \dfrac{\sqrt{x + 5} - 3}{x - 4}$

48. $\lim\limits_{x \to 3} \dfrac{\sqrt{x + 1} - 2}{x - 3}$

49. $\lim\limits_{x \to 0} \dfrac{\sqrt{x + 5} - \sqrt{5}}{x}$

50. $\lim\limits_{x \to 0} \dfrac{\sqrt{x + 2} - \sqrt{2}}{x}$

51. $\lim\limits_{x \to 2} f(x)$, where $f(x) = \begin{cases} 4 - x, & x \neq 2 \\ 0, & x = 2 \end{cases}$

52. $\lim\limits_{x \to 1} f(x)$, where $f(x) = \begin{cases} x^2 + 2, & x \neq 1 \\ 1, & x = 1 \end{cases}$

53. $\lim\limits_{x \to 3} f(x)$, where $f(x) = \begin{cases} \frac{1}{3}x - 5, & x \leq 3 \\ -3x + 7, & x > 3 \end{cases}$

54. $\lim\limits_{s \to 4} f(s)$, where $f(s) = \begin{cases} 3s - 4, & s \leq 4 \\ 5 - \frac{1}{2}s, & s > 4 \end{cases}$

55. $\lim\limits_{x \to -4} \dfrac{2}{x + 4}$

56. $\lim\limits_{x \to 5} \dfrac{4}{x - 5}$

57. $\lim\limits_{x \to 2} \dfrac{x - 2}{x^2 - 4x + 4}$

58. $\lim\limits_{t \to -6} \dfrac{t + 6}{t^2 + 12t + 36}$

 Finding Limits Graphically In Exercises 59–64, use the graph to find the limit (if it exists).

(a) $\lim\limits_{x \to c^+} f(x)$ (b) $\lim\limits_{x \to c^-} f(x)$ (c) $\lim\limits_{x \to c} f(x)$

59.

60.

61.

62.

63.

64.

 Finding One-Sided Limits In Exercises 65 and 66, use a graph to find the limit from the left and the limit from the right. *See Example 8.*

65. $\lim\limits_{x \to -3^-} \dfrac{|x + 3|}{x + 3}$

$\lim\limits_{x \to -3^+} \dfrac{|x + 3|}{x + 3}$

66. $\lim\limits_{x \to 6^-} \dfrac{|x - 6|}{x - 6}$

$\lim\limits_{x \to 6^+} \dfrac{|x - 6|}{x - 6}$

One-Sided Limits of Unbounded Functions In Exercises 67–70, use a graphing utility to graph the function and estimate the limit (if it exists). Use a table to reinforce your conclusion. If the limit does not exist, explain why the limit fails to exist.

67. $\lim\limits_{x \to 2^+} \dfrac{3}{x^2 - 4}$

68. $\lim\limits_{x \to -1^+} \dfrac{6}{x + 1}$

69. $\lim\limits_{x \to -2^-} \dfrac{1}{x + 2}$

70. $\lim\limits_{x \to 0^-} \dfrac{x + 1}{x}$

Estimating Limits In Exercises 71–74, use a graphing utility to estimate the limit (if it exists).

71. $\lim\limits_{x \to 2} \dfrac{x^2 - 5x + 6}{x^2 - 4x + 4}$

72. $\lim\limits_{x \to 1} \dfrac{x^2 + 6x - 7}{x^3 - x^2 + 2x - 2}$

73. $\lim\limits_{x \to -4} \dfrac{x^3 + 4x^2 + x + 4}{2x^2 + 7x - 4}$

74. $\lim\limits_{x \to -2} \dfrac{4x^3 + 7x^2 + x + 6}{3x^2 - x - 14}$

75. Environment The cost C (in dollars) of removing $p\%$ of the pollutants from the water in a small lake is given by

$$C = \dfrac{25{,}000p}{100 - p}, \quad 0 \le p < 100.$$

(a) Find the cost of removing 50% of the pollutants.

(b) What percent of the pollutants can be removed for $100,000?

(c) Evaluate $\lim\limits_{p \to 100^-} C$. Explain your results.

76. **HOW DO YOU SEE IT?** The graph shows the cost C (in dollars) of making x photocopies at a copy shop.

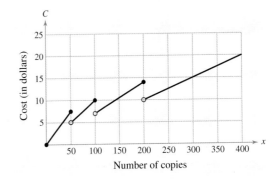

(a) Does $\lim\limits_{x \to 50} C$ exist? Explain your reasoning.

(b) Does $\lim\limits_{x \to 150} C$ exist? Explain your reasoning.

(c) You have to make 200 photocopies. Would it be better to make 200 or 201? Explain your reasoning.

77. Compound Interest Consider a certificate of deposit that pays 10% (annual percentage rate) on an initial deposit of $1000. The balance A after 10 years is $A = 1000(1 + 0.1x)^{10/x}$, where x is the length of the compounding period (in years).

(a) Use a graphing utility to graph A, where $0 \le x \le 1$.

(b) Use the *zoom* and *trace* features to estimate the balance for quarterly compounding and daily compounding.

(c) Use the *zoom* and *trace* features to estimate $\lim\limits_{x \to 0^+} A$. What do you think this limit represents? Explain your reasoning.

1.6 Continuity

■ Determine the continuity of functions.
■ Determine the continuity of functions on a closed interval.
■ Use the greatest integer function to model and solve real-life problems.
■ Use compound interest models to solve real-life problems.

Continuity

In mathematics, the term "continuous" has much the same meaning as it has in everyday use. To say that a function is continuous at

$$x = c$$

means that there is no interruption in the graph of f at c. That is, the graph of f

1. is unbroken at c.

2. has no holes, jumps, or gaps.

As simple as this concept may seem, its precise definition eluded mathematicians for many years. In fact, it was not until the early 1800s that a precise definition was finally developed.

Before looking at this definition, consider the function whose graph is shown in Figure 1.53. This figure identifies three values of x at which the function f is not continuous.

1. At $x = c_1$, $f(c_1)$ is not defined.

2. At $x = c_2$, $\lim_{x \to c_2} f(x)$ does not exist.

3. At $x = c_3$, $f(c_3) \neq \lim_{x \to c_3} f(x)$.

At all other points in the interval (a, b), the graph of f is uninterrupted, which implies that the function f is continuous at all other points in the interval (a, b).

In Exercise 67 on page 69, you will examine the continuity of a function that represents an account balance.

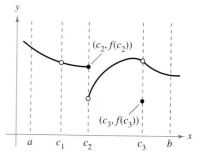

f is not continuous at $x = c_1, c_2, c_3$.
FIGURE 1.53

Definition of Continuity

Let c be a real number in the interval (a, b), and let f be a function whose domain contains the interval (a, b). The function f is **continuous at c** when these conditions are true.

1. $f(c)$ is defined.

2. $\lim_{x \to c} f(x)$ exists.

3. $\lim_{x \to c} f(x) = f(c)$

If f is continuous at every point in the interval (a, b), then f is **continuous on the open interval (a, b).**

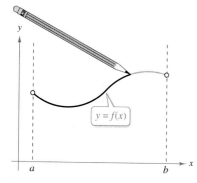

On the interval (a, b), the graph of f can be traced with a pencil.
FIGURE 1.54

Informally, you can say that a function is continuous on an interval when its graph on the interval can be traced using a pencil and paper without lifting the pencil from the paper, as shown in Figure 1.54.

Continuity of Polynomial and Rational Functions

1. A polynomial function is continuous at every real number.

2. A rational function is continuous at every number in its domain.

EXAMPLE 1   Determining Continuity of Polynomial Functions

Discuss the continuity of each function.

a. $f(x) = x^2 - 2x + 3$ **b.** $f(x) = x^3 - x$ **c.** $f(x) = x^4 - 2x^2 + 1$

SOLUTION   Each of these functions is a *polynomial function*. So, each is continuous on the entire real number line, as indicated in Figure 1.55.

(a)

(b)

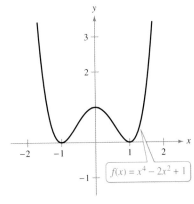

(c)

All three functions are continuous on $(-\infty, \infty)$.
FIGURE 1.55

✓ **Checkpoint 1**   *Worked-out solution available at LarsonAppliedCalculus.com*

Discuss the continuity of each function.

a. $f(x) = x^2 + x + 1$ **b.** $f(x) = x^3 + x$ **c.** $f(x) = x^4$ ■

Polynomial functions are one of the most important types of functions used in calculus. Be sure you see from Example 1 that the graph of a polynomial function is continuous on the entire real number line and therefore has no holes, jumps, or gaps. Rational functions, on the other hand, are not always continuous on the entire real number line, as shown in Example 2.

EXAMPLE 2   Determining Continuity of Rational Functions

Discuss the continuity of each function.

a. $f(x) = \dfrac{1}{x}$ **b.** $f(x) = \dfrac{x^2 - 1}{x - 1}$ **c.** $f(x) = \dfrac{1}{x^2 + 1}$

SOLUTION   Each of these functions is a rational function and is therefore continuous at every number in its domain.

a. The domain of $f(x) = 1/x$ consists of all real numbers except $x = 0$. So, this function is continuous on the intervals $(-\infty, 0)$ and $(0, \infty)$. [See Figure 1.56(a).]

b. The domain of $f(x) = (x^2 - 1)/(x - 1)$ consists of all real numbers except $x = 1$. So, this function is continuous on the intervals $(-\infty, 1)$ and $(1, \infty)$. [See Figure 1.56(b).]

c. The domain of $f(x) = 1/(x^2 + 1)$ consists of all real numbers. So, this function is continuous on the entire real number line. [See Figure 1.56(c).]

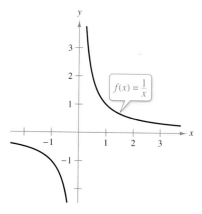

(a) Continuous on $(-\infty, 0)$ and $(0, \infty)$

FIGURE 1.56

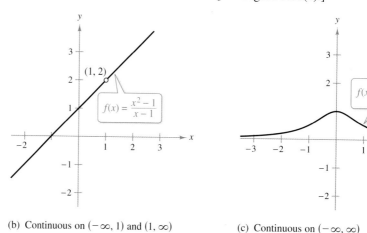

(b) Continuous on $(-\infty, 1)$ and $(1, \infty)$

(c) Continuous on $(-\infty, \infty)$

✓ **Checkpoint 2**   *Worked-out solution available at LarsonAppliedCalculus.com*

Discuss the continuity of each function.

a. $f(x) = \dfrac{1}{x - 1}$

b. $f(x) = \dfrac{x^2 - 4}{x - 2}$

c. $f(x) = \dfrac{1}{x^2 + 2}$ ■

TECH TUTOR

A graphing utility can give misleading information about the continuity of a function. For instance, try graphing the function from Example 2(b), $f(x) = (x^2 - 1)/(x - 1)$, in a standard viewing window. On most graphing utilities, the graph appears to be continuous at every real number. However, because $x = 1$ is not in the domain of f, you know that f is not continuous at $x = 1$. You can verify this on a graphing utility using the *trace* or *table* feature.

Consider an open interval I that contains a real number c. If a function f is defined on I (except possibly at c), and f is not continuous at c, then f is said to have a **discontinuity** at c. Discontinuities fall into two categories: **removable** and **nonremovable**. A discontinuity at c is called removable when f can be made continuous by appropriately defining (or redefining) $f(c)$. For instance, the function in Example 2(b) has a removable discontinuity at $(1, 2)$. To remove the discontinuity, all you need to do is redefine the function so that $f(1) = 2$.

A discontinuity at $x = c$ is nonremovable when the function cannot be made continuous at $x = c$ by defining or redefining the function at $x = c$. For instance, the function in Example 2(a) has a nonremovable discontinuity at $x = 0$.

Continuity on a Closed Interval

The intervals discussed in Examples 1 and 2 are open. To discuss continuity on a closed interval, you can use the concept of one-sided limits, as defined in Section 1.5.

Definition of Continuity on a Closed Interval

Let f be defined on a closed interval $[a, b]$. If f is continuous on the open interval (a, b) and

$$\lim_{x \to a^+} f(x) = f(a) \quad \text{and} \quad \lim_{x \to b^-} f(x) = f(b)$$

then f is **continuous on the closed interval** $[a, b]$. Moreover, f is **continuous from the right** at a and **continuous from the left** at b.

Similar definitions can be made to cover continuity on intervals of the form $(a, b]$ and $[a, b)$, or on infinite intervals. For instance, the function

$$f(x) = \sqrt{x}$$

is continuous on the infinite interval $[0, \infty)$.

EXAMPLE 3   Examining Continuity at an Endpoint

Discuss the continuity of

$$f(x) = \sqrt{3 - x}.$$

SOLUTION   Notice that the domain of f is the interval $(-\infty, 3]$. Moreover, f is continuous from the left at $x = 3$ because

$$\lim_{x \to 3^-} f(x) = \lim_{x \to 3^-} \sqrt{3 - x}$$
$$= \sqrt{3 - 3}$$
$$= 0$$
$$= f(3).$$

For all $x < 3$, the function f satisfies the three conditions for continuity. So, you can conclude that f is continuous on the interval $(-\infty, 3]$, as shown in Figure 1.57.

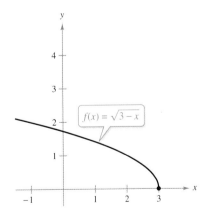

$f(x) = \sqrt{3 - x}$

FIGURE 1.57

✓ *Checkpoint 3*   *Worked-out solution available at LarsonAppliedCalculus.com*

Discuss the continuity of

$$f(x) = \sqrt{x - 2}.$$

STUDY TIP

When working with radical functions of the form

$$f(x) = \sqrt{g(x)}$$

remember that the domain of f coincides with the solution of $g(x) \geq 0$. (For a review of solving inequalities, see Appendix A.1.)

EXAMPLE 4   Examining Continuity on a Closed Interval

Discuss the continuity of

$$g(x) = \begin{cases} 5 - x, & -1 \le x \le 2 \\ x^2 - 1, & 2 < x \le 3 \end{cases}.$$

SOLUTION   The polynomial functions

$$5 - x$$

and

$$x^2 - 1$$

are continuous on the intervals $[-1, 2]$ and $(2, 3]$, respectively. So, to conclude that g is continuous on the entire interval

$$[-1, 3]$$

you need only to check the behavior of g when $x = 2$. You can do this by taking the one-sided limits when $x = 2$.

$$\lim_{x \to 2^-} g(x) = \lim_{x \to 2^-} (5 - x) = 5 - 2 = 3 \qquad \text{Limit from the left}$$

and

$$\lim_{x \to 2^+} g(x) = \lim_{x \to 2^+} (x^2 - 1) = 2^2 - 1 = 3 \qquad \text{Limit from the right}$$

Because these two limits are equal,

$$\lim_{x \to 2} g(x) = g(2) = 3.$$

So, g is continuous at $x = 2$ and, consequently, it is continuous on the entire interval

$$[-1, 3].$$

The graph of g is shown in Figure 1.58.

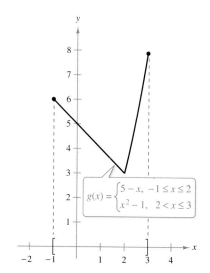

FIGURE 1.58

✓ *Checkpoint 4*   *Worked-out solution available at LarsonAppliedCalculus.com*

Discuss the continuity of

$$f(x) = \begin{cases} x + 2, & -1 \le x < 3 \\ 14 - x^2, & 3 \le x \le 5 \end{cases}.$$ ■

TECH TUTOR

Most graphing utilities can graph a piecewise-defined function. For instance, an *equation editor* entry for the function in Example 4 and its graph are shown below. (In some cases, it may be helpful to set your graphing utility to *dot* mode before graphing such a function.) For instructions on how to enter a piecewise-defined function into your graphing utility, consult your user's manual.

The Greatest Integer Function

Some functions that are used in business applications are **step functions.** For instance, the function in Example 10 in Section 1.5 is a step function. The **greatest integer function** is another example of a step function. This function is denoted by

$$[\![x]\!] = \text{greatest integer less than or equal to } x.$$

For example,

$$[\![-2.1]\!] = \text{greatest integer less than or equal to } -2.1 = -3$$
$$[\![-2]\!] = \text{greatest integer less than or equal to } -2 = -2$$
$$[\![1.5]\!] = \text{greatest integer less than or equal to } 1.5 = 1.$$

Note that the graph of the greatest integer function (Figure 1.59) jumps up one unit at each integer. This implies that the function is not continuous at each integer.

In real-life applications, the domain of the greatest integer function is often restricted to nonnegative values of x. In such cases, this function serves the purpose of **truncating** the decimal portion of x. For example, 1.345 is truncated to 1 and 3.57 is truncated to 3. That is,

$$[\![1.345]\!] = 1 \quad \text{and} \quad [\![3.57]\!] = 3.$$

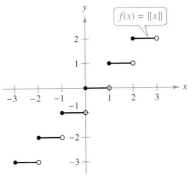

Greatest Integer Function

FIGURE 1.59

EXAMPLE 5 Modeling a Cost Function

Each day, a company produces 10,000 units in an eight-hour shift. The fixed cost *per shift* amounts to $5000, and the unit cost is $3. Using the greatest integer function, you can write the cost of producing x units as

$$C = 5000\left(1 + \left[\!\!\left[\frac{x-1}{10{,}000}\right]\!\!\right]\right) + 3x.$$

Sketch the graph of this cost function.

SOLUTION Note that during the first eight-hour shift,

$$\left[\!\!\left[\frac{x-1}{10{,}000}\right]\!\!\right] = 0, \quad 1 \le x \le 10{,}000$$

which implies

$$C = 5000\left(1 + \left[\!\!\left[\frac{x-1}{10{,}000}\right]\!\!\right]\right) + 3x = 5000 + 3x.$$

During the second eight-hour shift,

$$\left[\!\!\left[\frac{x-1}{10{,}000}\right]\!\!\right] = 1, \quad 10{,}001 \le x \le 20{,}000$$

which implies

$$C = 5000\left(1 + \left[\!\!\left[\frac{x-1}{10{,}000}\right]\!\!\right]\right) + 3x$$
$$= 10{,}000 + 3x.$$

In Figure 1.60, the graph of C is shown for three shifts. Note that the graph has discontinuities at $x = 10{,}000$, $x = 20{,}000$, and so on.

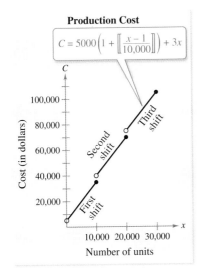

Production Cost

$$C = 5000\left(1 + \left[\!\!\left[\frac{x-1}{10{,}000}\right]\!\!\right]\right) + 3x$$

FIGURE 1.60

✓ **Checkpoint 5**   *Worked-out solution available at LarsonAppliedCalculus.com*

Use a graphing utility to graph the cost function in Example 5. ■

TECH TUTOR

You can use a spreadsheet or the *table* feature of a graphing utility to create a table. Try doing this for the data shown at the right. (Consult the user's manual of a spreadsheet software program for specific instructions on how to create a table.)

Extended Application: Compound Interest

Banks and other financial institutions differ on how interest is paid to an account. If the interest is added to the account so that future interest is paid on previously earned interest, then the interest is said to be **compounded.** For instance, a deposit of \$10,000 is made in an account that pays 6% interest, compounded quarterly. Because 6% is the annual interest rate, the quarterly rate is $\frac{1}{4}(0.06) = 0.015$, or 1.5%. The balances during the first five quarters are shown below.

Quarter	Balance
1st	\$10,000.00
2nd	$10{,}000.00 + (0.015)(10{,}000.00) = \$10{,}150.00$
3rd	$10{,}150.00 + (0.015)(10{,}150.00) = \$10{,}302.25$
4th	$10{,}302.25 + (0.015)(10{,}302.25) = \$10{,}456.78$
5th	$10{,}456.78 + (0.015)(10{,}456.78) = \$10{,}613.63$

EXAMPLE 6   Graphing Compound Interest

Sketch the graph of the balance in the account described above.

SOLUTION Let A represent the balance in the account and let t represent the time, in years. You can use the greatest integer function to represent the balance, as shown.

$$A = 10{,}000(1 + 0.015)^{[\![4t]\!]}$$

In the figure, notice that the graph of the function has a discontinuity at each quarter. That is, A has discontinuities at $t = \frac{1}{4}$, $t = \frac{1}{2}$, $t = \frac{3}{4}$, $t = 1$, $t = \frac{5}{4}$, and so on. Try to confirm this by entering the function A into a graphing utility and then using the *trace* or *table* feature to investigate the discontinuity of the function at each quarter.

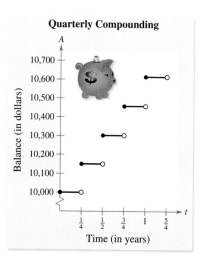

Quarterly Compounding

✓ *Checkpoint 6*   *Worked-out solution available at LarsonAppliedCalculus.com*

Write an equation that gives the balance of the account in Example 6 when the annual interest rate is 3%. Then sketch the graph of the equation. ■

SUMMARIZE (Section 1.6)

1. State the definition of continuity (*page 60*). For an example of a function that is continuous at every real number, see Example 1.

2. Describe the continuity of polynomial functions and rational functions (*page 61*). For examples of determining the continuity of polynomial functions and rational functions, see Examples 1 and 2.

3. State the definition of continuity on a closed interval (*page 63*). For an example of a function that is continuous on a closed interval, see Example 4.

4. State the definition of the greatest integer function (*page 65*). For real-life examples of the greatest integer function, see Examples 5 and 6.

SKILLS WARM UP 1.6

The following warm-up exercises involve skills that were covered in a previous course or in earlier sections. You will use these skills in the exercise set for this section. For additional help, review Appendices A.4 and A.5, and Section 1.5.

In Exercises 1–4, simplify the expression.

1. $\dfrac{x^2 + 6x + 8}{x^2 - 6x - 16}$

2. $\dfrac{x^2 - 5x - 6}{x^2 - 9x + 18}$

3. $\dfrac{2x^2 - 2x - 12}{4x^2 - 24x + 36}$

4. $\dfrac{x^3 - 16x}{x^3 + 2x^2 - 8x}$

In Exercises 5–8, solve for x.

5. $x^2 + 7x = 0$

6. $x^2 + 4x - 5 = 0$

7. $3x^2 + 8x + 4 = 0$

8. $3x^3 - x^2 - 24x = 0$

In Exercises 9 and 10, find the limit.

9. $\displaystyle\lim_{x \to 3} (2x^2 - 3x + 4)$

10. $\displaystyle\lim_{x \to -2} \sqrt{x^2 - x + 3}$

Exercises 1.6

See *CalcChat.com* for tutorial help and worked-out solutions to odd-numbered exercises.

 Determining Continuity In Exercises 1–10, determine whether the function is continuous on the entire real number line. Explain your reasoning. *See Examples 1 and 2.*

1. $f(x) = 5x^3 - x^2 + 2$

2. $f(x) = (x^2 - 1)^3$

3. $f(x) = \dfrac{3}{x^2 - 16}$

4. $f(x) = \dfrac{1}{9 - x^2}$

5. $f(x) = \dfrac{1}{4 + x^2}$

6. $f(x) = \dfrac{5x}{x^2 + 8}$

7. $f(x) = \dfrac{2x - 1}{x^2 - 8x + 15}$

8. $f(x) = \dfrac{x + 4}{x^2 - 6x + 5}$

9. $g(x) = \dfrac{x^2 - 8x + 12}{x^2 - 36}$

10. $g(x) = \dfrac{x^2 - 11x + 30}{x^2 - 25}$

 Determining Continuity In Exercises 11–40, describe the interval(s) on which the function is continuous. Explain why the function is continuous on the interval(s). If the function has a discontinuity, identify the conditions of continuity that are not satisfied. *See Examples 1, 2, 3, 4, and 5.*

11. $f(x) = \dfrac{x^2 - 1}{x}$

12. $f(x) = \dfrac{1}{x^2 - 4}$

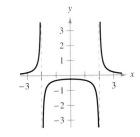

13. $f(x) = \dfrac{x^2 - 1}{x + 1}$

14. $f(x) = \dfrac{x^3 - 27}{x - 3}$

15. $f(x) = x^2 - 9x + 14$

16. $f(x) = 3 - 2x - x^2$

17. $f(x) = \dfrac{x}{x^2 - 1}$

18. $f(x) = \dfrac{x - 3}{x^2 - 9}$

19. $f(x) = \dfrac{7x}{x^2 + 5}$

20. $f(x) = \dfrac{6}{x^2 + 3}$

21. $f(x) = \dfrac{x - 5}{x^2 - 9x + 20}$

22. $f(x) = \dfrac{x - 1}{x^2 + x - 2}$

23. $f(x) = \sqrt{4 - x}$

24. $f(x) = \sqrt{x - 1}$

25. $f(x) = \sqrt{x} + 2$

26. $f(x) = 3 - \sqrt{x}$

27. $f(x) = \begin{cases} -2x + 3, & -1 \le x \le 1 \\ x^2, & 1 < x \le 3 \end{cases}$

28. $f(x) = \begin{cases} \frac{1}{2}x + 1, & -3 \le x \le 2 \\ 3 - x, & 2 < x \le 4 \end{cases}$

29. $f(x) = \begin{cases} 4 - 2x, & x \le 2 \\ x^2 - 3, & x > 2 \end{cases}$

30. $f(x) = \begin{cases} x^2 - 2, & x \le -1 \\ 3x + 2, & x > -1 \end{cases}$

31. $f(x) = \dfrac{|x + 1|}{x + 1}$

32. $f(x) = \dfrac{|4 - x|}{4 - x}$

33. $f(x) = x\sqrt{x + 3}$

34. $f(x) = \dfrac{x + 1}{\sqrt{x}}$

35. $f(x) = [\![2x]\!] + 1$

36. $f(x) = \dfrac{[\![x]\!]}{2} + x$

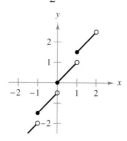

37. $f(x) = [\![x - 1]\!]$

38. $f(x) = x - [\![x]\!]$

39. $h(x) = f(g(x)), \quad f(x) = \dfrac{1}{\sqrt{x}}, \quad g(x) = x - 1, x > 1$

40. $h(x) = f(g(x)), \quad f(x) = \dfrac{1}{x - 1}, \quad g(x) = x^2 + 5$

Determining Continuity In Exercises 41–46, sketch the graph of the function and describe the interval(s) on which the function is continuous. If there are any discontinuities, determine whether they are removable.

41. $f(x) = \dfrac{x^2 - 16}{x - 4}$

42. $f(x) = \dfrac{2x^2 + x}{x}$

43. $f(x) = \dfrac{x + 4}{3x^2 - 12}$

44. $f(x) = \dfrac{-x}{x^3 - x}$

45. $f(x) = \begin{cases} x^2 + 1, & x < 0 \\ x - 1, & x \ge 0 \end{cases}$

46. $f(x) = \begin{cases} x^2 - 3, & x \le 0 \\ 2x + 3, & x > 0 \end{cases}$

Determining Continuity on a Closed Interval In Exercises 47–50, discuss the continuity of the function on the closed interval. If there are any discontinuities, determine whether they are removable.

Function	Interval
47. $f(x) = x^2 - 4x - 5$	$[-1, 5]$
48. $f(x) = \dfrac{5}{x^2 + 1}$	$[-2, 2]$
49. $f(x) = \dfrac{1}{x - 2}$	$[1, 4]$
50. $f(x) = \dfrac{x - 1}{x^2 - 4x + 3}$	$[0, 4]$

Finding Discontinuities In Exercises 51–56, use a graphing utility to graph the function. Use the graph to determine any x-value(s) at which the function is not continuous. Explain why the function is not continuous at the x-value(s).

51. $h(x) = \dfrac{1}{x^2 - x - 2}$

52. $k(x) = \dfrac{4 - x}{x^2 + x - 12}$

53. $f(x) = \begin{cases} 2x - 4, & x \le 3 \\ x^2 - 2x, & x > 3 \end{cases}$

54. $f(x) = \begin{cases} 3x - 2, & x \le 2 \\ x + 1, & x > 2 \end{cases}$

55. $f(x) = x - 2[\![x]\!]$

56. $f(x) = [\![2x - 1]\!]$

Making a Function Continuous In Exercises 57 and 58, find the constant a (Exercise 57) and the constants a and b (Exercise 58) such that the function is continuous on the entire real number line.

57. $f(x) = \begin{cases} x^3, & x \le 2 \\ ax^2, & x > 2 \end{cases}$

58. $f(x) = \begin{cases} 2, & x \le -1 \\ ax + b, & -1 < x < 3 \\ -2, & x \ge 3 \end{cases}$

Writing In Exercises 59 and 60, use a graphing utility to graph the function on the interval $[-4, 4]$. Does the graph of the function appear to be continuous on this interval? Is the function in fact continuous on $[-4, 4]$? Write a short paragraph about the importance of examining a function analytically as well as graphically.

59. $f(x) = \dfrac{x^2 + x}{x}$

60. $f(x) = \dfrac{x^3 - 8}{x - 2}$

61. Environmental Cost The cost C (in millions of dollars) of removing x percent of the pollutants emitted from the smokestack of a factory can be modeled by

$$C = \frac{2x}{100 - x}.$$

(a) What is the implied domain of C? Explain your reasoning.

(b) Use a graphing utility to graph the cost function. Is the function continuous on its domain? Explain your reasoning.

(c) Find the cost of removing 75% of the pollutants from the smokestack.

62. **HOW DO YOU SEE IT?**   The graph shows the number of gallons G of gasoline in a person's car after t days.

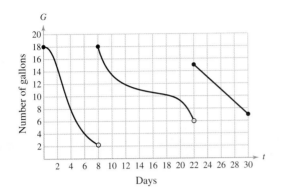

(a) On what days is the graph not continuous?

(b) What do you think happens on these days?

63. Biology   The gestation period of rabbits is about 29 to 35 days. Therefore, the population of a form (rabbits' home) can increase dramatically in a short period of time. The table gives the population of a form, where t is the time in months and N is the rabbit population.

t	0	1	2	3	4	5	6
N	2	8	10	14	10	15	12

Spreadsheet at LarsonAppliedCalculus.com

Graph the population as a function of time. Find any points of discontinuity in the function. Explain your reasoning.

64. Owning a Franchise   You have purchased a franchise. You have determined a linear model for your revenue as a function of time. Is the model a continuous function? Would your actual revenue be a continuous function of time? Explain your reasoning.

65. Consumer Awareness   The United States Postal Service first class mail rates for sending a letter are $0.49 for the first ounce and $0.21 for each additional ounce or fraction thereof up to 3.5 ounces. A model for the cost C (in dollars) of a first class mailing that weighs 3.5 ounces or less is given below. *(Source: United States Postal Service)*

$$C(x) = \begin{cases} 0.49, & 0 \le x \le 1 \\ 0.70, & 1 < x \le 2 \\ 0.91, & 2 < x \le 3 \\ 1.12, & 3 < x \le 3.5 \end{cases}$$

(a) Use a graphing utility to graph the function and then discuss its continuity. At what values is the function not continuous? Explain your reasoning.

(b) Find the cost of mailing a 2.5-ounce letter.

66. Salary Contract   A union contract guarantees an 11% yearly salary increase for 5 years. For a current salary of $45,300, the salaries S for the next 5 years are given by

$$S = 45{,}300(1.11)^{[\![t]\!]}$$

where $t = 0$ represents the present year.

(a) Use the greatest integer function of a graphing utility to graph the salary function and then discuss its continuity.

(b) Find the salary during the fifth year (when $t = 5$).

67. Compound Interest   A deposit of $7500 is made in an account that pays 6% interest, compounded quarterly. The amount A in the account after t years is

$$A = 7500(1.015)^{[\![4t]\!]}, \quad t \ge 0.$$

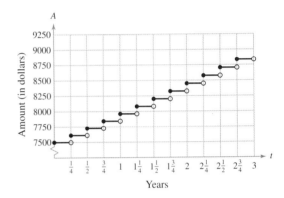

(a) Is the graph continuous? Explain your reasoning.

(b) What is the balance after 2 years?

(c) What is the balance after 7 years?

68. Inventory Management   The number of units in inventory in a small company is

$$N = 25\left(2\left[\![\frac{t+2}{2}\right]\!\right] - t\right)$$

where t is the time in months.

(a) Use the greatest integer function of a graphing utility to graph this function and then discuss its continuity.

(b) Find the number of units in inventory during the seventh month.

(c) How often must the company replenish its inventory?

69. Health Food Store   A co-op health food store charges $3.50 for the first pound of organically grown peanuts and $1.90 for each additional pound or fraction thereof.

(a) Use the greatest integer function to create a model for the cost C for x pounds of organically grown peanuts.

(b) Use a graphing utility to graph the function and then discuss its continuity.

ALGEBRA TUTOR

Order of Operations

Much of the algebra in this chapter involves evaluation of algebraic expressions. When you evaluate an algebraic expression, you need to know the priorities assigned to different operations. These priorities are called the *order of operations*.

1. Perform operations inside *symbols of grouping or absolute value symbols*, starting with the innermost symbol.

2. Evaluate all *exponential* expressions.

3. Perform all *multiplications* and *divisions* from left to right.

4. Perform all *additions* and *subtractions* from left to right.

EXAMPLE 1   Using Order of Operations

Evaluate each expression.

a. $20 - 2 \cdot 3^2$ **b.** $3 + 8 \div 2 \cdot 2$ **c.** $7 - [(5 \cdot 3) + 2^3]$

d. $[36 \div (3^2 \cdot 2)] + 6$ **e.** $36 - [3^2 \cdot (2 \div 6)]$ **f.** $10 - 2(8 + |5 - 7|)$

SOLUTION

a. $20 - 2 \cdot 3^2 = 20 - 2 \cdot 9$ Evaluate exponential expression.

$\qquad\qquad\quad = 20 - 18$ Multiply.

$\qquad\qquad\quad = 2$ Subtract.

b. $3 + 8 \div 2 \cdot 2 = 3 + 4 \cdot 2$ Divide.

$\qquad\qquad\quad = 3 + 8$ Multiply.

$\qquad\qquad\quad = 11$ Add.

c. $7 - [(5 \cdot 3) + 2^3] = 7 - [15 + 2^3]$ Multiply inside parentheses.

$\qquad\qquad\qquad = 7 - [15 + 8]$ Evaluate exponential expression.

$\qquad\qquad\qquad = 7 - 23$ Add inside brackets.

$\qquad\qquad\qquad = -16$ Subtract.

d. $[36 \div (3^2 \cdot 2)] + 6 = [36 \div (9 \cdot 2)] + 6$ Evaluate exponential expression inside parentheses.

$\qquad\qquad\qquad = [36 \div 18] + 6$ Multiply inside parentheses.

$\qquad\qquad\qquad = 2 + 6$ Divide inside brackets.

$\qquad\qquad\qquad = 8$ Add.

e. $36 - [3^2 \cdot (2 \div 6)] = 36 - \left[3^2 \cdot \frac{1}{3}\right]$ Divide inside parentheses.

$\qquad\qquad\qquad = 36 - \left[9 \cdot \frac{1}{3}\right]$ Evaluate exponential expression.

$\qquad\qquad\qquad = 36 - 3$ Multiply inside brackets.

$\qquad\qquad\qquad = 33$ Subtract.

f. $10 - 2(8 + |5 - 7|) = 10 - 2(8 + |-2|)$ Subtract inside absolute value symbols.

$\qquad\qquad\qquad = 10 - 2(8 + 2)$ Evaluate absolute value.

$\qquad\qquad\qquad = 10 - 2(10)$ Add inside parentheses.

$\qquad\qquad\qquad = 10 - 20$ Multiply.

$\qquad\qquad\qquad = -10$ Subtract.

TECH TUTOR

Most scientific and graphing calculators use the same order of operations listed above. Try entering the expressions in Example 1 into your calculator. Do you get the same results?

Solving Equations

A second algebraic skill used in this chapter is solving an equation in one variable.

1. To solve a *linear equation,* you can add or subtract the same quantity to or from each side of the equation. You can also multiply or divide each side of the equation by the same *nonzero* quantity.

2. To solve a *quadratic equation,* you can take the square root of each side, use factoring, or use the Quadratic Formula. (For a review of factoring and of the Quadratic Formula, see Appendix A.4.)

3. To solve a *radical equation,* isolate the radical on one side of the equation and square each side of the equation.

4. To solve an *absolute value equation,* use the definition of absolute value to rewrite the equation as two equations.

EXAMPLE 2 Solving Equations

Solve each equation.

a. $3x - 3 = 5x - 7$ **b.** $2x^2 = 10$

c. $2x^2 + 5x - 6 = 6$ **d.** $\sqrt{2x - 7} = 5$

e. $|3x + 6| = 9$

SOLUTION

a. $3x - 3 = 5x - 7$	Write original (linear) equation.
$-3 = 2x - 7$	Subtract $3x$ from each side.
$4 = 2x$	Add 7 to each side.
$2 = x$	Divide each side by 2.
b. $2x^2 = 10$	Write original (quadratic) equation.
$x^2 = 5$	Divide each side by 2.
$x = \pm\sqrt{5}$	Take the square root of each side.
c. $2x^2 + 5x - 6 = 6$	Write original (quadratic) equation.
$2x^2 + 5x - 12 = 0$	Write in general form.
$(2x - 3)(x + 4) = 0$	Factor.
$2x - 3 = 0 \implies x = \frac{3}{2}$	Set first factor equal to zero.
$x + 4 = 0 \implies x = -4$	Set second factor equal to zero.
d. $\sqrt{2x - 7} = 5$	Write original (radical) equation.
$2x - 7 = 25$	Square each side.
$2x = 32$	Add 7 to each side.
$x = 16$	Divide each side by 2.
e. $|3x + 6| = 9$	Write original (absolute value) equation.
$3x + 6 = -9$ or $3x + 6 = 9$	Rewrite equivalent equations.
$3x = -15$ or $3x = 3$	Subtract 6 from each side.
$x = -5$ or $x = 1$	Divide each side by 3.

STUDY TIP

Solving radical equations can sometimes lead to *extraneous solutions* (those that do not satisfy the original equation). For instance, squaring each side of the following equation yields two possible solutions, one of which is extraneous.

$$\sqrt{x} = x - 2$$
$$x = x^2 - 4x + 4$$
$$0 = x^2 - 5x + 4$$
$$0 = (x - 4)(x - 1)$$
$$x - 4 = 0 \implies x = 4$$
(solution)
$$x - 1 = 0 \implies x = 1$$
(extraneous)

SUMMARY AND STUDY STRATEGIES

After studying this chapter, you should have acquired the following skills.
The exercise numbers are keyed to the Review Exercises that begin on page 74.
Answers to odd-numbered Review Exercises are given in the back of the text.*

Section 1.1	Review Exercises
■ Plot points in a coordinate plane.	*1, 2*
■ Find the distance between two points in a coordinate plane.	*3–8*
$d = \sqrt{(x_2 - x_1)^2 + (y_2 - y_1)^2}$	
■ Find the midpoint of a line segment connecting two points.	*9–14*
Midpoint $= \left(\dfrac{x_1 + x_2}{2}, \dfrac{y_1 + y_2}{2} \right)$	
■ Interpret real-life data that are presented graphically.	*15, 16*
■ Translate points in a coordinate plane.	*17, 18*

Section 1.2	
■ Sketch graphs of equations by hand.	*19–28*
■ Find the x- and y-intercepts of graphs of equations.	*29–32*
■ Find the standard forms of equations of circles.	*33–36*
$(x - h)^2 + (y - k)^2 = r^2$	
■ Find the points of intersection of two graphs.	*37–40*
■ Find the break-even point for a business.	*41, 42*
The break-even point occurs when the revenue R is equal to the cost C.	
■ Find the equilibrium points of supply equations and demand equations.	*43*
The equilibrium point is the point of intersection of the graphs of the supply and demand equations.	
■ Use mathematical models to model and solve real-life problems.	*44*

Section 1.3	
■ Use the slope-intercept form of a linear equation to sketch graphs of lines.	*45–52*
$y = mx + b$	
■ Find slopes of lines passing through two points.	*53–56*
$m = \dfrac{y_2 - y_1}{x_2 - x_1}$	
■ Use the point-slope form to write equations of lines.	*57–64*
$y - y_1 = m(x - x_1)$	

---

* Several study aids are available to help you master the material in this chapter. The *Student Solutions Manual* and *CalcChat.com* have solutions to all odd-numbered exercises, and *CalcView.com* has video solutions for selected exercises. *LarsonAppliedCalculus.com* offers algebra help, data spreadsheets, and much more. A *Graphing Technology Guide* with step-by-step commands for a variety of graphing calculators is at *CengageBrain.com.*

Section 1.3 (continued)

Review Exercises

■ Find equations of parallel and perpendicular lines.

Parallel lines: $m_1 = m_2$

Perpendicular lines: $m_1 = -\dfrac{1}{m_2}$

65, 66

■ Use linear equations to model and solve real-life problems.

67, 68

Section 1.4

■ Use the Vertical Line Test to decide whether the relationship between two variables is a function.

69–72

If every vertical line intersects the graph of an equation at most once, then the equation defines y as a function of x.

■ Find the domains and ranges of functions.

73–78

■ Use function notation and evaluate functions.

79, 80

■ Combine functions to create other functions.

81, 82

■ Use the Horizontal Line Test to determine whether functions have inverse functions. If they do, find the inverse functions.

83–88

A function is one-to-one when every horizontal line intersects the graph of the function at most once. For a function to have an inverse function, it must be one-to-one.

Section 1.5

■ Use a table to estimate limits.

89–92

■ Use different analytic techniques to evaluate limits of functions.

93–110

Section 1.6

■ Determine the continuity of functions.

111–122

■ Determine the constant such that f is continuous.

123, 124

■ Use piecewise-defined functions and the greatest integer function to model and solve real-life problems.

125–128

Study Strategies

■ **Use a Graphing Utility**   A graphing calculator or graphing software for a computer can help you in this course in two important ways. As an *exploratory device,* a graphing utility allows you to focus on learning concepts. For instance, sketching the graphs of $y = x^2$, $y = x^2 + 1$, and $y = x^2 - 1$ helps confirm that adding (or subtracting) a constant to (or from) a function shifts the graph of the function vertically. As a *problem-solving tool,* a graphing utility frees you of some of the drudgery of sketching complicated graphs by hand. The time that you save can be spent using mathematics to solve real-life problems.

■ **Use the Skills Warm-Up Exercises**   Each exercise set in this text begins with a set of skills warm-up exercises. You should begin each homework session by quickly working all of these exercises. (All are answered in the back of the text.) The "old" skills covered in these exercises are needed to master the "new" skills in the section exercise set. The skills warm-up exercises remind you that mathematics is cumulative—to be successful in this course, you must retain "old" skills.

■ **Use the Additional Study Aids**   The additional study aids were prepared specifically to help you master the concepts discussed in the text. They are the student websites at *LarsonAppliedCalculus.com, CalcChat.com,* and *CalcView.com,* the *Student Solutions Manual,* and the *Graphing Technology Guide.*

Review Exercises   See *CalcChat.com* for tutorial help and worked-out solutions to odd-numbered exercises.

Plotting Points in the Cartesian Plane   In Exercises 1 and 2, plot the points in the Cartesian plane.

1. $(2, 3), (0, 6), (-5, 1), (4, -3), (-3, -1)$

2. $(-3, 4), (-1, -2), (6, -5), (-2, 0), (5, 5)$

Finding a Distance   In Exercises 3–8, find the distance between the two points.

3. $(0, 0), (5, 2)$

4. $(0, 2), (3, 4)$

5. $(-1, 3), (-4, 6)$

6. $(-8, 5), (-2, 7)$

7. $\left(\frac{1}{4}, -8\right), \left(\frac{3}{4}, -6\right)$

8. $(-0.6, 3), (4, -1.8)$

Finding the Midpoint of a Line Segment   In Exercises 9–14, find the midpoint of the line segment joining the two points.

9. $(5, 6), (9, 2)$

10. $(-7, 0), (3, 6)$

11. $(-10, 4), (-6, 8)$

12. $(7, -9), (-3, 5)$

13. $(-2, 0.1), (4.3, -3)$

14. $\left(\frac{1}{2}, \frac{5}{2}\right), \left(1, -\frac{3}{4}\right)$

Revenues, Costs, and Profits   In Exercises 15 and 16, use the graph below, which gives the revenues, costs, and profits for Google from 2009 through 2013. *(Source: Google Inc.)*

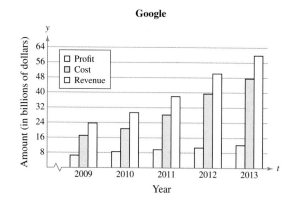

15. Write an equation that relates the revenue R, cost C, and profit P. Explain the relationship between the heights of the bars and the equation.

16. Estimate the revenue, cost, and profit for Google for each year.

Translating Points in the Plane   In Exercises 17 and 18, use the translation and the graph to find the vertices of the figure after it has been translated.

17. 2 units right and 3 units up

18. 5 units left and 2 units down

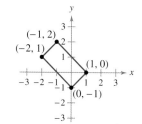

Sketching the Graph of an Equation   In Exercises 19–28, sketch the graph of the equation. Use a graphing utility to verify your results.

19. $y = 4x - 12$

20. $y = 4 - 3x$

21. $y = x^2 + 5$

22. $y = 1 - x^2$

23. $y = |4 - x|$

24. $y = |2x - 3|$

25. $y = 2 - x^3$

26. $y = 2x^3 - 1$

27. $y = \sqrt{4x + 1}$

28. $y = \sqrt{2x}$

Finding x- and y-Intercepts   In Exercises 29–32, find the x- and y-intercepts of the graph of the equation.

29. $4x + y + 3 = 0$

30. $3x - y + 6 = 0$

31. $y = x^2 + 2x - 8$

32. $y = (x - 1)^3 + 2(x - 1)^2$

Finding the Equation of a Circle   In Exercises 33–36, find the standard form of the equation of the circle with the given characteristics and sketch its graph.

33. Center: $(0, 0)$; radius: 8

34. Center: $(-5, -2)$; radius: 6

35. Center: $(2, 0)$; solution point: $\left(4, \sqrt{5}\right)$

36. Center: $(3, -4)$; solution point: $(-1, -1)$

Finding Points of Intersection   In Exercises 37–40, find the point(s) of intersection (if any) of the graphs of the equations. Use a graphing utility to check your results.

37. $y = 2x + 13, \quad y = -5x - 1$

38. $y = x^2 + 3, \quad y = 9 - x$

39. $y = x^3, \quad y = x$

40. $y = x^3 + 4x^2 - 3, \quad y = -2x^2 + 27x - 3$

41. Break-Even Analysis A student organization wants to raise money by having a T-shirt sale. Each shirt costs $8. The silk screening costs $200 for the design, plus $2 per shirt. Each shirt will sell for $14.

(a) Find equations for the total cost C and the total revenue R for x shirts.

(b) Find the break-even point.

(c) How many units would yield a profit of $600?

42. Break-Even Analysis You are starting a part-time business. You make an initial investment of $6000. The unit cost of the product is $6.50, and the selling price is $13.90.

(a) Find equations for the total cost C and the total revenue R for x units.

(b) Find the break-even point.

(c) How many units would yield a profit of $1500?

43. Supply and Demand The demand and supply equations for a cordless screwdriver are given by

$p = 91.4 - 0.009x$ ⠀⠀Demand equation

$p = 6.4 + 0.008x$ ⠀⠀Supply equation

where p is the price in dollars and x represents the number of units. Find the equilibrium point for this market.

44. Wind Energy The table shows the annual amounts of U.S. consumption W (in trillion Btu) of wind energy for the years 2008 through 2013. *(Source: U.S. Energy Information Administration)*

Year	2008	2009	2010
Consumption	546	721	923

Year	2011	2012	2013
Consumption	1168	1340	1595

Spreadsheet at LarsonAppliedCalculus.com

A mathematical model for the data is given by

$W = -0.630t^3 + 24.83t^2 - 100.2t + 77$

where t represents the year, with $t = 8$ corresponding to 2008.

(a) Use a graph and a table to compare the actual consumptions with the model.

(b) Use the model to predict the consumption in 2019.

Graphing Linear Equations In Exercises 45–52, find the slope and y-intercept (if possible) of the equation of the line. Then sketch the graph of the equation.

45. $y = -x + 12$ ⠀⠀⠀**46.** $y = 2x - 5$

47. $4x - y = 6$ ⠀⠀⠀**48.** $2x - 4y = -8$

49. $y = \frac{7}{4}$ ⠀⠀⠀⠀⠀⠀**50.** $x = -3$

51. $-2x - 5y - 5 = 0$ ⠀⠀**52.** $3.2x - 0.8y + 5.6 = 0$

Finding Slopes of Lines In Exercises 53–56, find the slope of the line passing through the pair of points.

53. $(0, 0), (7, 6)$ ⠀⠀⠀⠀**54.** $(-1, 5), (-5, 7)$

55. $(5, 29), (-3, 29)$

56. $(-6, -11), (-6, -1)$

Using the Point-Slope Form In Exercises 57–60, find an equation of the line that passes through the given point and has the given slope. Then sketch the line.

	Point	*Slope*
57.	$(3, -1)$	$m = -2$
58.	$(-3, -3)$	$m = \frac{1}{2}$
59.	$(1.5, -4)$	$m = 0$
60.	$\left(\frac{4}{5}, 2\right)$	$m = -\frac{3}{4}$

Writing an Equation of a Line In Exercises 61–64, find an equation of the line that passes through the points. Then sketch the line.

61. $(1, -7), (7, 5)$

62. $(-4, 7), (8, -9)$

63. $(5, 7), (5, 14)$

64. $(4, -3), (-2, -3)$

Finding Parallel and Perpendicular Lines In Exercises 65 and 66, find an equation of the line passing through the given point and satisfying the given condition.

65. Point: $(-3, 6)$

(a) Parallel to the line $y = \frac{5}{8}x$

(b) Perpendicular to the line $y = -5x - 3$

(c) Parallel to the line $4x + 2y = 7$

(d) Perpendicular to the line $3x - 2y = 2$

66. Point: $(1, -3)$

(a) Parallel to the x-axis

(b) Perpendicular to the x-axis

(c) Parallel to the line $-4x + 5y = -3$

(d) Perpendicular to the line $5x - 2y = 3$

67. Demand When a wholesaler sold a product at $32 per unit, sales were 750 units per week. After a price increase of $5 per unit, the sales dropped to 700 units per week. Assume that the relationship between the price p and the units sold per week x is linear.

(a) Write a linear equation expressing x in terms of p.

(b) Predict the number of units sold per week at a price of $34.50 per unit.

(c) Predict the number of units sold per week at a price of $42 per unit.

68. Linear Depreciation A printing company purchases an advanced color copier/printer for $117,000. After 9 years, the equipment will be obsolete and have no value.

(a) Write a linear equation giving the value v (in dollars) of the equipment in terms of the time t (in years), $0 \leq t \leq 9$.

(b) Use a graphing utility to graph the equation.

(c) Move the cursor along the graph and estimate (to two-decimal-place accuracy) the value of the equipment after 4 years.

(d) Move the cursor along the graph and estimate (to two-decimal-place accuracy) the time when the value of the equipment will be $84,000.

Vertical Line Test In Exercises 69–72, use the Vertical Line Test to determine whether y is a function of x.

69. $y = -x^2 + 2$

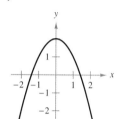

70. $x = y^2 - 2$

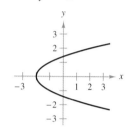

71. $y^2 - \frac{1}{4}x^2 = 4$

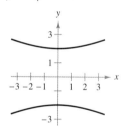

72. $y = |x + 4|$

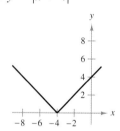

Finding the Domain and Range of a Function In Exercises 73–78, find the domain and range of the function. Use a graphing utility to verify your results.

73. $f(x) = x^3 + 2x^2 - x + 2$ **74.** $f(x) = 2$

75. $f(x) = \sqrt{x + 5}$

76. $f(x) = -|x| + 3$

77. $f(x) = \dfrac{x - 1}{x^2 - 1}$

78. $f(x) = \begin{cases} 6 - x, & x < 2 \\ 3x - 2, & x \geq 2 \end{cases}$

Evaluating a Function In Exercises 79 and 80, evaluate the function at the specified values of the independent variable. Simplify the results.

79. $f(x) = 3x + 4$

(a) $f(1)$ (b) $f(-5)$ (c) $f(x + 1)$

80. $f(x) = x^2 + 4x + 3$

(a) $f(0)$ (b) $f(3)$ (c) $f(x - 1)$

Combinations of Functions In Exercises 81 and 82, find (a) $f(x) + g(x)$, (b) $f(x) - g(x)$, (c) $f(x) \cdot g(x)$, (d) $f(x)/g(x)$, (e) $f(g(x))$, and (f) $g(f(x))$, if defined.

81. $f(x) = 6 + x^2$, $g(x) = 3x - 5$

82. $f(x) = 2x - 3$, $g(x) = \sqrt{x + 1}$

Determining Whether a Function Is One-to-One In Exercises 83–88, use a graphing utility to graph the function. Then use the Horizontal Line Test to determine whether the function is one-to-one. If it is, find its inverse function.

83. $f(x) = 4x - 3$

84. $f(x) = \frac{3}{2}x$

85. $f(x) = \sqrt{x - \frac{1}{2}}$

86. $f(x) = x^3 - 1$

87. $f(x) = |x + 1|$

88. $f(x) = 6$

Finding Limits Numerically In Exercises 89–92, complete the table and use the result to estimate the limit. Use a graphing utility to graph the function to confirm your result.

89. $\lim\limits_{x \to 1} (4x - 3)$

x	0.9	0.99	0.999	1	1.001	1.01	1.1
$f(x)$				?			

90. $\lim\limits_{x \to 3} \dfrac{x - 3}{x^2 - 2x - 3}$

x	2.9	2.99	2.999	3	3.001	3.01	3.1
$f(x)$				?			

91. $\lim\limits_{x \to 0} \dfrac{\sqrt{x + 6} - \sqrt{6}}{x}$

x	-0.1	-0.01	-0.001	0	0.001	0.01	0.1
$f(x)$				?			

92. $\lim\limits_{x \to 7} \dfrac{\dfrac{1}{x - 7} - \dfrac{1}{7}}{x}$

x	6.9	6.99	6.999	7	7.001	7.01	7.1
$f(x)$				?			

Finding Limits　In Exercises 93–110, find the limit (if it exists).

93. $\lim_{x\to3} 8$

94. $\lim_{x\to6} x^4$

95. $\lim_{x\to2} (5x - 3)$

96. $\lim_{x\to5} (3x^2 + 4)$

97. $\lim_{x\to-1} \dfrac{x + 3}{6x + 1}$

98. $\lim_{t\to-3} \dfrac{6t + 5}{t + 8}$

99. $\lim_{t\to0} \dfrac{t^2 + 1}{t}$

100. $\lim_{t\to2} \dfrac{t + 1}{t - 2}$

101. $\lim_{x\to-2} \dfrac{x + 2}{x^2 - 4}$

102. $\lim_{x\to4} \dfrac{x^2 - 6x + 8}{x - 4}$

103. $\lim_{x\to0^+} \left(x - \dfrac{1}{x}\right)$

104. $\lim_{x\to(1/2)^-} \dfrac{-x}{6x - 3}$

105. $\lim_{x\to0} \dfrac{\sqrt{3} - \sqrt{x + 3}}{x}$

106. $\lim_{s\to0} \dfrac{\left(1/\sqrt{1 + s}\right) - 1}{s}$

107. $\lim_{x\to0} f(x)$, where $f(x) = \begin{cases} x + 5, & x \neq 0 \\ 3, & x = 0 \end{cases}$

108. $\lim_{x\to-2} f(x)$, where $f(x) = \begin{cases} \frac{1}{2}x + 5, & x < -2 \\ -x + 2, & x \geq -2 \end{cases}$

109. $\lim_{\Delta x\to0} \dfrac{(x + \Delta x)^3 - (x + \Delta x) - (x^3 - x)}{\Delta x}$

110. $\lim_{\Delta x\to0} \dfrac{1 - (x + \Delta x)^2 - (1 - x^2)}{\Delta x}$

Determining Continuity　In Exercises 111–122, describe the interval(s) on which the function is continuous. Explain why the function is continuous on the interval(s). If the function has a discontinuity, identify the conditions of continuity that are not satisfied.

111. $f(x) = x + 6$

112. $f(x) = x^2 + 3x + 2$

113. $f(x) = \dfrac{1}{(x + 4)^2}$

114. $f(x) = \dfrac{x + 2}{x}$

115. $f(x) = \dfrac{3}{x + 1}$

116. $f(x) = \dfrac{x + 1}{2x + 2}$

117. $f(x) = \sqrt{x - 8}$

118. $f(x) = \sqrt{5 - x}$

119. $f(x) = [\![x + 3]\!]$

120. $f(x) = [\![x]\!] - 2$

121. $f(x) = \begin{cases} x, & x \leq 0 \\ x + 1, & x > 0 \end{cases}$

122. $f(x) = \begin{cases} x, & x \leq 0 \\ x^2, & x > 0 \end{cases}$

Making a Function Continuous　In Exercises 123 and 124, find the constant a such that the function is continuous on the entire real number line.

123. $f(x) = \begin{cases} -x + 1, & x \leq 3 \\ ax - 8, & x > 3 \end{cases}$

124. $f(x) = \begin{cases} x + 1, & x < 1 \\ 2x + a, & x \geq 1 \end{cases}$

125. Consumer Awareness　The cost C (in dollars) of purchasing x bottles of vitamins at an organic food store is shown below.

$$C(x) = \begin{cases} 5.99x, & 0 < x \leq 5 \\ 4.99x, & 5 < x \leq 10 \\ 3.99x, & 10 < x \leq 15 \\ 2.99x, & x > 15 \end{cases}$$

(a) Use a graphing utility to graph the function and then discuss its continuity. At what values is the function not continuous? Explain your reasoning.

(b) Find the cost of purchasing 10 bottles.

126. Salary Contract　A union contract guarantees a 3.5% yearly salary increase for 2 years. For a current salary of \$40,000, the salaries S for the next 3 years are given by

$$S(t) = \begin{cases} 40,000, & 0 < t \leq 1 \\ 41,400, & 1 < t \leq 2 \\ 42,849, & 2 < t \leq 3 \end{cases}$$

where $t = 0$ represents the present year. Does the limit of S exist as t approaches 2? Explain your reasoning.

127. Recycling　A recycling center pays \$0.50 for each pound of aluminum cans. Twenty-four aluminum cans weigh one pound. A mathematical model for the amount A paid by the recycling center is

$$A = \dfrac{1}{2} \left[\!\left[\dfrac{x}{24} \right]\!\right]$$

where x is the number of cans.

(a) Use a graphing utility to graph the function and then discuss its continuity.

(b) How much does the recycling center pay out for 1500 cans?

128. Consumer Awareness　The cost of sending an overnight package from New York to Atlanta is \$32.30 for the first pound and \$2.90 for each additional pound or fraction thereof.

(a) Use the greatest integer function to create a model for the cost C of overnight delivery of a package weighing x pounds.

(b) Use a graphing utility to graph the function and then discuss its continuity.

TEST YOURSELF

See *CalcChat.com* for tutorial help and worked-out solutions to odd-numbered exercises.

Take this test as you would take a test in class. When you are done, check your work against the answers given in the back of the book.

In Exercises 1–3, (a) find the distance between the points, (b) find the midpoint of the line segment joining the points, (c) find the slope of the line passing through the points, (d) find an equation of the line passing through the points, and (e) sketch the graph of the equation.

1. $(1, -1), (-4, 4)$ **2.** $\left(\frac{5}{2}, 2\right), (0, 2)$ **3.** $(2, 3), (-4, 1)$

4. The demand and supply equations for a product are $p = 65 - 2.1x$ and $p = 43 + 1.9x$, respectively, where p is the price (in dollars) and x represents the number of units (in thousands). Find the equilibrium point for this market.

In Exercises 5–7, find the slope and y-intercept (if possible) of the equation of the line. Then sketch the graph of the linear equation.

5. $y = \frac{1}{5}x - 2$ **6.** $x - \frac{9}{2} = 0$ **7.** $-x - 0.4y + 2.5 = 0$

8. Write an equation of the line that passes through the point $(-3, -1)$ and is perpendicular to the line $-6x + y = 3$.

9. Write an equation of the line that passes through the point $(2, 1)$ and is parallel to the line $5x - 2y = 8$.

In Exercises 10–12, (a) graph the function and label the intercepts, (b) find the domain and range of the function, (c) find the values of the function when x is -3, -2, and 3, and (d) determine whether the function is one-to-one.

10. $f(x) = 2x + 5$ **11.** $f(x) = x^2 - x - 2$ **12.** $f(x) = \sqrt{x + 9}$

In Exercises 13 and 14, find the inverse function of f.

13. $f(x) = 4x + 6$ **14.** $f(x) = \sqrt[3]{8 - 3x}$

In Exercises 15–18, find the limit (if it exists).

15. $\lim\limits_{x \to 0} \dfrac{x - 2}{x + 2}$ **16.** $\lim\limits_{x \to 5} \dfrac{x + 5}{x - 5}$

17. $\lim\limits_{x \to -3} \dfrac{x^2 + 2x - 3}{x^2 + 4x + 3}$ **18.** $\lim\limits_{x \to 0} \dfrac{\sqrt{x + 9} - 3}{x}$

In Exercises 19–21, describe the interval(s) on which the function is continuous. Explain why the function is continuous on the interval(s). If the function has a discontinuity, identify the conditions of continuity that are not satisfied.

19. $f(x) = \dfrac{x^2 - 36}{x - 6}$ **20.** $f(x) = \sqrt{5 - x}$ **21.** $f(x) = \begin{cases} 1 - x, & x < 1 \\ x - x^2, & x \geq 1 \end{cases}$

Year	2008	2009	2010
y	8924	14,265	14,825

Year	2011	2012	2013
y	13,747	12,506	11,460

Spreadsheet at LarsonAppliedCalculus.com

Table for 22

22. The table lists the numbers of unemployed workers y (in thousands) in the United States from 2008 through 2013. A mathematical model for the data is given by $y = 271.343t^3 - 9246.20t^2 + 103,234.1t - 364,018$, where t represents the year, with $t = 8$ corresponding to 2008. *(Source: U.S. Bureau of Labor Statistics)*

(a) Use a graph and a table to compare the actual numbers of unemployed workers with the model.

(b) Use the model to predict the number of unemployed workers in 2018. Do you think this prediction is valid?

Revenue for Buffalo Wild Wings

Slope ≈ 191.3

Year (8 ↔ 2008)

Example 11 on page 99 shows how differentiation can be used to find the rate of change in a company's revenue.

2 Differentiation

Monkey Business Images/Shutterstock.com

2.1 The Derivative and the Slope of a Graph

■ Identify tangent lines to a graph at a point.
■ Approximate the slopes of tangent lines to graphs at points.
■ Use the limit definition to find the slopes of graphs at points.
■ Use the limit definition to find the derivatives of functions.
■ Describe the relationship between differentiability and continuity.

Under Armour

In Exercise 13 on page 89, you will estimate and interpret the slope of the graph of a revenue function.

Tangent Line to a Graph

Calculus is a branch of mathematics that studies rates of change of functions. In this course, you will learn that rates of change have many applications in real life. In Section 1.3, you learned how the slope of a line indicates the rate at which the line rises or falls. For a line, this rate (or slope) is the same at every point on the line. For graphs other than lines, the rate at which the graph rises or falls changes from point to point. For instance, in Figure 2.1, the parabola is rising more quickly at the point (x_1, y_1) than it is at the point (x_2, y_2). At the vertex (x_3, y_3), the graph levels off, and at the point (x_4, y_4), the graph is falling.

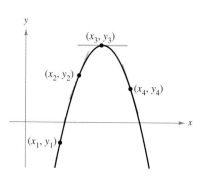

The slope of a nonlinear graph changes from one point to another.
FIGURE 2.1

To determine the rate at which a graph rises or falls at a *single point*, you can find the slope of the **tangent line** at the point. In simple terms, the tangent line to the graph of a function f at a point $P(x_1, y_1)$ is the line that best approximates the slope of the graph at that point, as shown in Figure 2.1. Figure 2.2 shows other examples of tangent lines.

  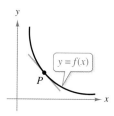

Tangent Line to a Graph at a Point
FIGURE 2.2

When Isaac Newton (1642–1727) was working on the "tangent line problem," he realized that it is difficult to define precisely what is meant by a tangent to a general curve. From geometry, you know that a line is tangent to a circle when the line intersects the circle at only one point, as shown in Figure 2.3. Tangent lines to a noncircular graph, however, can intersect the graph at more than one point. For instance, in the second graph in Figure 2.2, when the tangent line is extended, it intersects the graph at a point other than the point of tangency. In this section, you will see how the notion of a limit can be used to define a general tangent line.

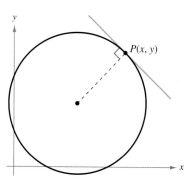

Tangent Line to a Circle
FIGURE 2.3

Slope of a Graph

Because a tangent line approximates the slope of a graph at a point, the problem of finding the slope of the graph at the point becomes one of finding the slope of the tangent line at the point.

EXAMPLE 1   Approximating the Slope of a Graph

Use Figure 2.4 to approximate the slope of the graph of

$$f(x) = x^2$$

at the point $(1, 1)$.

SOLUTION   From the graph of

$$f(x) = x^2$$

you can see that the tangent line at $(1, 1)$ rises approximately two units for each unit change in x. So, the slope of the tangent line at $(1, 1)$ is given by

$$\text{Slope} = \frac{\Delta y}{\Delta x} = \frac{\text{change in } y}{\text{change in } x} \approx \frac{2}{1} = 2.$$

Because the tangent line at the point $(1, 1)$ has a slope of about 2, you can conclude that the graph has a slope of about 2 at the point $(1, 1)$.

✓ **Checkpoint 1**   *Worked-out solution available at LarsonAppliedCalculus.com*

Use Figure 2.5 to approximate the slope of the graph of

$$f(x) = x^3$$

at the point $(1, 1)$.

When visually approximating the slope of a graph, note that the scales on the horizontal and vertical axes may differ. When this happens (as it frequently does in applications), the slope of the tangent line is distorted, and you must be careful to account for the difference in scales. (See Example 2).

EXAMPLE 2   Interpreting Slope

Figure 2.6 graphically depicts the average monthly temperature (in degrees Fahrenheit) in Duluth, Minnesota. Estimate the slope of this graph at the indicated point and give a physical interpretation of the result.   *(Source: National Oceanic and Atmospheric Administration)*

SOLUTION   From the graph, you can see that the tangent line at the given point falls approximately 28 units for each two-unit change in x. So, you can estimate the slope at the given point to be

$$\text{Slope} = \frac{\Delta y}{\Delta x} = \frac{\text{change in } y}{\text{change in } x} \approx \frac{-28}{2} = -14 \text{ degrees per month.}$$

This means that you can expect the average daily temperatures in November to be about 14 degrees *lower* than the corresponding temperatures in October.

✓ **Checkpoint 2**   *Worked-out solution available at LarsonAppliedCalculus.com*

In Figure 2.6, for which months do the slopes of the tangent lines appear to be positive? Negative? Interpret these slopes in the context of the problem.

FIGURE 2.4

FIGURE 2.5

FIGURE 2.6

Slope and the Limit Process

In Examples 1 and 2, you approximated the slope of a graph at a point by making a careful graph and then "eyeballing" the tangent line at the point of tangency. A more precise method of approximating the slope of a tangent line makes use of a **secant line** through the point of tangency and a second point on the graph, as shown in the figure at the right. If $(x, f(x))$ is the point of tangency and

$$(x + \Delta x, f(x + \Delta x))$$

is a second point on the graph of f, then the slope of the secant line through the two points is

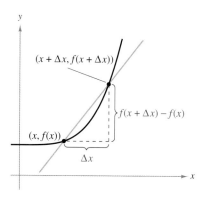

The Secant Line Through the Two Points $(x, f(x))$ and $(x + \Delta x, f(x + \Delta x))$

$$m = \frac{y_2 - y_1}{x_2 - x_1} \qquad \text{Formula for slope}$$

$$m_{\text{sec}} = \frac{f(x + \Delta x) - f(x)}{(x + \Delta x) - x} \qquad \frac{\text{Change in } y}{\text{Change in } x}$$

$$m_{\text{sec}} = \frac{f(x + \Delta x) - f(x)}{\Delta x}. \qquad \text{Slope of secant line}$$

The right side of this equation is called the **difference quotient.** The denominator Δx is the **change in x,** and the numerator is the **change in y.** The beauty of this procedure is that you obtain more and more accurate approximations of the slope of the tangent line by choosing points closer and closer to the point of tangency, as shown in Figure 2.7. Using the limit process, you can find the *exact* slope of the tangent line at $(x, f(x))$, which is also the slope of the graph of f at $(x, f(x))$.

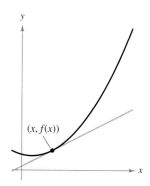

FIGURE 2.7   As Δx approaches 0, the secant lines approach the tangent line.

Definition of the Slope of a Graph

The **slope** m of the graph of f at the point

$$(x, f(x))$$

is equal to the slope of the tangent line to the graph of f at $(x, f(x))$, and is given by

$$m = \lim_{\Delta x \to 0} m_{\text{sec}} = \lim_{\Delta x \to 0} \frac{f(x + \Delta x) - f(x)}{\Delta x}$$

provided this limit exists.

EXAMPLE 3   Finding Slope by the Limit Process

Find the slope of the graph of

$$f(x) = x^2$$

at the point $(-2, 4)$.

SOLUTION   Begin by finding an expression that represents the slope of a secant line at the point $(-2, 4)$.

$$m_{sec} = \frac{f(-2 + \Delta x) - f(-2)}{\Delta x} \qquad \text{Set up difference quotient.}$$

$$= \frac{(-2 + \Delta x)^2 - (-2)^2}{\Delta x} \qquad \text{Use } f(x) = x^2.$$

$$= \frac{4 - 4\Delta x + (\Delta x)^2 - 4}{\Delta x} \qquad \text{Expand terms.}$$

$$= \frac{-4\Delta x + (\Delta x)^2}{\Delta x} \qquad \text{Simplify.}$$

$$= \frac{\Delta x(-4 + \Delta x)}{\Delta x} \qquad \text{Factor and divide out.}$$

$$= -4 + \Delta x, \quad \Delta x \neq 0 \qquad \text{Simplify.}$$

Next, take the limit of m_{sec} as $\Delta x \to 0$.

$$m = \lim_{\Delta x \to 0} m_{sec} = \lim_{\Delta x \to 0} (-4 + \Delta x) = -4 + 0 = -4$$

So, the graph of f has a slope of -4 at the point $(-2, 4)$, as shown in Figure 2.8.

✓ *Checkpoint 3*   *Worked-out solution available at LarsonAppliedCalculus.com*

Find the slope of the graph of

$$f(x) = x^2$$

at the point $(2, 4)$.

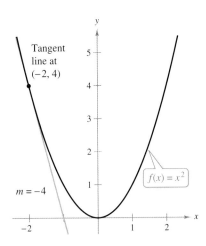

Tangent line at $(-2, 4)$

$m = -4$

$f(x) = x^2$

FIGURE 2.8

EXAMPLE 4   Finding the Slope of a Graph

Find the slope of the graph of $f(x) = -2x + 4$.

SOLUTION   You know from your study of linear functions that the line given by $f(x) = -2x + 4$ has a slope of -2, as shown in Figure 2.9. This conclusion is consistent with the limit definition of slope.

$$m = \lim_{\Delta x \to 0} \frac{f(x + \Delta x) - f(x)}{\Delta x}$$

$$= \lim_{\Delta x \to 0} \frac{[-2(x + \Delta x) + 4] - (-2x + 4)}{\Delta x}$$

$$= \lim_{\Delta x \to 0} \frac{-2x - 2\Delta x + 4 + 2x - 4}{\Delta x}$$

$$= \lim_{\Delta x \to 0} \frac{-2\Delta x}{\Delta x}$$

$$= -2$$

✓ *Checkpoint 4*   *Worked-out solution available at LarsonAppliedCalculus.com*

Find the slope of the graph of $f(x) = 2x + 5$.

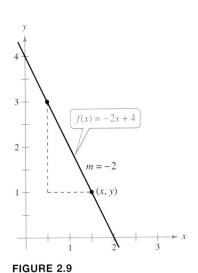

$f(x) = -2x + 4$

$m = -2$

(x, y)

FIGURE 2.9

EXAMPLE 5   Finding a Formula for the Slope of a Graph

Find a formula for the slope of the graph of $f(x) = x^2 + 1$. What are the slopes at the points $(-1, 2)$ and $(2, 5)$?

SOLUTION   Begin by finding an expression that represents the slope of a secant line at *any* point $(x, f(x))$.

$$
\begin{aligned}
m_{\text{sec}} &= \frac{f(x + \Delta x) - f(x)}{\Delta x} && \text{Set up difference quotient.} \\
&= \frac{[(x + \Delta x)^2 + 1] - (x^2 + 1)}{\Delta x} && \text{Use } f(x) = x^2 + 1. \\
&= \frac{x^2 + 2x\Delta x + (\Delta x)^2 + 1 - x^2 - 1}{\Delta x} && \text{Expand terms.} \\
&= \frac{2x\Delta x + (\Delta x)^2}{\Delta x} && \text{Simplify.} \\
&= \frac{\cancel{\Delta x}(2x + \Delta x)}{\cancel{\Delta x}} && \text{Factor and divide out.} \\
&= 2x + \Delta x, \quad \Delta x \neq 0 && \text{Simplify.}
\end{aligned}
$$

Next, take the limit of m_{sec} as $\Delta x \to 0$.

$$
\begin{aligned}
m &= \lim_{\Delta x \to 0} m_{\text{sec}} \\
&= \lim_{\Delta x \to 0} (2x + \Delta x) \\
&= 2x + 0 \\
&= 2x
\end{aligned}
$$

Using the formula $m = 2x$, you can find the slopes at the specified points. At $(-1, 2)$, the slope is $m = 2(-1) = -2$, and at $(2, 5)$, the slope is $m = 2(2) = 4$. The graph of f is shown in Figure 2.10.

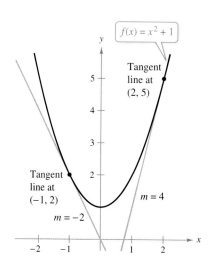

FIGURE 2.10

✔ **Checkpoint 5**   *Worked-out solution available at LarsonAppliedCalculus.com*

Find a formula for the slope of the graph of

$$f(x) = 4x^2 + 1.$$

What are the slopes at the points $(0, 1)$ and $(1, 5)$?

The Derivative of a Function

In Example 5, you started with the function

$$f(x) = x^2 + 1$$

and used the limit process to derive another function, $m = 2x$, that represents the slope of the graph of f at the point $(x, f(x))$. This derived function is called the **derivative** of f at x. It is denoted by $f'(x)$, which is read as "f prime of x."

Definition of the Derivative

The **derivative of f at x** is given by

$$f'(x) = \lim_{\Delta x \to 0} \frac{f(x + \Delta x) - f(x)}{\Delta x}$$

provided this limit exists. A function is **differentiable** at x when its derivative exists at x. The process of finding derivatives is called **differentiation.**

In addition to $f'(x)$, other notations can be used to denote the derivative of $y = f(x)$. The most common are

$$\frac{dy}{dx}, \quad y', \quad \frac{d}{dx}[f(x)], \quad \text{and} \quad D_x[y].$$

EXAMPLE 6   Finding a Derivative

Find the derivative of

$$f(x) = 3x^2 - 2x.$$

SOLUTION

$$
\begin{aligned}
f'(x) &= \lim_{\Delta x \to 0} \frac{f(x + \Delta x) - f(x)}{\Delta x} \\
&= \lim_{\Delta x \to 0} \frac{[3(x + \Delta x)^2 - 2(x + \Delta x)] - (3x^2 - 2x)}{\Delta x} \\
&= \lim_{\Delta x \to 0} \frac{3x^2 + 6x\Delta x + 3(\Delta x)^2 - 2x - 2\Delta x - 3x^2 + 2x}{\Delta x} \\
&= \lim_{\Delta x \to 0} \frac{6x\Delta x + 3(\Delta x)^2 - 2\Delta x}{\Delta x} \\
&= \lim_{\Delta x \to 0} \frac{\Delta x(6x + 3\Delta x - 2)}{\Delta x} \\
&= \lim_{\Delta x \to 0} (6x + 3\Delta x - 2) \\
&= 6x + 3(0) - 2 \\
&= 6x - 2
\end{aligned}
$$

So, the derivative of $f(x) = 3x^2 - 2x$ is

$$f'(x) = 6x - 2.$$

✓ **Checkpoint 6**   Worked-out solution available at LarsonAppliedCalculus.com

Find the derivative of

$$f(x) = x^2 - 5x.$$

STUDY TIP

The notation dy/dx is read as "the derivative of y with respect to x" or simply "dy, dx." Using limit notation, you can write

$$
\begin{aligned}
\frac{dy}{dx} &= \lim_{\Delta x \to 0} \frac{\Delta y}{\Delta x} \\
&= \lim_{\Delta x \to 0} \frac{f(x + \Delta x) - f(x)}{\Delta x} \\
&= f'(x).
\end{aligned}
$$

In many applications, it is convenient to use a variable other than x as the independent variable. Example 7 shows a function that uses t as the independent variable.

TECH TUTOR

You can use a graphing utility to confirm the result in Example 7. One way to do this is to choose a point on the graph of $y = 2/t$, such as $(1, 2)$, and find the equation of the tangent line at that point. Using the derivative found in the example, you know that the slope of the tangent line when $t = 1$ is $m = -2$. This means that the tangent line at the point $(1, 2)$ is

$$y - y_1 = m(t - t_1)$$
$$y - 2 = -2(t - 1)$$
$$y = -2t + 4.$$

By graphing $y = 2/t$ and $y = -2t + 4$ in the same viewing window, as shown below, you can confirm that the line is tangent to the graph at the point $(1, 2)$.

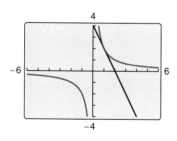

EXAMPLE 7   Finding a Derivative

Find the derivative of y with respect to t for the function

$$y = \frac{2}{t}.$$

SOLUTION   Consider $y = f(t)$, and use the limit process as shown.

$$\frac{dy}{dt} = \lim_{\Delta t \to 0} \frac{f(t + \Delta t) - f(t)}{\Delta t} \qquad \text{Set up difference quotient.}$$

$$= \lim_{\Delta t \to 0} \frac{\dfrac{2}{t + \Delta t} - \dfrac{2}{t}}{\Delta t} \qquad \text{Use } f(t) = 2/t.$$

$$= \lim_{\Delta t \to 0} \frac{\dfrac{2t - 2(t + \Delta t)}{t(t + \Delta t)}}{\Delta t} \qquad \text{Combine fractions in numerator.}$$

$$= \lim_{\Delta t \to 0} \frac{2t - 2t - 2\Delta t}{\Delta t(t)(t + \Delta t)} \qquad \text{Expand terms in numerator.}$$

$$= \lim_{\Delta t \to 0} \frac{-2\Delta t}{\Delta t(t)(t + \Delta t)} \qquad \text{Factor and divide out.}$$

$$= \lim_{\Delta t \to 0} \frac{-2}{t(t + \Delta t)} \qquad \text{Simplify.}$$

$$= \frac{-2}{t(t + 0)} \qquad \text{Direct substitution}$$

$$= -\frac{2}{t^2} \qquad \text{Simplify.}$$

So, the derivative of y with respect to t is

$$\frac{dy}{dt} = -\frac{2}{t^2}.$$

✓ **Checkpoint 7**   *Worked-out solution available at LarsonAppliedCalculus.com*

Find the derivative of y with respect to t for the function $y = 4/t$.   ▪

Remember that the derivative of a function gives you a formula for finding the slope of the tangent line at any point on the graph of the function. For instance, in Example 7 the slope of the tangent line to the graph of f at the point $(1, 2)$ is given by

$$f'(1) = -\frac{2}{1^2} = -2.$$

To find the slopes of the graph at other points, substitute the t-coordinate of the point into the derivative, as shown below.

Point	t-Coordinate	Slope
$(2, 1)$	$t = 2$	$m = f'(2) = -\dfrac{2}{2^2} = -\dfrac{1}{2}$
$(-2, -1)$	$t = -2$	$m = f'(-2) = -\dfrac{2}{(-2)^2} = -\dfrac{1}{2}$

Differentiability and Continuity

Not every function is differentiable. Figure 2.11 shows some common situations in which a function will not be differentiable at a point—vertical tangent lines, discontinuities, and sharp turns in the graph. Each of the functions shown in Figure 2.11 is differentiable at every value of *x except x* = 0.

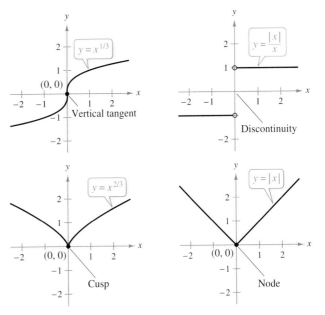

Functions That Are Not Differentiable at *x* = 0

FIGURE 2.11

In Figure 2.11, you can see that all but one of the functions are continuous at *x* = 0 but none are differentiable there. This shows that continuity is not a strong enough condition to guarantee differentiability. On the other hand, if a function is differentiable at a point, then it must be continuous at that point. This important result is stated in the following theorem.

Differentiability Implies Continuity

If a function f is differentiable at $x = c$, then f is continuous at $x = c$.

SUMMARIZE   (Section 2.1)

1. Describe a tangent line and how it can be used to approximate the slope of a graph at a point *(page 80)*. For an example of a tangent line, see Example 1.

2. State the definition of the slope of a graph using the limit process *(page 82)*. For examples of finding the slopes of graphs at points using the limit process, see Examples 3, 4, and 5.

3. State the definition of the derivative of a function *(page 85)*. For examples of finding derivatives of functions, see Examples 6 and 7.

4. Describe the relationship between differentiability and continuity *(page 87)*. For an example showing that continuity does not guarantee differentiability, see Figure 2.11.

SKILLS WARM UP 2.1

The following warm-up exercises involve skills that were covered in earlier sections. You will use these skills in the exercise set for this section. For additional help, review Sections 1.3, 1.4, and 1.5.

In Exercises 1–4, find an equation of the line containing P and Q.

1. $P(3, 1)$, $Q(3, 6)$

2. $P(2, 2)$, $Q(-5, 2)$

3. $P(1, 5)$, $Q(4, -1)$

4. $P(3, 5)$, $Q(-1, -7)$

In Exercises 5–8, find the limit.

5. $\displaystyle\lim_{\Delta x\to 0}\frac{2x\Delta x + (\Delta x)^2}{\Delta x}$

6. $\displaystyle\lim_{\Delta x\to 0}\frac{3x^2\Delta x + 3x(\Delta x)^2 + (\Delta x)^3}{\Delta x}$

7. $\displaystyle\lim_{\Delta x\to 0}\frac{1}{x(x + \Delta x)}$

8. $\displaystyle\lim_{\Delta x\to 0}\frac{(x + \Delta x)^2 - x^2}{\Delta x}$

In Exercises 9–12, find the domain of the function.

9. $f(x) = 3x$

10. $f(x) = \dfrac{1}{x - 1}$

11. $f(x) = \dfrac{1}{5}x^3 - 2x^2 + \dfrac{1}{3}x - 1$

12. $f(x) = \dfrac{6x}{x^3 + x}$

Exercises 2.1

See *CalcChat.com* for tutorial help and worked-out solutions to odd-numbered exercises.

Sketching Tangent Lines In Exercises 1–6, trace the graph and sketch the tangent lines at (x_1, y_1) and (x_2, y_2).

1.

2.

3.

4.

5.

6.
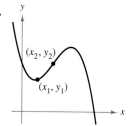

Approximating the Slope of a Graph In Exercises 7–12, approximate the slope of the graph at the point (x, y). (Each square on the grid is 1 unit by 1 unit.) See Example 1.

7.

8.

9.

10.

11.

12.
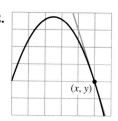

13. Revenue The graph represents the revenue R (in millions of dollars) for Under Armour from 2008 through 2013, where t represents the year, with $t = 8$ corresponding to 2008. Estimate and interpret the slopes of the graph for the years 2009 and 2011. *(Source: Under Armour, Inc.)*

Under Armour

Year (8 ↔ 2008)

14. Sales The graph represents the sales S (in millions of dollars) for Fossil from 2007 through 2013, where t represents the year, with $t = 7$ corresponding to 2007. Estimate and interpret the slopes of the graph for the years 2010 and 2012. *(Source: Fossil, Group)*

Fossil

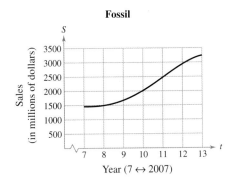

Year (7 ↔ 2007)

15. Temperature The graph represents the average monthly temperature F (in degrees Fahrenheit) in Bland, Virginia, for one year, where t represents the month, with $t = 1$ corresponding to January, $t = 2$ corresponding to February, and so on. Estimate and interpret the slopes of the graph at $t = 3$, 7, and 10. *(Source: National Oceanic and Atmospheric Administration)*

Average Temperature in Bland

Month (1 ↔ January)

16. HOW DO YOU SEE IT? Two long distance runners starting out side by side begin a 10,000-meter run. Their distances are given by $s = f(t)$ and $s = g(t)$, where s is measured in thousands of meters and t is measured in minutes.

10,000-Meter Run

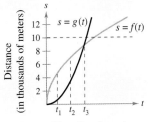

Time (in minutes)

(a) Which runner is running faster at t_1?

(b) What conclusion can you make regarding their rates at t_2?

(c) What conclusion can you make regarding their rates at t_3?

(d) Which runner finishes the race first? Explain.

Finding the Slope of a Graph In Exercises 17–26, use the limit definition to find the slope of the graph of f at the given point. *See Examples 3, 4, and 5.*

17. $f(x) = -1;\ (0, -1)$

18. $f(x) = 6;\ (-2, 6)$

19. $f(x) = 13 - 4x;\ (3, 1)$

20. $f(x) = 6x + 3;\ (1, 9)$

21. $f(x) = 2x^2 - 3;\ (2, 5)$

22. $f(x) = 11 - x^2;\ (3, 2)$

23. $f(x) = x^3 - 4x;\ (-1, 3)$

24. $f(x) = 7x - x^3;\ (-3, 6)$

25. $f(x) = 2\sqrt{x};\ (4, 4)$

26. $f(x) = \sqrt{x + 1};\ (8, 3)$

Finding a Derivative In Exercises 27–40, use the limit definition to find the derivative of the function. *See Examples 6 and 7.*

27. $f(x) = 3$

28. $f(x) = -2$

29. $f(x) = -5x$

30. $f(x) = 4x + 1$

31. $g(s) = \frac{1}{3}s + 2$

32. $h(t) = 6 - \frac{1}{2}t$

33. $f(x) = 4x^2 - 5x$

34. $f(x) = 2x^2 + 7x$

35. $h(t) = \sqrt{t - 3}$

36. $f(x) = \sqrt{x + 2}$

37. $f(t) = t^3 - 12t$

38. $f(t) = t^3 + t^2$

39. $f(x) = \dfrac{1}{x + 2}$

40. $g(s) = \dfrac{1}{s - 4}$

Finding an Equation of a Tangent Line In Exercises 41–48, find an equation of the tangent line to the graph of f at the given point. Then verify your results by using a graphing utility to graph the function and its tangent line at the point.

41. $f(x) = \frac{1}{2}x^2$; $(2, 2)$ **42.** $f(x) = -\frac{1}{8}x^2$; $(-4, -2)$

43. $f(x) = (x - 1)^2$; $(-2, 9)$ **44.** $f(x) = 2x^2 - 5$; $(-1, -3)$

45. $f(x) = \sqrt{x} + 1$; $(4, 3)$ **46.** $f(x) = \sqrt{x + 3}$; $(6, 3)$

47. $f(x) = \frac{1}{5x}$; $\left(-\frac{1}{5}, -1\right)$ **48.** $f(x) = \frac{1}{x - 3}$; $(2, -1)$

Finding an Equation of a Tangent Line In Exercises 49–52, find an equation(s) of the line(s) that is tangent to the graph of f and parallel to the given line.

	Function	Line
49.	$f(x) = -\frac{1}{4}x^2$	$x + y = 0$
50.	$f(x) = x^2 - 7$	$2x + y = 0$
51.	$f(x) = -\frac{1}{3}x^3$	$9x + y - 6 = 0$
52.	$f(x) = x^3 + 2$	$3x - y - 4 = 0$

Determining Differentiability In Exercises 53–58, describe the x-values at which the function is differentiable. Explain your reasoning.

53. $y = |x + 3|$ **54.** $y = |x^2 - 9|$

55. $y = -\sqrt[5]{2x + 1}$ **56.** $y = \sqrt{x - 1}$

57. $y = \dfrac{x^2}{x^2 - 4}$ **58.** $y = \begin{cases} x^3 + 3, & x < 0 \\ x^3 - 3, & x \geq 0 \end{cases}$

Writing a Function Using Derivatives In Exercises 59 and 60, write a function f that has the given characteristics. Then sketch the function.

59. $f(0) = 2$; $f'(x) = -3$ for $-\infty < x < \infty$

60. $f(-2) = f(4) = 0$; $f'(1) = 0$; $f'(x) < 0$ for $x < 1$; $f'(x) > 0$ for $x > 1$

Graphical, Numerical, and Analytic Analysis In Exercises 61–64, use a graphing utility to graph f on the interval $[-2, 2]$. Complete the table by graphically estimating the slopes of the graph at the given points. Then evaluate the slopes analytically and compare your results with those obtained graphically.

x	-2	$-\frac{3}{2}$	-1	$-\frac{1}{2}$	0	$\frac{1}{2}$	1	$\frac{3}{2}$	2
$f(x)$									
$f'(x)$									

61. $f(x) = \frac{1}{4}x^2$ **62.** $f(x) = \frac{3}{4}x^3$

63. $f(x) = -\frac{1}{2}x^3$ **64.** $f(x) = -\frac{3}{2}x^2$

Graphing a Function and Its Derivative In Exercises 65–68, find the derivative of f. Then use a graphing utility to graph f and its derivative in the same viewing window. What does the x-intercept of the derivative indicate about the graph of f?

65. $f(x) = x^2 - 4x$ **66.** $f(x) = 2 + 6x - x^2$

67. $f(x) = x^3 - 3x$ **68.** $f(x) = x^3 - 6x^2$

True or False? In Exercises 69–72, determine whether the statement is true or false. If it is false, explain why or give an example that shows it is false.

69. The slope of the graph of $y = x^2$ is different at every point on the graph of f.

70. If a function is continuous at a point, then it is differentiable at that point.

71. If a function is differentiable at a point, then it is continuous at that point.

72. A tangent line to a graph can intersect the graph at more than one point.

73. Writing Use a graphing utility to graph the two functions

$$f(x) = x^2 + 1 \quad \text{and} \quad g(x) = |x| + 1$$

in the same viewing window. Use the *zoom* and *trace* features to analyze the graphs near the point $(0, 1)$. What do you observe? Which function is differentiable at this point? Write a short paragraph describing the geometric significance of differentiability at a point.

2.2 Some Rules for Differentiation

■ Find the derivatives of functions using the Constant Rule.
■ Find the derivatives of functions using the Power Rule.
■ Find the derivatives of functions using the Constant Multiple Rule.
■ Find the derivatives of functions using the Sum and Difference Rules.
■ Use derivatives to answer questions about real-life situations.

The Constant Rule

In Section 2.1, you found derivatives by the limit process. This process is tedious, even for simple functions, but fortunately there are rules that greatly simplify differentiation. These rules allow you to calculate derivatives without the *direct* use of limits.

> **The Constant Rule**
>
> The derivative of a constant function is zero. That is,
>
> $$\frac{d}{dx}[c] = 0, \quad c \text{ is a constant.}$$

In Exercise 76 on page 102, you will use differentiation to find the rate of change in a company's sales.

PROOF Let $f(x) = c$. Then, by the limit definition of the derivative, you can write

$$f'(x) = \lim_{\Delta x \to 0} \frac{f(x + \Delta x) - f(x)}{\Delta x} = \lim_{\Delta x \to 0} \frac{c - c}{\Delta x} = \lim_{\Delta x \to 0} 0 = 0.$$

So, $\dfrac{d}{dx}[c] = 0$.

Note in Figure 2.12 that the Constant Rule is equivalent to saying that the slope of a horizontal line is zero. An interpretation of the Constant Rule says that the tangent line to a constant function is the function itself. For instance, the equation of the tangent line to $f(x) = 4$ at $x = -1$ is $y = 4$.

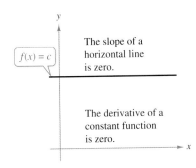

$f(x) = c$

The slope of a horizontal line is zero.

The derivative of a constant function is zero.

FIGURE 2.12

EXAMPLE 1   Finding Derivatives of Constant Functions

Original Function	Derivative
a. $y = 7$	$\dfrac{dy}{dx} = 0$
b. $f(x) = 0$	$f'(x) = 0$
c. $y = 2$	$y' = 0$
d. $g(t) = -\frac{3}{2}$	$g'(t) = 0$

✓ **Checkpoint 1**   Worked-out solution available at LarsonAppliedCalculus.com

Find the derivative of each function.

a. $f(x) = -2$

b. $y = \pi$

c. $g(w) = \sqrt{5}$

d. $s(t) = 320.5$

The Power Rule

The binomial expansion process is used in proving a special case of the Power Rule.

$$(x + \Delta x)^2 = x^2 + 2x\Delta x + (\Delta x)^2$$
$$(x + \Delta x)^3 = x^3 + 3x^2\Delta x + 3x(\Delta x)^2 + (\Delta x)^3$$
$$(x + \Delta x)^n = x^n + nx^{n-1}\Delta x + \frac{n(n-1)x^{n-2}}{2}(\Delta x)^2 + \cdots + (\Delta x)^n$$

$(\Delta x)^2$ is a factor of these terms.

The (Simple) Power Rule

$$\frac{d}{dx}[x^n] = nx^{n-1}, \quad n \text{ is a real number.}$$

PROOF   This proof is limited to the case in which n is a positive integer. Let $f(x) = x^n$. Using the binomial expansion, you can write

$$f'(x) = \lim_{\Delta x \to 0} \frac{f(x + \Delta x) - f(x)}{\Delta x} \qquad \text{Definition of derivative}$$

$$= \lim_{\Delta x \to 0} \frac{(x + \Delta x)^n - x^n}{\Delta x}$$

$$= \lim_{\Delta x \to 0} \frac{x^n + nx^{n-1}\Delta x + \frac{n(n-1)x^{n-2}}{2}(\Delta x)^2 + \cdots + (\Delta x)^n - x^n}{\Delta x}$$

$$= \lim_{\Delta x \to 0} \left[nx^{n-1} + \frac{n(n-1)x^{n-2}}{2}(\Delta x) + \cdots + (\Delta x)^{n-1} \right]$$

$$= nx^{n-1} + 0 + \cdots + 0$$

$$= nx^{n-1}.$$

For the Power Rule, the case in which $n = 1$ is worth remembering as a separate differentiation rule. That is,

$$\frac{d}{dx}[x] = 1. \qquad \text{The derivative of } x \text{ is 1.}$$

This rule is consistent with the fact that the slope of the line $y = x$ is 1. (See Figure 2.13.)

The slope of the line $y = x$ is 1.

FIGURE 2.13

EXAMPLE 2   Applying the Power Rule

Original Function		*Derivative*
a. $f(x) = x^3$		$f'(x) = 3x^2$
b. $y = \dfrac{1}{x^2} = x^{-2}$		$\dfrac{dy}{dx} = (-2)x^{-3} = -\dfrac{2}{x^3}$
c. $g(t) = t$		$g'(t) = 1$

✔ **Checkpoint 2**   *Worked-out solution available at LarsonAppliedCalculus.com*

Find the derivative of each function.

a. $f(x) = x^4$ **b.** $y = \dfrac{1}{x^3}$ **c.** $g(w) = w^2$

In Example 2(b), note that *before* differentiating, you should rewrite $1/x^2$ as x^{-2}. Rewriting is the first step in *many* differentiation problems.

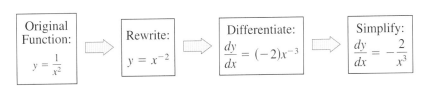

Original Function:
$$y = \frac{1}{x^2}$$

Rewrite:
$$y = x^{-2}$$

Differentiate:
$$\frac{dy}{dx} = (-2)x^{-3}$$

Simplify:
$$\frac{dy}{dx} = -\frac{2}{x^3}$$

Remember that the derivative of a function f is another function that gives the slope of the graph of f at any point at which f is differentiable. So, you can use the derivative to find slopes, as shown in Example 3.

EXAMPLE 3   Finding the Slope of a Graph

Find the slopes of the graph of

$$f(x) = x^2$$

at $x = -2, -1, 0, 1,$ and 2.

SOLUTION   Begin by using the Power Rule to find the derivative of f.

$$f'(x) = 2x \qquad \text{Derivative}$$

You can use the derivative to find the slopes of the graph of f, as shown.

x-Value	Slope of Graph of f
$x = -2$	$m = f'(-2) = 2(-2) = -4$
$x = -1$	$m = f'(-1) = 2(-1) = -2$
$x = 0$	$m = f'(0) = 2(0) = 0$
$x = 1$	$m = f'(1) = 2(1) = 2$
$x = 2$	$m = f'(2) = 2(2) = 4$

The graph of f is shown in Figure 2.14.

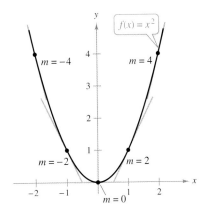

FIGURE 2.14

✓ *Checkpoint 3*   Worked-out solution available at LarsonAppliedCalculus.com

Find the slopes of the graph of

$$f(x) = x^3$$

at $x = -1, 0,$ and 1.

TECH TUTOR

You can use a graphing utility to confirm the slopes found in Example 3. One way to do this is to use the *derivative* feature. For instance, as shown below, at $x = -2$, the slope is -4, which is the same result found in Example 3. (Consult the user's manual of a graphing utility for specific instructions.)

The Constant Multiple Rule

To prove the Constant Multiple Rule, the following property of limits is used.

$$\lim_{x \to a} cg(x) = c\left[\lim_{x \to a} g(x)\right]$$

The Constant Multiple Rule

If f is a differentiable function of x and c is a real number, then

$$\frac{d}{dx}[cf(x)] = cf'(x), \quad c \text{ is a constant.}$$

PROOF   Apply the definition of the derivative to produce

$$\frac{d}{dx}[cf(x)] = \lim_{\Delta x \to 0} \frac{cf(x + \Delta x) - cf(x)}{\Delta x} \qquad \text{Definition of derivative}$$

$$= \lim_{\Delta x \to 0} c\left[\frac{f(x + \Delta x) - f(x)}{\Delta x}\right]$$

$$= c\left[\lim_{\Delta x \to 0} \frac{f(x + \Delta x) - f(x)}{\Delta x}\right]$$

$$= cf'(x).$$

Informally, the Constant Multiple Rule states that constants can be factored out of the differentiation process.

$$\frac{d}{dx}[cf(x)] = c\frac{d}{dx}[f(x)] = cf'(x)$$

The usefulness of this rule is often overlooked, especially when the constant appears in the denominator, as shown below.

$$\frac{d}{dx}\left[\frac{f(x)}{c}\right] = \frac{d}{dx}\left[\frac{1}{c}f(x)\right]$$

$$= \frac{1}{c}\left(\frac{d}{dx}[f(x)]\right)$$

$$= \frac{1}{c}f'(x)$$

To use the Constant Multiple Rule efficiently, look for constants that can be factored out *before* differentiating. For example,

$$\frac{d}{dx}[5x^2] = 5\frac{d}{dx}[x^2] \qquad \text{Factor out 5.}$$

$$= 5(2x) \qquad \text{Differentiate.}$$

$$= 10x \qquad \text{Simplify.}$$

and

$$\frac{d}{dx}\left[\frac{x^2}{5}\right] = \frac{1}{5}\left(\frac{d}{dx}[x^2]\right) \qquad \text{Factor out } \tfrac{1}{5}.$$

$$= \frac{1}{5}(2x) \qquad \text{Differentiate.}$$

$$= \frac{2}{5}x. \qquad \text{Simplify.}$$

EXAMPLE 4   Using the Power and Constant Multiple Rules

Find the derivative of (a) $y = 2x^{1/2}$ and (b) $f(t) = \dfrac{4t^2}{5}$.

SOLUTION

a. Using the Constant Multiple Rule and the Power Rule, you can write

$$\frac{dy}{dx} = \frac{d}{dx}[2x^{1/2}] = \underbrace{2\frac{d}{dx}[x^{1/2}]}_{\substack{\text{Constant} \\ \text{Multiple Rule}}} = \underbrace{2\left(\frac{1}{2}x^{-1/2}\right)}_{\substack{\text{Power} \\ \text{Rule}}} = x^{-1/2} = \frac{1}{\sqrt{x}}.$$

b. Begin by rewriting $f(t)$ as

$$f(t) = \frac{4t^2}{5} = \frac{4}{5}t^2.$$

Then, use the Constant Multiple Rule and the Power Rule to obtain

$$f'(t) = \frac{d}{dt}\left[\frac{4}{5}t^2\right] = \frac{4}{5}\left(\frac{d}{dt}[t^2]\right) = \frac{4}{5}(2t) = \frac{8}{5}t.$$

✓ **Checkpoint 4**   *Worked-out solution available at LarsonAppliedCalculus.com*

Find the derivative of (a) $y = 4x^2$ and (b)   ■

You may find it helpful to combine the Constant Multiple Rule and the Power Rule into one combined rule.

$$\frac{d}{dx}[cx^n] = cnx^{n-1}, \quad n \text{ is a real number, } c \text{ is a constant.}$$

For instance, in Example 4(b), you can apply this combined rule to obtain

$$\frac{d}{dt}\left[\frac{4}{5}t^2\right] = \left(\frac{4}{5}\right)(2)(t) = \frac{8}{5}t.$$

The three functions in the next example are simple, yet errors are frequently made in differentiating functions involving constant multiples of the first power of x. Keep in mind that

$$\frac{d}{dx}[cx] = c, \quad c \text{ is a constant.}$$

EXAMPLE 5   Applying the Constant Multiple Rule

Original Function	*Derivative*
a. $y = -\dfrac{3x}{2}$	$y' = -\dfrac{3}{2}$
b. $y = 3\pi x$	$y' = 3\pi$
c. $y = -\dfrac{x}{2}$	$y' = -\dfrac{1}{2}$

✓ **Checkpoint 5**   *Worked-out solution available at LarsonAppliedCalculus.com*

Find the derivative of (a) $y = \dfrac{t}{4}$ and (b) $y = -\dfrac{2x}{5}$.   ■

Parentheses can play an important role in the use of the Constant Multiple Rule and the Power Rule. In Example 6, be sure you understand the mathematical conventions involving the use of parentheses.

EXAMPLE 6 Using Parentheses When Differentiating

Original Function	Rewrite	Differentiate	Simplify
a. $y = \dfrac{5}{2x^3}$	$y = \dfrac{5}{2}(x^{-3})$	$y' = \dfrac{5}{2}(-3x^{-4})$	$y' = -\dfrac{15}{2x^4}$
b. $y = \dfrac{5}{(2x)^3}$	$y = \dfrac{5}{8}(x^{-3})$	$y' = \dfrac{5}{8}(-3x^{-4})$	$y' = -\dfrac{15}{8x^4}$
c. $y = \dfrac{7}{3x^{-2}}$	$y = \dfrac{7}{3}(x^2)$	$y' = \dfrac{7}{3}(2x)$	$y' = \dfrac{14x}{3}$
d. $y = \dfrac{7}{(3x)^{-2}}$	$y = 63(x^2)$	$y' = 63(2x)$	$y' = 126x$

✓ **Checkpoint 6** *Worked-out solution available at LarsonAppliedCalculus.com*

Find the derivative of each function.

a. $y = \dfrac{9}{4x^2}$

b. $y = \dfrac{9}{(4x)^2}$ ■

When differentiating functions involving radicals, you should rewrite the function with rational exponents. For instance, you should rewrite

$$y = \sqrt[3]{x} \quad \text{as} \quad y = x^{1/3}$$

and you should rewrite

$$y = \frac{1}{\sqrt[3]{x^4}} \quad \text{as} \quad y = x^{-4/3}.$$

EXAMPLE 7 Differentiating Radical Functions

Original Function	Rewrite	Differentiate	Simplify
a. $y = \sqrt{x}$	$y = x^{1/2}$	$y' = \left(\dfrac{1}{2}\right)x^{-1/2}$	$y' = \dfrac{1}{2\sqrt{x}}$
b. $y = \dfrac{1}{2\sqrt[3]{x^2}}$	$y = \dfrac{1}{2}x^{-2/3}$	$y' = \dfrac{1}{2}\left(-\dfrac{2}{3}\right)x^{-5/3}$	$y' = -\dfrac{1}{3x^{5/3}}$
c. $y = \sqrt{2x}$	$y = \sqrt{2}(x^{1/2})$	$y' = \sqrt{2}\left(\dfrac{1}{2}\right)x^{-1/2}$	$y' = \dfrac{1}{\sqrt{2x}}$

✓ **Checkpoint 7** *Worked-out solution available at LarsonAppliedCalculus.com*

Find the derivative of each function.

a. $y = \sqrt{5x}$

b. $y = \sqrt[4]{x}$ ■

The Sum and Difference Rules

To differentiate $y = 3x + 2x^3$, you would probably write

$$y' = 3 + 6x^2$$

without questioning your answer. The validity of differentiating a sum or difference of functions term by term is given by the Sum and Difference Rules.

The Sum and Difference Rules

The derivative of the sum or difference of two differentiable functions is the sum or difference of their derivatives.

$$\frac{d}{dx}[f(x) + g(x)] = f'(x) + g'(x) \qquad \text{Sum Rule}$$

$$\frac{d}{dx}[f(x) - g(x)] = f'(x) - g'(x) \qquad \text{Difference Rule}$$

PROOF Let $h(x) = f(x) + g(x)$. Then, you can prove the Sum Rule as shown.

$$h'(x) = \lim_{\Delta x \to 0} \frac{h(x + \Delta x) - h(x)}{\Delta x} \qquad \text{Definition of derivative}$$

$$= \lim_{\Delta x \to 0} \frac{f(x + \Delta x) + g(x + \Delta x) - f(x) - g(x)}{\Delta x}$$

$$= \lim_{\Delta x \to 0} \frac{f(x + \Delta x) - f(x) + g(x + \Delta x) - g(x)}{\Delta x}$$

$$= \lim_{\Delta x \to 0} \left[\frac{f(x + \Delta x) - f(x)}{\Delta x} + \frac{g(x + \Delta x) - g(x)}{\Delta x} \right]$$

$$= \lim_{\Delta x \to 0} \frac{f(x + \Delta x) - f(x)}{\Delta x} + \lim_{\Delta x \to 0} \frac{g(x + \Delta x) - g(x)}{\Delta x}$$

$$= f'(x) + g'(x)$$

So,

$$\frac{d}{dx}[f(x) + g(x)] = f'(x) + g'(x).$$

The Difference Rule can be proved in a similar manner.

The Sum and Difference Rules can be extended to the sum or difference of any finite number of functions. For instance, if $y = f(x) + g(x) + h(x)$, then $y' = f'(x) + g'(x) + h'(x)$.

STUDY TIP

Look back at Example 6 on page 85. Notice that the example asks for the derivative of the difference of two functions. Compare the result with the one obtained in Example 8(b) at the right.

EXAMPLE 8   Using the Sum and Difference Rules

Original Function	*Derivative*
a. $y = x^3 + 4x^2$	$y' = 3x^2 + 8x$
b. $f(x) = 3x^2 - 2x$	$f'(x) = 6x - 2$
c. $y = \dfrac{3x^2 - x + 1}{x} = 3x - 1 + \dfrac{1}{x}$	$y' = 3 - \dfrac{1}{x^2} = \dfrac{3x^2 - 1}{x^2}$

✓ *Checkpoint 8*   *Worked-out solution available at LarsonAppliedCalculus.com*

Find the derivative of each function.

a. $f(x) = 2x^2 + 5x$ **b.** $y = x^4 - 2x$ **c.** $y = \dfrac{x^3 + 5x - 1}{x}$

With the differentiation rules listed in this section, you can differentiate any polynomial function.

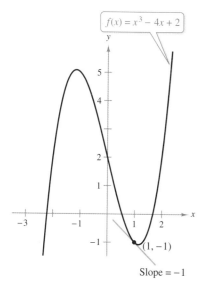

$f(x) = x^3 - 4x + 2$

Slope = −1

FIGURE 2.15

EXAMPLE 9   Finding the Slope of a Graph

Find the slope of the graph of $f(x) = x^3 - 4x + 2$ at the point $(1, -1)$.

SOLUTION   The derivative of $f(x)$ is

$$f'(x) = 3x^2 - 4.$$

So, the slope of the graph of f at $(1, -1)$ is

$$\text{Slope} = f'(1) = 3(1)^2 - 4 = 3 - 4 = -1$$

as shown in Figure 2.15.

✓ *Checkpoint 9*   *Worked-out solution available at LarsonAppliedCalculus.com*

Find the slope of the graph of $f(x) = x^2 - 5x + 1$ at the point $(2, -5)$.   ■

Example 9 illustrates the use of the derivative for determining the shape of a graph. A rough sketch of the graph of $f(x) = x^3 - 4x + 2$ might lead you to think that the point $(1, -1)$ is a minimum point of the graph. After finding the slope at this point to be -1, however, you can conclude that the minimum point (where the slope is 0) is farther to the right. (You will study techniques for finding minimum and maximum points in Section 3.2.)

EXAMPLE 10   Finding an Equation of a Tangent Line

Find an equation of the tangent line to the graph of

$$g(x) = -\frac{1}{2}x^4 + 3x^3 - 2x$$

at the point $\left(-1, -\frac{3}{2}\right)$.

SOLUTION   The derivative of $g(x)$ is $g'(x) = -2x^3 + 9x^2 - 2$. So, the slope of the graph of g at $\left(-1, -\frac{3}{2}\right)$ is

$$\begin{aligned}
\text{Slope} &= g'(-1) \\
&= -2(-1)^3 + 9(-1)^2 - 2 \\
&= 2 + 9 - 2 \\
&= 9
\end{aligned}$$

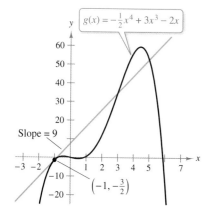

$g(x) = -\frac{1}{2}x^4 + 3x^3 - 2x$

Slope = 9

$\left(-1, -\frac{3}{2}\right)$

FIGURE 2.16

as shown in Figure 2.16. Using the point-slope form, you can write the equation of the tangent line at $\left(-1, -\frac{3}{2}\right)$ as shown.

$$y - \left(-\frac{3}{2}\right) = 9[x - (-1)] \qquad \text{Point-slope form}$$

$$y + \frac{3}{2} = 9x + 9 \qquad \text{Simplify.}$$

$$y = 9x + \frac{15}{2} \qquad \text{Equation of tangent line}$$

✓ *Checkpoint 10*   *Worked-out solution available at LarsonAppliedCalculus.com*

Find an equation of the tangent line to the graph of $f(x) = -x^2 + 3x - 2$ at the point $(2, 0)$.   ■

Application

There are many applications of the derivative that you will study in this textbook. In Example 11, you will use a derivative to find the rate of change of a company's revenue with respect to time.

EXAMPLE 11   Modeling Revenue

From 2008 through 2013, the revenue R (in millions of dollars) for Buffalo Wild Wings can be modeled by

$$R = 22.766t^2 - 309.59t + 1452.0, \quad 8 \le t \le 13$$

where t represents the year, with $t = 8$ corresponding to 2008. At what rate was the revenue for Buffalo Wild Wings changing in 2011?   *(Source: Buffalo Wild Wings, Inc.)*

SOLUTION   One way to answer this question is to find the derivative of the revenue model with respect to time.

$$\frac{dR}{dt} = 45.532t - 309.59, \quad 8 \le t \le 13$$

In 2011 (at $t = 11$), the rate of change of the revenue with respect to time is given by

$$\frac{dR}{dt} = 45.532(11) - 309.59 \approx 191.3.$$

Because R is measured in millions of dollars and t is measured in years, it follows that the derivative dR/dt is measured in millions of dollars per year. So, at the end of 2011, the revenue for Buffalo Wild Wings was increasing at a rate of about \$191.3 million per year, as shown in Figure 2.17.

Revenue for Buffalo Wild Wings

Slope ≈ 191.3

Year (8 ↔ 2008)

FIGURE 2.17

✔ **Checkpoint 11**   *Worked-out solution available at LarsonAppliedCalculus.com*

From 2001 through 2013, the sales per share S (in dollars) for Wal-Mart can be modeled by

$$S = 0.1655t^2 + 6.129t + 42.04, \quad 1 \le t \le 13$$

where t represents the year, with $t = 1$ corresponding to 2001. At what rate was Wal-Mart's sales per share changing in 2010?   *(Source: Wal-Mart Stores, Inc.)* ■

SUMMARIZE   (Section 2.2)

1. State the Constant Rule *(page 91)*. For an example of the Constant Rule, see Example 1.

2. State the Power Rule *(page 92)*. For examples of the Power Rule, see Examples 2 and 3.

3. State the Constant Multiple Rule *(page 94)*. For examples of the Constant Multiple Rule, see Examples 4, 5, 6, and 7.

4. State the Sum Rule *(page 97)*. For an example of the Sum Rule, see Example 8.

5. State the Difference Rule *(page 97)*. For an example of the Difference Rule, see Example 8.

6. Describe a real-life example of how differentiation can be used to analyze the rate of change of a company's revenue *(page 99, Example 11)*.

SKILLS WARM UP 2.2

The following warm-up exercises involve skills that were covered in a previous course. You will use these skills in the exercise set for this section. For additional help, review Appendices A.3 and A.4.

In Exercises 1 and 2, evaluate each expression when $x = 2$.

1. (a) $2x^2$ (b) $(5x)^2$ (c) $6x^{-2}$

2. (a) $\dfrac{1}{(3x)^2}$ (b) $\dfrac{1}{4x^3}$ (c) $\dfrac{(2x)^{-3}}{4x^{-2}}$

In Exercises 3–6, simplify the expression.

3. $4(3)x^3 + 2(2)x$

4. $\frac{1}{2}(3)x^2 - \frac{3}{2}x^{1/2}$

5. $\left(\frac{1}{4}\right)x^{-3/4}$

6. $\frac{1}{3}(3)x^2 - 2\left(\frac{1}{2}\right)x^{-1/2} + \frac{1}{3}x^{-2/3}$

In Exercises 7–10, solve the equation.

7. $3x^2 + 2x = 0$

8. $x^3 - x = 0$

9. $x^2 + 8x - 20 = 0$

10. $3x^2 - 10x + 8 = 0$

Exercises 2.2

See *CalcChat.com* for tutorial help and worked-out solutions to odd-numbered exercises.

 Finding Derivatives In Exercises 1–24, find the derivative of the function. *See Examples 1, 2, 4, 5, and 8.*

1. $y = 3$

2. $f(x) = -8$

3. $y = x^5$

4. $f(x) = \dfrac{1}{x^6}$

5. $h(x) = 3x^3$

6. $h(x) = 6x^5$

7. $y = \dfrac{5x^4}{6}$

8. $g(t) = \dfrac{3t^2}{4}$

9. $f(x) = 4x$

10. $g(x) = \dfrac{x}{3}$

11. $y = 8 - x^3$

12. $y = t^2 - 6$

13. $f(x) = 4x^2 - 3x$

14. $g(x) = 3x^2 + 5x^3$

15. $f(t) = -3t^2 + 2t - 4$

16. $y = 7x^3 - 9x^2 + 8$

17. $s(t) = 4t^4 - 2t^2 + t + 3$

18. $y = 2x^3 - x^2 + 3x - 1$

19. $g(x) = x^{2/3}$

20. $h(x) = x^{5/2}$

21. $y = 4t^{4/3}$

22. $f(x) = 12x^{1/6}$

23. $y = 4x^{-2} + 2x^2$

24. $s(t) = 8t^{-4} + t$

 Using Parentheses When Differentiating In Exercises 25–30, find the derivative of the function. *See Example 6.*

Function	Rewrite	Differentiate	Simplify
25. $y = \dfrac{2}{7x^4}$			
26. $y = \dfrac{2}{3x^2}$			
27. $y = \dfrac{1}{(4x)^3}$			
28. $y = \dfrac{\pi}{(2x)^6}$			
29. $y = \dfrac{4}{(2x)^{-5}}$			
30. $y = \dfrac{4x}{x^{-3}}$			

 Differentiating Radical Functions In Exercises 31–36, find the derivative of the function. *See Example 7.*

Function	Rewrite	Differentiate	Simplify
31. $y = 6\sqrt{x}$			
32. $y = \dfrac{3\sqrt{x}}{4}$			
33. $y = \dfrac{1}{7\sqrt[6]{x}}$			
34. $y = \dfrac{3}{2\sqrt[4]{x^3}}$			
35. $y = \sqrt[5]{8x}$			
36. $y = \sqrt[3]{6x^2}$			

Finding the Slope of a Graph In Exercises 37–44, find the slope of the graph of the function at the given point. Use the *derivative* feature of a graphing utility to confirm your results. *See Examples 3 and 9.*

37. $y = x^{3/2}$

38. $y = x^{-1}$

 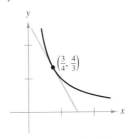

39. $f(t) = t^{-4};\ \left(\frac{1}{2}, 16\right)$ **40.** $f(x) = x^{-1/3};\ \left(8, \frac{1}{2}\right)$

41. $f(x) = 2x^3 + 8x^2 - x - 4;\ (-1, 3)$

42. $f(x) = x^4 - 2x^3 + 5x^2 - 7x;\ (-1, 15)$

43. $f(x) = -\frac{1}{2}x(1 + x^2);\ (1, -1)$

44. $f(x) = 3(5 - x)^2;\ (5, 0)$

 Finding an Equation of a Tangent Line In Exercises 45–50, (a) find an equation of the tangent line to the graph of the function at the given point, (b) use a graphing utility to graph the function and its tangent line at the point, and (c) use the *tangent* feature of a graphing utility to confirm your results. *See Example 10.*

45. $y = -2x^4 + 5x^2 - 3;\ (1, 0)$

46. $y = x^3 + x + 4;\ (-2, -6)$

47. $f(x) = \sqrt[3]{x} + \sqrt[5]{x};\ (1, 2)$ **48.** $f(x) = \dfrac{1}{\sqrt[3]{x^2}} - x;\ (-1, 2)$

49. $y = 3x\left(x^2 - \dfrac{2}{x}\right);\ (2, 18)$ **50.** $y = (2x + 1)^2;\ (0, 1)$

Finding Derivatives In Exercises 51–62, find $f'(x)$.

51. $f(x) = x^2 - \dfrac{4}{x} - 3x^{-2}$

52. $f(x) = 6x^2 - 5x^{-2} + 7x^{-3}$

53. $f(x) = x^2 - 2x - \dfrac{2}{x^4}$ **54.** $f(x) = x^2 + 4x + \dfrac{1}{x}$

55. $f(x) = x^{4/5} + x$ **56.** $f(x) = x^{1/3} - 1$

57. $f(x) = x(x^2 + 1)$ **58.** $f(x) = (x^2 + 2x)(x + 1)$

59. $f(x) = \dfrac{2x^3 - 4x^2 + 3}{x^2}$ **60.** $f(x) = \dfrac{2x^2 - 3x + 1}{x}$

61. $f(x) = \dfrac{4x^3 - 3x^2 + 2x + 5}{x^2}$

62. $f(x) = \dfrac{-6x^3 + 3x^2 - 2x + 1}{x}$

 Finding Horizontal Tangent Lines In Exercises 63–66, determine the point(s), if any, at which the graph of the function has a horizontal tangent line.

63. $y = x^4 - 2x^2 + 3$ **64.** $y = x^3 + 3x^2$

65. $y = \frac{1}{2}x^2 + 5x$ **66.** $y = x^2 + 2x$

Using the Derivative In Exercises 67 and 68, determine the point(s), if any, at which the graph of the function has a tangent line with the given slope.

	Function	*Slope*
67.	$y = x^2 + 3$	$m = 4$
68.	$y = x^2 + 2x$	$m = -10$

 Exploring Relationships In Exercises 69 and 70, (a) sketch the graphs of f and g, (b) find $f'(1)$ and $g'(1)$, (c) sketch the tangent line to each graph at $x = 1$, and (d) explain the relationship between f' and g'.

69. $f(x) = x^3$ **70.** $f(x) = x^2$
     $g(x) = x^3 + 3$ $g(x) = 3x^2$

Exploring Relationships In Exercises 71–74, the relationship between f and g is given. Explain the relationship between f' and g'.

71. $g(x) = f(x) + 6$ **72.** $g(x) = 2f(x)$

73. $g(x) = -5f(x)$ **74.** $g(x) = 3f(x) - 1$

75. Revenue The revenue R (in millions of dollars) for Under Armour from 2008 through 2013 can be modeled by

$$R = -4.1685t^3 + 175.037t^2 - 1950.88t + 7265.3$$

where t is the year, with $t = 8$ corresponding to 2008. *(Source: Under Armour, Inc.)*

Under Armour

(a) Find the slopes of the graph for the years 2009 and 2011.

(b) Compare your results with those obtained in Exercise 13 in Section 2.1.

(c) Interpret the slope of the graph in the context of the problem.

76. Sales The sales S (in millions of dollars) for Fossil from 2007 through 2013 can be modeled by

$$S = -2.67538t^4 + 94.0568t^3 - 1155.203t^2$$
$$+ 6002.42t - 9794.2$$

where t is the year, with $t = 7$ corresponding to 2007. *(Source: Fossil, Group)*

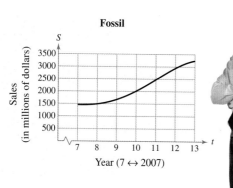

Fossil

(a) Find the slopes of the graph for the years 2010 and 2012.

(b) Compare your results with those obtained in Exercise 14 in Section 2.1.

(c) Interpret the slope of the graph in the context of the problem.

77. Psychology: Migraine Prevalence The graph illustrates the prevalence of migraine headaches in males and females in selected income groups. *(Source: Adapted from Sue/Sue/Sue, Understanding Abnormal Behavior, Seventh Edition)*

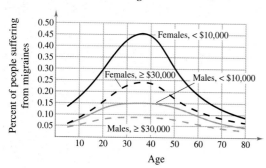

Prevalence of Migraine Headaches

(a) Write a short paragraph describing your general observations about the prevalence of migraines in females and males with respect to age group and income bracket.

(b) Describe the graph of the derivative of each curve, and explain the significance of each derivative. Include an explanation of the units of the derivatives, and indicate the time intervals in which the derivatives would be positive and negative.

78. HOW DO YOU SEE IT? The attendance for four high school basketball games is given by $s = f(t)$, and the attendance for four high school football games is given by $s = g(t)$, where $t = 1$ corresponds to the first game.

Attendance of High School Sports

(a) Which attendance rate, f' or g', is greater at game 1?

(b) What conclusion can you make regarding the attendance rates, f' and g', at game 3?

(c) What conclusion can you make regarding the attendance rates, f' and g', at game 4?

(d) Which sport do you think would have a greater attendance for game 5? Explain your reasoning.

79. Cost The marginal cost for manufacturing an electrical component is $7.75 per unit, and the fixed cost is $500. Write the cost C as a function of x, the number of units produced. Show that the derivative of this cost function is a constant and is equal to the marginal cost.

80. Political Fundraiser A politician raises funds by selling tickets to a dinner for $500. The politician pays $150 for each dinner and has fixed costs of $7000 to rent a dining hall and wait staff. Write the profit P as a function of x, the number of dinners sold. Show that the derivative of the profit function is a constant and is equal to the increase in profit from each dinner sold.

Finding Horizontal Tangent Lines In Exercises 81 and 82, use a graphing utility to graph f and f' over the given interval. Determine any points at which the graph of f has a horizontal tangent line.

Function	*Interval*
81. $f(x) = 4.1x^3 - 12x^2 + 2.5x$	$[0, 3]$
82. $f(x) = x^3 - 1.4x^2 - 0.96x + 1.44$	$[-2, 2]$

True or False? In Exercises 83 and 84, determine whether the statement is true or false. If it is false, explain why or give an example that shows it is false.

83. If $f'(x) = g'(x)$, then $f(x) = g(x)$.

84. If $f(x) = g(x) + c$, then $f'(x) = g'(x)$.

2.3 Rates of Change: Velocity and Marginals

■ Find the average rates of change of functions over intervals.
■ Find the instantaneous rates of change of functions at points.
■ Find the marginal revenues, marginal costs, and marginal profits for products.

Average Rate of Change

In Sections 2.1 and 2.2, you studied the two primary applications of derivatives.

1. Slope The derivative of f is a function that gives the slope of the graph of f at a point $(x, f(x))$.

2. Rate of Change The derivative of f is a function that gives the rate of change of f with respect to x at the point $(x, f(x))$.

In this section, you will see that there are many real-life applications of rates of change. A few are velocity, acceleration, population growth rates, unemployment rates, production rates, and water flow rates. Although rates of change often involve change with respect to time, you can investigate the rate of change of one variable with respect to any other related variable.

When determining the rate of change of one variable with respect to another, you must be careful to distinguish between *average* and *instantaneous* rates of change. The distinction between these two rates of change is comparable to the distinction between the slope of the secant line through two points on a graph and the slope of the tangent line at one point on the graph.

Definition of Average Rate of Change

If $y = f(x)$, then the **average rate of change** of y with respect to x on the interval $[a, b]$ is

$$\text{Average rate of change} = \frac{f(b) - f(a)}{b - a} = \frac{\Delta y}{\Delta x}.$$

Note that $f(a)$ is the value of the function at the *left* endpoint of the interval, $f(b)$ is the value of the function at the *right* endpoint of the interval, and $b - a$ is the width of the interval, as shown in Figure 2.18.

Visitors to a National Park

In Exercise 13 on page 114, you will use the graph of a function to estimate the rate of change of the number of visitors to a national park.

STUDY TIP

In real-life problems, it is important to list the units of measure for a rate of change. The units for $\Delta y / \Delta x$ are "y-units" per "x-units." For example, if y is measured in miles and x is measured in hours, then $\Delta y / \Delta x$ is measured in *miles per hour*.

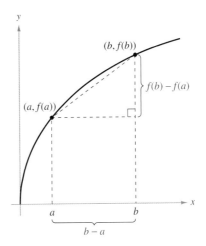

FIGURE 2.18

EXAMPLE 1   Medicine

The concentration C (in milligrams per milliliter) of a drug in a patient's bloodstream is monitored over 10-minute intervals for 2 hours, where t is measured in minutes, as shown in the table.

DATA	t	0	10	20	30	40	50	60	70	80	90	100	110	120
	C	0	2	17	37	55	73	89	103	111	113	113	103	68

Spreadsheet at LarsonAppliedCalculus.com

Find the average rate of change of C over each interval.

a. $[0, 10]$

b. $[0, 20]$

c. $[100, 110]$

SOLUTION

a. For the interval $[0, 10]$, the average rate of change is

Value of C at right endpoint
Value of C at left endpoint

$$\frac{\Delta C}{\Delta t} = \frac{2 - 0}{10 - 0} = \frac{2}{10} = 0.2 \text{ milligram per millimeter per minute.}$$

Width of interval

b. For the interval $[0, 20]$, the average rate of change is

$$\frac{\Delta C}{\Delta t} = \frac{17 - 0}{20 - 0} = \frac{17}{20} = 0.85 \text{ milligram per milliliter per minute.}$$

c. For the interval $[100, 110]$, the average rate of change is

$$\frac{\Delta C}{\Delta t} = \frac{103 - 113}{110 - 100} = \frac{-10}{10} = -1 \text{ milligram per milliliter per minute.}$$

Notice in Figure 2.19 that the average rate of change is positive when the concentration increases and negative when the concentration decreases.

Drug Concentration in Bloodstream

Concentration (in mg/mL)

Time (in minutes)

FIGURE 2.19

✓ **Checkpoint 1**   *Worked-out solution available at LarsonAppliedCalculus.com*

Use the table in Example 1 to find the average rate of change of C over each interval.

a. $[0, 120]$ **b.** $[90, 100]$ **c.** $[90, 120]$ ■

The rates of change in Example 1 are in milligrams per milliliter per minute because the concentration is measured in milligrams per milliliter and the time is measured in minutes.

Concentration is measured in milligrams per milliliter.

Rate of change is measured in milligrams per milliliter per minute.

$$\frac{\Delta C}{\Delta t} = \frac{2 - 0}{10 - 0} = \frac{2}{10} = 0.2 \text{ milligram per milliliter per minute}$$

Time is measured in minutes.

A common application of an average rate of change is to find the **average velocity** of an object that is moving in a straight line. That is,

$$\text{Average velocity} = \frac{\text{change in distance}}{\text{change in time}}.$$

This formula is demonstrated in Example 2.

EXAMPLE 2   **Finding an Average Velocity**

A free-falling object is dropped from a height of 100 feet. *Neglecting air resistance,* the height s (in feet) of the object at time t (in seconds) is given by

$$s = -16t^2 + 100.$$

Find the average velocity of the object over each interval. (See Figure 2.20.)

a. $[1, 2]$ **b.** $[1, 1.5]$ **c.** $[1, 1.1]$

SOLUTION   You can use the position equation $s = -16t^2 + 100$ to determine the heights at

$$t = 1, 1.1, 1.5, \text{ and } 2$$

as shown in the table.

t (in seconds)	0	1	1.1	1.5	2
s (in feet)	100	84	80.64	64	36

a. For the interval $[1, 2]$, the object falls from a height of 84 feet to a height of 36 feet. So, the average velocity is

$$\frac{\Delta s}{\Delta t} = \frac{36 - 84}{2 - 1} = \frac{-48}{1} = -48 \text{ feet per second.}$$

b. For the interval $[1, 1.5]$, the average velocity is

$$\frac{\Delta s}{\Delta t} = \frac{64 - 84}{1.5 - 1} = \frac{-20}{0.5} = -40 \text{ feet per second.}$$

c. For the interval $[1, 1.1]$, the average velocity is

$$\frac{\Delta s}{\Delta t} = \frac{80.64 - 84}{1.1 - 1} = \frac{-3.36}{0.1} = -33.6 \text{ feet per second.}$$

✓ *Checkpoint 2*   *Worked-out solution available at LarsonAppliedCalculus.com*

The height s (in feet) of a free-falling object at time t (in seconds) is given by

$$s = -16t^2 + 180.$$

Find the average velocity of the object over each interval.

a. $[0, 1]$ **b.** $[1, 2]$ **c.** $[2, 3]$ ■

Some falling objects have considerable air resistance. Other falling objects have negligible air resistance. When modeling a falling-body problem, you must decide whether to account for air resistance or neglect it.

FIGURE 2.20

t = 0
t = 1
t = 1.1
t = 1.5
100 ft
t = 2
Falling object

STUDY TIP

In Example 2, the average velocities are negative because the object is moving downward.

Instantaneous Rate of Change and Velocity

Suppose in Example 2 you wanted to find the rate of change of s at the instant $t = 1$ second. Such a rate is called an **instantaneous rate of change.** You can approximate the instantaneous rate of change at $t = 1$ by calculating the average rate of change over smaller and smaller intervals of the form $[1, 1 + \Delta t]$, as shown in the table. From the table, it seems reasonable to conclude that the instantaneous rate of change of the height at $t = 1$ is -32 feet per second.

Δt approaches 0.

Δt	1	0.5	0.1	0.01	0.001	0.0001	0
$\dfrac{\Delta s}{\Delta t}$	-48	-40	-33.6	-32.16	-32.016	-32.0016	-32

$\dfrac{\Delta s}{\Delta t}$ approaches -32.

STUDY TIP

The limit in the definition of instantaneous rate of change is the same as the limit in the definition of the derivative of f at x. This is the second major interpretation of the derivative—as an *instantaneous rate of change in one variable with respect to another.* Recall that the first interpretation of the derivative is as the slope of the graph of f at x.

Definition of Instantaneous Rate of Change

The **instantaneous rate of change** (or simply **rate of change**) of $y = f(x)$ at x is the limit of the average rate of change on the interval

$$[x, x + \Delta x]$$

as Δx approaches 0.

$$\lim_{\Delta x \to 0} \frac{\Delta y}{\Delta x} = \lim_{\Delta x \to 0} \frac{f(x + \Delta x) - f(x)}{\Delta x}$$

If y is a distance and x is time, then the rate of change is a **velocity.**

EXAMPLE 3   Finding an Instantaneous Rate of Change

Find the velocity of the object in Example 2 at $t = 1$.

SOLUTION From Example 2, you know that the height of the falling object is given by

$$s = -16t^2 + 100. \qquad \text{Position function}$$

By taking the derivative of this position function, you obtain the velocity function.

$$s' = -32t \qquad \text{Velocity function}$$

The velocity function gives the velocity at *any* time. So, at $t = 1$, the velocity is

$$s' = -32(1)$$
$$= -32 \text{ feet per second.}$$

✓ **Checkpoint 3**   *Worked-out solution available at LarsonAppliedCalculus.com*

The height of the object in Checkpoint 2 is given by

$$s = -16t^2 + 180.$$

Find the velocities of the object at

a. $t = 1.75$.

b. $t = 2$.

The general **position function** for a free-falling object, neglecting air resistance, is

$$s = -16t^2 + v_0 t + s_0 \qquad \text{Position function}$$

where s is the height (in feet), t is the time (in seconds), v_0 is the initial velocity (in feet per second), and s_0 is the initial height (in feet). Remember that the model assumes that positive velocities indicate upward motion and negative velocities indicate downward motion. The derivative of the position function is the **velocity function.**

$$s' = -32t + v_0 \qquad \text{Velocity function}$$

The absolute value of the velocity is the **speed** of the object.

EXAMPLE 4   Finding the Velocity of a Diver

At time $t = 0$, a diver jumps from a diving board that is 32 feet high, as shown in Figure 2.21. The diver's initial velocity is 16 feet per second, so the position of the diver is given by

$$s = -16t^2 + 16t + 32. \qquad \text{Position function}$$

a. When does the diver hit the water?

b. What is the diver's velocity at impact?

SOLUTION

a. To find the time when the diver hits the water, let $s = 0$ and solve for t.

$$-16t^2 + 16t + 32 = 0 \qquad \text{Set } s \text{ equal to 0.}$$
$$-16(t^2 - t - 2) = 0 \qquad \text{Factor out common factor.}$$
$$-16(t + 1)(t - 2) = 0 \qquad \text{Factor.}$$
$$t = -1 \text{ or } t = 2 \qquad \text{Solve for } t.$$

The solution $t = -1$ does not make sense in the problem because it would mean that the diver hits the water 1 second before jumping. So, you can conclude that the diver hits the water at $t = 2$ seconds.

b. The velocity at time t is given by the derivative

$$s' = -32t + 16. \qquad \text{Velocity function}$$

So, the velocity at time $t = 2$ is

$$s' = -32(2) + 16 = -48 \text{ feet per second.}$$

FIGURE 2.21

✓ **Checkpoint 4**   *Worked-out solution available at LarsonAppliedCalculus.com*

At time $t = 0$, a diver jumps from a diving board that is 12 feet high with an initial velocity of 16 feet per second. The diver's position function is $s = -16t^2 + 16t + 12$.

a. When does the diver hit the water?

b. What is the diver's velocity at impact?   ■

In Example 4, note that the diver's initial velocity is $v_0 = 16$ feet per second (upward) and the diver's initial height is $s_0 = 32$ feet.

Initial velocity is 16 feet per second.
Initial height is 32 feet.

$$s = -16t^2 + 16t + 32$$

Rates of Change in Economics: Marginals

Another important use of rates of change is in the field of economics. Economists refer to *marginal profit, marginal revenue,* and *marginal cost* as the rates of change of the profit, revenue, and cost with respect to x, the number of units produced or sold. An equation that relates these three quantities is

$$P = R - C$$

where P, R, and C represent the following quantities.

$$P = \text{total profit}, \quad R = \text{total revenue}, \quad \text{and} \quad C = \text{total cost}$$

The derivatives of these quantities are called the **marginal profit, marginal revenue,** and **marginal cost,** respectively.

$$\frac{dP}{dx} = \text{marginal profit}$$

$$\frac{dR}{dx} = \text{marginal revenue}$$

$$\frac{dC}{dx} = \text{marginal cost}$$

In many business and economics problems, the number of units produced or sold is restricted to nonnegative integer values, as indicated in Figure 2.22. (Of course, it could happen that a sale involves half or quarter units, but it is hard to conceive of a sale involving $\sqrt{2}$ units.) The variable that denotes such units is called a **discrete variable.**

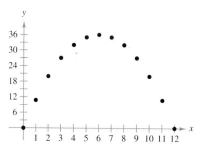

Function of a Discrete Variable

FIGURE 2.22

To analyze a function of a discrete variable x, you can temporarily assume that x is a **continuous variable** and is able to take on any real value in a given interval, as indicated in Figure 2.23. Then, you can use the methods of calculus to find the x-value that corresponds to the marginal revenue, maximum profit, minimum cost, or whatever is called for. Finally, you should round the solution to the nearest sensible x-value—cents, dollars, units, or days, depending on the context of the problem.

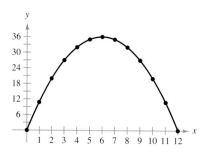

Function of a Continuous Variable

FIGURE 2.23

EXAMPLE 5   Finding the Marginal Profit

The profit (in dollars) derived from selling x units of an alarm clock is given by

$$P = 0.0002x^3 + 10x.$$

a. Find the marginal profit for a production level of 50 units.

b. Compare the marginal profit with the actual gain in profit obtained by increasing the production level from 50 to 51 units.

SOLUTION

a. The profit is $P = 0.0002x^3 + 10x$. The marginal profit is given by the derivative

$$\frac{dP}{dx} = 0.0006x^2 + 10.$$

When $x = 50$, the marginal profit is

$$\frac{dP}{dx} = 0.0006(50)^2 + 10 \qquad \text{Substitute 50 for } x.$$
$$= 0.0006(2500) + 10$$
$$= 1.5 + 10$$
$$= \$11.50 \text{ per unit.} \qquad \text{Marginal profit for } x = 50$$

b. For $x = 50$, the actual profit is

$$P = 0.0002(50)^3 + 10(50) \qquad \text{Substitute 50 for } x.$$
$$= 0.0002(125,000) + 500$$
$$= 25 + 500$$
$$= \$525.00 \qquad \text{Actual profit for } x = 50$$

and for $x = 51$, the actual profit is

$$P = 0.0002(51)^3 + 10(51) \qquad \text{Substitute 51 for } x.$$
$$= 0.0002(132,651) + 510$$
$$\approx 26.53 + 510$$
$$= \$536.53. \qquad \text{Actual profit for } x = 51$$

So, the additional profit obtained by increasing production from 50 to 51 units is

$$536.53 - 525.00 = \$11.53. \qquad \text{Extra profit for one unit}$$

Note that the actual profit increase of \$11.53 (when x increases from 50 to 51 units) can be approximated by the marginal profit of \$11.50 per unit (when $x = 50$), as shown in Figure 2.24.

Marginal Profit

FIGURE 2.24

✓ **Checkpoint 5**   *Worked-out solution available at LarsonAppliedCalculus.com*

Find the marginal profit in Example 5 for a production level of 100 units. Compare this with the actual gain in profit by increasing production from 100 to 101 units.   ∎

STUDY TIP

In Example 5, the marginal profit gives a good approximation of the actual change in profit because the graph of P is nearly straight over the interval $50 \le x \le 51$. You will study more about the use of marginals to approximate actual changes in Section 3.8.

The profit function in Example 5 is unusual in that the profit continues to increase as long as the number of units sold increases. In practice, it is more common to encounter situations in which sales can be increased only by lowering the price per item. Such reductions in price will ultimately cause the profit to decline.

The number of units x that consumers are willing to purchase at a given price per unit p is given by the **demand function**

$$p = f(x). \qquad \text{Demand function}$$

The total revenue R is then related to the price per unit and the quantity demanded (or sold) by the equation

$$R = xp. \qquad \text{Revenue function}$$

EXAMPLE 6   Finding a Demand Function

The table shows the numbers x (in millions) of smartphones sold by a company and the average unit prices p (in dollars) from 2010 through 2014. Use this information to find the demand function and the total revenue function.

Year	2010	2011	2012	2013	2014
x	38	54	72	96	127
p	474	443	407	372	332

SOLUTION   Begin by making a scatter plot of the data using the ordered pairs (x, p), as shown in the figure. From the graph, it appears that a linear model would be a good fit for the data. To find a linear model for the demand function, use any two points, such as

$$(38, 474) \quad \text{and} \quad (127, 332).$$

The slope of the line through these two points is

$$m = \frac{332 - 474}{127 - 38}$$
$$= \frac{-142}{89}$$
$$\approx -1.6.$$

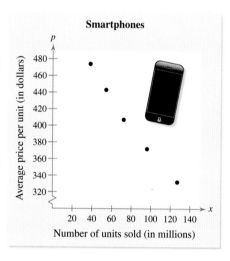

Smartphones

Average price per unit (in dollars) — vertical axis (p) with values 320, 340, 360, 380, 400, 420, 440, 460, 480

Number of units sold (in millions) — horizontal axis (x) with values 20, 40, 60, 80, 100, 120, 140

Using the point-slope form of a line, you can approximate the equation of the demand function to be

$$p = -1.6x + 535. \qquad \text{Demand function}$$

So, the total revenue function for this company's smartphones is

$$R = xp = x(-1.6x + 535) = -1.6x^2 + 535x. \qquad \text{Revenue function}$$

✓ **Checkpoint 6**   *Worked-out solution available at LarsonAppliedCalculus.com*

Use the *regression* feature of a graphing utility or a spreadsheet to find the demand function in Example 6. Then find the total revenue function. Compare your results with those in Example 6. ■

TECH TUTOR

Another way to find a linear model for the demand function in Example 6 is to use the *linear regression* feature of a graphing utility or a spreadsheet software program. (See Checkpoint 6.) You will learn more about linear regression in Section 7.7. (Consult the user's manual of a graphing utility or a spreadsheet software program for specific instructions.)

Demand Function

As the price decreases, more hamburgers are sold.

FIGURE 2.25

EXAMPLE 7   Finding the Marginal Revenue

A fast-food restaurant has determined that the monthly demand for its hamburgers is given by

$$p = \frac{60,000 - x}{20,000}.$$

Figure 2.25 shows that as the price decreases, the quantity demanded increases. The table shows the demands for hamburgers at various prices.

x	60,000	50,000	40,000	30,000	20,000	10,000	0
p	$0.00	$0.50	$1.00	$1.50	$2.00	$2.50	$3.00

Spreadsheet at LarsonAppliedCalculus.com

Find the increase in revenue per hamburger for monthly sales of 20,000 hamburgers. In other words, find the marginal revenue when $x = 20,000$.

SOLUTION   Because the demand is given by

$$p = \frac{60,000 - x}{20,000}$$

and the revenue is given by $R = xp$, you have

$$R = xp \qquad \text{Formula for revenue}$$

$$= x\left(\frac{60,000 - x}{20,000}\right) \qquad \text{Substitute for } p.$$

$$= \frac{1}{20,000}(60,000x - x^2). \qquad \text{Revenue function}$$

By differentiating, you can find the marginal revenue to be

$$\frac{dR}{dx} = \frac{1}{20,000}(60,000 - 2x).$$

So, at $x = 20,000$, the marginal revenue is

$$\frac{dR}{dx} = \frac{1}{20,000}(60,000 - 2x) \qquad \text{Marginal revenue}$$

$$= \frac{1}{20,000}[60,000 - 2(20,000)] \qquad \text{Substitute 20,000 for } x.$$

$$= \frac{1}{20,000}(60,000 - 40,000) \qquad \text{Multiply.}$$

$$= \frac{1}{20,000}(20,000) \qquad \text{Subtract.}$$

$$= \$1 \text{ per unit.} \qquad \text{Marginal revenue when } x = 20,000$$

So, for monthly sales of 20,000 hamburgers, you can conclude that the increase in revenue per hamburger is $1.

✓ **Checkpoint 7**   *Worked-out solution available at LarsonAppliedCalculus.com*

Find the revenue function and marginal revenue for a demand function of

$$p = 2000 - 4x.$$

Find the marginal revenue when $x = 250$.

STUDY TIP

Writing a demand function in the form $p = f(x)$ is a convention used in economics. From a consumer's point of view, it might seem more reasonable to think that the quantity demanded is a function of the price. Mathematically, however, the two points of view are equivalent because a typical demand function is one-to-one and so has an inverse function. For instance, in Example 7, you could write the demand function as $x = 60,000 - 20,000p$.

EXAMPLE 8   Finding the Marginal Profit

For the fast-food restaurant in Example 7, the cost of producing x hamburgers is

$$C = 5000 + 0.56x, \quad 0 \le x \le 50,000.$$

Find the profit and the marginal profit for each production level.

a. $x = 20,000$

b. $x = 24,400$

c. $x = 30,000$

SOLUTION   From Example 7, you know that the total revenue from selling x hamburgers is

$$R = \frac{1}{20,000}(60,000x - x^2).$$

Because the total profit is given by $P = R - C$, you have

$$P = \frac{1}{20,000}(60,000x - x^2) - (5000 + 0.56x)$$

$$= 3x - \frac{x^2}{20,000} - 5000 - 0.56x$$

$$= 2.44x - \frac{x^2}{20,000} - 5000. \qquad \text{See Figure 2.26.}$$

So, the marginal profit is

$$\frac{dP}{dx} = 2.44 - \frac{x}{10,000}.$$

Using these formulas, you can compute the profit and marginal profit.

Production	Profit	Marginal Profit
a. $x = 20,000$	$P = \$23,800$	$2.44 - \dfrac{20,000}{10,000} = \$0.44$ per unit
b. $x = 24,400$	$P = \$24,768$	$2.44 - \dfrac{24,400}{10,000} = \$0.00$ per unit
c. $x = 30,000$	$P = \$23,200$	$2.44 - \dfrac{30,000}{10,000} = -\$0.56$ per unit

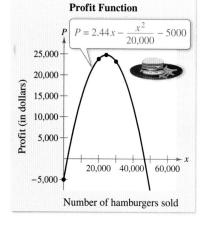

Profit Function

$P = 2.44x - \dfrac{x^2}{20,000} - 5000$

FIGURE 2.26

✓ **Checkpoint 8**   *Worked-out solution available at LarsonAppliedCalculus.com*

From Example 8, compare the marginal profit when 10,000 units are produced with the actual increase in profit from 10,000 units to 10,001 units.   ■

SUMMARIZE   (Section 2.3)

1. State the definition of average rate of change *(page 103)*. For examples of average rates of change, see Examples 1 and 2.

2. State the definition of instantaneous rate of change *(page 106)*. For examples of instantaneous rates of change, see Examples 3 and 4.

3. Describe a real-life example of how rates of change can be used in the field of economics *(pages 109–112, Examples 5, 6, 7, and 8)*.

The following warm-up exercises involve skills that were covered in a previous course or earlier sections. You will use these skills in the exercise set for this section. For additional help, review Sections 2.1 and 2.2.

In Exercises 1–4, evaluate the expression.

1. $\dfrac{-63 - (-105)}{21 - 7}$

2. $\dfrac{-43 - 35}{6 - (-7)}$

3. $\dfrac{24 - 33}{9 - 6}$

4. $\dfrac{40 - 16}{18 - 8}$

In Exercises 5–12, find the derivative of the function.

5. $y = 4x^2 - 2x + 7$

6. $s = -2t^3 + 8t^2 - 7t$

7. $s = -16t^2 + 24t + 30$

8. $y = -16x^2 + 54x + 70$

9. $A = \frac{1}{10}(-2r^3 + 3r^2 + 5r)$

10. $y = \frac{1}{9}(6x^3 - 18x^2 + 63x - 15)$

11. $y = 12x - \dfrac{x^2}{5000}$

12. $y = 138 + 74x - \dfrac{x^3}{10,000}$

Exercises 2.3

See *CalcChat.com* for tutorial help and worked-out solutions to odd-numbered exercises.

1. Research and Development The table shows the amounts A (in billions of dollars) spent on Research and Development in the United States from 1980 through 2012, where t is the year, with $t = 0$ corresponding to 1980. Approximate the average rate of change of A during each period. *(Source: U.S. National Science Foundation)*

(a) 1980–1986   (b) 1986–1992

(c) 1992–1998   (d) 1998–2004

(e) 2004–2010   (f) 1980–2012

(g) 1990–2012   (h) 2000–2012

t	0	1	2	3	4	5	6	7
A	63	72	81	90	102	115	120	126

t	8	9	10	11	12	13	14
A	134	142	152	161	165	166	169

t	15	16	17	18	19	20	21
A	184	197	212	226	245	269	280

t	22	23	24	25	26	27	28
A	279	293	305	327	353	380	407

t	29	30	31	32
A	405	408	428	453

2. Trade Deficit The graph shows the values I (in billions of dollars) of goods imported to the United States and the values E (in billions of dollars) of goods exported from the United States from 1980 through 2013. Approximate the average rates of change of I and E during each period. *(Source: U.S. Census Bureau)*

(a) Imports: 1980–1990

(b) Exports: 1980–1990

(c) Imports: 1990–2000

(d) Exports: 1990–2000

(e) Imports: 2000–2010

(f) Exports: 2000–2010

(g) Imports: 1980–2013

(h) Exports: 1980–2013

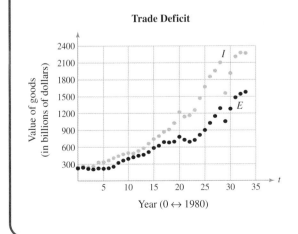

Trade Deficit

Finding Rates of Change In Exercises 3–12, find the average rate of change of the function over the given interval. Compare this rate with the instantaneous rates of change at the endpoints of the interval.

3. $f(t) = 3t + 5;\ [1, 2]$ 4. $h(x) = 7 - 2x;\ [1, 3]$

5. $h(x) = x^2 - 4x + 2;\ [-2, 2]$

6. $f(x) = -x^2 - 6x - 5;\ [-3, 1]$

7. $f(x) = 3x^{4/3};\ [1, 8]$ 8. $f(x) = x^{3/2};\ [1, 4]$

9. $f(x) = \dfrac{1}{x};\ [1, 5]$ 10. $f(x) = \dfrac{1}{\sqrt{x}};\ [1, 9]$

11. $f(t) = t^4 - 2t^2;\ [-2, -1]$

12. $g(x) = x^3 - 1;\ [-1, 1]$

13. **Consumer Trends** The graph shows the number of visitors V (in thousands) to a national park during a one-year period, where $t = 1$ represents January.

Visitors to a National Park

(a) Estimate the average rate of change of V over the interval $[9, 12]$ and explain your results.

(b) Over what interval is the average rate of change approximately equal to the rate of change at $t = 8$? Explain your reasoning.

14. **Medicine** The graph shows the estimated number of milligrams of a pain medication M in the bloodstream t hours after a 1000-milligram dose of the drug has been given.

Pain Medication in Bloodstream

(a) Estimate the average rate of change of M over the interval $[1, 3]$ and explain your results.

(b) Over what interval is the average rate of change approximately equal to the rate of change at $t = 4$? Explain your reasoning.

15. **Velocity** The height s (in feet) at time t (in seconds) of a ball thrown upward from the top of a building is given by $s = -16t^2 + 30t + 250$. Find the average velocity over each indicated interval and compare this velocity with the instantaneous velocities at the endpoints of the interval.

(a) $[0, 1]$ (b) $[1, 2]$ (c) $[2, 3]$ (d) $[3, 4]$

16. **Chemistry: Wind Chill** At $0°$ Celsius, the heat loss H (in kilocalories per square meter per hour) from a person's body can be modeled by

$$H = 33\left(10\sqrt{v} - v + 10.45\right)$$

where v is the wind speed (in meters per second).

(a) Find $\dfrac{dH}{dv}$ and interpret its meaning in this situation.

(b) Find the instantaneous rates of change of H when $v = 2$ and $v = 5$.

17. **Velocity** The height s (in feet) at time t (in seconds) of a silver dollar dropped from the top of a building is given by $s = -16t^2 + 555$.

(a) Find the average velocity over the interval $[2, 3]$.

(b) Find the instantaneous velocities when $t = 2$ and $t = 3$.

(c) How long will it take the coin to hit the ground?

(d) Find the velocity of the coin when it hits the ground.

18. **Velocity** A ball is thrown straight down from the top of a 210-foot building with an initial velocity of -18 feet per second.

(a) Find the position and velocity functions for the ball.

(b) Find the average velocity over the interval $[1, 2]$.

(c) Find the instantaneous velocities when $t = 1$ and $t = 2$.

(d) How long will it take the ball to hit the ground?

(e) Find the velocity of the ball when it hits the ground.

 Marginal Cost In Exercises 19–22, find the marginal cost for producing x units. (The cost is measured in dollars.)

19. $C = 205{,}000 + 9800x$

20. $C = 150{,}000 + 7x^3$

21. $C = 55{,}000 + 470x - 0.25x^2,\ 0 \le x \le 940$

22. $C = 100\left(9 + 3\sqrt{x}\right)$

 Marginal Revenue In Exercises 23–26, find the marginal revenue for selling x units. (The revenue is measured in dollars.)

23. $R = 50x - 0.5x^2$ 24. $R = 30x - x^2$

25. $R = -6x^3 + 8x^2 + 200x$ 26. $R = 50(20x - x^{3/2})$

Marginal Profit In Exercises 27–30, find the marginal profit for selling x units. (The profit is measured in dollars.)

27. $P = -2x^2 + 72x - 145$

28. $P = -0.25x^2 + 2000x - 1,250,000$

29. $P = 0.0013x^3 + 12x$

30. $P = -0.5x^3 + 30x^2 - 164.25x - 1000$

31. **Marginal Cost** The cost C (in dollars) of producing x units of a product is given by

$$C = 3.6\sqrt{x} + 500.$$

(a) Find the marginal cost when $x = 9$.

(b) Find the additional cost when the production increases from 9 to 10 units.

(c) Compare the results of parts (a) and (b).

32. **Marginal Revenue** The revenue R (in dollars) from renting x apartments can be modeled by

$$R = 2x(900 + 32x - x^2).$$

(a) Find the marginal revenue when $x = 14$.

(b) Find the additional revenue when the number of rentals is increased from 14 to 15.

(c) Compare the results of parts (a) and (b).

33. **Marginal Profit** The profit P (in dollars) from selling x tablet computers is given by

$$P = -0.04x^2 + 25x - 1500.$$

(a) Find the marginal profit when $x = 150$.

(b) Find the additional profit when the sales increase from 150 to 151 units.

(c) Compare the results of parts (a) and (b).

34. **Marginal Profit** The profit P (in dollars) from selling x units of a product is given by

$$P = 36,000 + 2048\sqrt{x} - \frac{1}{8x^2}, \quad 150 \le x \le 275.$$

Find the marginal profit for each of the following sales.

(a) $x = 150$ (b) $x = 175$ (c) $x = 200$

(d) $x = 225$ (e) $x = 250$ (f) $x = 275$

35. **Population Growth** The population P (in thousands) of Australia from 1990 through 2013 can be modeled by

$$P = 1.73t^2 + 190.6t + 16,994$$

where t is the year, with $t = 0$ corresponding to 1990. *(Source: U.S. Census Bureau)*

(a) Evaluate P for $t = 0, 3, 6, 9, 12, 15, 18,$ and 21. Explain these values.

(b) Find dP/dt and explain its meaning in this situation.

(c) Evaluate dP/dt for the same values as in part (a). Explain your results.

36. **Health** The temperature T (in degrees Fahrenheit) of a person during an illness can be modeled by the equation

$$T = -0.0375t^2 + 0.3t + 100.4$$

where t is time in hours since the person started to show signs of a fever.

(a) Use a graphing utility to graph the function. Be sure to choose an appropriate window.

(b) Do the slopes of the tangent lines appear to be positive or negative? What does this tell you?

(c) Evaluate the function for $t = 0, 4, 8,$ and 12.

(d) Find dT/dt and explain its meaning in this situation.

(e) Evaluate dT/dt for $t = 0, 4, 8,$ and 12. Explain your results.

37. **Economics** Use the information in the table to find the models and answer the questions below.

DATA Quantity produced and sold (Q)	Price (p)	Total revenue (TR)	Marginal revenue (MR)
0	160	0	—
2	140	280	130
4	120	480	90
6	100	600	50
8	80	640	10
10	60	600	-30

Spreadsheet at LarsonAppliedCalculus.com

(a) Use the *regression* feature of a graphing utility to find a quadratic model that relates the total revenue (TR) to the quantity produced and sold (Q).

(b) Using derivatives, find a model for marginal revenue from the model you found in part (a).

(c) Calculate the marginal revenue for all values of Q using your model in part (b), and compare these values with the actual values given. How good is your model? *(Source: Adapted from Taylor, Economics, Fifth Edition)*

38. **Profit** The monthly demand function p and cost function C for x magazines downloaded from a digital store are $p = 5 - 0.001x$ and $C = 35 + 1.5x$.

(a) Find the monthly revenue R as a function of x.

(b) Find the monthly profit P as a function of x.

(c) Complete the table.

x	600	1200	1800	2400	3000
dR/dx					
dP/dx					
P					

39. Marginal Profit When the price of a glass of lemonade at a lemonade stand was $1.75, 400 glasses were sold. When the price was lowered to $1.50, 500 glasses were sold. Assume that the demand function is linear and that the marginal and fixed costs are $0.10 and $25, respectively.

(a) Find the profit P as a function of x, the number of glasses of lemonade sold.

(b) Use a graphing utility to graph P, and comment about the slopes of P when $x = 300$, $x = 530$, and $x = 700$.

(c) Find the marginal profits when 300 glasses of lemonade are sold, when 530 glasses of lemonade are sold, and when 700 glasses of lemonade are sold.

40. Marginal Profit When the admission price for a baseball game was $30 per ticket, 36,000 tickets were sold. When the price was raised to $35, only 32,000 tickets were sold. Assume that the demand function is linear and that the marginal and fixed costs for the ballpark owners are $5 and $700,000, respectively.

(a) Find the profit P as a function of x, the number of tickets sold.

(b) Use a graphing utility to graph P, and comment about the slopes of P when $x = 18,000$, $x = 28,000$, and $x = 36,000$.

(c) Find the marginal profits when 18,000 tickets are sold, when 28,000 tickets are sold, and when 36,000 tickets are sold.

41. Fuel Cost A car is driven 15,000 miles a year and gets x miles per gallon. Assume that the average fuel cost is $2.60 per gallon.

(a) Find the annual cost of fuel C as a function of x.

(b) Find dC/dx and explain its meaning in this situation.

(c) Use the functions to complete the table.

x	10	15	20	25	30	35	40
C							
dC/dx							

(d) Who would benefit more from a 1 mile per gallon increase in fuel efficiency—the driver who gets 15 miles per gallon or the driver who gets 35 miles per gallon? Explain.

42. Gasoline Sales The number N of gallons of regular unleaded gasoline sold by a gasoline station at a price of p dollars per gallon is given by $N = f(p)$.

(a) Describe the meaning of $f'(2.959)$.

(b) Is $f'(2.959)$ usually positive or negative? Explain.

43. Dow Jones Industrial Average The table shows the year-end closing prices p of the Dow Jones Industrial Average (DJIA) from 2000 through 2013, where t is the year, with $t = 0$ corresponding to 2000. *(Source: Dow Jones Industrial Average)*

t	0	1	2	3
p	10,786.85	10,021.50	8341.63	10,453.92

t	4	5	6	7
p	10,783.01	10,717.50	12,463.15	13,264.82

t	8	9	10	11
p	8776.39	10,428.05	11,577.51	12,217.56

t	12	13
p	13,104.14	16,576.66

Spreadsheet at LarsonAppliedCalculus.com

(a) Determine the average rate of change in the value of the DJIA from 2000 through 2013.

(b) Estimate the instantaneous rate of change in 2005 by finding the average rate of change from 2003 to 2007.

(c) Estimate the instantaneous rate of change in 2005 by finding the average rate of change from 2004 to 2006.

(d) Compare your answers for parts (b) and (c). Which interval do you think produced the best estimate for the instantaneous rate of change in 2005?

44. HOW DO YOU SEE IT? Many populations in nature exhibit logistic growth, which consists of four phases, as shown in the figure. Describe the rate of growth of the population in each phase, and give possible reasons as to why the rates might be changing from phase to phase. *(Source: Adapted from Levine/Miller, Biology: Discovering Life, Second Edition)*

2.4 The Product and Quotient Rules

■ Find the derivatives of functions using the Product Rule.
■ Find the derivatives of functions using the Quotient Rule.
■ Use derivatives to answer questions about real-life situations.

The Product Rule

In Section 2.2, you saw that the derivative of a sum or difference of two functions is simply the sum or difference of their derivatives. The rules for the derivative of a product or quotient of two functions are not as simple.

Population of Bacteria

In Exercise 65 on page 125, you will use the Quotient Rule to find the rate of change of a population of bacteria.

The Product Rule

The derivative of the product of two differentiable functions is equal to the first function times the derivative of the second plus the second function times the derivative of the first.

$$\frac{d}{dx}[f(x)g(x)] = f(x)g'(x) + g(x)f'(x)$$

PROOF Some mathematical proofs, such as the proof of the Sum Rule, are straightforward. Others involve clever steps that may not appear to follow clearly from a prior step. This proof involves such a step—subtracting and adding the same quantity—which is shown in color below. Let $F(x) = f(x)g(x)$.

$$F'(x) = \lim_{\Delta x \to 0} \frac{F(x + \Delta x) - F(x)}{\Delta x}$$

$$= \lim_{\Delta x \to 0} \frac{f(x + \Delta x)g(x + \Delta x) - f(x)g(x)}{\Delta x}$$

$$= \lim_{\Delta x \to 0} \frac{f(x + \Delta x)g(x + \Delta x) - f(x + \Delta x)g(x) + f(x + \Delta x)g(x) - f(x)g(x)}{\Delta x}$$

$$= \lim_{\Delta x \to 0} \left[f(x + \Delta x)\frac{g(x + \Delta x) - g(x)}{\Delta x} + g(x)\frac{f(x + \Delta x) - f(x)}{\Delta x} \right]$$

$$= \lim_{\Delta x \to 0} \left[f(x + \Delta x)\frac{g(x + \Delta x) - g(x)}{\Delta x} \right] + \lim_{\Delta x \to 0} \left[g(x)\frac{f(x + \Delta x) - f(x)}{\Delta x} \right]$$

$$= \lim_{\Delta x \to 0} f(x + \Delta x) \cdot \lim_{\Delta x \to 0} \frac{g(x + \Delta x) - g(x)}{\Delta x} + \lim_{\Delta x \to 0} g(x) \cdot \lim_{\Delta x \to 0} \frac{f(x + \Delta x) - f(x)}{\Delta x}$$

$$= f(x)g'(x) + g(x)f'(x)$$

STUDY TIP

Rather than trying to remember the formula for the Product Rule, it may be more helpful to remember its verbal statement:

the first function times the derivative of the second plus the second function times the derivative of the first.

EXAMPLE 1   Using the Product Rule

Find the derivative of $y = (3x - 2x^2)(5 + 4x)$.

SOLUTION   Using the Product Rule, you can write

$$\frac{dy}{dx} = \overbrace{(3x - 2x^2)}^{\text{First}} \overbrace{\frac{d}{dx}[5 + 4x]}^{\text{Derivative of second}} + \overbrace{(5 + 4x)}^{\text{Second}} \overbrace{\frac{d}{dx}[3x - 2x^2]}^{\text{Derivative of first}}$$

$$= (3x - 2x^2)(4) + (5 + 4x)(3 - 4x)$$

$$= (12x - 8x^2) + (15 - 8x - 16x^2)$$

$$= 15 + 4x - 24x^2.$$

✓ *Checkpoint 1*   *Worked-out solution available at LarsonAppliedCalculus.com*

Find the derivative of $y = (4x + 3x^2)(6 - 3x)$.   ■

In general, the derivative of the product of two functions is not equal to the product of the derivatives of the two functions. To see this, compare the product of the derivatives of

$$f(x) = 3x - 2x^2 \quad \text{and} \quad g(x) = 5 + 4x$$

with the derivative found in Example 1.

In the next example, notice that the first step in differentiating is *rewriting the original function.*

TECH TUTOR

If you have access to a symbolic differentiation utility, try using it to confirm several of the derivatives in this section.

EXAMPLE 2   Using the Product Rule

Find the derivative of $f(x) = \left(\dfrac{1}{x} + 1\right)(x - 1)$.

SOLUTION   Begin by rewriting the function.

$$f(x) = \left(\frac{1}{x} + 1\right)(x - 1) \qquad \text{Write original function.}$$

$$= (x^{-1} + 1)(x - 1) \qquad \text{Rewrite function.}$$

Next, apply the Product Rule.

$$f'(x) = (x^{-1} + 1)\frac{d}{dx}[x - 1] + (x - 1)\frac{d}{dx}[x^{-1} + 1] \quad \text{Product Rule}$$

$$= (x^{-1} + 1)(1) + (x - 1)(-x^{-2})$$

$$= \frac{1}{x} + 1 - \frac{x - 1}{x^2}$$

$$= \frac{x + x^2 - x + 1}{x^2} \qquad \text{Write with common denominator.}$$

$$= \frac{x^2 + 1}{x^2} \qquad \text{Simplify.}$$

✓ *Checkpoint 2*   *Worked-out solution available at LarsonAppliedCalculus.com*

Find the derivative of

$$f(x) = \left(\frac{1}{x} + 1\right)(2x + 1).$$ ■

You now have two differentiation rules that deal with products—the Constant Multiple Rule and the Product Rule. The difference between these two rules is that the Constant Multiple Rule is used when one of the factors is a constant

$$F(x) = cf(x)$$ Use Constant Multiple Rule.

whereas the Product Rule is used when both of the factors are variable quantities

$$F(x) = f(x)g(x).$$ Use Product Rule.

The next example compares these two rules.

EXAMPLE 3 **Comparing Differentiation Rules**

Find the derivative of each function.

a. $y = 2x(x^2 + 3x)$

b. $y = 2(x^2 + 3x)$

SOLUTION

a. Because both factors are variable quantities, use the Product Rule.

$$y = 2x(x^2 + 3x)$$

$$\frac{dy}{dx} = (2x)\frac{d}{dx}[x^2 + 3x] + (x^2 + 3x)\frac{d}{dx}[2x]$$ Product Rule

$$= (2x)(2x + 3) + (x^2 + 3x)(2)$$

$$= 4x^2 + 6x + 2x^2 + 6x$$

$$= 6x^2 + 12x$$

b. Because one of the factors is a constant, use the Constant Multiple Rule.

$$y = 2(x^2 + 3x)$$

$$\frac{dy}{dx} = 2\frac{d}{dx}[x^2 + 3x]$$ Constant Multiple Rule

$$= 2(2x + 3)$$

$$= 4x + 6$$

✓ **Checkpoint 3**   *Worked-out solution available at LarsonAppliedCalculus.com*

Find the derivative of each function.

a. $y = 3x(2x^2 + 5x)$

b. $y = 3(2x^2 + 5x)$

The Product Rule can be extended to products that have more than two factors. For example, if f, g, and h are differentiable functions of x, then

$$\frac{d}{dx}[f(x)g(x)h(x)] = f'(x)g(x)h(x) + f(x)g'(x)h(x) + f(x)g(x)h'(x).$$

STUDY TIP

You could calculate the derivative in Example 3(a) without the Product Rule. For instance,

$$y = 2x(x^2 + 3x) = 2x^3 + 6x^2$$

and

$$\frac{dy}{dx} = 6x^2 + 12x.$$

The Quotient Rule

In Section 2.2, you saw that by using the Constant Rule, the Power Rule, the Constant Multiple Rule, and the Sum and Difference Rules, you were able to differentiate any polynomial function. By combining these rules with the Quotient Rule, you can now differentiate any *rational* function.

STUDY TIP

As suggested for the Product Rule, it may be more helpful to remember the verbal statement of the Quotient Rule rather than trying to remember the formula for the rule.

The Quotient Rule

The derivative of the quotient of two differentiable functions is equal to the denominator times the derivative of the numerator minus the numerator times the derivative of the denominator, all divided by the square of the denominator.

$$\frac{d}{dx}\left[\frac{f(x)}{g(x)}\right] = \frac{g(x)f'(x) - f(x)g'(x)}{[g(x)]^2}, \quad g(x) \neq 0$$

PROOF   Begin by letting

$$F(x) = \frac{f(x)}{g(x)}.$$

As in the proof of the Product Rule, a key step in this proof is subtracting and adding the same quantity—which is shown in color below.

$$F'(x) = \lim_{\Delta x \to 0} \frac{F(x + \Delta x) - F(x)}{\Delta x}$$

$$= \lim_{\Delta x \to 0} \frac{\dfrac{f(x + \Delta x)}{g(x + \Delta x)} - \dfrac{f(x)}{g(x)}}{\Delta x}$$

$$= \lim_{\Delta x \to 0} \left[\left(\frac{g(x)f(x + \Delta x)}{g(x)g(x + \Delta x)} - \frac{f(x)g(x + \Delta x)}{g(x)g(x + \Delta x)}\right) \div \Delta x\right]$$

$$= \lim_{\Delta x \to 0} \left[\frac{g(x)f(x + \Delta x) - f(x)g(x + \Delta x)}{g(x)g(x + \Delta x)} \cdot \frac{1}{\Delta x}\right]$$

$$= \lim_{\Delta x \to 0} \frac{g(x)f(x + \Delta x) - f(x)g(x + \Delta x)}{\Delta x g(x)g(x + \Delta x)}$$

$$= \lim_{\Delta x \to 0} \frac{g(x)f(x + \Delta x) - f(x)g(x) + f(x)g(x) - f(x)g(x + \Delta x)}{\Delta x g(x)g(x + \Delta x)}$$

$$= \frac{\displaystyle\lim_{\Delta x \to 0} \frac{g(x)[f(x + \Delta x) - f(x)]}{\Delta x} - \lim_{\Delta x \to 0} \frac{f(x)[g(x + \Delta x) - g(x)]}{\Delta x}}{\displaystyle\lim_{\Delta x \to 0} [g(x)g(x + \Delta x)]}$$

$$= \frac{g(x)\left[\displaystyle\lim_{\Delta x \to 0} \frac{f(x + \Delta x) - f(x)}{\Delta x}\right] - f(x)\left[\displaystyle\lim_{\Delta x \to 0} \frac{g(x + \Delta x) - g(x)}{\Delta x}\right]}{\displaystyle\lim_{\Delta x \to 0} [g(x)g(x + \Delta x)]}$$

$$= \frac{g(x)f'(x) - f(x)g'(x)}{[g(x)]^2}$$

From the Quotient Rule, you can see that the derivative of a quotient is not, in general, the quotient of the derivatives. That is,

$$\frac{d}{dx}\left[\frac{f(x)}{g(x)}\right] \neq \frac{f'(x)}{g'(x)}.$$

ALGEBRA TUTOR

When applying the Quotient Rule, it is suggested that you enclose all factors and derivatives in symbols of grouping, such as parentheses. Also, pay special attention to the subtraction required in the numerator. For help in simplifying expressions like the one in Example 4, see the *Chapter 2 Algebra Tutor* on pages 158 and 159.

EXAMPLE 4   Using the Quotient Rule

Find the derivative of $y = \dfrac{x-1}{2x+3}$.

SOLUTION   Apply the Quotient Rule, as shown.

$$\frac{dy}{dx} = \frac{(2x+3)\dfrac{d}{dx}[x-1] - (x-1)\dfrac{d}{dx}[2x+3]}{(2x+3)^2}$$

$$= \frac{(2x+3)(1) - (x-1)(2)}{(2x+3)^2}$$

$$= \frac{2x+3 - 2x+2}{(2x+3)^2}$$

$$= \frac{5}{(2x+3)^2}$$

✓ **Checkpoint 4**   *Worked-out solution available at LarsonAppliedCalculus.com*

Find the derivative of $y = \dfrac{x+4}{5x-2}$.

EXAMPLE 5   Finding an Equation of a Tangent Line

Find an equation of the tangent line to the graph of

$$y = \frac{2x^2 - 4x + 3}{2 - 3x}$$

at the point $(1, -1)$.

SOLUTION   Apply the Quotient Rule, as shown.

$$\frac{dy}{dx} = \frac{(2-3x)\dfrac{d}{dx}[2x^2 - 4x + 3] - (2x^2 - 4x + 3)\dfrac{d}{dx}[2 - 3x]}{(2-3x)^2}$$

$$= \frac{(2-3x)(4x-4) - (2x^2 - 4x + 3)(-3)}{(2-3x)^2}$$

$$= \frac{-12x^2 + 20x - 8 - (-6x^2 + 12x - 9)}{(2-3x)^2}$$

$$= \frac{-12x^2 + 20x - 8 + 6x^2 - 12x + 9}{(2-3x)^2}$$

$$= \frac{-6x^2 + 8x + 1}{(2-3x)^2}$$

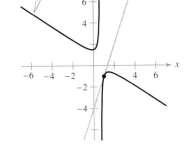

FIGURE 2.27

To find the slope at $(1, -1)$, evaluate dy/dx when $x = 1$.

$$\frac{dy}{dx} = \frac{-6(1)^2 + 8(1) + 1}{[2 - 3(1)]^2} = 3 \qquad \text{Slope of graph at } (1, -1)$$

Using the point-slope form of a line, you can find the equation of the tangent line to be $y = 3x - 4$. The graph of the function and the tangent line is shown in Figure 2.27.

✓ **Checkpoint 5**   *Worked-out solution available at LarsonAppliedCalculus.com*

Find an equation of the tangent line to the graph of $y = \dfrac{x^2 - 4}{2x + 5}$ at the point $\left(0, -\dfrac{4}{5}\right)$. ■

EXAMPLE 6   Rewriting Before Differentiating

Find the derivative of

$$y = \frac{3 - (1/x)}{x + 5}.$$

SOLUTION   Begin by rewriting the function. Then apply the Quotient Rule and simplify the result.

$$y = \frac{3 - (1/x)}{x + 5} \qquad \text{Write original function.}$$

$$= \frac{x[3 - (1/x)]}{x(x + 5)} \qquad \text{Multiply numerator and denominator by } x.$$

$$= \frac{3x - 1}{x^2 + 5x} \qquad \text{Rewrite.}$$

$$\frac{dy}{dx} = \frac{(x^2 + 5x)(3) - (3x - 1)(2x + 5)}{(x^2 + 5x)^2} \qquad \text{Apply Quotient Rule.}$$

$$= \frac{(3x^2 + 15x) - (6x^2 + 13x - 5)}{(x^2 + 5x)^2}$$

$$= \frac{-3x^2 + 2x + 5}{(x^2 + 5x)^2} \qquad \text{Simplify.}$$

✓ **Checkpoint 6**   *Worked-out solution available at LarsonAppliedCalculus.com*

Find the derivative of

$$y = \frac{3 - (2/x)}{x + 4}.$$ ■

Not every quotient needs to be differentiated by the Quotient Rule. For instance, each quotient in the next example can be considered as the product of a constant and a function of *x*. In such cases, the Constant Multiple Rule is more efficient than the Quotient Rule.

EXAMPLE 7   Using the Constant Multiple Rule

	Original Function	*Rewrite*	*Differentiate*	*Simplify*
a.	$y = \dfrac{x^2 + 3x}{6}$	$y = \dfrac{1}{6}(x^2 + 3x)$	$y' = \dfrac{1}{6}(2x + 3)$	$y' = \dfrac{1}{3}x + \dfrac{1}{2}$
b.	$y = \dfrac{5x^4}{8}$	$y = \dfrac{5}{8}x^4$	$y' = \dfrac{5}{8}(4x^3)$	$y' = \dfrac{5}{2}x^3$
c.	$y = \dfrac{-3(3x - 2x^2)}{7x}$	$y = -\dfrac{3}{7}(3 - 2x)$	$y' = -\dfrac{3}{7}(-2)$	$y' = \dfrac{6}{7}$
d.	$y = \dfrac{9}{5x^2}$	$y = \dfrac{9}{5}(x^{-2})$	$y' = \dfrac{9}{5}(-2x^{-3})$	$y' = -\dfrac{18}{5x^3}$

✓ **Checkpoint 7**   *Worked-out solution available at LarsonAppliedCalculus.com*

Find the derivative of each function.

a. $y = \dfrac{x^2 + 4x}{5}$ **b.** $y = \dfrac{3x^4}{4}$ ■

Application

 EXAMPLE 8   Rate of Change of Systolic Blood Pressure

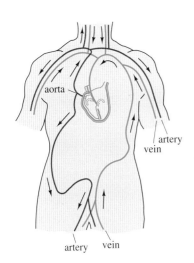

aorta

artery
vein

artery vein

As blood moves from the heart through the major arteries out to the capillaries and back through the veins, the systolic blood pressure continuously drops. Consider a person whose systolic blood pressure P (in millimeters of mercury) is given by

$$P = \frac{25t^2 + 125}{t^2 + 1}, \quad 0 \le t \le 10$$

where t is measured in seconds. At what rate is the blood pressure changing 5 seconds after blood leaves the heart?

SOLUTION   Begin by applying the Quotient Rule.

$$P = \frac{25t^2 + 125}{t^2 + 1} \qquad \text{Write original function.}$$

$$\frac{dP}{dt} = \frac{(t^2 + 1)(50t) - (25t^2 + 125)(2t)}{(t^2 + 1)^2} \qquad \text{Quotient Rule}$$

$$= \frac{50t^3 + 50t - 50t^3 - 250t}{(t^2 + 1)^2}$$

$$= -\frac{200t}{(t^2 + 1)^2} \qquad \text{Simplify.}$$

When $t = 5$, the rate of change is

$$\frac{dP}{dt} = -\frac{200(5)}{26^2} \approx -1.48 \text{ millimeters of mercury per second.}$$

So, the pressure is *dropping* at a rate of about 1.48 millimeters of mercury per second at $t = 5$ seconds.

✔ ***Checkpoint 8***   *Worked-out solution available at LarsonAppliedCalculus.com*

In Example 8, find the rate at which systolic blood pressure is changing at each time shown in the table below. Describe the changes in blood pressure as the blood moves away from the heart.

t	0	1	2	3	4	5	6	7
$\dfrac{dP}{dt}$								

SUMMARIZE   (Section 2.4)

1. State the Product Rule *(page 117)*. For examples of the Product Rule, see Examples 1, 2, and 3.

2. State the Quotient Rule *(page 120)*. For examples of the Quotient Rule, see Examples 4, 5, and 6.

3. Describe a real-life example of how the Quotient Rule can be used to analyze the rate of change of systolic blood pressure *(page 123, Example 8)*.

SKILLS WARM UP 2.4

The following warm-up exercises involve skills that were covered in a previous course or earlier sections. You will use these skills in the exercise set for this section. For additional help, review Appendices A.3 and A.5, and Section 2.2.

In Exercises 1–10, simplify the expression.

1. $(x^2 + 1)(2) + (2x + 7)(2x)$

2. $(2x - x^3)(8x) + (4x^2)(2 - 3x^2)$

3. $x(4)(x^2 + 2)^3(2x) + (x^2 + 4)(1)$

4. $x^2(2)(2x + 1)(2) + (2x + 1)^4(2x)$

5. $\dfrac{(2x + 7)(5) - (5x + 6)(2)}{(2x + 7)^2}$

6. $\dfrac{(x^2 - 4)(2x + 1) - (x^2 + x)(2x)}{(x^2 - 4)^2}$

7. $\dfrac{(x^2 + 1)(2) - (2x + 1)(2x)}{(x^2 + 1)^2}$

8. $\dfrac{(1 - x^4)(4) - (4x - 1)(-4x^3)}{(1 - x^4)^2}$

9. $(x^{-1} + x)(2) + (2x - 3)(-x^{-2} + 1)$

10. $\dfrac{(1 - x^{-1})(1) - (x - 4)(x^{-2})}{(1 - x^{-1})^2}$

In Exercises 11–14, find $f'(2)$.

11. $f(x) = 3x^2 - x + 4$

12. $f(x) = -x^3 + x^2 + 8x$

13. $f(x) = \dfrac{2}{7x}$

14. $f(x) = x^2 - \dfrac{1}{x^2}$

Exercises 2.4

See *CalcChat.com* for tutorial help and worked-out solutions to odd-numbered exercises.

 Using the Product Rule In Exercises 1–10, use the Product Rule to find the derivative of the function. *See Examples 1, 2, and 3.*

1. $f(x) = (2x - 3)(1 - 5x)$

2. $g(x) = (4x - 7)(3x + 1)$

3. $f(x) = (6x - x^2)(4 + 3x)$

4. $f(x) = (5x - x^3)(2x + 9)$

5. $f(x) = x(x^2 + 3)$

6. $f(x) = x^2(3x^3 - 1)$

7. $h(x) = \left(\dfrac{2}{x} - 3\right)(x^2 + 7)$

8. $f(x) = (3 - x)\left(\dfrac{4}{x^2} - 5\right)$

9. $g(x) = (x^2 - 4x + 3)(x - 2)$

10. $g(x) = (x^2 - 2x + 1)(x^3 - 1)$

 Using the Quotient Rule In Exercises 11–20, use the Quotient Rule to find the derivative of the function. *See Examples 4 and 6.*

11. $h(x) = \dfrac{x}{x - 5}$

12. $h(x) = \dfrac{x^2}{x + 3}$

13. $f(t) = \dfrac{2t^2 - 3}{3t + 1}$

14. $f(x) = \dfrac{7x + 3}{4x - 9}$

15. $f(t) = \dfrac{t + 6}{t^2 - 8}$

16. $g(x) = \dfrac{4x - 5}{x^2 - 1}$

17. $f(x) = \dfrac{x^2 + 6x + 5}{2x - 1}$

18. $f(x) = \dfrac{4x^2 - x + 2}{3 - 4x}$

19. $f(x) = \dfrac{6 + (2/x)}{3x - 1}$

20. $f(x) = \dfrac{5 - (1/x^2)}{x + 2}$

 Using the Constant Multiple Rule In Exercises 21–30, find the derivative of the function. *See Example 7.*

Original Function	*Rewrite*	*Differentiate*	*Simplify*

21. $f(x) = \dfrac{x^3 + 6x}{3}$

22. $f(x) = \dfrac{x^3 + 2x^2}{10}$

23. $y = \dfrac{7x^2}{5}$

24. $y = \dfrac{2x^4}{9}$

25. $y = \dfrac{7}{3x^3}$

26. $y = \dfrac{4}{5x^2}$

27. $y = \dfrac{4x^2 - 3x}{8\sqrt{x}}$

28. $y = \dfrac{5(3x^2 + 2x)}{6\sqrt[3]{x}}$

29. $y = \dfrac{x^2 - 4x + 3}{2(x - 1)}$

30. $y = \dfrac{x^2 - 4}{(x + 2)}$

Finding Derivatives In Exercises 31–46, find the derivative of the function. State which differentiation rule(s) you used to find the derivative.

31. $f(x) = (x^3 - 3x)(2x^2 + 3x + 5)$

32. $h(t) = (t^5 - 1)(4t^2 - 7t - 3)$

33. $h(t) = \frac{1}{3}(6t - 4)$

34. $f(x) = \frac{1}{2}(3x - 8)$

35. $f(x) = \frac{x^3 + 3x + 2}{x^2 - 1}$

36. $f(x) = \frac{2x^3 - 4x^2 + 9}{x^3 - 5}$

37. $f(x) = \frac{x^2 - x - 20}{x + 4}$

38. $h(t) = \frac{3t^2 + 22t + 7}{t + 7}$

39. $g(t) = (2t^3 - 1)^2$

40. $f(x) = (4x^3 - 2x - 3)^2$

41. $g(s) = \frac{s^2 - 2s + 5}{\sqrt{s}}$

42. $f(x) = \frac{x^3 - 5x^2 - 6x}{\sqrt{x}}$

43. $f(x) = \frac{(x - 2)(3x + 1)}{4x + 2}$

44. $f(x) = \frac{(x + 1)(2x - 7)}{2x + 1}$

45. $g(x) = (x + 4)(2x + 9)(x - 3)$

46. $f(x) = (3x^3 + 4x)(x - 5)(x + 1)$

 Finding an Equation of a Tangent Line In Exercises 47–54, find an equation of the tangent line to the graph of the function at the given point. Then use a graphing utility to graph the function and the tangent line in the same viewing window. *See Example 5.*

Function	*Point*
47. $f(x) = (5x + 2)(x^2 + x)$	$(-1, 0)$
48. $h(x) = (x^2 - 1)(x^3 - 3x)$	$(-2, -6)$
49. $f(x) = x^3(x^2 - 4)$	$(1, -3)$
50. $f(x) = \sqrt{x}(x - 3)$	$(9, 18)$
51. $f(x) = \frac{3x - 2}{x + 1}$	$(4, 2)$
52. $f(x) = \frac{2x + 1}{x - 1}$	$(2, 5)$
53. $f(x) = \frac{(3x - 2)(6x + 5)}{2x - 3}$	$(-1, -1)$
54. $g(x) = \frac{(x + 2)(x^2 + x)}{x - 4}$	$(1, -2)$

 Finding Horizontal Tangent Lines In Exercises 55–58, find the point(s), if any, at which the graph of f has a horizontal tangent line.

55. $f(x) = \frac{x^2}{x - 1}$ **56.** $f(x) = \frac{x^2}{x^2 + 1}$

57. $f(x) = \frac{x^4}{x^3 + 1}$

58. $f(x) = \frac{x^4 + 3}{x^2 + 1}$

Graphing a Function and Its Derivative In Exercises 59–62, use a graphing utility to graph f and f' on the interval $[-2, 2]$.

59. $f(x) = x(x + 1)$

60. $f(x) = x^2(x + 1)$

61. $f(x) = x(x + 1)(x - 1)$

62. $f(x) = x^2(x + 1)(x - 1)$

Demand In Exercises 63 and 64, use the demand function to find the rate of change in the demand x for the given price p.

63. $x = 275\left(1 - \frac{3p}{5p + 1}\right)$, $p = \$4$

64. $x = 300 - p - \frac{2p}{p + 1}$, $p = \$3$

65. Population Growth A population of bacteria is introduced into a culture. The number of bacteria P can be modeled by

$$P = 500\left(1 + \frac{4t}{50 + t^2}\right)$$

where t is the time (in hours). Find the rate of change of the population at $t = 2$.

66. Quality Control The percent P of defective parts produced by a new employee t days after the employee starts work can be modeled by

$$P = \frac{t + 1750}{50(t + 2)}.$$

Find the rate of change of P at (a) $t = 1$ and (b) $t = 10$.

67. Environment The model

$$P = \frac{t^2 - t + 1}{t^2 + 1}$$

measures the percent P (in decimal form) of the normal level of oxygen in a pond, where t is the time (in weeks) after organic waste is dumped into the pond. Find the rate of change of P at (a) $t = 0.5$, (b) $t = 2$, and (c) $t = 8$. Interpret the meaning of these values.

68. Physical Science The temperature T (in degrees Fahrenheit) of food placed in a refrigerator is modeled by

$$T = 10\left(\frac{4t^2 + 16t + 75}{t^2 + 4t + 10}\right)$$

where t is the time (in hours). What is the initial temperature of the food? Find the rate of change of T with respect to t at (a) $t = 1$, (b) $t = 3$, (c) $t = 5$, and (d) $t = 10$. Interpret the meaning of these values.

 69. Cost The cost C of producing x units of a product is given by $C = x^3 - 15x^2 + 87x - 73$ for $4 \leq x \leq 9$.

(a) Use a graphing utility to graph the marginal cost function and the average cost function, C/x, in the same viewing window.

(b) Find the point of intersection of the graphs of dC/dx and C/x. Does this point have any significance?

70. HOW DO YOU SEE IT? The advertising manager for a new product determines that P percent of the potential market is aware of the product t weeks after the advertising campaign begins.

Market Awareness

(a) What happens to the percent of people who are aware of the product in the long run?

(b) What happens to the rate of change of the percent of people who are aware of the product in the long run?

71. Inventory Replenishment The ordering and transportation cost C (in thousands of dollars) of the components used in manufacturing a product is given by

$$C = 100\left(\frac{200}{x^2} + \frac{x}{x + 30}\right), \quad x \geq 1$$

where x is the order size (in hundreds). Find the rate of change of C with respect to x when (a) $x = 10$, (b) $x = 15$, and (c) $x = 20$. What do these rates of change imply about increasing the size of an order? Of the given order sizes, which would you choose? Explain.

72. Managing a Store You are managing a store and have been adjusting the price of an item. You have found that you make a profit of $50 when 10 units are sold, $60 when 12 units are sold, and $65 when 14 units are sold.

(a) Use the *regression* feature of a graphing utility to find a quadratic model that relates the profit P to the number of units sold x.

(b) Use a graphing utility to graph P.

(c) Find the point on the graph at which the marginal profit is zero. Interpret this point in the context of the problem.

Using Relationships In Exercises 73–76, use the given information to find $f'(2)$.

$$g(2) = 3 \quad \text{and} \quad g'(2) = -2$$
$$h(2) = -1 \quad \text{and} \quad h'(2) = 4$$

73. $f(x) = 2g(x) + h(x)$ **74.** $f(x) = 3 - g(x)$

75. $f(x) = g(x)h(x)$ **76.** $f(x) = \dfrac{g(x)}{h(x)}$

Business Capsule

In October 2011, Melissa and Rick Hinnant created Grace & Lace, originally a boot sock company, out of their home. They hired stay-at-home moms and local seamstresses to sew and ship their socks. By 2013, they expanded to women's apparel and sales reached $2.8 million. The heart of the company's mission is to make a life-changing difference in the world. They have built two orphanages in India and one Freedom Home in Nepal, rescuing over 100 women and children off the street. They plan to fund five more homes by 2016.

77. Research Project Use your school's library, the Internet, or some other reference source to find information on a company that is noted for its philanthropy and community commitment. (One such business is described above.) Write a short paper about the company.

QUIZ YOURSELF   See *CalcChat.com* for tutorial help and worked-out solutions to odd-numbered exercises.

Take this quiz as you would take a quiz in class. When you are done, check your work against the answers given in the back of the book.

In Exercises 1–3, use the limit definition to find the derivative of the function. Then find the slope of the graph of *f* at the given point.

1. $f(x) = 5x + 3; (-2, -7)$

2. $f(x) = \sqrt{x + 3}; (1, 2)$

3. $f(x) = 3x - x^2; (4, -4)$

In Exercises 4–13, find the derivative of the function.

4. $f(x) = 12$

5. $f(x) = 19x + 9$

6. $f(x) = x^4 - 3x^3 - 5x^2 + 8$

7. $f(x) = 12x^{1/4}$

8. $f(x) = 4x^{-2}$

9. $f(x) = 10x^{-1/5} + x^{-3}$

10. $f(x) = \dfrac{2x + 3}{3x + 2}$

11. $f(x) = (x^2 + 1)(-2x + 4)$

12. $f(x) = (x^2 + 3x + 4)(5x - 2)$

13. $f(x) = \dfrac{4x}{x^2 + 3}$

In Exercises 14–17, find the average rate of change of the function over the given interval. Compare this rate with the instantaneous rates of change at the endpoints of the interval.

14. $f(x) = x^2 - 3x + 1; [0, 3]$

15. $f(x) = 2x^3 + x^2 - x + 4; [-1, 1]$

16. $f(x) = \dfrac{1}{3x}; [-5, -2]$

17. $f(x) = \sqrt[3]{x}; [8, 27]$

18. The profit *P* (in dollars) from selling *x* units of a product is given by

$$P = -0.0125x^2 + 16x - 600.$$

(a) Find the marginal profit when $x = 175$.

(b) Find the additional profit when sales increase from 175 to 176 units.

(c) Compare the results of parts (a) and (b).

In Exercises 19–22, find an equation of the tangent line to the graph of *f* at the given point. Then use a graphing utility to graph the function and the tangent line in the same viewing window.

19. $f(x) = 5x^2 + 6x - 1; (-1, -2)$

20. $f(x) = \dfrac{8}{\sqrt{x^3}}; (4, 1)$

21. $f(x) = (x^2 + 1)(4x - 3); (1, 2)$

22. $f(x) = \dfrac{5x + 4}{2 - 3x}; (1, -9)$

23. From 2007 through 2013, the sales per share *S* (in dollars) for the Hershey Company can be modeled by

$$S = -0.01722t^3 + 0.7333t^2 - 7.657t + 45.47, \quad 7 \le t \le 13$$

where *t* represents the year, with $t = 7$ corresponding to 2007.   *(Source: Hershey Company)*

(a) Find the rate of change of the sales per share with respect to the year.

(b) At what rate was the sales per share changing in 2008? in 2011? in 2012?

2.5 The Chain Rule

■ Find the derivatives of functions using the Chain Rule.
■ Find the derivatives of functions using the General Power Rule.
■ Write derivatives in simplified form.
■ Use derivatives to answer questions about real-life situations.
■ Review the basic differentiation rules for algebraic functions.

The Chain Rule

In this section, you will study one of the most powerful rules of differential calculus—the **Chain Rule.** This differentiation rule deals with composite functions and adds versatility to the rules presented in Sections 2.2 and 2.4. For example, compare the functions below. Those on the left can be differentiated without the Chain Rule, whereas those on the right are best differentiated with the Chain Rule.

Without the Chain Rule	*With the Chain Rule*
$y = x^2 + 1$	$y = \sqrt{x^2 + 1}$
$y = x + 1$	$y = (x + 1)^{-1/2}$
$y = 3x + 2$	$y = (3x + 2)^5$
$y = \dfrac{x + 5}{x^2 + 2}$	$y = \left(\dfrac{x + 5}{x^2 + 2}\right)^2$
$y = \dfrac{x + 1}{x}$	$y = \sqrt{\dfrac{x + 1}{x}}$

Balance in an Account

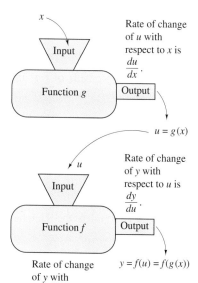

In Exercise 69 on page 136, you will use the General Power Rule to find the rate of change of the balance in an account.

The Chain Rule

If $y = f(u)$ is a differentiable function of u and $u = g(x)$ is a differentiable function of x, then $y = f(g(x))$ is a differentiable function of x and

$$\frac{dy}{dx} = \frac{dy}{du} \cdot \frac{du}{dx}$$

or, equivalently,

$$\frac{d}{dx}[f(g(x))] = f'(g(x))g'(x).$$

Basically, the Chain Rule states that if y changes dy/du times as fast as u, and u changes du/dx times as fast as x, then y changes

$$\frac{dy}{du} \cdot \frac{du}{dx}$$

times as fast as x, as illustrated in Figure 2.28. One advantage of the

$$\frac{dy}{dx}$$

notation for derivatives is that it helps you remember differentiation rules, such as the Chain Rule. For instance, in the formula

$$\frac{dy}{dx} = \frac{dy}{du} \cdot \frac{du}{dx}$$

you can imagine that the du's divide out.

x

Rate of change of u with respect to x is $\dfrac{du}{dx}$.

Input

Function g Output

$u = g(x)$

u

Rate of change of y with respect to u is $\dfrac{dy}{du}$.

Input

Function f Output

Rate of change of y with respect to x is $\dfrac{dy}{dx} = \dfrac{dy}{du} \cdot \dfrac{du}{dx}.$

$y = f(u) = f(g(x))$

FIGURE 2.28

When applying the Chain Rule, it is helpful to think of the composite function $y = f(g(x))$ or $y = f(u)$ as having two parts—an *inside* and an *outside*—as illustrated below.

$$y = f(g(x)) = f(u)$$

The Chain Rule tells you that the derivative of $y = f(u)$ is the derivative of the outer function (at the inner function u) *times* the derivative of the inner function. That is,

$$y' = f'(u) \cdot u'.$$

EXAMPLE 1   Decomposing Composite Functions

Write each function as the composition of two functions.

a. $y = \dfrac{1}{x + 1}$ **b.** $y = \sqrt{3x^2 - x + 1}$

SOLUTION   There is more than one correct way to decompose each function. One way for each is shown below.

$y = f(g(x))$	$u = g(x)$ *(inside)*	$y = f(u)$ *(outside)*
a. $y = \dfrac{1}{x+1}$	$u = x + 1$	$y = \dfrac{1}{u}$
b. $y = \sqrt{3x^2 - x + 1}$	$u = 3x^2 - x + 1$	$y = \sqrt{u}$

✓ **Checkpoint 1**   *Worked-out solution available at LarsonAppliedCalculus.com*

Write each function as the composition of two functions, where $y = f(g(x))$.

a. $y = \dfrac{1}{\sqrt{x + 1}}$ **b.** $y = (x^2 + 2x + 5)^3$

STUDY TIP

You can also solve Example 2 without using the Chain Rule by expanding the function to obtain

$$y = x^6 + 3x^4 + 3x^2 + 1$$

and then finding the derivative

$$y' = 6x^5 + 12x^3 + 6x.$$

Verify that this is the same as the derivative in Example 2. Which method would you use to find

$$\frac{d}{dx}[(x^2 + 1)^{50}]?$$

EXAMPLE 2   Using the Chain Rule

Find the derivative of $y = (x^2 + 1)^3$.

SOLUTION   To apply the Chain Rule, you need to identify the inside function u.

$$y = \overbrace{(x^2 + 1)}^{u}{}^3 = u^3$$

The inside function is $u = x^2 + 1$. By the Chain Rule, you can write the derivative as shown.

$$\frac{dy}{dx} = \overbrace{3(x^2 + 1)^2}^{\frac{dy}{du}}\overbrace{(2x)}^{\frac{du}{dx}} = 6x(x^2 + 1)^2$$

✓ **Checkpoint 2**   *Worked-out solution available at LarsonAppliedCalculus.com*

Find the derivative of $y = (x^3 + 1)^2$.

The General Power Rule

The function in Example 2 illustrates one of the most common types of composite functions—a power function of the form

$$y = [u(x)]^n.$$

The rule for differentiating such functions is called the **General Power Rule,** and it is a special case of the Chain Rule.

The General Power Rule

If $y = [u(x)]^n$, where u is a differentiable function of x and n is a real number, then

$$\frac{dy}{dx} = n[u(x)]^{n-1}\frac{du}{dx}$$

or, equivalently,

$$\frac{d}{dx}[u^n] = nu^{n-1}u'.$$

PROOF   Apply the Chain Rule and the Simple Power Rule as shown.

$$\frac{dy}{dx} = \frac{dy}{du} \cdot \frac{du}{dx}$$

$$= \frac{d}{du}[u^n]\frac{du}{dx}$$

$$= nu^{n-1}\frac{du}{dx}$$

TECH TUTOR

If you have access to a symbolic differentiation utility, try using it to confirm the result of Example 3.

EXAMPLE 3   Using the General Power Rule

Find the derivative of

$$y = (3x - 2x^2)^3.$$

SOLUTION   To apply the General Power Rule, you need to identify the inside function u.

$$y = \overbrace{(3x - 2x^2)}^{u}{}^3 = u^3$$

The inside function is

$$u = 3x - 2x^2.$$

So, by the General Power Rule,

$$\frac{dy}{dx} = \overset{n}{3}(\overbrace{3x - 2x^2)^2}^{u^{n-1}}\overbrace{\frac{d}{dx}[3x - 2x^2]}^{u'}$$

$$= 3(3x - 2x^2)^2(3 - 4x).$$

✓ **Checkpoint 3**   Worked-out solution available at LarsonAppliedCalculus.com

Find the derivative of

$$y = (x^2 + 3x)^4.$$

EXAMPLE 4   Finding an Equation of a Tangent Line

Find an equation of the tangent line to the graph of

$$y = \sqrt[3]{(x^2 + 4)^2}$$

at $x = 2$.

SOLUTION   Begin by rewriting the function in rational exponent form.

$$y = (x^2 + 4)^{2/3} \qquad \text{Rewrite original function.}$$

Then, using the inside function, $u = x^2 + 4$, apply the General Power Rule.

$$\frac{dy}{dx} = \frac{2}{3}(x^2 + 4)^{-1/3}(2x) \qquad \text{Apply General Power Rule.}$$

$$= \frac{4x(x^2 + 4)^{-1/3}}{3}$$

$$= \frac{4x}{3\sqrt[3]{x^2 + 4}} \qquad \text{Write in radical form.}$$

When $x = 2$, $y = 4$ and the slope of the line tangent to the graph at $(2, 4)$ is $\frac{4}{3}$. Using the point-slope form, you can find the equation of the tangent line to be $y = \frac{4}{3}x + \frac{4}{3}$. The graph of the function and the tangent line is shown in Figure 2.29.

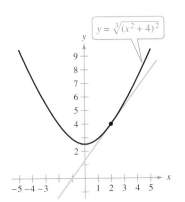

FIGURE 2.29

✓ **Checkpoint 4**   Worked-out solution available at LarsonAppliedCalculus.com

Find an equation of the tangent line to the graph of $y = \sqrt[3]{(x + 4)^2}$ at $x = 4$.

EXAMPLE 5   Differentiating a Quotient with a Constant Numerator

Find the derivative of

$$y = \frac{5}{(4x - 3)^2}.$$

SOLUTION   Begin by rewriting the function in rational exponent form.

$$y = 5(4x - 3)^{-2} \qquad \text{Rewrite original function.}$$

Then, using the inside function, $u = 4x - 3$, apply the General Power Rule.

$$\frac{dy}{dx} = 5(-2)(4x - 3)^{-3}(4) \qquad \text{Apply General Power Rule.}$$

Constant
Multiple Rule

$$= -40(4x - 3)^{-3} \qquad \text{Simplify.}$$

$$= -\frac{40}{(4x - 3)^3} \qquad \text{Write with positive exponent.}$$

✓ **Checkpoint 5**   Worked-out solution available at LarsonAppliedCalculus.com

Find the derivative of

$$y = \frac{4}{2x + 1}.$$

Simplification Techniques

Throughout this chapter, writing derivatives in simplified form has been emphasized. The reason for this is that most applications of derivatives require a simplified form. The next two examples illustrate some useful simplification techniques.

ALGEBRA TUTOR

In Example 6, note that you subtract exponents when factoring. That is, when $(1 - x^2)^{-1/2}$ is factored out of $(1 - x^2)^{1/2}$, the *remaining* factor has an exponent of $\frac{1}{2} - \left(-\frac{1}{2}\right) = 1$. So, $(1 - x^2)^{1/2}$ is equal to the product of $(1 - x^2)^{-1/2}$ and $(1 - x^2)^1$. For help in simplifying expressions like the one in Example 6, see the *Chapter 2 Algebra Tutor* on pages 158 and 159.

EXAMPLE 6   Simplifying by Factoring Out Least Powers

Find the derivative of $y = x^2 \sqrt{1 - x^2}$.

SOLUTION

$$
\begin{aligned}
y &= x^2\sqrt{1 - x^2} && \text{Write original function.} \\
&= x^2(1 - x^2)^{1/2} && \text{Rewrite function.} \\
y' &= x^2\frac{d}{dx}[(1 - x^2)^{1/2}] + (1 - x^2)^{1/2}\frac{d}{dx}[x^2] && \text{Product Rule} \\
&= x^2\left[\frac{1}{2}(1 - x^2)^{-1/2}(-2x)\right] + (1 - x^2)^{1/2}(2x) && \text{General Power Rule} \\
&= -x^3(1 - x^2)^{-1/2} + 2x(1 - x^2)^{1/2} && \text{Simplify.} \\
&= x(1 - x^2)^{-1/2}[-x^2(1) + 2(1 - x^2)] && \text{Factor.} \\
&= x(1 - x^2)^{-1/2}(2 - 3x^2) && \text{Simplify.} \\
&= \frac{x(2 - 3x^2)}{\sqrt{1 - x^2}} && \text{Write in radical form.}
\end{aligned}
$$

✓ **Checkpoint 6**   *Worked-out solution available at LarsonAppliedCalculus.com*

Find the derivative of

$$y = x^2\sqrt{x^2 + 1}.$$

STUDY TIP

In Example 7, try to find $f'(x)$ by applying the Quotient Rule to

$$f(x) = \frac{(3x - 1)^2}{(x^2 + 3)^2}.$$

Which method do you prefer?

EXAMPLE 7   Differentiating a Quotient Raised to a Power

Find the derivative of

$$f(x) = \left(\frac{3x - 1}{x^2 + 3}\right)^2.$$

SOLUTION

$$
\begin{aligned}
f'(x) &= 2\overset{n}{\left(\frac{3x - 1}{x^2 + 3}\right)}\overset{u^{n-1}}{\frac{d}{dx}}\overset{u'}{\left[\frac{3x - 1}{x^2 + 3}\right]} && \text{General Power Rule} \\
&= \left[\frac{2(3x - 1)}{x^2 + 3}\right]\left[\frac{(x^2 + 3)(3) - (3x - 1)(2x)}{(x^2 + 3)^2}\right] && \text{Quotient Rule} \\
&= \frac{2(3x - 1)(3x^2 + 9 - 6x^2 + 2x)}{(x^2 + 3)^3} && \text{Multiply.} \\
&= \frac{2(3x - 1)(-3x^2 + 2x + 9)}{(x^2 + 3)^3} && \text{Simplify.}
\end{aligned}
$$

✓ **Checkpoint 7**   *Worked-out solution available at LarsonAppliedCalculus.com*

Find the derivative of

$$f(x) = \left(\frac{x + 1}{x - 5}\right)^2.$$

Application

EXAMPLE 8   **Finding Rates of Change**

From 2004 through 2013, the sales per share S (in dollars) for Apple can be modeled by

$$S = (0.0351t^2 - 0.134t + 1.19)^2, \quad 4 \le t \le 13$$

where t is the year, with $t = 4$ corresponding to 2004. Use the model to approximate the rates of change in the sales per share in 2006, 2009, and 2012. Would Apple stockholders have been satisfied with the performance of this stock from 2004 through 2013?   *(Source: Apple, Inc.)*

SOLUTION   The rate of change in S is given by the derivative dS/dt. You can use the General Power Rule to find the derivative.

$$\frac{dS}{dt} = 2(0.0351t^2 - 0.134t + 1.19)(0.0702t - 0.134)$$

$$= (0.1404t - 0.268)(0.0351t^2 - 0.134t + 1.19)$$

In 2006, the sales per share was changing at a rate of

$$[0.1404(6) - 0.268][0.0351(6)^2 - 0.134(6) + 1.19] \approx \$0.95 \text{ per year.}$$

In 2009, the sales per share was changing at a rate of

$$[0.1404(9) - 0.268][0.0351(9)^2 - 0.134(9) + 1.19] \approx \$2.81 \text{ per year.}$$

In 2012, the sales per share was changing at a rate of

$$[0.1404(12) - 0.268][0.0351(12)^2 - 0.134(12) + 1.19] \approx \$6.57 \text{ per year.}$$

The graph of the sales per share function S is shown in Figure 2.30. So, most stockholders would have been satisfied with the performance of this stock.

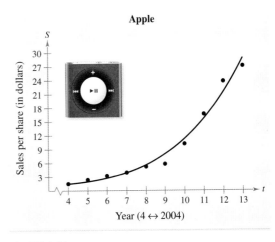

Apple

Sales per share (in dollars)

Year (4 ↔ 2004)

FIGURE 2.30

✓ **Checkpoint 8**   *Worked-out solution available at LarsonAppliedCalculus.com*

From 2004 through 2013, the sales per share S (in dollars) for Dollar Tree can be modeled by

$$S = (0.0120t^2 + 0.140t + 2.30)^2, \quad 4 \le t \le 13$$

where t is the year, with $t = 4$ corresponding to 2004. Use the model to approximate the rate of change in the sales per share in 2010.   *(Source: Dollar Tree, Inc.)*   ■

Review of Basic Differentiation Rules

You now have all the rules you need to differentiate *any* algebraic function. For your convenience, they are summarized below.

Summary of Basic Differentiation Rules

Let u and v be differentiable functions of x.

1. **Constant Rule**
$$\frac{d}{dx}[c] = 0, \quad c \text{ is a constant.}$$

2. **Constant Multiple Rule**
$$\frac{d}{dx}[cu] = c\frac{du}{dx}, \quad c \text{ is a constant.}$$

3. **Sum and Difference Rules**
$$\frac{d}{dx}[u \pm v] = \frac{du}{dx} \pm \frac{dv}{dx}$$

4. **Product Rule**
$$\frac{d}{dx}[uv] = u\frac{dv}{dx} + v\frac{du}{dx}$$

5. **Quotient Rule**
$$\frac{d}{dx}\left[\frac{u}{v}\right] = \frac{v\dfrac{du}{dx} - u\dfrac{dv}{dx}}{v^2}$$

6. **Power Rules**
$$\frac{d}{dx}[x^n] = nx^{n-1}, \quad n \text{ is a real number.}$$
$$\frac{d}{dx}[u^n] = nu^{n-1}\frac{du}{dx}, \quad n \text{ is a real number.}$$

7. **Chain Rule**
If $y = f(u)$ is a differentiable function of u and $u = g(x)$ is a differentiable function of x, then $y = f(g(x))$ is a differentiable function of x and
$$\frac{dy}{dx} = \frac{dy}{du} \cdot \frac{du}{dx}.$$

SUMMARIZE (Section 2.5)

1. State the Chain Rule *(page 128)*. For an example of the Chain Rule, see Example 2.

2. State the General Power Rule *(page 130)*. For examples of the General Power Rule, see Examples 3, 4, and 5.

3. Describe a real-life example of how the General Power Rule can be used to analyze the rate of change of a company's sales per share *(page 133, Example 8)*.

4. Use the Summary of Basic Differentiation Rules to identify the differentiation rules illustrated by (a)–(f) below *(page 134)*.

(a) $\dfrac{d}{dx}[2x] = 2\dfrac{d}{dx}[x]$ (b) $\dfrac{d}{dx}[x^4] = 4x^3$ (c) $\dfrac{d}{dx}[8] = 0$

(d) $\dfrac{d}{dx}[x^2 + x] = \dfrac{d}{dx}[x^2] + \dfrac{d}{dx}[x]$ (e) $\dfrac{d}{dx}[x - x^3] = \dfrac{d}{dx}[x] - \dfrac{d}{dx}[x^3]$

(f) $\dfrac{d}{dx}[x(x + 1)] = (x)\dfrac{d}{dx}[x + 1] + (x + 1)\dfrac{d}{dx}[x]$

SKILLS WARM UP 2.5
The following warm-up exercises involve skills that were covered in a previous course. You will use these skills in the exercise set for this section. For additional help, review Appendices A.3 and A.4.

In Exercises 1–6, rewrite the expression with rational exponents.

1. $\sqrt[5]{(1-5x)^2}$

2. $\sqrt[4]{(2x-1)^3}$

3. $\dfrac{1}{\sqrt{4x^2+1}}$

4. $\dfrac{1}{\sqrt[6]{2x^3+9}}$

5. $\dfrac{\sqrt{x}}{\sqrt[3]{1-2x}}$

6. $\dfrac{\sqrt{(3-7x)^3}}{2x}$

In Exercises 7–10, factor the expression.

7. $3x^3 - 6x^2 + 5x - 10$

8. $5x\sqrt{x} - x - 5\sqrt{x} + 1$

9. $4(x^2+1)^2 - x(x^2+1)^3$

10. $-x^5 + 6x^3 + 7x^2 - 42$

Exercises 2.5
See *CalcChat.com* for tutorial help and worked-out solutions to odd-numbered exercises.

 Decomposing Composite Functions In Exercises 1–6, identify the inside function, $u = g(x)$, and the outside function, $y = f(u)$. *See Example 1.*

$$y = f(g(x)) \qquad u = g(x) \qquad y = f(u)$$

1. $y = (6x-5)^4$

2. $y = (x^2 - 2x + 3)^3$

3. $y = \sqrt{5x-2}$

4. $y = \sqrt[3]{9-x^2}$

5. $y = \dfrac{1}{3x+1}$

6. $y = \dfrac{1}{\sqrt{x^2-3}}$

 Using the Chain Rule In Exercises 7–12, use the Chain Rule to find the derivative of the function. *See Example 2.*

7. $y = (4x+7)^2$

8. $y = (3x^2-2)^3$

9. $y = \sqrt{3-x^2}$

10. $y = 4\sqrt[4]{6x+5}$

11. $y = (5x^4 - 2x)^{2/3}$

12. $y = (x^3 + 2x^2)^{-1}$

Choosing a Differentiation Rule In Exercises 13–18, choose the rule that you would use to *most efficiently* find the derivative of the function.

(a) Simple Power Rule   (b) Constant Rule

(c) General Power Rule   (d) Quotient Rule

13. $f(x) = \dfrac{2}{1-x^3}$

14. $f(x) = \dfrac{7}{(1-x)^3}$

15. $f(x) = \sqrt[3]{8^2}$

16. $f(x) = \sqrt[3]{x^2}$

17. $f(x) = \dfrac{x^2+9}{x^3+4x^2-6}$

18. $f(x) = \dfrac{\sqrt{x}}{x^3+2x-5}$

 Using the General Power Rule In Exercises 19–30, use the General Power Rule to find the derivative of the function. *See Examples 3 and 5.*

19. $y = (2x-7)^3$

20. $g(x) = (3-5x)^4$

21. $h(x) = (6x - x^3)^2$

22. $f(x) = (2x^3 - 6x)^{4/3}$

23. $f(t) = \sqrt{t+1}$

24. $g(x) = \sqrt{5-3x}$

25. $s(t) = \sqrt{2t^2 + 5t + 2}$

26. $y = 9\sqrt[3]{4x^2+3}$

27. $f(x) = \dfrac{2}{(2-9x)^3}$

28. $g(x) = \dfrac{3}{(7x^2+6x)^5}$

29. $f(x) = \dfrac{1}{\sqrt{(x^2+11)^7}}$

30. $y = \dfrac{1}{\sqrt[3]{(4-x^3)^4}}$

 Finding an Equation of a Tangent Line In Exercises 31–36, find an equation of the tangent line to the graph of f at the point $(2, f(2))$. Then use a graphing utility to graph the function and the tangent line in the same viewing window. *See Example 4.*

31. $f(x) = 2(x^2-1)^3$

32. $f(x) = 3(9x-4)^4$

33. $f(x) = \sqrt{4x^2-7}$

34. $f(x) = x\sqrt{x^2+5}$

35. $f(x) = \sqrt{x^2-2x+1}$

36. $f(x) = (4 - 3x^2)^{-2/3}$

Using Technology In Exercises 37–40, use a symbolic differentiation utility to find the derivative of the function. Graph the function and its derivative in the same viewing window. Describe the behavior of the function when the derivative is zero.

37. $f(x) = \dfrac{\sqrt{x}+1}{x^2+1}$

38. $f(x) = \sqrt{\dfrac{2x}{x+1}}$

39. $f(x) = \sqrt{\dfrac{x+1}{x}}$

40. $f(x) = \sqrt{x}(2-x^2)$

Finding Derivatives In Exercises 41–56, find the derivative of the function. State which differentiation rule(s) you used to find the derivative.

41. $y = \dfrac{1}{4 - x^2}$

42. $s(t) = \dfrac{1}{t^2 + 3t - 1}$

43. $y = -\dfrac{5t}{(t + 8)^2}$

44. $f(x) = \dfrac{3x}{(x^3 - 4)^2}$

45. $y = (2x - 1)(9 - 3x^2)$

46. $y = (7x + 4)(x^3 - 2x^2)$

47. $y = \dfrac{1}{\sqrt{x + 2}}$

48. $g(x) = \dfrac{3}{\sqrt[3]{x^3 - 1}}$

49. $f(x) = x(3x - 9)^3$

50. $f(x) = x^3(x - 4)^2$

51. $y = x\sqrt{2x + 3}$

52. $y = 2t\sqrt{t + 6}$

53. $y = t^2\sqrt{t - 2}$

54. $y = \sqrt{x}(x - 2)^2$

55. $y = \left(\dfrac{6 - 5x}{x^2 - 1}\right)^2$

56. $y = \left(\dfrac{4x^2 - 5}{2 - x}\right)^3$

Finding an Equation of a Tangent Line In Exercises 57–64, find an equation of the tangent line to the graph of the function at the given point. Then use a graphing utility to graph the function and the tangent line in the same viewing window.

57. $y = (x^3 - 2x^2 - x + 1)^2;\ (1, 1)$

58. $y = \sqrt[5]{3x^3 + 4x};\ (2, 2)$

59. $f(t) = \dfrac{36}{(3 - t)^2};\ (0, 4)$

60. $s(x) = \dfrac{1}{\sqrt{x^2 - 3x + 4}};\ \left(3, \tfrac{1}{2}\right)$

61. $f(t) = (t^2 - 9)\sqrt{t + 2};\ (-1, -8)$

62. $y = -\dfrac{2x}{\sqrt{1 - x}};\ (-3, 3)$

63. $f(x) = \dfrac{x + 1}{\sqrt{2x - 3}};\ (2, 3)$ **64.** $y = \dfrac{x}{\sqrt{25 + x^2}};\ (0, 0)$

Finding Horizontal Tangent Lines In Exercises 65–68, find the point(s), if any, at which the graph of f has a horizontal tangent line.

65. $f(x) = \sqrt[3]{x^2 + 4}$

66. $f(x) = \sqrt{5x^2 + x - 3}$

67. $f(x) = \dfrac{x}{\sqrt{2x - 1}}$

68. $f(x) = \dfrac{5x}{\sqrt{3x - 2}}$

69. Compound Interest You deposit $1000 in an account with an annual interest rate of r (in decimal form) compounded monthly. At the end of 5 years, the balance A is

$$A = 1000\left(1 + \frac{r}{12}\right)^{60}.$$

Find the rate of change of A with respect to r when (a) $r = 0.08$, (b) $r = 0.10$, and (c) $r = 0.12$.

70. Biology The number N of bacteria in a culture after t days is modeled by

$$N = 400\left[1 - \frac{3}{(t^2 + 2)^2}\right].$$

Find the rate of change of N with respect to t when (a) $t = 0$, (b) $t = 1$, (c) $t = 2$, (d) $t = 3$, and (e) $t = 4$. (f) What can you conclude?

71. Depreciation The value V of a machine t years after it is purchased is inversely proportional to the square root of $t + 1$. The initial value of the machine is $10,000. Write V as a function of t. Then find the rates of depreciation when $t = 1$ and $t = 3$.

72. HOW DO YOU SEE IT? The cost C (in dollars) of producing x units of a product is $C = 60x + 1350$. For one week, management determined that the number of units produced x at the end of t hours was $x = -1.6t^3 + 19t^2 - 0.5t - 1$. The graph shows the cost C in terms of the time t.

Cost of Producing a Product

(a) Using the graph, which is greater, the rate of change of the cost after 1 hour or the rate of change of the cost after 4 hours?

(b) Explain why the cost function is not increasing at a constant rate during the eight-hour shift.

73. Credit Card Rate The average annual rate r (in percent form) for commercial bank credit cards from 2008 through 2013 can be modeled by

$$r = \sqrt{0.3017t^4 - 9.657t^3 + 97.35t^2 - 266.8t - 242}$$

where t represents the year, with $t = 8$ corresponding to 2008. *(Source: Board of Governors of the Federal Reserve System)*

(a) Find the derivative of this model. Which differentiation rule(s) did you use?

(b) Use a graphing utility to graph the derivative on the interval $8 \le t \le 13$.

(c) Use the *trace* feature to find the year(s) during which the finance rate was changing the most and the least.

2.6 Higher-Order Derivatives

■ Find higher-order derivatives.

■ Find and use a position function to determine the velocity and acceleration of a moving object.

Second, Third, and Higher-Order Derivatives

The "standard" derivative f' is often called the **first derivative** of f. The derivative of f' is the **second derivative** of f and is denoted by f''.

$$\frac{d}{dx}[f'(x)] = f''(x) \qquad \text{Second derivative}$$

The derivative of f'' is the **third derivative** of f and is denoted by f'''.

$$\frac{d}{dx}[f''(x)] = f'''(x) \qquad \text{Third derivative}$$

By continuing this process, you obtain **higher-order derivatives** of f. Higher-order derivatives are denoted as follows.

Position of a Ball

Height (in feet) vs. Time (in seconds)

In Exercise 35 on page 142, you will use derivatives to find the velocity function and the acceleration function of a ball.

Notation for Higher-Order Derivatives

1. 1st derivative: y', $f'(x)$, $\dfrac{dy}{dx}$, $\dfrac{d}{dx}[f(x)]$, $D_x[y]$

2. 2nd derivative: y'', $f''(x)$, $\dfrac{d^2y}{dx^2}$, $\dfrac{d^2}{dx^2}[f(x)]$, $D_x^2[y]$

3. 3rd derivative: y''', $f'''(x)$, $\dfrac{d^3y}{dx^3}$, $\dfrac{d^3}{dx^3}[f(x)]$, $D_x^3[y]$

4. 4th derivative: $y^{(4)}$, $f^{(4)}(x)$, $\dfrac{d^4y}{dx^4}$, $\dfrac{d^4}{dx^4}[f(x)]$, $D_x^4[y]$

5. nth derivative: $y^{(n)}$, $f^{(n)}(x)$, $\dfrac{d^ny}{dx^n}$, $\dfrac{d^n}{dx^n}[f(x)]$, $D_x^n[y]$

EXAMPLE 1   Finding Higher-Order Derivatives

Find the first five derivatives of

$$f(x) = 2x^4 - 3x^2.$$

SOLUTION

$$\begin{aligned}
f(x) &= 2x^4 - 3x^2 &&\text{Write original function.}\\
f'(x) &= 8x^3 - 6x &&\text{First derivative}\\
f''(x) &= 24x^2 - 6 &&\text{Second derivative}\\
f'''(x) &= 48x &&\text{Third derivative}\\
f^{(4)}(x) &= 48 &&\text{Fourth derivative}\\
f^{(5)}(x) &= 0 &&\text{Fifth derivative}
\end{aligned}$$

✓ **Checkpoint 1**   Worked-out solution available at LarsonAppliedCalculus.com

Find the first four derivatives of $f(x) = 6x^3 - 2x^2 + 1$.

EXAMPLE 2   Finding Higher-Order Derivatives

Find $g'''(2)$ for the function

$$g(t) = -t^4 + 2t^3 + t + 4.$$

SOLUTION   Begin by differentiating three times.

$g'(t) = -4t^3 + 6t^2 + 1$	First derivative
$g''(t) = -12t^2 + 12t$	Second derivative
$g'''(t) = -24t + 12$	Third derivative

Then, evaluate the third derivative of g at $t = 2$.

$$g'''(2) = -24(2) + 12$$
$$= -36 \qquad \text{Value of third derivative at } t = 2$$

✓ **Checkpoint 2**   *Worked-out solution available at LarsonAppliedCalculus.com*

Find $g'''(1)$ for $g(x) = x^4 - x^3 + 2x$.   ◼

Examples 1 and 2 show how to find higher-order derivatives of *polynomial* functions. Note that with each successive differentiation, the degree of the polynomial drops by one. Eventually, higher-order derivatives of polynomial functions degenerate to a constant function. Specifically, the nth-order derivative of an nth-degree polynomial function

$$f(x) = a_n x^n + a_{n-1}x^{n-1} + \cdots + a_1 x + a_0$$

is the constant function

$$f^{(n)}(x) = n! a_n$$

where $n! = 1 \cdot 2 \cdot 3 \cdots n$. Each derivative of order higher than n is the zero function.

EXAMPLE 3   Finding Higher-Order Derivatives

Find the first four derivatives of $y = x^{-1}$.

SOLUTION

$y = x^{-1} = \dfrac{1}{x}$	Write original function.
$y' = (-1)x^{-2} = -\dfrac{1}{x^2}$	First derivative
$y'' = (-1)(-2)x^{-3} = \dfrac{2}{x^3}$	Second derivative
$y''' = (-1)(-2)(-3)x^{-4} = -\dfrac{6}{x^4}$	Third derivative
$y^{(4)} = (-1)(-2)(-3)(-4)x^{-5} = \dfrac{24}{x^5}$	Fourth derivative

✓ **Checkpoint 3**   *Worked-out solution available at LarsonAppliedCalculus.com*

Find the fourth derivative of

$$y = \frac{1}{x^2}.$$ ◼

Acceleration

In Section 2.3, you saw that the velocity of a free-falling object (neglecting air resistance) is given by the derivative of its position function. In other words, the rate of change of the position with respect to time is defined to be the velocity. In a similar way, the rate of change of the velocity with respect to time is defined to be the **acceleration** of the object.

$$s = f(t) \qquad \text{Position function}$$

$$\frac{ds}{dt} = f'(t) \qquad \text{Velocity function}$$

$$\frac{d^2s}{dt^2} = f''(t) \qquad \text{Acceleration function}$$

To find the position, velocity, or acceleration at a particular time t, substitute the value of t into the appropriate function, as shown in Example 4.

EXAMPLE 4   Finding Acceleration

A ball is thrown upward from the top of a 160-foot cliff, as shown in the figure. The initial velocity of the ball is 48 feet per second. The position function is

$$s = -16t^2 + 48t + 160$$

where the time t is measured in seconds. Find the height, velocity, and acceleration of the ball at $t = 3$.

160 ft

Not drawn to scale

SOLUTION   Begin by differentiating to find the velocity and acceleration functions.

$$s = -16t^2 + 48t + 160 \qquad \text{Position function}$$

$$\frac{ds}{dt} = -32t + 48 \qquad \text{Velocity function}$$

$$\frac{d^2s}{dt^2} = -32 \qquad \text{Acceleration function}$$

To find the height, velocity, and acceleration at $t = 3$, substitute $t = 3$ into each of the functions above.

$$\text{Height} = -16(3)^2 + 48(3) + 160 = 160 \text{ feet}$$

$$\text{Velocity} = -32(3) + 48 = -48 \text{ feet per second}$$

$$\text{Acceleration} = -32 \text{ feet per second squared}$$

✔ *Checkpoint 4*   *Worked-out solution available at LarsonAppliedCalculus.com*

A ball is thrown upward from the top of an 80-foot cliff with an initial velocity of 64 feet per second. The position function is

$$s = -16t^2 + 64t + 80$$

where the time t is measured in seconds. Find the height, velocity, and acceleration of the ball at $t = 2$.

The acceleration due to gravity on the surface of the moon is only about one-sixth that exerted on the surface of Earth.

In Example 4, notice that the acceleration of the ball is -32 feet per second squared at any time t. This constant acceleration is due to the gravitational force of Earth and is called the **acceleration due to gravity.** Note that the negative value indicates that the ball is being pulled *down*—toward Earth.

Although the acceleration exerted on a falling object is relatively constant near Earth's surface, it varies greatly throughout our solar system. Large planets exert a much greater gravitational pull than small planets or moons. The next example describes the motion of a free-falling object on the moon.

EXAMPLE 5   Finding Acceleration on the Moon

An astronaut standing on the surface of the moon throws a rock upward. The height s (in feet) of the rock is given by

$$s = -\frac{27}{10}t^2 + 27t + 6$$

where t is measured in seconds. How does the acceleration due to gravity on the moon compare with that on Earth?

SOLUTION

$$s = -\frac{27}{10}t^2 + 27t + 6 \qquad \text{Position function}$$

$$\frac{ds}{dt} = -\frac{27}{5}t + 27 \qquad \text{Velocity function}$$

$$\frac{d^2s}{dt^2} = -\frac{27}{5} \qquad \text{Acceleration function}$$

So, the acceleration at any time is

$$-\frac{27}{5} = -5.4 \text{ feet per second squared}$$

which is about one-sixth of the acceleration due to gravity on Earth.

✓ *Checkpoint 5*   *Worked-out solution available at LarsonAppliedCalculus.com*

The position function on Earth, where s is measured in meters, t is measured in seconds, v_0 is the initial velocity in meters per second, and h_0 is the initial height in meters, is

$$s = -4.9t^2 + v_0t + h_0.$$

An object is thrown upward with an initial velocity of 2.2 meters per second from an initial height of 3.6 meters. What is the acceleration due to gravity on Earth in meters per second squared? ∎

The position function described in Example 5 neglects air resistance, which is appropriate because the moon has no atmosphere—and *no air resistance*. This means that the position function for any free-falling object on the moon is given by

$$s = -\frac{27}{10}t^2 + v_0t + h_0$$

where s is the height (in feet), t is the time (in seconds), v_0 is the initial velocity (in feet per second), and h_0 is the initial height (in feet). For instance, the rock in Example 5 was thrown upward with an initial velocity of 27 feet per second and had an initial height of 6 feet. This position function is valid for all objects, whether heavy ones such as hammers or light ones such as feathers.

EXAMPLE 6   Finding Velocity and Acceleration

The velocity v (in feet per second) of a certain automobile starting from rest is

$$v = \frac{80t}{t + 5}$$ Velocity function

where t is the time (in seconds). Find the velocity and acceleration of the automobile at 10-second intervals from $t = 0$ to $t = 60$.

SOLUTION   To find the acceleration function, differentiate the velocity function.

$$\frac{dv}{dt} = \frac{(t + 5)(80) - (80t)(1)}{(t + 5)^2}$$ Apply Quotient Rule.

$$= \frac{400}{(t + 5)^2}$$ Acceleration function

t (seconds)	0	10	20	30	40	50	60
v (ft/sec)	0	53.3	64	68.6	71.1	72.7	73.8
$\frac{dv}{dt}$ (ft/sec²)	16	1.78	0.64	0.33	0.20	0.13	0.09

In the table, note that the acceleration approaches zero as the velocity levels off. This observation should agree with your experience—when riding in an accelerating automobile, you do not feel the velocity, but you do feel the acceleration. In other words, you feel changes in velocity.

 Checkpoint 6   *Worked-out solution available at LarsonAppliedCalculus.com*

Use a graphing utility to graph the velocity function and acceleration function in Example 6 in the same viewing window. Compare the graphs with the table in Example 6. As the velocity levels off, what does the acceleration approach?   ■

SUMMARIZE   (Section 2.6)

1. State the meaning of each derivative listed below *(page 137)*. For examples of higher-order derivatives, see Examples 1, 2, and 3.

   (a) y'' (b) $f^{(4)}(x)$ (c) $\dfrac{d^3y}{dx^3}$

2. Match the notation with its meaning *(page 139)*. For examples of position, velocity, and acceleration functions, see Examples 4, 5, and 6.

   Notation   *Meaning*

   (i) $s = f(t)$ (a) velocity function

   (ii) $\dfrac{ds}{dt} = f'(t)$ (b) acceleration function

   (iii) $\dfrac{d^2s}{dt^2} = f''(t)$ (c) position function

3. Describe a real-life example of how higher-order derivatives can be used to analyze the velocity and acceleration of an object *(page 139, Examples 4, 5, and 6).*

SKILLS WARM UP 2.6

The following warm-up exercises involve skills that were covered in earlier sections. You will use these skills in the exercise set for this section. For additional help, review Appendix A.4 and Sections 1.4 and 2.4.

In Exercises 1–4, solve the equation.

1. $-16t^2 + 292 = 0$

2. $-16t^2 + 88t = 0$

3. $-16t^2 + 128t + 320 = 0$

4. $-16t^2 + 9t + 1440 = 0$

In Exercises 5–8, find dy/dx.

5. $y = x^2(2x + 7)$

6. $y = (x^2 + 3x)(2x^2 - 5)$

7. $y = \dfrac{x^2}{2x + 7}$

8. $y = \dfrac{x^2 + 3x}{2x^2 - 5}$

In Exercises 9 and 10, find the domain and range of f.

9. $f(x) = x^2 - 4$

10. $f(x) = \sqrt{x - 7}$

Exercises 2.6

See *CalcChat.com* for tutorial help and worked-out solutions to odd-numbered exercises.

 Finding Higher-Order Derivatives In Exercises 1–12, find the second derivative of the function. *See Examples 1 and 3.*

1. $f(x) = 9 - 2x$

2. $f(x) = 4x + 15$

3. $f(x) = x^2 + 7x - 4$

4. $f(x) = 3x^2 + 4x$

5. $g(t) = \frac{1}{3}t^3 - 4t^2 + 2t$

6. $f(x) = -\frac{5}{4}x^4 + 3x^2 - 6x$

7. $f(t) = \dfrac{2}{t^3}$

8. $g(t) = \dfrac{5}{6t^4}$

9. $f(x) = 3(2 - x^2)^3$

10. $y = 4(x^2 + 5x)^3$

11. $f(x) = \dfrac{x + 1}{x - 1}$

12. $g(x) = \dfrac{1 - 4x}{x - 3}$

 Finding Higher-Order Derivatives In Exercises 13–18, find the third derivative of the function. *See Examples 1 and 3.*

13. $f(x) = x^5 - 3x^4$

14. $f(x) = x^4 - 2x^3$

15. $f(x) = 5x(x + 4)^3$

16. $f(x) = (x^3 - 6)^4$

17. $f(x) = \dfrac{3}{8x^4}$

18. $f(x) = -\dfrac{2}{25x^5}$

 Finding Higher-Order Derivatives In Exercises 19–24, find the given value. *See Example 2.*

Function	*Value*
19. $g(t) = 5t^4 + 10t^2 + 3$	$g''(2)$
20. $f(x) = 9 - x^2$	$f''(-\sqrt{5})$
21. $f(x) = \sqrt{4 - x}$	$f'''(-5)$
22. $f(t) = \sqrt{2t + 3}$	$f'''(\frac{1}{2})$
23. $f(x) = (x^3 - 2x)^3$	$f''(1)$
24. $g(x) = (x^2 + 3x)^4$	$g''(-1)$

 Finding Higher-Order Derivatives In Exercises 25–30, find the higher-order derivative. *See Examples 1 and 3.*

Given	*Derivative*
25. $f'(x) = 2x^2$	$f''(x)$
26. $f''(x) = 20x^3 - 36x^2$	$f'''(x)$
27. $f'''(x) = 4x^{-4}$	$f^{(4)}(x)$
28. $f''(x) = 4\sqrt{x - 2}$	$f^{(5)}(x)$
29. $f^{(4)}(x) = (x^2 + 1)^2$	$f^{(6)}(x)$
30. $f''(x) = 2x^2 + 7x - 12$	$f^{(5)}(x)$

 Using Derivatives In Exercises 31–34, find the second derivative of the function and solve the equation $f''(x) = 0$.

31. $f(x) = x^3 - 9x^2 + 27x - 27$

32. $f(x) = (x + 2)(x - 2)(x + 3)(x - 3)$

33. $f(x) = x\sqrt{x^2 - 1}$

34. $f(x) = \dfrac{x}{x^2 + 3}$

35. Velocity and Acceleration A ball is propelled straight upward from ground level with an initial velocity of 144 feet per second.

(a) Write the position, velocity, and acceleration functions of the ball.

(b) Find the height, velocity, and acceleration at $t = 3$.

(c) When is the ball at its highest point? How high is this point?

(d) How fast is the ball traveling when it hits the ground? How is this speed related to the initial velocity?

36. Velocity and Acceleration A brick becomes dislodged from the top of the Empire State Building (at a height of 1250 feet) and falls to the sidewalk below.

(a) Write the position, velocity, and acceleration functions of the brick.

(b) How long does it take the brick to hit the sidewalk?

(c) How fast is the brick traveling when it hits the sidewalk?

37. Velocity and Acceleration The velocity v (in feet per second) of an automobile starting from rest is modeled by $v = 90t/(t + 10)$, where t is the time (in seconds). Create a table showing the velocity and acceleration at 10-second intervals during the first minute of travel. What can you conclude?

38. Stopping Distance A car is traveling at a rate of 66 feet per second (45 miles per hour) when the brakes are applied. The position function for the car is given by $s = -8.25t^2 + 66t$, where s is measured in feet and t is measured in seconds. Use this function to complete the table showing the position, velocity, and acceleration for each given value of t. What can you conclude?

t	0	1	2	3	4
s					
ds/dt					
d^2s/dt^2					

39. Derivatives of Polynomial Functions Consider the function $f(x) = x^2 - 6x + 6$.

 (a) Use a graphing utility to graph f, f', and f'' in the same viewing window.

(b) What is the relationship among the degree of f and the degrees of its successive derivatives?

(c) Repeat parts (a) and (b) for $f(x) = 3x^3 - 9x$.

(d) In general, what is the relationship among the degree of a polynomial function and the degrees of its successive derivatives?

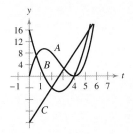

40. HOW DO YOU SEE IT? The graph shows the position, velocity, and acceleration functions of a particle. Identify each function. Explain your reasoning.

41. Modeling Data The table shows the sales y (in millions of dollars) for PetSmart from 2008 to 2013, where t is the year, with $t = 8$ corresponding to 2008. *(Source: PetSmart, Inc.)*

t	8	9	10	11	12	13
y	5065	5336	5694	6113	6758	6917

Spreadsheet at LarsonAppliedCalculus.com

(a) Use a graphing utility to find a cubic model for the sales $y(t)$ of PetSmart.

(b) Find the first and second derivatives of the function.

(c) Show that the sales of PetSmart were increasing from 2008 to 2013.

(d) Find the year when the sales were increasing at the greatest rate by solving $y''(t) = 0$.

42. Finding a Pattern Develop a general rule for $[xf(x)]^n$, where f is a differentiable function of x.

True or False? In Exercises 43 and 44, determine whether the statement is true or false. If it is false, explain why or give an example that shows it is false.

43. If $y = (x + 1)(x + 2)(x + 3)(x + 4)$, then $\dfrac{d^5y}{dx^5} = 0$.

44. The second derivative represents the rate of change of the first derivative.

45. Project: Cell Phone Subscribers in U.S. For a project analyzing the number of cell phone subscribers in the United States from 2000 to 2013, visit this text's website at *LarsonAppliedCalculus.com.* *(Source: CTIA–The Wireless Association)*

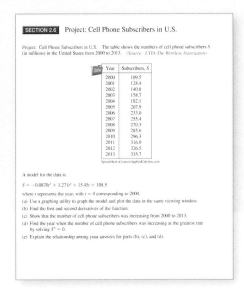

2.7 Implicit Differentiation

- Find derivatives explicitly.
- Find derivatives implicitly.
- Use implicit differentiation to answer questions about real-life situations.

Demand of a Product

In Exercise 43 on page 150, you will use implicit differentiation to find the rate of change for a demand function.

Explicit and Implicit Functions

So far in this text, most functions involving two variables have been expressed in the **explicit form**

$$y = f(x). \qquad \text{Explicit form}$$

That is, one of the two variables has been explicitly given in terms of the other. For example, in the equation

$$y = 3x - 5$$

the variable y is explicitly written as a function of x. Some functions, however, are not given explicitly and are only implied by a given equation, as shown in Example 1.

EXAMPLE 1   Finding a Derivative Explicitly

Find dy/dx for the equation

$$xy = 1.$$

SOLUTION   In this equation, y is **implicitly** defined as a function of x. One way to find dy/dx is first to solve the equation for y, then differentiate as usual.

$xy = 1$	Write original equation.
$y = \dfrac{1}{x}$	Solve for y.
$= x^{-1}$	Rewrite.
$\dfrac{dy}{dx} = -x^{-2}$	Differentiate with respect to x.
$= -\dfrac{1}{x^2}$	Simplify.

✓ **Checkpoint 1**   *Worked-out solution available at LarsonAppliedCalculus.com*

Find dy/dx for the equation

$$x^2 y = 1.$$

The procedure shown in Example 1 works whenever you can write the function explicitly. You cannot, however, use this procedure when you are unable to solve for y as a function of x. For instance, how would you find dy/dx for the equations

$$x^2 - 2y^3 + 4y = 2$$

and

$$x^2 + 2xy - y^3 = 5?$$

For either equation, it is difficult to express y as a function of x explicitly. To differentiate such equations, you can use a procedure called **implicit differentiation.**

Implicit Differentiation

To understand how to find dy/dx implicitly, you must realize that the differentiation is taking place *with respect to x*. This means that when you differentiate terms involving *x* alone, you can differentiate as usual. *But* when you differentiate terms involving *y*, you must apply the Chain Rule because you are assuming that *y* is defined implicitly as a differentiable function of *x*. Study the next example carefully. Note in particular how the Chain Rule is used to introduce the dy/dx factors in Examples 2(b), 2(c), and 2(d).

EXAMPLE 2   Applying the Chain Rule

Differentiate each expression with respect to *x*.

a. $3x^2$ **b.** $2y^3$

c. $x + 3y$ **d.** xy^2

SOLUTION

a. The only variable in this expression is *x*. So, to differentiate with respect to *x*, you can use the Simple Power Rule and the Constant Multiple Rule to obtain

$$\frac{d}{dx}[3x^2] = 6x.$$

b. This case is different. The variable in the expression is *y*, and yet you are asked to differentiate with respect to *x*. To do this, assume that *y* is a differentiable function of *x*. Then you can use the Constant Multiple Rule and the Chain Rule.

$$\frac{d}{dx}[2y^3] = \overset{c}{2} \quad \overset{n}{(3)} \quad \overset{u^{n-1}}{y^2} \quad \overset{u'}{\frac{dy}{dx}} \qquad \text{Chain Rule}$$

$$= 6y^2\frac{dy}{dx}$$

c. This expression involves both *x* and *y*. By the Sum Rule, the Constant Multiple Rule, and the Chain Rule, you can write

$$\frac{d}{dx}[x + 3y] = 1 + 3\frac{dy}{dx}. \qquad \text{Chain Rule: } \frac{d}{dx}[3y] = 3\frac{dy}{dx}$$

d. By the Product Rule and the Chain Rule, you can write

$$\frac{d}{dx}[xy^2] = x\frac{d}{dx}[y^2] + y^2\frac{d}{dx}[x] \qquad \text{Product Rule}$$

$$= x\left(2y\frac{dy}{dx}\right) + y^2(1) \qquad \text{Chain Rule}$$

$$= 2xy\frac{dy}{dx} + y^2.$$

✓ **Checkpoint 2**   *Worked-out solution available at LarsonAppliedCalculus.com*

Differentiate each expression with respect to *x*.

a. $4x^3$

b. $3y^2$

c. $x + 5y$

d. xy^3

Guidelines for Implicit Differentiation

Consider an equation involving x and y in which y is a differentiable function of x. You can use the steps below to find dy/dx.

1. Differentiate both sides of the equation *with respect to x*.

2. Collect all terms involving dy/dx on the left side of the equation and move all other terms to the right side of the equation.

3. Factor dy/dx out of the left side of the equation.

4. Solve for dy/dx by dividing both sides of the equation by the left-hand factor that does not contain dy/dx.

In Example 3, note that implicit differentiation can produce an expression for dy/dx that contains both x and y.

EXAMPLE 3   Using Implicit Differentiation

Find dy/dx for the equation $y^3 + y^2 - 5y - x^2 = -4$.

SOLUTION

1. Differentiate both sides of the equation with respect to x.

$$\frac{d}{dx}[y^3 + y^2 - 5y - x^2] = \frac{d}{dx}[-4]$$

$$\frac{d}{dx}[y^3] + \frac{d}{dx}[y^2] - \frac{d}{dx}[5y] - \frac{d}{dx}[x^2] = \frac{d}{dx}[-4]$$

$$3y^2\frac{dy}{dx} + 2y\frac{dy}{dx} - 5\frac{dy}{dx} - 2x = 0$$

2. Collect the dy/dx terms on the left side of the equation and move all other terms to the right side of the equation.

$$3y^2\frac{dy}{dx} + 2y\frac{dy}{dx} - 5\frac{dy}{dx} = 2x$$

3. Factor dy/dx out of the left side of the equation.

$$\frac{dy}{dx}(3y^2 + 2y - 5) = 2x$$

4. Solve for dy/dx by dividing by $(3y^2 + 2y - 5)$.

$$\frac{dy}{dx} = \frac{2x}{3y^2 + 2y - 5}$$

✓ **Checkpoint 3**   *Worked-out solution available at LarsonAppliedCalculus.com*

Find dy/dx for the equation $y^2 + x^2 - 2y - 4x = 4$.   ■

To see how you can use an implicit derivative, consider the graph shown in Figure 2.31. The derivative found in Example 3 gives a formula for the slope of the tangent line at a point on this graph. For instance, the slope at the point $(1, -3)$ is

$$\frac{dy}{dx} = \frac{2(1)}{3(-3)^2 + 2(-3) - 5} = \frac{1}{8}.$$

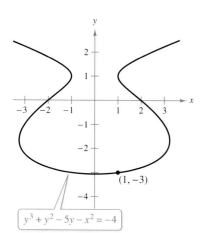

$(1, -3)$

$y^3 + y^2 - 5y - x^2 = -4$

FIGURE 2.31

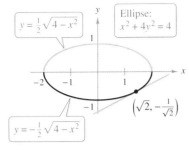

$y = \frac{1}{2}\sqrt{4 - x^2}$

Ellipse: $x^2 + 4y^2 = 4$

$y = -\frac{1}{2}\sqrt{4 - x^2}$

$\left(\sqrt{2}, -\frac{1}{\sqrt{2}}\right)$

Slope of tangent line is $\frac{1}{2}$.

FIGURE 2.32

STUDY TIP

To see the benefit of implicit differentiation, try reworking Example 4 using the explicit function

$$y = -\frac{1}{2}\sqrt{4 - x^2}.$$

The graph of this function is the lower half of the ellipse.

EXAMPLE 4   Finding the Slope of a Graph Implicitly

Find the slope of the tangent line to the ellipse given by $x^2 + 4y^2 = 4$ at the point $\left(\sqrt{2}, -1/\sqrt{2}\right)$, as shown in Figure 2.32.

SOLUTION   Begin by finding dy/dx implicitly.

$x^2 + 4y^2 = 4$	Write original equation.
$\dfrac{d}{dx}[x^2 + 4y^2] = \dfrac{d}{dx}[4]$	Differentiate with respect to x.
$2x + 8y\dfrac{dy}{dx} = 0$	Implicit differentiation
$8y\dfrac{dy}{dx} = -2x$	Subtract $2x$ from each side.
$\dfrac{dy}{dx} = \dfrac{-2x}{8y}$	Divide each side by $8y$.
$\dfrac{dy}{dx} = -\dfrac{x}{4y}$	Simplify.

To find the slope at the given point, substitute $x = \sqrt{2}$ and $y = -1/\sqrt{2}$ into the derivative, as shown below.

$$\frac{dy}{dx} = -\frac{\sqrt{2}}{4(-1/\sqrt{2})}$$

$$= \frac{1}{2}$$

✓ *Checkpoint 4*   *Worked-out solution available at LarsonAppliedCalculus.com*

Find the slope of the tangent line to the circle $x^2 + y^2 = 25$ at the point $(3, -4)$.

EXAMPLE 5   Finding the Slope of a Graph Implicitly

Find the slope of the graph of $2x^2 - y^2 = 1$ at the point $(1, 1)$.

SOLUTION   Begin by finding dy/dx implicitly.

$2x^2 - y^2 = 1$	Write original equation.
$4x - 2y\dfrac{dy}{dx} = 0$	Differentiate with respect to x.
$-2y\dfrac{dy}{dx} = -4x$	Subtract $4x$ from each side.
$\dfrac{dy}{dx} = \dfrac{2x}{y}$	Divide each side by $-2y$.

At the point $(1, 1)$, the slope of the graph is

$$\frac{dy}{dx} = \frac{2(1)}{1}$$

$$= 2$$

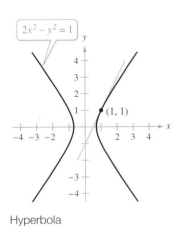

$2x^2 - y^2 = 1$

$(1, 1)$

Hyperbola

FIGURE 2.33

as shown in Figure 2.33. The graph is called a **hyperbola.**

✓ *Checkpoint 5*   *Worked-out solution available at LarsonAppliedCalculus.com*

Find the slope of the graph of $x^2 - 9y^2 = 16$ at the point $(5, 1)$.

Demand Function

FIGURE 2.34

Application

EXAMPLE 6   Using a Demand Function

The demand function for a product is modeled by

$$p = \frac{3}{0.000001x^3 + 0.01x + 1}$$

where p is measured in dollars and x is measured in thousands of units, as shown in Figure 2.34. Find the rate of change of the demand x with respect to the price p when $x = 100$.

SOLUTION   To simplify the differentiation, begin by rewriting the function. Then, differentiate *with respect to p*.

$$p = \frac{3}{0.000001x^3 + 0.01x + 1}$$

$$0.000001x^3 + 0.01x + 1 = \frac{3}{p}$$

$$0.000003x^2 \frac{dx}{dp} + 0.01 \frac{dx}{dp} = -\frac{3}{p^2}$$

$$(0.000003x^2 + 0.01) \frac{dx}{dp} = -\frac{3}{p^2}$$

$$\frac{dx}{dp} = -\frac{3}{p^2(0.000003x^2 + 0.01)}$$

When $x = 100$, the price is

$$p = \frac{3}{0.000001(100)^3 + 0.01(100) + 1} = \$1.$$

So, when $x = 100$ and $p = 1$, the rate of change of the demand with respect to the price is

$$\frac{dx}{dp} = -\frac{3}{(1)^2[0.000003(100)^2 + 0.01]} = -75.$$

This means that when $x = 100$, the demand is dropping at the rate of 75 thousand units for each dollar increase in price.

✓ *Checkpoint 6*   Worked-out solution available at LarsonAppliedCalculus.com

The demand function for a product is given by

$$p = \frac{2}{0.001x^2 + x + 1}.$$

Find dx/dp implicitly.   ■

SUMMARIZE   (Section 2.7)

1. State the guidelines for implicit differentiation *(page 146)*. For examples of implicit differentiation, see Examples 2, 3, 4, and 5.

2. Describe a real-life example of how implicit differentiation can be used to analyze the rate of change of a product's demand *(page 148, Example 6)*.

SKILLS WARM UP 2.7

The following warm-up exercises involve skills that were covered in a previous course. You will use these skills in the exercise set for this section. For additional help, review Appendix A.3.

In Exercises 1–6, solve the equation for y.

1. $x - \dfrac{y}{x} = 2$

2. $\dfrac{4}{x-3} = \dfrac{1}{y}$

3. $xy - x + 6y = 6$

4. $7 + 4y = 3x^2 + x^2y$

5. $x^2 + y^2 = 5$

6. $x = \pm\sqrt{6 - y^2}$

In Exercises 7–9, evaluate the expression at the given point.

7. $\dfrac{3x^2 - 4}{3y^2}$, $(2, 1)$

8. $\dfrac{x^2 - 2}{1 - y}$, $(0, -3)$

9. $\dfrac{7x}{4y^2 + 13y + 3}$, $\left(-\dfrac{1}{7}, -2\right)$

Exercises 2.7

See *CalcChat.com* for tutorial help and worked-out solutions to odd-numbered exercises.

 Finding Derivatives In Exercises 1–12, find dy/dx. *See Examples 1 and 3.*

1. $x^3y = 6$

2. $3x^2 - y = 8x$

3. $y^2 = 1 - x^2, 0 \le x \le 1$

4. $y^3 = 5x^3 + 8x$

5. $y^4 - y^2 + 7y - 6x = 9$

6. $4y^3 + 5y^2 - y - 3x^3 = 8x$

7. $xy^2 + 4xy = 10$

8. $2xy^3 - x^2y = 2$

9. $\dfrac{2x + y}{x - 5y} = 1$

10. $\dfrac{xy - y^2}{y - x} = 1$

11. $\dfrac{2y}{y^2 + 3} = 4x$

12. $\dfrac{4y^2}{y^2 - 9} = x^2$

 Finding the Slope of a Graph Implicitly In Exercises 13–26, find the slope of the graph of the equation at the given point. *See Examples 4 and 5.*

Equation	*Point*
13. $x^2 + y^2 = 16$	$(0, 4)$
14. $x^2 - y^2 = 25$	$(5, 0)$
15. $y + xy = 4$	$(-5, -1)$
16. $xy - 3y^2 = 2$	$(7, 2)$
17. $x^2 - xy + y^2 = 3$	$(-2, -1)$
18. $x^2y + y^3x = -6$	$(2, -1)$
19. $xy - x = y$	$\left(\dfrac{3}{2}, 3\right)$
20. $x^3 + y^3 = 6xy$	$\left(\dfrac{4}{3}, \dfrac{8}{3}\right)$
21. $x^{1/2} + y^{1/2} = 9$	$(16, 25)$
22. $x^{2/3} + y^{2/3} = 5$	$(8, 1)$
23. $\sqrt{xy} = x - 2y$	$(4, 1)$
24. $(x + y)^3 = x^3 + y^3$	$(-1, 1)$

Equation	*Point*
25. $y^2(x^2 + y^2) = 2x^2$	$(1, 1)$
26. $(x^2 + y^2)^2 = 8x^2y$	$(2, 2)$

 Finding the Slope of a Graph Implicitly In Exercises 27–32, find the slope of the graph at the given point. *See Examples 4 and 5.*

27. $3x^2 - 2y + 5 = 0$

28. $4x^2 + 2y - 1 = 0$

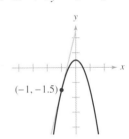

29. $x^2 + y^2 = 4$

30. $4x^2 + 9y^2 = 36$

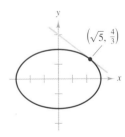

31. $x^2 - y^3 = 0$

32. $(4 - x)y^2 = x^3$

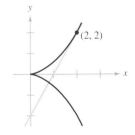

Finding Derivatives Implicitly and Explicitly In Exercises 33 and 34, find dy/dx implicitly and explicitly (the explicit functions are shown on the graph) and show that the results are equivalent. Use the graph to estimate the slope of the tangent line at the labeled point. Then verify your result analytically by evaluating dy/dx at the point.

33. $x - y^2 - 1 = 0$　　　**34.** $4y^2 - x^2 = 7$

 Finding Equations of Tangent Lines In Exercises 35–42, find equations of the tangent lines to the graph of the equation at the given points. Use a graphing utility to graph the equation and the tangent lines in the same viewing window.

Equation	Points
35. $x^2 + y^2 = 100$	$(8, 6)$ and $(-6, 8)$
36. $x^2 + y^2 = 9$	$(0, 3)$ and $\left(2, \sqrt{5}\right)$
37. $y^2 = 5x^3$	$\left(1, \sqrt{5}\right)$ and $\left(1, -\sqrt{5}\right)$
38. $4xy + x^2 = 5$	$(1, 1)$ and $(5, -1)$
39. $x^3 + y^3 = 8$	$(0, 2)$ and $(2, 0)$
40. $x^2y - 8 = -4y$	$(-2, 1)$ and $\left(6, \frac{1}{5}\right)$
41. $y^2 = \dfrac{x^3}{4 - x}$	$(2, 2)$ and $(2, -2)$
42. $x + y^3 = 6xy^3 - 1$	$(-1, 0)$ and $(0, -1)$

Demand In Exercises 43–46, find the rate of change of x with respect to p. *See Example 6.*

43. $p = \dfrac{2}{0.00001x^3 + 0.1x}, \quad x \geq 0$

44. $p = \dfrac{4}{0.000001x^2 + 0.05x + 1}, \quad x \geq 0$

45. $p = \sqrt{\dfrac{200 - x}{2x}}, \quad 0 < x \leq 200$

46. $p = \sqrt{\dfrac{500 - x}{2x}}, \quad 0 < x \leq 500$

47. Production Let x represent the units of labor and y the capital invested in a manufacturing process. When 135,540 units are produced, the relationship between labor and capital can be modeled by $100x^{0.75}y^{0.25} = 135,540$.

(a) Find the rate of change of y with respect to x when $x = 1500$ and $y = 1000$.

 (b) The model used in this problem is called the *Cobb-Douglas production function*. Graph the model on a graphing utility and describe the relationship between labor and capital.

48. HOW DO YOU SEE IT? The graph shows the demand function for a product.

Demand Function

(a) What happens to the demand as the price increases?

(b) Over what interval is the rate of change of the demand with respect to the price decreasing?

49. Health: U.S. Chickenpox Decline The numbers (in thousands) of cases y of chickenpox reported in the years 2008 through 2012 can be modeled by

$$y^2 - 35,892.5 = -27.0021t^3 + 888.789t^2 - 9753.25t$$

where t represents the year, with $t = 8$ corresponding to 2008. *(Source: U.S. Centers for Disease Control and Prevention)*

(a) Use a graphing utility to graph the model and describe the results.

(b) Use the graph to estimate the year during which the number of reported cases was decreasing at the greatest rate.

(c) Complete the table to estimate the year during which the number of reported cases was decreasing at the greatest rate. Compare this estimate with your answer in part (b).

t	8	9	10	11	12
y					
y'					

2.8 Related Rates

■ Examine related variables.
■ Solve related-rate problems.

Related Variables

In this section, you will study problems involving variables that are changing with respect to time. If two or more such variables are related to each other, then their rates of change with respect to time are also related.

For instance, suppose that x and y are related by the equation

$$y = 2x.$$

If both variables are changing with respect to time, then their rates of change will also be related.

x and y are related.	The rates of change of x and y are related.
$y = 2x$	$\dfrac{dy}{dt} = 2\dfrac{dx}{dt}$

In this simple example, you can see that because y always has twice the value of x, it follows that the rate of change of y with respect to time is always twice the rate of change of x with respect to time.

In Exercise 25 on page 157, you will use related rates to find the rate of change of the sales for a product.

EXAMPLE 1 **Examining Two Rates That Are Related**

The variables x and y are differentiable functions of t and are related by the equation

$$y = x^2 + 3.$$

When $x = 1$, $dx/dt = 2$. Find dy/dt when $x = 1$.

SOLUTION Use the Chain Rule to differentiate both sides of the equation with respect to t.

$$y = x^2 + 3 \qquad \text{Write original equation.}$$

$$\frac{d}{dt}[y] = \frac{d}{dt}[x^2 + 3] \qquad \text{Differentiate with respect to } t.$$

$$\frac{dy}{dt} = 2x\frac{dx}{dt} \qquad \text{Apply Chain Rule.}$$

When $x = 1$ and $dx/dt = 2$, you have

$$\frac{dy}{dt} = 2(1)(2)$$

$$= 4.$$

✓ **Checkpoint 1** *Worked-out solution available at LarsonAppliedCalculus.com*

The variables x and y are differentiable functions of t and are related by the equation

$$y = x^3 + 2.$$

When $x = 1$, $dx/dt = 3$. Find dy/dt when $x = 1$.

Solving Related-Rate Problems

In Example 1, you were *given* the mathematical model.

> *Given equation:* $y = x^2 + 3$
>
> *Given rate:* $\dfrac{dx}{dt} = 2$ when $x = 1$
>
> *Find:* $\dfrac{dy}{dt}$ when $x = 1$

In the next example, you must *create* a mathematical model from a verbal description.

EXAMPLE 2   Changing Area

A pebble is dropped into a calm pool of water, causing ripples in the form of concentric circles, as shown in the photo. The radius r of the outer ripple is increasing at a constant rate of 1 foot per second. When the radius is 4 feet, at what rate is the total area A of the disturbed water changing?

SOLUTION   The variables r and A are related by the equation for the area of a circle, $A = \pi r^2$. To solve this problem, use the fact that the rate of change of the radius is constant at 1 foot per second. So, $dr/dt = 1$.

> *Equation:* $A = \pi r^2$
>
> *Given rate:* $\dfrac{dr}{dt} = 1$ when $r = 4$
>
> *Find:* $\dfrac{dA}{dt}$ when $r = 4$

Total area increases as the outer radius increases.

Using this information, you can proceed as in Example 1.

$A = \pi r^2$	Write original equation.
$\dfrac{d}{dt}[A] = \dfrac{d}{dt}[\pi r^2]$	Differentiate with respect to t.
$\dfrac{dA}{dt} = 2\pi r \dfrac{dr}{dt}$	Apply Chain Rule.

When $r = 4$ and $dr/dt = 1$, you have

$$\frac{dA}{dt} = 2\pi(4)(1) = 8\pi. \qquad \text{Substitute 4 for } r \text{ and 1 for } dr/dt.$$

When the radius is 4 feet, the area is changing at a rate of 8π square feet per second.

✓ **Checkpoint 2**   *Worked-out solution available at LarsonAppliedCalculus.com*

As in Example 2, a pebble is dropped into the pool, but this time the radius r of the outer ripple is increasing at a rate of 2 feet per second. At what rate is the total area of the disturbed water changing when the radius is 3 feet?   ■

In Example 2, note that the radius changes at a *constant* rate ($dr/dt = 1$ for all t), but the area changes at a *nonconstant* rate, as shown in the table.

r (ft)	1	2	3	4
$\dfrac{dA}{dt}$ (ft²/sec)	2π	4π	6π	8π

The solution shown in Example 2 illustrates the steps for solving a related-rate problem.

Guidelines for Solving a Related-Rate Problem

1. Identify all *given* quantities and all quantities *to be determined*. If possible, make a sketch and label the quantities.

2. Write an equation that relates all variables whose rates of change are either given or to be determined.

3. Use the Chain Rule to implicitly differentiate both sides of the equation *with respect to time*.

4. *After* completing Step 3, substitute into the resulting equation all known values for the variables and their rates of change. Then solve for the required rate of change.

STUDY TIP

Be sure you notice the order of Steps 3 and 4 in the guidelines. Do not substitute the known values for the variables until after you have differentiated.

In Step 2 of the guidelines, note that you must write an equation that relates the given variables. To help you with this step, reference tables that summarize many common formulas are included in the appendices. For instance, the volume of a sphere of radius r is given by the formula

$$V = \frac{4}{3}\pi r^3$$

as listed in Appendix D.

The table below lists examples of the mathematical models for some common rates of change that can be used in the first step of the solution of a related-rate problem.

Verbal statement	Mathematical model
The velocity of a car after traveling for 1 hour is 50 miles per hour.	x = distance traveled $\frac{dx}{dt} = 50 \text{ mi/h when } t = 1$
Water is being pumped into a swimming pool at a rate of 10 cubic feet per minute.	V = volume of water in pool $\frac{dV}{dt} = 10 \text{ ft}^3/\text{min}$
A population of bacteria is increasing at a rate of 2000 per hour.	x = number in population $\frac{dx}{dt} = 2000 \text{ bacteria per hour}$
Revenue is increasing at a rate of $4000 per month.	R = revenue $\frac{dR}{dt} = 4000 \text{ dollars per month}$
Profit is decreasing at a rate of $2500 per day.	P = profit $\frac{dP}{dt} = -2500 \text{ dollars per day}$

EXAMPLE 3   Analyzing a Profit Function

A company's profit P (in dollars) from selling x units of a product can be modeled by

$$P = 500x - \frac{1}{4}x^2. \qquad \text{Model for profit}$$

The sales are increasing at a rate of 10 units per day. Find the rate of change of the profit with respect to time (in days) when 500 units have been sold.

SOLUTION   Because the sales are increasing at a rate of 10 units per day, you know that at time t the rate of change is $dx/dt = 10$. So, the problem can be stated as shown.

$$\text{Given rate: } \frac{dx}{dt} = 10$$

$$\text{Find: } \frac{dP}{dt} \text{ when } x = 500$$

To find the rate of change of the profit, use the model for profit that relates the profit P and the units of the product sold x.

$$\text{Equation: } P = 500x - \frac{1}{4}x^2$$

By differentiating both sides of the equation with respect to t, you obtain

$$\frac{d}{dt}[P] = \frac{d}{dt}\left[500x - \frac{1}{4}x^2\right] \qquad \text{Differentiate with respect to } t.$$

$$\frac{dP}{dt} = \left(500 - \frac{1}{2}x\right)\frac{dx}{dt}. \qquad \text{Apply Chain Rule.}$$

When $x = 500$ units and $dx/dt = 10$, the rate of change of the profit is

$$\frac{dP}{dt} = \left[500 - \frac{1}{2}(500)\right](10) = (500 - 250)(10) = 250(10) = \$2500 \text{ per day.}$$

The graph of the profit function (in terms of x) is shown in Figure 2.35.

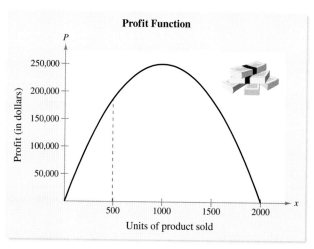

Profit Function

FIGURE 2.35

✓ *Checkpoint 3*   *Worked-out solution available at LarsonAppliedCalculus.com*

Find the rate of change of the profit (in dollars) with respect to time (in days) when 50 units have been sold, sales have increased at a rate of 10 units per day, and $P = 200x - \frac{1}{2}x^2$.

EXAMPLE 4   Increasing Production

A company is increasing the production of a product at the rate of 200 units per week. The weekly demand function is modeled by

$$p = 100 - 0.001x$$

where p is the price per unit (in dollars) and x is the number of units produced in a week. Find the rate of change of the revenue (in dollars) with respect to time (in weeks) when the weekly production is 2000 units. Is the rate of change of the revenue greater than $20,000 per week?

SOLUTION   Because production is increasing at a rate of 200 units per week, you know that at time t the rate of change is $dx/dt = 200$. So, the problem can be stated as shown.

Given rate: $\dfrac{dx}{dt} = 200$

Find: $\dfrac{dR}{dt}$ when $x = 2000$

To find the rate of change of the revenue, you must find an equation that relates the revenue R and the number of units produced x.

Equation: $R = xp = x(100 - 0.001x) = 100x - 0.001x^2$

By differentiating both sides of the equation with respect to t, you obtain

$R = 100x - 0.001x^2$	Write original equation.
$\dfrac{d}{dt}[R] = \dfrac{d}{dt}[100x - 0.001x^2]$	Differentiate with respect to t.
$\dfrac{dR}{dt} = (100 - 0.002x)\dfrac{dx}{dt}.$	Apply Chain Rule.

When $x = 2000$ and $dx/dt = 200$, the rate of change of the revenue is

$$\frac{dR}{dt} = [100 - 0.002(2000)](200)$$

$$= \$19{,}200 \text{ per week.}$$

So, the rate of change of the revenue is not greater than $20,000 per week.

 ✓ *Checkpoint 4*   *Worked-out solution available at LarsonAppliedCalculus.com*

Find the rate of change of the revenue with respect to time for the company in Example 4 when the weekly demand function is

$$p = 150 - 0.002x.$$ ■

SUMMARIZE   (Section 2.8)

1. Give a description of related variables *(page 151)*. For an example of two related variables, see Example 1.

2. State the guidelines for solving a related-rate problem *(page 153)*. For examples of solving related-rate problems, see Examples 2, 3, and 4.

3. Describe a real-life example of how related rates can be used to analyze the rate of change of a company's revenue *(page 155, Example 4)*.

SKILLS WARM UP 2.8

The following warm-up exercises involve skills that were covered in earlier sections. You will use these skills in the exercise set for this section. For additional help, review Section 2.7.

In Exercises 1–6, write a formula for the given quantity.

1. Area of a circle

2. Volume of a sphere

3. Surface area of a cube

4. Volume of a cube

5. Volume of a cone

6. Area of a triangle

In Exercises 7–10, find dy/dx by implicit differentiation.

7. $x^2 + y^2 = 9$ 8. $3xy - x^2 = 6$ 9. $x^3 - 4y + 2xy = 12x$ 10. $x + xy^2 - y^2 = xy$

Exercises 2.8

See *CalcChat.com* for tutorial help and worked-out solutions to odd-numbered exercises.

Using Related Rates In Exercises 1–4, assume that x and y are both differentiable functions of t. Use the given values to find (a) dy/dt and (b) dx/dt. *See Example 1.*

Equation	Find	Given
1. $y = \sqrt{x}$	(a) $\dfrac{dy}{dt}$ when $x = 4$,	$\dfrac{dx}{dt} = 3$
	(b) $\dfrac{dx}{dt}$ when $x = 25$,	$\dfrac{dy}{dt} = 2$
2. $y = 3x^2 - 5x$	(a) $\dfrac{dy}{dt}$ when $x = 3$,	$\dfrac{dx}{dt} = 2$
	(b) $\dfrac{dx}{dt}$ when $x = 2$,	$\dfrac{dy}{dt} = 4$
3. $xy = 4$	(a) $\dfrac{dy}{dt}$ when $x = 8$,	$\dfrac{dx}{dt} = 10$
	(b) $\dfrac{dx}{dt}$ when $x = 1$,	$\dfrac{dy}{dt} = -6$
4. $x^2 + y^2 = 25$	(a) $\dfrac{dy}{dt}$ when $x = 3, y = 4$,	$\dfrac{dx}{dt} = 8$
	(b) $\dfrac{dx}{dt}$ when $x = 4, y = 3$,	$\dfrac{dy}{dt} = -2$

5. Area The radius r of a circle is increasing at a rate of 3 inches per minute. Find the rate of change of the area when (a) $r = 6$ inches and (b) $r = 24$ inches.

6. Volume The radius r of a sphere is increasing at a rate of 3 inches per minute. Find the rate of change of the volume when (a) $r = 9$ inches and (b) $r = 16$ inches.

7. Area Let A be the area of a circle of radius r that is changing with respect to time. If dr/dt is constant, is dA/dt constant? Explain your reasoning.

8. Volume Let V be the volume of a sphere of radius r that is changing with respect to time. If dr/dt is constant, is dV/dt constant? Explain your reasoning.

9. Volume A spherical balloon is inflated with gas at a rate of 10 cubic feet per minute. How fast is the radius of the balloon changing at the instant the radius is (a) 1 foot and (b) 2 feet?

10. Volume The radius r of a right circular cone is increasing at a rate of 2 inches per minute. The height h of the cone is related to the radius by $h = 3r$. Find the rate of change of the volume when (a) $r = 6$ inches and (b) $r = 24$ inches.

11. Cost, Revenue, and Profit A company that manufactures sport supplements calculates that its cost C and revenue R can be modeled by the equations

$$C = 125,000 + 0.75x \quad \text{and} \quad R = 250x - \tfrac{1}{10}x^2$$

where x is the number of units of sport supplements produced in 1 week. Production during one particular week is 1000 units and is increasing at a rate of 150 units per week. Find the rate of change of the (a) cost, (b) revenue, and (c) profit.

12. Cost, Revenue, and Profit A company that manufactures pet toys calculates that its cost C and revenue R can be modeled by the equations

$$C = 75,000 + 1.05x \quad \text{and} \quad R = 500x - \frac{x^2}{25}$$

where x is the number of toys produced in 1 week. Production during one particular week is 5000 toys and is increasing at a rate of 250 toys per week. Find the rate of change of the (a) cost, (b) revenue, and (c) profit.

13. Revenue The revenue R from selling x units of a product is given by

$$R = 1200x - x^2.$$

The sales are increasing at a rate of 23 units per day. Find the rate of change of the revenue when (a) $x = 300$ units and (b) $x = 450$ units.

14. Profit The profit P from selling x units of a product is given by $P = 510x - 0.3x^2$. The sales are increasing at a rate of 9 units per day. Find the rate of change of the profit when (a) $x = 400$ units and (b) $x = 600$ units.

15. Volume All edges of a cube are expanding at a rate of 6 centimeters per second. How fast is the volume changing when each edge is (a) 2 centimeters and (b) 10 centimeters?

16. Surface Area All edges of a cube are expanding at a rate of 6 centimeters per second. How fast is the surface area changing when each edge is (a) 2 centimeters and (b) 10 centimeters?

17. Boating A boat is pulled by a winch on a dock, and the winch is 12 feet above the deck of the boat (see figure). The winch pulls the rope at a rate of 4 feet per second. Find the speed of the boat when 13 feet of rope is out. What happens to the speed of the boat as it gets closer and closer to the dock?

Figure for 17

18. Shadow Length A man 6 feet tall walks at a rate of 5 feet per second away from a light that is 15 feet above the ground (see figure).

(a) When he is 10 feet from the base of the light, at what rate is the tip of his shadow moving?

(b) When he is 10 feet from the base of the light, at what rate is the length of his shadow changing?

19. Air Traffic Control An airplane flying at an altitude of 6 miles passes directly over a radar antenna (see figure). When the airplane is 10 miles away ($s = 10$), the radar detects that the distance s is changing at a rate of 240 miles per hour. What is the speed of the airplane?

Figure for 19

20. Baseball A (square) baseball diamond has sides that are 90 feet long (see figure). A player running from second base to third base at a speed of 30 feet per second is 26 feet from third base. At what rate is the player's distance from home plate changing?

21. Air Traffic Control An air traffic controller spots two airplanes at the same altitude converging to a point as they fly at right angles to each other. One airplane is 150 miles from the point and has a speed of 450 miles per hour. The other is 200 miles from the point and has a speed of 600 miles per hour.

(a) At what rate is the distance between the planes changing?

(b) How much time does the controller have to get one of the airplanes on a different flight path?

22. Advertising Costs A retail sporting goods store estimates that weekly sales S and weekly advertising costs x are related by the equation $S = 2250 + 50x + 0.35x^2$. The current weekly advertising costs are $1500, and these costs are increasing at a rate of $125 per week. Find the current rate of change of the weekly sales with respect to time.

23. Environment An accident at an oil drilling platform is causing a circular oil slick. The slick is 0.08 foot thick, and when the radius of the slick is 150 feet, the radius is increasing at the rate of 0.5 foot per minute. At what rate (in cubic feet per minute) is oil flowing from the site of the accident?

24. Profit A company is increasing the production of a product at the rate of 25 units per week. The demand and cost functions for the product are given by $p = 50 - 0.01x$ and $C = 4000 + 40x - 0.02x^2$, where x is the number of units produced per week. Find the rate of change of the profit with respect to time when the weekly sales are $x = 800$ units.

25. Sales The profit for a product is increasing at a rate of $5600 per week. The demand and cost functions for the product are given by $p = 6000 - 25x$ and $C = 2400x + 5200$, where x is the number of units produced per week. Find the rate of change of the sales with respect to time when the weekly sales are $x = 44$ units.

26. **HOW DO YOU SEE IT?** The graph shows the demand and supply equations for a product, where x represents the number of units (in thousands) and p is the price (in dollars). Using the graph, (a) determine whether dp/dt is positive or negative given that dx/dt is negative, and (b) determine whether dx/dt is positive or negative given that dp/dt is positive. Explain.

Figure for 20

ALGEBRA TUTOR

Simplifying Algebraic Expressions

To be successful in using derivatives, you must be good at simplifying algebraic expressions. Here are some helpful simplification techniques.

1. Combine *like terms*. This may involve expanding an expression by multiplying factors.

2. Divide out *common factors* in the numerator and denominator of an expression.

3. Factor an expression.

4. Rationalize a denominator.

5. Add, subtract, multiply, or divide fractions.

TECH TUTOR

Symbolic algebra systems can simplify algebraic expressions. If you have access to such a system, try using it to simplify the expressions in this Algebra Tutor.

EXAMPLE 1   Simplifying Fractional Expressions

a. $\dfrac{[3(x + \Delta x) + 5] - (3x + 5)}{\Delta x} = \dfrac{3x + 3\Delta x + 5 - 3x - 5}{\Delta x}$ Multiply factors and remove parentheses.

$= \dfrac{3\Delta x}{\Delta x}$ Combine like terms.

$= 3, \quad \Delta x \neq 0$ Divide out common factor.

b. $\dfrac{(x + \Delta x)^2 - x^2}{\Delta x} = \dfrac{x^2 + 2x(\Delta x) + (\Delta x)^2 - x^2}{\Delta x}$ Expand terms.

$= \dfrac{2x(\Delta x) + (\Delta x)^2}{\Delta x}$ Combine like terms.

$= \dfrac{\Delta x(2x + \Delta x)}{\Delta x}$ Factor.

$= 2x + \Delta x, \quad \Delta x \neq 0$ Divide out common factor.

c. $\dfrac{(x^2 - 1)(-2 - 2x) - (3 - 2x - x^2)(2)}{(x^2 - 1)^2}$

$= \dfrac{(-2x^2 - 2x^3 + 2 + 2x) - (6 - 4x - 2x^2)}{(x^2 - 1)^2}$ Multiply factors.

$= \dfrac{-2x^2 - 2x^3 + 2 + 2x - 6 + 4x + 2x^2}{(x^2 - 1)^2}$ Remove parentheses.

$= \dfrac{-2x^3 + 6x - 4}{(x^2 - 1)^2}$ Combine like terms.

d. $2\left(\dfrac{2x + 1}{3x}\right)\left[\dfrac{3x(2) - (2x + 1)(3)}{(3x)^2}\right]$

$= 2\left(\dfrac{2x + 1}{3x}\right)\left[\dfrac{6x - (6x + 3)}{(3x)^2}\right]$ Multiply factors.

$= \dfrac{2(2x + 1)(6x - 6x - 3)}{(3x)^3}$ Multiply fractions and remove parentheses.

$= \dfrac{2(2x + 1)(-3)}{3(9)x^3}$ Combine like terms and factor.

$= \dfrac{-2(2x + 1)}{9x^3}$ Divide out common factor. ■

EXAMPLE 2   Simplifying Expressions with Powers

Simplify each expression.

a. $(2x + 1)^2(6x + 1) + (3x^2 + x)(2)(2x + 1)(2)$

b. $(-1)(3x^2 - 2x)^{-2}(6x - 2)$

c. $(x)\left(\frac{1}{2}\right)(2x + 3)^{-1/2}(2) + (2x + 3)^{1/2}(1)$

d. $\dfrac{x^2\left(\frac{1}{2}\right)(x^2 + 1)^{-1/2}(2x) - (x^2 + 1)^{1/2}(2x)}{x^4}$

SOLUTION

a. $(2x + 1)^2(6x + 1) + (3x^2 + x)(2)(2x + 1)(2)$

$\quad = (2x + 1)[(2x + 1)(6x + 1) + (3x^2 + x)(2)(2)]$ Factor.

$\quad = (2x + 1)[12x^2 + 8x + 1 + (12x^2 + 4x)]$ Multiply factors.

$\quad = (2x + 1)(12x^2 + 8x + 1 + 12x^2 + 4x)$ Remove parentheses.

$\quad = (2x + 1)(24x^2 + 12x + 1)$ Combine like terms.

b. $(-1)(3x^2 - 2x)^{-2}(6x - 2)$

$\quad = \dfrac{(-1)(6x - 2)}{(3x^2 - 2x)^2}$ Rewrite as a fraction.

$\quad = \dfrac{(-1)(2)(3x - 1)}{(3x^2 - 2x)^2}$ Factor.

$\quad = \dfrac{-2(3x - 1)}{(3x^2 - 2x)^2}$ Multiply factors.

c. $(x)\left(\frac{1}{2}\right)(2x + 3)^{-1/2}(2) + (2x + 3)^{1/2}(1)$

$\quad = (2x + 3)^{-1/2}[x + (2x + 3)]$ Factor.

$\quad = \dfrac{x + 2x + 3}{(2x + 3)^{1/2}}$ Rewrite as a fraction.

$\quad = \dfrac{3(x + 1)}{(2x + 3)^{1/2}}$ Combine like terms and factor.

d. $\dfrac{x^2\left(\frac{1}{2}\right)(x^2 + 1)^{-1/2}(2x) - (x^2 + 1)^{1/2}(2x)}{x^4}$

$\quad = \dfrac{(x^3)(x^2 + 1)^{-1/2} - (x^2 + 1)^{1/2}(2x)}{x^4}$ Multiply factors.

$\quad = \dfrac{(x^2 + 1)^{-1/2}(x)[x^2 - (x^2 + 1)(2)]}{x^4}$ Factor.

$\quad = \dfrac{x[x^2 - (2x^2 + 2)]}{(x^2 + 1)^{1/2}x^4}$ Write with positive exponents.

$\quad = \dfrac{x^2 - 2x^2 - 2}{(x^2 + 1)^{1/2}x^3}$ Divide out common factor and remove parentheses.

$\quad = \dfrac{-x^2 - 2}{(x^2 + 1)^{1/2}x^3}$ Combine like terms. ■

STUDY TIP

All of the expressions in this Algebra Tutor are derivatives. Can you see what the original function is for each expression? Explain your reasoning.

SUMMARY AND STUDY STRATEGIES

After studying this chapter, you should have acquired the following skills.
The exercise numbers are keyed to the Review Exercises that begin on page 162.
Answers to odd-numbered Review Exercises are given in the back of the text.*

Section 2.1   Review Exercises

- Approximate the slope of the tangent line to a graph at a point.   *1–4*
- Interpret the slope of a graph in a real-life setting.   *5–8*
- Use the limit definition to find the slope of a graph at a point and the derivative *9–24*
  of a function.

$$f'(x) = \lim_{\Delta x \to 0} \frac{f(x + \Delta x) - f(x)}{\Delta x}$$

- Use the graph of a function to recognize points at which the function is   *25–28*
  not differentiable.

Section 2.2

- Use the Constant Rule for differentiation.   *29, 30*

  $$\frac{d}{dx}[c] = 0, \ c \text{ is a constant.}$$

- Use the Power Rule for differentiation.   *31, 32*

  $$\frac{d}{dx}[x^n] = nx^{n-1}, \ n \text{ is a real number.}$$

- Use the Constant Multiple Rule for differentiation.   *33–36*

  $$\frac{d}{dx}[cf(x)] = cf'(x), \ c \text{ is a constant.}$$

- Use the Sum and Difference Rules for differentiation.   *37–40*

  $$\frac{d}{dx}[f(x) \pm g(x)] = f'(x) \pm g'(x)$$

- Use derivatives to find the slope of a graph at a point.   *41–44*
- Use derivatives to write equations of tangent lines.   *45–48*
- Use derivatives to answer questions about real-life situations.   *49, 50*

Section 2.3

- Find the average rate of change of a function over an interval and the instantaneous *51–54*
  rate of change at a point.

  Average rate of change: $\dfrac{f(b) - f(a)}{b - a}$;   Instantaneous rate of change: $\lim\limits_{\Delta x \to 0} \dfrac{f(x + \Delta x) - f(x)}{\Delta x}$

- Use derivatives to find the velocities of objects.   *55, 56*
- Find the marginal revenues, marginal costs, and marginal profits for products.   *57–66*
- Use derivatives to answer questions about real-life situations.   *67, 68*

---

* Several study aids are available to help you master the material in this chapter. The *Student Solutions Manual* and *Calchat.com* have solutions to all odd-numbered exercises, and *CalcView.com* has video solutions for selected exercises. *LarsonAppliedCalculus.com* offers algebra help, data spreadsheets, and much more. A *Graphing Technology Guide* with step-by-step commands for a variety of graphing calculators is at *CengageBrain.com*.

Section 2.4

- Use the Product Rule for differentiation.

$$\frac{d}{dx}[f(x)g(x)] = f(x)g'(x) + g(x)f'(x)$$

- Use the Quotient Rule for differentiation.

$$\frac{d}{dx}\left[\frac{f(x)}{g(x)}\right] = \frac{g(x)f'(x) - f(x)g'(x)}{[g(x)]^2}, \quad g(x) \neq 0$$

Section 2.5

- Use the General Power Rule for differentiation.

$$\frac{d}{dx}[u^n] = nu^{n-1}u', \quad n \text{ is a real number.}$$

- Use differentiation rules efficiently to find the derivative of any algebraic function, then simplify the result.

- Use derivatives to answer questions about real-life situations.   (Sections 2.1–2.5)

Section 2.6

- Find higher-order derivatives.

- Find and use a position function to determine the velocity and acceleration of a moving object.

Section 2.7

- Find derivatives implicitly.

- Use implicit differentiation to write equations of tangent lines.

Section 2.8

- Solve related-rate problems.

Study Strategies

- **Simplify Your Derivatives**   You may ask if you have to simplify your derivatives. The answer is "Yes, if you expect to use them." In the next chapter, you will see that almost all applications of derivatives require that the derivatives be written in simplified form. It is not difficult to see the advantage of a derivative in simplified form. Consider, for instance, the derivative of

$$f(x) = \frac{x}{\sqrt{x^2 + 1}}.$$

The "raw form" produced by the Quotient and Chain Rules

$$f'(x) = \frac{(x^2 + 1)^{1/2}(1) - (x)(\frac{1}{2})(x^2 + 1)^{-1/2}(2x)}{\left(\sqrt{x^2 + 1}\right)^2}$$

is obviously much more difficult to use than the simplified form

$$f'(x) = \frac{1}{(x^2 + 1)^{3/2}}.$$

- **List Units of Measure in Applied Problems**   When using derivatives in real-life applications, be sure to list the units of measure for each variable. For instance, if R is measured in dollars and t is measured in years, then the derivative dR/dt is measured in dollars per year.

Review Exercises
See *CalcChat.com* for tutorial help and worked-out solutions to odd-numbered exercises.

Approximating the Slope of a Graph In Exercises 1–4, approximate the slope of the graph at the point (x, y). (Each square on the grid is 1 unit by 1 unit.)

1.

2.

3.

4.
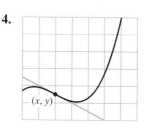

5. Revenue The graph represents the revenue R (in millions of dollars) for Chipotle Mexican Grill from 2007 through 2013, where t represents the year, with $t = 7$ corresponding to 2007. Estimate and interpret the slopes of the graph for the years 2008 and 2010. *(Source: Chipotle Mexican Grill, Inc.)*

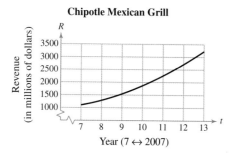

Chipotle Mexican Grill

6. Farms The graph represents the number of farms N (in thousands) in the United States from 2008 through 2013, where t represents the year, with $t = 8$ corresponding to 2008. Estimate and interpret the slopes of the graph for the years 2010 and 2012. *(Source: U.S. Department of Agriculture)*

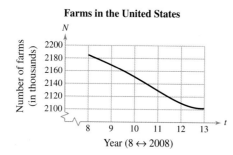

Farms in the United States

7. Consumer Trends The graph shows the number of visitors V (in thousands) to a national park during a one-year period, where $t = 1$ corresponds to January. Estimate and interpret the slopes of the graph at $t = 1$, 8, and 12.

Visitors to a National Park

8. White-Water Rafting Two white-water rafters leave a campsite simultaneously and start downstream on a 9-mile trip. Their distances from the campsite are given by $s = f(t)$ and $s = g(t)$, where s is measured in miles and t is measured in hours.

White-Water Rafting

(a) Which rafter is traveling at a greater rate at t_1?

(b) What can you conclude about their rates at t_2?

(c) What can you conclude about their rates at t_3?

(d) Which rafter finishes the trip first? Explain your reasoning.

Finding the Slope of a Graph In Exercises 9–16, use the limit definition to find the slope of the graph of f at the given point.

9. $f(x) = -3x - 5; (-2, 1)$

10. $f(x) = 7x + 3; (-1, -4)$

11. $f(x) = x^2 + 9; (3, 18)$

12. $f(x) = x^2 - 7x; (1, -6)$

13. $f(x) = \sqrt{x + 9}; (-5, 2)$

14. $f(x) = \sqrt{x - 1}; (10, 3)$

15. $f(x) = \dfrac{1}{x - 5}; (6, 1)$ **16.** $f(x) = \dfrac{1}{x + 6}; \left(-2, \dfrac{1}{4}\right)$

Finding a Derivative   In Exercises 17–24, use the limit definition to find the derivative of the function.

17. $f(x) = 9x + 1$

18. $f(x) = 1 - 4x$

19. $f(x) = -\frac{1}{2}x^2 + 2x$

20. $f(x) = 3x^2 - \frac{1}{4}x$

21. $f(x) = \sqrt{x - 5}$

22. $f(x) = \sqrt{x} + 3$

23. $f(x) = \frac{5}{x}$

24. $f(x) = \frac{1}{x + 4}$

Determining Differentiability   In Exercises 25–28, describe the x-values at which the function is differentiable. Explain your reasoning.

25. $y = \sqrt[3]{3x + 3}$

26. $y = -|x| + 3$

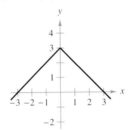

27. $y = \begin{cases} -x - 2, & x \le 0 \\ x^3 + 2, & x > 0 \end{cases}$

28. $y = (x + 1)^{2/3}$

Finding Derivatives   In Exercises 29–40, find the derivative of the function.

29. $y = -6$

30. $f(x) = 5$

31. $f(x) = x^7$

32. $h(x) = \frac{1}{x^4}$

33. $f(x) = 4x^2$

34. $g(t) = 8t^6$

35. $f(x) = \frac{5x^3}{4}$

36. $y = 3x^{2/3}$

37. $g(x) = 2x^4 + 3x^2$

38. $f(x) = 6x^2 - 4x$

39. $y = x^2 + 6x - 7$

40. $y = 2x^4 - 3x^3 + x$

Finding the Slope of a Graph   In Exercises 41–44, find the slope of the graph of the function at the given point. Use the *derivative* feature of a graphing utility to confirm your results.

41. $f(x) = 2x^{-1/2}$; $(4, 1)$

42. $y = \frac{3}{2x} + 3$; $\left(\frac{1}{2}, 6\right)$

43. $g(x) = x^3 - 4x^2 - 6x + 8$; $(-1, 9)$

44. $y = 2x^4 - 5x^3 + 6x^2 - x$; $(1, 2)$

Finding an Equation of a Tangent Line   In Exercises 45–48, (a) find an equation of the tangent line to the graph of the function at the given point, (b) use a graphing utility to graph the function and its tangent line at the point, and (c) use the *tangent* feature of a graphing utility to confirm your results.

45. $f(x) = 2x^2 - 3x + 1$; $(2, 3)$

46. $y = 11x^4 - 5x^2 + 1$; $(-1, 7)$

47. $f(x) = \sqrt{x} - \frac{1}{\sqrt{x}}$; $(1, 0)$

48. $f(x) = \sqrt[3]{x} - x$; $(-8, 6)$

49. Revenue   The revenue R (in millions of dollars) for Chipotle Mexican Grill from 2007 through 2013 can be modeled by

$$R = -0.5972t^3 + 51.187t^2 - 485.54t + 2199.0$$

where t is the year, with $t = 7$ corresponding to 2007. *(Source: Chipotle Mexican Grill, Inc.)*

(a) Find the slopes of the graph for the years 2008 and 2010.

(b) Compare your results with those obtained in Exercise 5.

(c) Interpret the slope of the graph in the context of the problem.

50. Farms   The number of farms N (in thousands) in the United States from 2008 through 2013 can be modeled by

$$N = 0.2083t^4 - 7.954t^3 + 111.96t^2 - 706.5t + 3891$$

where t is the year, with $t = 8$ corresponding to 2008. *(Source: U.S. Department of Agriculture)*

(a) Find the slopes of the graph for the years 2010 and 2012.

(b) Compare your results with those obtained in Exercise 6.

(c) Interpret the slope of the graph in the context of the problem.

Finding Rates of Change In Exercises 51–54, find the average rate of change of the function over the given interval. Compare this rate with the instantaneous rates of change at the endpoints of the interval.

51. $f(t) = 4t + 3; [-3, 1]$

52. $f(x) = x^2 + 3x - 4; [0, 1]$

53. $f(x) = x^{2/3}; [1, 8]$

54. $f(x) = x^3 - x^2 + 3; [-2, 2]$

55. Velocity The height s (in feet) at time t (in seconds) of a ball thrown straight downward from the top of a 600-foot building with an initial velocity of -30 feet per second is given by

$$s = -16t^2 - 30t + 600.$$

(a) Find the average velocity over the interval $[1, 3]$.

(b) Find the instantaneous velocities when $t = 1$ and $t = 3$.

(c) How long will it take the ball to hit the ground?

(d) Find the velocity of the ball when it hits the ground.

56. Velocity A rock is dropped from a tower on the Brooklyn Bridge, 276 feet above the East River. Let t represent the time in seconds.

(a) Find the position and velocity functions for the rock.

(b) Find the average velocity over the interval $[0, 2]$.

(c) Find the instantaneous velocities at $t = 2$ and $t = 3$.

(d) How long will it take the rock to hit the water?

(e) Find the velocity of the rock when it hits the water.

Marginal Cost In Exercises 57–60, find the marginal cost for producing x units. (The cost is measured in dollars.)

57. $C = 2500 + 320x$

58. $C = 24{,}000 + 450x - x^2, 0 \le x \le 225$

59. $C = 370 + 2.55\sqrt{x}$

60. $C = 475 + 5.25x^{2/3}$

Marginal Revenue In Exercises 61–64, find the marginal revenue for selling x units. (The revenue is measured in dollars.)

61. $R = 150x - 0.6x^2$ **62.** $R = 150x - \frac{3}{4}x^2$

63. $R = -4x^3 + 2x^2 + 100x$

64. $R = 4x + 10\sqrt{x}$

Marginal Profit In Exercises 65 and 66, find the marginal profit for selling x units. (The profit is measured in dollars.)

65. $P = -0.0002x^3 + 6x^2 - x - 2000$

66. $P = -\frac{1}{15}x^3 + 4000x^2 - 120x - 144{,}000$

67. Marginal Profit The profit P (in dollars) from selling x units of a product is given by

$$P = -0.05x^2 + 20x - 1000.$$

(a) Find the marginal profit when $x = 100$ units.

(b) Find the additional profit when the sales increase from 100 to 101 units.

(c) Compare the results of parts (a) and (b).

68. Population Growth The population P (in millions) of Brazil from 1990 through 2013 can be modeled by

$$P = -0.021t^2 + 2.77t + 148.9$$

where t represents the year, with $t = 0$ corresponding to 1990. *(Source: U.S. Census Bureau)*

(a) Evaluate P for $t = 0, 4, 8, 12, 16, 20,$ and 23. Explain these values.

(b) Find dP/dt and explain its meaning in this situation.

(c) Evaluate dP/dt for the same values as in part (a). Explain your results.

Finding Derivatives In Exercises 69–90, find the derivative of the function. State which differentiation rule(s) you used to find the derivative.

69. $f(x) = x^3(5 - 3x^2)$

70. $y = 4x^2(2x^2 - 5)$

71. $y = (4x - 3)(x^3 - 2x^2)$

72. $s = \left(4 - \frac{1}{t^2}\right)(t^2 - 3t)$

73. $g(x) = \dfrac{x}{x + 3}$

74. $f(x) = \dfrac{2 - 5x}{3x + 1}$

75. $f(x) = \dfrac{6x - 5}{x^2 + 1}$

76. $f(x) = \dfrac{x^2 + x - 1}{x^2 - 1}$

77. $f(x) = (5x^2 + 2)^3$

78. $f(x) = \sqrt[3]{x^2 - 1}$

79. $h(x) = \dfrac{2}{\sqrt{x + 1}}$ **80.** $g(x) = \dfrac{6}{(3x^2 - 5x)^4}$

81. $g(x) = x\sqrt{x^2 + 1}$ **82.** $g(t) = \dfrac{t}{(1 - t)^3}$

83. $f(x) = x(1 - 4x^2)^2$ **84.** $f(x) = \left(x^2 + \dfrac{1}{x}\right)^5$

85. $h(x) = [x^2(2x + 3)]^3$ **86.** $f(x) = [(x - 2)(x + 4)]^2$

87. $f(x) = x^2(x - 7)^{6/5}$ **88.** $f(s) = s^3(s^2 - 1)^{5/2}$

89. $h(t) = \dfrac{\sqrt{3t + 1}}{(1 - 3t)^2}$ **90.** $g(x) = \left(\dfrac{3x + 1}{x^2 + 1}\right)^2$

91. Physical Science The temperature T (in degrees Fahrenheit) of food placed in a freezer can be modeled by

$$T = \frac{1300}{t^2 + 2t + 25}$$

where t is the time (in hours).

(a) Find the rates of change of T at $t = 1, 3, 5,$ and 10.

(b) Graph the model on a graphing utility and describe the rate at which the temperature is changing over time.

92. Forestry According to the *Doyle Log Rule*, the volume V (in board-feet) of a log of length L (in feet) and diameter D (in inches) at the small end is

$$V = \left(\frac{D - 4}{4}\right)^2 L.$$

Find the rate at which the volume is changing with respect to D for a 12-foot-long log whose smallest diameter is (a) 8 inches, (b) 16 inches, (c) 24 inches, and (d) 36 inches.

Finding Higher-Order Derivatives In Exercises 93–100, find the higher-order derivative.

Given	Derivative
93. $f(x) = 3x^2 + 7x + 1$	$f''(x)$
94. $f'(x) = 5x^4 - 6x^2 + 2x$	$f'''(x)$
95. $f'''(x) = -\dfrac{3}{x^4}$	$f^{(6)}(x)$
96. $f(x) = \sqrt{x}$	$f^{(4)}(x)$
97. $f'(x) = 8x^{5/2}$	$f^{(4)}(x)$
98. $f''(x) = 9\sqrt[3]{x}$	$f^{(5)}(x)$
99. $f(x) = x^2 + \dfrac{3}{x}$	$f''(x)$
100. $f'''(x) = 20x^4 - \dfrac{2}{x^3}$	$f^{(5)}(x)$

101. Athletics A person dives from a 30-foot platform with an initial velocity of 5 feet per second (upward).

(a) Write the position, velocity, and acceleration functions of the diver.

(b) How long does it take the diver to hit the water?

(c) What is the diver's velocity at impact?

(d) What is the diver's acceleration at impact?

102. Velocity and Acceleration The position function of a particle is given by

$$s = \frac{1}{t^2 + 2t + 1}$$

where s is the height (in feet) and t is the time (in seconds). Find the velocity and acceleration functions.

Finding Derivatives In Exercises 103–106, use implicit differentiation to find dy/dx.

103. $x^2 + 3xy + y^3 = 10$

104. $x^2 + 9xy + y^2 = 0$

105. $y^2 - x^2 + 8x - 9y - 1 = 0$

106. $y^2 + x^2 - 6y - 2x - 5 = 0$

Finding an Equation of a Tangent Line In Exercises 107–110, use implicit differentiation to find an equation of the tangent line to the graph of the equation at the given point.

Equation	Point
107. $y^2 = x - y$	$(2, 1)$
108. $2\sqrt[3]{x} + 3\sqrt{y} = 10$	$(8, 4)$
109. $y^2 - 2x = xy$	$(1, 2)$
110. $y^3 - 2x^2y + 3xy^2 = -1$	$(0, -1)$

111. Area The radius r of a circle is increasing at a rate of 2 inches per minute. Find the rate of change of the area when (a) $r = 3$ inches and (b) $r = 10$ inches.

112. Profit The profit P from selling x units of a product is given by

$$P = 375x - 1.5x^2.$$

The sales are increasing at a rate of 2 units per day. Find the rate of change of the profit when (a) $x = 50$ units and (b) $x = 100$ units.

113. Water Level A swimming pool is 40 feet long, 20 feet wide, 4 feet deep at the shallow end, and 9 feet deep at the deep end (see figure). Water is being pumped into the pool at the rate of 10 cubic feet per minute. How fast is the water level rising when there is 4 feet of water in the deep end?

114. Profit A company is increasing the production of a product at a rate of 15 units per week. The demand and cost functions for the product can be modeled by

$$p = 211 - 0.002x$$

and

$$C = 30x + 1,500,000$$

where x is the number of units produced per week. Find the rate of change of the profit with respect to time when the weekly sales are $x = 1600$ units.

TEST YOURSELF   See *CalcChat.com* for tutorial help and worked-out solutions to odd-numbered exercises.

Take this test as you would take a test in class. When you are done, check your work against the answers given in the back of the book.

In Exercises 1 and 2, use the limit definition to find the derivative of the function. Then find the slope of the graph of *f* at the given point.

1. $f(x) = x^2 + 3; (3, 12)$ **2.** $f(x) = \sqrt{x} - 2; (4, 0)$

In Exercises 3–11, find the derivative of the function.

3. $f(t) = t^3 + 2t$ **4.** $f(x) = 4x^2 - 8x + 1$ **5.** $f(x) = x^{3/2} + 6x^{1/2}$

6. $f(x) = 5x^2 - \dfrac{3}{x^3}$ **7.** $f(x) = (x + 3)(x^2 + 2x)$ **8.** $f(x) = \sqrt{x}(5 + x)$

9. $f(x) = (3x^2 + 4)^2$ **10.** $f(x) = \sqrt{1 - 2x}$ **11.** $f(x) = \dfrac{(5x - 1)^3}{x}$

12. Find an equation of the tangent line to the graph of

$$f(x) = x - \frac{1}{x}$$

at the point $(1, 0)$. Then use a graphing utility to graph the function and the tangent line in the same viewing window.

13. The sales *S* (in billions of dollars) of CVS Health from 2009 through 2013 can be modeled by

$$S = -2.1083t^3 + 70.811t^2 - 777.05t + 2893.6$$

where *t* represents the year, with $t = 9$ corresponding to 2009. *(Source: CVS Health Corporation)*

(a) Find the average rate of change from 2010 through 2012.

(b) Find the instantaneous rates of change for 2010 and 2012.

(c) Interpret the results of parts (a) and (b) in the context of the problem.

14. The monthly demand function *p* and cost function *C* for *x* units of a product are given by

$$p = 1700 - 0.016x \quad \text{and} \quad C = 715{,}000 + 240x.$$

(a) Write the profit function for this product.

(b) Find the marginal profit when the monthly sales are $x = 700$ units.

In Exercises 15–17, find the third derivative of the function.

15. $f(x) = 2x^2 + 3x + 1$ **16.** $f(x) = \sqrt{3 - x}$ **17.** $f(x) = \dfrac{2x + 1}{2x - 1}$

18. A ball is thrown straight upward from a height of 75 feet above the ground with an initial velocity of 30 feet per second. Write the position, velocity, and acceleration functions of the ball. Find the height, velocity, and acceleration when $t = 2$.

In Exercises 19–21, use implicit differentiation to find *dy/dx*.

19. $x + xy = 6$ **20.** $y^2 + 2x - 2y + 1 = 0$ **21.** $4x^2 - 3y^2 + x^3y = 5$

22. The radius *r* of a right circular cylinder is increasing at a rate of 0.25 centimeter per minute. The height *h* of the cylinder is related to the radius by $h = 20r$. Find the rate of change of the volume when (a) $r = 0.5$ centimeter and (b) $r = 1$ centimeter.

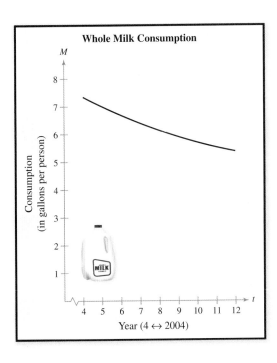

Whole Milk Consumption

Example 2 on page 169 shows how the derivative can be used to show that whole milk consumption decreased in the United States from 2004 to 2012.

3 Applications of the Derivative

Nyul | Dreamstime.com

3.1 Increasing and Decreasing Functions

■ Test for increasing and decreasing functions.

■ Find the critical numbers of functions and find the open intervals on which functions are increasing or decreasing.

■ Use increasing and decreasing functions to model and solve real-life problems.

Increasing and Decreasing Functions

A function is **increasing** when its graph moves up as x moves to the right and **decreasing** when its graph moves down as x moves to the right. The following definition states this more formally.

In Exercise 54 on page 176, you will use derivatives and critical numbers to find the intervals on which the profit from selling popcorn is increasing and decreasing.

Definitions of Increasing and Decreasing Functions

A function f is **increasing** on an interval when, for any two numbers x_1 and x_2 in the interval,

$$x_2 > x_1 \quad \text{implies} \quad f(x_2) > f(x_1).$$

A function f is **decreasing** on an interval when, for any two numbers x_1 and x_2 in the interval,

$$x_2 > x_1 \quad \text{implies} \quad f(x_2) < f(x_1).$$

The function in Figure 3.1 is decreasing on the interval $(-\infty, a)$, constant on the interval (a, b), and increasing on the interval (b, ∞). Actually, from the definitions of increasing and decreasing functions, the function shown in Figure 3.1 is decreasing on the interval $(-\infty, a]$ and increasing on the interval $[b, \infty)$. This text restricts the discussion to finding *open* intervals on which a function is increasing or decreasing.

The derivative of a function can be used to determine whether the function is increasing or decreasing on an interval.

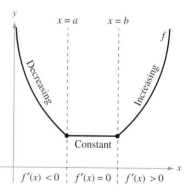

FIGURE 3.1

Test for Increasing and Decreasing Functions

Let f be differentiable on the interval (a, b).

1. If $f'(x) > 0$ for all x in (a, b), then f is increasing on (a, b).

2. If $f'(x) < 0$ for all x in (a, b), then f is decreasing on (a, b).

3. If $f'(x) = 0$ for all x in (a, b), then f is constant on (a, b).

STUDY TIP

The conclusions in the first two cases of testing for increasing and decreasing functions are valid even when $f'(x) = 0$ at a finite number of x-values in (a, b).

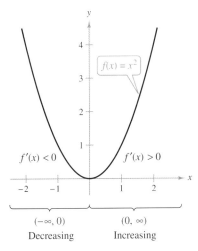

$f(x) = x^2$

$f'(x) < 0$ $f'(x) > 0$

$(-\infty, 0)$ $(0, \infty)$
Decreasing   Increasing

FIGURE 3.2

EXAMPLE 1   **Testing for Increasing and Decreasing Functions**

Show that the function $f(x) = x^2$ is decreasing on the open interval $(-\infty, 0)$ and increasing on the open interval $(0, \infty)$.

SOLUTION   The derivative of f is

$$f'(x) = 2x.$$

On the open interval $(-\infty, 0)$, the fact that x is negative implies that $f'(x) = 2x$ is also negative. So, by the test for a decreasing function, you can conclude that f is *decreasing* on this interval. Similarly, on the open interval $(0, \infty)$, the fact that x is positive implies that $f'(x) = 2x$ is also positive. So, it follows that f is *increasing* on this interval, as shown in Figure 3.2.

✓ *Checkpoint 1*   Worked-out solution available at LarsonAppliedCalculus.com

Show that the function $f(x) = x^4$ is decreasing on the open interval $(-\infty, 0)$ and increasing on the open interval $(0, \infty)$.

EXAMPLE 2   **Modeling Consumption**

From 2004 through 2012, the consumption M (in gallons per person) of whole milk in the United States can be modeled by

$$M = 0.014t^2 - 0.47t + 9.0, \quad 4 \le t \le 12$$

where $t = 4$ corresponds to 2004 (see Figure 3.3). Show that the consumption of whole milk was decreasing from 2004 to 2012.   *(Source: U.S. Department of Agriculture)*

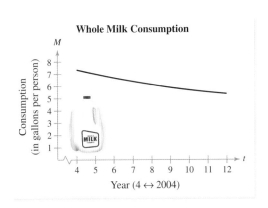

Whole Milk Consumption

Consumption (in gallons per person)

Year (4 ↔ 2004)

FIGURE 3.3

SOLUTION   The derivative of this model is $dM/dt = 0.028t - 0.47$. Note that "from 2004 to 2012" corresponds to the open interval $(4, 12)$. For this interval, the derivative is negative. So, the function is decreasing, which implies that the consumption of whole milk was decreasing from 2004 to 2012.

✓ *Checkpoint 2*   Worked-out solution available at LarsonAppliedCalculus.com

From 2007 through 2012, the consumption B (in pounds per person) of blueberries in the United States can be modeled by

$$B = -0.0179t^2 + 0.485t - 2.03, \quad 7 \le t \le 12$$

where $t = 7$ corresponds to 2007. Show that the consumption of blueberries was increasing from 2007 to 2012.   *(Source: U.S. Department of Agriculture)*

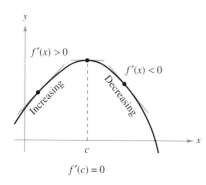

$f'(x) > 0$

Increasing

$f'(x) < 0$

Decreasing

c

$f'(c) = 0$

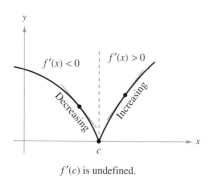

$f'(x) < 0$ $f'(x) > 0$

Decreasing

Increasing

c

$f'(c)$ is undefined.

FIGURE 3.4

Critical Numbers and Their Use

In Example 1, you were given two intervals: one on which the function was decreasing and one on which it was increasing. Suppose you had been asked to determine these intervals. To do this, you could have used the fact that for a continuous function, $f'(x)$ can change signs only at x-values for which $f'(x) = 0$ or at x-values for which $f'(x)$ is undefined, as shown in Figure 3.4. These two types of numbers are called the **critical numbers** of f.

Definition of a Critical Number

If f is defined at c, then c is a **critical number** of f when $f'(c) = 0$ or when $f'(c)$ is undefined.

Be sure you understand that the definition of a critical number requires that a critical number be in the domain of the function. For instance, $x = 0$ is not a critical number of the function $f(x) = 1/x$ even though $f'(x) = -1/x^2$ is undefined at $x = 0$.

EXAMPLE 3   Finding Critical Numbers

Find the critical numbers of $f(x) = 2x^3 - 9x^2$.

SOLUTION   Begin by differentiating the function.

$$f(x) = 2x^3 - 9x^2 \qquad \text{Write original function.}$$
$$f'(x) = 6x^2 - 18x \qquad \text{Differentiate.}$$

To find the critical numbers of f, you must find all x-values for which $f'(x) = 0$ and all x-values for which $f'(x)$ is undefined.

$$6x^2 - 18x = 0 \qquad \text{Set } f'(x) \text{ equal to 0.}$$
$$6x(x - 3) = 0 \qquad \text{Factor.}$$
$$x = 0, x = 3 \qquad \text{Critical numbers}$$

Because there are no x-values for which $f'(x)$ is undefined, you can conclude that $x = 0$ and $x = 3$ are the only critical numbers of f.

✓ **Checkpoint 3**   Worked-out solution available at LarsonAppliedCalculus.com

Find the critical numbers of $f(x) = x^2 - x$. ■

To determine the intervals on which a continuous function is increasing or decreasing, you can use the guidelines below.

Guidelines for Applying the Increasing/Decreasing Test

1. Find the derivative of f.

2. Locate the critical numbers of f and use these numbers to determine test intervals. That is, find all x for which $f'(x) = 0$ or $f'(x)$ is undefined.

3. Determine the sign of $f'(x)$ at one test value in each of the intervals.

4. Use the test for increasing and decreasing functions to decide whether f is increasing or decreasing on each interval.

EXAMPLE 4   Intervals on Which f Is Increasing or Decreasing

Find the open intervals on which the function is increasing or decreasing.

$$f(x) = x^3 - \frac{3}{2}x^2$$

SOLUTION   Begin by finding the derivative of f. Then set the derivative equal to zero and solve for the critical numbers.

$f'(x) = 3x^2 - 3x$	Differentiate original function.
$3x^2 - 3x = 0$	Set derivative equal to 0.
$3x(x - 1) = 0$	Factor.
$x = 0, x = 1$	Critical numbers

Because there are no x-values for which $f'(x)$ is undefined, it follows that $x = 0$ and $x = 1$ are the only critical numbers. So, the intervals that need to be tested are

$(-\infty, 0), (0, 1),$ and $(1, \infty).$ Test intervals

The table summarizes the testing of these three intervals.

Interval	$-\infty < x < 0$	$0 < x < 1$	$1 < x < \infty$
Test value	$x = -1$	$x = \frac{1}{2}$	$x = 2$
Sign of $f'(x)$	$f'(-1) = 6 > 0$	$f'(\frac{1}{2}) = -\frac{3}{4} < 0$	$f'(2) = 6 > 0$
Conclusion	Increasing	Decreasing	Increasing

So, f is increasing on the intervals $(-\infty, 0)$ and $(1, \infty)$ and decreasing on the interval $(0, 1)$, as shown in Figure 3.5. Note that the test values in the intervals were chosen for convenience—other x-values could have been used.

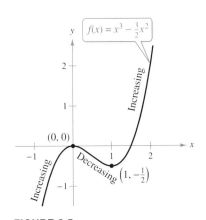

FIGURE 3.5

✓ *Checkpoint 4*   *Worked-out solution available at LarsonAppliedCalculus.com*

Find the open intervals on which $f(x) = x^3 - 12x$ is increasing or decreasing.   ■

TECH TUTOR

You can use the *trace* feature of a graphing utility to confirm the result of Example 4. Begin by graphing the function, as shown below. Then use the *trace* feature and move the cursor from left to right. On intervals in which the function is increasing, note that the y-values increase as the x-values increase, whereas on intervals in which the function is decreasing, the y-values decrease as the x-values increase.

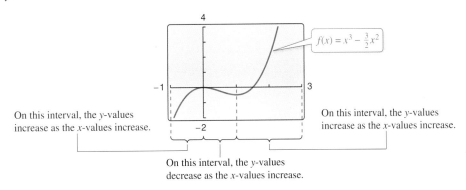

On this interval, the y-values increase as the x-values increase.

On this interval, the y-values increase as the x-values increase.

On this interval, the y-values decrease as the x-values increase.

Not only is the function in Example 4 continuous on the entire real number line, it is also differentiable there. For such functions, the only critical numbers are those for which $f'(x) = 0$. The next example considers a continuous function that has *both* types of critical numbers—those for which $f'(x) = 0$ and those for which $f'(x)$ is undefined.

ALGEBRA TUTOR

For help on the algebra in Example 5, see Example 2(d) in the *Chapter 3 Algebra Tutor*, on page 243.

EXAMPLE 5   Intervals on Which *f* Is Increasing or Decreasing

Find the open intervals on which

$$f(x) = (x^2 - 4)^{2/3}$$

is increasing or decreasing.

SOLUTION   Begin by finding the derivative of the function.

$$f'(x) = \frac{2}{3}(x^2 - 4)^{-1/3}(2x) \qquad \text{Differentiate.}$$

$$= \frac{4x}{3(x^2 - 4)^{1/3}} \qquad \text{Simplify.}$$

From this, you can see that the derivative is zero when $x = 0$ and the derivative is undefined when $x = \pm 2$. So, the critical numbers are

$$x = -2, \quad x = 0, \quad \text{and} \quad x = 2. \qquad \text{Critical numbers}$$

This implies that the test intervals are

$$(-\infty, -2), \quad (-2, 0), \quad (0, 2), \quad \text{and} \quad (2, \infty). \qquad \text{Test intervals}$$

The table summarizes the testing of these four intervals.

Interval	$-\infty < x < -2$	$-2 < x < 0$	$0 < x < 2$	$2 < x < \infty$
Test value	$x = -3$	$x = -1$	$x = 1$	$x = 3$
Sign of $f'(x)$	$f'(-3) < 0$	$f'(-1) > 0$	$f'(1) < 0$	$f'(3) > 0$
Conclusion	Decreasing	Increasing	Decreasing	Increasing

So, f is increasing on the intervals $(-2, 0)$ and $(2, \infty)$ and decreasing on the intervals $(-\infty, -2)$ and $(0, 2)$, as shown in Figure 3.6.

STUDY TIP

To test the intervals in the table in Example 5, it is not necessary to *evaluate* $f'(x)$ at each test value—you only need to determine its sign. For instance, you can determine the sign of $f'(-3)$ as shown.

$$f'(-3) = \frac{4(-3)}{3(9 - 4)^{1/3}}$$

$$= \frac{\text{negative}}{\text{positive}}$$

$$= \text{negative}$$

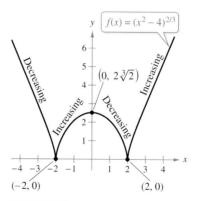

FIGURE 3.6

✓ *Checkpoint 5*   *Worked-out solution available at LarsonAppliedCalculus.com*

Find the open intervals on which $f(x) = x^{2/3}$ is increasing or decreasing. ■

The functions in Examples 1 through 5 are continuous on the entire real number line. If there are isolated x-values at which a function is not continuous, then these x-values should be used along with the critical numbers to determine the test intervals.

EXAMPLE 6   Testing a Function That Is Not Continuous

The function

$$f(x) = \frac{x^4 + 1}{x^2}$$

is not continuous at $x = 0$. Also, the derivative of f

$$f'(x) = \frac{2(x^4 - 1)}{x^3}$$

is zero at $x = \pm 1$. So, use these numbers to determine the test intervals.

$x = -1,\ x = 1$ Critical numbers

$x = 0$ Discontinuity

After testing $f'(x)$, you can determine that f is decreasing on the intervals $(-\infty, -1)$ and $(0, 1)$, and increasing on the intervals $(-1, 0)$ and $(1, \infty)$, as shown in Figure 3.7.

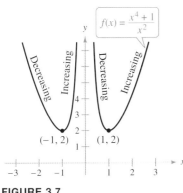

FIGURE 3.7

✔ *Checkpoint 6*

Find the open invervals on which

$$f(x) = \frac{x^2 + 1}{x}$$

is increasing or decreasing.

The converse of the test for increasing and decreasing functions is *not* true. For instance, it is possible for a function to be increasing on an interval even though its derivative is not positive at every point in the interval.

EXAMPLE 7   Testing an Increasing Function

Show that $f(x) = x^3 - 3x^2 + 3x$ is increasing on the entire real number line.

SOLUTION   From the derivative of f

$$f'(x) = 3x^2 - 6x + 3 = 3(x - 1)^2$$

you can see that the only critical number is $x = 1$. So, the test intervals are $(-\infty, 1)$ and $(1, \infty)$. The table summarizes the testing of these two intervals. From Figure 3.8, you can see that f is increasing on the entire real number line, even though $f'(1) = 0$. To convince yourself of this, look back at the definition of an increasing function.

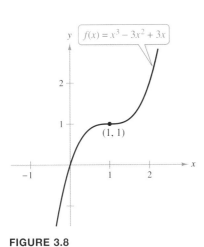

FIGURE 3.8

Interval	$-\infty < x < 1$	$1 < x < \infty$
Test value	$x = 0$	$x = 2$
Sign of $f'(x)$	$f'(0) = 3(-1)^2 > 0$	$f'(2) = 3(1)^2 > 0$
Conclusion	Increasing	Increasing

✔ *Checkpoint 7*

Show that $f(x) = -x^3 + 2$ is decreasing on the entire real number line.

Application

EXAMPLE 8   Profit Analysis

A national toy distributor determines the cost and revenue functions for one of its games.

$$C = 2.4x - 0.0002x^2, \quad 0 \le x \le 6000$$
$$R = 7.2x - 0.001x^2, \quad 0 \le x \le 6000$$

Determine the open interval on which the profit function is increasing.

SOLUTION   The profit for producing x games is

$$P = R - C$$
$$= (7.2x - 0.001x^2) - (2.4x - 0.0002x^2)$$
$$= 4.8x - 0.0008x^2.$$

To find the interval on which the profit is increasing, set the marginal profit P' equal to zero and solve for x.

$$P' = 4.8 - 0.0016x \qquad \text{Differentiate profit function.}$$
$$4.8 - 0.0016x = 0 \qquad \text{Set } P' \text{ equal to 0.}$$
$$-0.0016x = -4.8 \qquad \text{Subtract 4.8 from each side.}$$
$$x = \frac{-4.8}{-0.0016} \qquad \text{Divide each side by } -0.0016.$$
$$x = 3000 \text{ games} \qquad \text{Simplify.}$$

On the interval $(0, 3000)$, $P'(x)$ is positive and the profit is *increasing*. On the interval $(3000, 6000)$, $P'(x)$ is negative and the profit is *decreasing*. The graphs of the cost, revenue, and profit functions are shown in Figure 3.9.

Profit Analysis

FIGURE 3.9

✓ *Checkpoint 8*   *Worked-out solution available at LarsonAppliedCalculus.com*

A national distributor of pet toys determines the cost and revenue functions for one of its toys.

$$C = 1.2x - 0.0001x^2, \quad 0 \le x \le 6000$$
$$R = 3.6x - 0.0005x^2, \quad 0 \le x \le 6000$$

Determine the open interval on which the profit function is increasing. ■

SUMMARIZE (Section 3.1)

1. State the test for increasing and decreasing functions *(page 168)*. For an example of testing for increasing and decreasing functions, see Example 1.

2. State the definition of a critical number *(page 170)*. For an example of finding critical numbers, see Example 3.

3. State the guidelines for determining the intervals on which a continuous function is increasing or decreasing *(page 170)*. For examples of finding the intervals on which functions are increasing or decreasing, see Examples 4, 5, and 7.

4. Describe a real-life example of how testing for increasing and decreasing functions can be used to analyze the profit of a company *(page 174, Example 8)*.

The following warm-up exercises involve skills that were covered in a previous course or earlier sections. You will use these skills in the exercise set for this section. For additional help, review Appendix A.3 and Section 1.4.

In Exercises 1–4, solve the equation.

1. $x^2 = 8x$

2. $9x = \dfrac{5}{3}x^2$

3. $\dfrac{x^2 - 25}{x^3} = 0$

4. $\dfrac{2x}{\sqrt{1 - x^2}} = 0$

In Exercises 5–8, find the domain of the function.

5. $y = \dfrac{x + 3}{x - 3}$

6. $y = \dfrac{4}{\sqrt{5 - 2x}}$

7. $y = \dfrac{2x + 1}{x^2 - 3x - 10}$

8. $y = \dfrac{3x}{\sqrt{9 - 3x^2}}$

In Exercises 9–12, evaluate the expression when $x = -2, 0,$ and 2.

9. $-2(x + 1)(x - 1)$

10. $4(2x + 1)(2x - 1)$

11. $\dfrac{2x + 1}{(x - 1)^2}$

12. $\dfrac{-2(x + 1)}{(x - 4)^2}$

Exercises 3.1

See *CalcChat.com* for tutorial help and worked-out solutions to odd-numbered exercises.

Using Graphs In Exercises 1–4, use the graph to estimate the open intervals on which the function is increasing or decreasing.

1. $f(x) = -(x + 1)^2$

2. $f(x) = \dfrac{x^3}{4} - 3x$

3. $f(x) = x^4 - 2x^2$

4. $f(x) = -(x^2 - 9)^{2/3}$

 Testing for Increasing and Decreasing Functions In Exercises 5–8, show that the function is increasing and decreasing on the given open intervals. *See Example 1.*

5. $y = -x^2$; increasing on $(-\infty, 0)$; decreasing on $(0, \infty)$

6. $y = x^4 + 1$; decreasing on $(-\infty, 0)$; increasing on $(0, \infty)$

7. $y = (x - 2)^2$; decreasing on $(-\infty, 2)$; increasing on $(2, \infty)$

8. $y = x^3$; increasing on $(-\infty, \infty)$

 Finding Critical Numbers In Exercises 9–14, find the critical numbers of the function. *See Example 3.*

9. $f(x) = 4x^2 - 6x$

10. $f(x) = 28x - 2x^2$

11. $y = x^4 + 4x^3 + 8$

12. $g(x) = x^4 - 8x^2 + 1$

13. $f(x) = \sqrt{x^2 - 25}$

14. $y = \dfrac{x}{x^2 + 16}$

 Intervals on Which f Is Increasing or Decreasing In Exercises 15–38, find the critical numbers and the open intervals on which the function is increasing or decreasing. Use a graphing utility to verify your results. *See Examples 4, 5, and 7.*

15. $f(x) = 2x - 3$

16. $f(x) = 5 - 3x$

17. $y = x^2 - 6x$

18. $y = -2x^2 + 8x$

19. $f(x) = -2x^2 + 4x + 3$

20. $f(x) = \frac{1}{2}x^2 + 6x + 9$

21. $y = 3x^3 + 12x^2 + 15x$

22. $y = x^3 - 3x + 2$

23. $f(x) = x^4 - 2x^3$

24. $f(x) = \frac{1}{4}x^4 - \frac{1}{3}x^3 - x^2$

25. $g(x) = (x + 2)^2$

26. $y = (x - 2)^3$

27. $g(x) = 2(5x - 1)^3$

28. $y = -3(8x - 7)^4$

29. $y = x^{1/3} + 1$

30. $y = x^{2/3} - 4$

31. $f(x) = \sqrt[3]{x^2 - 1}$

32. $f(x) = \sqrt{9 - x^2}$

33. $g(x) = (x + 2)^{1/3}$

34. $g(x) = (x - 1)^{2/3}$

35. $f(x) = x\sqrt{x + 1}$

36. $h(x) = x\sqrt[3]{x - 1}$

37. $f(x) = \dfrac{x}{x^2 + 9}$

38. $f(x) = \dfrac{x^2}{x^2 + 4}$

Intervals on Which f Is Increasing or Decreasing In Exercises 39–48, find the critical numbers and the open intervals on which the function is increasing or decreasing. (*Hint:* Check for discontinuities.) Sketch the graph of the function to verify your results. See Example 6.

39. $f(x) = \dfrac{x + 4}{x - 5}$

40. $f(x) = \dfrac{2x + 1}{x + 2}$

41. $f(x) = \dfrac{x^3 + 2}{x}$

42. $f(x) = \dfrac{x^4 + 16}{x^2}$

43. $f(x) = \dfrac{2x}{16 - x^2}$

44. $f(x) = \dfrac{x^2}{x^2 - 9}$

45. $y = \begin{cases} 4 - x^2, & x \le 0 \\ -2x, & x > 0 \end{cases}$

46. $y = \begin{cases} 3x + 1, & x \le 1 \\ 5 - x^2, & x > 1 \end{cases}$

47. $y = \begin{cases} 2x + 1, & x \le -1 \\ x^2 - 2, & x > -1 \end{cases}$

48. $y = \begin{cases} -x^3 + 1, & x \le 0 \\ -x^2 + 2x, & x > 0 \end{cases}$

49. Sales The sales S (in billions of dollars) for Macy's from 2009 through 2013 can be modeled by

$$S = -0.193t^2 + 5.39t - 9.5, \quad 9 \le t \le 13$$

where t is the year, with $t = 9$ corresponding to 2009. Show that the sales were increasing from 2009 through 2013. *(Source: Macy's, Inc.)*

50. HOW DO YOU SEE IT? Plots of the relative numbers of N_2 (nitrogen) molecules that have a given velocity at each of three temperatures (in degrees Kelvin) are shown in the figure. Identify the differences in the average velocities (indicated by the peaks of the curves) for the three temperatures, and describe the open intervals on which the velocity is increasing and decreasing for each of the three temperatures. *(Source: Adapted from Zumdahl, Chemistry, Seventh Edition)*

Molecular Velocity

51. Bachelor's Degrees The number y (in thousands) of biological and biomedical science bachelor's degrees conferred in the United States from 1998 through 2012 can be modeled by

$$y = -0.0253t^3 + 1.431t^2 - 22.63t + 170.1, \quad 8 \le t \le 22$$

where t is the year, with $t = 8$ corresponding to 1998. *(Source: U.S. National Center for Education Statistics)*

(a) Use a graphing utility to graph the model. Then graphically estimate the years during which the model is increasing and the years during which it is decreasing.

(b) Use the test for increasing and decreasing functions to verify the result of part (a).

52. Cost The ordering and transportation cost C (in hundreds of dollars) for an automobile dealership is modeled by

$$C = 10\left(\dfrac{1}{x} + \dfrac{x}{x + 3}\right), \quad x \ge 1$$

where x is the number of automobiles ordered.

(a) Find the open intervals on which C is increasing or decreasing.

(b) Use a graphing utility to graph the cost function.

(c) Use the *trace* feature to determine the order sizes for which the cost is \$900. Assuming that the revenue function is increasing for $x \ge 0$, which order size would you use? Explain your reasoning.

53. Profit Analysis A fast-food restaurant determines the cost and revenue models for its hamburgers.

$$C = 0.6x + 7500, \quad 0 \le x \le 50{,}000$$

$$R = \dfrac{1}{20{,}000}(65{,}000x - x^2), \quad 0 \le x \le 50{,}000$$

(a) Write the profit function for this situation.

(b) Determine the open intervals on which the profit function is increasing and decreasing.

(c) Determine how many hamburgers the restaurant needs to sell to obtain a maximum profit. Explain your reasoning.

54. Profit The profit P (in dollars) made by a cinema from selling x bags of popcorn can be modeled by

$$P = 2.36x - \dfrac{x^2}{25{,}000} - 3500, \quad 0 \le x \le 50{,}000.$$

(a) Find the open intervals on which P is increasing and decreasing.

(b) If you owned the cinema, what price would you charge to obtain a maximum profit from popcorn sales? Explain your reasoning.

3.2 Extrema and the First-Derivative Test

- Recognize the occurrence of relative extrema of functions.
- Use the First-Derivative Test to find the relative extrema of functions.
- Find absolute extrema of continuous functions on a closed interval.
- Find minimum and maximum values of real-life models and interpret the results in context.

Profit from Selling Soft Drinks

In Exercise 49 on page 185, you will use the First-Derivative Test to find the price of a soft drink that yields a maximum profit.

Relative Extrema

You have used the derivative to determine the intervals on which a function is increasing or decreasing. In this section, you will examine the points at which a function changes from increasing to decreasing, or vice versa. At such a point, the function has a **relative extremum.** (The plural of extremum is *extrema*.) The **relative extrema** of a function include the **relative minima** and **relative maxima** of the function. For instance, the function shown in Figure 3.10 has a relative maximum at the left point and a relative minimum at the right point.

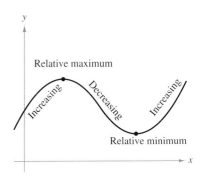

FIGURE 3.10

Definition of Relative Extrema

Let f be a function defined at c.

1. $f(c)$ is a **relative maximum** of f when there exists an interval (a, b) containing c such that $f(x) \leq f(c)$ for all x in (a, b).
2. $f(c)$ is a **relative minimum** of f when there exists an interval (a, b) containing c such that $f(x) \geq f(c)$ for all x in (a, b).

If $f(c)$ is a relative extremum of f, then the relative extremum is said to occur at $x = c$.

For a continuous function, the relative extrema must occur at critical numbers of the function, as shown in Figure 3.11.

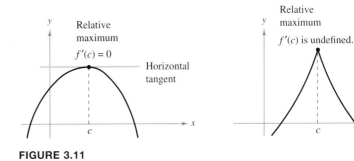

FIGURE 3.11

Occurrences of Relative Extrema

If f has a relative minimum or relative maximum at $x = c$, then c is a critical number of f. That is, either $f'(c) = 0$ or $f'(c)$ is undefined.

The First-Derivative Test

The discussion on the preceding page implies that in your search for relative extrema of a continuous function, you need to test only the critical numbers of the function. Once you have determined that c is a critical number of a function f, the **First-Derivative Test** for relative extrema enables you to classify $f(c)$ as a relative minimum, a relative maximum, or neither.

First-Derivative Test for Relative Extrema

Let f be continuous on the interval (a, b) in which c is the only critical number. If f is differentiable on the interval (except possibly at c), then $f(c)$ can be classified as a relative minimum, a relative maximum, or neither, as shown.

1. On the interval (a, b), if $f'(x)$ is negative to the left of $x = c$ and positive to the right of $x = c$, then $f(c)$ is a relative minimum.

2. On the interval (a, b), if $f'(x)$ is positive to the left of $x = c$ and negative to the right of $x = c$, then $f(c)$ is a relative maximum.

3. On the interval (a, b), if $f'(x)$ is positive on both sides of $x = c$ or negative on both sides of $x = c$, then $f(c)$ is neither a relative minimum nor a relative maximum.

A graphical interpretation of the First-Derivative Test is shown in Figure 3.12.

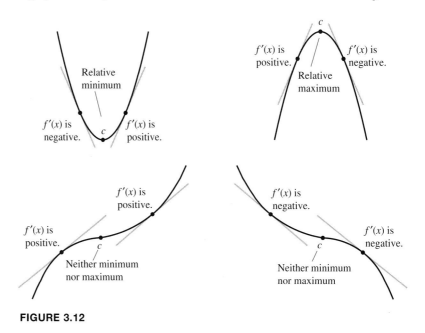

FIGURE 3.12

Guidelines for Finding Relative Extrema

1. Find the derivative of f.

2. Locate the critical numbers of f and use these numbers to determine the test intervals.

3. Determine the sign of $f'(x)$ at an arbitrary number in each of the test intervals.

4. For each critical number c, use the First-Derivative Test to decide whether $f(c)$ is a relative minimum, a relative maximum, or neither.

EXAMPLE 1 Finding Relative Extrema

Find all relative extrema of the function

$$f(x) = 2x^3 - 3x^2 - 36x + 14.$$

SOLUTION Begin by finding the derivative of f.

$$f'(x) = 6x^2 - 6x - 36 \qquad \text{Differentiate.}$$

Next, find the critical numbers of f.

$$6x^2 - 6x - 36 = 0 \qquad \text{Set derivative equal to 0.}$$
$$6(x^2 - x - 6) = 0 \qquad \text{Factor out common factor.}$$
$$6(x - 3)(x + 2) = 0 \qquad \text{Factor.}$$
$$x = -2, x = 3 \qquad \text{Critical numbers}$$

Because $f'(x)$ is defined for all x, the only critical numbers of f are

$$x = -2 \quad \text{and} \quad x = 3. \qquad \text{Critical numbers}$$

Using these numbers, you can form the three test intervals

$$(-\infty, -2), \quad (-2, 3), \quad \text{and} \quad (3, \infty). \qquad \text{Test intervals}$$

The testing of the three intervals is shown in the table.

Interval	$-\infty < x < -2$	$-2 < x < 3$	$3 < x < \infty$
Test value	$x = -3$	$x = 0$	$x = 4$
Sign of $f'(x)$	$f'(-3) = 36 > 0$	$f'(0) = -36 < 0$	$f'(4) = 36 > 0$
Conclusion	Increasing	Decreasing	Increasing

Using the First-Derivative Test, you can conclude that the critical number $x = -2$ yields a relative maximum [$f'(x)$ changes sign from positive to negative], and the critical number $x = 3$ yields a relative minimum [$f'(x)$ changes sign from negative to positive]. The graph of f is shown in Figure 3.13. The relative maximum is

$$f(-2) = 58$$

and the relative minimum is

$$f(3) = -67.$$

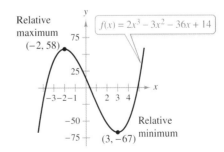

Relative maximum $(-2, 58)$

Relative minimum $(3, -67)$

$f(x) = 2x^3 - 3x^2 - 36x + 14$

FIGURE 3.13

✓ **Checkpoint 1** *Worked-out solution available at LarsonAppliedCalculus.com*

Find all relative extrema of

$$f(x) = 2x^3 - 6x + 1.$$

TECH TUTOR

Some graphing calculators have a special feature that allows you to find the minimum or maximum of a function on an interval. Consult the user's manual for information on the *minimum* and *maximum* features of your graphing utility.

In Example 1, both critical numbers yielded relative extrema. In the next example, only one of the two critical numbers yields a relative extremum.

ALGEBRA TUTOR

For help on the algebra in Example 2, see Example 2(c) in the *Chapter 3 Algebra Tutor*, on page 243.

EXAMPLE 2   Finding Relative Extrema

Find all relative extrema of the function $f(x) = x^4 - x^3$.

SOLUTION   From the derivative of the function

$$f'(x) = 4x^3 - 3x^2 = x^2(4x - 3)$$

you can see that the function has only two critical numbers: $x = 0$ and $x = \frac{3}{4}$. These numbers produce the test intervals $(-\infty, 0)$, $\left(0, \frac{3}{4}\right)$, and $\left(\frac{3}{4}, \infty\right)$, which are tested in the table.

Interval	$-\infty < x < 0$	$0 < x < \frac{3}{4}$	$\frac{3}{4} < x < \infty$
Test value	$x = -1$	$x = \frac{1}{2}$	$x = 1$
Sign of $f'(x)$	$f'(-1) = -7 < 0$	$f'\left(\frac{1}{2}\right) = -\frac{1}{4} < 0$	$f'(1) = 1 > 0$
Conclusion	Decreasing	Decreasing	Increasing

By the First-Derivative Test, it follows that f has a relative minimum at $x = \frac{3}{4}$, as shown in Figure 3.14. The relative minimum is

$$f\left(\frac{3}{4}\right) = -\frac{27}{256}.$$

Note that the critical number $x = 0$ does not yield a relative extremum because $f'(x)$ is negative on both sides of $x = 0$.

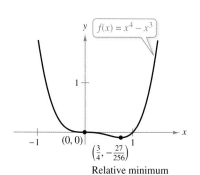

FIGURE 3.14

✓ *Checkpoint 2*   Worked-out solution available at LarsonAppliedCalculus.com

Find all relative extrema of $f(x) = x^4 - 4x^3$.

EXAMPLE 3   Finding Relative Extrema

Find all relative extrema of the function

$$f(x) = 2x - 3x^{2/3}.$$

SOLUTION   From the derivative of the function

$$f'(x) = 2 - \frac{2}{x^{1/3}} = \frac{2(x^{1/3} - 1)}{x^{1/3}}$$

you can see that $f'(1) = 0$ and $f'(x)$ is undefined at $x = 0$. So, the function has two critical numbers: $x = 1$ and $x = 0$. These numbers produce the test intervals $(-\infty, 0)$, $(0, 1)$, and $(1, \infty)$. By testing these intervals, you can conclude that f has a relative maximum at $(0, 0)$ and a relative minimum at $(1, -1)$, as shown in Figure 3.15.

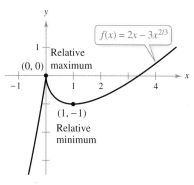

FIGURE 3.15

✓ *Checkpoint 3*   Worked-out solution available at LarsonAppliedCalculus.com

Find all relative extrema of $f(x) = 3x^{2/3} - 2x$.

Absolute Extrema

The terms *relative minimum* and *relative maximum* describe the *local* behavior of a function. To describe the *global* behavior of the function on an entire interval, you can use the terms **absolute maximum** and **absolute minimum.**

Definition of Absolute Extrema

Let f be defined on an interval I containing c.

1. $f(c)$ is an **absolute minimum of** f on I when $f(c) \leq f(x)$ for every x in I.

2. $f(c)$ is an **absolute maximum of** f on I when $f(c) \geq f(x)$ for every x in I.

The absolute minimum and absolute maximum values of a function on an interval are sometimes called simply the **minimum** and **maximum** of f on I.

Be sure that you understand the distinction between relative extrema and absolute extrema. For instance, in Figure 3.16, the function has a relative minimum that also happens to be an absolute minimum on the interval $[a, b]$. The relative maximum of f, however, is not the absolute maximum on the interval $[a, b]$. The next theorem points out that if a continuous function has a closed interval as its domain, then it *must* have both an absolute minimum and an absolute maximum on the interval. From Figure 3.16, note that these extrema can occur at the endpoints of the interval.

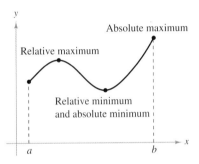

FIGURE 3.16

Extreme Value Theorem

If f is continuous on a closed interval $[a, b]$, then f has both a minimum value and a maximum value on $[a, b]$.

Although a continuous function has just one minimum and one maximum value on a closed interval, either of these values can occur for more than one x-value. For instance, on the interval $[-3, 3]$, the function

$$f(x) = 9 - x^2$$

has a minimum value of zero at $x = -3$ *and* at $x = 3$, as shown in Figure 3.17.

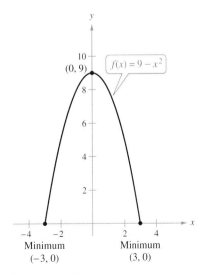

FIGURE 3.17

When looking for extrema of a function on a *closed* interval, remember that you must consider the values of the function at the endpoints as well as at the critical numbers of the function. You can use the guidelines below to find extrema on a closed interval.

Guidelines for Finding Extrema on a Closed Interval

To find the extrema of a continuous function f on a closed interval $[a, b]$, use the following steps.

1. Find the critical numbers of f in the open interval (a, b).

2. Evaluate f at each of its critical numbers in (a, b).

3. Evaluate f at each endpoint, a and b.

4. The least of these values is the minimum, and the greatest is the maximum.

EXAMPLE 4   Finding Extrema on a Closed Interval

Find the minimum and maximum values of

$$f(x) = x^2 - 6x + 2$$

on the interval $[0, 5]$.

SOLUTION   Begin by differentiating the function.

$$f(x) = x^2 - 6x + 2 \qquad \text{Write original function.}$$
$$f'(x) = 2x - 6 \qquad \text{Differentiate.}$$

Next, find the critical numbers of f.

$$2x - 6 = 0 \qquad \text{Set derivative equal to 0.}$$
$$2x = 6 \qquad \text{Add 6 to each side.}$$
$$x = 3 \qquad \text{Solve for } x.$$

Because $f'(x)$ is defined for all x, you can conclude that the only critical number of f is $x = 3$. Because this number lies in the interval $[0, 5]$, you should evaluate f at this number *and* at the endpoints of the interval, as shown in the table.

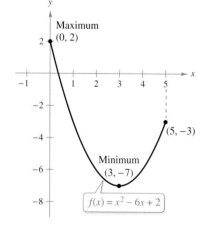

FIGURE 3.18

x-value	Endpoint: $x = 0$	Critical number: $x = 3$	Endpoint: $x = 5$
$f(x)$	$f(0) = 2$	$f(3) = -7$	$f(5) = -3$
Conclusion	Maximum is 2.	Minimum is -7.	Neither maximum nor minimum

From the table, you can see that the minimum of f on the interval $[0, 5]$ is $f(3) = -7$. Moreover, the maximum of f on the interval $[0, 5]$ is $f(0) = 2$. This is confirmed by the graph of f, as shown in Figure 3.18.

✓ Checkpoint 4   *Worked-out solution available at LarsonAppliedCalculus.com*

Find the minimum and maximum values of

$$f(x) = x^2 - 8x + 10$$

on the interval $[0, 7]$. Sketch the graph of f and label the minimum and maximum values.

Application

Finding the minimum and maximum values of a function is one of the most common applications of calculus.

EXAMPLE 5   **Finding the Maximum Profit**

Recall the fast-food restaurant in Examples 7 and 8 in Section 2.3. The restaurant's profit function from selling x hamburgers is given by

$$P = 2.44x - \frac{x^2}{20,000} - 5000, \quad 0 \le x \le 50,000.$$

Find the sales level that yields a maximum profit.

SOLUTION   To begin, find an equation for marginal profit.

$$\frac{dP}{dx} = 2.44 - \frac{x}{10,000} \qquad \text{Find marginal profit.}$$

Next, set the marginal profit equal to zero and solve for x.

$$2.44 - \frac{x}{10,000} = 0 \qquad \text{Set marginal profit equal to 0.}$$

$$-\frac{x}{10,000} = -2.44 \qquad \text{Subtract 2.44 from each side.}$$

$$x = 24,400 \text{ hamburgers} \qquad \text{Critical number}$$

In Figure 3.19, you can see that the critical number $x = 24,400$ corresponds to the sales level that yields a maximum profit. To find the maximum profit, substitute $x = 24,400$ into the profit function.

$$P = 2.44x - \frac{x^2}{20,000} - 5000$$

$$= 2.44(24,400) - \frac{(24,400)^2}{20,000} - 5000$$

$$= \$24,768$$

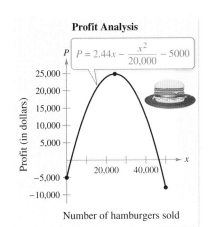

Profit Analysis

$$P = 2.44x - \frac{x^2}{20,000} - 5000$$

Profit (in dollars)

Number of hamburgers sold

FIGURE 3.19

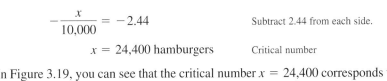

✔ *Checkpoint 5*   *Worked-out solution available at LarsonAppliedCalculus.com*

Verify the results of Example 5 by completing the table.

x	24,000	24,200	24,300	24,400	24,500	24,600	24,800
P							

SUMMARIZE   (Section 3.2)

1. State the First-Derivative Test *(page 178)*. For examples in which the First-Derivative Test is used, see Examples 1, 2, and 3.

2. State the guidelines for finding extrema on a closed interval *(page 182)*. For an example of finding the extrema of a function on a closed interval, see Example 4.

3. Describe a real-life example of how the First-Derivative Test can be used to find the sales level that yields a maximum profit for a company *(page 183, Example 5)*.

SKILLS WARM UP 3.2

The following warm-up exercises involve skills that were covered in earlier sections. You will use these skills in the exercise set for this section. For additional help, review Sections 2.2, 2.4, and 3.1.

In Exercises 1–6, solve the equation $f'(x) = 0$.

1. $f(x) = x^4 - 18x^2 + 7$

2. $f(x) = \frac{1}{3}x^3 - \frac{3}{2}x^2 - 10x$

3. $f(x) = 5x^{4/5} - 4x$

4. $f(x) = \frac{1}{2}x^2 - 3x^{5/3}$

5. $f(x) = \dfrac{x + 4}{x^2 + 1}$

6. $f(x) = \dfrac{3x + 1}{x^2 - 2}$

In Exercises 7–10, use $g(x) = -x^5 - 2x^4 + 4x^3 + 2x - 1$ to determine the sign of the derivative.

7. $g'(-4)$

8. $g'(0)$

9. $g'(1)$

10. $g'(3)$

In Exercises 11 and 12, decide whether the function is increasing or decreasing on the given interval.

11. $f(x) = 2x^2 - 11x - 6$, $(3, 6)$

12. $f(x) = x^3 + 2x^2 - 4x - 8$, $(-2, 0)$

Exercises 3.2

See *CalcChat.com* for tutorial help and worked-out solutions to odd-numbered exercises.

 Finding Relative Extrema In Exercises 1–12, find all relative extrema of the function. *See Examples 1, 2, and 3.*

1. $f(x) = -2x^2 + 4x + 3$
2. $f(x) = x^2 + 8x + 10$
3. $f(x) = x^2 - 6x$
4. $f(x) = -4x^2 + 4x + 1$
5. $f(x) = x^4 - 12x^3$
6. $g(x) = \frac{1}{3}x^3 - x$
7. $h(x) = -(x + 4)^3$
8. $h(x) = 2(x - 3)^3$
9. $f(x) = 5x^3 - 10x^2 + 3$
10. $f(x) = x^4 - 32x + 4$
11. $f(x) = 6x^{2/3} + 4x$
12. $f(x) = 3x - 36x^{1/3}$

 Finding Relative Extrema In Exercises 13–18, use a graphing utility to find all relative extrema of the function.

13. $f(x) = 2x - 6x^{2/3}$
14. $f(t) = (t - 1)^{1/3}$

15. $g(t) = t - \dfrac{1}{2t^2}$
16. $f(x) = 9x + \dfrac{4}{x}$

17. $f(x) = \dfrac{x}{x + 1}$
18. $h(x) = \dfrac{6}{x^2 + 2}$

Finding Extrema on a Closed Interval In Exercises 19–30, find the absolute extrema of the function on the closed interval. Use a graphing utility to verify your results. *See Example 4.*

19. $f(x) = 3(1 - x)$, $[-4, -1]$
20. $f(x) = \frac{1}{4}(3x + 5)$, $[-9, -3]$
21. $f(x) = 5 - 2x^2$, $[-3, 1]$
22. $f(x) = x^2 - 2x - 3$, $[0, 1]$

23. $f(x) = x^3 - 6x^2$, $[0, 6]$
24. $f(x) = x^3 - 12x$, $[0, 4]$
25. $h(s) = \dfrac{1}{3 - s}$, $[0, 2]$

26. $h(t) = \dfrac{t}{t - 2}$, $[3, 5]$

27. $g(t) = \dfrac{1}{x} - \dfrac{2}{x^2}$, $[-2, 1]$

28. $g(x) = 4\left(1 + \dfrac{1}{x} + \dfrac{1}{x^2}\right)$, $[-4, 5]$

29. $h(t) = (t - 1)^{2/3}$, $[-7, 2]$
30. $g(x) = (x^2 - 4)^{2/3}$, $[-6, 3]$

Determining Types of Extrema In Exercises 31–34, approximate the critical numbers of the function on the open interval shown in the graph. Determine whether the function has a relative maximum, a relative minimum, an absolute maximum, an absolute minimum, or none of these at each critical number.

31.

32.

33.

34.

 Finding Extrema on a Closed Interval In Exercises 35–38, use a graphing utility to find the absolute extrema of the function on the closed interval.

35. $f(x) = 0.4x^3 - 1.8x^2 + x - 3$, $[0, 5]$

36. $f(x) = 3.2x^5 + 5x^3 - 3.5x$, $[0, 1]$

37. $f(x) = \frac{4}{3}x\sqrt{3 - x}$, $[0, 3]$

38. $f(x) = 4\sqrt{x} - 2x + 1$, $[0, 6]$

 Finding Absolute Extrema In Exercises 39–42, find the absolute extrema of the function on the interval $[0, \infty)$.

39. $f(x) = \dfrac{8}{x + 1}$

40. $f(x) = 5 - \dfrac{2}{3x + 4}$

41. $f(x) = \dfrac{2x}{x^2 + 4}$

42. $f(x) = 8 - \dfrac{4x}{x^2 + 1}$

Creating the Graph of a Function In Exercises 43 and 44, graph a function on the closed interval $[-2, 5]$ having the given characteristics. (There are many correct answers.)

43. Absolute maximum at $x = -2$

Absolute minimum at $x = 1$

Relative maximum at $x = 3$

44. Relative minimum at $x = -1$

Critical number at $x = 0$, but no extrema

Absolute maximum at $x = 2$

Absolute minimum at $x = 5$

45. Population The resident population P (in millions) of the United States from 2000 through 2013 can be modeled by $P = -0.0023t^3 + 0.015t^2 + 2.83t + 281.8$, $0 \le t \le 13$, where $t = 0$ corresponds to 2000. *(Source: U.S. Census Bureau)*

(a) Make a conjecture about the maximum and minimum populations of the United States from 2000 to 2013.

(b) Analytically find the maximum and minimum populations over the interval.

(c) Write a brief paragraph comparing your conjecture with your results in part (b).

 46. **HOW DO YOU SEE IT?** The graph of the United States fertility rate shows the number of births per 1000 women in their lifetime according to the birth rate in that particular year. *(Source: U.S. National Center for Health Statistics)*

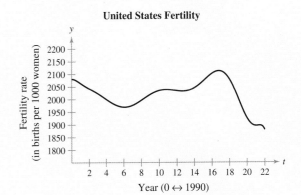

(a) Around what year was the fertility rate the highest, and to how many births per 1000 women did this rate correspond?

(b) During which time period was the fertility rate increasing most rapidly? most slowly?

(c) During which time period was the fertility rate decreasing most rapidly? most slowly?

(d) Give some possible real-life reasons for fluctuations in the fertility rate.

47. Cost A retailer determines that the cost C of ordering and storing x units of a product can be modeled by

$$C = 3x + \frac{20{,}000}{x}, \quad 0 < x \le 200.$$

The delivery truck can carry at most 200 units per order. Find the order size that will minimize the cost. Use a graphing utility to verify your result.

48. Medical Science Coughing forces the trachea (windpipe) to contract, which in turn affects the velocity of the air passing through the trachea. The velocity of the air during coughing can be modeled by

$$v = k(R - r)r^2, \quad 0 \le r < R$$

where k is a positive constant, R is the normal radius of the trachea, and r is the radius during coughing. What radius r will produce the maximum air velocity?

49. Profit When soft drinks are sold for $1.00 per bottle at football games, approximately 6000 bottles are sold. When the price is raised to $1.20 per bottle, the quantity demanded drops to 5600. The initial cost is $5000 and the cost per unit is $0.50. Assuming that the demand function is linear, what number of bottles and what price per bottle will yield a maximum profit?

3.3 Concavity and the Second-Derivative Test

Production Cost

In Exercise 69 on page 194, you will use the Second-Derivative Test to find the production level that will minimize the average cost per unit.

■ Determine the intervals on which the graphs of functions are concave upward or concave downward.
■ Find the points of inflection of the graphs of functions.
■ Use the Second-Derivative Test to find the relative extrema of functions.
■ Find the points of diminishing returns of input-output models.

Concavity

You already know that locating the intervals over which a function f increases or decreases helps to describe its graph. In this section, you will see that locating the intervals on which f' increases or decreases can determine where the graph of f is curving upward or curving downward. This property of curving upward or downward is defined formally as the **concavity** of the graph of the function.

> **Definition of Concavity**
>
> Let f be differentiable on an open interval I. The graph of f is
>
> 1. **concave upward** on I when f' is increasing on the interval.
>
> 2. **concave downward** on I when f' is decreasing on the interval.

In Figure 3.20, you can observe the following graphical interpretation of concavity.

1. A curve that is concave upward lies *above* its tangent lines. [See Figure 3.20(a).]

2. A curve that is concave downward lies *below* its tangent lines. [See Figure 3.20(b).]

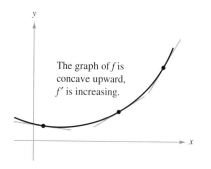

The graph of f is concave upward, f' is increasing.

The graph of f is concave downward, f' is decreasing.

(a) The graph of f lies above its tangent lines.

(b) The graph of f lies below its tangent lines.

FIGURE 3.20

To find the open intervals on which the graph of a function is concave upward or concave downward, you can use the second derivative of the function as follows.

> **Test for Concavity**
>
> Let f be a function whose second derivative exists on an open interval I.
>
> 1. If $f''(x) > 0$ for all x in I, then the graph of f is concave upward on I.
>
> 2. If $f''(x) < 0$ for all x in I, then the graph of f is concave downward on I.

EXAMPLE 1   Determining Concavity

Find the second derivative of f and discuss the concavity of its graph.

a. $f(x) = x^2$ **b.** $f(x) = \sqrt{x}$

SOLUTION

a. The graph of the function

$$f(x) = x^2 \qquad \text{Original function}$$

is concave upward on the entire real number line because its second derivative

$$f''(x) = 2 \qquad \text{Second derivative}$$

is positive for all x. (See Figure 3.21.)

b. The graph of the function

$$f(x) = \sqrt{x} \qquad \text{Original function}$$

is concave downward for $x > 0$ because its second derivative

$$f''(x) = -\frac{1}{4}x^{-3/2} \qquad \text{Second derivative}$$

is negative for all $x > 0$. (See Figure 3.22.)

Concave Upward

FIGURE 3.21

Concave Downward

FIGURE 3.22

 Checkpoint 1   *Worked-out solution available at LarsonAppliedCalculus.com*

Find the second derivative of f and discuss the concavity of its graph.

a. $f(x) = -2x^2$ **b.** $f(x) = -2\sqrt{x}$ ■

For a *continuous* function f, you can find the open intervals on which the graph of f is concave upward and concave downward using the guidelines below. [When there are x-values at which the function is not continuous, these values should be used, along with the points at which $f''(x) = 0$ or $f''(x)$ is undefined, to form the test intervals.]

Guidelines for Applying the Concavity Test

1. Locate the x-values at which $f''(x) = 0$ or $f''(x)$ is undefined.

2. Use these x-values to determine the test intervals.

3. Determine the sign of $f''(x)$ in each test interval.

ALGEBRA TUTOR

For help on the algebra in Example 2, see Example 1(a) in the *Chapter 3 Algebra Tutor*, on page 242.

EXAMPLE 2   Applying the Test for Concavity

Determine the open intervals on which the graph of

$$f(x) = \frac{6}{x^2 + 3}$$

is concave upward or concave downward.

SOLUTION   Begin by finding the second derivative of f.

$$f(x) = 6(x^2 + 3)^{-1} \qquad \text{Rewrite original function.}$$

$$f'(x) = 6(-1)(x^2 + 3)^{-2}(2x) \qquad \text{Chain Rule}$$

$$= \frac{-12x}{(x^2 + 3)^2} \qquad \text{Simplify.}$$

$$f''(x) = \frac{(x^2 + 3)^2(-12) - (-12x)(2)(x^2 + 3)(2x)}{(x^2 + 3)^4} \qquad \text{Quotient Rule}$$

$$= \frac{-12(x^2 + 3) + 48x^2}{(x^2 + 3)^3} \qquad \text{Simplify.}$$

$$= \frac{36(x^2 - 1)}{(x^2 + 3)^3} \qquad \text{Simplify.}$$

From this, you can see that $f''(x)$ is defined for all real numbers and $f''(x) = 0$ when $x = \pm 1$. So, you can test the concavity of f by testing the intervals

$$(-\infty, -1), \quad (-1, 1), \quad \text{and} \quad (1, \infty). \qquad \text{Test intervals}$$

The results are shown in the table and in Figure 3.23.

Interval	$-\infty < x < -1$	$-1 < x < 1$	$1 < x < \infty$
Test value	$x = -2$	$x = 0$	$x = 2$
Sign of $f''(x)$	$f''(-2) > 0$	$f''(0) < 0$	$f''(2) > 0$
Conclusion	Concave upward	Concave downward	Concave upward

STUDY TIP

In Example 2, f' is increasing on the interval $(1, \infty)$ even though f is decreasing there. Be sure you see that the increasing or decreasing of f' does not necessarily correspond to the increasing or decreasing of f.

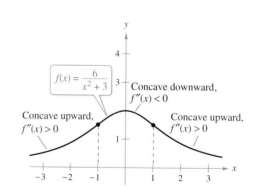

$$f(x) = \frac{6}{x^2 + 3}$$

Concave downward, $f''(x) < 0$

Concave upward, $f''(x) > 0$

Concave upward, $f''(x) > 0$

FIGURE 3.23

✓ *Checkpoint 2*   Worked-out solution available at LarsonAppliedCalculus.com

Determine the open intervals on which the graph of

$$f(x) = \frac{12}{x^2 + 4}$$

is concave upward or concave downward.

Points of Inflection

If the tangent line to a graph exists at a point at which the concavity changes, then the point is a **point of inflection.** Three examples of inflection points are shown in Figure 3.24. (Note that the third graph has a vertical tangent line at its point of inflection.)

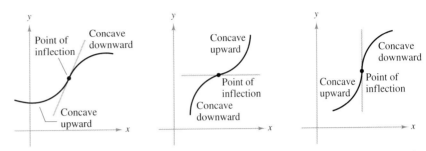

The concavity of the graph of a function changes at a point of inflection.

FIGURE 3.24

Definition of Point of Inflection

If the graph of a continuous function has a tangent line at a point where its concavity changes from upward to downward (or downward to upward), then the point is a **point of inflection.**

Because a point of inflection occurs where the concavity of a graph changes, it must be true that at such points the sign of $f''(x)$ changes. So, to locate possible points of inflection, you need to determine only the values of x for which $f''(x) = 0$ or for which $f''(x)$ is undefined. This parallels the procedure for locating the relative extrema of f by determining the critical numbers of f.

Property of Points of Inflection

If $(c, f(c))$ is a point of inflection of the graph of f, then either $f''(c) = 0$ or $f''(c)$ is undefined.

EXAMPLE 3   Finding a Point of Inflection

Discuss the concavity of the graph of $f(x) = 2x^3 + 1$ and find its point of inflection.

SOLUTION   Differentiating twice produces the following.

$$f(x) = 2x^3 + 1 \qquad \text{Write original function.}$$
$$f'(x) = 6x^2 \qquad \text{Find first derivative.}$$
$$f''(x) = 12x \qquad \text{Find second derivative.}$$

Setting $f''(x) = 0$, you can determine that the only possible point of inflection occurs at $x = 0$. After testing the intervals $(-\infty, 0)$ and $(0, \infty)$, you can determine that the graph is concave downward on $(-\infty, 0)$ and concave upward on $(0, \infty)$. Because the concavity changes at $x = 0$, you can conclude that the graph of f has a point of inflection at $(0, 1)$, as shown in Figure 3.25.

$f(x) = 2x^3 + 1$

FIGURE 3.25

✓ *Checkpoint 3*   *Worked-out solution available at LarsonAppliedCalculus.com*

Discuss the concavity of the graph of $f(x) = -x^3$ and find its point of inflection. ■

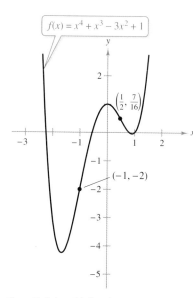

Two Points of Inflection
FIGURE 3.26

EXAMPLE 4   Finding Points of Inflection

Discuss the concavity of the graph of

$$f(x) = x^4 + x^3 - 3x^2 + 1$$

and find its points of inflection.

SOLUTION   Begin by finding the second derivative of f.

$$f(x) = x^4 + x^3 - 3x^2 + 1 \qquad \text{Write original function.}$$
$$f'(x) = 4x^3 + 3x^2 - 6x \qquad \text{Find first derivative.}$$
$$f''(x) = 12x^2 + 6x - 6 \qquad \text{Find second derivative.}$$
$$= 6(2x - 1)(x + 1) \qquad \text{Factor.}$$

From this, you can see that the possible points of inflection occur at $x = \frac{1}{2}$ and $x = -1$. After testing the intervals $(-\infty, -1)$, $\left(-1, \frac{1}{2}\right)$, and $\left(\frac{1}{2}, \infty\right)$, you can determine that the graph is concave upward on $(-\infty, -1)$, concave downward on $\left(-1, \frac{1}{2}\right)$, and concave upward on $\left(\frac{1}{2}, \infty\right)$. Because the concavity changes at $x = -1$ and $x = \frac{1}{2}$, you can conclude that the graph of f has points of inflection at these x-values, as shown in Figure 3.26. The points of inflection are

$$(-1, -2) \quad \text{and} \quad \left(\frac{1}{2}, \frac{7}{16}\right).$$

✓ **Checkpoint 4**   *Worked-out solution available at LarsonAppliedCalculus.com*

Discuss the concavity of the graph of

$$f(x) = x^4 - 2x^3 + 1$$

and find its points of inflection. ■

It is possible for the second derivative to be zero at a point that is *not* a point of inflection. For example, compare the graphs of

$$f(x) = x^3 \quad \text{and} \quad g(x) = x^4$$

as shown in Figure 3.27. Both second derivatives are zero when $x = 0$, but only the graph of f has a point of inflection at $x = 0$. This shows that before concluding that a point of inflection exists at a value of x for which $f''(x) = 0$, you must test to be certain that the concavity actually changes at that point.

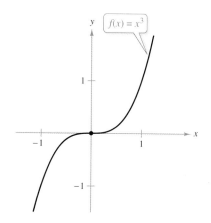

$f''(0) = 0$, and $(0, 0)$ is a point of inflection.

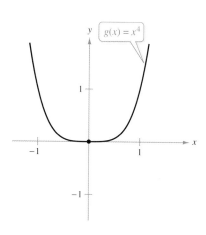

$g''(0) = 0$, but $(0, 0)$ is not a point of inflection.

FIGURE 3.27

The Second-Derivative Test

In addition to testing for concavity, the second derivative can be used to perform a simple test for relative minima and relative maxima. If f is a function such that $f'(c) = 0$ and the graph of f is concave upward at $x = c$, then $f(c)$ is a relative minimum of f. Similarly, if f is a function such that $f'(c) = 0$ and the graph of f is concave downward at $x = c$, then $f(c)$ is a relative maximum of f, as shown in Figure 3.28.

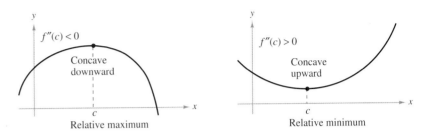

FIGURE 3.28

Second-Derivative Test

Let $f'(c) = 0$, and let f'' exist on an open interval containing c.

1. If $f''(c) > 0$, then $f(c)$ is a relative minimum.

2. If $f''(c) < 0$, then $f(c)$ is a relative maximum.

3. If $f''(c) = 0$, then the test fails. In such cases, you can use the First-Derivative Test to determine whether $f(c)$ is a relative minimum, a relative maximum, or neither.

EXAMPLE 5 **Using the Second-Derivative Test**

Find all relative extrema of $f(x) = -3x^5 + 5x^3$.

SOLUTION Begin by finding the first derivative of f.

$$f'(x) = -15x^4 + 15x^2 = 15x^2(1 - x^2)$$

From this derivative, you can see that $x = 0$, $x = -1$, and $x = 1$ are the only critical numbers of f. Using the second derivative

$$f''(x) = -60x^3 + 30x = 30x(1 - 2x^2)$$

you can apply the Second-Derivative Test, as shown.

Point	$(-1, -2)$	$(0, 0)$	$(1, 2)$
Sign of $f''(x)$	$f''(-1) > 0$	$f''(0) = 0$	$f''(1) < 0$
Conclusion	Relative minimum	Test fails.	Relative maximum

Because the Second-Derivative Test fails at $(0, 0)$, you can use the First-Derivative Test and observe that $f'(x)$ is positive on both sides of $x = 0$. So, $(0, 0)$ is neither a relative minimum nor a relative maximum. A test for concavity would show that $(0, 0)$ is a point of inflection. The graph of f is shown in Figure 3.29.

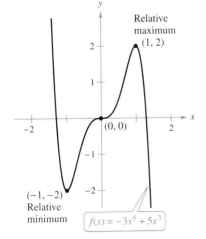

FIGURE 3.29

✓ *Checkpoint 5* *Worked-out solution available at LarsonAppliedCalculus.com*

Find all relative extrema of $f(x) = x^4 - 4x^3 + 1$.

FIGURE 3.30

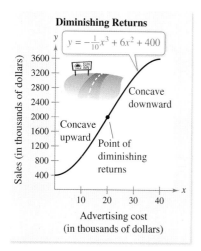

FIGURE 3.31

Extended Application: Diminishing Returns

In economics, the notion of concavity is related to the concept of **diminishing returns.** Consider a function

Output ⟶ ⟵ Input

$$y = f(x)$$

where x measures input (in dollars) and y measures output (in dollars). In Figure 3.30, notice that the graph of this function is concave upward on the interval (a, c) and is concave downward on the interval (c, b). On the interval (a, c), each additional dollar of input returns more than the previous input dollar. By contrast, on the interval (c, b), each additional dollar of input returns less than the previous input dollar. The point $(c, f(c))$ is called the **point of diminishing returns.** An increased investment beyond this point is usually considered a poor use of capital.

EXAMPLE 6   Exploring Diminishing Returns

By increasing its advertising cost x (in thousands of dollars) for a product, a company discovers that it can increase the sales y (in thousands of dollars) according to the model

$$y = -\frac{1}{10}x^3 + 6x^2 + 400, \quad 0 \le x \le 40.$$

Find the point of diminishing returns for this product.

SOLUTION   Begin by finding the first and second derivatives.

$$y' = 12x - \frac{3x^2}{10} \qquad \text{First derivative}$$

$$y'' = 12 - \frac{3x}{5} \qquad \text{Second derivative}$$

The second derivative is zero only when $x = 20$. By testing for concavity on the intervals $(0, 20)$ and $(20, 40)$, you can conclude that the graph has a point of diminishing returns when $x = 20$, as shown in Figure 3.31. So, the point of diminishing returns for this product occurs when \$20,000 is spent on advertising.

✓ *Checkpoint 6*   *Worked-out solution available at LarsonAppliedCalculus.com*

Find the point of diminishing returns for the model below, where R is the revenue (in thousands of dollars) and x is the advertising cost (in thousands of dollars).

$$R = \frac{1}{20,000}(450x^2 - x^3), \quad 0 \le x \le 300$$

SUMMARIZE   (Section 3.3)

1. State the test for concavity *(page 186)*. For examples of applying the test for concavity, see Examples 1 and 2.

2. State the definition of point of inflection *(page 189)*. For examples of finding points of inflection, see Examples 3 and 4.

3. State the Second-Derivative Test *(page 191)*. For an example of using the Second-Derivative Test, see Example 5.

4. Describe a real-life example of how the second derivative can be used to find the point of diminishing returns for a product *(page 192, Example 6)*.

SKILLS WARM UP 3.3

The following warm-up exercises involve skills that were covered in earlier sections. You will use these skills in the exercise set for this section. For additional help, review Sections 2.2, 2.4, 2.5, 2.6, and 3.1.

In Exercises 1–6, find the second derivative of the function.

1. $f(x) = 4x^4 - 9x^3 + 5x - 1$

2. $g(s) = (s^2 - 1)(s^2 - 3s + 2)$

3. $g(x) = (x^2 + 1)^4$

4. $f(x) = (3x - 7)^{4/3}$

5. $h(x) = \dfrac{4x + 3}{5x - 1}$

6. $f(x) = \dfrac{2x - 1}{3x + 2}$

In Exercises 7–10, find the critical numbers of the function.

7. $f(x) = 6x^3 - 3x + 11$

8. $f(x) = x^4 - 4x^3 - 10$

9. $g(t) = \dfrac{16 + t^2}{t}$

10. $h(x) = \dfrac{x^4 - 50x^2}{8}$

Exercises 3.3

See *CalcChat.com* for tutorial help and worked-out solutions to odd-numbered exercises.

 Using Graphs In Exercises 1–4, state the signs of $f'(x)$ and $f''(x)$ on the interval $(0, 2)$.

1.

2.

3.

4.

 Determining Concavity In Exercises 5–12, determine the open intervals on which the graph of the function is concave upward or concave downward. *See Examples 1 and 2.*

5. $f(x) = -3x^2$

6. $f(x) = -5\sqrt{x}$

7. $y = -x^3 + 3x^2 - 2$

8. $y = 2x^3 + 6x^2 - x - 1$

9. $f(x) = \dfrac{x^2 + 1}{x^2 - 1}$

10. $f(x) = \dfrac{x^2 + 4}{4 - x^2}$

11. $f(x) = \dfrac{24}{x^2 + 12}$

12. $f(x) = \dfrac{x^2}{x^2 + 1}$

 Finding Points of Inflection In Exercises 13–20, discuss the concavity of the graph of the function and find the points of inflection. *See Examples 3 and 4.*

13. $f(x) = x^3 - 9x^2 + 24x - 18$

14. $f(x) = -4x^3 - 8x^2 + 32$

15. $f(x) = 2x^3 - 3x^2 - 12x + 5$

16. $f(x) = -\frac{1}{3}x^3 - 4x^2 - 5x - 9$

17. $g(x) = 4 - x - 3x^4$

18. $g(x) = 2x^4 - 8x^3 + 12x^2 + 12x$

19. $f(x) = x(x - 4)^3$

20. $f(x) = (x - 1)^3(x - 5)$

 Using the Second-Derivative Test In Exercises 21–34, find all relative extrema of the function. Use the Second-Derivative Test when applicable. *See Example 5.*

21. $f(x) = 6x - x^2$

22. $f(x) = x^2 + 3x - 8$

23. $f(x) = x^4 - 4x^3 + 2$

24. $f(x) = x^3 - 5x^2 + 7x$

25. $f(x) = x^{2/3} - 3$

26. $f(x) = x + \dfrac{4}{x}$

27. $f(x) = \sqrt{x^2 + 1}$

28. $f(x) = \sqrt{2x^2 + 6}$

29. $f(x) = \sqrt{9 - x^2}$

30. $f(x) = \sqrt{4 - x^2}$

31. $f(x) = \dfrac{8}{x^2 + 2}$

32. $f(x) = \dfrac{x}{x^2 + 16}$

33. $f(x) = \dfrac{x}{x - 1}$

34. $f(x) = \dfrac{x}{x^2 - 1}$

Finding Relative Extrema In Exercises 35–38, use a graphing utility to find all relative extrema of the function.

35. $f(x) = 5 + 3x^2 - x^3$ **36.** $f(x) = x^3 - 6x^2 + 7$

37. $f(x) = \frac{1}{2}x^4 - \frac{1}{3}x^3 - \frac{1}{2}x^2$

38. $f(x) = x - \frac{3}{2}x^2 - \frac{1}{3}x^4$

Using the Second-Derivative Test In Exercises 39–50, find all relative extrema and points of inflection. Then use a graphing utility to graph the function.

39. $f(x) = x^3 - 12x$ **40.** $f(x) = x^3 - 3x$

41. $g(x) = \sqrt{x} + \dfrac{4}{\sqrt{x}}$ **42.** $f(x) = 2\sqrt{x} + \dfrac{3}{\sqrt{x}}$

43. $f(x) = \frac{1}{4}x^4 - 2x^2$ **44.** $f(x) = 2 + 3x - \frac{3}{2}x^4$

45. $g(x) = (x + 2)(x + 5)^2$ **46.** $g(x) = (x - 6)(x + 2)^3$

47. $g(x) = x\sqrt{x + 3}$ **48.** $g(x) = x\sqrt{9 - x}$

49. $f(x) = \dfrac{4}{1 + x^2}$ **50.** $f(x) = \dfrac{2}{x^2 - 1}$

Creating a Function In Exercises 51–54, sketch a graph of a function f having the given characteristics. (There are many correct answers.)

51. $f(0) = f(2) = 0$
$f'(x) > 0$ if $x < 1$
$f'(1) = 0$
$f'(x) < 0$ if $x > 1$
$f''(x) < 0$

52. $f(0) = f(2) = 0$
$f'(x) < 0$ if $x < 1$
$f'(1) = 0$
$f'(x) > 0$ if $x > 1$
$f''(x) > 0$

53. $f(2) = f(4) = 0$
$f'(x) > 0$ if $x < 3$
$f'(3)$ is undefined.
$f'(x) < 0$ if $x > 3$
$f''(x) > 0,\ x \neq 3$

54. $f(2) = f(4) = 0$
$f'(x) < 0$ if $x < 3$
$f'(3)$ is undefined.
$f'(x) > 0$ if $x > 3$
$f''(x) < 0,\ x \neq 3$

Using Graphs In Exercises 55 and 56, use the graph of f to sketch the graph of f'. Find the open intervals on which (a) $f'(x)$ is positive, (b) $f'(x)$ is negative, (c) f' is increasing, and (d) f' is decreasing. For each of these intervals, describe the corresponding behavior of f.

55.

56.

Using the First Derivative In Exercises 57–60, you are given f'. Find the open intervals on which (a) $f'(x)$ is increasing or decreasing and (b) the graph of f is concave upward or concave downward. (c) Find the x-values of the relative extrema and points of inflection of f.

57. $f'(x) = 2x + 5$ **58.** $f'(x) = 9x^2 - 1$

59. $f'(x) = -x^2 + 2x - 1$ **60.** $f'(x) = x^2 + x - 6$

Point of Diminishing Returns In Exercises 61 and 62, find the point of diminishing returns for the function, where R is the revenue (in thousands of dollars) and x is the advertising cost (in thousands of dollars). Use a graphing utility to verify your result. *See Example 6.*

61. $R = \dfrac{1}{50,000}(600x^2 - x^3),\quad 0 \leq x \leq 400$

62. $R = -\frac{4}{9}x^3 + 4x^2 + 12,\quad 0 \leq x \leq 5$

Productivity In Exercises 63 and 64, consider a college student who works from 7 P.M. to 11 P.M. assembling mechanical components. The number N of components assembled after t hours is given by the function. At what time is the student assembling components at the greatest rate?

63. $N = -0.12t^3 + 0.54t^2 + 8.22t,\quad 0 \leq t \leq 4$

64. $N = \dfrac{20t^2}{4 + t^2},\quad 0 \leq t \leq 4$

Comparing a Function and Its Derivatives In Exercises 65–68, use a graphing utility to graph f, f', and f'' in the same viewing window. Graphically locate the relative extrema and points of inflection of the graph of f. State the relationship between the behavior of f and the signs of f' and f''.

65. $f(x) = \frac{1}{2}x^3 - x^2 + 3x - 5,\quad [0, 3]$

66. $f(x) = -\frac{1}{20}x^5 - \frac{1}{12}x^2 - \frac{1}{3}x + 1,\quad [-2, 2]$

67. $f(x) = \dfrac{2}{x^2 + 1},\quad [-3, 3]$

68. $f(x) = \dfrac{x^2}{x^2 + 1},\quad [-3, 3]$

69. Average Cost The total cost C (in dollars) of operating a factory is $C = 0.5x^2 + 10x + 7200$, where x is the number of units produced. At what level of production will the average cost per unit be minimized? (The average cost per unit is C/x.)

70. Inventory Cost The cost C (in dollars) of ordering and storing x units is $C = 2x + (300,000/x)$. What order size will produce a minimum cost?

71. Home Sales The median sales price p (in thousands of dollars) of new single-family houses sold in the United States from 2005 through 2013 can be modeled by

$p = -0.1238t^4 + 5.105t^3 - 73.61t^2 + 438.7t - 674,$

for $5 \le t \le 13$, where t is the year, with $t = 5$ corresponding to 2005. *(Source: U.S. Census Bureau)*

(a) Use a graphing utility to graph the model on the interval $[5, 13]$.

(b) Use the graph in part (a) to estimate the year corresponding to the absolute minimum sales price.

(c) Use the graph in part (a) to estimate the year corresponding to the absolute maximum sales price.

(d) During approximately which year was the rate of increase of the sales price the greatest? the least?

72. HOW DO YOU SEE IT? The graph shows the Dow Jones Industrial Average y on Black Monday, October 19, 1987, where $t = 0$ corresponds to 9:30 A.M., when the market opened, and $t = 6.5$ corresponds to 4 P.M., when the market closed.
(Source: Wall Street Journal)

(a) Estimate the relative extrema and absolute extrema of the graph. Interpret your results in the context of the problem.

(b) Estimate the point of inflection of the graph on the interval $[1, 3]$. Interpret your result in the context of the problem.

73. Poverty Level From 2005 through 2013, the population p (in millions) living below poverty in the United States can be modeled by

$p = 0.01469t^4 - 0.6364t^3 + 9.677t^2 - 59.63t + 163.8$

for $5 \le t \le 13$, where t is the year, with $t = 5$ corresponding to 2005. *(Source: U.S. Census Bureau)*

(a) Use a graphing utility to graph the model on the interval $[5, 13]$.

(b) Use the second derivative to determine the concavity of the graph of p.

(c) Find the point(s) of inflection of the graph of p.

(d) Interpret the meaning of the point(s) of inflection from part (c).

74. Think About It Let S represent monthly sales of Bluetooth headphones. Write a statement describing S' and S'' for each of the following.

(a) The rate of change of sales is increasing.

(b) Sales are increasing, but at a greater rate.

(c) The rate of change of sales is steady.

(d) Sales are steady.

(e) Sales are declining, but at a lower rate.

(f) Sales have bottomed out and have begun to rise.

75. Sketching a Graph Sketch the graph of a function f that does *not* have a point of inflection at $(c, f(c))$ even though $f''(c) = 0$.

Business Capsule

While working in New York City in 2004, Matthew Corrin noticed an abundance of fresh food bars and decided that if someone could successfully brand one, that person could create the "Starbucks of the fresh food business." With $275,000, he opened his first Freshii store in Toronto in 2005, and soon began developments for more. Now, Freshii opens two new stores each week and has over 160 locations in more than 70 cities and 13 countries, with over 400 new locations under development. The mission of this eco-friendly chain is "to eliminate the excuse of people not eating fresh food because it isn't convenient," Corrin said.

76. Research Project Use your school's library, the Internet, or some other reference source to research the financial history of a fast-growing company like the one discussed above. Gather data on the company's costs and revenues over a period of time, and use a graphing utility to graph a scatter plot of the data. Fit models to the data. Do the models appear to be concave upward or concave downward? Do they appear to be increasing or decreasing? Discuss the implications of your answers.

3.4 Optimization Problems

■ Solve real-life optimization problems.

Solving Optimization Problems

One of the most common applications of calculus is the determination of optimum (minimum or maximum) values. Before learning a general method for solving optimization problems, consider the next example.

Cost of a Storage Box

Cost (in dollars) vs Side length (in centimeters)

In Exercise 13 on page 202, you will use primary equations, secondary equations, and derivatives to find the dimensions of a box that will minimize the cost of making the box.

ALGEBRA TUTOR

For help on the algebra in Example 1, see Example 1(c) in the *Chapter 3 Algebra Tutor*, on page 242.

EXAMPLE 1   Finding the Maximum Volume

A manufacturer wants to design an open box that has a square base and a surface area S of 108 square inches, as shown in Figure 3.32. What dimensions will produce a box with a maximum volume?

SOLUTION   Because the base of the box is square, the volume is

$$V = x^2h.$$ Primary equation

This equation is called the **primary equation** because it gives a formula for the quantity to be optimized. The surface area of the box is

$$S = (\text{area of base}) + (\text{area of four sides})$$
$$108 = x^2 + 4xh.$$ Secondary equation

Open Box with Square Base:
$S = x^2 + 4xh = 108$

FIGURE 3.32

Because V is to be optimized, it helps to express V as a function of just one variable. To do this, solve the secondary equation for h in terms of x to obtain

$$h = \frac{108 - x^2}{4x}$$

and substitute for h in the primary equation.

$$V = x^2h = x^2\left(\frac{108 - x^2}{4x}\right) = 27x - \frac{1}{4}x^3$$ Function of one variable

Before finding which x-value yields a maximum value of V, you need to determine the *feasible domain* of the function. That is, what values of x make sense in this problem? Because x must be nonnegative and the area of the base ($A = x^2$) is at most 108, you can conclude that the feasible domain is

$$0 \le x \le \sqrt{108}.$$ Feasible domain

Using the techniques described in the first three sections of this chapter, you can determine that $\left(\text{on the interval } 0 \le x \le \sqrt{108}\right)$ this function has an absolute maximum at $x = 6$ inches and $h = 3$ inches.

✓ **Checkpoint 1**   *Worked-out solution available at LarsonAppliedCalculus.com*

Use a graphing utility to graph the volume function

$$V = 27x - \tfrac{1}{4}x^3$$

from Example 1 on $0 \le x \le \sqrt{108}$. Verify that the function has an absolute maximum at $x = 6$. What is the maximum volume? ■

In Example 1, be sure that you understand the basic question that it asks. Remember that you are not ready to begin solving an optimization problem until you have clearly identified the problem. Once you are sure you understand what is being asked, you are ready to begin considering a method for solving the problem.

For instance, in Example 1, you should realize that there are infinitely many open boxes having 108 square inches of surface area. To begin solving this problem, you might ask yourself which basic shape would seem to yield a maximum volume. Should the box be tall, squat, or nearly cubical? You might even try calculating a few volumes, as shown in Figure 3.33, to see if you can get a better feeling for what the optimum dimensions should be.

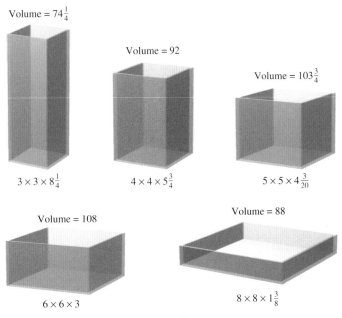

Volume = $74\frac{1}{4}$

$3 \times 3 \times 8\frac{1}{4}$

Volume = 92

$4 \times 4 \times 5\frac{3}{4}$

Volume = $103\frac{3}{4}$

$5 \times 5 \times 4\frac{3}{20}$

Volume = 108

$6 \times 6 \times 3$

Volume = 88

$8 \times 8 \times 1\frac{3}{8}$

Which box has the greatest volume?

FIGURE 3.33

There are several steps in the solution to Example 1. The first step is to sketch a diagram and identify all *known* quantities and all quantities *to be determined*. The second step is to write a primary equation for the quantity to be optimized. Then, a secondary equation is used to rewrite the primary equation as a function of one variable. Finally, calculus is used to determine the optimum value. These steps are summarized below.

Guidelines for Solving Optimization Problems

1. Identify all given quantities and all quantities to be determined. If possible, make a sketch.

2. Write a **primary equation** for the quantity that is to be maximized or minimized. (A summary of several common formulas is given in Appendix D.)

3. Reduce the primary equation to one having a single independent variable. This may involve the use of a **secondary equation** that relates the independent variables of the primary equation.

4. Determine the feasible domain of the primary equation. That is, determine the values for which the stated problem makes sense.

5. Determine the desired maximum or minimum value by the calculus techniques discussed in Sections 3.1 through 3.3.

EXAMPLE 2 Finding a Minimum Distance

Find the points on the graph of $y = 4 - x^2$ that are closest to $(0, 2)$.

SOLUTION

1. Figure 3.34 shows that there are two points at a minimum distance from $(0, 2)$.

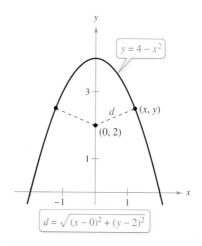

$$d = \sqrt{(x - 0)^2 + (y - 2)^2}$$

FIGURE 3.34

ALGEBRA TUTOR

For help on the algebra in Example 2, see Example 1(b) in the *Chapter 3 Algebra Tutor*, on page 242.

2. You are asked to minimize the distance d. So, you can use the Distance Formula to obtain a primary equation.

$$d = \sqrt{(x - 0)^2 + (y - 2)^2} \qquad \text{Primary equation}$$

3. Using the secondary equation $y = 4 - x^2$, you can rewrite the primary equation as a function of a single variable.

$$
\begin{aligned}
d &= \sqrt{x^2 + (4 - x^2 - 2)^2} & \text{Substitute } 4 - x^2 \text{ for } y. \\
&= \sqrt{x^2 + (2 - x^2)^2} & \text{Simplify.} \\
&= \sqrt{x^2 + 4 - 4x^2 + x^4} & \text{Expand binomial.} \\
&= \sqrt{x^4 - 3x^2 + 4} & \text{Combine like terms.}
\end{aligned}
$$

Because d is smallest when the expression under the radical is smallest, you simplify the problem by finding the minimum value of $f(x) = x^4 - 3x^2 + 4$.

4. The domain of f is the entire real number line.

5. To find the minimum value of $f(x)$, first find the critical numbers of f.

$$
\begin{aligned}
f'(x) &= 4x^3 - 6x & \text{Find derivative of } f. \\
0 &= 4x^3 - 6x & \text{Set derivative equal to 0.} \\
0 &= 2x(2x^2 - 3) & \text{Factor.} \\
x &= 0, \; x = \sqrt{\tfrac{3}{2}}, \; x = -\sqrt{\tfrac{3}{2}} & \text{Critical numbers}
\end{aligned}
$$

The First-Derivative Test verifies that $x = 0$ yields a relative maximum, whereas both $x = \sqrt{3/2}$ and $x = -\sqrt{3/2}$ yield a minimum. So, the points closest to $(0, 2)$ are

$$\left(\sqrt{\tfrac{3}{2}}, \tfrac{5}{2} \right) \quad \text{and} \quad \left(-\sqrt{\tfrac{3}{2}}, \tfrac{5}{2} \right).$$

✓ **Checkpoint 2**   *Worked-out solution available at LarsonAppliedCalculus.com*

Find the points on the graph of $y = 4 - x^2$ that are closest to $(0, 3)$.

EXAMPLE 3   Finding a Minimum Area

A rectangular page will contain 24 square inches of print. The margins at the top and bottom of the page are $1\frac{1}{2}$ inches wide. The margins on each side are 1 inch wide. What should the dimensions of the page be to minimize the amount of paper used?

SOLUTION

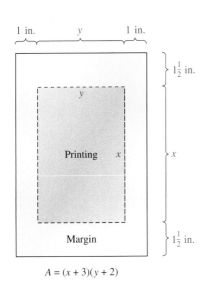

$A = (x + 3)(y + 2)$

FIGURE 3.35

1. A diagram of the page is shown in Figure 3.35.

2. Letting A be the area to be minimized, the primary equation is

$$A = (x + 3)(y + 2). \qquad \text{Primary equation}$$

3. The printed area inside the margins is given by

$$24 = xy. \qquad \text{Secondary equation}$$

Solving this equation for y produces

$$y = \frac{24}{x}.$$

By substituting this result into the primary equation, you obtain

$$A = (x + 3)\left(\frac{24}{x} + 2\right) \qquad \text{Write as a function of one variable.}$$

$$= (x + 3)\left(\frac{24 + 2x}{x}\right) \qquad \text{Rewrite second factor as a single fraction.}$$

$$= \frac{2x^2}{x} + \frac{30x}{x} + \frac{72}{x} \qquad \text{Multiply and separate into terms.}$$

$$= 2x + 30 + \frac{72}{x}. \qquad \text{Simplify.}$$

4. Because x must be positive, the feasible domain is $x > 0$.

5. To find the minimum area, begin by finding the critical numbers of A.

$$\frac{dA}{dx} = 2 - \frac{72}{x^2} \qquad \text{Find derivative of } A.$$

$$0 = 2 - \frac{72}{x^2} \qquad \text{Set derivative equal to 0.}$$

$$-2 = -\frac{72}{x^2} \qquad \text{Subtract 2 from each side.}$$

$$x^2 = 36 \qquad \text{Simplify.}$$

$$x = \pm 6 \qquad \text{Critical numbers}$$

Because $x = -6$ is not in the feasible domain, you need to consider only the critical number $x = 6$. Using the First-Derivative Test, it follows that A is a minimum when $x = 6$. So, the dimensions of the page should be

$$x + 3 = 6 + 3 = 9 \text{ inches} \quad \text{by} \quad y + 2 = \frac{24}{6} + 2 = 6 \text{ inches.}$$

✓ *Checkpoint 3*   *Worked-out solution available at LarsonAppliedCalculus.com*

A rectangular page will contain 54 square inches of print. The margins at the top and bottom of the page are $1\frac{1}{2}$ inches wide. The margins on each side are 1 inch wide. What should the dimensions of the page be to minimize the amount of paper used?

As applications go, the examples described in this section are fairly simple, and yet the resulting primary equations are quite complicated. Real-life applications often involve equations that are at least as complex as the ones in the examples. Remember that one of the main goals of this course is to enable you to use the power of calculus to analyze equations that at first glance seem formidable.

Also remember that once you have found the primary equation, you can use the graph of the equation to help solve the problem. For instance, the graphs of the primary equations in Examples 1 through 3 are shown in Figure 3.36.

Example 1

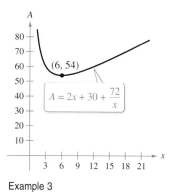

Example 2

Example 3

FIGURE 3.36

SUMMARIZE   (Section 3.4)

1. State what is meant by the primary equation of an optimization problem *(page 196)*. For examples of primary equations in optimization problems, see Examples 1, 2, and 3.

2. State what is meant by the feasible domain of a function *(page 196)*. For examples of feasible domains, see Examples 1, 2, and 3.

3. State what is meant by the secondary equation of an optimization problem *(page 197)*. For examples of secondary equations in optimization problems, see Examples 1, 2, and 3.

4. State the guidelines for solving optimization problems *(page 197)*. For examples of solving optimization problems, see Examples 2 and 3.

5. Describe a real-life example of how solving an optimization problem can be used to determine the dimensions of a page so that the amount of paper used is minimized *(page 199, Example 3)*.

SKILLS WARM UP 3.4

The following warm-up exercises involve skills that were covered in earlier sections. You will use these skills in the exercise set for this section. For additional help, review Section 3.1.

In Exercises 1–4, write a formula for the written statement.

1. The sum of one number and half a second number is 12.

2. The product of one number and three times a second number is 36.

3. The area of a rectangle is 24 square units.

4. The distance between two points is 10 units.

In Exercises 5–10, find the critical numbers of the function.

5. $y = 3x^2 + 12x - 8$

6. $y = 2x^3 - x^2 - 4x$

7. $y = 5x + \dfrac{125}{x}$

8. $y = 3x + \dfrac{96}{x^2}$

9. $y = \dfrac{x^2 + 1}{x}$

10. $y = \dfrac{x}{x^2 + 9}$

Exercises 3.4

See *CalcChat.com* for tutorial help and worked-out solutions to odd-numbered exercises.

 Maximum Area In Exercises 1 and 2, find the length and width of a rectangle that has the given perimeter and a maximum area.

1. Perimeter: 80 meters   **2.** Perimeter: P units

 Minimum Perimeter In Exercises 3 and 4, find the length and width of a rectangle that has the given area and a minimum perimeter.

3. Area: 49 square feet   **4.** Area: A square centimeters

5. Maximum Area An animal shelter has 160 feet of fencing to enclose two adjacent rectangular playpen areas for dogs (see figure). What dimensions should be used so that the enclosed area will be a maximum?

6. Minimum Dimensions A campground owner plans to enclose a rectangular field adjacent to a river. The owner wants the field to contain 180,000 square meters. No fencing is required along the river. What dimensions will use the least amount of fencing?

7. Maximum Volume

(a) Verify that each of the rectangular solids shown in the figure has a surface area of 150 square inches.

(b) Find the volume of each solid.

(c) Determine the dimensions of a rectangular solid (with a square base) of maximum volume when its surface area is 150 square inches.

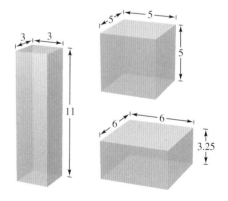

8. Maximum Volume A rectangular solid with a square base has a surface area of 337.5 square centimeters.

(a) Determine the dimensions that yield the maximum volume.

(b) Find the maximum volume.

9. Minimum Surface Area A rectangular solid with a square base has a volume of 8000 cubic inches.

(a) Determine the dimensions that yield the minimum surface area.

(b) Find the minimum surface area.

10. **HOW DO YOU SEE IT?**   The graph shows the profit P (in thousands of dollars) of a company in terms of its advertising cost x (in thousands of dollars).

Profit of a Company

(a) Estimate the interval on which the profit is increasing.

(b) Estimate the interval on which the profit is decreasing.

(c) Estimate the amount of money the company should spend on advertising in order to yield a maximum profit.

(d) Estimate the point of diminishing returns.

11. Maximum Area   An indoor physical fitness room consists of a rectangular region with a semicircle on each end. The perimeter of the room is to be a 200-meter running track. Find the dimensions that will make the area of the rectangular region as large as possible.

12. Maximum Area   A Norman window is constructed by adjoining a semicircle to the top of an ordinary rectangular window (see figure). Find the dimensions of a Norman window of maximum area when the total perimeter is 16 feet.

13. Minimum Cost   A storage box with a square base must have a volume of 80 cubic centimeters. The top and bottom cost $0.20 per square centimeter and the sides cost $0.10 per square centimeter. Find the dimensions that will minimize cost.

14. Minimum Surface Area   A net enclosure for golf practice is open at one end (see figure). The volume of the enclosure is $83\frac{1}{3}$ cubic meters. Find the dimensions that require the least amount of netting.

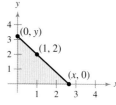

Figure for 14   Figure for 15

15. Maximum Volume   An open box is to be made from a six-inch by six-inch square piece of material by cutting equal squares from the corners and turning up the sides (see figure). Find the volume of the largest box that can be made.

16. Maximum Volume   An open box is to be made from a three-foot by eight-foot rectangular piece of material by cutting equal squares from the corners and turning up the sides. Find the volume of the largest box that can be made.

17. Minimum Area   A rectangular page will contain 36 square inches of print. The margins at the top and bottom and on each side are $1\frac{1}{2}$ inches wide. Find the dimensions of the page that will minimize the amount of paper used.

18. Minimum Area   A rectangular page will contain 50 square inches of print. The margins at the top and bottom of the page are 2 inches wide. The margins on each side are 1 inch wide. Find the dimensions of the page that will minimize the amount of paper used.

19. Maximum Area   A rectangle is bounded by the x- and y-axes and the graph of

$$y = \tfrac{1}{2}(6 - x)$$

(see figure). What length and width should the rectangle have so that its area is a maximum?

Figure for 19   Figure for 20

20. Minimum Length and Minimum Area   A right triangle is formed in the first quadrant by the x- and y-axes and a line through the point $(1, 2)$ (see figure).

(a) Write the length L of the hypotenuse as a function of x.

(b) Use a graphing utility to approximate x graphically such that the length of the hypotenuse is a minimum.

(c) Find the vertices of the triangle such that its area is a minimum.

21. Maximum Area A rectangle is bounded by the x-axis and the semicircle $y = \sqrt{25 - x^2}$ (see figure). What length and width should the rectangle have so that its area is a maximum?

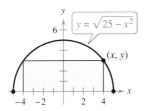

22. Maximum Area Find the dimensions of the largest rectangle that can be inscribed in a semicircle of radius r. (See Exercise 21.)

23. Minimum Surface Area You are designing a juice container that has the shape of a right circular cylinder. The container is supposed to hold 8 fluid ounces (1 fluid ounce is approximately 1.80469 cubic inches). Find the dimensions that will use a minimum amount of construction material.

24. Minimum Cost An energy drink container of the shape described in Exercise 23 must have a volume of 16 fluid ounces. The cost per square inch of constructing the top and bottom is twice the cost of constructing the lateral side. Find the dimensions that will minimize cost.

 Finding a Minimum Distance In Exercises 25–28, find the point on the graph of the function that is closest to the given point. *See Example 2.*

25. $f(x) = x^2$, $\left(2, \frac{1}{2}\right)$ **26.** $f(x) = (x - 1)^2$, $(-5, 3)$

27. $f(x) = \sqrt{x}$, $(4, 0)$ **28.** $f(x) = \sqrt{x - 8}$, $(12, 0)$

29. Maximum Volume A rectangular package to be sent by a postal service can have a maximum combined length and girth (perimeter of a cross section) of 108 inches. Find the dimensions of the package with maximum volume. Assume that the package's dimensions are x by x by y (see figure).

30. Minimum Surface Area A solid is formed by adjoining two hemispheres to the ends of a right circular cylinder. The total volume of the solid is 14 cubic centimeters. Find the radius of the cylinder that produces the minimum surface area.

31. Minimum Cost An industrial tank of the shape described in Exercise 30 must have a volume of 4000 cubic feet. The hemispherical ends cost twice as much per square foot of surface area as the lateral side. Find the dimensions that will minimize cost.

32. Minimum Time You are in a boat 2 miles from the nearest point on the coast. You are to go to point Q, located 3 miles down the coast and 1 mile inland (see figure). You can row at a rate of 2 miles per hour and you can walk at a rate of 4 miles per hour. Toward what point on the coast should you row in order to reach point Q in the least amount of time?

33. Minimum Area The sum of the circumference of a circle and the perimeter of a square is 16. Find the dimensions of the circle and square that produce a minimum total area.

34. Minimum Area The sum of the perimeters of an equilateral triangle and a square is 10. Find the dimensions of the triangle and square that produce a minimum total area.

35. Area Four feet of wire is to be used to form a square and a circle.

(a) Express the sum of the areas of the square and the circle as a function A of a side of the square x.

(b) What is the domain of A?

(c) Use a graphing utility to graph A on its domain.

(d) How much wire should be used for the square and how much for the circle in order to enclose the least total area? the greatest total area?

36. Maximum Yield A home gardener estimates that 16 apple trees will produce an average yield of 80 apples per tree. But because of the size of the garden, for each additional tree planted, the yield will decrease by four apples per tree. How many trees should be planted to maximize the total yield of apples? What is the maximum yield?

37. Farming A strawberry farmer will receive $30 per bushel of strawberries during the first week of harvesting. Each week after that, the value will drop $0.80 per bushel. The farmer estimates that there are approximately 120 bushels of strawberries in the fields, and that the crop is increasing at a rate of four bushels per week. When should the farmer harvest the strawberries to maximize their value? How many bushels of strawberries will yield the maximum value? What is the maximum value of the strawberries?

QUIZ YOURSELF

See *CalcChat.com* for tutorial help and worked-out solutions to odd-numbered exercises.

Take this quiz as you would take a quiz in class. When you are done, check your work against the answers given in the back of the book.

In Exercises 1–3, find the critical numbers and the open intervals on which the function is increasing or decreasing. Use a graphing utility to verify your results.

1. $f(x) = x^2 - 8x + 6$

2. $f(x) = 2x^3 + 12x^2$

3. $f(x) = \dfrac{x}{x^2 + 25}$

In Exercises 4–6, find all relative extrema of the function.

4. $f(x) = x^3 + 3x^2 - 5$

5. $f(x) = x^4 - 8x^2 + 3$

6. $f(x) = 2x^{2/3}$

In Exercises 7–9, find the absolute extrema of the function on the closed interval. Use a graphing utility to verify your results.

7. $f(x) = x^2 + 4x - 12, \quad [-5, -1]$

8. $f(x) = x^3 - 27x, \quad [-4, 4]$

9. $f(x) = \dfrac{x}{x^2 + 1}, \quad [0, 2]$

10. The resident population P (in thousands) of West Virginia from 2000 through 2014 can be modeled by

$$P = -0.077t^3 + 1.47t^2 - 2.3t + 1805, \quad 0 \le t \le 14$$

where t is the year, with $t = 0$ corresponding to 2000.   *(Source: U.S. Census Bureau)*

(a) During which year(s) was the population increasing? decreasing?

(b) During which year, from 2000 through 2014, was the population the greatest? the least?

In Exercises 11 and 12, discuss the concavity of the graph of the function and find the points of inflection.

11. $f(x) = x^3 - 6x^2 + 7x$

12. $f(x) = \frac{1}{12}x^4 - \frac{9}{2}x^2$

In Exercises 13 and 14, use the Second-Derivative Test to find all relative extrema of the function.

13. $f(x) = 2x^3 + 3x^2 - 12x + 16$

14. $f(x) = 2x + \dfrac{18}{x}$

15. By increasing its advertising cost x (in thousands of dollars) for a product, a company discovers that it can increase the sales S (in thousands of dollars) according to the model

$$S = \frac{1}{3600}(360x^2 - x^3), \quad 0 \le x \le 240.$$

Find the point of diminishing returns for this product.

16. A gardener has 200 feet of fencing to enclose a rectangular garden adjacent to a river (see figure). No fencing is needed along the river. What dimensions should be used so that the area of the garden will be a maximum?

17. A rectangular page will contain 36 square inches of print. The margins at the top and bottom and on each side are 1 inch wide. Find the dimensions of the page that will minimize the amount of paper used.

Figure for 16

3.5 Business and Economics Applications

Maximum Profit

In Exercise 15 on page 212, you will use derivatives to find the price that yields a maximum profit.

■ Solve business and economics optimization problems.
■ Find the price elasticity of demand for demand functions.
■ Recognize basic business terms and formulas.

Optimization in Business and Economics

The problems in this section are primarily optimization problems. So, the five-step procedure used in Section 3.4 is an appropriate strategy to follow.

EXAMPLE 1   Finding the Maximum Revenue

A company has determined that its total revenue (in dollars) for a product can be modeled by

$$R = -x^3 + 450x^2 + 52,500x$$

where x is the number of units produced (and sold). What production level will yield a maximum revenue?

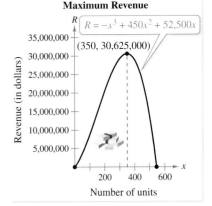

Maximum revenue occurs when $dR/dx = 0$.

FIGURE 3.37

SOLUTION

1. A sketch of the revenue function is shown in Figure 3.37.

2. The primary equation is the given revenue function.

$$R = -x^3 + 450x^2 + 52,500x$$

3. Because R is a function of one variable, you do not need a secondary equation.

4. The feasible domain of the primary equation is $0 \le x \le 546$. This is determined by finding the x-intercepts of the revenue function, as shown in Figure 3.37.

5. To maximize the revenue, begin by finding the derivative of R.

$$\frac{dR}{dx} = -3x^2 + 900x + 52,500 \qquad \text{Find derivative of } R.$$

Next, find the critical numbers of R.

$$-3x^2 + 900x + 52,500 = 0 \qquad \text{Set derivative equal to 0.}$$
$$-3(x - 350)(x + 50) = 0 \qquad \text{Factor.}$$
$$x = 350, \ x = -50 \qquad \text{Critical numbers}$$

The only critical number in the feasible domain is $x = 350$. From the graph of the function, you can see that the production level of 350 units corresponds to a maximum revenue.

✓ **Checkpoint 1**   *Worked-out solution available at LarsonAppliedCalculus.com*

Find the number of units that must be produced to maximize the revenue function $R = -x^3 + 150x^2 + 9375x$, where R is the total revenue (in dollars) and x is the number of units produced (and sold). What is the maximum revenue?   ■

To study the effects of production levels on cost, one method economists use is the **average cost function** $\overline{C}$, which is defined as

$$\overline{C} = \frac{C}{x}$$

Average cost function

where $C = f(x)$ is the total cost function and x is the number of units produced.

EXAMPLE 2   Finding the Minimum Average Cost

A company estimates that the cost (in dollars) of producing x units of a product can be modeled by

$$C = 800 + 0.04x + 0.0002x^2.$$

Find the production level that minimizes the average cost per unit.

SOLUTION

1. C represents the total cost, x represents the number of units produced, and $\overline{C}$ represents the average cost per unit.

2. The primary equation is

$$\overline{C} = \frac{C}{x}.$$

Primary equation

3. Substituting the given equation for C produces

$$\overline{C} = \frac{800 + 0.04x + 0.0002x^2}{x}$$

Substitute for C.

$$= \frac{800}{x} + 0.04 + 0.0002x.$$

Function of one variable

4. The feasible domain of $\overline{C}$ is $x > 0$ because the company cannot produce a negative number of units.

5. To minimize the average cost, begin by finding the derivative of $\overline{C}$.

$$\frac{d\overline{C}}{dx} = -\frac{800}{x^2} + 0.0002$$

Find derivative of C.

Next, find the critical numbers of $\overline{C}$.

$$-\frac{800}{x^2} + 0.0002 = 0$$

Set derivative equal to 0.

$$0.0002 = \frac{800}{x^2}$$

$$x^2 = \frac{800}{0.0002}$$

Multiply each side by x^2 and divide each side by 0.0002.

$$x^2 = 4{,}000{,}000$$

$$x = \pm 2000$$

Critical numbers

By choosing the positive value of x and sketching the graph of $\overline{C}$, as shown in Figure 3.38, you can see that a production level of 2000 units minimizes the average cost per unit.

✓ Checkpoint 2   *Worked-out solution available at LarsonAppliedCalculus.com*

Find the production level that minimizes the average cost per unit for the cost function $C = 400 + 0.05x + 0.0025x^2$, where C is the cost (in dollars) of producing x units of a product.

STUDY TIP

To see that $x = 2000$ corresponds to the minimum average cost in Example 2, try evaluating $\overline{C}$ for several values of x. For instance, when $x = 400$, the average cost per unit is $\overline{C} = \$2.12$, but when $x = 2000$, the average cost per unit is $\overline{C} = \$0.84$.

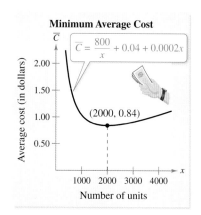

Minimum Average Cost

Minimum average cost occurs when $d\overline{C}/dx = 0$.

FIGURE 3.38

EXAMPLE 3   Finding the Maximum Revenue

A business sells 2000 units of a product per month at a price of $10 each. It can sell 250 more items per month for each $0.25 reduction in price. What price per unit will maximize the monthly revenue?

SOLUTION

1. Let x represent the number of units sold in a month, let p represent the price per unit, and let R represent the monthly revenue.

2. Because the revenue is to be maximized, the primary equation is $R = xp$.

3. A price of $p = $10 corresponds to $x = 2000$, and a price of $p = $9.75 corresponds to $x = 2250$. Using this information, you can use the two-point form to write the demand equation.

$$p - 10 = \frac{10 - 9.75}{2000 - 2250}(x - 2000) \qquad \text{Two-point form}$$

$$p - 10 = -0.001(x - 2000) \qquad \text{Simplify.}$$

$$p = -0.001x + 12 \qquad \text{Secondary equation}$$

Substituting this result into the revenue equation produces

$$R = x(-0.001x + 12) \qquad \text{Substitute for } p.$$

$$= -0.001x^2 + 12x. \qquad \text{Function of one variable}$$

4. The feasible domain of the revenue function is $0 \le x \le 12{,}000$. This is determined by finding the x-intercepts of the revenue function.

5. To maximize the revenue, begin by finding the derivative of R.

$$\frac{dR}{dx} = 12 - 0.002x \qquad \text{Find derivative of } R.$$

Next, find the critical numbers of R.

$$12 - 0.002x = 0 \qquad \text{Set derivative equal to 0.}$$

$$-0.002x = -12$$

$$x = 6000 \qquad \text{Critical number}$$

From the graph of R in Figure 3.39, you can see that this production level yields a maximum revenue. The price that corresponds to this production level is

$$p = 12 - 0.001x \qquad \text{Demand function}$$

$$= 12 - 0.001(6000) \qquad \text{Substitute 6000 for } x.$$

$$= \$6. \qquad \text{Price per unit}$$

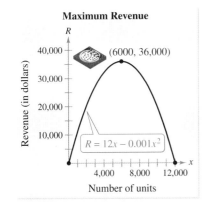

Maximum Revenue

(6000, 36,000)

$R = 12x - 0.001x^2$

Revenue (in dollars)

Number of units

FIGURE 3.39

✓ **Checkpoint 3**   *Worked-out solution available at LarsonAppliedCalculus.com*

Find the price per unit that will maximize the monthly revenue for the business in Example 3 when it can sell only 200 more items per month for each $0.25 reduction in price. ■

In Example 3, the revenue function was written as a function of x. It could also have been written as a function of p. That is,

$$R = 1000(12p - p^2).$$

By finding the critical numbers of this function, you can determine that the maximum revenue occurs at $p = 6$.

ALGEBRA TUTOR

For help on the algebra in Example 4, see Example 2(b) in the *Chapter 3 Algebra Tutor*, on page 243.

EXAMPLE 4   **Finding the Maximum Profit**

The marketing department of a business has determined that the demand for a product can be modeled by $p = 50/\sqrt{x}$, where p is the price per unit (in dollars) and x is the number of units. The cost (in dollars) of producing x units is given by $C = 0.5x + 500$. What price will yield a maximum profit?

SOLUTION

1. Let R represent the revenue, P the profit, p the price per unit, x the number of units, and C the total cost of producing x units.

2. Because you are maximizing the profit, the primary equation is $P = R - C$.

3. Because the revenue is $R = xp$, you can write the profit function as

$$P = R - C$$
$$= xp - (0.5x + 500) \qquad \text{Substitute for } R \text{ and } C.$$
$$= x\left(\frac{50}{\sqrt{x}}\right) - 0.5x - 500 \qquad \text{Substitute for } p.$$
$$= 50\sqrt{x} - 0.5x - 500. \qquad \text{Function of one variable}$$

4. The feasible domain of the function is $127 < x \le 7872$. (When x is less than or equal to 127 or greater than 7872, the profit is negative. So, these values should be excluded from the domain of P.)

5. To maximize the profit, begin by finding the derivative of P.

$$\frac{dP}{dx} = \frac{25}{\sqrt{x}} - 0.5 \qquad \text{Find derivative of } P.$$

Next, find the critical numbers of P.

$$\frac{25}{\sqrt{x}} - 0.5 = 0 \qquad \text{Set derivative equal to 0.}$$
$$\frac{25}{\sqrt{x}} = 0.5 \qquad \text{Add 0.5 to each side.}$$
$$50 = \sqrt{x} \qquad \text{Isolate } x\text{-term on one side.}$$
$$2500 = x \qquad \text{Critical number}$$

Maximum Profit

$P = 50\sqrt{x} - 0.5x - 500$

$(2500, 750)$

$1 ea.

Profit (in dollars)

Number of units

FIGURE 3.40

From the graph of the profit function shown in Figure 3.40, you can see that a maximum profit occurs at $x = 2500$. The price that corresponds to $x = 2500$ is

$$p = \frac{50}{\sqrt{x}} = \frac{50}{\sqrt{2500}} = \frac{50}{50} = \$1. \qquad \text{Price per unit}$$

✓ *Checkpoint 4*   *Worked-out solution available at LarsonAppliedCalculus.com*

Find the price that will maximize profit for the demand and cost functions $p = 40/\sqrt{x}$ and $C = 2x + 50$, where p is the price per unit (in dollars), x is the number of units, and C is the cost (in dollars). ∎

To find the maximum profit in Example 4, the equation $P = R - C$ was differentiated and set equal to zero. From the equation

$$\frac{dP}{dx} = \frac{dR}{dx} - \frac{dC}{dx} = 0$$

it follows that the maximum profit occurs when the marginal revenue is equal to the marginal cost, as shown in Figure 3.41.

Marginal Revenue and Marginal Cost

$R = 50\sqrt{x}$

Maximum profit: $\frac{dR}{dx} = \frac{dC}{dx}$

$C = 0.5x + 500$

Revenue and cost (in dollars)

Number of units

FIGURE 3.41

Price Elasticity of Demand

STUDY TIP

In the discussion of price elasticity of demand, the price is assumed to decrease as the quantity demanded increases. So, the demand function $p = f(x)$ is decreasing and dp/dx is negative.

One way in which economists measure the responsiveness of consumers to a change in the price of a product is with **price elasticity of demand.** For example, a drop in the price of fresh tomatoes might result in a much greater demand for fresh tomatoes; such a demand is called **elastic.** On the other hand, the demand for items such as coffee and gasoline is relatively unresponsive to changes in price; the demand for such items is called **inelastic.**

More formally, the elasticity of demand is the percent change of a quantity demanded x, divided by the percent change in its price p. You can develop a formula for price elasticity of demand using the approximation

$$\frac{\Delta p}{\Delta x} \approx \frac{dp}{dx}$$

which is based on the definition of the derivative. Using this approximation, you can write

$$
\begin{aligned}
\text{Price elasticity of demand} &= \frac{\text{rate of change in demand}}{\text{rate of change in price}} \\
&= \frac{\Delta x/x}{\Delta p/p} \\
&= \frac{p/x}{\Delta p/\Delta x} \\
&\approx \frac{p/x}{dp/dx}.
\end{aligned}
$$

Definition of Price Elasticity of Demand

If $p = f(x)$ is a differentiable function, then the **price elasticity of demand** is given by

$$\eta = \frac{p/x}{dp/dx}$$

where η is the lowercase Greek letter eta. For a given price, the demand is **elastic** when $|\eta| > 1$, the demand is **inelastic** when $|\eta| < 1$, and the demand has **unit elasticity** when $|\eta| = 1$.

Price elasticity of demand is related to the total revenue function, as indicated in Figure 3.42 and the list below.

1. If the demand is *elastic*, then a decrease in price is accompanied by an increase in unit sales sufficient to increase the total revenue.

2. If the demand is *inelastic*, then a decrease in price is not accompanied by an increase in unit sales sufficient to increase the total revenue.

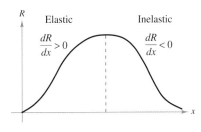

Revenue Curve
FIGURE 3.42

Demand Function of a Product

$p = 24 - 2\sqrt{x}$

FIGURE 3.43

EXAMPLE 5   Comparing Elasticity and Revenue

The demand function for a product is modeled by

$$p = 24 - 2\sqrt{x}, \quad 0 \le x \le 144$$

where p is the price per unit (in dollars) and x is the number of units. (See Figure 3.43.)

a. Determine when the demand is elastic, inelastic, and of unit elasticity.

b. Use the result of part (a) to describe the behavior of the revenue function.

SOLUTION

a. The price elasticity of demand is given by

$$\eta = \frac{p/x}{dp/dx} \qquad\qquad \text{Formula for price elasticity of demand}$$

$$= \frac{\dfrac{24 - 2\sqrt{x}}{x}}{\dfrac{-1}{\sqrt{x}}} \qquad\qquad \text{Substitute for } \quad \text{and } dp/dx.$$

$$= \frac{\left(\dfrac{24 - 2\sqrt{x}}{x}\right)\left(-\sqrt{x}\right)}{\left(\dfrac{-1}{\sqrt{x}}\right)\left(-\sqrt{x}\right)} \qquad\qquad \text{Multiply numerator and denominator by } -\sqrt{x}.$$

$$= \frac{-24\sqrt{x} + 2x}{x} \qquad\qquad \text{Simplify.}$$

$$= -\frac{24}{\sqrt{x}} + 2. \qquad\qquad \text{Rewrite as two fractions and simplify.}$$

The demand is of unit elasticity when $|\eta| = 1$. In the interval $[0, 144]$, the only solution of the equation

$$\left| -\frac{24}{\sqrt{x}} + 2 \right| = 1 \qquad\qquad \text{Unit elasticity } (|\eta| = 1)$$

is $x = 64$. So, the demand is of unit elasticity when $x = 64$. For x-values in the interval $(0, 64)$,

$$\left| -\frac{24}{\sqrt{x}} + 2 \right| > 1, \quad 0 < x < 64 \qquad\qquad \text{Elastic } (|\eta| > 1)$$

which implies that the demand is elastic when $0 < x < 64$. For x-values in the interval $(64, 144)$,

$$\left| -\frac{24}{\sqrt{x}} + 2 \right| < 1, \quad 64 < x < 144 \qquad\qquad \text{Inelastic } (|\eta| < 1)$$

which implies that the demand is inelastic when $64 < x < 144$.

b. From part (a), you can conclude that the revenue function R is increasing on the open interval $(0, 64)$, is decreasing on the open interval $(64, 144)$, and is a maximum when $x = 64$, as indicated in Figure 3.44.

Revenue Function of a Product

$R = xp = x(24 - 2\sqrt{x})$

(64, 512)

FIGURE 3.44

✓ *Checkpoint 5*   *Worked-out solution available at LarsonAppliedCalculus.com*

The demand function for a product is modeled by $p = 36 - 2\sqrt{x}$, $0 \le x \le 324$, where p is the price per unit (in dollars) and x is the number of units. Determine when the demand is elastic, inelastic, and of unit elasticity. ■

Business Terms and Formulas

This section concludes with a summary of the basic business terms and formulas used in this section. A summary of the graphs of the demand, revenue, cost, and profit functions is shown in Figure 3.45.

Summary of Business Terms and Formulas

x = number of units produced (or sold)	η = price elasticity of demand
p = price per unit	$\quad = \dfrac{p/x}{dp/dx}$
R = total revenue from selling x units = xp	$\dfrac{dR}{dx}$ = marginal revenue
C = total cost of producing x units	$\dfrac{dC}{dx}$ = marginal cost
P = total profit from selling x units = $R - C$	$\dfrac{dP}{dx}$ = marginal profit
$\overline{C}$ = average cost per unit = $\dfrac{C}{x}$	

Demand Function
Quantity demanded increases
as price decreases.

Revenue Function
The low prices required to sell
more units eventually result in
a decreasing revenue.

Cost Function
The total cost to produce
x units includes the fixed cost.

FIGURE 3.45

Profit Function
The break-even point occurs
when $R = C$.

SUMMARIZE (Section 3.5)

1. Describe a real-life example of how optimization can be used to find the maximum revenue for a product *(page 205, Example 1)*.

2. State the definition of the average cost function *(page 206)*. For an example of an average cost function, see Example 2.

3. State the definition of price elasticity of demand *(page 209)*. For an example of price elasticity of demand, see Example 5.

SKILLS WARM UP 3.5

The following warm-up exercises involve skills that were covered in a previous course or earlier sections. You will use these skills in the exercise set for this section. For additional help, review Appendices A.2 and A.3, and Section 2.3.

In Exercises 1–4, evaluate the expression for $x = 150$.

1. $\left| -\dfrac{300}{x} + 3 \right|$ **2.** $\left| -\dfrac{600}{5x} + 2 \right|$ **3.** $\left| \dfrac{(20x^{-1/2})/x}{-10x^{-3/2}} \right|$ **4.** $\left| \dfrac{(4000/x^2)/x}{-8000x^{-3}} \right|$

In Exercises 5–10, find the marginal revenue, marginal cost, or marginal profit.

5. $C = 650 + 1.2x + 0.003x^2$ **6.** $P = 0.01x^2 + 11x$ **7.** $P = -0.3x^3 + 8x^2 - 4x + 11$

8. $C = 1200 + x + 0.03x^3$ **9.** $R = 14x - \dfrac{x^2}{2000}$ **10.** $R = 3.4x - \dfrac{x^2}{1500}$

Exercises 3.5

See *CalcChat.com* for tutorial help and worked-out solutions to odd-numbered exercises.

Finding the Maximum Revenue In Exercises 1–4, find the number of units x that maximizes the revenue R. *See Example 1.*

1. $R = 800x - 0.2x^2$ **2.** $R = 48x^2 - 0.02x^3$
3. $R = 900x + 222x^2 - x^3$ **4.** $R = 30x^{2/3} - 2x$

Finding the Minimum Average Cost In Exercises 5–8, find the number of units x that minimizes the average cost per unit $\overline{C}$. *See Example 2.*

5. $C = 0.125x^2 + 20x + 5000$
6. $C = 2x^2 + 348x + 7200$
7. $C = 0.001x^3 + 5x + 250$
8. $C = 0.02x^3 + 55x^2 + 1380$

Finding the Maximum Profit In Exercises 9–12, find the price that will maximize profit for the demand and cost functions, where p is the price, x is the number of units, and C is the cost. *See Example 4.*

Demand Function	Cost Function
9. $p = 90 - x$	$C = 100 + 30x$
10. $p = 70 - 0.01x$	$C = 8000 + 50x + 0.03x^2$
11. $p = 50 - 0.1\sqrt{x}$	$C = 35x + 500$
12. $p = \dfrac{24}{\sqrt{x}}$	$C = 0.4x + 600$

Average Cost In Exercises 13 and 14, use the cost function to find the production level at which the average cost is a minimum. For this production level, show that the marginal cost and average cost are equal. Use a graphing utility to graph the average cost function and verify your results.

13. $C = 2x^2 + 5x + 18$ **14.** $C = x^3 - 8x^2 + 20x$

15. Maximum Profit A commodity has a demand function modeled by

$$p = 80 - 0.2x$$

and a total cost function modeled by

$$C = 30x + 40$$

where x is the number of units.

(a) What price yields a maximum profit?

(b) When the profit is maximized, what is the average cost per unit?

16. Maximum Profit A commodity has a demand function modeled by $p = 100 - 0.5x$ and a total cost function modeled by $C = 50x + 37.5$, where x is the number of units.

(a) What price yields a maximum profit?

(b) When the profit is maximized, what is the average cost per unit?

Maximum Profit In Exercises 17 and 18, find the amount s spent on advertising (in thousands of dollars) that maximizes the profit P (in thousands of dollars). Then find the point of diminishing returns.

17. $P = -2s^3 + 35s^2 - 100s + 200$
18. $P = -0.1s^3 + 6s^2 + 400$

19. Maximum Profit The cost per unit of producing an MP3 player is $90. The manufacturer charges $150 per player for orders of 100 or less. To encourage large orders, however, the manufacturer reduces the charge by $0.10 per player for orders in excess of 100 players. For instance, an order of 101 players would be $149.90 per player, an order of 102 players would be $149.80 per player, and so on. Find the largest order the manufacturer should allow to obtain a maximum profit.

20. Maximum Profit A real estate office handles a 50-unit apartment complex. When the rent is $580 per month, all units are occupied. For each $40 increase in rent, however, an average of one unit becomes vacant. Each occupied unit requires an average of $45 per month for service and repairs. What rent should be charged to obtain a maximum profit?

21. Maximum Revenue When a wholesaler sold a product at $40 per unit, sales were 300 units per week. After a price increase of $5, however, the average number of units sold dropped to 275 per week. Assuming that the demand function is linear, what price per unit will maximize the weekly revenue?

22. Maximum Revenue A small theater has a seating capacity of 2000. When the ticket price is $20, attendance is 1500. After a price decrease of $2, however, the attendance increased to 1700. Assuming that the demand function is linear, what ticket price will maximize the revenue?

23. Minimum Cost A power station is on one side of a river that is 0.5 mile wide, and a factory is 6 miles downstream on the other side of the river (see figure). It costs $18 per foot to run overland power lines and $25 per foot to run underwater power lines. Write a cost function for running the power lines from the power station to the factory. Use a graphing utility to graph your function. Estimate the value of x that minimizes the cost. Explain your results.

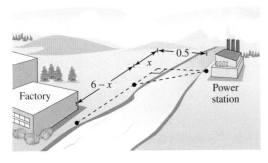

24. Minimum Cost An offshore oil well is 1 mile off the coast. The oil refinery is 2 miles down the coast. Laying pipe in the ocean is twice as expensive as laying it on land. Find the most economical path for the pipe from the well to the oil refinery.

Minimum Cost In Exercises 25 and 26, find the speed v, in miles per hour, that will minimize costs on a 110-mile delivery trip. The cost per hour for fuel is C dollars, and the driver is paid W dollars per hour. (Assume there are no costs other than wages and fuel.)

25. Fuel cost: $C = \dfrac{v^2}{310}$; Driver: $W = \$13$

26. Fuel cost: $C = \dfrac{v^2}{250}$; Driver: $W = \$18.50$

Elasticity In Exercises 27–32, find the price elasticity of demand for the demand function at the indicated x-value. Is the demand elastic, inelastic, or of unit elasticity at the indicated x-value? Use a graphing utility to graph the revenue function, and identify the intervals of elasticity and inelasticity.

Demand Function	Quantity Demanded
27. $p = 600 - 5x$	$x = 60$
28. $p = 400 - 3x$	$x = 20$
29. $p = 5 - 0.03x$	$x = 100$
30. $p = 15 - 0.0003x$	$x = 25{,}000$
31. $p = \dfrac{500}{x + 2}$	$x = 23$
32. $p = \dfrac{500}{x^2} + 5$	$x = 5$

33. Elasticity The demand function for a product is modeled by

$$p = 20 - 0.02x, \quad 0 \le x \le 1000$$

where p is the price per unit (in dollars) and x is the number of units.

(a) Determine when the demand is elastic, inelastic, and of unit elasticity.

(b) Use the result of part (a) to describe the behavior of the revenue function.

34. Elasticity The demand function for a product is modeled by $p = 800 - 4x$, $0 \le x \le 200$, where p is the price per unit (in dollars) and x is the number of units.

(a) Determine when the demand is elastic, inelastic, and of unit elasticity.

(b) Use the result of part (a) to describe the behavior of the revenue function.

35. Minimum Cost The shipping and handling cost C (in thousands of dollars) of a manufactured product is modeled by

$$C = 4\left(\frac{25}{x^2} - \frac{x}{x - 10}\right), \quad 0 < x < 10$$

where x is the number of units shipped (in hundreds). Use the *minimum* feature of a graphing utility to find the shipment size that minimizes the cost.

36. Minimum Cost The ordering and transportation cost C (in thousands of dollars) of the components used in manufacturing a product is modeled by

$$C = 8\left(\frac{2500}{x^2} - \frac{x}{x - 100}\right), \quad 0 < x < 100$$

where x is the order size (in hundreds). Use the *minimum* feature of a graphing utility to find the order size that minimizes the cost.

37. Revenue The demand for a car wash is $x = 900 - 45p$, where the current price is \$8. Can revenue be increased by lowering the price and thus attracting more customers? Use price elasticity of demand to determine your answer.

38. HOW DO YOU SEE IT? Match each graph with the function it best represents—a demand function, a revenue function, a cost function, or a profit function. Explain your reasoning. (The graphs are labeled *a–d*.)

39. Sales The sales S (in billions of dollars) for Lockheed Martin for the years 2004 through 2013 can be modeled by

$$S = -0.0227t^3 + 0.401t^2 - 0.32t + 31.8, \ 4 \le t \le 13$$

where t represents the year, with $t = 4$ corresponding to 2004. *(Source: Lockheed Martin)*

(a) Find the years, from 2004 through 2013, when the maximum and minimum sales occurred.

(b) During which year were the company's sales increasing most rapidly?

(c) During which year were the company's sales decreasing most rapidly?

(d) Use a graphing utility to graph the sales function and verify the results in parts (a), (b), and (c).

40. Sales The sales S (in billions of dollars) for Kohl's Corporation from 2004 through 2013 can be modeled by

$$S = \frac{26.287t - 41.07}{1.122t + 1}, \ \ 4 \le t \le 13$$

where t represents the year, with $t = 4$ corresponding to 2004. *(Source: Kohl's Corporation)*

(a) Find the years, from 2004 through 2013, when the maximum and minimum sales occurred.

(b) During which year were the sales increasing at the greatest rate?

(c) Use a graphing utility to graph the sales function and verify the results in parts (a) and (b).

41. Demand A demand function is modeled by

$$x = \frac{a}{p^m}$$

where a is a constant and $m > 1$. Show that $\eta = -m$. In other words, show that a 1% increase in price results in an m% decrease in the quantity demanded.

42. Think About It Throughout this text, it is assumed that demand functions are decreasing. Can you think of a product that has an increasing demand function? That is, can you think of a product that becomes more in demand as its price increases? Explain your reasoning, and sketch a graph of the function.

Business Capsule

The website vWorker.com is a marketplace that links freelance workers with employers looking to outsource jobs. The site's founder, Ian Ippolito, noticed that more and more companies were outsourcing rather than hiring full-time employees. He saw the potential in this trend, borrowed \$5000 from his parents, and turned the capital into an online business in 2001. Today his company is known as vWorker.com—for virtual worker—and it connects more than 150,000 employers with over 300,000 workers worldwide. Posting and bidding are free, but vWorker takes a percentage of the workers' final earnings. Annual revenues reached \$11.1 million in 2012. Ippolito sold the company at the end of 2012 to Freelancer.com.

43. Research Project Choose a company with an innovative product or service like the one described above. Use your school's library, the Internet, or some other reference source to research the history of the company. Collect data about the revenue that the product or service has generated, and find a mathematical model for the data. Summarize your findings.

3.6 Asymptotes

■ Find the vertical asymptotes of functions and find infinite limits.
■ Find the horizontal asymptotes of functions and find limits at infinity.
■ Use asymptotes to answer questions about real-life situations.

Vertical Asymptotes and Infinite Limits

In the first three sections of this chapter, you studied ways in which you can use calculus to help analyze the graph of a function. In this section, you will study another valuable aid to curve sketching: the determination of vertical and horizontal asymptotes.

Recall from Section 1.5, Example 11, that the function

$$f(x) = \frac{3}{x-2}$$

is unbounded as x approaches 2 (see Figure 3.46).

Average Cost

In Exercise 63 on page 225, you will use limits at infinity to find the limit of an average cost function as the number of units produced increases.

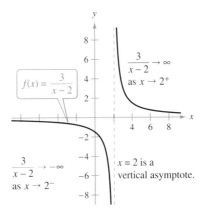

$$f(x) = \frac{3}{x-2}$$

$$\frac{3}{x-2} \to \infty$$
as $x \to 2^+$

$$\frac{3}{x-2} \to -\infty$$
as $x \to 2^-$

$x = 2$ is a vertical asymptote.

FIGURE 3.46

This type of behavior is described by saying that the line

$$x = 2 \qquad \text{Vertical asymptote}$$

is a **vertical asymptote** of the graph of f. A limit in which $f(x)$ approaches infinity (or negative infinity) as x approaches c from the left or from the right is an **infinite limit.** The infinite limits for the function $f(x) = 3/(x-2)$ can be written as

$$\lim_{x \to 2^-} \frac{3}{x-2} = -\infty \qquad \text{$f(x)$ decreases without bound as x approaches 2 from the left.}$$

and

$$\lim_{x \to 2^+} \frac{3}{x-2} = \infty. \qquad \text{$f(x)$ increases without bound as x approaches 2 from the right.}$$

Definition of Vertical Asymptote

If $f(x)$ approaches infinity (or negative infinity) as x approaches c from the right or from the left, then the line

$$x = c$$

is a **vertical asymptote** of the graph of f.

One of the most common instances of a vertical asymptote is the graph of a *rational function*—that is, a function of the form $f(x) = p(x)/q(x)$, where $p(x)$ and $q(x)$ are polynomials. If c is a real number such that $q(c) = 0$ and $p(c) \neq 0$, then the graph of f has a vertical asymptote at $x = c$. Example 1 shows four cases.

TECH TUTOR

Use a spreadsheet or table to verify the results shown in Example 1. (Consult the user's manual of a spreadsheet software program for specific instructions on how to create a table.) For instance, in Example 1(a), notice that the values of $f(x) = 1/(x - 1)$ decrease and increase without bound as x gets closer and closer to 1 from the left and the right.

x Approaches 1 from the Left

x	$f(x) = 1/(x - 1)$
0	-1
0.9	-10
0.99	-100
0.999	-1000
0.9999	$-10,000$

x Approaches 1 from the Right

x	$f(x) = 1/(x - 1)$
2	1
1.1	10
1.01	100
1.001	1000
1.0001	10,000

EXAMPLE 1   Finding Infinite Limits

Limit from the left   *Limit from the right*

a. $\displaystyle\lim_{x \to 1^-} \frac{1}{x - 1} = -\infty$ $\displaystyle\lim_{x \to 1^+} \frac{1}{x - 1} = \infty$ See Figure 3.47(a).

b. $\displaystyle\lim_{x \to 1^-} \frac{-1}{x - 1} = \infty$ $\displaystyle\lim_{x \to 1^+} \frac{-1}{x - 1} = -\infty$ See Figure 3.47(b).

c. $\displaystyle\lim_{x \to 1^-} \frac{-1}{(x - 1)^2} = -\infty$ $\displaystyle\lim_{x \to 1^+} \frac{-1}{(x - 1)^2} = -\infty$ See Figure 3.47(c).

d. $\displaystyle\lim_{x \to 1^-} \frac{1}{(x - 1)^2} = \infty$ $\displaystyle\lim_{x \to 1^+} \frac{1}{(x - 1)^2} = \infty$ See Figure 3.47(d).

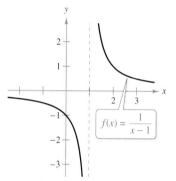

$$\lim_{x \to 1^-} \frac{1}{x - 1} = -\infty \qquad \lim_{x \to 1^+} \frac{1}{x - 1} = \infty$$

(a)

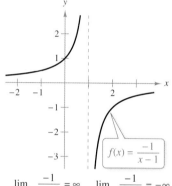

$$\lim_{x \to 1^-} \frac{-1}{x - 1} = \infty \qquad \lim_{x \to 1^+} \frac{-1}{x - 1} = -\infty$$

(b)

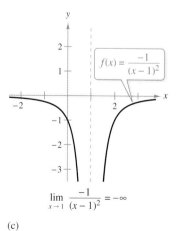

$$\lim_{x \to 1} \frac{-1}{(x - 1)^2} = -\infty$$

(c)

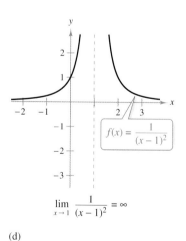

$$\lim_{x \to 1} \frac{1}{(x - 1)^2} = \infty$$

(d)

FIGURE 3.47

✔ ***Checkpoint 1***   *Worked-out solution available at LarsonAppliedCalculus.com*

Find each limit.

a. $\displaystyle\lim_{x \to 2^-} \frac{1}{x - 2}$ **b.** $\displaystyle\lim_{x \to 2^+} \frac{1}{x - 2}$ **c.** $\displaystyle\lim_{x \to -3^-} \frac{1}{x + 3}$ **d.** $\displaystyle\lim_{x \to -3^+} \frac{1}{x + 3}$ ■

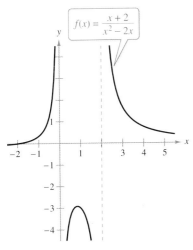

Vertical Asymptotes at $x = 0$ and $x = 2$

FIGURE 3.48

Each of the graphs in Example 1 has only one vertical asymptote. As shown in the next example, the graph of a rational function can have more than one vertical asymptote.

EXAMPLE 2   **Finding Vertical Asymptotes**

Determine all vertical asymptotes of the graph of

$$f(x) = \frac{x + 2}{x^2 - 2x}.$$

SOLUTION   The possible vertical asymptotes correspond to the x-values for which the denominator is zero.

$x^2 - 2x = 0$	Set denominator equal to 0.
$x(x - 2) = 0$	Factor.
$x = 0, x = 2$	Zeros of denominator

Because the numerator of f is not zero at either of these x-values, you can conclude that the graph of f has two vertical asymptotes—one at $x = 0$ and one at $x = 2$, as shown in Figure 3.48.

✓ **Checkpoint 2**   *Worked-out solution available at LarsonAppliedCalculus.com*

Determine all vertical asymptotes of the graph of

$$f(x) = \frac{x + 4}{x^2 - 4x}.$$

EXAMPLE 3   **Finding Vertical Asymptotes**

Determine all vertical asymptotes of the graph of

$$f(x) = \frac{x^2 + 2x - 8}{x^2 - 4}.$$

SOLUTION   First factor the numerator and denominator. Then divide out common factors.

$f(x) = \dfrac{x^2 + 2x - 8}{x^2 - 4}$	Write original function.
$= \dfrac{(x + 4)(x - 2)}{(x + 2)(x - 2)}$	Factor numerator and denominator.
$= \dfrac{(x + 4)(x - 2)}{(x + 2)(x - 2)}$	Divide out common factor.
$= \dfrac{x + 4}{x + 2},\ x \neq 2$	Simplify.

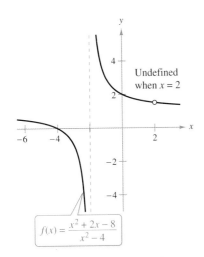

$f(x) = \dfrac{x^2 + 2x - 8}{x^2 - 4}$

Vertical Asymptote at $x = -2$

FIGURE 3.49

At all x-values other than $x = 2$, the graph of f coincides with the graph of the function $g(x) = (x + 4)/(x + 2)$. The only zero of the denominator is $x = -2$. Because the numerator is not zero when $x = -2$, you can conclude that the graph of f has a vertical asymptote at $x = -2$, as shown in Figure 3.49. Note that $x = 2$ is *not* a vertical asymptote, but there is a hole in the graph at $x = 2$ because $f(2)$ is undefined.

✓ **Checkpoint 3**   *Worked-out solution available at LarsonAppliedCalculus.com*

Determine all vertical asymptotes of the graph of

$$f(x) = \frac{x^2 + 4x + 3}{x^2 - 9}.$$

TECH TUTOR

When you use a graphing utility to graph a function that has a vertical asymptote, the utility may try to connect separate branches of the graph. For instance, the figure below shows the graph of

$$f(x) = \frac{3}{x - 2}$$

on a graphing calculator.

This line is not part of the graph of the function.

The graph of the function has two branches.

From Example 3, you know that the graph of

$$f(x) = \frac{x^2 + 2x - 8}{x^2 - 4}$$

has a vertical asymptote at $x = -2$. This implies that the limit of $f(x)$ as $x \to -2$ from the right (or from the left) is either ∞ or $-\infty$. But without looking at the graph, how can you determine that the limit from the left is *negative* infinity and the limit from the right is *positive* infinity? That is, why is the limit from the left

$$\lim_{x \to -2^-} \frac{x^2 + 2x - 8}{x^2 - 4} = -\infty \qquad \text{Limit from the left}$$

and why is the limit from the right

$$\lim_{x \to -2^+} \frac{x^2 + 2x - 8}{x^2 - 4} = \infty? \qquad \text{Limit from the right}$$

It is cumbersome to determine these limits analytically, and you may find the graphical method shown in Example 4 to be more efficient.

EXAMPLE 4   Determining Infinite Limits

Find the limits.

$$\lim_{x \to 1^-} \frac{x^2 - 3x}{x - 1} \quad \text{and} \quad \lim_{x \to 1^+} \frac{x^2 - 3x}{x - 1}$$

SOLUTION   Begin by considering the function

$$f(x) = \frac{x^2 - 3x}{x - 1}.$$

Because the denominator is zero when $x = 1$ and the numerator is not zero when $x = 1$, it follows that the graph of the function has a vertical asymptote at $x = 1$. This implies that each of the given limits is either ∞ or $-\infty$. To determine which, use a graphing utility to graph the function, as shown in Figure 3.50. From the graph, you can see that the limit from the left is positive infinity and the limit from the right is negative infinity. That is,

From the left, $f(x)$ approaches positive infinity.

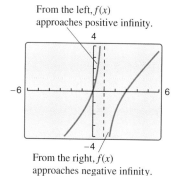

From the right, $f(x)$ approaches negative infinity.

FIGURE 3.50

$$\lim_{x \to 1^-} \frac{x^2 - 3x}{x - 1} = \infty \qquad \text{Limit from the left}$$

and

$$\lim_{x \to 1^+} \frac{x^2 - 3x}{x - 1} = -\infty. \qquad \text{Limit from the right}$$

✓ **Checkpoint 4**   *Worked-out solution available at LarsonAppliedCalculus.com*

Find the limits.

$$\lim_{x \to 2^-} \frac{x^2 - 4x}{x - 2} \quad \text{and} \quad \lim_{x \to 2^+} \frac{x^2 - 4x}{x - 2} \qquad \blacksquare$$

In Example 4, try evaluating f at x-values that are just barely to the left of 1. You will find that you can make $f(x)$ arbitrarily large by choosing x sufficiently close to 1. For instance, $f(0.99999) \approx 199,999$.

Horizontal Asymptotes and Limits at Infinity

Another type of limit, called a **limit at infinity,** specifies a finite value approached by a function as x increases (or decreases) without bound.

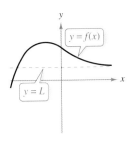

FIGURE 3.51

Definition of Horizontal Asymptote

If f is a function and L_1 and L_2 are real numbers, then the statements

$$\lim_{x \to \infty} f(x) = L_1 \quad \text{and} \quad \lim_{x \to -\infty} f(x) = L_2$$

denote **limits at infinity.** The lines $y = L_1$ and $y = L_2$ are **horizontal asymptotes** of the graph of f.

Figure 3.51 shows two ways in which the graph of a function can approach one or more horizontal asymptotes. Note that it is possible for the graph of a function to cross its horizontal asymptote.

Limits at infinity share many of the properties of limits discussed in Section 1.5. When finding horizontal asymptotes, you can use the property that

$$\lim_{x \to \infty} \frac{1}{x^r} = 0, \quad r > 0 \quad \text{and} \quad \lim_{x \to -\infty} \frac{1}{x^r} = 0, \quad r > 0.$$

(The second limit assumes that x^r is defined when $x < 0$.)

EXAMPLE 5   Finding Limits at Infinity

Find the limit: $\lim\limits_{x \to \infty} \left(5 - \dfrac{2}{x^2} \right)$.

SOLUTION

$$\lim_{x \to \infty} \left(5 - \frac{2}{x^2} \right) = \lim_{x \to \infty} 5 - \lim_{x \to \infty} \frac{2}{x^2} \qquad \lim_{x \to \infty} [f(x) - g(x)] = \lim_{x \to \infty} f(x) - \lim_{x \to \infty} g(x)$$

$$= \lim_{x \to \infty} 5 - 2 \left(\lim_{x \to \infty} \frac{1}{x^2} \right) \qquad \lim_{x \to \infty} cf(x) = c \lim_{x \to \infty} f(x)$$

$$= 5 - 2(0)$$

$$= 5$$

You can verify this limit by sketching the graph of

$$f(x) = 5 - \frac{2}{x^2}$$

as shown in the figure. Note that the graph has $y = 5$ as a horizontal asymptote to the right. By evaluating the limit

$$\lim_{x \to -\infty} \left(5 - \frac{2}{x^2} \right)$$

you can show that $y = 5$ is also a horizontal asymptote to the left.

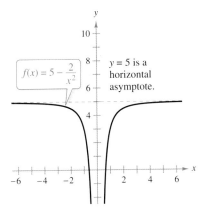

$y = 5$ is a horizontal asymptote.

$f(x) = 5 - \dfrac{2}{x^2}$

✓ **Checkpoint 5**

Find the limit: $\lim\limits_{x \to \infty} \left(2 + \dfrac{5}{x^2} \right)$.

There is an easy way to determine whether the graph of a *rational* function has a horizontal asymptote. This shortcut is based on a comparison of the degrees of the numerator and denominator of the rational function.

Horizontal Asymptotes of Rational Functions

Let $f(x) = p(x)/q(x)$ be a rational function.

1. If the degree of the numerator is less than the degree of the denominator, then $y = 0$ is a horizontal asymptote of the graph of f (to the left and to the right).

2. If the degree of the numerator is equal to the degree of the denominator, then $y = a/b$ is a horizontal asymptote of the graph of f (to the left and to the right), where a and b are the leading coefficients of $p(x)$ and $q(x)$, respectively.

3. If the degree of the numerator is greater than the degree of the denominator, then the graph of f has no horizontal asymptote.

EXAMPLE 6   Finding Horizontal Asymptotes

Find the horizontal asymptote of the graph of each function.

a. $y = \dfrac{-2x + 3}{3x^2 + 1}$ **b.** $y = \dfrac{-2x^2 + 3}{3x^2 + 1}$ **c.** $y = \dfrac{-2x^3 + 3}{3x^2 + 1}$

SOLUTION

a. Because the degree of the numerator is less than the degree of the denominator, $y = 0$ is a horizontal asymptote. [See Figure 3.52(a).]

b. Because the degree of the numerator is equal to the degree of the denominator, the line $y = -\frac{2}{3}$ is a horizontal asymptote. [See Figure 3.52(b).]

c. Because the degree of the numerator is greater than the degree of the denominator, the graph has no horizontal asymptote. [See Figure 3.52(c).]

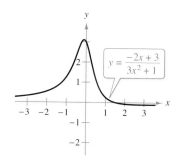
(a) $y = 0$ is a horizontal asymptote.

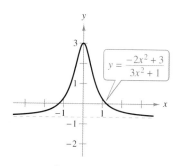
(b) $y = -\frac{2}{3}$ is a horizontal asymptote.

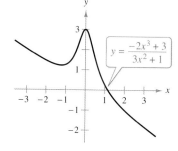
(c) No horizontal asymptote

FIGURE 3.52

✓ *Checkpoint 6*   *Worked-out solution available at LarsonAppliedCalculus.com*

Find the horizontal asymptote of the graph of each function.

a. $y = \dfrac{2x + 1}{4x^2 + 5}$ **b.** $y = \dfrac{2x^2 + 1}{4x^2 + 5}$ **c.** $y = \dfrac{2x^3 + 1}{4x^2 + 5}$

Some functions have two horizontal asymptotes: one to the right and one to the left (see Exercises 61 and 62).

Applications

There are many examples of asymptotic behavior in real life. For instance, Example 7 describes the asymptotic behavior of an average cost function.

EXAMPLE 7   Modeling Average Cost

A small business invests $5000 in a new product. In addition to this initial investment, the product will cost $0.50 per unit to produce.

a. Find the average cost per unit when 1000 units are produced.

b. Find the average cost per unit when 10,000 units are produced.

c. Find the average cost per unit when 100,000 units are produced.

d. What is the limit of the average cost as the number of units produced increases?

SOLUTION   From the given information, you can model the total cost C (in dollars) by

$$C = 0.5x + 5000 \qquad \text{Total cost function}$$

where x is the number of units produced. This implies that the average cost function is

$$\overline{C} = \frac{C}{x} = 0.5 + \frac{5000}{x}. \qquad \text{Average cost function}$$

a. When only 1000 units are produced, the average cost per unit is

$$\overline{C} = 0.5 + \frac{5000}{1000} \qquad \text{Substitute 1000 for } x.$$

$$= \$5.50. \qquad \text{Average cost for 1000 units}$$

b. When 10,000 units are produced, the average cost per unit is

$$\overline{C} = 0.5 + \frac{5000}{10,000} \qquad \text{Substitute 10,000 for } x.$$

$$= \$1.00. \qquad \text{Average cost for 10,000 units}$$

c. When 100,000 units are produced, the average cost per unit is

$$\overline{C} = 0.5 + \frac{5000}{100,000} \qquad \text{Substitute 100,000 for } x.$$

$$= \$0.55. \qquad \text{Average cost for 100,000 units}$$

d. As x approaches infinity, the limiting average cost per unit is

$$\lim_{x \to \infty} \left(0.5 + \frac{5000}{x} \right) = \$0.50.$$

Average Cost

$$\overline{C} = \frac{C}{x} = 0.5 + \frac{5000}{x}$$

Number of units

As $x \to \infty$, the average cost per unit approaches $0.50.

FIGURE 3.53

As shown in Figure 3.53, this example points out one of the major problems faced by small businesses. That is, it is difficult to have competitively low prices when the production level is low.

✓ *Checkpoint 7*   *Worked-out solution available at LarsonAppliedCalculus.com*

A small business invests $25,000 in a new product. In addition, the product will cost $0.75 per unit to produce. Find the cost function and the average cost function. What is the limit of the average cost function as the number of units produced increases?   ■

In Example 7, suppose that the small business had made an initial investment of $50,000. How would this change the answers to the questions? Would it change the average cost of producing x units? Would it change the limiting average cost per unit?

EXAMPLE 8   Modeling Smokestack Emission

A manufacturing plant has determined that the cost C (in dollars) of removing $p\%$ of the smokestack pollutants of its main smokestack is modeled by

$$C = \frac{80,000p}{100 - p}, \quad 0 \leq p < 100.$$

What is the vertical asymptote of this function? What does the vertical asymptote mean to the plant owners?

SOLUTION   The graph of the cost function is shown in Figure 3.54. From the graph, you can see that $p = 100$ is the vertical asymptote. This means that as the plant attempts to remove higher and higher percents of the pollutants, the cost increases dramatically. For instance, the cost of removing 85% of the pollutants is

$$C = \frac{80,000(85)}{100 - 85} \approx \$453,333 \qquad \text{Cost for 85\% removal}$$

but the cost of removing 90% is

$$C = \frac{80,000(90)}{100 - 90} = \$720,000. \qquad \text{Cost for 90\% removal}$$

FIGURE 3.54

✓ *Checkpoint 8*   Worked-out solution available at LarsonAppliedCalculus.com

According to the cost function in Example 8, is it possible to remove 100% of the smokestack pollutants? Why or why not?   ■

SUMMARIZE   (Section 3.6)

1. State the definition of vertical asymptote *(page 215)*. For examples of vertical asymptotes, see Examples 1, 2, and 3.

2. State the definition of horizontal asymptote *(page 219)*. For examples of horizontal asymptotes, see Example 6.

3. Describe a real-life example of how asymptotic behavior can be used to analyze the average cost for a new product *(page 221, Example 7)*.

SKILLS WARM UP 3.6

The following warm-up exercises involve skills that were covered in earlier sections. You will use these skills in the exercise set for this section. For additional help, review Sections 1.5, 2.3, and 3.5.

In Exercises 1–8, find the limit.

1. $\lim\limits_{x \to 2} (x + 1)$

2. $\lim\limits_{x \to -1} (3x + 4)$

3. $\lim\limits_{x \to -3} \dfrac{2x^2 + x - 15}{x + 3}$

4. $\lim\limits_{x \to 3} \dfrac{4x^2 - 9x - 9}{x - 3}$

5. $\lim\limits_{x \to 4^-} \dfrac{x^2 - 12x + 32}{x^2 - 16}$

6. $\lim\limits_{x \to 1^-} \dfrac{x^2 - 6x + 5}{x^2 - 1}$

7. $\lim\limits_{x \to 0^+} \sqrt{x}$

8. $\lim\limits_{x \to 1^+} \left(x + \sqrt{x - 1}\right)$

In Exercises 9–12, find the average cost function and the marginal cost function.

9. $C = 150 + 3x$

10. $C = 1900 + 1.7x + 0.002x^2$

11. $C = 0.005x^2 + 0.5x + 1375$

12. $C = 760 + 0.05x$

Exercises 3.6

See *CalcChat.com* for tutorial help and worked-out solutions to odd-numbered exercises.

Vertical and Horizontal Asymptotes In Exercises 1–6, find the vertical and horizontal asymptotes.

1. $f(x) = \dfrac{x^2 + 1}{x^2}$

2. $f(x) = \dfrac{4}{(x - 2)^3}$

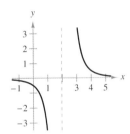

3. $f(x) = \dfrac{x^2 - 2}{x^2 - x - 2}$

4. $f(x) = \dfrac{x^2 - 1}{2x^2 - 8}$

5. $f(x) = \dfrac{3x^2}{2(x^2 + 1)}$

6. $f(x) = \dfrac{-4x}{x^2 + 4}$

Finding Infinite Limits In Exercises 7–12, find the limit. *See Example 1.*

7. $\lim\limits_{x \to 4^+} \dfrac{1}{x - 4}$

8. $\lim\limits_{x \to -5^-} \dfrac{1}{x + 5}$

9. $\lim\limits_{x \to -6^+} \dfrac{-4}{x + 6}$

10. $\lim\limits_{x \to 3^-} \dfrac{-0.5}{x - 3}$

11. $\lim\limits_{x \to 2^-} \dfrac{1}{(x - 2)^2}$

12. $\lim\limits_{x \to -4^+} \dfrac{1}{(x + 4)^2}$

Finding Vertical Asymptotes In Exercises 13–18, determine all vertical asymptotes of the graph of the function. *See Examples 2 and 3.*

13. $f(x) = \dfrac{x - 3}{x^2 + 3x}$

14. $f(x) = \dfrac{x + 8}{x^2 + 7x}$

15. $f(x) = \dfrac{x^2 - 8x + 15}{x^2 - 9}$

16. $f(x) = \dfrac{x^2 + 2x - 35}{x^2 - 25}$

17. $f(x) = \dfrac{2x^2 - x - 3}{2x^2 - 11x + 12}$

18. $f(x) = \dfrac{x^2 + x - 30}{4x^2 - 17x - 15}$

Determining Infinite Limits In Exercises 19–24, use a graphing utility to find the limit. *See Example 4.*

19. $\lim\limits_{x \to 6^+} \dfrac{1}{(x - 6)^2}$

20. $\lim\limits_{x \to -2^-} \dfrac{1}{x + 2}$

21. $\lim\limits_{x \to 3^+} \dfrac{x - 4}{x - 3}$

22. $\lim\limits_{x \to 1^+} \dfrac{2 + x}{1 - x}$

23. $\lim\limits_{x \to -1^-} \dfrac{x^2 + 1}{x^2 - 1}$

24. $\lim\limits_{x \to 5^+} \dfrac{2x - 3}{x^2 - 25}$

Finding Limits at Infinity In Exercises 25–28, find the limit. *See Example 5.*

25. $\lim\limits_{x \to \infty} \left(1 + \dfrac{1}{x}\right)$

26. $\lim\limits_{x \to -\infty} \left(3 + \dfrac{4}{x^2}\right)$

27. $\lim\limits_{x \to -\infty} \left(\dfrac{5}{x^3} - 8\right)$

28. $\lim\limits_{x \to \infty} \left(\dfrac{3}{x^4} - 9\right)$

Finding Horizontal Asymptotes In Exercises 29–36, find the horizontal asymptote of the graph of the function. *See Example 6.*

29. $f(x) = \dfrac{4x - 3}{2x + 1}$

30. $f(x) = \dfrac{5x^2 + 1}{10x^3 - 3x^2 + 7}$

31. $f(x) = \dfrac{3x - 7}{4 - x - 2x^2}$

32. $f(x) = \dfrac{2x^2 - 5x - 12}{1 - 6x - 8x^2}$

33. $f(x) = \dfrac{5x^2}{x + 3}$

34. $f(x) = \dfrac{x^3 - 2x^2 + 3x + 1}{x^2 - 3x + 2}$

35. $f(x) = \dfrac{2x}{x - 1} + \dfrac{3x}{x + 1}$

36. $f(x) = \dfrac{4x^2}{x^3 - 1} - \dfrac{3x}{2x + 1}$

Using Horizontal Asymptotes In Exercises 37–40, match the function with its graph. Use horizontal asymptotes as an aid. [The graphs are labeled (a)–(d).]

(a)

(b)

(c)

(d)

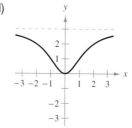

37. $f(x) = \dfrac{3x^2}{x^2 + 2}$

38. $f(x) = \dfrac{x}{x^2 + 2}$

39. $f(x) = 2 + \dfrac{x^2}{x^4 + 1}$

40. $f(x) = 5 - \dfrac{1}{x^2 + 1}$

Finding Limits at Infinity In Exercises 41 and 42, find $\lim\limits_{x \to \infty} h(x)$, if possible.

41. $f(x) = 5x^3 - 3$

  (a) $h(x) = \dfrac{f(x)}{x^2}$ (b) $h(x) = \dfrac{f(x)}{x^3}$ (c) $h(x) = \dfrac{f(x)}{x^4}$

42. $f(x) = -4x^2 + 2x - 5$

  (a) $h(x) = \dfrac{f(x)}{x}$ (b) $h(x) = \dfrac{f(x)}{x^2}$ (c) $h(x) = \dfrac{f(x)}{x^3}$

Finding Limits at Infinity In Exercises 43 and 44, find each limit, if possible.

43. (a) $\lim\limits_{x \to \infty} \dfrac{x^2 + 2}{x^3 - 1}$ (b) $\lim\limits_{x \to \infty} \dfrac{x^2 + 2}{x^2 - 1}$ (c) $\lim\limits_{x \to \infty} \dfrac{x^2 + 2}{x - 1}$

44. (a) $\lim\limits_{x \to \infty} \dfrac{4 - 5x}{2x^3 + 6}$ (b) $\lim\limits_{x \to \infty} \dfrac{4 - 5x}{2x + 6}$ (c) $\lim\limits_{x \to \infty} \dfrac{4 - 5x^2}{2x + 6}$

Estimating Limits at Infinity In Exercises 45–48, use a graphing utility or spreadsheet software program to complete the table. Then use the result to estimate the limit of $f(x)$ as x approaches infinity.

x	10^0	10^1	10^2	10^3	10^4	10^5	10^6
$f(x)$							

45. $f(x) = \sqrt{x^3 + 6} - 2x$

46. $f(x) = x - \sqrt{x(x - 1)}$

47. $f(x) = \dfrac{x + 1}{x\sqrt{x}}$

48. $f(x) = \dfrac{\sqrt{x}}{x^2 + 3}$

Sketching Graphs In Exercises 49–62, sketch the graph of the equation. Use intercepts, extrema, and asymptotes as sketching aids.

49. $y = \dfrac{3x}{1 - x}$

50. $y = \dfrac{x - 3}{x - 2}$

51. $f(x) = \dfrac{x^2}{x^2 + 9}$

52. $f(x) = \dfrac{x}{x^2 + 4}$

53. $g(x) = \dfrac{x^2}{x^2 - 16}$

54. $g(x) = \dfrac{x}{x^2 - 36}$

55. $y = 1 - \dfrac{3}{x^2}$

56. $y = 1 + \dfrac{1}{x}$

57. $g(x) = \dfrac{x^2 - x - 2}{x - 2}$

58. $g(x) = \dfrac{x^2 - 9}{x + 3}$

59. $y = \dfrac{2x^2 - 6}{(x - 1)^2}$

60. $y = \dfrac{x}{(x + 1)^2}$

61. $y = \dfrac{x}{\sqrt{x^2 + 1}}$

62. $y = \dfrac{2x}{\sqrt{x^2 + 4}}$

63. Average Cost The cost C (in dollars) of producing x units of a product is $C = 1.15x + 6000$.

(a) Find the average cost function $\overline{C}$.

(b) Find $\overline{C}$ when $x = 600$ and when $x = 6000$.

(c) Determine the limit of the average cost function as x approaches infinity. Interpret the limit in the context of the problem.

64. Average Cost The cost C (in dollars) for a company to recycle x tons of material is $C = 1.25x + 10{,}500$.

(a) Find the average cost function $\overline{C}$.

(b) Find $\overline{C}$ when $x = 100$ and when $x = 1000$.

(c) Determine the limit of the average cost function as x approaches infinity. Interpret the limit in the context of the problem.

65. Average Profit The cost C and revenue R functions (in dollars) for producing and selling x units of a product are $C = 34.5x + 15{,}000$ and $R = 69.9x$.

(a) Find the average profit function

$$\overline{P} = \frac{R - C}{x}.$$

(b) Find the average profits when x is 1000, 10,000, and 100,000.

(c) What is the limit of the average profit function as x approaches infinity? Explain your reasoning.

66. HOW DO YOU SEE IT? The graph shows the temperature T (in degrees Fahrenheit) of an apple pie t seconds after it is removed from an oven.

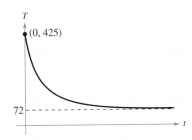

(a) Find $\lim\limits_{t \to 0^+} T$. What does this limit represent?

(b) Find $\lim\limits_{t \to \infty} T$. What does this limit represent?

67. Seizing Drugs The cost C (in millions of dollars) for the federal government to seize $p\%$ of an illegal drug as it enters the country is modeled by

$$C = \frac{528p}{100 - p}, \quad 0 \le p < 100.$$

(a) Find the costs of seizing 25%, 50%, and 75%.

(b) Find the limit of C as $p \to 100^-$. Interpret the limit in the context of the problem. Use a graphing utility to verify your result.

68. Recycling Bins The cost C (in dollars) of supplying recycling bins to $p\%$ of the population in a rural township is modeled by

$$C = \frac{26{,}000p}{100 - p}, \quad 0 \le p < 100.$$

(a) Find the costs of supplying bins to 20%, 50%, and 95% of the population.

(b) Find the limit of C as $p \to 100^-$. Interpret the limit in the context of the problem. Use a graphing utility to verify your result.

69. Learning Curve Psychologists have developed mathematical models to predict performance P (the percent of correct responses in decimal form) as a function of n, the number of times a task is performed. One such model is

$$P = \frac{0.5 + 0.9(n - 1)}{1 + 0.9(n - 1)}, \quad n > 0.$$

(a) Use a spreadsheet software program to complete the table for the model.

n	1	2	3	4	5	6	7	8	9	10
P										

(b) Find the limit of P as n approaches infinity.

(c) Use a graphing utility to graph this learning curve, and interpret the graph in the context of the problem.

70. Project: Alternative-Fueled Vehicles For a project analyzing the number of alternative-fueled vehicles in use in the United States, visit this text's website at *LarsonAppliedCalculus.com* (Source: U.S. Energy Information Administration)

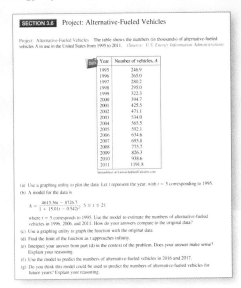

3.7 Curve Sketching: A Summary

■ Analyze and sketch the graphs of functions.
■ Recognize the graphs of simple polynomial functions.

Summary of Curve-Sketching Techniques

It would be difficult to overstate the importance of using graphs in mathematics. Descartes's introduction of analytic geometry contributed significantly to the rapid advances in calculus that began during the mid-seventeenth century.

So far, you have studied several concepts that are useful in analyzing the graph of a function.

In Exercise 45 on page 234, you will analyze the graph of the average monthly Social Security benefits to determine whether the model is a good fit for the data.

• x-intercepts and y-intercepts	(Section 1.2)
• Domain and range	(Section 1.4)
• Continuity	(Section 1.6)
• Differentiability	(Section 2.1)
• Relative extrema	(Section 3.2)
• Concavity	(Section 3.3)
• Points of inflection	(Section 3.3)
• Vertical asymptotes	(Section 3.6)
• Horizontal asymptotes	(Section 3.6)

When you are sketching the graph of a function, either by hand or with a graphing utility, remember that you cannot normally show the *entire* graph. The decision as to which part of the graph to show is crucial. For instance, which of the viewing windows in Figure 3.55 better represents the graph of

$$f(x) = x^3 - 25x^2 + 74x - 20?$$

Figure 3.55(a) gives a more complete view of the graph, but the context of the problem might indicate that Figure 3.55(b) is better.

(a)

(b)

FIGURE 3.55

Guidelines for Analyzing the Graph of a Function

1. Determine the domain and range of the function. When the function models a real-life situation, consider the context.

2. Determine the intercepts and asymptotes of the graph.

3. Locate the x-values at which $f'(x)$ and $f''(x)$ are zero or undefined. Use the results to determine where the relative extrema and the points of inflection occur.

Analyze and sketch the graph of

$$f(x) = x^3 + 3x^2 - 9x + 5.$$

SOLUTION   The domain of f is all real numbers, and its range is all real numbers. The y-intercept occurs at $(0, 5)$. Because this function factors as

$$f(x) = (x - 1)^2(x + 5) \qquad \text{Factored form}$$

the x-intercepts occur at $(-5, 0)$ and $(1, 0)$. The first derivative is

$$f'(x) = 3x^2 + 6x - 9 \qquad \text{First derivative}$$
$$= 3(x - 1)(x + 3). \qquad \text{Factored form}$$

So, the critical numbers of f are $x = 1$ and $x = -3$. The second derivative of f is

$$f''(x) = 6x + 6 \qquad \text{Second derivative}$$
$$= 6(x + 1). \qquad \text{Factored form}$$

which implies that the second derivative is zero when $x = -1$. By testing $f'(x)$ and $f''(x)$, as shown in the table, you can see that f has one relative minimum, one relative maximum, and one point of inflection. The graph of f is shown in Figure 3.56.

	$f(x)$	$f'(x)$	$f''(x)$	Characteristics of graph
x in $(-\infty, -3)$		$+$	$-$	Increasing, concave downward
$x = -3$	32	0	$-$	Relative maximum
x in $(-3, -1)$		$-$	$-$	Decreasing, concave downward
$x = -1$	16	$-$	0	Point of inflection
x in $(-1, 1)$		$-$	$+$	Decreasing, concave upward
$x = 1$	0	0	$+$	Relative minimum
x in $(1, \infty)$		$+$	$+$	Increasing, concave upward

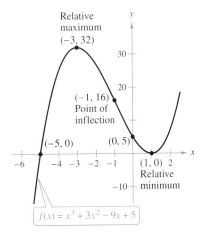

FIGURE 3.56

TECH TUTOR

In Example 1, you are able to find the zeros of f, f', and f'' algebraically (by factoring). When this is not feasible, you can use a graphing utility to find the zeros. For instance, the function

$$g(x) = x^3 + 3x^2 - 9x + 6$$

is similar to the function in the example, but it does not factor with integer coefficients. Using a graphing utility, you can determine that the function has only one x-intercept, $x \approx -5.0275$.

✓ **Checkpoint 1**   *Worked-out solution available at LarsonAppliedCalculus.com*

Analyze and sketch the graph of $f(x) = -x^3 + 3x^2 + 9x - 27$. ■

EXAMPLE 2   Analyzing a Graph

Analyze and sketch the graph of

$$f(x) = x^4 - 12x^3 + 48x^2 - 64x.$$

SOLUTION   The domain of f is all real numbers, and its range is $[-27, \infty)$. One of the intercepts occurs at $(0, 0)$. Because this function factors as

$$f(x) = x(x^3 - 12x^2 + 48x - 64)$$
$$= x(x - 4)^3 \qquad \text{Factored form}$$

a second x-intercept occurs at $(4, 0)$. The first derivative is

$$f'(x) = 4x^3 - 36x^2 + 96x - 64 \qquad \text{First derivative}$$
$$= 4(x - 1)(x - 4)^2. \qquad \text{Factored form}$$

So, the critical numbers of f are $x = 1$ and $x = 4$. The second derivative is

$$f''(x) = 12x^2 - 72x + 96 \qquad \text{Second derivative}$$
$$= 12(x - 4)(x - 2) \qquad \text{Factored form}$$

which implies that the second derivative is zero when $x = 2$ and $x = 4$. By testing $f'(x)$ and $f''(x)$, as shown in the table, you can see that f has one relative minimum and two points of inflection. The graph of f is shown in Figure 3.57.

	$f(x)$	$f'(x)$	$f''(x)$	Characteristics of graph
x in $(-\infty, 1)$		$-$	$+$	Decreasing, concave upward
$x = 1$	-27	0	$+$	Relative minimum
x in $(1, 2)$		$+$	$+$	Increasing, concave upward
$x = 2$	-16	$+$	0	Point of inflection
x in $(2, 4)$		$+$	$-$	Increasing, concave downward
$x = 4$	0	0	0	Point of inflection
x in $(4, \infty)$		$+$	$+$	Increasing, concave upward

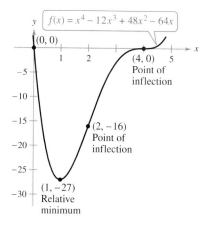

FIGURE 3.57

✓ **Checkpoint 2**   *Worked-out solution available at LarsonAppliedCalculus.com*

Analyze and sketch the graph of $f(x) = x^4 - 4x^3 + 5$.

The fourth-degree polynomial function in Example 2 has one relative minimum and no relative maxima. In general, a polynomial function of degree n can have *at most* $n - 1$ relative extrema, and *at most* $n - 2$ points of inflection. Moreover, polynomial functions of even degree must have at least one relative extremum.

EXAMPLE 3   Analyzing a Graph

Analyze and sketch the graph of

$$f(x) = \frac{x^2 - 2x + 4}{x - 2}.$$

SOLUTION   The domain of f is all real numbers except $x = 2$. The range of f is $(-\infty, -2] \cup [4, \infty)$. The y-intercept occurs at $(0, -2)$. Using the Quadratic Formula on the numerator, you can see that there are no x-intercepts. Because the denominator is zero when $x = 2$ (and the numerator is not zero when $x = 2$), it follows that $x = 2$ is a vertical asymptote of the graph. There are no horizontal asymptotes because the degree of the numerator is greater than the degree of the denominator. The first derivative is

$$f'(x) = \frac{(x - 2)(2x - 2) - (x^2 - 2x + 4)}{(x - 2)^2} \qquad \text{First derivative}$$

$$= \frac{x(x - 4)}{(x - 2)^2}. \qquad \text{Factored form}$$

So, the critical numbers of f are $x = 0$ and $x = 4$. The second derivative is

$$f''(x) = \frac{(x - 2)^2(2x - 4) - (x^2 - 4x)(2)(x - 2)}{(x - 2)^4} \qquad \text{Second derivative}$$

$$= \frac{(x - 2)(2x^2 - 8x + 8 - 2x^2 + 8x)}{(x - 2)^4}$$

$$= \frac{8}{(x - 2)^3}. \qquad \text{Factored form}$$

Because the second derivative has no zeros and because $x = 2$ is not in the domain of the function, you can conclude that the graph has no points of inflection. By testing $f'(x)$ and $f''(x)$, as shown in the table, you can see that f has one relative minimum and one relative maximum. The graph of f is shown in Figure 3.58.

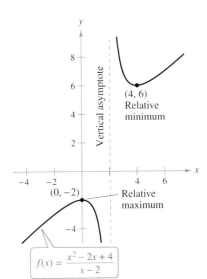

$$f(x) = \frac{x^2 - 2x + 4}{x - 2}$$

FIGURE 3.58

	$f(x)$	$f'(x)$	$f''(x)$	Characteristics of graph
x in $(-\infty, 0)$		$+$	$-$	Increasing, concave downward
$x = 0$	-2	0	$-$	Relative maximum
x in $(0, 2)$		$-$	$-$	Decreasing, concave downward
$x = 2$	Undef.	Undef.	Undef.	Vertical asymptote
x in $(2, 4)$		$-$	$+$	Decreasing, concave upward
$x = 4$	6	0	$+$	Relative minimum
x in $(4, \infty)$		$+$	$+$	Increasing, concave upward

✔ **Checkpoint 3**   *Worked-out solution available at LarsonAppliedCalculus.com*

Analyze and sketch the graph of

$$f(x) = \frac{x^2}{x - 1}.$$

EXAMPLE 4   Analyzing a Graph

Analyze and sketch the graph of

$$f(x) = \frac{2(x^2 - 9)}{x^2 - 4}.$$

SOLUTION   Begin by writing the function in factored form.

$$f(x) = \frac{2(x - 3)(x + 3)}{(x - 2)(x + 2)} \qquad \text{Factored form}$$

The domain of f is all real numbers except $x = -2$ and $x = 2$. The range of f is $(-\infty, 2) \cup \left[\frac{9}{2}, \infty\right)$. The y-intercept is $\left(0, \frac{9}{2}\right)$, and the x-intercepts are $(-3, 0)$ and $(3, 0)$. The graph of f has vertical asymptotes at $x = \pm 2$ and a horizontal asymptote at $y = 2$. The first derivative is

$$f'(x) = \frac{2[(x^2 - 4)(2x) - (x^2 - 9)(2x)]}{(x^2 - 4)^2} \qquad \text{First derivative}$$

$$= \frac{2(2x^3 - 8x - 2x^3 + 18x)}{(x^2 - 4)^2} \qquad \text{Multiply.}$$

$$= \frac{20x}{(x^2 - 4)^2}. \qquad \text{Factored form}$$

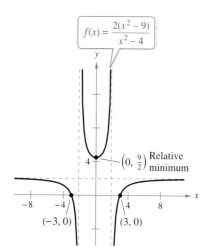

$f(x) = \dfrac{2(x^2 - 9)}{x^2 - 4}$

$\left(0, \frac{9}{2}\right)$ Relative minimum

$(-3, 0)$ $(3, 0)$

FIGURE 3.59

So, the critical number of f is $x = 0$. The second derivative is

$$f''(x) = \frac{(x^2 - 4)^2(20) - (20x)(2)(x^2 - 4)(2x)}{(x^2 - 4)^4} \qquad \text{Second derivative}$$

$$= \frac{20(x^2 - 4)(x^2 - 4 - 4x^2)}{(x^2 - 4)^4}$$

$$= \frac{20(3x^2 + 4)}{(x^2 - 4)^3}. \qquad \text{Factored form}$$

Because the second derivative has no zeros and $x = \pm 2$ are not in the domain of the function, you can conclude that the graph has no points of inflection. By testing $f'(x)$ and $f''(x)$, as shown in the table, you can see that f has one relative minimum. The graph of f is shown in Figure 3.59.

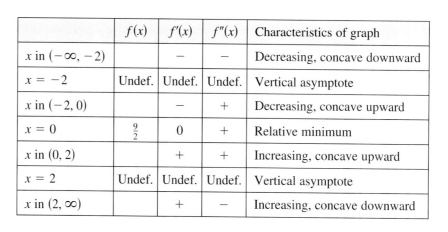

	$f(x)$	$f'(x)$	$f''(x)$	Characteristics of graph
x in $(-\infty, -2)$		−	−	Decreasing, concave downward
$x = -2$	Undef.	Undef.	Undef.	Vertical asymptote
x in $(-2, 0)$		−	+	Decreasing, concave upward
$x = 0$	$\frac{9}{2}$	0	+	Relative minimum
x in $(0, 2)$		+	+	Increasing, concave upward
$x = 2$	Undef.	Undef.	Undef.	Vertical asymptote
x in $(2, \infty)$		+	−	Increasing, concave downward

✓ **Checkpoint 4**   *Worked-out solution available at LarsonAppliedCalculus.com*

Analyze and sketch the graph of

$$f(x) = \frac{x^2 + 1}{x^2 - 1}.$$

EXAMPLE 5   Analyzing a Graph

Analyze and sketch the graph of

$$f(x) = 2x^{5/3} - 5x^{4/3}.$$

SOLUTION   Begin by writing the function in factored form.

$$f(x) = x^{4/3}(2x^{1/3} - 5) \qquad \text{Factored form}$$

The domain of f is all real numbers, and its range is all real numbers. One of the intercepts is $(0, 0)$. A second x-intercept occurs when $2x^{1/3} - 5 = 0$.

$$2x^{1/3} - 5 = 0$$
$$2x^{1/3} = 5$$
$$x^{1/3} = \frac{5}{2}$$
$$x = \left(\frac{5}{2}\right)^3$$
$$x = \frac{125}{8}$$

So, the second x-intercept occurs at $\left(\frac{125}{8}, 0\right)$. The first derivative is

$$f'(x) = \frac{10}{3}x^{2/3} - \frac{20}{3}x^{1/3} \qquad \text{First derivative}$$
$$= \frac{10}{3}x^{1/3}(x^{1/3} - 2). \qquad \text{Factored form}$$

So, the critical numbers of f are $x = 0$ and $x = 8$. The second derivative is

$$f''(x) = \frac{20}{9}x^{-1/3} - \frac{20}{9}x^{-2/3} \qquad \text{Second derivative}$$
$$= \frac{20}{9}x^{-2/3}(x^{1/3} - 1)$$
$$= \frac{20(x^{1/3} - 1)}{9x^{2/3}}. \qquad \text{Factored form}$$

So, possible points of inflection occur at $x = 1$ and $x = 0$. By testing $f'(x)$ and $f''(x)$, as shown in the table, you can see that f has one relative maximum, one relative minimum, and one point of inflection. The graph of f is shown in Figure 3.60.

	$f(x)$	$f'(x)$	$f''(x)$	Characteristics of graph
x in $(-\infty, 0)$		+	−	Increasing, concave downward
$x = 0$	0	0	Undef.	Relative maximum
x in $(0, 1)$		−	−	Decreasing, concave downward
$x = 1$	−3	−	0	Point of inflection
x in $(1, 8)$		−	+	Decreasing, concave upward
$x = 8$	−16	0	+	Relative minimum
x in $(8, \infty)$		+	+	Increasing, concave upward

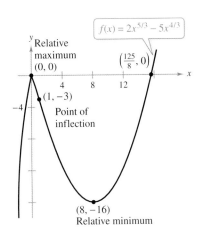

FIGURE 3.60

✓ **Checkpoint 5**   *Worked-out solution available at LarsonAppliedCalculus.com*

Analyze and sketch the graph of $f(x) = 2x^{3/2} - 6x^{1/2}$.

Summary of Simple Polynomial Graphs

A summary of the graphs of polynomial functions of degrees 0, 1, 2, and 3 is shown in Figure 3.61. Because of their simplicity, lower-degree polynomial functions are commonly used as mathematical models.

Constant function (degree 0): $y = a$

Horizontal line

Linear function (degree 1): $y = ax + b$

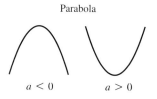

Line of slope a

$a < 0$ $a > 0$

Quadratic function (degree 2): $y = ax^2 + bx + c$

Parabola

$a < 0$ $a > 0$

Cubic function (degree 3): $y = ax^3 + bx^2 + cx + d$

Cubic curve

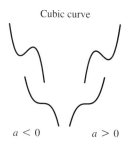

$a < 0$ $a > 0$

FIGURE 3.61

SUMMARIZE   (Section 3.7)

1. List the concepts you have learned that are useful in analyzing the graph of a function *(page 226)*. For an example that uses some of these concepts to analyze the graph of a function, see Example 1.

2. State the guidelines for analyzing the graph of a function *(page 226)*. For examples that use these guidelines, see Examples 3, 4, and 5.

3. State a general rule relating the degree n of a polynomial function with (a) the number of relative extrema and (b) the number of points of inflection *(page 229)*. For an example where this rule can be used to analyze the graph of a polynomial function, see Example 2.

SKILLS WARM UP 3.7

The following warm-up exercises involve skills that were covered in earlier sections. You will use these skills in the exercise set for this section. For additional help, review Sections 3.1 and 3.6.

In Exercises 1–4, find the vertical and horizontal asymptotes of the graph of the function.

1. $f(x) = \dfrac{1}{x^2}$ **2.** $f(x) = \dfrac{11}{(x-4)^2}$ **3.** $f(x) = \dfrac{40x}{x+3}$ **4.** $f(x) = \dfrac{x^2-3}{x^2-4x+3}$

In Exercises 5–10, find the open intervals on which the function is increasing or decreasing.

5. $f(x) = x^2 + 4x + 2$ **6.** $f(x) = -x^2 - 8x + 1$

7. $f(x) = x^3 - 3x + 1$ **8.** $f(x) = \dfrac{-x^3 + x^2 - 1}{x^2}$

9. $f(x) = \dfrac{x-1}{2x-5}$ **10.** $f(x) = -x^3 - 4x^2 + 3x + 2$

Exercises 3.7

See *CalcChat.com* for tutorial help and worked-out solutions to odd-numbered exercises.

Analyzing a Graph In Exercises 1–22, analyze and sketch the graph of the function. Label any intercepts, relative extrema, points of inflection, and asymptotes. *See Examples 1, 2, 3, 4, and 5.*

1. $y = -x^2 - 2x + 3$ **2.** $y = 2x^2 - 4x + 1$

3. $y = x^3 - 3x^2 + 4$ **4.** $y = -x^3 + 3x - 2$

5. $y = 2 - x - x^3$ **6.** $y = x^3 + 3x^2 + 3x + 2$

7. $y = 3x^4 + 4x^3$ **8.** $y = x^4 - 4x^2$

9. $y = x^4 - 8x^3 + 18x^2 - 16x + 5$

10. $y = x^4 - 4x^3 + 16x - 16$

11. $y = \dfrac{x^2 + 1}{x}$ **12.** $y = \dfrac{x + 2}{x}$

13. $y = \dfrac{x^2 - 6x + 12}{x - 4}$ **14.** $y = \dfrac{x^2 + 4x + 7}{x + 3}$

15. $y = \dfrac{x^2 + 1}{x^2 - 9}$ **16.** $y = \dfrac{2x}{x^2 - 1}$

17. $y = 3x^{2/3} - x^2$ **18.** $y = x^{5/3} - 5x^{2/3}$

19. $y = x\sqrt{9 - x}$ **20.** $y = x\sqrt{4 - x^2}$

21. $y = \begin{cases} x^2 + 1, & x \le 0 \\ 1 - 2x, & x > 0 \end{cases}$

22. $y = \begin{cases} x^2 + 4, & x < 0 \\ 4 - x, & x \ge 0 \end{cases}$

 Graphing a Function In Exercises 23–36, use a graphing utility to graph the function. Choose a window that allows all relative extrema and points of inflection to be identified on the graph.

23. $y = 3x^3 - 9x + 1$ **24.** $y = -4x^3 + 6x^2$

25. $y = -(x + 2)^4$ **26.** $y = (x - 1)^5$

27. $y = \dfrac{5 - 3x}{x - 2}$ **28.** $y = \dfrac{4x}{x^2 + 1}$

29. $y = 1 - x^{2/3}$ **30.** $y = (1 - x)^{2/3}$

31. $y = x^{4/3}$ **32.** $y = x^{-1/3}$

33. $y = \dfrac{x}{\sqrt{x^2 - 4}}$

34. $y = \dfrac{x - 3}{x}$

35. $y = \dfrac{x^3}{x^3 - 1}$

36. $y = \dfrac{x^4}{x^4 - 1}$

 Interpreting a Graph In Exercises 37–40, use the graph of f' or f'' to sketch the graph of f. (There are many correct answers.)

37.

38.

39.

40.

Sketching a Function In Exercises 41 and 42, sketch a graph of a function f having the given characteristics. (There are many correct answers.)

41. $f(-2) = f(0) = 0$

$f'(x) > 0$ if $x < -1$

$f'(x) < 0$ if $-1 < x < 0$

$f'(x) > 0$ if $x > 0$

$f'(-1) = f'(0) = 0$

42. $f(-1) = f(3) = 0$

$f'(1)$ is undefined.

$f'(x) < 0$ if $x < 1$

$f'(x) > 0$ if $x > 1$

$f''(x) < 0,\ x \neq 1$

$\lim\limits_{x \to \infty} f(x) = 4$

Creating a Function In Exercises 43 and 44, create a function whose graph has the given characteristics. (There are many correct answers.)

43. Vertical asymptote: $x = 3$

Horizontal asymptote: $y = 0$

44. Vertical asymptote: $x = -4$

Horizontal asymptote: None

45. Social Security The table lists the average monthly Social Security benefits B (in dollars) for retired workers aged 62 and over from 2009 through 2014, where $t = 9$ corresponds to 2009. A model for the data is

$$B = \frac{2057.91 - 64.007t + 40.6404t^2}{1 + 0.217t + 0.0144t^2},\quad 9 \le t \le 14.$$

(Source: U.S. Social Security Administration)

DATA	t	9	10	11	12	13	14
	B	1164	1175	1229	1262	1294	1329

Spreadsheet at LarsonAppliedCalculus.com

(a) Use a graphing utility to create a scatter plot of the data and graph the model in the same viewing window. How well does the model fit the data?

(b) Use the model to predict the average monthly benefit in 2018.

(c) Should this model be used to predict the average monthly Social Security benefits in future years? Why or why not?

46. Cost An employee of a delivery company earns $14 per hour driving a delivery van in an area where gasoline costs $2.33 per gallon. When the van is driven at a constant speed s (in miles per hour, with $40 \le s \le 70$), the van gets $700/s$ miles per gallon.

(a) Find the cost C as a function of s for a 100-mile trip on an interstate highway.

(b) Use a graphing utility to graph the function found in part (a) and determine the most economical speed.

47. Meteorology The monthly average high temperature T (in degrees Fahrenheit) for Boston, Massachusetts, can be modeled by

$$T = \frac{31.58 - 3.616t + 0.2243t^2}{1 - 0.211t + 0.0141t^2},\quad 1 \le t \le 12$$

where t is the month, with $t = 1$ corresponding to January. Use a graphing utility to graph the model and find all absolute extrema. Interpret the meaning of these values in the context of the problem. *(Source: National Climatic Data Center)*

48. **HOW DO YOU SEE IT?** The graph shows a company's profits P for the years 1985 through 2015, where t is the year, with $t = 0$ corresponding to 2000.

(a) For which values of t is P' zero? positive? negative? Interpret the meanings of these values in the context of the problem.

(b) For which values of t is P'' zero? positive? negative? Interpret the meanings of these values in the context of the problem.

Writing In Exercises 49 and 50, use a graphing utility to graph the function. Explain why there is no vertical asymptote when a superficial examination of the function may indicate that there should be one.

49. $h(x) = \dfrac{6 - 2x}{3 - x}$

50. $g(x) = \dfrac{x^2 + x - 2}{x - 1}$

51. Discovery Consider the function

$$f(x) = \frac{x^2 - 2x + 4}{x - 2}.$$

(a) Show that f can be rewritten as

$$f(x) = x + \frac{4}{x - 2}.$$

(b) Use a graphing utility to graph f and the line $y = x$. How do the two graphs compare as you zoom out?

(c) Use the results of part (b) to describe what is meant by a "slant asymptote."

3.8 Differentials and Marginal Analysis

■ Find the differentials of functions.
■ Use differentials in economics to approximate changes in revenue, cost, and profit.
■ Find the differential of a function using differentiation formulas.

Differentials

When the derivative was defined in Section 2.1 as the limit of the ratio $\Delta y/\Delta x$, it seemed natural to retain the quotient symbolism for the limit itself. So, the derivative of y with respect to x was denoted by

$$\frac{dy}{dx} = \lim_{\Delta x \to 0} \frac{\Delta y}{\Delta x}$$

even though dy/dx was not interpreted as the quotient of two separate quantities. In this section, you will see that the quantities dy and dx can be assigned meanings in such a way that their quotient, when $dx \neq 0$, is equal to the derivative of y with respect to x.

In Exercise 35 on page 241, you will use differentials to approximate the change in revenue for a one-unit increase in sales of a product.

Definition of Differentials

Let $y = f(x)$ represent a differentiable function. The **differential of x** (denoted by dx) is any nonzero real number. The **differential of y** (denoted by dy) is $dy = f'(x)\,dx$.

In the definition of differentials, dx can have any nonzero value. In most applications, however, dx is chosen to be small, and this choice is denoted by $dx = \Delta x$.

One use of differentials is in approximating the change in y that corresponds to a change in x, as shown in Figure 3.62. This change is denoted by

$$\Delta y = f(x + \Delta x) - f(x). \qquad \text{Change in } y$$

In Figure 3.62, notice that as Δx gets smaller and smaller, the values of dy and Δy get closer and closer. That is, when Δx is small, $dy \approx \Delta y$. This **tangent line approximation** is the basis for most applications of differentials.

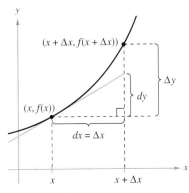

FIGURE 3.62

Note in Figure 3.62 that near the point of tangency, the graph of f is very close to the tangent line. This is the essence of the approximations used in this section. In other words, near the point of tangency, $dy \approx \Delta y$.

EXAMPLE 1   Comparing Δy and dy

Consider the function given by

$$f(x) = x^2.$$

Find the value of dy when $x = 1$ and $dx = 0.01$. Compare this with the value of Δy when $x = 1$ and $\Delta x = 0.01$.

SOLUTION   Begin by finding the derivative of f.

$$f'(x) = 2x \qquad \text{Derivative of } f$$

When $x = 1$ and $dx = 0.01$, the value of the differential dy is

$$
\begin{aligned}
dy &= f'(x)\, dx && \text{Differential of } y \\
&= f'(1)(0.01) && \text{Substitute 1 for } x \text{ and 0.01 for } dx. \\
&= 2(1)(0.01) && \text{Use } f'(x) = 2x. \\
&= 0.02. && \text{Simplify.}
\end{aligned}
$$

When $x = 1$ and $\Delta x = 0.01$, the value of Δy is

$$
\begin{aligned}
\Delta y &= f(x + \Delta x) - f(x) && \text{Change in } y \\
&= f(1.01) - f(1) && \text{Substitute 1 for } x \text{ and 0.01 for } \Delta x. \\
&= (1.01)^2 - (1)^2 \\
&= 1.0201 - 1 \\
&= 0.0201. && \text{Simplify.}
\end{aligned}
$$

Note that $dy \approx \Delta y$, as shown in Figure 3.63.

$y = 2x - 1$

$f(x) = x^2$

Δy

dy

$f(1.01)$

$f(1)$

$(1, 1)$

$\Delta x = 0.01$

FIGURE 3.63

✓ Checkpoint 1   *Worked-out solution available at LarsonAppliedCalculus.com*

Find the value of dy when $x = 2$ and $dx = 0.01$ for $f(x) = x^4$. Compare this with the value of Δy when $x = 2$ and $\Delta x = 0.01$.   ■

In Example 1, the tangent line to the graph of $f(x) = x^2$ at $x = 1$ is $y = 2x - 1$. For x-values near 1, this line is close to the graph of f, as shown in Figure 3.63 and in the table.

x	0.5	0.9	0.99	1	1.01	1.1	1.5
$f(x) = x^2$	0.25	0.81	0.9801	1	1.0201	1.21	2.25
$y = 2x - 1$	0	0.8	0.98	1	1.02	1.2	2

The validity of the approximation $dy \approx \Delta y$, $dx \ne 0$, stems from the definition of the derivative. That is, the existence of the limit

$$f'(x) = \lim_{\Delta x \to 0} \frac{f(x + \Delta x) - f(x)}{\Delta x}$$

implies that when Δx is close to zero, then $f'(x)$ is close to the difference quotient. So, you can write

$$\frac{f(x + \Delta x) - f(x)}{\Delta x} \approx f'(x)$$

$$f(x + \Delta x) - f(x) \approx f'(x)\Delta x$$

$$\Delta y \approx f'(x)\Delta x.$$

Substituting dx for Δx and dy for $f'(x)\, dx$ produces $\Delta y \approx dy$.

Marginal Analysis

Differentials are used in economics to approximate changes in revenue, cost, and profit. Let $R = f(x)$ be the total revenue for selling x units of a product. When the number of units increases by 1, the change in x is $\Delta x = 1$, and the change in R is

$$\Delta R = f(x + \Delta x) - f(x) \approx dR = \frac{dR}{dx}\, dx.$$

In other words, you can use the differential dR to approximate the change in revenue that accompanies the sale of one additional unit. Similarly, the differentials dC and dP can be used to approximate the changes in cost and profit that accompany the sale (or production) of one additional unit.

EXAMPLE 2   Using Marginal Analysis

The demand function for a product is modeled by

$$p = 400 - x, \quad 0 \le x \le 400$$

where p is the price per unit (in dollars) and x is the number of units. Use differentials to approximate the change in revenue as sales increase from 149 units to 150 units. Compare this with the actual change in revenue.

SOLUTION   Begin by finding the revenue function. Because the demand is given by $p = 400 - x$, the revenue is

$R = xp$	Formula for revenue
$\quad = x(400 - x)$	Use $p = 400 - x$.
$\quad = 400x - x^2.$	Multiply.

Next, find the marginal revenue, dR/dx.

$$\frac{dR}{dx} = 400 - 2x \qquad \text{Power Rule}$$

When $x = 149$ and $dx = \Delta x = 1$, the approximate change in the revenue is

$$\Delta R \approx dR$$
$$\quad = \frac{dR}{dx}\, dx$$
$$\quad = (400 - 2x)\, dx$$
$$\quad = [400 - 2(149)](1)$$
$$\quad = \$102.$$

When x increases from 149 to 150 and $R = f(x) = 400x - x^2$, the actual change in revenue is

$$\Delta R = f(x + \Delta x) - f(x)$$
$$\quad = [400(150) - 150^2] - [400(149) - 149^2]$$
$$\quad = 37{,}500 - 37{,}399$$
$$\quad = \$101.$$

✓ *Checkpoint 2*   *Worked-out solution available at LarsonAppliedCalculus.com*

The demand function for a product is modeled by $p = 200 - x$, $0 \le x \le 200$, where p is the price per unit (in dollars) and x is the number of units. Use differentials to approximate the change in revenue as sales increase from 89 to 90 units. Compare this with the actual change in revenue.   ■

EXAMPLE 3   Using Marginal Analysis

The profit (in dollars) derived from selling x units of an alarm clock is modeled by

$$P = 0.0002x^3 + 10x.$$

Use the differential dP to approximate the change in profit when the production level changes from 50 to 51 units. Compare this with the actual gain in profit obtained by increasing the production level from 50 to 51 units.

SOLUTION   The marginal profit is

$$\frac{dP}{dx} = 0.0006x^2 + 10.$$

When $x = 50$ and $dx = \Delta x = 1$, the approximate change in profit is

$$\Delta P \approx dP$$
$$= \frac{dP}{dx}\, dx$$
$$= (0.0006x^2 + 10)\, dx$$
$$= [0.0006(50)^2 + 10](1)$$
$$= \$11.50.$$

When x changes from 50 to 51 units and $P = f(x) = 0.0002x^3 + 10x$, the actual change in profit is

$$\Delta P = f(x + \Delta x) - f(x)$$
$$= [(0.0002)(51)^3 + 10(51)] - [(0.0002)(50)^3 + 10(50)]$$
$$\approx 536.53 - 525.00$$
$$= \$11.53.$$

These values are shown graphically in the figure.

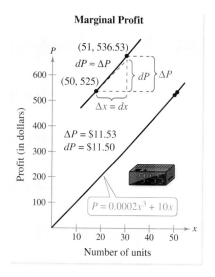

Marginal Profit

✓ **Checkpoint 3**   *Worked-out solution available at LarsonAppliedCalculus.com*

Use the differential dP to approximate the change in profit for the profit function in Example 3 when the production level changes from 40 to 41 units. Compare this with the actual gain in profit obtained by increasing the production level from 40 to 41 units.

Formulas for Differentials

You can use the definition of differentials to rewrite each differentiation rule in **differential form.**

Differential Forms of Differentiation Rules

Constant Multiple Rule: $d[cu] = c\, du$

Sum or Difference Rule: $d[u \pm v] = du \pm dv$

Product Rule: $d[uv] = u\, dv + v\, du$

Quotient Rule: $d\left[\dfrac{u}{v}\right] = \dfrac{v\, du - u\, dv}{v^2}$

Constant Rule: $d[c] = 0$

Power Rule: $d[x^n] = nx^{n-1}\, dx$

The next example compares the derivatives and differentials of several simple functions.

EXAMPLE 4   Finding Differentials

Function	Derivative	Differential
a. $y = x^2$	$\dfrac{dy}{dx} = 2x$	$dy = 2x\, dx$
b. $y = \dfrac{3x + 2}{5}$	$\dfrac{dy}{dx} = \dfrac{3}{5}$	$dy = \dfrac{3}{5}\, dx$
c. $y = 2x^2 - 3x$	$\dfrac{dy}{dx} = 4x - 3$	$dy = (4x - 3)\, dx$
d. $y = \dfrac{1}{x}$	$\dfrac{dy}{dx} = -\dfrac{1}{x^2}$	$dy = -\dfrac{1}{x^2}\, dx$

 Checkpoint 4   *Worked-out solution available at LarsonAppliedCalculus.com*

Find the differential dy of each function.

a. $y = 4x^3$ **b.** $y = \dfrac{2x + 1}{3}$

c. $y = 3x^2 - 2x$ **d.** $y = \dfrac{1}{x^2}$

SUMMARIZE (Section 3.8)

1. State the definition of differentials *(page 235)*. For an example of a differential, see Example 1.

2. Explain what is meant by marginal analysis *(page 237)*. For examples of marginal analysis, see Examples 2 and 3.

3. State the differential forms of the differential rules *(page 239)*. For an example that uses the differential forms of the differential rules, see Example 4.

SKILLS WARM UP 3.8

The following warm-up exercises involve skills that were covered in earlier sections. You will use these skills in the exercise set for this section. For additional help, review Sections 2.2 and 2.4.

In Exercises 1–12, find the derivative.

1. $C = 44 + 0.09x^2$

2. $C = 250 + 0.15x$

3. $R = x(1.25 + 0.02\sqrt{x})$

4. $R = x(15.5 - 1.55x)$

5. $P = -0.03x^{1/3} + 1.4x - 2250$

6. $P = -0.04x^3 + 950x - 786$

7. $A = \frac{1}{4}\sqrt{3}x^2$

8. $A = \frac{2}{9}x^2$

9. $C = 2\pi r$

10. $P = 4w$

11. $S = 4\pi r^2$

12. $P = 2x + \sqrt{2}x$

In Exercises 13–16, write a formula for the quantity.

13. Area A of a circle of radius r

14. Area A of a square of side x

15. Volume V of a cube of edge x

16. Volume V of a sphere of radius r

Exercises 3.8

See *CalcChat.com* for tutorial help and worked-out solutions to odd-numbered exercises.

 Comparing Δy and dy In Exercises 1–6, compare the values of dy and Δy for the function. *See Example 1.*

Function	x-Value	Differential of x
1. $f(x) = 0.5x^3$	$x = 1$	$\Delta x = dx = 0.1$
2. $f(x) = 0.2x^2$	$x = 3$	$\Delta x = dx = -0.1$
3. $f(x) = x^4 + 3$	$x = -2$	$\Delta x = dx = -0.01$
4. $f(x) = 2x + 1$	$x = 1$	$\Delta x = dx = 0.01$
5. $f(x) = 3\sqrt{x}$	$x = 4$	$\Delta x = dx = 0.1$
6. $f(x) = 6x^{4/3}$	$x = -1$	$\Delta x = dx = 0.01$

Finding Differentials In Exercises 7–12, let $x = 2$ and complete the table for the function.

$dx = \Delta x$	dy	Δy	$\Delta y - dy$	$\dfrac{dy}{\Delta y}$
1.000				
0.500				
0.100				
0.010				
0.001				

7. $y = x^3$

8. $y = x^5$

9. $y = \dfrac{1}{x^3}$

10. $y = \dfrac{3}{x^2}$

11. $y = \sqrt[4]{x}$

12. $y = \sqrt{x}$

 Marginal Analysis In Exercises 13–18, use differentials to approximate the change in cost, revenue, or profit corresponding to an increase in sales (or production) of one unit. (For instance, in Exercise 13, approximate the change in cost as x increases from 12 units to 13 units.) Then compare this with the actual change in cost, revenue, or profit. *See Examples 2 and 3.*

Function	x-Value
13. $C = 0.05x^2 + 4x + 10$	$x = 12$
14. $C = 0.025x^2 + 8x + 5$	$x = 10$
15. $R = 30x - 0.15x^2$	$x = 75$
16. $R = 50x - 1.5x^2$	$x = 15$
17. $P = -0.5x^3 + 2500x - 6000$	$x = 50$
18. $P = -0.1x^3 + 900x - 50$	$x = 60$

 Finding Differentials In Exercises 19–28, find the differential dy. *See Example 4.*

19. $y = 6x^4$

20. $y = \dfrac{8 - 4x}{3}$

21. $y = 5x^3 - 6x$

22. $y = \dfrac{9}{x^{2/3}}$

23. $y = \dfrac{x + 1}{2x - 1}$

24. $y = \dfrac{x}{x^2 + 1}$

25. $y = \sqrt{9 - x^2}$

26. $y = \sqrt[3]{6x^2}$

27. $y = x(3x - 7)^3$

28. $y = (x^2 + 3)(2x + 4)^2$

Finding an Equation of a Tangent Line In Exercises 29–32, find an equation of the tangent line to the function at the given point. Then find the function values and the tangent line values at $f(x + \Delta x)$ and $y(x + \Delta x)$ for $\Delta x = -0.01$ and 0.01.

29. $f(x) = 2x^3 - x^2 + 1$; $(-2, -19)$

30. $f(x) = 5 - 3x^2$; $(-1, 2)$

31. $f(x) = \dfrac{x}{x^2 + 1}$; $(0, 0)$

32. $f(x) = \sqrt{25 - x^2}$; $(3, 4)$

33. Profit The profit P for a company producing x units is
$$P = (500x - x^2) - \left(\tfrac{1}{2}x^2 - 77x + 3000\right).$$

(a) Use differentials to approximate the change in profit when the production level changes from 115 to 120 units.

(b) Compare this with the actual change in profit.

34. Revenue The revenue R for a company selling x units is $R = 900x - 0.1x^2$.

(a) Use differentials to approximate the change in revenue as the sales increase from 3000 units to 3100 units.

(b) Compare this with the actual change in revenue.

35. Demand The demand function for a product is modeled by $p = 68 - 0.2x$, where p is the price per unit (in dollars) and x is the number of units.

(a) Use differentials to approximate the change in revenue as sales increase from 7 units to 8 units. Compare this with the actual change in revenue.

(b) Repeat part (a) as sales increase from 70 units to 71 units.

36. HOW DO YOU SEE IT? The graph shows the profit P (in dollars) from selling x units of an item. Use the graph to determine which is greater, the change in profit when the production level changes from 400 to 401 units or the change in profit when the production level changes from 900 to 901 units. Explain your reasoning.

Profit

37. Biology: Wildlife Management A state game commission introduces 50 deer into newly acquired state game lands. The population N of the herd can be modeled by
$$N = \frac{10(5 + 3t)}{1 + 0.04t}$$
where t is the time in years. Use differentials to approximate the change in the herd size from $t = 5$ to $t = 6$.

38. Medical Science The concentration C (in milligrams per milliliter) of a drug in a patient's bloodstream t hours after injection into muscle tissue is modeled by
$$C = \frac{3t}{27 + t^3}.$$
Use differentials to approximate the change in the concentration when t changes from $t = 1$ to $t = 1.5$.

39. Marginal Analysis A retailer has determined that the monthly sales x of a watch are 150 units when the price is \$50, but decrease to 120 units when the price is \$60. Assume that the demand is a linear function of the price. Find the revenue R as a function of x and approximate the change in revenue for a one-unit increase in sales when $x = 141$. Make a sketch showing dR and ΔR.

40. Marginal Analysis The demand x for a web camera is 30,000 units per month when the price is \$25 and 40,000 units when the price is \$20. The initial investment is \$275,000 and the cost per unit is \$17. Assume that the demand is a linear function of the price. Find the profit P as a function of x and approximate the change in profit for a one-unit increase in sales when $x = 28,000$. Make a sketch showing dP and ΔP.

Error Propagation In Exercises 41 and 42, use the following information. Given the error in a measurement (Δx), the *propagated error* (Δy) can be approximated by the differential dy. The ratio dy/y is the *relative error*, which corresponds to a *percentage error* of $dy/y \times 100\%$.

41. Area The side of a square measures 6 inches, with a possible error of $\pm\tfrac{1}{16}$ inch. Estimate the propagated error and the percentage error in computing the area of the square.

42. Volume The radius of a sphere measures 6 inches, with a possible error of ±0.02 inch. Estimate the propagated error and the percentage error in computing the volume of the sphere.

True or False? In Exercises 43 and 44, determine whether the statement is true or false. If it is false, explain why or give an example that shows it is false.

43. If $y = x + c$, then $dy = dx$.

44. If $y = ax + b$, then $\Delta y/\Delta x = dy/dx$.

ALGEBRA TUTOR

Solving Equations

Much of the algebra in Chapter 3 involves simplifying algebraic expressions (see pages 158 and 159) and solving algebraic equations (see page 71). The Algebra Tutor on page 71 illustrates some of the basic techniques for solving equations. On these two pages, you can review some of the more complicated techniques for solving equations.

When solving an equation, remember that your basic goal is to isolate the variable on one side of the equation. To do this, you use inverse operations. For instance, to isolate x in

$$x - 2 = 0$$

you add 2 to each side of the equation, because *addition* is the inverse operation of *subtraction*. To isolate x in

$$\sqrt{x} = 2$$

you square each side of the equation, because *squaring* is the inverse operation of *taking the square root*.

EXAMPLE 1   Solving Equations

Solve each equation.

a. $\dfrac{36(x^2 - 1)}{(x^2 + 3)^3} = 0$ **b.** $0 = 2x(2x^2 - 3)$ **c.** $\dfrac{dV}{dx} = 0$, where $V = 27x - \dfrac{1}{4}x^3$

SOLUTION

a. $\dfrac{36(x^2 - 1)}{(x^2 + 3)^3} = 0$

    Example 2, page 188

$$36(x^2 - 1) = 0$$

    A fraction is equal to zero only when its numerator is zero.

$$x^2 - 1 = 0$$

    Divide each side by 36.

$$x^2 = 1$$

    Add 1 to each side.

$$x = \pm 1$$

    Take the square root of each side.

b. $0 = 2x(2x^2 - 3)$

    Example 2, page 198

$$2x = 0 \implies x = 0$$

    Set first factor equal to zero.

$$2x^2 - 3 = 0 \implies x = \pm\sqrt{\tfrac{3}{2}}$$

    Set second factor equal to zero.

c. $V = 27x - \dfrac{1}{4}x^3$

    Example 1, page 196

$$\frac{dV}{dx} = 27 - \frac{3}{4}x^2$$

    Find derivative of V.

$$0 = 27 - \frac{3}{4}x^2$$

    Set derivative equal to 0.

$$\frac{3}{4}x^2 = 27$$

    Add $\tfrac{3}{4}x^2$ to each side.

$$x^2 = 36$$

    Multiply each side by $\tfrac{4}{3}$.

$$x = \pm 6$$

    Take the square root of each side.

EXAMPLE 2   Solving Equations

Solve each equation.

a. $\dfrac{20(x^{1/3} - 1)}{9x^{2/3}} = 0$ **b.** $\dfrac{25}{\sqrt{x}} - 0.5 = 0$

c. $x^2(4x - 3) = 0$ **d.** $\dfrac{4x}{3(x^2 - 4)^{1/3}} = 0$

e. $g'(x) = 0$, where $g(x) = (x - 2)(x + 1)^2$

SOLUTION

a. $\dfrac{20(x^{1/3} - 1)}{9x^{2/3}} = 0$ Example 5, page 231

$20(x^{1/3} - 1) = 0$ A fraction is equal to zero only when its numerator is zero.

$x^{1/3} - 1 = 0$ Divide each side by 20.

$x^{1/3} = 1$ Add 1 to each side.

$x = 1$ Cube each side.

b. $\dfrac{25}{\sqrt{x}} - 0.5 = 0$ Example 4, page 208

$\dfrac{25}{\sqrt{x}} = 0.5$ Add 0.5 to each side.

$25 = 0.5\sqrt{x}$ Multiply each side by $\sqrt{x}$.

$50 = \sqrt{x}$ Divide each side by 0.5.

$2500 = x$ Square each side.

c. $x^2(4x - 3) = 0$ Example 2, page 180

$x^2 = 0 \implies x = 0$ Set first factor equal to zero.

$4x - 3 = 0 \implies x = \frac{3}{4}$ Set second factor equal to zero.

d. $\dfrac{4x}{3(x^2 - 4)^{1/3}} = 0$ Example 5, page 172

$4x = 0$ A fraction is equal to zero only when its numerator is zero.

$x = 0$ Divide each side by 4.

e. $g(x) = (x - 2)(x + 1)^2$ Original function

$g'(x) = (x - 2)(2)(x + 1)(1) + (x + 1)^2(1)$ Find derivative of g.

$(x - 2)(2)(x + 1) + (x + 1)^2 = 0$ Set derivative equal to zero.

$(x + 1)[2(x - 2) + (x + 1)] = 0$ Factor.

$(x + 1)(2x - 4 + x + 1) = 0$ Multiply factors.

$(x + 1)(3x - 3) = 0$ Combine like terms.

$x + 1 = 0 \implies x = -1$ Set first factor equal to zero.

$3x - 3 = 0 \implies x = 1$ Set second factor equal to zero.

SUMMARY AND STUDY STRATEGIES

After studying this chapter, you should have acquired the following skills. The exercise numbers are keyed to the Review Exercises that begin on page 246. Answers to odd-numbered Review Exercises are given in the back of the text.*

Section 3.1	Review Exercises
■ Find the critical numbers of a function.	*1–6*
c is a critical number of f when $f'(c) = 0$ or $f'(c)$ is undefined.	
■ Find the open intervals on which a function is increasing or decreasing.	*7–16*
f is increasing when $f'(x) > 0$.	
f is decreasing when $f'(x) < 0$.	
■ Find intervals on which a real-life model is increasing or decreasing.	*17, 18*

Section 3.2	
■ Use the First-Derivative Test to find the relative extrema of a function.	*19–28*
■ Find the absolute extrema of a continuous function on a closed interval.	*29–36*
■ Find minimum and maximum values of a real-life model and interpret the results in context.	*37, 38*

Section 3.3	
■ Find the open intervals on which the graph of a function is concave upward or concave downward.	*39–42*
f is concave upward when $f''(x) > 0$.	
f is concave downward when $f''(x) < 0$.	
■ Find the points of inflection of the graph of a function.	*43–46*
■ Use the Second-Derivative Test to find the relative extrema of a function.	*47–52*
■ Find the point of diminishing returns of an input-output model.	*53, 54*

Section 3.4	
■ Solve real-life optimization problems.	*55–58*

Section 3.5	
■ Solve business and economics optimization problems.	*59–64*
■ Find the price elasticity of demand for a demand function.	*65, 66*

---

* Several study aids are available to help you master the material in this chapter. The *Student Solutions Manual* and *CalcChat.com* have solutions to all odd-numbered exercises, and *CalcView.com* has video solutions for selected exercises. *LarsonAppliedCalculus.com* offers algebra help, data spreadsheets, and much more. A *Graphing Technology Guide* with step-by-step commands for a variety of graphing calculators is at *CengageBrain.com*.

Section 3.6 Review Exercises

- Find the vertical asymptotes of a function. *67–70*
- Find infinite limits. *71–74*
- Find the horizontal asymptotes of a function. *75–78*
- Use asymptotes to answer questions about real-life situations. *79–82*

Section 3.7

- Analyze the graph of a function. *83–96*

Section 3.8

- Use differentials to approximate changes in a function. *97–100*
- Use differentials in economics to approximate changes in cost, revenue, and profit. *101–106*
- Find the differential of a function using differentiation formulas. *107–112*
- Use differentials to approximate changes in real-life models. *113–115*

Study Strategies

- **Solve Problems Graphically, Analytically, and Numerically**   When analyzing the graph of a function, use a variety of problem-solving strategies. For instance, to analyze the graph of

$$f(x) = x^3 - 4x^2 + 5x - 4$$

  you could begin *graphically*. That is, you could use a graphing utility to find a viewing window that appears to show the important characteristics of the graph. From the graph shown below, the function appears to have one relative minimum, one relative maximum, and one point of inflection.

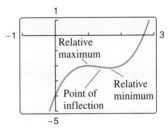

  Next, you could use calculus to *analyze* the graph. Because the derivative of f is

$$f'(x) = 3x^2 - 8x + 5 = (3x - 5)(x - 1)$$

  the critical numbers of f are $x = \frac{5}{3}$ and $x = 1$. By the First-Derivative Test, you can conclude that $x = \frac{5}{3}$ yields a relative minimum and $x = 1$ yields a relative maximum. The second derivative is

$$f''(x) = 6x - 8.$$

  By testing the intervals $\left(-\infty, \frac{4}{3}\right)$ and $\left(\frac{4}{3}, \infty\right)$, you can conclude that $x = \frac{4}{3}$ yields a point of inflection. Finally, you could analyze the graph *numerically*. For instance, you could construct a table of values and observe that f is increasing on the interval $(-\infty, 1)$, decreasing on the interval $\left(1, \frac{5}{3}\right)$, and increasing on the interval $\left(\frac{5}{3}, \infty\right)$.

- **Problem-Solving Strategies**   When you get stuck while trying to solve an optimization problem, consider the strategies below.

  1. *Draw a Diagram.*   If feasible, draw a diagram that represents the problem. Label all known values and unknown values on the diagram.

  2. *Solve a Simpler Problem.*   Simplify the problem, or write several simple examples of the problem. For instance, if you are asked to find the dimensions that will produce a maximum area, try calculating the areas of several examples.

  3. *Rewrite the Problem in Your Own Words.*   Rewriting a problem can help you understand it better.

  4. *Guess, Check, and Revise.*   Try guessing the answer, then check your guess in the statement of the original problem. By refining your guesses, you may be able to think of a general strategy for solving the problem.

Review Exercises   See *CalcChat.com* for tutorial help and worked-out solutions to odd-numbered exercises.

Finding Critical Numbers In Exercises 1–6, find the critical numbers of the function.

1. $f(x) = -x^2 - 5x + 9$

2. $y = 3x^2 + 18x$

3. $y = 4x^3 - 108x$

4. $f(x) = x^4 - 8x^2 + 13$

5. $g(x) = (x - 1)^2(x - 3)$

6. $h(x) = \sqrt{x}(x - 3)$

Intervals on Which f Is Increasing or Decreasing In Exercises 7–16, find the critical numbers and the open intervals on which the function is increasing or decreasing. Use a graphing utility to verify your results.

7. $f(x) = x^2 + x - 2$

8. $g(x) = (x + 5)^4$

9. $f(x) = -x^3 + 6x^2 - 2$

10. $y = x^3 - 5x^2 + 8x$

11. $y = (x - 6)^{2/3}$

12. $y = 2x^{1/3} - 3$

13. $g(x) = \sqrt{25 - x^2}$

14. $y = x\sqrt[3]{x^2 - 64}$

15. $y = \dfrac{x}{x^2 + 49}$

16. $f(x) = \dfrac{x^2}{x^2 + 16}$

17. **Sales** The sales S (in millions of dollars) of Polaris Industries from 2009 through 2013 can be modeled by

$$S = 12.236t^2 + 294.92t - 2109.0, \quad 9 \le t \le 13$$

where t is the year, with $t = 9$ corresponding to 2009. Show that the sales were increasing from 2009 through 2013. *(Source: Polaris Industries, Inc.)*

18. **Revenue** The revenue R (in billions of dollars) of Cintas from 2009 through 2014 can be modeled by

$$R = -0.02454t^3 + 0.8838t^2 - 10.284t + 42.60$$

for $9 \le t \le 14$, where t is the year, with $t = 9$ corresponding to 2009. *(Source: Cintas Corporation)*

(a) Use a graphing utility to graph the model. Then graphically estimate the years during which the revenue was increasing and the years during which the revenue was decreasing.

(b) Use the test for increasing and decreasing functions to verify the result of part (a).

Finding Relative Extrema In Exercises 19–28, use the First-Derivative Test to find all relative extrema of the function. Use a graphing utility to verify your results.

19. $f(x) = 4x^3 - 6x^2 - 2$

20. $f(x) = \frac{1}{4}x^4 - 8x$

21. $h(x) = 4 + 10x - x^2$

22. $g(x) = 2x^3 + 15x^2 - 8$

23. $h(x) = 2x^2 - x^4$

24. $s(x) = \frac{1}{2}x^4 - 9x^2 + 5$

25. $f(x) = \dfrac{6}{x^2 + 1}$

26. $f(x) = \dfrac{7x}{x^2 + 4}$

27. $h(x) = x - 3x^{1/3}$

28. $g(x) = x - 6\sqrt{x}, \quad x > 0$

Finding Extrema on a Closed Interval In Exercises 29–36, find the absolute extrema of the function on the closed interval. Use a graphing utility to verify your results.

29. $f(x) = x^2 + 5x + 6; \quad [-3, 0]$

30. $f(x) = x^4 - 2x^3; \quad [0, 2]$

31. $f(x) = x^3 - 12x + 1; \quad [-4, 4]$

32. $f(x) = x^3 + 2x^2 - 3x + 4; \quad [-3, 2]$

33. $f(x) = 2\sqrt{x} - x; \quad [0, 9]$

34. $f(x) = \dfrac{x}{\sqrt{x^2 + 1}}; \quad [0, 2]$

35. $f(x) = \dfrac{2x}{x^2 + 1}; \quad [-1, 2]$

36. $f(x) = \dfrac{8}{x} + x; \quad [1, 4]$

37. **Surface Area** A right circular cylinder of radius r and height h has a volume of 25 cubic inches (see figure). The total surface area of the cylinder in terms of r is given by

$$S = 2\pi r\left(r + \frac{25}{\pi r^2}\right).$$

Find the radius that will minimize the surface area. Use a graphing utility to verify your result.

38. Profit The profit P (in dollars) made by a company from selling x tablet computers can be modeled by

$$P = 1.64x - \frac{x^2}{15,000} - 2500.$$

Find the number of units sold that will yield a maximum profit. What is the maximum profit?

Determining Concavity In Exercises 39–42, determine the open intervals on which the graph of the function is concave upward or concave downward.

39. $f(x) = (x - 2)^3$

40. $h(x) = x^5 - 10x^2$

41. $g(x) = \frac{1}{4}(-x^4 + 8x^2 - 12)$

42. $h(x) = \dfrac{18}{x^2 + 9}$

Finding Points of Inflection In Exercises 43–46, discuss the concavity of the graph of the function and find the points of inflection.

43. $f(x) = \frac{1}{2}x^4 - 4x^3$

44. $f(x) = \frac{1}{12}x^4 - 2x^2 - x + 3$

45. $f(x) = x(2x - 3)^2$

46. $f(x) = (x - 1)^2(x - 3)$

Using the Second-Derivative Test In Exercises 47–52, find all relative extrema of the function. Use the Second-Derivative Test when applicable.

47. $f(x) = x^3 - 6x^2 + 12x$

48. $f(x) = x^4 - 32x^2 + 12$

49. $f(x) = x^5 - 5x^3$

50. $f(x) = x(x^2 - 3x - 9)$

51. $f(x) = 2x^2(1 - x^2)$

52. $f(x) = x - 4\sqrt{x + 1}$

Point of Diminishing Returns In Exercises 53 and 54, find the point of diminishing returns for the function, where R is the revenue (in thousands of dollars) and x is the advertising cost (in thousands of dollars). Use a graphing utility to verify your result.

53. $R = \frac{1}{1500}(150x^2 - x^3)$, $0 \le x \le 100$

54. $R = -\frac{2}{3}(x^3 - 12x^2 - 6)$, $0 \le x \le 8$

55. Minimum Perimeter Find the length and width of a rectangle that has an area of 225 square meters and a minimum perimeter.

56. Maximum Volume A rectangular solid with a square base has a surface area of 432 square centimeters.

(a) Determine the dimensions that yield the maximum volume.

(b) Find the maximum volume.

57. Maximum Volume An open box is to be made from a 10-inch by 16-inch rectangular piece of material by cutting equal squares from the corners and turning up the sides (see figure). Find the volume of the largest box that can be made.

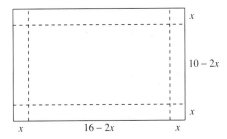

58. Minimum Area A rectangular page will contain 108 square inches of print. The margins at the top and bottom of the page are $\frac{3}{4}$ inch wide. The margins on each side are 1 inch wide. Find the dimensions of the page that will minimize the amount of paper used.

Finding the Maximum Revenue In Exercises 59 and 60, find the number of units x that maximizes the revenue R.

59. $R = 450x - 0.25x^2$

60. $R = 36x^2 - 0.05x^3$

Finding the Minimum Average Cost In Exercises 61 and 62, find the number of units x that minimizes the average cost per unit $\overline{C}$.

61. $C = 0.2x^2 + 10x + 4500$

62. $C = 0.03x^3 + 30x + 3840$

63. Maximum Profit A commodity has a demand function modeled by

$$p = 36 - 4x$$

and a total cost function modeled by

$$C = 2x^2 + 6$$

where x is the number of units.

(a) What price yields a maximum profit?

(b) When the profit is maximized, what is the average cost per unit?

64. Maximum Profit The profit P (in thousands of dollars) for a company in terms of the amount s spent on advertising (in thousands of dollars) can be modeled by

$$P = -4s^3 + 60s^2 - 108s + 300.$$

(a) Find the amount spent on advertising that maximizes the profit.

(b) Find the point of diminishing returns.

65. Elasticity The demand function for a product is modeled by

$$p = 60 - 0.04x, \quad 0 \le x \le 1500$$

where p is the price per unit (in dollars) and x is the number of units.

(a) Determine when the demand is elastic, inelastic, and of unit elasticity.

(b) Use the result of part (a) to describe the behavior of the revenue function.

66. Elasticity The demand function for a product is modeled by

$$p = 960 - x, \quad 0 \le x \le 960$$

where p is the price per unit (in dollars) and x is the number of units.

(a) Determine when the demand is elastic, inelastic, and of unit elasticity.

(b) Use the result of part (a) to describe the behavior of the revenue function.

Finding Vertical Asymptotes In Exercises 67–70, determine all vertical asymptotes of the graph of the function.

67. $f(x) = \dfrac{x + 7}{x^2 + 9x}$

68. $f(x) = \dfrac{x - 3}{x^2 - 36}$

69. $f(x) = \dfrac{x^2 - 16}{2x^2 + 9x + 4}$

70. $f(x) = \dfrac{x^2 + 6x + 9}{x^2 - 5x - 24}$

Determining Infinite Limits In Exercises 71–74, use a graphing utility to find the limit.

71. $\displaystyle \lim_{x \to 0^+} \left(x - \frac{1}{x^3} \right)$

72. $\displaystyle \lim_{x \to -8^-} \frac{2x - 3}{x + 8}$

73. $\displaystyle \lim_{x \to -1^+} \frac{x^2 - 2x + 1}{x + 1}$

74. $\displaystyle \lim_{x \to 3^-} \frac{3x^2 + 1}{x^2 - 9}$

Finding Horizontal Asymptotes In Exercises 75–78, find the horizontal asymptote of the graph of the function.

75. $f(x) = \dfrac{3x^2}{5 - 4x^2}$

76. $f(x) = \dfrac{3x^2 - 2x + 3}{x + 1}$

77. $f(x) = \dfrac{3x}{x^2 + 1}$

78. $f(x) = \dfrac{x}{x - 2} + \dfrac{2x}{x + 2}$

79. Average Cost The cost C (in dollars) of producing x units of a product is

$$C = 0.75x + 4000.$$

(a) Find the average cost function $\overline{C}$.

(b) Find $\overline{C}$ when $x = 100$ and when $x = 1000$.

(c) Determine the limit of the average cost function as x approaches infinity. Interpret the limit in the context of the problem.

80. Average Cost The cost C (in dollars) of producing x units of a product is

$$C = 1.50x + 8000.$$

(a) Find the average cost function $\overline{C}$.

(b) Find $\overline{C}$ when $x = 1000$ and when $x = 10,000$.

(c) Determine the limit of the average cost function as x approaches infinity. Interpret the limit in the context of the problem.

81. Seizing Drugs The cost C (in millions of dollars) for the federal government to seize $p\%$ of an illegal drug as it enters the country is modeled by

$$C = \frac{250p}{100 - p}, \quad 0 \le p < 100.$$

(a) Find the costs of seizing 20%, 50%, and 90%.

(b) Find the limit of C as $p \to 100^-$. Interpret the limit in the context of the problem. Use a graphing utility to verify your result.

82. Removing Pollutants The cost C (in dollars) of removing $p\%$ of the air pollutants in the stack emission of a utility company that burns coal is modeled by

$$C = \frac{160,000p}{100 - p}, \quad 0 \le p < 100.$$

(a) Find the costs of removing 25%, 50%, and 75%.

(b) Find the limit of C as $p \to 100^-$. Interpret the limit in the context of the problem. Use a graphing utility to verify your result.

Analyzing a Graph In Exercises 83–94, analyze and sketch the graph of the function. Label any intercepts, relative extrema, points of inflection, and asymptotes.

83. $f(x) = 4x - x^2$

84. $f(x) = 4x^3 - 2x^4$

85. $f(x) = x^3 - 6x^2 + 3x + 10$

86. $f(x) = -x^3 + 3x^2 + 9x - 2$

87. $f(x) = x^4 - 4x^3 + 16x - 16$

88. $f(x) = x^5 + 1$

89. $f(x) = \dfrac{x + 1}{x - 1}$

90. $f(x) = \dfrac{4x}{x^2 + 9}$

91. $f(x) = 3x^{2/3} - 2x$

92. $f(x) = x^{4/5}$

93. $f(x) = x\sqrt{6 - x}$

94. $f(x) = x\sqrt{16 - x^2}$

95. Bacteria The data in the table show the number N of bacteria in a culture at time t, where t is measured in days.

t	1	2	3	4
N	25	200	804	1756

t	5	6	7	8
N	2296	2434	2467	2473

Spreadsheet at LarsonAppliedCalculus.com

A model for these data is

$$N = \frac{24{,}670 - 35{,}153t + 13{,}250t^2}{100 - 39t + 7t^2}, \quad 1 \le t \le 8.$$

(a) Use a graphing utility to create a scatter plot of the data and graph the model in the same viewing window. How well does the model fit the data?

(b) Use the model to predict the number of bacteria in the culture after 10 days.

(c) Should this model be used to predict the number of bacteria in the culture after a few months? Why or why not?

96. Meteorology The monthly average high temperature T (in degrees Fahrenheit) in New York City can be modeled by

$$T = \frac{31.887 - 1.7686t + 0.09031t^2}{1 - 0.1924t + 0.01286t^2}, \quad 1 \le t \le 12$$

where t is the month, with $t = 1$ corresponding to January. Use a graphing utility to graph the model and find all absolute extrema. Interpret the meaning of these values in the context of the problem. (*Source: National Climatic Data Center*)

Comparing Δy and dy In Exercises 97–100, compare the values of dy and Δy for the function.

Function	x-Value	Differential of x
97. $f(x) = 2x^2$	$x = 2$	$\Delta x = dx = 0.01$
98. $f(x) = x^4 - 4$	$x = -1$	$\Delta x = dx = 0.1$
99. $f(x) = 6x - x^3$	$x = 3$	$\Delta x = dx = 0.1$
100. $f(x) = 5x^{3/2}$	$x = 9$	$\Delta x = dx = 0.01$

Marginal Analysis In Exercises 101–106, use differentials to approximate the change in cost, revenue, or profit corresponding to an increase in sales (or production) of one unit. (For instance, in Exercise 101, approximate the change in cost as x increases from 10 units to 11 units.) Then compare this with the actual change in cost, revenue, or profit.

Function	x-Value
101. $C = 40x^2 + 1225$	$x = 10$
102. $C = 1.5\sqrt[3]{x} + 500$	$x = 125$
103. $R = 6.25x + 0.4x^{3/2}$	$x = 225$
104. $R = 80x - 0.35x^2$	$x = 80$
105. $P = 0.003x^2 + 0.019x - 1200$	$x = 750$
106. $P = -0.2x^3 + 3000x - 7500$	$x = 50$

Finding Differentials In Exercises 107–112, find the differential dy.

107. $y = 0.8x^3$

108. $y = 7x^4 + 2x^2$

109. $y = (3x^2 - 2)^3$

110. $y = \sqrt{36 - x^2}$

111. $y = \dfrac{2 - x}{x + 5}$

112. $y = \dfrac{3x^2}{x - 4}$

113. Profit The profit P (in dollars) for a company producing x units is

$$P = -0.8x^2 + 324x - 2000.$$

(a) Use differentials to approximate the change in profit when the production level changes from 100 to 101 units.

(b) Compare this with the actual change in profit.

114. Demand The demand function for a product is

$$p = 108 - 0.2x$$

where p is the price per unit (in dollars) and x is the number of units.

(a) Use differentials to approximate the change in revenue as sales increase from 20 units to 21 units. Compare this with the actual change in revenue.

(b) Repeat part (a) when sales increase from 40 units to 41 units.

115. Physiology: Body Surface Area The body surface area (BSA) of a 180-centimeter-tall (about six-foot-tall) person is modeled by

$$B = 0.1\sqrt{5w}$$

where B is the BSA (in square meters) and w is the weight (in kilograms). Use differentials to approximate the change in the person's BSA when the person's weight changes from 90 kilograms to 95 kilograms.

TEST YOURSELF

See *CalcChat.com* for tutorial help and worked-out solutions to odd-numbered exercises.

Take this test as you would take a test in class. When you are done, check your work against the answers given in the back of the book.

In Exercises 1–3, find the critical numbers and the open intervals on which the function is increasing or decreasing .

1. $f(x) = 3x^2 - 4$ **2.** $f(x) = 27x - x^3$ **3.** $f(x) = (x - 6)^4$

In Exercises 4–6, use the First-Derivative Test to find all relative extrema of the function.

4. $f(x) = \frac{1}{3}x^3 - 9x + 4$ **5.** $f(x) = 2x^4 - 4x^2 - 5$ **6.** $f(x) = \frac{8}{x^2 + 3}$

In Exercises 7–9, find the absolute extrema of the function on the closed interval. Use a graphing utility to verify your results.

7. $f(x) = x^2 + 6x + 8, \ [-4, 0]$ **8.** $f(x) = 12\sqrt{x} - 4x, \ [0, 5]$

9. $f(x) = \frac{6}{x} + \frac{x}{2}, \ [1, 6]$

In Exercises 10–13, discuss the concavity of the graph of the function and find the points of inflection.

10. $f(x) = x^5 - 80x^2$ **11.** $f(x) = \frac{20}{3x^2 + 8}$

12. $f(x) = x^4 + 6$ **13.** $f(x) = x^4 - 54x^2 + 230$

In Exercises 14 and 15, find all relative extrema of the function. Use the Second-Derivative Test when applicable.

14. $f(x) = x^3 - 6x^2 - 36x + 50$ **15.** $f(x) = \frac{3}{5}x^5 - 9x^3$

In Exercises 16–18, find the vertical and horizontal asymptotes of the graph of the function.

16. $f(x) = \frac{3x + 2}{x - 5}$ **17.** $f(x) = \frac{9x + 4}{x^2 + 8}$ **18.** $f(x) = \frac{2x^2 - 5}{x - 1}$

In Exercises 19–21, analyze and sketch the graph of the function. Label any intercepts, relative extrema, points of inflection, and asymptotes.

19. $y = -x^4 - 4x^3 - 3$ **20.** $y = x^5 - 5x$ **21.** $y = \frac{x}{x^2 - 4}$

22. Find the value of dy when $x = 3$ and $dx = 0.01$ for $f(x) = 5x^2 - 3$. Compare this with the value of Δy when $x = 3$ and $\Delta x = 0.01$.

23. A commodity has a demand function modeled by $p = 280 - 0.4x$, and a total cost function modeled by $C = 80x + 120$, where x is the number of units.

(a) What price yields a maximum profit?

(b) Find the average cost per unit when $x = 50$ and $x = 650$.

(c) Determine when the demand is elastic, inelastic, and of unit elasticity.

(d) Use differentials to approximate the change in revenue as sales increase from 210 units to 220 units. Compare this with the actual change in revenue.

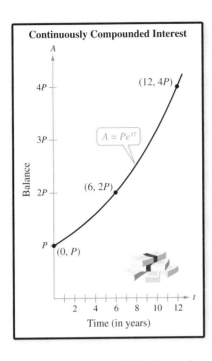

Continuously Compounded Interest

$A = Pe^{rt}$

Points shown: $(0, P)$, $(6, 2P)$, $(12, 4P)$

Balance vs. Time (in years)

Example 3 on page 297 shows how an exponential growth model can be used to find the annual interest rate of an account.

4 Exponential and Logarithmic Functions

4.1 Exponential Functions

Radioactive Decay

In Exercise 5 on page 256, you will evaluate an exponential function to find the remaining amount of a radioactive material.

■ Use the properties of exponents to evaluate and simplify exponential expressions.
■ Sketch the graphs of exponential functions.

Exponential Functions

You are already familiar with the behavior of algebraic functions such as

$$f(x) = x^2, \quad g(x) = \sqrt{x} = x^{1/2}, \quad \text{and} \quad h(x) = \frac{1}{x} = x^{-1}$$

each of which involves a variable raised to a constant power. By interchanging roles and raising a constant to a variable power, you obtain another important class of functions called **exponential functions.** Some examples are

$$f(x) = 2^x, \quad g(x) = \left(\frac{1}{10}\right)^x = \frac{1}{10^x}, \quad \text{and} \quad h(x) = 3^{2x} = 9^x.$$

In general, you can use any positive number $a \neq 1$ as the base of an exponential function.

Definition of Exponential Function

If $a > 0$ and $a \neq 1$, then the **exponential function** with base a is $f(x) = a^x$.

In the definition of an exponential function, the base $a = 1$ is excluded because it yields $f(x) = 1^x = 1$. This is a constant function, not an exponential function.

When working with exponential functions, the properties of exponents, shown below, are useful.

Properties of Exponents

Let a and b be positive real numbers, and let x and y be real numbers.

1. $a^0 = 1$ **2.** $a^x a^y = a^{x+y}$ **3.** $\dfrac{a^x}{a^y} = a^{x-y}$ **4.** $(a^x)^y = a^{xy}$

5. $(ab)^x = a^x b^x$ **6.** $\left(\dfrac{a}{b}\right)^x = \dfrac{a^x}{b^x}$ **7.** $a^{-x} = \dfrac{1}{a^x}$

EXAMPLE 1 Applying Properties of Exponents

a. $(2^2)(2^3) = 2^{2+3} = 2^5 = 32$ Apply Property 2.

b. $(2^2)(2^{-3}) = 2^{2-3} = 2^{-1} = \dfrac{1}{2}$ Apply Properties 2 and 7.

c. $(3^2)^3 = 3^{2(3)} = 3^6 = 729$ Apply Property 4.

✓ **Checkpoint 1**   *Worked-out solution available at LarsonAppliedCalculus.com*

Use the properties of exponents to simplify each expression.

a. $(3^2)(3^3)$ **b.** $(3^2)(3^{-1})$ **c.** $(2^3)^2$

EXAMPLE 2 Applying Properties of Exponents

a. $\left(\dfrac{1}{3}\right)^{-2} = \left(\dfrac{3}{1}\right)^{2} = \dfrac{3^2}{1^2} = 9$ Apply Properties 7 and 6.

b. $\dfrac{3^2}{3^3} = 3^{2-3} = 3^{-1} = \dfrac{1}{3}$ Apply Properties 3 and 7.

c. $(2^{1/2})(3^{1/2}) = [(2)(3)]^{1/2} = 6^{1/2} = \sqrt{6}$ Apply Property 5.

✓ **Checkpoint 2** Worked-out solution available at LarsonAppliedCalculus.com

Use the properties of exponents to simplify each expression.

a. $\left(\dfrac{1}{2}\right)^{-3}$ **b.** $\dfrac{2^2}{2^3}$ **c.** $(2^{1/2})(5^{1/2})$

Although Examples 1 and 2 demonstrate the properties of exponents with integer and rational exponents, it is important to realize that the properties hold for *all* real number exponents. With a calculator, you can evaluate a^x for any positive number a and any real number x. Here are some examples.

$$2^{-0.6} \approx 0.660, \qquad \pi^{0.75} \approx 2.360, \qquad (1.56)^{\sqrt{2}} \approx 1.876$$

EXAMPLE 3 Dating Organic Material

Organic Material

FIGURE 4.1

In living organic material, the ratio of radioactive carbon isotopes to the total number of carbon atoms is about 1 to 10^{12}. When organic material dies, its radioactive carbon isotopes begin to decay, with a half-life of about 5715 years. This means that after 5715 years, the ratio of isotopes to atoms will have decreased to one-half of the original ratio; after a second 5715 years, the ratio will have decreased to one-fourth of the original; and so on. Figure 4.1 shows this decreasing ratio. The formula for the ratio R of carbon isotopes to carbon atoms is

$$R = \left(\dfrac{1}{10^{12}}\right)\left(\dfrac{1}{2}\right)^{t/5715}$$

where t is the time in years. Find the value of R for each period of time.

a. 10,000 years

b. 20,000 years

c. 25,000 years

SOLUTION

a. $R = \left(\dfrac{1}{10^{12}}\right)\left(\dfrac{1}{2}\right)^{10,000/5715} \approx 2.973 \times 10^{-13}$ Ratio for 10,000 years

b. $R = \left(\dfrac{1}{10^{12}}\right)\left(\dfrac{1}{2}\right)^{20,000/5715} \approx 8.842 \times 10^{-14}$ Ratio for 20,000 years

c. $R = \left(\dfrac{1}{10^{12}}\right)\left(\dfrac{1}{2}\right)^{25,000/5715} \approx 4.821 \times 10^{-14}$ Ratio for 25,000 years

✓ **Checkpoint 3** Worked-out solution available at LarsonAppliedCalculus.com

Use the formula for the ratio of carbon isotopes to carbon atoms in Example 3 to find the value of R for each period of time.

a. 5000 years   **b.** 15,000 years   **c.** 30,000 years

Graphs of Exponential Functions

The basic nature of the graph of an exponential function can be determined by the point-plotting method or by using a graphing utility.

EXAMPLE 4   Graphing Exponential Functions

Sketch the graph of each exponential function.

a. $f(x) = 2^x$

b. $g(x) = \left(\dfrac{1}{2}\right)^x = 2^{-x}$

c. $h(x) = 3^x$

SOLUTION   To sketch these functions by hand, you can begin by constructing a table of values, as shown below.

x	-3	-2	-1	0	1	2	3	4
$f(x) = 2^x$	$\frac{1}{8}$	$\frac{1}{4}$	$\frac{1}{2}$	1	2	4	8	16
$g(x) = 2^{-x}$	8	4	2	1	$\frac{1}{2}$	$\frac{1}{4}$	$\frac{1}{8}$	$\frac{1}{16}$
$h(x) = 3^x$	$\frac{1}{27}$	$\frac{1}{9}$	$\frac{1}{3}$	1	3	9	27	81

The graphs of the three functions are shown in Figure 4.2. Note that the graphs of $f(x) = 2^x$ and $h(x) = 3^x$ are increasing, whereas the graph of $g(x) = 2^{-x}$ is decreasing.

(a)

(b)

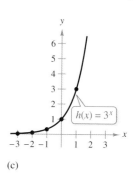

(c)

FIGURE 4.2

✓ *Checkpoint 4*   *Worked-out solution available at LarsonAppliedCalculus.com*

Sketch the graph of

$$f(x) = 5^x.$$

TECH TUTOR

Try graphing the functions

$$f(x) = 2^x \quad \text{and} \quad h(x) = 3^x$$

in the same viewing window, as shown at the right. From the display, you can see that the graph of h is increasing more rapidly than the graph of f.

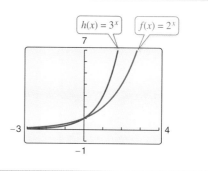

The forms of the graphs in Figure 4.2 are typical of the graphs of the exponential functions $y = a^{-x}$ and $y = a^x$, where $a > 1$. The basic characteristics of such graphs are summarized in Figure 4.3.

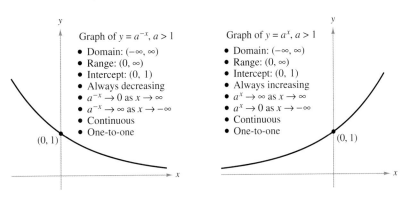

Graph of $y = a^{-x}$, $a > 1$
- Domain: $(-\infty, \infty)$
- Range: $(0, \infty)$
- Intercept: $(0, 1)$
- Always decreasing
- $a^{-x} \to 0$ as $x \to \infty$
- $a^{-x} \to \infty$ as $x \to -\infty$
- Continuous
- One-to-one

Graph of $y = a^x$, $a > 1$
- Domain: $(-\infty, \infty)$
- Range: $(0, \infty)$
- Intercept: $(0, 1)$
- Always increasing
- $a^x \to \infty$ as $x \to \infty$
- $a^x \to 0$ as $x \to -\infty$
- Continuous
- One-to-one

Characteristics of the Exponential Functions $y = a^{-x}$ and $y = a^x$, where $a > 1$

FIGURE 4.3

EXAMPLE 5   **Graphing an Exponential Function**

Sketch the graph of $f(x) = 3^{-x} - 1$.

SOLUTION   Begin by creating a table of values, as shown below.

x	-2	-1	0	1	2
$f(x)$	$3^2 - 1 = 8$	$3^1 - 1 = 2$	$3^0 - 1 = 0$	$3^{-1} - 1 = -\frac{2}{3}$	$3^{-2} - 1 = -\frac{8}{9}$

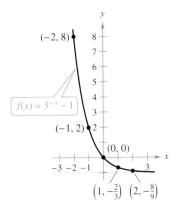

FIGURE 4.4

From the limit

$$\lim_{x \to \infty} (3^{-x} - 1) = \lim_{x \to \infty} 3^{-x} - \lim_{x \to \infty} 1$$
$$= \lim_{x \to \infty} \frac{1}{3^x} - \lim_{x \to \infty} 1$$
$$= 0 - 1$$
$$= -1$$

you can see that $y = -1$ is a horizontal asymptote of the graph. The graph is shown in Figure 4.4.

 Checkpoint 5   Worked-out solution available at LarsonAppliedCalculus.com

Sketch the graph of $f(x) = 2^{-x} + 1$.   ■

SUMMARIZE   (Section 4.1)

1. State the definition of an exponential function *(page 252)*. For examples of exponential functions, see Example 4.

2. State the properties of exponents *(page 252)*. For examples of using the properties of exponents to simplify expressions, see Examples 1 and 2.

3. State the basic characteristics of the graphs of the exponential functions $y = a^{-x}$ and $y = a^x$, where $a > 1$ *(page 255)*. For an example of the graph of an exponential function, see Example 5.

SKILLS WARM UP 4.1

The following warm-up exercises involve skills that were covered in earlier sections. You will use these skills in the exercise set for this section. For additional help, review Section 1.4.

In Exercises 1–6, describe how the graph of g is related to the graph of f.

1. $g(x) = f(x + 2)$

2. $g(x) = -f(x)$

3. $g(x) = f(x) - 3$

4. $g(x) = f(-x)$

5. $g(x) = f(x - 1)$

6. $g(x) = f(x) + 2$

In Exercises 7–12, evaluate each expression.

7. $25^{3/2}$

8. $16^{3/4}$

9. $27^{2/3}$

10. $\left(\dfrac{1}{5}\right)^{3}$

11. $\left(\dfrac{1}{8}\right)^{1/3}$

12. $\left(\dfrac{5}{8}\right)^{2}$

In Exercises 13–18, solve for x.

13. $2x - 6 = 4$

14. $3x + 1 = 5$

15. $(x + 4)^2 = 25$

16. $(x - 2)^2 = 8$

17. $x^2 + 4x - 5 = 0$

18. $2x^2 - 3x + 1 = 0$

Exercises 4.1

See *CalcChat.com* for tutorial help and worked-out solutions to odd-numbered exercises.

 Applying Properties of Exponents In Exercises 1–4, use the properties of exponents to simplify each expression. *See Examples 1 and 2.*

1. (a) $(6^2)(6^4)$ (b) $(9^3)(9^{-4})$

   (c) $(5^2)^4$ (d) 5^{-3}

2. (a) $\dfrac{6^4}{6^8}$ (b) $\left(\dfrac{1}{5}\right)^{-2}$

   (c) $(8^{1/2})(2^{1/2})$ (d) $(7^{-2})\left(\dfrac{1}{2}\right)^{3}(2^0)$

3. (a) $\dfrac{7^5}{49^3}$ (b) $(9^{2/3})(3)(3^{2/3})$

   (c) $[(25^{1/2})(5^2)]^{1/3}$ (d) $\left(\dfrac{1}{8}\right)^{-3}(2^{-3})(4^0)$

4. (a) $(4^3)(4^2)$ (b) $\left(\dfrac{1}{4}\right)^{2}(4^2)$

   (c) $(4^6)^{1/2}$ (d) $[(8^{-1})(8^{2/3})]^3$

5. Radioactive Decay Beginning with 16 grams of a radioactive element whose half-life is 30 years, the mass y (in grams) remaining after t years is given by

$$y = 16\left(\frac{1}{2}\right)^{t/30}, \quad t \geq 0.$$

How much of the initial mass remains after 90 years?

6. Radioactive Decay Beginning with 23 grams of a radioactive element whose half-life is 45 years, the mass y (in grams) remaining after t years is given by

$$y = 23\left(\frac{1}{2}\right)^{t/45}, \quad t \geq 0.$$

How much of the initial mass remains after 150 years?

 Graphing Exponential Functions In Exercises 7–18, sketch the graph of the function. *See Examples 4 and 5.*

7. $f(x) = 6^x$

8. $f(x) = 4^x$

9. $f(x) = \left(\dfrac{1}{5}\right)^{x} = 5^{-x}$

10. $f(x) = \left(\dfrac{1}{4}\right)^{x} = 4^{-x}$

11. $y = 2^x - 3$

12. $y = 3^{x+2}$

13. $y = -2^x$

14. $y = -5^x$

15. $y = 3^{-x^2}$

16. $y = 2^{-x^2}$

17. $s(t) = \frac{1}{4}(3^{-t})$

18. $s(t) = 2^{-t} + 3$

19. Population Growth The resident populations P (in millions) of Texas from 2005 through 2014 can be modeled by the exponential function $P(t) = 20.922(1.0186)^t$, where t is the time in years, with $t = 5$ corresponding to 2005. Use the model to estimate the populations in the years (a) 2017 and (b) 2022. *(Source: U.S. Census Bureau)*

20. Sales The sales S (in billions of dollars) for Starbucks from 2009 through 2014 can be modeled by the exponential function

$$S(t) = 3.71(1.112)^t$$

where t is the time in years, with $t = 9$ corresponding to 2009. *(Source: Starbucks Corporation)*

(a) Use the model to estimate the sales in 2016.

(b) Use the model to estimate the sales in 2020.

21. Property Value A piece of property sells for $64,000. The value of the property doubles every 15 years. A model for the value V of the property t years after the date of purchase is

$$V(t) = 64,000(2)^{t/15}.$$

Use the model to approximate the value of the property (a) 5 years and (b) 20 years after it is purchased.

22. HOW DO YOU SEE IT? Match the exponential function with its graph. Explain your reasoning. [The graphs are labeled (i), (ii), (iii), (iv), (v), and (vi).]

(i)

(ii)

(iii)

(iv)

(v)

(vi)

(a) $f(x) = 3^x$

(b) $f(x) = 3^{-x/2}$

(c) $f(x) = -3^x$

(d) $f(x) = 3^{x-2}$

(e) $f(x) = 3^{-x} - 1$

(f) $f(x) = 3^x + 2$

23. Inflation Rate With an annual rate of inflation of 4% over the next 10 years, the approximate cost C of goods or services during any year in the decade is given by

$$C(t) = P(1.04)^t, \ 0 \leq t \leq 10$$

where t is the time (in years) and P is the present cost. The price of an oil change for a car is presently $24.95. Estimate the price 10 years from now.

24. Inflation Rate Repeat Exercise 23 using an annual rate of inflation of 1.5% over the next 10 years. The approximate cost C of goods or services is given by

$$C(t) = P(1.015)^t, \ 0 \leq t \leq 10.$$

25. Depreciation A car sells for $30,000. The car depreciates such that each year it is worth $\frac{4}{5}$ of its value from the previous year. Find a model for the value V of the car after t years. Sketch a graph of the model and determine the value of the car 5 years after it is purchased.

26. Drug Concentration Immediately following an injection, the concentration of a drug in the bloodstream is 300 milligrams per milliliter. After t hours, the concentration is 75% of the level of the previous hour. Find a model for $C(t)$, the concentration of the drug after t hours. Sketch a graph of the model and determine the concentration of the drug after 8 hours.

27. Pharmacists For the years 2008 through 2013, the average salaries y (in dollars) of pharmacists in the United States are shown in the table. *(Source: U.S. Bureau of Labor Statistics)*

Year	2008	2009	2010
Salary	104,260	106,630	109,380

Year	2011	2012	2013
Salary	112,160	114,950	116,500

Spreadsheet at LarsonAppliedCalculus.com

A model for these data is given by

$$y = 86,827(1.0233)^t$$

where t represents the year, with $t = 8$ corresponding to 2008.

(a) Compare the actual salaries with those given by the model. How well does the model fit the data? Explain your reasoning.

(b) Use a graphing utility to graph the model.

(c) Use the *zoom* and *trace* features of the graphing utility to predict the year during which the average salary of pharmacists will reach $130,000.

4.2 Natural Exponential Functions

■ Evaluate and graph functions involving the natural exponential function.
■ Solve compound interest problems.
■ Solve present value problems.

Natural Exponential Functions

In Section 4.1, exponential functions were introduced using an unspecified base *a*. In calculus, the most convenient (or natural) choice for a base is the irrational number *e*, whose decimal approximation is

$$e \approx 2.71828182846.$$

Although this choice of base may seem unusual, its convenience will become apparent as the rules for differentiating exponential functions are developed in Section 4.3. In that development, you will encounter the limit used in the definition of *e*.

Zambia

Limit Definition of e

The irrational number *e* is defined to be the limit of $(1 + x)^{1/x}$ as $x \to 0$. That is,

$$\lim_{x \to 0} (1 + x)^{1/x} = e.$$

EXAMPLE 1 Graphing the Natural Exponential Function

Complete the table of values for $f(x) = e^x$. Then sketch the graph of f.

x	-2	-1	0	1	2
$f(x)$					

SOLUTION Begin by completing the table as shown.

x	-2	-1	0	1	2
$f(x)$	$e^{-2} \approx 0.135$	$e^{-1} \approx 0.368$	$e^0 = 1$	$e^1 \approx 2.718$	$e^2 \approx 7.389$

Then use the point-plotting method to sketch the graph of f, as shown in Figure 4.5. Note that e^x is positive for all values of x. Moreover, the graph has the x-axis as a horizontal asymptote to the left. That is,

$$\lim_{x \to -\infty} e^x = 0.$$

In Exercise 46 on page 266, you will evaluate a natural exponential function to find the population of Zambia for several years.

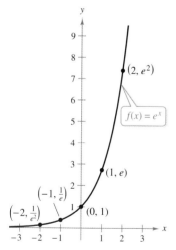

FIGURE 4.5

✓ **Checkpoint 1** *Worked-out solution available at LarsonAppliedCalculus.com*

Complete the table of values for $g(x) = e^{-x}$. Then sketch the graph of g.

x	-2	-1	0	1	2
$g(x)$					

Exponential functions are often used to model the growth of a quantity or a population. When the quantity's growth *is not* restricted, an exponential model is often used. When the quantity's growth *is* restricted, the best model is often a **logistic growth model** of the form

$$f(t) = \frac{a}{1 + be^{-kt}}.$$

Graphs of both types of population growth models are shown in Figure 4.6. The graph of a logistic growth model is called a *logistic curve*.

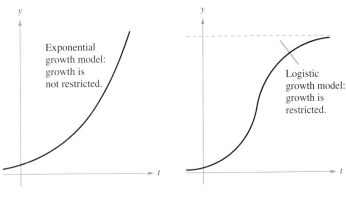

Exponential growth model: growth is not restricted.

Logistic growth model: growth is restricted.

FIGURE 4.6

EXAMPLE 2   Modeling a Population

A bacterial culture is growing according to the logistic growth model

$$y = \frac{1.25}{1 + 0.25e^{-0.4t}}, \quad t \geq 0$$

where y is the culture weight (in grams) and t is the time (in hours). Find the weight of the culture after 0 hours, 1 hour, and 10 hours. What is the limit of the model as t increases without bound?

SOLUTION   The graph of the model is shown in Figure 4.7.

$$y = \frac{1.25}{1 + 0.25e^{-0.4(0)}} \approx 1 \text{ gram} \qquad \text{Weight when } t = 0$$

$$y = \frac{1.25}{1 + 0.25e^{-0.4(1)}} \approx 1.071 \text{ grams} \qquad \text{Weight when } t = 1$$

$$y = \frac{1.25}{1 + 0.25e^{-0.4(10)}} \approx 1.244 \text{ grams} \qquad \text{Weight when } t = 10$$

As t approaches infinity, the limit of y is

$$\lim_{t \to \infty} \frac{1.25}{1 + 0.25e^{-0.4t}} = \lim_{t \to \infty} \frac{1.25}{1 + (0.25/e^{0.4t})} = \frac{1.25}{1 + 0} = 1.25.$$

So, as t increases without bound, the weight of the culture approaches 1.25 grams.

Growth of Bacterial Culture

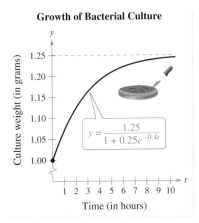

$$y = \frac{1.25}{1 + 0.25e^{-0.4t}}$$

Time (in hours)

When a culture is grown in a dish, the size of the dish and the available food limit the culture's growth.

FIGURE 4.7

✓ **Checkpoint 2**   *Worked-out solution available at LarsonAppliedCalculus.com*

A bacterial culture is growing according to the model $y = 1.50/(1 + 0.2e^{-0.5t})$, $t \geq 0$, where y is the culture weight (in grams) and t is the time (in hours). Find the weight of the culture after 0 hours, 1 hour, and 10 hours. What is the limit of the model as t increases without bound?

Extended Application: Compound Interest

An amount of P dollars is deposited in an account at an annual interest rate of r (in decimal form). What is the balance after 1 year? The answer depends on the number of times the interest is compounded, according to the formula

$$A = P\left(1 + \frac{r}{n}\right)^n$$

where n is the number of compoundings per year. The balances after 1 year for a deposit of \$1000 at 8%, for various compounding periods, are shown in the table.

Number of times compounded per year, n	Balance (in dollars), A
Annually, $n = 1$	$A = 1000\left(1 + \frac{0.08}{1}\right)^1 = \$1080.00$
Semiannually, $n = 2$	$A = 1000\left(1 + \frac{0.08}{2}\right)^2 = \$1081.60$
Quarterly, $n = 4$	$A = 1000\left(1 + \frac{0.08}{4}\right)^4 = \$1082.43$
Monthly, $n = 12$	$A = 1000\left(1 + \frac{0.08}{12}\right)^{12} = \$1083.00$
Daily, $n = 365$	$A = 1000\left(1 + \frac{0.08}{365}\right)^{365} = \$1083.28$

You may be surprised to discover that as n increases, the balance A approaches a limit, as indicated in the following development. In this development, let

$$x = \frac{r}{n} \quad \text{or} \quad \frac{1}{x} = \frac{n}{r}.$$

Then $x \to 0$ as $n \to \infty$, and you have

$$A = \lim_{n \to \infty} P\left(1 + \frac{r}{n}\right)^n \qquad \text{Take limit as } n \to \infty.$$

$$= P \lim_{n \to \infty} \left(1 + \frac{r}{n}\right)^n \qquad \text{Apply property of limits.}$$

$$= P \lim_{n \to \infty} \left[\left(1 + \frac{r}{n}\right)^{n/r}\right]^r \qquad \text{Rewrite.}$$

$$= P\left[\lim_{x \to 0} (1 + x)^{1/x}\right]^r \qquad \text{Let } x = r/n. \text{ Then } x \to 0 \text{ as } n \to \infty.$$

$$= Pe^r. \qquad \text{Apply limit definition of } e.$$

This limit is the balance after 1 year of **continuous compounding.** So, for a deposit of \$1000 at 8%, compounded continuously, the balance at the end of the year would be

$$A = 1000e^{0.08} = \$1083.29.$$

To find the balance in an account after t years, use the appropriate formula from the summary.

STUDY TIP

The interest rate r in the formulas for compound interest should be written in decimal form. For instance, an interest rate of 2.5% would be written as $r = 0.025$.

Summary of Compound Interest Formulas

Let P be the amount deposited, t the number of years, A the balance, and r the annual interest rate (in decimal form).

1. Compounded n times per year: $A = P\left(1 + \frac{r}{n}\right)^{nt}$

2. Compounded continuously: $A = Pe^{rt}$

The average interest rates paid by banks on savings accounts have varied greatly during the past several decades. At times savings accounts have earned as much as 12% annual interest, and at times they have earned less than 1%. The next example shows how the annual interest rate can affect the balance of an account.

EXAMPLE 3   Finding Account Balances

You are creating a trust fund for your newborn nephew. You plan to deposit $12,000 in an account, with instructions that the account be turned over to your nephew on his 25th birthday. Compare the balances in the account for each situation. Which account should you choose?

a. 7%, compounded continuously

b. 7%, compounded quarterly

c. 11%, compounded continuously

d. 11%, compounded quarterly

SOLUTION

a. $A = 12{,}000e^{0.07(25)} = \$69{,}055.23$ 7%, compounded continuously

b. $A = 12{,}000\left(1 + \dfrac{0.07}{4}\right)^{4(25)} = \$68{,}017.87$ 7%, compounded quarterly

c. $A = 12{,}000e^{0.11(25)} = \$187{,}711.58$ 11%, compounded continuously

d. $A = 12{,}000\left(1 + \dfrac{0.11}{4}\right)^{4(25)} = \$180{,}869.07$ 11%, compounded quarterly

The balances in the accounts for parts (a) and (c) are shown in Figure 4.8. Notice the dramatic difference between the balances at 7% and 11%. You should choose the account described in part (c) because it earns more money than the other accounts.

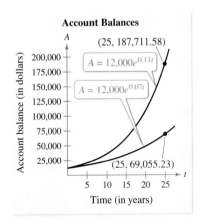

Account Balances

FIGURE 4.8

✓ **Checkpoint 3**   *Worked-out solution available at LarsonAppliedCalculus.com*

Find the balance in an account when $2000 is deposited for 10 years at an interest rate of 9%, compounded as follows. Compare the results and make a general statement about compounding.

a. quarterly   **b.** monthly

c. daily   **d.** continuously

In Example 3, note that the interest earned depends on the frequency with which the interest is compounded. The annual interest rate is called the **stated rate** or **nominal rate.** However, the nominal rate does not reflect the actual rate at which interest is earned, which means that the compounding produced an **effective rate** that is greater than the nominal rate. In general, the effective rate corresponding to a nominal rate of r that is compounded n times per year is

$$\text{Effective rate} = r_{eff} = \left(1 + \frac{r}{n}\right)^n - 1.$$

EXAMPLE 4   Finding the Effective Rate of Interest

Find the effective rate of interest corresponding to a nominal rate of 6% per year compounded (a) annually, (b) semiannually, (c) quarterly, and (d) monthly.

SOLUTION

a. $r_{eff} = \left(1 + \dfrac{r}{n}\right)^n - 1$ Formula for effective rate of interest

$\qquad = \left(1 + \dfrac{0.06}{1}\right)^1 - 1$ Substitute for r and n.

$\qquad = 1.06 - 1$ Simplify.

$\qquad = 0.06$

So, the effective rate is 6% per year.

b. $r_{eff} = \left(1 + \dfrac{r}{n}\right)^n - 1$ Formula for effective rate of interest

$\qquad = \left(1 + \dfrac{0.06}{2}\right)^2 - 1$ Substitute for r and n.

$\qquad = (1.03)^2 - 1$ Simplify.

$\qquad = 0.0609$

So, the effective rate is 6.09% per year.

c. $r_{eff} = \left(1 + \dfrac{r}{n}\right)^n - 1$ Formula for effective rate of interest

$\qquad = \left(1 + \dfrac{0.06}{4}\right)^4 - 1$ Substitute for r and n.

$\qquad = (1.015)^4 - 1$ Simplify.

$\qquad \approx 0.0614$

So, the effective rate is about 6.14% per year.

d. $r_{eff} = \left(1 + \dfrac{r}{n}\right)^n - 1$ Formula for effective rate of interest

$\qquad = \left(1 + \dfrac{0.06}{12}\right)^{12} - 1$ Substitute for r and n.

$\qquad = (1.005)^{12} - 1$ Simplify.

$\qquad \approx 0.0617$

So, the effective rate is about 6.17% per year.

✓ *Checkpoint 4*   *Worked-out solution available at LarsonAppliedCalculus.com*

Repeat Example 4 using a nominal rate of 7%.

Present Value

In planning for the future, this problem often arises: "How much money P should be deposited now, at a fixed rate of interest r, in order to have a balance of A, t years from now?" The answer to this question is given by the **present value** of A.

To find the present value of a future investment, use the formula for compound interest as shown.

$$A = P\left(1 + \frac{r}{n}\right)^{nt} \qquad \text{Formula for compound interest}$$

Solving for P gives a present value of

$$P = \frac{A}{\left(1 + \dfrac{r}{n}\right)^{nt}} \quad \text{or} \quad P = \frac{A}{(1 + i)^N}$$

where $i = r/n$ is the interest rate per compounding period and $N = nt$ is the total number of compounding periods. You will learn another way to find the present value of a future investment in Section 6.1.

EXAMPLE 5 **Finding Present Value**

An investor is purchasing a 10-year certificate of deposit that pays an annual interest rate of 8%, compounded monthly. How much should the person invest in order to obtain a balance of $15,000 at maturity?

SOLUTION Here, $A = 15,000$, $r = 0.08$, $n = 12$, and $t = 10$. Using the formula for present value, you obtain

$$P = \frac{15,000}{\left(1 + \dfrac{0.08}{12}\right)^{12(10)}} \qquad \text{Substitute for } A, r, n, \text{ and } t.$$

$$\approx 6757.85. \qquad \text{Simplify.}$$

So, the person should invest $6757.85 in the certificate of deposit.

✓ **Checkpoint 5**   *Worked-out solution available at LarsonAppliedCalculus.com*

How much money should be deposited in an account paying 6% interest compounded monthly in order to have a balance of $20,000 after 3 years? ■

SUMMARIZE (Section 4.2)

1. State the limit definition of e *(page 258)*. For an example of a graph of a natural exponential function, see Example 1.

2. Describe a real-life example of how an exponential function can be used to model a population *(page 259, Example 2)*.

3. State the compound interest formulas for n compoundings per year and for continuous compounding *(page 260)*. For applications of these formulas, see Example 3.

4. State the formula for finding the effective rate of interest *(page 262)*. For an application of this formula, see Example 4.

5. State the formula for present value *(page 263)*. For an application of this formula, see Example 5.

SKILLS WARM UP 4.2

The following warm-up exercises involve skills that were covered in earlier sections. You will use these skills in the exercise set for this section. For additional help, review Sections 1.6 and 3.6.

In Exercises 1–4, discuss the continuity of the function.

1. $f(x) = \dfrac{3x^2 + 2x + 1}{x^2 + 1}$

2. $f(x) = \dfrac{x + 1}{x^2 - 4}$

3. $f(x) = \dfrac{x^2 - 6x + 5}{x^2 - 3}$

4. $g(x) = \dfrac{x^2 - 9x + 20}{x - 4}$

In Exercises 5–12, find the horizontal asymptote of the graph of the function.

5. $f(x) = \dfrac{25}{1 + 4x}$

6. $f(x) = \dfrac{16x}{3 + x^2}$

7. $f(x) = \dfrac{8x^3 + 2}{2x^3 + x}$

8. $f(x) = \dfrac{x}{2x}$

9. $f(x) = \dfrac{3}{2 + (1/x)}$

10. $f(x) = \dfrac{6}{1 + x^{-2}}$

11. $f(x) = 8^{-x}$

12. $f(x) = \dfrac{3x^2 + 8}{2x + 11}$

Exercises 4.2

See *CalcChat.com* for tutorial help and worked-out solutions to odd-numbered exercises.

 Applying Properties of Exponents In Exercises 1–4, use the properties of exponents to simplify each expression.

1. (a) $(e^3)(e^4)$　　(b) $(e^3)^4$

　(c) $(e^3)^{-2}$　　(d) e^0

2. (a) $\left(\dfrac{1}{e}\right)^{-2}$　　(b) $\left(\dfrac{e^5}{e^2}\right)^{-1}$

　(c) $(e^5)(e^3)$　　(d) $\dfrac{e^0}{e^{-4}}$

3. (a) $(e^2)^{5/2}$　　(b) $(e^2)(e^{1/2})$

　(c) $(e^{-2})^{-3}$　　(d) $\dfrac{e^5}{e^{-2}}$

4. (a) $(e^{-3})^{2/3}$　　(b) $\dfrac{e^4}{e^{-1/2}}$

　(c) $(e^{-2})^{-4}$　　(d) $(e^{-4})(e^{-3/2})$

Matching In Exercises 5–10, match the function with its graph. [The graphs are labeled (a)–(f).]

(a)

(b)

(c)

(d)

(e)

(f)

5. $f(x) = e^{2x+1}$

6. $f(x) = e^{-x/2}$

7. $f(x) = e^{x^2}$

8. $f(x) = e^{-1/x}$

9. $f(x) = e^{\sqrt{x}}$

10. $f(x) = -e^x + 1$

 Graphing Natural Exponential Functions In Exercises 11–16, sketch the graph of the function. *See Example 1.*

11. $f(x) = e^{-x/3}$

12. $f(x) = e^{2x}$

13. $g(x) = e^x - 2$

14. $h(x) = e^{-x} + 5$

15. $g(x) = e^{1-x}$

16. $j(x) = e^{x+2}$

Graphing Functions In Exercises 17–24, use a graphing utility to graph the function. Determine whether the function has any horizontal asymptotes and discuss the continuity of the function.

17. $N(t) = 600e^{-0.4t}$

18. $A(t) = 500e^{0.15t}$

19. $g(x) = \dfrac{4}{1 + e^{x^3}}$

20. $g(x) = \dfrac{10}{1 + e^{-x}}$

21. $f(x) = \dfrac{e^x + e^{-x}}{2}$

22. $f(x) = \dfrac{e^x - e^{-x}}{2}$

23. $f(x) = \dfrac{2}{1 + e^{1/x}}$

24. $f(x) = \dfrac{2}{1 + 2e^{-0.2x}}$

25. Graphing Exponential Functions Use a graphing utility to graph $f(x) = e^x$ and the given function in the same viewing window. How are the two graphs related?

(a) $g(x) = e^{x-2}$ (b) $h(x) = -\frac{1}{2}e^x$

(c) $q(x) = e^x + 3$

26. Graphing Logistic Growth Functions Use a graphing utility to graph the function. Describe the shape of the graph for very large and very small values of x.

(a) $f(x) = \dfrac{8}{1 + e^{-0.5x}}$ (b) $g(x) = \dfrac{8}{1 + e^{-0.5/x}}$

 Finding Account Balances In Exercises 27–30, complete the table to determine the balance A for $3000 invested at rate r for t years, compounded n times per year. *See Example 3.*

n	1	2	4	12	365	Continuous compounding
A						

27. $r = 3\%, t = 10$ years

28. $r = 2.5\%, t = 20$ years

29. $r = 4.5\%, t = 30$ years

30. $r = 5\%, t = 40$ years

 Finding Present Value In Exercises 31–34, complete the table to determine the amount of money P that should be invested at rate r to produce a final balance of $100,000 in t years. *See Example 5.*

t	1	10	20	30	40	50
P						

31. $r = 4\%$, compounded continuously

32. $r = 3\%$, compounded continuously

33. $r = 5\%$, compounded monthly

34. $r = 6\%$, compounded daily

35. Trust Fund You deposit $15,000 in a trust fund for your newborn niece that pays 8% interest, compounded continuously. The account is turned over to your niece on her 30th birthday. Determine the balance in the account when your niece turns 30.

36. Trust Fund A deposit of $10,000 is made in a trust fund that pays 7% interest, compounded continuously. It is specified that the balance will be given to the college from which the donor graduated after the money has earned interest for 50 years. How much will the college receive?

37. Effective Rate Find the effective rate of interest corresponding to a nominal rate of 9% per year compounded (a) annually, (b) semiannually, (c) quarterly, and (d) monthly.

38. Effective Rate Find the effective rate of interest corresponding to a nominal rate of 7.5% per year compounded (a) annually, (b) semiannually, (c) quarterly, and (d) monthly.

39. Present Value How much should be deposited in an account paying 7.2% interest compounded monthly in order to have a balance of $8000 after 3 years?

40. Present Value How much should be deposited in an account paying 7.8% interest compounded monthly in order to have a balance of $21,000 after 4 years?

41. Demand The demand function for a product is modeled by

$$p = 5000\left(1 - \frac{4}{4 + e^{-0.002x}}\right).$$

Find the price p (in dollars) of the product when the quantity demanded is (a) $x = 100$ units and (b) $x = 500$ units. (c) What is the limit of the price as x increases without bound?

42. Demand The demand function for a product is modeled by

$$p = 10,000\left(1 - \frac{3}{3 + e^{-0.001x}}\right).$$

Find the price p (in dollars) of the product when the quantity demanded is (a) $x = 1000$ units and (b) $x = 1500$ units. (c) What is the limit of the price as x increases without bound?

43. Probability The average time between incoming calls at a switchboard is 3 minutes. If a call has just come in, then the probability that the next call will come within the next t minutes is

$$P(t) = 1 - e^{-t/3}.$$

Find the probability of each situation.

(a) A call comes in within $\frac{1}{2}$ minute.

(b) A call comes in within 2 minutes.

(c) A call comes in within 5 minutes.

44. Consumer Awareness An automobile gets 28 miles per gallon at speeds up to and including 50 miles per hour. At speeds greater than 50 miles per hour, the number of miles per gallon drops at the rate of 12% for each 10 miles per hour. If s is the speed (in miles per hour) and y is the number of miles per gallon, then

$$y = 28e^{0.6 - 0.012s}, \ s > 50.$$

Use this information and a spreadsheet to create a table showing the miles per gallon for $s = 50, 55, 60, 65,$ and 70. What can you conclude?

45. Revenue The revenues R (in billions of dollars) for Netflix from 2007 through 2014 are shown in the table. *(Source: Netflix, Inc.)*

Year	2007	2008	2009	2010
Revenue	1.21	1.36	1.67	2.16

Year	2011	2012	2013	2014
Revenue	3.20	3.61	4.37	5.50

Spreadsheet at LarsonAppliedCalculus.com

A model for these data is given by $R = 0.23e^{0.228t}$, where t represents the year, with $t = 7$ corresponding to 2007.

(a) How well does the model fit the data?

(b) Find a linear model for the data. How well does the linear model fit the data? Which model, exponential or linear, is a better fit?

(c) Use both models to predict the year in which the revenue of Netflix will exceed 20 billion dollars.

46. Population The populations P (in millions) of Zambia from 1950 through 2013 can be modeled by $P = 2.49e^{0.0276t}$, where t is the time in years, with $t = 0$ corresponding to 1950. *(Source: U.S. Census Bureau)*

(a) Find the populations in 1950, 1960, 1980, 1990, 2010, and 2013.

(b) Explain why the change in population from 1950 to 1960 is not the same as the change in population from 1980 to 1990.

(c) Use the model to estimate the population in 2023.

47. Certificate of Deposit You want to invest $5000 in a certificate of deposit for 12 months. You are given the options below. Which would you choose? Explain.

(a) $r = 5.25\%$, quarterly compounding

(b) $r = 5\%$, monthly compounding

(c) $r = 4.75\%$, continuous compounding

48. HOW DO YOU SEE IT? The figure shows the graphs of $y = 2^x$, $y = e^x$, $y = 10^x$, $y = 2^{-x}$, $y = e^{-x}$, and $y = 10^{-x}$. Match each function with its graph. (The graphs are labeled a–f.) Explain your reasoning.

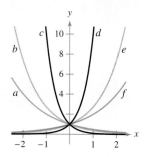

49. Learning Theory In a learning theory project, the proportion P of correct responses after n trials can be modeled by $P = 0.83/(1 + e^{-0.2n})$.

(a) Find the proportion of correct responses after 3 trials.

(b) Find the proportion of correct responses after 7 trials.

(c) Use a graphing utility to graph the model. Find the number of trials required for the proportion of correct responses to be 0.75.

(d) Does the proportion of correct responses have a limit as n increases without bound? Explain your reasoning.

50. Learning Theory In a typing class, the average number N of words per minute typed after t weeks of lessons can be modeled by $N = 95/(1 + 8.5e^{-0.12t})$.

(a) Find the average number of words per minute typed after 10 weeks.

(b) Find the average number of words per minute typed after 20 weeks.

(c) Use a graphing utility to graph the model. Find the number of weeks required to achieve an average of 70 words per minute.

(d) Does the number of words per minute have a limit as t increases without bound? Explain your reasoning.

51. Biology The population y of a bacterial culture is modeled by the logistic growth function $y = 925/(1 + e^{-0.3t})$, where t is the time in days.

(a) Use a graphing utility to graph the model. Find the number of days it takes for the population to reach 711.

(b) Does the population have a limit as t increases without bound? Explain your answer.

(c) How does the limit change when the model is $y = 1000/(1 + e^{-0.3t})$? Explain your answer. Draw some conclusions about this type of model.

4.3 Derivatives of Exponential Functions

■ Find the derivatives of natural exponential functions.

■ Use calculus to analyze the graphs of real-life functions that involve the natural exponential function.

■ Explore normal probability density functions.

Derivatives of Exponential Functions

In Section 4.2, it was stated that the most convenient base for exponential functions is the irrational number e. The convenience of this base stems primarily from the fact that the function

$$f(x) = e^x$$

is its own derivative. You will see that this is not true of other exponential functions of the form

$$y = a^x$$

where $a \neq e$. To verify that $f(x) = e^x$ is its own derivative, notice that the limit

$$\lim_{\Delta x \to 0} (1 + \Delta x)^{1/\Delta x} = e$$

implies that for small values of Δx,

$$e \approx (1 + \Delta x)^{1/\Delta x}$$

or

$$e^{\Delta x} \approx 1 + \Delta x.$$

This approximation is used in the following derivation.

$$f'(x) = \lim_{\Delta x \to 0} \frac{f(x + \Delta x) - f(x)}{\Delta x} \qquad \text{Definition of derivative}$$

$$= \lim_{\Delta x \to 0} \frac{e^{x + \Delta x} - e^x}{\Delta x} \qquad \text{Use } f(x) = e^x.$$

$$= \lim_{\Delta x \to 0} \frac{e^x(e^{\Delta x} - 1)}{\Delta x} \qquad \text{Factor numerator.}$$

$$= \lim_{\Delta x \to 0} \frac{e^x[(1 + \Delta x) - 1]}{\Delta x} \qquad \text{Substitute } 1 + \Delta x \text{ for } e^{\Delta x}.$$

$$= \lim_{\Delta x \to 0} \frac{e^x(\Delta x)}{\Delta x} \qquad \text{Divide out common factor.}$$

$$= \lim_{\Delta x \to 0} e^x \qquad \text{Simplify.}$$

$$= e^x \qquad \text{Evaluate limit.}$$

When u is a differentiable function of x, you can apply the Chain Rule to obtain the derivative of e^u with respect to x. Both formulas are summarized below.

Derivative of the Natural Exponential Function

Let u be a differentiable function of x.

1. $\dfrac{d}{dx}[e^x] = e^x$ **2.** $\dfrac{d}{dx}[e^u] = e^u \dfrac{du}{dx}$

Average Typing Speed

In Exercise 48 on page 274, you will use the derivative of an exponential function to find the rate of change of the average typing speed after 5, 10, and 30 weeks of lessons.

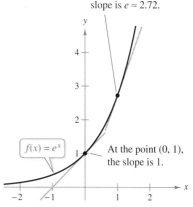

At the point $(1, e)$, the slope is $e \approx 2.72$.

At the point $(0, 1)$, the slope is 1.

$f(x) = e^x$

FIGURE 4.9

EXAMPLE 1   Finding Slopes of Tangent Lines

Find the slopes of the tangent lines to the graph of

$$f(x) = e^x$$

at the points $(0, 1)$ and $(1, e)$. What conclusion can you make?

SOLUTION   Because the derivative of f is

$$f'(x) = e^x \qquad \text{Derivative}$$

it follows that the slope of the tangent line to the graph of f is

$$f'(0) = e^0 = 1 \qquad \text{Slope at point } (0, 1)$$

at the point $(0, 1)$ and

$$f'(1) = e^1 = e \qquad \text{Slope at point } (1, e)$$

at the point $(1, e)$, as shown in Figure 4.9. From this pattern, you can see that the slope of the tangent line to the graph of $f(x) = e^x$ at any point (x, e^x) is equal to the y-coordinate of the point.

✓ *Checkpoint 1*   *Worked-out solution available at LarsonAppliedCalculus.com*

Find the slopes of the tangent lines to the graph of $f(x) = 2e^x$ at the points $(0, 2)$ and $(1, 2e)$.

STUDY TIP

In Example 2, notice that when you differentiate an exponential function, the exponent does not change. For instance, the derivative of $f(x) = e^{3x}$ is $f'(x) = 3e^{3x}$. In both f and f', the exponent is $3x$.

EXAMPLE 2   Differentiating Exponential Functions

Differentiate each function.

a. $f(x) = e^{2x}$ **b.** $f(x) = e^{-3x^2}$

c. $f(x) = 6e^{x^3}$ **d.** $f(x) = e^{-x}$

SOLUTION

a. Let $u = 2x$. Then $du/dx = 2$, and you can apply the Chain Rule.

$$f'(x) = e^u \frac{du}{dx} = e^{2x}(2) = 2e^{2x}$$

b. Let $u = -3x^2$. Then $du/dx = -6x$, and you can apply the Chain Rule.

$$f'(x) = e^u \frac{du}{dx} = e^{-3x^2}(-6x) = -6xe^{-3x^2}$$

c. Let $u = x^3$. Then $du/dx = 3x^2$, and you can apply the Chain Rule.

$$f'(x) = 6e^u \frac{du}{dx} = 6e^{x^3}(3x^2) = 18x^2 e^{x^3}$$

d. Let $u = -x$. Then $du/dx = -1$, and you can apply the Chain Rule.

$$f'(x) = e^u \frac{du}{dx} = e^{-x}(-1) = -e^{-x}$$

✓ *Checkpoint 2*   *Worked-out solution available at LarsonAppliedCalculus.com*

Differentiate each function.

a. $f(x) = e^{3x}$ **b.** $f(x) = e^{-2x^3}$

c. $f(x) = 4e^{x^2}$ **d.** $f(x) = e^{-2x}$

The differentiation rules that you studied in Chapter 2 can be used with exponential functions, as shown in Example 3.

EXAMPLE 3 Differentiating Exponential Functions

Differentiate each function.

a. $f(x) = 4e$ **b.** $f(x) = e^{2x-1}$

c. $f(x) = xe^x$ **d.** $f(x) = \dfrac{e^x - e^{-x}}{2}$

e. $f(x) = \dfrac{e^x}{x}$ **f.** $f(x) = xe^x - e^x$

TECH TUTOR

If you have access to a symbolic differentiation utility, try using it to find the derivatives of the functions in Example 3.

SOLUTION

a. $f(x) = 4e$ Write original function.

    $f'(x) = 0$ Constant Rule

b. $f(x) = e^{2x-1}$ Write original function.

    $f'(x) = (e^{2x-1})(2)$ Chain Rule

        $= 2e^{2x-1}$ Simplify.

c. $f(x) = xe^x$ Write original function.

    $f'(x) = xe^x + e^x(1)$ Product Rule

        $= xe^x + e^x$ Simplify.

d. $f(x) = \dfrac{e^x - e^{-x}}{2}$ Write original function.

        $= \dfrac{1}{2}(e^x - e^{-x})$ Rewrite.

    $f'(x) = \dfrac{1}{2}[e^x - e^{-x}(-1)]$ Constant Multiple and Chain Rules

        $= \dfrac{1}{2}(e^x + e^{-x})$ Simplify.

e. $f(x) = \dfrac{e^x}{x}$ Write original function.

    $f'(x) = \dfrac{xe^x - e^x(1)}{x^2}$ Quotient Rule

        $= \dfrac{e^x(x - 1)}{x^2}$ Simplify.

f. $f(x) = xe^x - e^x$ Write original function.

    $f'(x) = [xe^x + e^x(1)] - e^x$ Product and Difference Rules

        $= xe^x + e^x - e^x$

        $= xe^x$ Simplify.

✓ **Checkpoint 3**   *Worked-out solution available at LarsonAppliedCalculus.com*

Differentiate each function.

a. $f(x) = 9e$ **b.** $f(x) = e^{3x+1}$ **c.** $f(x) = x^2 e^x$

d. $f(x) = \dfrac{e^x + e^{-x}}{2}$ **e.** $f(x) = \dfrac{e^x}{x^2}$ **f.** $f(x) = x^2 e^x - e^x$ ■

Applications

In Chapter 3, you learned how to use derivatives to analyze the graphs of functions. The next example applies those techniques to a function composed of exponential functions. In the example, notice that $e^a = e^b$ implies that $a = b$.

EXAMPLE 4   Analyzing a Catenary

When a telephone wire is hung between two poles, the wire forms a U-shaped curve called a **catenary.** For instance, the function

$$y = 30(e^{x/60} + e^{-x/60}), \quad -30 \le x \le 30$$

models the shape of a telephone wire strung between two poles that are 60 feet apart (x and y are measured in feet). Show that the lowest point on the wire is midway between the two poles. How much does the wire sag between the two poles?

SOLUTION   First, find the derivative of the function.

$$y = 30(e^{x/60} + e^{-x/60}) \qquad \text{Write original function.}$$

$$y' = 30\left[e^{x/60}\left(\frac{1}{60}\right) + e^{-x/60}\left(-\frac{1}{60}\right)\right] \qquad \text{Derivative}$$

$$= 30\left(\frac{1}{60}\right)(e^{x/60} - e^{-x/60}) \qquad \text{Factor out } \frac{1}{60}.$$

$$= \frac{1}{2}(e^{x/60} - e^{-x/60}) \qquad \text{Simplify.}$$

To find the critical numbers, set the derivative equal to zero.

$$\frac{1}{2}(e^{x/60} - e^{-x/60}) = 0 \qquad \text{Set derivative equal to 0.}$$

$$e^{x/60} - e^{-x/60} = 0 \qquad \text{Multiply each side by 2.}$$

$$e^{x/60} = e^{-x/60} \qquad \text{Add } e^{-x/60} \text{ to each side.}$$

$$\frac{x}{60} = -\frac{x}{60} \qquad \text{If } e^a = e^b, \text{ then } a = b.$$

$$x = -x \qquad \text{Multiply each side by 60.}$$

$$2x = 0 \qquad \text{Add } x \text{ to each side.}$$

$$x = 0 \qquad \text{Divide each side by 2.}$$

Using the First-Derivative Test, you can determine that the critical number $x = 0$ yields a relative minimum of the function. From the graph in Figure 4.10, you can see that this relative minimum is actually a minimum on the interval $[-30, 30]$. So, you can conclude that the lowest point on the wire lies midway between the two poles. To find how much the wire sags between the two poles, you can compare its height at each pole with its height at the relative minimum.

$$y = 30(e^{-30/60} + e^{-(-30)/60}) \approx 67.7 \text{ feet} \qquad \text{Height at left pole}$$

$$y = 30(e^{0/60} + e^{-(0)/60}) = 60 \text{ feet} \qquad \text{Height at relative minimum}$$

$$y = 30(e^{30/60} + e^{-(30)/60}) \approx 67.7 \text{ feet} \qquad \text{Height at right pole}$$

From this, you can see that the wire sags about 7.7 feet.

FIGURE 4.10

✓ *Checkpoint 4*   Worked-out solution available at LarsonAppliedCalculus.com

Use a graphing utility to graph the function in Example 4. Verify the minimum value. Use the information in the example to choose an appropriate viewing window. ■

EXAMPLE 5 Finding a Maximum Revenue

The demand function for a product is modeled by

$$p = 56e^{-0.000012x} \qquad \text{Demand function}$$

where p is the price per unit (in dollars) and x is the number of units. What price will yield a maximum revenue? What is the maximum revenue at this price?

SOLUTION The revenue function is

$$R = xp = 56xe^{-0.000012x}. \qquad \text{Revenue function}$$

To find the maximum revenue *analytically*, you would first find the marginal revenue

$$\frac{dR}{dx} = 56xe^{-0.000012x}(-0.000012) + e^{-0.000012x}(56).$$

You would then set dR/dx equal to zero

$$56xe^{-0.000012x}(-0.000012) + e^{-0.000012x}(56) = 0$$

and solve for x. At this point, you can see that the analytical approach is rather cumbersome. In this problem, it is easier to use a *graphical* approach. After experimenting to find a reasonable viewing window, you can obtain a graph of R that is similar to that shown in Figure 4.11. Using the *maximum* feature, you can conclude that the maximum revenue occurs when x is about 83,333 units. To find the price that corresponds to this production level, substitute $x \approx 83,333$ into the demand function.

$$p \approx 56e^{-0.000012(83,333)} \approx \$20.60$$

So, a price of about $20.60 will yield a maximum revenue of

$$R \approx 56(83,333)e^{-0.000012(83,333)}$$

$$\approx \$1,716,771. \qquad \text{Maximum revenue}$$

Use the *maximum* feature to approximate the x-value that corresponds to the maximum revenue.

FIGURE 4.11

✓ *Checkpoint 5*   *Worked-out solution available at LarsonAppliedCalculus.com*

The demand function for a product is modeled by

$$p = 50e^{-0.0000125x}$$

where p is the price per unit (in dollars) and x is the number of units. What price will yield a maximum revenue? What is the maximum revenue at this price? ■

Try solving Example 5 analytically. When you do this, you must solve the equation

$$56xe^{-0.000012x}(-0.000012) + e^{-0.000012x}(56) = 0.$$

Explain how you would solve this equation. What is the solution?

Normal Probability Density Functions

If you take a course in statistics or quantitative business analysis, you will spend quite a bit of time studying the characteristics and use of the **normal probability density function** given by

$$f(x) = \frac{1}{\sigma\sqrt{2\pi}}e^{-(x-\mu)^2/(2\sigma^2)}$$

where σ is the lowercase Greek letter sigma and μ is the lowercase Greek letter mu. In this formula, σ represents the *standard deviation* of the probability distribution, and μ represents the *mean* of the probability distribution.

EXAMPLE 6   Exploring a Probability Density Function

Consider a normal probability density function where $\mu = 0$ and $\sigma = 1$

$$f(x) = \frac{1}{\sqrt{2\pi}}e^{-x^2/2}$$

which is called the *standard normal probability density function*. Show that the graph of this function has points of inflection at $x = \pm 1$. (See Figure 4.12.)

SOLUTION   Begin by finding the second derivative of the function.

$$f'(x) = \frac{1}{\sqrt{2\pi}}(-x)e^{-x^2/2} \qquad \text{First derivative}$$

$$f''(x) = \frac{1}{\sqrt{2\pi}}[(-x)(-x)e^{-x^2/2} + (-1)e^{-x^2/2}] \qquad \text{Second derivative}$$

$$= \frac{1}{\sqrt{2\pi}}(e^{-x^2/2})(x^2 - 1) \qquad \text{Simplify.}$$

By setting the second derivative equal to 0, you can determine that $x = \pm 1$. By testing the concavity of the graph, you can then conclude that these x-values yield points of inflection, as shown in Figure 4.12.

✓ *Checkpoint 6*   *Worked-out solution available at LarsonAppliedCalculus.com*

Graph the normal probability density function

$$f(x) = \frac{1}{4\sqrt{2\pi}}e^{-x^2/32}$$

and approximate the points of inflection.

Two points of inflection

$$f(x) = \frac{1}{\sqrt{2\pi}}e^{-x^2/2}$$

The graph of the standard normal probability density function is bell-shaped.

FIGURE 4.12

SUMMARIZE   (Section 4.3)

1. State the derivative of the natural exponential function *(page 267)*. For examples of finding derivatives of natural exponential functions, see Examples 2 and 3.

2. Describe a real-life example of how a natural exponential function can be used to analyze the graph of a catenary *(page 270, Example 4)*.

3. Describe a real-life example of how a natural exponential function can be used to analyze a company's maximum revenue *(page 271, Example 5)*.

4. Describe a use of the natural exponential function in statistics *(page 272)*. For an example of the natural exponential function in statistics, see Example 6.

SKILLS WARM UP 4.3

The following warm-up exercises involve skills that were covered in a previous course or in earlier sections. You will use these skills in the exercise set for this section. For additional help, review Sections 2.2, 2.4, and 3.2.

In Exercises 1–4, factor the expression.

1. $x^2 e^x - \frac{1}{2} e^x$ **2.** $(xe^{-x})^{-1} + e^x$ **3.** $xe^x - e^{2x}$ **4.** $8e^x - x^3 e^{-x}$

In Exercises 5–8, find the derivative of the function.

5. $f(x) = \dfrac{3}{7x^2}$ **6.** $g(x) = 3x^2 - \dfrac{x}{6}$ **7.** $f(x) = (4x - 3)(x^2 + 9)$ **8.** $f(t) = \dfrac{t - 2}{\sqrt{t + 5}}$

In Exercises 9 and 10, find all relative extrema of the function.

9. $f(x) = \frac{1}{8} x^3 - 2x$ **10.** $f(x) = x^4 - 2x^2 + 5$

Exercises 4.3

See *CalcChat.com* for tutorial help and worked-out solutions to odd-numbered exercises.

Differentiating Exponential Functions In Exercises 1–16, find the derivative of the function. *See Examples 2 and 3.*

1. $f(x) = 3e$

2. $f(x) = -5e$

3. $y = e^{5x}$

4. $y = e^{-3x^4}$

5. $y = e^{3-x}$

6. $f(x) = e^{x^2 + 7}$

7. $f(x) = 7e^{-1/x^2}$

8. $g(x) = 8e^{\sqrt{x}}$

9. $y = 4x^3 e^{-x}$

10. $f(x) = (x^2 + 1)e^{4x}$

11. $f(x) = \dfrac{2}{(e^x + e^{-x})^3}$

12. $f(x) = \dfrac{(e^x + e^{-x})^4}{2}$

13. $f(x) = \dfrac{e^x + 1}{e^x - 1}$

14. $f(x) = \dfrac{e^{2x}}{e^{2x} + 1}$

15. $y = xe^x - 4e^{-x}$

16. $y = x^2 e^x - 2xe^x + 2e^x$

Finding the Slope of a Tangent Line In Exercises 17–20, find the slope of the tangent line to the graph of the exponential function at the point (0, 1).

17. $y = e^{4x}$

18. $y = e^{-3x}$

19. $y = e^{-x/2}$

20. $y = e^{x/2}$

Finding an Equation of a Tangent Line In Exercises 21–26, find an equation of the tangent line to the graph of the function at the given point.

21. $y = e^{3x - x^2}$, $(3, 1)$ **22.** $g(x) = e^{x^3 - 4x}$, $(-1, e^3)$

23. $y = x^2 e^{-x}$, $\left(2, \dfrac{4}{e^2}\right)$ **24.** $y = \dfrac{3x}{e^x}$, $\left(1, \dfrac{3}{e}\right)$

25. $y = (e^{2x} + 1)^3$, $(0, 8)$ **26.** $y = (e^{4x} - 2)^2$, $(0, 1)$

Finding Derivatives Implicitly In Exercises 27–30, find dy/dx implicitly.

27. $xe^y - 10x + 3y = 0$

28. $x^2 y - e^y - 4 = 0$

29. $x^2 e^{-x} + 2y^2 - xy = 0$

30. $e^{xy} + x^2 - y^2 = 10$

Finding Second Derivatives In Exercises 31–34, find the second derivative of the function.

31. $f(x) = 2e^{3x} + 3e^{-2x}$

32. $f(x) = 5e^{-x} - 2e^{-5x}$

33. $f(x) = (1 + 2x)e^{4x}$

34. $f(x) = (3 + 2x)e^{-3x}$

Analyzing a Graph In Exercises 35–38, analyze and sketch the graph of the function. Label any relative extrema, points of inflection, and asymptotes.

35. $f(x) = \dfrac{1}{2 - e^{-x}}$ **36.** $f(x) = \dfrac{e^x - e^{-x}}{2}$

37. $f(x) = x^2 e^{-x}$ **38.** $f(x) = xe^{-x}$

Solving Equations   In Exercises 39–42, solve the equation for x.

39. $e^{-3x} = e$

40. $e^x = 1$

41. $e^{\sqrt{x}} = e^3$

42. $e^{-1/x} = e^{1/2}$

43. Catenary   A power cable that is suspended between two towers that are 50 feet apart can be modeled by

$$y = 25(e^{x/50} + e^{-x/50}), \quad -25 \le x \le 25$$

where x and y are measured in feet.

(a) Show that the lowest point on the wire is midway between the two towers.

(b) Use a graphing utility to graph the function and verify your result in part (a).

(c) How much does the cable sag between the two towers?

44. Maximum Revenue   The demand function for a product is modeled by

$$p = 73e^{-0.000025x}$$

where p is the price per unit (in dollars) and x is the number of units.

(a) What price will yield a maximum revenue?

(b) What is the maximum revenue at the price found in part (a)?

45. Depreciation   The value V (in dollars) of an item after t years is given by $V = 15,000e^{-0.6286t}$.

(a) Sketch the function over the interval [0, 10]. Use a graphing utility to verify your graph.

(b) Find the rate of change of V after 1 year.

(c) Find the rate of change of V after 6 years.

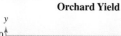

46.   HOW DO YOU SEE IT?   The yield y (in pounds per acre) of an orchard at age t (in years) is modeled by $y = 7955.6e^{-0.0458/t}$. The graph is shown below.

Orchard Yield

(a) What happens to the yield in the long run?

(b) What happens to the rate of change of the yield in the long run?

47. Compound Interest   The balance A (in dollars) in a savings account is given by $A = 5000e^{0.08t}$, where t is measured in years. Find the rate at which the balance is changing when (a) $t = 1$ year, (b) $t = 10$ years, and (c) $t = 50$ years.

48. Learning Theory   The average typing speed N (in words per minute) after t weeks of lessons is modeled by

$$N = \frac{95}{1 + 8.5e^{-0.12t}}.$$

Find the rate at which the typing speed is changing when (a) $t = 5$ weeks, (b) $t = 10$ weeks, and (c) $t = 30$ weeks.

49. Probability   In a recent year, the mean SAT score for college-bound seniors on the mathematics portion was 513, with a standard deviation of 120.   *(Source: The College Board)*

(a) Assuming the data can be modeled by a normal probability density function, find a model for these data.

(b) Use a graphing utility to graph the model. Be sure to choose an appropriate viewing window.

(c) Find the derivative of the model.

(d) Show that $f' > 0$ for $x < \mu$ and $f' < 0$ for $x > \mu$.

50. Probability   A survey of a college freshman class has determined that the mean height of females in the class is 64 inches, with a standard deviation of 3.2 inches.

(a) Assuming the data can be modeled by a normal probability density function, find a model for these data.

(b) Use a graphing utility to graph the model. Be sure to choose an appropriate viewing window.

(c) Find the derivative of the model.

(d) Show that $f' > 0$ for $x < \mu$ and $f' < 0$ for $x > \mu$.

51. Normal Probability Density Function   Use a graphing utility to graph the normal probability density functions with $\mu = 0$ and $\sigma = 2$, 3, and 4 in the same viewing window. What effect does the standard deviation σ have on the function? Explain your reasoning.

52. Normal Probability Density Function   Use a graphing utility to graph the normal probability density functions with $\sigma = 1$ and $\mu = -2$, 1, and 3 in the same viewing window. What effect does the mean μ have on the function? Explain your reasoning.

53. Normal Probability Density Function   Use Example 6 as a model to show that the graph of the normal probability density function with $\mu = 0$,

$$f(x) = \frac{1}{\sigma\sqrt{2\pi}}e^{-x^2/(2\sigma^2)}$$

has points of inflection at $x = \pm\sigma$. What is the maximum value of the function? Use a graphing utility to verify your answer by graphing the function for several values of σ.

QUIZ YOURSELF

See *CalcChat.com* for tutorial help and worked-out solutions to odd-numbered exercises.

Take this quiz as you would take a quiz in class. When you are done, check your work against the answers given in the back of the book.

In Exercises 1–8, use the properties of exponents to simplify the expression.

1. $(4^5)(4^3)$

2. $\left(\dfrac{1}{6}\right)^{-3}$

3. $\dfrac{3^8}{3^5}$

4. $(5^{1/2})(3^{1/2})$

5. $(e^2)(e^5)$

6. $(e^{3/4})(e^2)$

7. $\dfrac{e^2}{e^{-4}}$

8. $(e^{-1})^{-3}$

In Exercises 9–14, sketch the graph of the function.

9. $f(x) = 3^x - 2$

10. $f(x) = 5^{-x} + 2$

11. $f(x) = 6^{x-3}$

12. $f(x) = e^{x+4}$

13. $f(x) = e^x - 3$

14. $f(x) = e^{-2x} + 1$

15. After t years, the remaining mass y (in grams) of an initial mass of 35 grams of a radioactive element whose half-life is 80 years is given by

$$y = 35\left(\dfrac{1}{2}\right)^{t/80}, \ t \ge 0.$$

How much of the initial mass remains after 50 years?

16. With an annual rate of inflation of 4.5% over the next 10 years, the approximate cost C of goods or services during any year in the decade is given by

$$C(t) = P(1.045)^t, \ 0 \le t \le 10$$

where t is the time (in years) and P is the present cost. The price of a baseball game ticket is presently $20. Estimate the price 10 years from now.

17. For $P = \$2000$, $r = 3.5\%$, and $t = 6$ years, find the balance in an account when interest is compounded (a) quarterly, (b) monthly, and (c) continuously.

18. How much should be deposited in an account paying 6% interest compounded monthly in order to have a balance of $14,000 after 5 years?

In Exercises 19–22, find the derivative of the function.

19. $y = e^{7x}$

20. $y = e^{x-4}$

21. $y = 5e^{x+2}$

22. $y = 3e^x - xe^x$

23. Find an equation of the tangent line to the graph of $y = e^{-2x}$ at the point $(0, 1)$.

24. Analyze and sketch the graph of $f(x) = 0.5x^2 e^{-0.5x}$. Label any relative extrema, points of inflection, and asymptotes.

25. The demand function for a product is modeled by $p = 54e^{-0.00002x}$, where p is the price per unit (in dollars) and x is the number of units.

(a) What price will yield a maximum revenue?

(b) What is the maximum revenue at the price found in part (a)?

4.4 Logarithmic Functions

Orlando, Florida

■ Understand the definition of the natural logarithmic function, and sketch the graphs of natural logarithmic functions.

■ Use properties of logarithms to simplify, expand, and condense logarithmic expressions.

■ Use inverse properties of exponential and logarithmic functions to solve exponential and logarithmic equations.

■ Use properties of natural logarithms to answer questions about real-life situations.

The Natural Logarithmic Function

From your previous algebra courses, you should be somewhat familiar with logarithms. For instance, the **common logarithm** $\log_{10} x$ is defined as

$$\log_{10} x = b \quad \text{if and only if} \quad 10^b = x.$$

The base of common logarithms is 10. In calculus, the most useful base for logarithms is the number e.

Definition of the Natural Logarithmic Function

The **natural logarithmic function,** denoted by $\ln x$, is defined as

$$\ln x = b \quad \text{if and only if} \quad e^b = x.$$

The notation $\ln x$ is read as "el en of x" or as "the natural log of x."

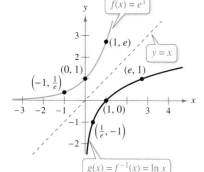

In Exercise 77 on page 283, you will solve a natural exponential equation to predict when the population of Orlando, Florida, will reach 310,000.

This definition implies that the natural logarithmic function and the natural exponential function are inverse functions. So, every logarithmic equation can be written in an equivalent exponential form, and every exponential equation can be written in logarithmic form. Here are some examples.

Logarithmic form:	*Exponential form:*
$\ln 1 = 0$	$e^0 = 1$
$\ln e = 1$	$e^1 = e$
$\ln \dfrac{1}{e} = -1$	$e^{-1} = \dfrac{1}{e}$
$\ln 2 \approx 0.693$	$e^{0.693} \approx 2$
$\ln 0.1 \approx -2.303$	$e^{-2.303} \approx 0.1$

Because the functions $f(x) = e^x$ and $g(x) = \ln x$ are inverse functions, their graphs are reflections of each other in the line $y = x$. This reflective property is illustrated in Figure 4.13. Here are several properties of the natural logarithmic function.

Properties of the Natural Logarithmic Function

Let $g(x) = \ln x$.

1. The domain of g is $(0, \infty)$ and the range of g is $(-\infty, \infty)$.

2. The x-intercept of the graph of g is $(1, 0)$.

3. The function g is continuous, increasing, and one-to-one.

4. The limit of $g(x)$ as $x \to 0^+$ is $-\infty$ and the limit of $g(x)$ as $x \to \infty$ is ∞.

FIGURE 4.13

Notice that the domain of the natural logarithmic function is the set of *positive real numbers*—be sure you see that $\ln x$ is not defined for zero or for negative numbers. You can test this on your calculator. When you try evaluating $\ln(-1)$ or $\ln 0$, your calculator should indicate that the value is not a real number.

EXAMPLE 1   Graphing Logarithmic Functions

Sketch the graph of each function.

a. $f(x) = \ln(x + 1)$

b. $f(x) = 2 \ln(x - 2)$

SOLUTION

a. Because the natural logarithmic function is defined only for positive values, the domain of the function is $x + 1 > 0$, or $x > -1$. To sketch the graph, begin by constructing a table of values, as shown below. Then plot the points in the table and connect them with a smooth curve, as shown in Figure 4.14(a).

x	-0.5	0	0.5	1	1.5	2
$\ln(x + 1)$	-0.693	0	0.405	0.693	0.916	1.099

b. The domain of this function is $x - 2 > 0$, or $x > 2$. A table of values for the function is shown below, and its graph is shown in Figure 4.14(b).

x	2.5	3	3.5	4	4.5	5
$2 \ln(x - 2)$	-1.386	0	0.811	1.386	1.833	2.197

(a)

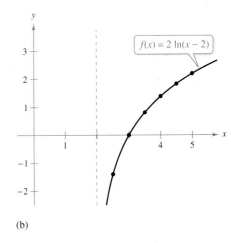

(b)

FIGURE 4.14

✓ **Checkpoint 1**   *Worked-out solution available at LarsonAppliedCalculus.com*

Complete the table and sketch the graph of

$$f(x) = \ln(x + 2).$$

x	-1.5	-1	-0.5	0	0.5	1
$f(x)$						

Properties of Logarithmic Functions

Recall from Section 1.4 that inverse functions have the property that

$$f(f^{-1}(x)) = x \quad \text{and} \quad f^{-1}(f(x)) = x.$$

The properties listed below follow from the fact that the natural logarithmic function and the natural exponential function are inverse functions.

Inverse Properties of Logarithms and Exponents

1. $\ln e^x = x$ 2. $e^{\ln x} = x$

EXAMPLE 2   Applying Inverse Properties

Simplify each expression.

a. $\ln e^{\sqrt{2}}$ **b.** $e^{\ln 3x}$

SOLUTION

a. Because $\ln e^x = x$, it follows that

$$\ln e^{\sqrt{2}} = \sqrt{2}.$$

b. Because $e^{\ln x} = x$, it follows that

$$e^{\ln 3x} = 3x.$$

✓ **Checkpoint 2**   *Worked-out solution available at LarsonAppliedCalculus.com*

Simplify each expression.

a. $\ln e^3$ **b.** $e^{\ln(x+1)}$ ■

Most of the properties of exponential functions can be rewritten in terms of logarithmic functions. For instance, the property

$$e^x e^y = e^{x+y}$$

states that you can multiply two exponential expressions by adding their exponents. In terms of logarithms, this property becomes

$$\ln xy = \ln x + \ln y.$$

This property and two other properties of logarithms are summarized below.

Properties of Logarithms

Let x, y, and n be real numbers, where $x > 0$ and $y > 0$.

1. $\ln xy = \ln x + \ln y$ **2.** $\ln \dfrac{x}{y} = \ln x - \ln y$ **3.** $\ln x^n = n \ln x$

STUDY TIP

There is no general property that can be used to rewrite $\ln(x + y)$. Specifically, $\ln(x + y)$ is not equal to $\ln x + \ln y$.

Rewriting a logarithm of a single quantity as the sum, difference, or multiple of logarithms is called *expanding* the logarithmic expression. The reverse procedure is called *condensing* a logarithmic expression.

TECH TUTOR

Try using a graphing utility to verify the result of Example 3(b). That is, try graphing the functions

$$y = \ln\sqrt{x^2 + 1}$$

and

$$y = \frac{1}{2}\ln(x^2 + 1).$$

Because these two functions are equivalent, their graphs should coincide.

EXAMPLE 3   Expanding Logarithmic Expressions

Use the properties of logarithms to rewrite each expression as a sum, difference, or multiple of logarithms. (Assume $x > 0$ and $y > 0$.)

a. $\ln\dfrac{10}{9}$ **b.** $\ln\sqrt{x^2 + 1}$

c. $\ln\dfrac{xy}{5}$ **d.** $\ln[x^2(x + 1)]$

SOLUTION

a. $\ln\dfrac{10}{9} = \ln 10 - \ln 9$ Property 2

b. $\ln\sqrt{x^2 + 1} = \ln(x^2 + 1)^{1/2}$ Rewrite with rational exponent.

$\qquad\qquad = \dfrac{1}{2}\ln(x^2 + 1)$ Property 3

c. $\ln\dfrac{xy}{5} = \ln xy - \ln 5$ Property 2

$\qquad\quad = \ln x + \ln y - \ln 5$ Property 1

d. $\ln[x^2(x + 1)] = \ln x^2 + \ln(x + 1)$ Property 1

$\qquad\qquad\quad = 2\ln x + \ln(x + 1)$ Property 3

✓ **Checkpoint 3**   Worked-out solution available at LarsonAppliedCalculus.com

Use the properties of logarithms to rewrite each expression as a sum, difference, or multiple of logarithms. (Assume $x > 0$ and $y > 0$.)

a. $\ln\dfrac{2}{5}$ **b.** $\ln\sqrt[3]{x + 2}$ **c.** $\ln\dfrac{x}{5y}$ **d.** $\ln x(x + 1)^2$

EXAMPLE 4   Condensing Logarithmic Expressions

Use the properties of logarithms to rewrite each expression as the logarithm of a single quantity. (Assume $x > 0$ and $y > 0$.)

a. $\ln x + 2\ln y$ **b.** $2\ln(x + 2) - 3\ln x$

SOLUTION

a. $\ln x + 2\ln y = \ln x + \ln y^2$ Property 3

$\qquad\qquad\quad = \ln xy^2$ Property 1

b. $2\ln(x + 2) - 3\ln x = \ln(x + 2)^2 - \ln x^3$ Property 3

$\qquad\qquad\qquad\quad = \ln\dfrac{(x + 2)^2}{x^3}$ Property 2

✓ **Checkpoint 4**   Worked-out solution available at LarsonAppliedCalculus.com

Use the properties of logarithms to rewrite each expression as the logarithm of a single quantity. (Assume $x > 0$ and $y > 0$.)

a. $4\ln x + 3\ln y$ **b.** $\ln(x + 1) - 2\ln(x + 3)$

Solving Exponential and Logarithmic Equations

To solve an exponential equation, first isolate the exponential expression. Then take the logarithm of each side of the equation and solve for the variable.

EXAMPLE 5   Solving Exponential Equations

Solve each equation.

a. $e^x = 5$ **b.** $10 + e^{0.1t} = 14$

SOLUTION

a. $e^x = 5$ Write original equation.

    $\ln e^x = \ln 5$ Take natural log of each side.

       $x = \ln 5$ Inverse property: $\ln e^x = x$

b. $10 + e^{0.1t} = 14$ Write original equation.

        $e^{0.1t} = 4$ Subtract 10 from each side.

    $\ln e^{0.1t} = \ln 4$ Take natural log of each side.

      $0.1t = \ln 4$ Inverse property: $\ln e^{0.1t} = 0.1t$

         $t = 10 \ln 4$ Multiply each side by 10.

✓ **Checkpoint 5**   *Worked-out solution available at LarsonAppliedCalculus.com*

Solve each equation.

a. $e^x = 6$ **b.** $5 + e^{0.2t} = 10$

To solve a logarithmic equation, first isolate the logarithmic expression. Then exponentiate each side and solve for the variable.

EXAMPLE 6   Solving Logarithmic Equations

Solve each equation.

a. $\ln x = 5$ **b.** $3 + 2 \ln x = 7$

SOLUTION

a. $\ln x = 5$ Write original equation.

   $e^{\ln x} = e^5$ Exponentiate each side.

      $x = e^5$ Inverse property: $e^{\ln x} = x$

b. $3 + 2 \ln x = 7$ Write original equation.

    $2 \ln x = 4$ Subtract 3 from each side.

     $\ln x = 2$ Divide each side by 2.

   $e^{\ln x} = e^2$ Exponentiate each side.

      $x = e^2$ Inverse property: $e^{\ln x} = x$

✓ **Checkpoint 6**   *Worked-out solution available at LarsonAppliedCalculus.com*

Solve each equation.

a. $\ln x = 4$ **b.** $4 + 5 \ln x = 19$

Because the domain of a logarithmic function generally does not include all real numbers, be sure to check for extraneous solutions of logarithmic equations.

TECH TUTOR

One way to solve an equation using a graphing utility is to graph the left- and right-hand sides of the equation and then use the *intersect* feature. For instance, to solve Example 6(b), enter $y_1 = 3 + 2 \ln x$ and $y_2 = 7$. The solution of the original equation is the x-value of each point of intersection (see figure). So, the solution of the original equation is $x \approx 7.389 \approx e^2$.

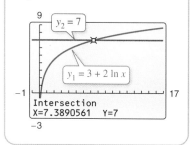

Application

EXAMPLE 7   Finding Doubling Time

You deposit P dollars in an account whose annual interest rate is r, compounded continuously. How long will it take for your balance to double?

SOLUTION   The balance in the account after t years is $A = Pe^{rt}$. So, the balance will have doubled when $Pe^{rt} = 2P$. To find the "doubling time," solve this equation for t.

$Pe^{rt} = 2P$	Balance in account has doubled.
$e^{rt} = 2$	Divide each side by P.
$\ln e^{rt} = \ln 2$	Take natural log of each side.
$rt = \ln 2$	Inverse property: $\ln e^{rt} = rt$
$t = \dfrac{1}{r}\ln 2$	Divide each side by r.

From this result, you can see that the time it takes for the balance to double is inversely proportional to the interest rate r. The table shows the doubling times for several interest rates. Notice that the doubling time decreases as the rate increases. The relationship between doubling time and the interest rate is shown graphically in Figure 4.15.

Doubling Account Balances

$t = \dfrac{1}{r}\ln 2$

FIGURE 4.15

r	3%	4%	5%	6%	7%	8%	9%	10%	11%	12%
t	23.1	17.3	13.9	11.6	9.9	8.7	7.7	6.9	6.3	5.8

Spreadsheet at LarsonAppliedCalculus.com

✓ **Checkpoint 7**   *Worked-out solution available at LarsonAppliedCalculus.com*

Use the equation found in Example 7 to determine the amount of time it would take for your balance to double at an interest rate of 8.75%.

SUMMARIZE   (Section 4.4)

1. State the definition of the natural logarithmic function *(page 276)*. For an example of graphing logarithmic functions, see Example 1.

2. State the inverse properties of logarithms and exponents *(page 278)*. For an example of applying these properties, see Example 2.

3. State the properties of logarithms *(page 278)*. For examples of using these properties to expand and condense logarithmic expressions, see Examples 3 and 4.

4. Identify the properties of logarithms and exponents used to solve the exponential and logarithmic equations in Examples 5 and 6 *(page 280)*.

5. Describe a real-life example of how a logarithm is used to determine the time it will take for an account balance to double *(page 281, Example 7)*.

SKILLS WARM UP 4.4

The following warm-up exercises involve skills that were covered in a previous course or in earlier sections. You will use these skills in the exercise set for this section. For additional help, review Appendix A.1 and Sections 1.4 and 4.2.

In Exercises 1–4, find the inverse function of f.

1. $f(x) = 5x$ **2.** $f(x) = x - 8$ **3.** $f(x) = 3x + 2$ **4.** $f(x) = \frac{3}{4}x - 9$

In Exercises 5–8, solve the inequality.

5. $0 < x + 4$

7. $0 < \sqrt{x^2 - 1}$

6. $0 < x^2 + 1$

8. $0 < 2x - 5$

In Exercises 9 and 10, find the balance in the account after 10 years.

9. $P = \$1900$, $r = 6\%$, compounded continuously

10. $P = \$2500$, $r = 3\%$, compounded continuously

Exercises 4.4

See *CalcChat.com* for tutorial help and worked-out solutions to odd-numbered exercises.

 Logarithmic and Exponential Forms of Equations In Exercises 1–8, write the logarithmic equation as an exponential equation, or vice versa.

1. $\ln 5 = 1.6094\ldots$ **2.** $\ln 9 = 2.1972\ldots$

3. $\ln 0.2 = -1.6094\ldots$ **4.** $\ln 0.05 = -2.9957\ldots$

5. $e^3 = 20.0855\ldots$ **6.** $e^2 = 7.3891\ldots$

7. $e^{-3} = 0.0498\ldots$ **8.** $e^{0.25} = 1.2840\ldots$

Matching In Exercises 9–12, match the function with its graph. [The graphs are labeled (a)–(d).]

(a)

(b)

(c)

(d)
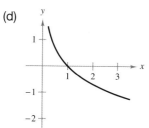

9. $f(x) = 2 + \ln x$

10. $f(x) = -\ln x$

11. $f(x) = \ln(x + 3)$

12. $f(x) = -\ln(x - 1)$

 Graphing Logarithmic Functions In Exercises 13–20, sketch the graph of the function. *See Example 1.*

13. $y = \ln(x - 1)$ **14.** $y = \ln(x + 4)$

15. $y = \ln 2x$ **16.** $y = 5 + \ln x$

17. $y = 3\ln(x + 1)$ **18.** $y = 4\ln(x - 3)$

19. $y = -2 + \frac{1}{3}\ln x$ **20.** $y = \frac{1}{2}\ln|2x|$

 Applying Inverse Properties In Exercises 21–26, apply the inverse properties of logarithms and exponents to simplify the expression. *See Example 2.*

21. $\ln e^{x^2}$ **22.** $\ln e^{2x-1}$

23. $e^{\ln(5x+2)}$ **24.** $e^{\ln\sqrt{x}}$

25. $-1 + \ln e^{2x}$ **26.** $-8 + e^{\ln x^3}$

 Expanding Logarithmic Expressions In Exercises 27–36, use the properties of logarithms to rewrite the expression as a sum, difference, or multiple of logarithms. (Assume all variables are positive.) *See Example 3.*

27. $\ln \frac{2}{3}$ **28.** $\ln \frac{1}{5}$

29. $\ln \frac{xy}{z}$ **30.** $\ln xyz$

31. $\ln \sqrt[3]{2x + 7}$ **32.** $\ln \sqrt[4]{\frac{5x^2}{x + 3}}$

33. $\ln z(z - 1)^2$ **34.** $\ln x\sqrt[3]{x^2 + 1}$

35. $\ln \frac{3x(x + 1)}{(2x + 1)^2}$

36. $\ln \frac{2x}{\sqrt{x^2 - 1}}$

Using Properties of Logarithms In Exercises 37 and 38, use the properties of logarithms and the fact that ln 2 ≈ 0.6931 and ln 3 ≈ 1.0986 to approximate the logarithm. Then use a calculator to confirm your approximation.

37. (a) $\ln 6$ (b) $\ln \frac{2}{3}$ (c) $\ln 81$ (d) $\ln \sqrt{3}$

38. (a) $\ln 0.25$ (b) $\ln 24$ (c) $\ln \sqrt[3]{12}$ (d) $\ln \frac{1}{72}$

Condensing Logarithmic Expressions In Exercises 39–46, use the properties of logarithms to rewrite the expression as the logarithm of a single quantity. (Assume all variables are positive.) *See Example 4.*

39. $\ln(x - 2) - \ln(x + 2)$ **40.** $\ln(2x + 1) + \ln(2x - 1)$

41. $3 \ln x + 2 \ln y - 4 \ln z$ **42.** $4 \ln x + 6 \ln y - \ln z$

43. $5 \ln(x - 6) + \frac{1}{2} \ln(4x + 1)$

44. $7 \ln(5x + 8) - \frac{3}{2} \ln(x - 9)$

45. $\frac{2}{3}[\ln x + \ln(x + 5) - 4 \ln(x^2 + 4)]$

46. $\frac{1}{3}[2 \ln(x + 3) + \ln x - \ln(x^2 - 1)]$

Solving Exponential and Logarithmic Equations In Exercises 47–68, solve for x or t. *See Examples 5 and 6.*

47. $e^{\ln x} = 4$ **48.** $e^{\ln x^2} - 9 = 0$

49. $e^{x + 2} = 8$ **50.** $e^{-0.5x} = 0.075$

51. $300e^{-0.2t} = 700$ **52.** $400e^{-0.0174t} = 1000$

53. $4e^{2x - 1} - 1 = 5$ **54.** $2e^{3x - 7} - 6 = 3$

55. $\ln x = 0$ **56.** $2 \ln x = 4$

57. $\ln 2x = 2.4$ **58.** $\ln 4x = 1.6$

59. $3 + 4 \ln x = 15$ **60.** $6 + 3 \ln x = 8$

61. $\ln x - \ln(x - 6) = 3$ **62.** $\ln x + \ln(x + 2) = 0$

63. $5^{2x} = 15$ **64.** $2^{1 - x} = 6$

65. $500(1.07)^t = 1000$ **66.** $400(1.06)^t = 1300$

67. $\left(1 + \frac{0.07}{12}\right)^{12t} = 3$ **68.** $\left(1 + \frac{0.06}{12}\right)^{12t} = 5$

Inverse Functions In Exercises 69–72, analytically show that the functions are inverse functions. Then use a graphing utility to show this graphically.

69. $f(x) = e^{2x}$
$g(x) = \ln \sqrt{x}$

70. $f(x) = e^x - 1$
$g(x) = \ln(x + 1)$

71. $f(x) = e^{2x - 1}$
$g(x) = \frac{1}{2} + \ln \sqrt{x}$

72. $f(x) = e^{x/3}$
$g(x) = \ln x^3$

Compound Interest In Exercises 73 and 74, $3000 is invested in an account at interest rate r, compounded continuously. Find the time required for the amount to (a) double and (b) triple.

73. $r = 8.5\%$ **74.** $r = 5.5\%$

75. Compound Interest A deposit of $1000 is made in an account that earns interest at an annual rate of 5%. How long will it take for the balance to double when the interest is compounded (a) annually, (b) monthly, (c) daily, and (d) continuously?

76. Compound Interest Complete the table to determine the time t necessary for P dollars to triple when the interest is compounded continuously at rate r.

r	2%	4%	6%	8%	10%	12%	14%
t							

77. Population Growth The population P (in thousands) of Orlando, Florida, from 1980 through 2013 can be modeled by $P = 130e^{0.0204t}$, where $t = 0$ corresponds to 1980. *(Source: U.S. Census Bureau)*

(a) What was the population of Orlando in 2013?

(b) In what year will the population of Orlando reach 310,000?

78. Population Growth The population P (in thousands) of Charlotte, North Carolina, from 1980 through 2013 can be modeled by

$$P = 321e^{0.0275t}$$

where $t = 0$ corresponds to 1980. *(Source: U.S. Census Bureau)*

(a) What was the population of Charlotte in 2013?

(b) In what year will the population of Charlotte reach 1,000,000?

Carbon Dating In Exercises 79–82, you are given the ratio of carbon isotopes to carbon atoms in a fossil. Use the information to estimate the age of the fossil. In living organic material, the ratio of radioactive carbon isotopes to the total number of carbon atoms is about 1 to 10^{12}. (See Example 3 in Section 4.1.) When organic material dies, its radioactive carbon isotopes begin to decay, with a half-life of about 5715 years. So, the ratio R of carbon isotopes to carbon-14 atoms is modeled by $R = 10^{-12}(\frac{1}{2})^{t/5715}$, where t is the time (in years) and $t = 0$ represents the time when the organic material died.

79. $R = 0.32 \times 10^{-12}$ **80.** $R = 0.27 \times 10^{-12}$

81. $R = 0.22 \times 10^{-12}$ **82.** $R = 0.13 \times 10^{-12}$

83. Learning Theory Students in a mathematics class were given an exam and then retested monthly with equivalent exams. The average scores S (on a 100-point scale) for the class can be modeled by $S = 80 - 14 \ln(t + 1), 0 \le t \le 12$, where t is the time in months.

(a) What was the average score on the original exam?

(b) What was the average score after 4 months?

(c) After how many months was the average score 46?

84. HOW DO YOU SEE IT? The graph shows the percents of American males and females ages 20 and over who are no more than x inches tall. *(Source: National Center for Health Statistics)*

Heights of Americans Ages 20 and Over

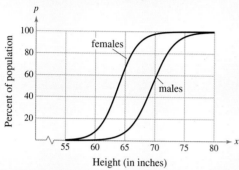

(a) Use the graph to determine the limit of each function as x approaches infinity. What do they mean?

(b) What is the median height of each sex?

85. Demand The demand function for a product is given by

$$p = 6000\left(1 - \frac{4}{4 + e^{-0.002x}}\right)$$

where p is the price per unit (in dollars) and x is the number of units sold. Find the number of units sold when (a) $p = \$200$ and (b) $p = \$700$. Use a graphing utility to verify your results in parts (a) and (b).

86. Demand The demand function for a product is given by

$$p = 12{,}000\left(1 - \frac{3}{3 + e^{-0.001x}}\right)$$

where p is the price per unit (in dollars) and x is the number of units sold. Find the number of units sold when (a) $p = \$500$ and (b) $p = \$1600$. Use a graphing utility to verify your results in parts (a) and (b).

87. Using a Property of Logarithms Demonstrate that

$$\frac{\ln x}{\ln y} \neq \ln\frac{x}{y} = \ln x - \ln y$$

by using a spreadsheet to complete the table.

x	y	$\dfrac{\ln x}{\ln y}$	$\ln\dfrac{x}{y}$	$\ln x - \ln y$
1	2			
3	4			
10	5			
4	0.5			

88. Finding Limits and Relative Extrema Use a spreadsheet to complete the table using

$$f(x) = \frac{\ln x}{x}.$$

x	1	5	10	10^2	10^4	10^6
$f(x)$						

(a) Use the table to estimate the limit: $\displaystyle\lim_{x\to\infty} f(x)$.

(b) Use a graphing utility to estimate the relative extrema of f.

Verifying Properties of Logarithms In Exercises 89 and 90, use a graphing utility to verify that the functions are equivalent for $x > 0$.

89. $f(x) = \ln\dfrac{x^2}{4}$

$g(x) = 2\ln x - \ln 4$

90. $f(x) = \ln\sqrt{x(x^2 + 1)}$

$g(x) = \frac{1}{2}[\ln x + \ln(x^2 + 1)]$

True or False? In Exercises 91–96, determine whether the statement is true or false given that $f(x) = \ln x$. If it is false, explain why or give an example that shows it is false.

91. $f(0) = 0$

92. $f(ax) = f(a) + f(x), \quad a > 0, x > 0$

93. $f(x - 2) = f(x) - f(2), \quad x > 2$

94. $\sqrt{f(x)} = \frac{1}{2}f(x)$

95. If $f(u) = 2f(v)$, then $v = u^2$.

96. If $f(x) < 0$, then $0 < x < 1$.

97. Finance You are investing P dollars at an annual interest rate of r, compounded continuously, for t years. Which of the following options would you choose to get the highest value of the investment? Explain your reasoning.

(a) Double the amount you invest.

(b) Double your interest rate.

(c) Double the number of years.

98. Think About It Are the times required for the investments in Exercises 73 and 74 to quadruple twice as long as the times for them to double? Give a reason for your answer and verify your answer algebraically.

99. Pursuit Curve Use a graphing utility to graph

$$y = 10\ln\left(\frac{10 + \sqrt{100 - x^2}}{10}\right) - \sqrt{100 - x^2}$$

over the interval $(0, 10]$. This graph is called a *tractrix* or *pursuit curve*. Use your school's library, the Internet, or some other reference source to find information about a tractrix. Explain how such a curve can arise in a real-life setting.

4.5 Derivatives of Logarithmic Functions

■ Find the derivatives of natural logarithmic functions.
■ Find the derivatives of exponential and logarithmic functions involving other bases.

Derivatives of Logarithmic Functions

In Exercise 73 on page 292, you will use the derivative of a logarithmic function to find the rate of change of a demand function.

Implicit differentiation (see Section 2.7) can be used to develop the derivative of the natural logarithmic function.

$y = \ln x$	Natural logarithmic function
$e^y = x$	Write in exponential form.
$\dfrac{d}{dx}[e^y] = \dfrac{d}{dx}[x]$	Differentiate with respect to x.
$e^y \dfrac{dy}{dx} = 1$	Chain Rule
$\dfrac{dy}{dx} = \dfrac{1}{e^y}$	Divide each side by e^y.
$\dfrac{dy}{dx} = \dfrac{1}{x}$	Substitute x for e^y.

This result and its Chain Rule version are summarized below.

Derivative of the Natural Logarithmic Function

Let u be a differentiable function of x.

1. $\dfrac{d}{dx}[\ln x] = \dfrac{1}{x}, \; x > 0$

2. $\dfrac{d}{dx}[\ln u] = \dfrac{1}{u}\dfrac{du}{dx}, \; u > 0$

EXAMPLE 1   Differentiating a Logarithmic Function

Find the derivative of

$$f(x) = \ln 2x.$$

SOLUTION   Let $u = 2x$. Then $du/dx = 2$, and you can apply the Chain Rule as shown.

$$f'(x) = \frac{1}{u}\frac{du}{dx} \qquad \text{Chain Rule}$$

$$= \frac{1}{2x}(2)$$

$$= \frac{1}{x} \qquad \text{Simplify.}$$

✓ **Checkpoint 1**   Worked-out solution available at LarsonAppliedCalculus.com

Find the derivative of

$$f(x) = \ln 5x.$$

EXAMPLE 2   Differentiating Logarithmic Functions

Find the derivative of each function.

a. $f(x) = \ln(2x^2 + 4)$ **b.** $f(x) = x \ln x$ **c.** $f(x) = \dfrac{\ln x}{x}$

SOLUTION

a. $f'(x) = \dfrac{1}{u}\dfrac{du}{dx}$ Chain Rule

$\quad\quad = \dfrac{1}{2x^2 + 4}(4x)$ $u = 2x^2 + 4,\ du/dx = 4x$

$\quad\quad = \dfrac{2x}{x^2 + 2}$ Simplify.

b. $f'(x) = x\dfrac{d}{dx}[\ln x] + (\ln x)\dfrac{d}{dx}[x]$ Product Rule

$\quad\quad = x\left(\dfrac{1}{x}\right) + (\ln x)(1)$

$\quad\quad = 1 + \ln x$ Simplify.

c. $f'(x) = \dfrac{x\dfrac{d}{dx}[\ln x] - (\ln x)\dfrac{d}{dx}[x]}{x^2}$ Quotient Rule

$\quad\quad = \dfrac{x\left(\dfrac{1}{x}\right) - (\ln x)(1)}{x^2}$

$\quad\quad = \dfrac{1 - \ln x}{x^2}$ Simplify.

✓ **Checkpoint 2**   Worked-out solution available at LarsonAppliedCalculus.com

Find the derivative of each function.

a. $f(x) = \ln(x^2 - 4)$ **b.** $f(x) = x^2 \ln x$ **c.** $f(x) = -\dfrac{\ln x}{x^2}$

EXAMPLE 3   Rewriting Before Differentiating

$\quad f(x) = \ln\sqrt{x + 1}$ Original function

$\quad\quad = \ln(x + 1)^{1/2}$ Rewrite with rational exponent.

$\quad\quad = \dfrac{1}{2}\ln(x + 1)$ Property of logarithms

$\quad f'(x) = \dfrac{1}{2}\left(\dfrac{1}{x + 1}\right)$ Differentiate.

$\quad\quad = \dfrac{1}{2(x + 1)}$ Simplify.

✓ **Checkpoint 3**   Worked-out solution available at LarsonAppliedCalculus.com

Find the derivative of

$\quad f(x) = \ln\sqrt[3]{x + 1}.$

EXAMPLE 4   Rewriting Before Differentiating

Find the derivative of

$$f(x) = \ln[x(x^2 + 1)^2].$$

SOLUTION

$$
\begin{aligned}
f(x) &= \ln[x(x^2 + 1)^2] && \text{Write original function.} \\
&= \ln x + \ln(x^2 + 1)^2 && \text{Property of logarithms} \\
&= \ln x + 2\ln(x^2 + 1) && \text{Property of logarithms} \\
f'(x) &= \frac{1}{x} + 2\left(\frac{2x}{x^2 + 1}\right) && \text{Differentiate.} \\
&= \frac{1}{x} + \frac{4x}{x^2 + 1} && \text{Simplify.}
\end{aligned}
$$

✓ *Checkpoint 4*   *Worked-out solution available at LarsonAppliedCalculus.com*

Find the derivative of $f(x) = \ln\left[x^2\sqrt{x^2 + 1}\right]$. ■

Finding the derivative of the function in Example 4 without first rewriting would be a formidable task.

$$f'(x) = \frac{1}{x(x^2 + 1)^2}\frac{d}{dx}[x(x^2 + 1)^2]$$

You might try showing that this yields the same result obtained in Example 4, but be careful—the algebra is messy.

EXAMPLE 5   Finding an Equation of a Tangent Line

Find an equation of the tangent line to the graph of $f(x) = 2 + 3x \ln x$ at the point $(1, 2)$.

SOLUTION   Begin by finding the derivative of f.

$$
\begin{aligned}
f(x) &= 2 + 3x \ln x && \text{Write original function.} \\
f'(x) &= 3x\left(\frac{1}{x}\right) + (\ln x)(3) && \text{Differentiate.} \\
&= 3 + 3\ln x && \text{Simplify.}
\end{aligned}
$$

The slope of the line tangent to the graph of f at $(1, 2)$ is

$$f'(1) = 3 + 3\ln 1 = 3 + 3(0) = 3.$$

Using the point-slope form of a line, you can find the equation of the tangent line to be

$$y = 3x - 1.$$

The graph of the function and the tangent line are shown in Figure 4.16.

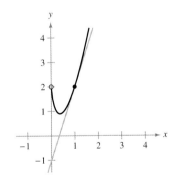

FIGURE 4.16

✓ *Checkpoint 5*   *Worked-out solution available at LarsonAppliedCalculus.com*

Find an equation of the tangent line to the graph of $f(x) = 4 \ln x$ at the point $(1, 0)$. ■

EXAMPLE 6   Finding Relative Extrema

Find all relative extrema of the function $f(x) = \dfrac{x^2}{2} - \ln x$.

FIGURE 4.17

SOLUTION   From Figure 4.17, it appears that the function has a minimum at $x = 1$. To find the minimum analytically, find the critical numbers by setting the derivative of f equal to zero and solving for x.

$$f(x) = \frac{x^2}{2} - \ln x \qquad \text{Write original function.}$$

$$f'(x) = x - \frac{1}{x} \qquad \text{Differentiate.}$$

$$x - \frac{1}{x} = 0 \qquad \text{Set derivative equal to 0.}$$

$$x = \frac{1}{x} \qquad \text{Add } 1/x \text{ to each side.}$$

$$x^2 = 1 \qquad \text{Multiply each side by } x.$$

$$x = \pm 1 \qquad \text{Take square root of each side.}$$

Of these two possible critical numbers, only the positive one lies in the domain of f. By applying the First-Derivative Test, you can confirm that the function has a relative minimum at $x = 1$.

✓ **Checkpoint 6**   *Worked-out solution available at LarsonAppliedCalculus.com*

Find all relative extrema of the function $f(x) = x - 2 \ln x$.

EXAMPLE 7   Finding a Rate of Change

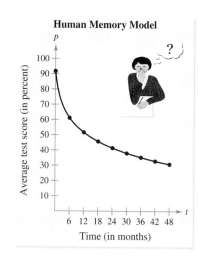

FIGURE 4.18

A group of 200 college students was tested every 6 months over a four-year period. The group was composed of students who took Spanish during the fall semester of their freshman year and did not take subsequent Spanish courses. The average test score p (in percent) is modeled by

$$p = 91.6 - 15.6 \ln(t + 1), \quad 0 \le t \le 48$$

where t is the time in months, as shown in Figure 4.18. At what rate was the average score changing after 1 year?

SOLUTION   The rate of change is

$$\frac{dp}{dt} = -\frac{15.6}{t + 1}.$$

The rate of change when $t = 12$ is

$$\frac{dp}{dt} = -\frac{15.6}{12 + 1} = -\frac{15.6}{13} = -1.2.$$

This means that the average score was decreasing at the rate of 1.2% per month after 1 year.

✓ **Checkpoint 7**   *Worked-out solution available at LarsonAppliedCalculus.com*

Suppose the average test score in Example 7 was modeled by

$$p = 92.3 - 16.9 \ln(t + 1), \quad 0 \le t \le 48$$

where t is the time in months. How would the rate at which the average test score was changing after 1 year compare with that of the model in Example 7?   ▪

Other Bases

This chapter began with a definition of a general exponential function

$$f(x) = a^x$$

where a is a positive number such that $a \neq 1$. The corresponding **logarithm to the base a** is defined by

$$\log_a x = b \quad \text{if and only if} \quad a^b = x.$$

As with the natural logarithmic function, the domain of the logarithmic function to the base a is the set of positive numbers.

EXAMPLE 8   Evaluating Logarithms

a. $\log_2 8 = 3$ $2^3 = 8$

b. $\log_{10} 100 = 2$ $10^2 = 100$

c. $\log_{10} \frac{1}{10} = -1$ $10^{-1} = \frac{1}{10}$

d. $\log_3 81 = 4$ $3^4 = 81$

✓ **Checkpoint 8**   *Worked-out solution available at LarsonAppliedCalculus.com*

Evaluate each logarithm without using a calculator.

a. $\log_2 16$ **b.** $\log_{10} \frac{1}{100}$ **c.** $\log_2 \frac{1}{32}$ **d.** $\log_5 125$ ∎

Most calculators have only two logarithm keys—a natural logarithm key denoted by (LN) and a common logarithm key denoted by (LOG). Logarithms to other bases can be evaluated with the following change-of-base formula.

$$\log_a x = \frac{\ln x}{\ln a} \qquad \text{Change-of-base formula}$$

EXAMPLE 9   Changing Bases to Evaluate Logarithms

Use the change-of-base formula and a calculator to evaluate each logarithm.

a. $\log_2 3$

b. $\log_3 6$

c. $\log_2(-1)$

SOLUTION   In each case, use the change-of-base formula and a calculator.

a. $\log_2 3 = \dfrac{\ln 3}{\ln 2} \approx 1.585$ $\log_a x = \dfrac{\ln x}{\ln a}$

b. $\log_3 6 = \dfrac{\ln 6}{\ln 3} \approx 1.631$ $\log_a x = \dfrac{\ln x}{\ln a}$

c. $\log_2(-1)$ is not defined.

✓ **Checkpoint 9**   *Worked-out solution available at LarsonAppliedCalculus.com*

Use the change-of-base formula and a calculator to evaluate each logarithm.

a. $\log_2 5$ **b.** $\log_3 18$ **c.** $\log_4 80$ **d.** $\log_{16} 0.25$ ∎

STUDY TIP

Remember that you can convert to base e using the formulas

$$a^x = e^{(\ln a)x}$$

and

$$\log_a x = \left(\frac{1}{\ln a}\right)\ln x.$$

To find derivatives of exponential or logarithmic functions to bases other than e, you can either convert to base e or use the differentiation rules shown below.

Other Bases and Differentiation

Let a be a positive real number $(a \neq 1)$ and let u be a differentiable function of x.

1. $\dfrac{d}{dx}[a^x] = (\ln a)a^x$ **2.** $\dfrac{d}{dx}[a^u] = (\ln a)a^u\dfrac{du}{dx}$

3. $\dfrac{d}{dx}[\log_a x] = \left(\dfrac{1}{\ln a}\right)\dfrac{1}{x}$ **4.** $\dfrac{d}{dx}[\log_a u] = \left(\dfrac{1}{\ln a}\right)\left(\dfrac{1}{u}\right)\dfrac{du}{dx}$

PROOF By definition, $a^x = e^{(\ln a)x}$. So, you can prove the first rule by letting $u = (\ln a)x$ and differentiating with base e to obtain

$$\frac{d}{dx}[a^x] = \frac{d}{dx}[e^{(\ln a)x}] = e^u\frac{du}{dx} = e^{(\ln a)x}(\ln a) = (\ln a)a^x.$$

The proofs of the other rules are similar.

EXAMPLE 10   **Finding a Rate of Change**

Radioactive carbon isotopes have a half-life of 5715 years. An object contains 1 gram of the isotopes. The amount A (in grams) that will be present after t years is

$$A = \left(\frac{1}{2}\right)^{t/5715}.$$

At what rate is the amount changing when $t = 10{,}000$ years?

SOLUTION   The derivative of A with respect to t is

$$\frac{dA}{dt} = \left(\ln\frac{1}{2}\right)\left(\frac{1}{2}\right)^{t/5715}\left(\frac{1}{5715}\right).$$

When $t = 10{,}000$, the rate at which the amount is changing is

$$\left(\ln\frac{1}{2}\right)\left(\frac{1}{2}\right)^{10,000/5715}\left(\frac{1}{5715}\right) \approx -0.000036$$

which implies that the amount of isotopes in the object is decreasing at the rate of 0.000036 gram per year.

✓ **Checkpoint 10**   *Worked-out solution available at LarsonAppliedCalculus.com*

Use a graphing utility to graph the model in Example 10. Describe what happens to the rate at which the amount is changing as time t increases.

SUMMARIZE   (Section 4.5)

1. State the derivative of the natural logarithmic function *(page 285)*. For examples of finding derivatives of natural logarithmic functions, see Examples 1, 2, 3, and 4.

2. State the derivatives of exponential and logarithmic functions to the base a *(page 290)*. For an example of finding the derivative of an exponential function with base a, see Example 10.

SKILLS WARM UP 4.5

The following warm-up exercises involve skills that were covered in earlier sections. You will use these skills in the exercise set for this section. For additional help, review Sections 2.6, 2.7, and 4.4.

In Exercises 1–6, expand the logarithmic expression. (Assume all variables are positive.)

1. $\ln(x + 1)^2$

2. $\ln x(x + 1)$

3. $\ln \dfrac{x}{x + 1}$

4. $\ln\left(\dfrac{x}{x - 3}\right)^3$

5. $\ln \dfrac{4x(x - 7)}{x^2}$

6. $\ln x^3 \sqrt{x + 1}$

In Exercises 7 and 8, find dy/dx implicitly.

7. $y^2 + xy = 7$

8. $x^2y - xy^2 = 3x$

In Exercises 9 and 10, find the second derivative of the function.

9. $f(x) = x^2(x + 1) - 3x^3$

10. $f(x) = -\dfrac{1}{6x^2}$

Exercises 4.5

See *CalcChat.com* for tutorial help and worked-out solutions to odd-numbered exercises.

 Differentiating Logarithmic Functions In Exercises 1–22, find the derivative of the function. *See Examples 1, 2, 3, and 4.*

1. $y = \ln x^8$

2. $f(x) = \ln 7x$

3. $y = \ln(x^2 + 3)$

4. $f(x) = \ln(8 - x^4)$

5. $y = \ln \sqrt[4]{3x - 5}$

6. $y = \ln \sqrt[5]{1 - 2x}$

7. $y = (\ln x)^4$

8. $y = (\ln x^2)^2$

9. $f(x) = 2x \ln x$

10. $y = \dfrac{\ln x}{x^3}$

11. $y = \ln\left(x\sqrt{x^6 - 2}\right)$

12. $y = \ln[x(2x + 3)^2]$

13. $y = \ln \dfrac{x}{x + 1}$

14. $y = \ln \dfrac{x}{x^2 + 1}$

15. $y = \ln \sqrt[3]{\dfrac{x - 1}{x + 1}}$

16. $y = \ln \sqrt{\dfrac{x + 1}{x - 1}}$

17. $y = \ln \dfrac{\sqrt{4 + x^2}}{x}$

18. $y = \ln \dfrac{(6 - x)^{3/2}}{x^{2/3}}$

19. $g(x) = e^{-x} \ln x$

20. $y = e^{x^2} \ln 4x^3$

21. $g(x) = \ln \dfrac{e^x + e^{-x}}{2}$

22. $f(x) = \ln \dfrac{1 + e^x}{1 - e^x}$

 Evaluating Logarithms In Exercises 23–28, evaluate the logarithm without using a calculator. *See Example 8.*

23. $\log_5 25$

24. $\log_7 49$

25. $\log_3 \frac{1}{27}$

26. $\log_6 \frac{1}{36}$

27. $\log_4 64$

28. $\log_8 512$

 Changing Bases to Evaluate Logarithms In Exercises 29–34, use the change-of-base formula and a calculator to evaluate the logarithm. *See Example 9.*

29. $\log_4 7$

30. $\log_6 10$

31. $\log_2 48$

32. $\log_5 12$

33. $\log_3 \frac{1}{2}$

34. $\log_7 \frac{2}{9}$

 Differentiating Functions of Other Bases In Exercises 35–44, find the derivative of the function.

35. $y = 3^x$

36. $y = \left(\frac{1}{4}\right)^x$

37. $f(x) = \log_2 x$

38. $g(x) = \log_5 x$

39. $h(x) = 4^{2x - 3}$

40. $f(x) = 7^{x^5}$

41. $y = \log_{10}(x^2 + 6x)$

42. $g(x) = \log_6(2x^3 - 5x^2)$

43. $y = x2^x$

44. $y = x3^{x+1}$

 Finding an Equation of a Tangent Line In Exercises 45–52, find an equation of the tangent line to the graph of the function at the given point. *See Example 5.*

45. $y = \ln x^3$; $(1, 0)$

46. $y = \ln x^{5/2}$; $(1, 0)$

47. $y = x^3 \ln x$; (e, e^3)

48. $y = \dfrac{\ln x}{x^4}$; $\left(e, \dfrac{1}{e^4}\right)$

49. $f(x) = \ln \dfrac{5(x + 2)}{x}$; $\left(-\dfrac{5}{2}, 0\right)$

50. $f(x) = \ln\left(x\sqrt{x + 3}\right)$; $\left(\frac{6}{5}, \frac{9}{10}\right)$

51. $y = \log_3 x$; $(27, 3)$

52. $g(x) = \log_8 4x$; $(2, 1)$

Finding Derivatives Implicitly In Exercises 53–56, find dy/dx implicitly.

53. $x^2 - 3 \ln y + y^2 = 10$

54. $\ln xy + 5x = 30$

55. $4x^3 + \ln y^2 + 2y = 2x$

56. $4xy + \ln(x^2y) = 7$

 Finding Second Derivatives In Exercises 57–62, find the second derivative of the function.

57. $f(x) = 2 + x^3 \ln x$

58. $f(x) = x \ln \sqrt{x} + 2x$

59. $f(x) = 5 + \dfrac{\ln x}{3x^2}$

60. $f(x) = \dfrac{\ln x}{x^3} + 4x^2$

61. $f(x) = 5^x$

62. $f(x) = \log_9 x$

63. Sound Intensity The relationship between the number of decibels β and the intensity of a sound I (in watts per square centimeter) is given by

$$\beta = 10 \log_{10}\left(\frac{I}{10^{-16}}\right).$$

Find the rate of change in the number of decibels when the intensity is 10^{-4} watt per square centimeter.

64. Chemistry The temperatures T (in °F) at which water boils at selected pressures p (in pounds per square inch) can be modeled by

$$T = 87.97 + 34.96 \ln p + 7.91 \sqrt{p}.$$

Find the rate of change of the temperature when the pressure is 60 pounds per square inch.

 Finding Relative Extrema In Exercises 65–68, find all relative extrema of the function. Use a graphing utility to verify your result. *See Example 6.*

65. $y = \ln x - x$

66. $y = \ln 2x - 2x^2$

67. $y = \ln(1 - 4x - x^2)$

68. $y = \ln(x^3 + 3x^2 + 2)$

 Analyzing a Graph In Exercises 69–72, analyze and sketch the graph of the function. Label any relative extrema, points of inflection, and asymptotes.

69. $y = \dfrac{x}{\ln x}$

70. $y = \dfrac{\ln 5x}{x^2}$

71. $y = x^2 \ln \dfrac{x}{4}$

72. $y = (\ln x)^2$

Demand In Exercises 73 and 74, find dx/dp for the demand function. Interpret this rate of change for a price of \$10.

73. $x = \ln \dfrac{1000}{p}$

74. $x = \dfrac{500}{\ln(p^2 + 1)}$

Price per unit (in dollars)

Price per unit (in dollars)

75. Demand Solve the demand function in Exercise 73 for p. Use the result to find dp/dx. Then find the rate of change when $p = \$10$. What is the relationship between this derivative and dx/dp?

76. Demand Solve the demand function in Exercise 74 for p. Use the result to find dp/dx. Then find the rate of change when $p = \$10$. What is the relationship between this derivative and dx/dp?

77. Minimum Average Cost The cost of producing x units of a product is modeled by

$$C = 500 + 300x - 300 \ln x, \quad x \ge 1.$$

(a) Find the average cost function $\overline{C}$.

(b) Find the minimum average cost analytically. Use a graphing utility to confirm your result.

78. Minimum Average Cost The cost of producing x units of a product is modeled by

$$C = 100 + 25x - 120 \ln x, \quad x \ge 1.$$

(a) Find the average cost function $\overline{C}$.

(b) Find the minimum average cost analytically. Use a graphing utility to confirm your result.

79. Consumer Trends The numbers of employees E (in thousands) at outpatient care centers from 2009 through 2013 are shown in the table.

Year	2009	2010	2011	2012	2013
Employees	558	600	621	649	682

The data can be modeled by $E = 323.3 \ln t - 151$, where $t = 9$ corresponds to 2009. *(Source: U.S. Bureau of Labor Statistics)*

(a) Use a graphing utility to plot the data and graph E over the interval $[9, 13]$.

(b) At what rate was the number of employees changing in 2011?

80. **HOW DO YOU SEE IT?** The graph shows the temperature T (in °C) of an object h hours after it is removed from a furnace.

Hours

(a) Find $\lim_{h \to \infty} T$. What does this limit represent?

(b) When is the temperature changing most rapidly?

81. Home Mortgage The term t (in years) of a $200,000 home mortgage at 7.5% interest can be approximated by

$$t = -13.375 \ln \frac{x - 1250}{x}, \quad x > 1250$$

where x is the monthly payment in dollars.

(a) Use a graphing utility to graph the model.

(b) Use the model to approximate the term of a home mortgage for which the monthly payment is $1398.43. What is the total amount paid?

(c) Use the model to approximate the term of a home mortgage for which the monthly payment is $1611.19. What is the total amount paid?

(d) Find the instantaneous rate of change of t with respect to x when $x = \$1398.43$ and $x = \$1611.19$.

(e) Write a short paragraph describing the benefit of the higher monthly payment.

82. Earthquake Intensity On the Richter scale, the magnitude R of an earthquake of intensity I is given by

$$R = \frac{\ln I - \ln I_0}{\ln 10}$$

where I_0 is the minimum intensity used for comparison. Assume $I_0 = 1$.

(a) Find the magnitude of an earthquake with an intensity of 39,811,000.

(b) Find the intensity of the January 23, 2015 earthquake in Vanuatu for which $R = 6.8$.

(c) Find the intensity of the April 1, 2014 earthquake in Chile for which $R = 8.2$.

(d) Find the factor by which the intensity is increased when the value of R is doubled.

(e) Find dR/dI.

83. Learning Theory Students in a learning theory study were given an exam and then retested monthly for 6 months with an equivalent exam. The data obtained in the study are shown in the table, where t is the time in months after the initial exam and s is the average score for the students.

DATA	t	1	2	3	4	5	6
	s	84.2	78.4	72.1	68.5	67.1	65.3

Spreadsheet at LarsonAppliedCalculus.com

(a) Use a graphing utility to find a logarithmic model for the average score s in terms of the time t.

(b) Use a graphing utility to plot the data and graph the model. How well does the model fit the data?

(c) Find the rates of change of s with respect to t when $t = 2$ and $t = 5$.

(d) Interpret the meaning of the rates of change found in part (c) in the context of the problem.

Business Capsule

Honeyfund, founded in 2006 by Sara and Josh Margulis, is a free, online crowd-gifting™ business, which allows wedding guests to fund honeymoons, down payments, and more for couples. As of 2015, more than 400,000 couples have used the popular wedding registry. The website gets 1.6 million page views per month and users have given more than $264 million in gifts. Sister-site Plumfund offers free crowd-gifting™ beyond the wedding.

84. Research Project Use your school's library, the Internet, or some other reference source to research information about an e-commerce company, such as the one discussed above. Collect data about the company (sales over a 10-year period, for example) and find a mathematical model that represents the data.

4.6 Exponential Growth and Decay

Population Growth of Bacteria

In Exercise 23 on page 300, you will use exponential growth to find the time it takes a population of bacteria to double.

■ Use exponential growth and decay to model real-life situations.

Exponential Growth and Decay

In this section, you will learn to create models of *exponential growth and decay*. Real-life situations that involve exponential growth and decay deal with a substance or population whose *rate of change at any time t is proportional to the amount of the substance present at that time*. For example, the rate of decomposition of a radioactive substance is proportional to the amount of radioactive substance at a given instant. In its simplest form, this relationship is represented by the equation

Rate of change of y ⌐ is ⌐ proportional to y.

$$\frac{dy}{dt} = ky.$$

In this equation, k is a constant and y is a function of t. The solution of this equation is shown below.

Exponential Growth and Decay

If y is a positive quantity whose rate of change with respect to time is proportional to the quantity present at any time t, then y is of the form

$$y = Ce^{kt}$$

where C is the **initial value** and k is the **constant of proportionality. Exponential growth** is indicated by $k > 0$ and **exponential decay** by $k < 0$.

PROOF   Because the rate of change of y is proportional to y, you can write

$$\frac{dy}{dt} = ky.$$

You can see that $y = Ce^{kt}$ is a solution of this equation by differentiating to obtain $dy/dt = kCe^{kt}$ and substituting.

$y = Ce^{kt}$	Original equation
$\dfrac{dy}{dt} = kCe^{kt}$	Differentiate.
$= k(Ce^{kt})$	Rewrite.
$= ky$	Substitute y for Ce^{kt}.

STUDY TIP

In the model $y = Ce^{kt}$, C is called the "initial value" because, when $t = 0$,

$$y = Ce^{k(0)}$$
$$= C(1)$$
$$= C.$$

Radioactive decay is measured in terms of **half-life,** the number of years required for half of the atoms in a sample of radioactive material to decay. The half-lives of some common radioactive isotopes are as shown.

Uranium (^{238}U)	4,470,000,000 years
Plutonium (^{239}Pu)	24,100 years
Carbon (^{14}C)	5,715 years
Radium (^{226}Ra)	1,599 years
Einsteinium (^{254}Es)	276 days
Nobelium (^{257}No)	25 seconds

EXAMPLE 1   **Modeling Radioactive Decay**

A sample contains 1 gram of radium. How much radium will remain after 1000 years?

SOLUTION   Let y represent the mass (in grams) of the radium in the sample. Because the rate of decay is proportional to y, you can conclude that y is of the form $y = Ce^{kt}$, where t is the time in years. From the given information, you know that $y = 1$ when $t = 0$. Substituting these values into the model produces

$$1 = Ce^{k(0)} \qquad \text{Substitute 1 for } y \text{ and 0 for } t.$$

which implies that $C = 1$. Because radium has a half-life of 1599 years, you know that $y = \frac{1}{2}$ when $t = 1599$. Substituting these values into the model allows you to solve for k.

Radioactive Half-Life of Radium

$y = e^{kt}$	Exponential decay model
$\frac{1}{2} = e^{k(1599)}$	Substitute $\frac{1}{2}$ for y and 1599 for t.
$\ln \frac{1}{2} = 1599k$	Take natural log of each side.
$\frac{1}{1599} \ln \frac{1}{2} = k$	Divide each side by 1599.

So, $k \approx -0.0004335$, and the exponential decay model is

$$y = e^{-0.0004335t}.$$

To find the amount of radium remaining in the sample after 1000 years, substitute $t = 1000$ into the model.

$$y = e^{-0.0004335(1000)} \approx 0.648 \text{ gram}$$

The graph of the model is shown in Figure 4.19.

FIGURE 4.19

✓ *Checkpoint 1*   Worked-out solution available at LarsonAppliedCalculus.com

Use the model in Example 1 to determine the number of years required for a 1-gram sample of radium to decay to 0.4 gram. ■

Instead of approximating the value of k in Example 1, you could leave the value exact and write

$$y = e^{[(1/1599)\ln(1/2)]t}$$
$$= e^{\ln[(1/2)^{(t/1599)}]}$$
$$= \left(\frac{1}{2}\right)^{t/1599}.$$

This version of the model clearly shows the "half-life." When $t = 1599$, the value of y is $\frac{1}{2}$; when $t = 2(1599)$, the value of y is $\frac{1}{4}$; and so on.

Guidelines for Modeling Exponential Growth and Decay

1. Use the given information to write *two* sets of conditions involving y and t.

2. Substitute the given conditions into the model $y = Ce^{kt}$ and use the results to solve for the constants C and k. (When one of the conditions involves $t = 0$, substitute that value first to solve for C.)

3. Use the model $y = Ce^{kt}$ to answer the question.

EXAMPLE 2   Modeling Population Growth

ALGEBRA TUTOR

For help with the algebra in Example 2, see Example 1(c) in the *Chapter 4 Algebra Tutor* on page 302.

In a research experiment, a population of fruit flies is increasing in accordance with the exponential growth model. After 2 days, there are 100 flies, and after 4 days, there are 300 flies. How many flies will there be after 5 days?

SOLUTION   Let y be the number of flies at time t. From the given information, you know that $y = 100$ when $t = 2$ and $y = 300$ when $t = 4$. Substituting this information into the model $y = Ce^{kt}$ produces

$$100 = Ce^{2k} \quad \text{and} \quad 300 = Ce^{4k}.$$

To solve for k, solve for C in the first equation and substitute the result into the second equation.

$300 = Ce^{4k}$	Second equation
$300 = \left(\dfrac{100}{e^{2k}}\right)e^{4k}$	Substitute $100/e^{2k}$ for C.
$\dfrac{300}{100} = e^{2k}$	Divide each side by 100.
$\ln 3 = 2k$	Take natural log of each side.
$\dfrac{1}{2}\ln 3 = k$	Solve for k.

Using $k = \frac{1}{2}\ln 3 \approx 0.5493$, you can determine that

$$C \approx \frac{100}{e^{2(0.5493)}}$$

$$\approx 33.$$

So, the exponential growth model is

$$y = 33e^{0.5493t}$$

as shown in Figure 4.20. This implies that the population after 5 days is

$$y = 33e^{0.5493(5)}$$

$$\approx 514 \text{ flies.}$$

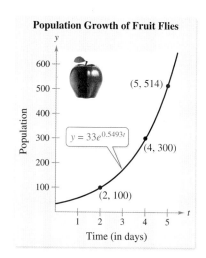

Population Growth of Fruit Flies

$y = 33e^{0.5493t}$

(5, 514)
(4, 300)
(2, 100)

Population
600
500
400
300
200
100

Time (in days)
1 2 3 4 5

FIGURE 4.20

✓ **Checkpoint 2**   *Worked-out solution available at LarsonAppliedCalculus.com*

Find the exponential growth model for a population of fruit flies for which there are 100 flies after 2 days and 400 flies after 4 days.

EXAMPLE 3   Modeling Compound Interest

Money is deposited in an account for which the interest is compounded continuously. The balance in the account doubles in 6 years. What is the annual interest rate?

SOLUTION   The balance A in an account with continuously compounded interest is given by the exponential growth model

$$A = Pe^{rt} \qquad \text{Exponential growth model}$$

where P is the original deposit, r is the annual interest rate (in decimal form), and t is the time (in years). From the given information, you know that

$$A = 2P$$

when $t = 6$, as shown in Figure 4.21. Use this information to solve for r.

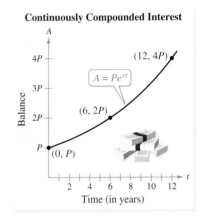

Continuously Compounded Interest

FIGURE 4.21

$A = Pe^{rt}$	Exponential growth model
$2P = Pe^{r(6)}$	Substitute $2P$ for A and 6 for t.
$2 = e^{6r}$	Divide each side by P.
$\ln 2 = 6r$	Take natural log of each side.
$\frac{1}{6}\ln 2 = r$	Divide each side by 6.

So, the annual interest rate is

$$r = \tfrac{1}{6}\ln 2$$
$$\approx 0.1155$$

or about 11.55%.

 Checkpoint 3   Worked-out solution available at LarsonAppliedCalculus.com

Find the annual interest rate for an account whose balance doubles in 8 years and for which the interest is compounded continuously.   ■

Each of the examples in this section uses the exponential growth model $y = Ce^{kt}$, in which the base is e. Exponential growth, however, can be modeled with any base. That is, the model

$$y = Ca^{bt}$$

also represents exponential growth. (To see this, note that the model can be written in the form $y = Ce^{(\ln a)bt}$.) In some real-life settings, bases other than e are more convenient. For instance, in Example 1, knowing that the half-life of radium is 1599 years, you can immediately write the exponential decay model as

$$y = \left(\frac{1}{2}\right)^{t/1599}.$$

Using this model, the amount of radium left in the sample after 1000 years is

$$y = \left(\frac{1}{2}\right)^{1000/1599}$$
$$\approx 0.648 \text{ gram}$$

which is the same answer obtained in Example 1.

EXAMPLE 4 **Modeling Sales**

Four months after discontinuing advertising on national television, a manufacturer notices that sales have dropped from 100,000 MP3 players per month to 80,000. Using an exponential pattern of decline, what will the sales be after another 4 months?

SOLUTION Let y represent the number of MP3 players, let t represent the time (in months), and consider the exponential decay model

$y = Ce^{kt}$. Exponential decay model

From the given information, you know that $y = 100{,}000$ when $t = 0$. Using this information, you have

$100{,}000 = Ce^0$

which implies that $C = 100{,}000$. To solve for k, use the fact that $y = 80{,}000$ when $t = 4$.

$y = 100{,}000e^{kt}$	Exponential decay model
$80{,}000 = 100{,}000e^{k(4)}$	Substitute 80,000 for y and 4 for t.
$0.8 = e^{4k}$	Divide each side by 100,000.
$\ln 0.8 = 4k$	Take natural log of each side.
$\frac{1}{4}\ln 0.8 = k$	Divide each side by 4.

So, $k = \frac{1}{4}\ln 0.8 \approx -0.0558$, which means that the model is

$y = 100{,}000e^{-0.0558t}$.

After four more months ($t = 8$), you can expect sales to drop to

$y = 100{,}000e^{-0.0558(8)}$

$\approx 64{,}000$ MP3 players

as shown in Figure 4.22.

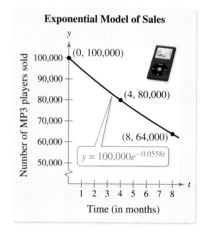

Exponential Model of Sales

(0, 100,000)

(4, 80,000)

(8, 64,000)

$y = 100{,}000e^{-0.0558t}$

Number of MP3 players sold

Time (in months)

FIGURE 4.22

✓ *Checkpoint 4* Worked-out solution available at LarsonAppliedCalculus.com

Use the model in Example 4 to determine when sales will drop to 50,000 MP3 players. ■

SUMMARIZE (Section 4.6)

1. State the model used for exponential growth and decay *(page 294)*. For examples of the use of this model, see Examples 1, 2, 3, and 4.

2. State the guidelines for modeling exponential growth and decay *(page 296)*. For examples of the use of these guidelines, see Examples 2, 3, and 4.

3. Describe a real-life example of an exponential decay model *(pages 295 and 298, Examples 1 and 4)*.

4. Describe a real-life example of an exponential growth model *(pages 296 and 297, Examples 2 and 3)*.

SKILLS WARM UP 4.6

The following warm-up exercises involve skills that were covered in earlier sections. You will use these skills in the exercise set for this section. For additional help, review Sections 4.3 and 4.4.

In Exercises 1–4, solve the equation for k.

1. $12 = 24e^{4k}$

2. $18 = 9e^{5k}$

3. $25 = 16e^{-0.01k}$

4. $22 = 32e^{-0.02k}$

In Exercises 5–8, find the derivative of the function.

5. $y = 32e^{0.23t}$

6. $y = 18e^{0.072t}$

7. $y = 24e^{-1.4t}$

8. $y = 25e^{-0.001t}$

In Exercises 9–12, simplify the expression.

9. $e^{\ln 4}$

10. $16e^{\ln 0.25}$

11. $e^{\ln(2x+1)}$

12. $e^{\ln(x^2+1)}$

Exercises 4.6

See *CalcChat.com* for tutorial help and worked-out solutions to odd-numbered exercises.

 Modeling Exponential Growth and Decay In Exercises 1–6, find the exponential function $y = Ce^{kt}$ that passes through the two given points.

1. $y = Ce^{kt}$

2. $y = Ce^{kt}$

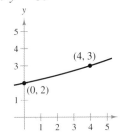

3. $y = Ce^{kt}$

4. $y = Ce^{kt}$

5. $y = Ce^{kt}$

6. $y = Ce^{kt}$

 Determining Exponential Growth and Decay In Exercises 7–10, use the given information to write an exponential equation for y. Does the function represent exponential growth or exponential decay?

7. $\dfrac{dy}{dt} = 2y, \quad y = 10$ when $t = 0$

8. $\dfrac{dy}{dt} = -\dfrac{2}{3}y, \quad y = 20$ when $t = 0$

9. $\dfrac{dy}{dt} = -4y, \quad y = 30$ when $t = 0$

10. $\dfrac{dy}{dt} = 5.2y, \quad y = 18$ when $t = 0$

 Modeling Radioactive Decay In Exercises 11–16, complete the table for each radioactive isotope. *See Example 1.*

Isotope	Half-life (in years)	Initial quantity	Amount after 1000 years	Amount after 10,000 years
11. ^{226}Ra	1599	10 grams		
12. ^{14}C	5715	3 grams		
13. ^{239}Pu	24,100		2.1 grams	
14. ^{63}Ni	96		1.5 grams	
15. ^{246}Cm	4730			2 grams
16. ^{229}Th	7340			0.3 gram

17. Radioactive Decay What percent of a present amount of radioactive radium (^{226}Ra) will remain after 900 years? (The half-life of ^{226}Ra is 1599 years.)

18. Radioactive Decay Find the half-life of a radioactive material for which 99.57% of the initial amount remains after 1 year.

19. Carbon Dating Carbon-14 (^{14}C) dating assumes that the carbon dioxide on the Earth today has the same radioactive content as it did centuries ago. If this is true, then the amount of ^{14}C absorbed by a tree that grew several centuries ago should be the same as the amount of ^{14}C absorbed by a similar tree today. A piece of ancient charcoal contains only 15% as much of the radioactive carbon as a piece of modern charcoal. How long ago was the tree burned to make the ancient charcoal? (The half-life of ^{14}C is 5715 years.)

20. Carbon Dating Repeat Exercise 19 for a piece of charcoal that contains 30% as much radioactive carbon as a modern piece.

Finding Exponential Models In Exercises 21 and 22, find exponential models $y_1 = Ce^{k_1 t}$ and $y_2 = C(2)^{k_2 t}$ that pass through the two given points. Compare the values of k_1 and k_2. Briefly explain your results.

21. $(0, 5), (12, 20)$　　　**22.** $\left(0, 8\right), \left(20, \frac{1}{2}\right)$

23. Population Growth The number of a certain type of bacteria increases continuously at a rate proportional to the number present. There are 150 bacteria at a given time and 450 bacteria 5 hours later.

(a) How many bacteria will there be 10 hours after the initial time?

(b) How long will it take for the population to double?

(c) Does the answer to part (b) depend on the starting time? Explain your reasoning.

24. HOW DO YOU SEE IT? The graph shows the populations (in millions) of Bulgaria and Tajikistan from 2006 through 2013 using exponential models. *(Source: U.S. Census Bureau)*

(a) Determine whether the population of each country is modeled by exponential growth or exponential decay. Explain your reasoning.

(b) Estimate the year when the two countries had the same population. What was this population?

Modeling Compound Interest In Exercises 25–32, complete the table for an account in which interest is compounded continuously. *See Example 3.*

Initial investment	Annual rate	Time to double	Amount after 10 years	Amount after 25 years
25. $1000	12%			
26. $20,000	$10\frac{1}{2}$%			
27. $750		8 years		
28. $10,000		10 years		
29. $500			$1292.85	
30. $2000				$6008.33
31.	4.5%			$15,000.00
32.	2%		$2000.00	

Finding Present Value In Exercises 33 and 34, determine the principal P that must be invested at interest rate r, compounded continuously, so that $1,000,000 will be available for retirement in t years.

33. $r = 7.5\%, t = 40$

34. $r = 8\%, t = 35$

35. Effective Rate The effective rate of interest r_{eff} is the annual rate that will produce the same interest per year as the nominal rate r.

(a) For a rate r (in decimal form) that is compounded n times per year, show that the effective rate r_{eff} (in decimal form) is

$$r_{eff} = \left(1 + \frac{r}{n}\right)^n - 1.$$

(b) For a rate r (in decimal form) that is compounded continuously, show that the effective rate r_{eff} (in decimal form) is $r_{eff} = e^r - 1$.

36. Effective Rate Use the results of Exercise 35 to complete the table showing the effective rates for nominal rates of (a) $r = 5\%$, (b) $r = 6\%$, and (c) $r = 7\frac{1}{2}\%$.

Number of compoundings per year	4	12	365	Continuous
Effective rate				

37. Investment: Rule of 70 Verify that the time necessary for an investment to double in value when compounded continuously is approximately $70/r$, where r is the annual interest rate entered as a percent.

38. Investment: Rule of 70 Use the Rule of 70 from Exercise 37 to approximate the times necessary for an investment to double in value when compounded continuously for (a) $r = 10\%$ and (b) $r = 7\%$.

39. Depreciation   A sports utility vehicle that costs $33,000 new has a book value of $20,856 after 3 years.

(a) Find a linear model for the value of the vehicle.

(b) Find an exponential model for the value of the vehicle.

(c) Find the book values of the vehicle after 1 year and after 4 years using each model.

(d) Use a graphing utility to graph the two models in the same viewing window. Which model depreciates faster in the first 3 years?

(e) Explain the advantages and disadvantages of using each model to a buyer and to a seller.

40. Population   The table shows the populations P (in thousands) of Lakewood, Colorado, from 2009 through 2013. *(Source: U.S. Census Bureau)*

Year	2009	2010	2011	2012	2013
Population, P	142	143	144	146	147

(a) Use the 2009 and 2010 data to find an exponential model P_1. Let $t = 9$ represent 2009.

(b) Use a graphing utility to find an exponential model P_2 for all of the data. Let $t = 9$ represent 2009.

(c) Use a graphing utility to plot the data and graph both models in the same viewing window. Compare the actual data with the estimates from the models. Which model is more accurate?

41. Sales   The cumulative sales S (in thousands of units) of a new product after it has been on the market for t years are modeled by

$$S = Ce^{k/t}.$$

During the first year, 5000 units were sold. The saturation point for the market is 30,000 units. That is, the limit of S as $t \to \infty$ is 30,000.

(a) Solve for C and k in the model.

(b) How many units will be sold after 5 years?

(c) Use a graphing utility to graph the sales function.

42. Sales   The cumulative sales S (in thousands of units) of a new product after it has been on the market for t years are modeled by

$$S = 50(1 - e^{kt}).$$

During the first year, 8000 units were sold.

(a) Solve for k in the model.

(b) What is the saturation point for this product? (The saturation point is the limit of S as $t \to \infty$.)

(c) How many units will be sold after 5 years?

(d) Use a graphing utility to graph the sales function.

43. Learning Curve   The management of a factory finds that the maximum number of units a worker can produce in a day is 30. The learning curve for the number of units N produced per day after a new employee has worked for t days is modeled by

$$N = 30(1 - e^{kt}).$$

After 20 days on the job, a worker is producing 19 units in a day. How many more days should pass before this worker is producing 25 units per day?

44. Learning Curve   The management in Exercise 43 requires that a new employee be producing at least 20 units per day after 30 days on the job.

(a) Find a learning curve model that describes this minimum requirement.

(b) Find the number of days before a minimal achiever is producing 25 units per day. How does this compare to the number of days found using the model from Exercise 43?

45. Revenue   A small business assumes that the demand function for one of its new products can be modeled by

$$p = Ce^{kx}.$$

When $p = \$45$, $x = 1000$ units, and when $p = \$40$, $x = 1200$ units.

(a) Solve for C and k in the model.

(b) Find the values of x and p that will maximize the revenue for this product.

46. Revenue   Repeat Exercise 45 given that when $p = \$5$, $x = 300$ units, and when $p = \$4$, $x = 400$ units.

47. Project: ATM Surcharge Fee   For a project analyzing the average ATM surcharge fee in the United States from 2002 to 2014, visit this text's website at *LarsonAppliedCalculus.com*. *(Source: Bankrate, Inc.)*

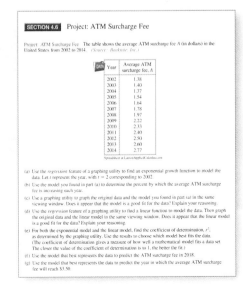

ALGEBRA TUTOR

Solving Exponential and Logarithmic Equations

To find the extrema or points of inflection of an exponential or logarithmic function, you must know how to solve exponential and logarithmic equations. A few examples are given on page 280. Some additional examples are presented in this Algebra Tutor.

As with all equations, remember that your basic goal is to isolate the variable on one side of the equation. To do this, you use inverse operations. For instance, to isolate x in

$$e^x = 7$$

take the natural log of each side of the equation and use the property $\ln e^x = x$. Similarly, to isolate x in

$$\ln x = 5$$

exponentiate each side of the equation and use the property $e^{\ln x} = x$.

EXAMPLE 1   Solving Exponential Equations

Solve each exponential equation.

a. $25 = 5e^{7t}$ **b.** $80{,}000 = 100{,}000e^{k(4)}$ **c.** $300 = \left(\dfrac{100}{e^{2k}}\right)e^{4k}$

SOLUTION

a.

$25 = 5e^{7t}$	Write original equation.
$5 = e^{7t}$	Divide each side by 5.
$\ln 5 = \ln e^{7t}$	Take natural log of each side.
$\ln 5 = 7t$	Apply the property $\ln e^a = a$.
$\dfrac{1}{7}\ln 5 = t$	Divide each side by 7.

b.

$80{,}000 = 100{,}000e^{k(4)}$	Example 4, page 298
$0.8 = e^{4k}$	Divide each side by 100,000.
$\ln 0.8 = \ln e^{4k}$	Take natural log of each side.
$\ln 0.8 = 4k$	Apply the property $\ln e^a = a$.
$\dfrac{1}{4}\ln 0.8 = k$	Divide each side by 4.

c.

$300 = \left(\dfrac{100}{e^{2k}}\right)e^{4k}$	Example 2, page 296
$300 = (100)\dfrac{e^{4k}}{e^{2k}}$	Rewrite product.
$300 = 100e^{4k-2k}$	To divide powers, subtract exponents.
$300 = 100e^{2k}$	Simplify.
$3 = e^{2k}$	Divide each side by 100.
$\ln 3 = \ln e^{2k}$	Take natural log of each side.
$\ln 3 = 2k$	Apply the property $\ln e^a = a$.
$\dfrac{1}{2}\ln 3 = k$	Divide each side by 2.

EXAMPLE 2 Solving Logarithmic Equations

Solve each logarithmic equation.

a. $\ln x = 2$　　　**b.** $5 + 2 \ln x = 4$

c. $2 \ln 3x = 4$　　**d.** $\ln x - \ln(x - 1) = 1$

SOLUTION

a. $\ln x = 2$　　　　　　　　　　Write original equation.

　　$e^{\ln x} = e^2$　　　　　　　　Exponentiate each side.

　　　$x = e^2$　　　　　　　　Apply the property $e^{\ln a} = a$.

b. $5 + 2 \ln x = 4$　　　　　　Write original equation.

　　　$2 \ln x = -1$　　　　　Subtract 5 from each side.

　　　　$\ln x = -\dfrac{1}{2}$　　　　Divide each side by 2.

　　　$e^{\ln x} = e^{-1/2}$　　　　Exponentiate each side.

　　　　$x = e^{-1/2}$　　　　Apply the property $e^{\ln a} = a$.

c. $2 \ln 3x = 4$　　　　　　　Write original equation.

　　　$\ln 3x = 2$　　　　　　Divide each side by 2.

　　　$e^{\ln 3x} = e^2$　　　　　　Exponentiate each side.

　　　　$3x = e^2$　　　　　　Apply the property $e^{\ln a} = a$.

　　　　　$x = \dfrac{1}{3}e^2$　　　　Divide each side by 3.

d. $\ln x - \ln(x - 1) = 1$　　　Write original equation.

　　　$\ln \dfrac{x}{x - 1} = 1$　　　　$\ln m - \ln n = \ln(m/n)$

　　　$e^{\ln[x/(x-1)]} = e^1$　　　　Exponentiate each side.

　　　　$\dfrac{x}{x - 1} = e^1$　　　　Apply the property $e^{\ln a} = a$.

　　　　　$x = ex - e$　　　Multiply each side by $x - 1$.

　　　$x - ex = -e$　　　Subtract ex from each side.

　　$x(1 - e) = -e$　　　Factor.

　　　　　$x = \dfrac{-e}{1 - e}$　　　Divide each side by $1 - e$.

　　　　　$x = \dfrac{e}{e - 1}$　　　Simplify.

STUDY TIP

Because the domain of a logarithmic function generally does not include all real numbers, be sure to check for extraneous solutions of logarithmic equations.

SUMMARY AND STUDY STRATEGIES

After studying this chapter, you should have acquired the following skills.
The exercise numbers are keyed to the Review Exercises that begin on page 306.
Answers to odd-numbered Review Exercises are given in the back of the text.*

Section 4.1

Review Exercises

- Use the properties of exponents to evaluate and simplify exponential expressions.   *1, 2*

$$a^0 = 1, \quad a^x a^y = a^{x+y}, \quad \frac{a^x}{a^y} = a^{x-y}, \quad (a^x)^y = a^{xy},$$

$$(ab)^x = a^x b^x, \quad \left(\frac{a}{b}\right)^x = \frac{a^x}{b^x}, \quad a^{-x} = \frac{1}{a^x}$$

- Sketch the graphs of exponential functions.   *3–8*
- Use properties of exponents to answer questions about real-life situations.   *9–12*

Section 4.2

- Use the properties of exponents to evaluate and simplify natural exponential expressions.   *13, 14*
- Sketch the graphs of natural exponential functions.   *15–18*
- Solve compound interest problems.   *19–24*

  $$A = P(1 + r/n)^{nt}, \quad A = Pe^{rt}$$

- Solve effective rate of interest problems.   *25, 26*

  $$r_{eff} = (1 + r/n)^n - 1$$

- Solve present value problems.   *27, 28*

  $$P = \frac{A}{(1 + r/n)^{nt}}$$

- Answer questions involving the natural exponential function as a real-life model.   *29–34*

Section 4.3

- Find the derivatives of natural exponential functions.   *35–40*

  $$\frac{d}{dx}[e^x] = e^x, \quad \frac{d}{dx}[e^u] = e^u \frac{du}{dx}$$

- Find equations of the tangent lines to the graphs of natural exponential functions.   *41–44*
- Use calculus to analyze the graphs of functions that involve the natural exponential function.   *45–48*

Section 4.4

- Use the definition of the natural logarithmic function to write exponential equations in logarithmic form, and vice versa.   *49–52*

  $\ln x = b$ if and only if $e^b = x$.

---

* Several study aids are available to help you master the material in this chapter. The *Student Solutions Manual* and *CalcChat.com* have solutions to all odd-numbered exercises, and *CalcView.com* has video solutions for selected exercises. *LarsonAppliedCalculus.com* offers algebra help, data spreadsheets, and much more. A *Graphing Technology Guide* with step-by-step commands for a variety of graphing calculators is at *CengageBrain.com*.

Section 4.4   (continued)   Review Exercises

- Sketch the graphs of natural logarithmic functions.   *53–56*
- Use properties of logarithms to simplify, expand, and condense logarithmic   *57–66*
  expressions.

  $$\ln xy = \ln x + \ln y, \quad \ln \frac{x}{y} = \ln x - \ln y, \quad \ln x^n = n \ln x$$

- Use inverse properties of exponential and logarithmic functions to solve   *67–80*
  exponential and logarithmic equations.

  $$\ln e^x = x, \quad e^{\ln x} = x$$

- Use properties of natural logarithms to answer questions about real-life situations.   *81–84*

Section 4.5

- Find the derivatives of natural logarithmic functions.   *85–98*

  $$\frac{d}{dx}[\ln x] = \frac{1}{x}, \quad \frac{d}{dx}[\ln u] = \frac{1}{u}\frac{du}{dx}$$

- Use the definition of logarithms to evaluate logarithmic expressions involving   *99–102*
  other bases.

  $$\log_a x = b \quad \text{if and only if} \quad a^b = x.$$

- Use the change-of-base formula to evaluate logarithmic expressions involving   *103–106*
  other bases.

  $$\log_a x = \frac{\ln x}{\ln a}$$

- Find the derivatives of exponential and logarithmic functions involving other bases.   *107–112*

  $$\frac{d}{dx}[a^x] = (\ln a)a^x, \quad \frac{d}{dx}[a^u] = (\ln a)a^u\frac{du}{dx}$$

  $$\frac{d}{dx}[\log_a x] = \left(\frac{1}{\ln a}\right)\frac{1}{x}, \quad \frac{d}{dx}[\log_a u] = \left(\frac{1}{\ln a}\right)\left(\frac{1}{u}\right)\frac{du}{dx}$$

- Use calculus to find the relative extrema of functions that involve the natural *113–116*
  logarithmic function.
- Use calculus to answer questions about real-life situations.   *117, 118*

Section 4.6

- Use exponential growth and decay to model real-life situations.   *119–132*

Study Strategies

- **Classifying Differentiation Rules**   Differentiation rules fall into two basic classes: (1) general rules that apply to all differentiable functions; and (2) specific rules that apply to special types of functions. At this point in the course, you have studied six general rules: the Constant Rule, the Constant Multiple Rule, the Sum Rule, the Difference Rule, the Product Rule, and the Quotient Rule. Although these rules were introduced in the context of algebraic functions, remember that they also can be used with exponential and logarithmic functions. You have also studied three specific rules: the Power Rule, the derivative of the natural exponential function, and the derivative of the natural logarithmic function. Each of these rules comes in two forms: the "simple" version, such as $D_x[e^x] = e^x$, and the Chain Rule version, such as $D_x[e^u] = e^u(du/dx)$.

- **To Memorize or Not to Memorize?**   When studying mathematics, you need to memorize some formulas and rules. Much of this will come from practice—the formulas that you use most often will be committed to memory. Some formulas, however, are used only infrequently. With these, it is helpful to be able to *derive* the formula from a *known* formula. For instance, knowing the Log Rule for differentiation and the change-of-base formula, $\log_a x = (\ln x)/(\ln a)$, allows you to derive the formula for the derivative of a logarithmic function to base a.

Review Exercises

See CalcChat.com for tutorial help and worked-out solutions to odd-numbered exercises.

Applying Properties of Exponents In Exercises 1 and 2, use the properties of exponents to simplify each expression.

1. (a) $(4^5)(4^2)$ (b) $(7^3)^3$

   (c) 2^{-4} (d) $\dfrac{9^8}{9^6}$

2. (a) $(5^4)(25^2)$ (b) $(9^{1/3})(3^{1/3})$

   (c) $\left(\dfrac{1}{3}\right)^{-3}$ (d) $(6^4)(6^{-5})$

Graphing Exponential Functions In Exercises 3–8, sketch the graph of the function.

3. $f(x) = 9^{x/2}$ **4.** $g(x) = 16^{3x/2}$

5. $f(t) = \left(\dfrac{1}{6}\right)^t$ **6.** $g(t) = \left(\dfrac{1}{7}\right)^{-t}$

7. $f(x) = \left(\dfrac{1}{2}\right)^{-2x} - 4$ **8.** $g(x) = \left(\dfrac{2}{3}\right)^{2x} + 1$

9. Population Growth The resident populations P (in thousands) of Virginia from 2005 through 2014 can be modeled by the exponential function

$$P(t) = 7195(1.0107)^t$$

where t is the time in years, with $t = 5$ corresponding to 2005. Use the model to estimate the populations in the years (a) 2016 and (b) 2025. *(Source: U.S. Census Bureau)*

10. Revenue The revenues R (in millions of dollars) for Panera Bread Company from 2006 through 2013 can be modeled by the exponential function

$$R(t) = 379(1.1536)^t$$

where t is the time in years, with $t = 6$ corresponding to 2006. Use the model to estimate the sales in the years (a) 2016 and (b) 2019. *(Source: Panera Bread Company)*

11. Property Value A piece of property sells for \$55,000. The value of the property doubles every 12 years. A model for the value V of the property t years after the date of purchase is

$$V(t) = 55{,}000(2)^{t/12}.$$

Use the model to approximate the value of the property (a) 4 years and (b) 25 years after it is purchased.

12. Inflation Rate With an annual rate of inflation of 2% over the next 10 years, the approximate cost C of goods or services during any year in the decade is given by

$$C(t) = P(1.02)^t, \ 0 \le t \le 10$$

where t is the time in years and P is the present cost. The price of a graphing calculator is presently \$120. Estimate the price 10 years from now.

Applying Properties of Exponents In Exercises 13 and 14, use the properties of exponents to simplify each expression.

13. (a) $(e^7)^3$ (b) $\dfrac{e^3}{e^5}$

   (c) $(e^3)(e^{5/3})$ (d) $(e^2)^{-4}$

14. (a) $(e^6)(e^{-3})$ (b) $(e^{-2})^{-5}$

   (c) $\left(\dfrac{e^6}{e^2}\right)^{-1}$ (d) $(e^3)^{4/3}$

Graphing Natural Exponential Functions In Exercises 15–18, sketch the graph of the function.

15. $f(x) = e^{-x} + 1$

16. $g(x) = e^{2x} - 1$

17. $f(x) = 1 - e^{x+3}$

18. $g(x) = 2 + e^{x-1}$

Finding Account Balances In Exercises 19–22, complete the table to determine the balance A for P dollars invested at rate r for t years, compounded n times per year.

n	1	2	4	12	365	Continuous compounding
A						

19. $P = \$1000, r = 4\%, t = 5$ years

20. $P = \$7000, r = 6\%, t = 20$ years

21. $P = \$3000, r = 3.5\%, t = 10$ years

22. $P = \$4500, r = 2\%, t = 25$ years

Comparing Account Balances In Exercises 23 and 24, \$2000 is deposited in an account. Decide which account, (a) or (b), will have the greater balance after 10 years.

23. (a) 5%, compounded continuously

   (b) 6%, compounded quarterly

24. (a) $6\frac{1}{2}\%$, compounded monthly

   (b) $6\frac{1}{4}\%$, compounded continuously

25. Effective Rate Find the effective rate of interest corresponding to a nominal rate of 4% per year compounded (a) annually, (b) semiannually, (c) quarterly, and (d) monthly.

26. Effective Rate Find the effective rate of interest corresponding to a nominal rate of 8.25% per year compounded (a) annually, (b) semiannually, (c) quarterly, and (d) monthly.

27. Present Value   How much should be deposited in an account paying 7% interest compounded quarterly in order to have a balance of $12,000 after 3 years?

28. Present Value   How much should be deposited in an account paying 8% interest compounded monthly in order to have a balance of $20,000 after 5 years?

29. Demand   The demand function for a product is modeled by

$$p = 12,500 - \frac{10,000}{2 + e^{-0.001x}}.$$

Find the price p (in dollars) of the product when the quantity demanded is (a) $x = 1000$ units and (b) $x = 2500$ units. (c) What is the limit of the price as x increases without bound?

30. Demand   The demand function for a product is modeled by

$$p = 8000\left(1 - \frac{5}{5 + e^{-0.002x}}\right).$$

Find the price p (in dollars) of the product when the quantity demanded is (a) $x = 1000$ units and (b) $x = 2500$ units. (c) What is the limit of the price as x increases without bound?

31. Profit   The net profits P (in millions of dollars) of Express Scripts from 2004 through 2013 are shown in the table.

Year	2004	2005	2006	2007	2008
Profit	300.1	388.1	468.7	620.6	779.6

Year	2009	2010	2011	2012	2013
Profit	826.5	1204.6	1275.8	1340.5	1898.2

Spreadsheet at LarsonAppliedCalculus.com

A model for these data is given by $P = 147.8e^{0.196t}$, where t represents the year, with $t = 4$ corresponding to 2004.   *(Source: Express Scripts Holding Co.)*

(a) How well does the model fit the data?

(b) Find a linear model for the data. How well does the linear model fit the data? Which model, exponential or linear, is a better fit?

(c) Use both models to predict the net profit in 2018.

32. Population   The populations P (in thousands) of Lincoln, Nebraska, from 2000 through 2013 can be modeled by $P = 227.5e^{0.0127t}$, where t is the time in years, with $t = 0$ corresponding to 2000.   *(Source: U.S. Census Bureau)*

(a) Find the populations in 2000, 2010, and 2013.

(b) Use the model to predict the population in 2022.

33. Biology   A lake is stocked with 500 fish, and the fish population P begins to increase according to the logistic growth model

$$P = \frac{10,000}{1 + 19e^{-t/5}}, \quad t \geq 0$$

where t is the time in months.

(a) Find the number of fish in the lake after 4 months.

(b) Use a graphing utility to graph the model. Find the number of months it takes for the population of fish to reach 4000.

(c) Does the population have a limit as t increases without bound? Explain your reasoning.

34. Medicine   On a college campus of 5000 students, the spread of a flu virus through the student body is modeled by

$$P = \frac{5000}{1 + 4999e^{-0.8t}}, \quad t \geq 0$$

where P is the total number of infected people and t is the time in days.

(a) Find the number of students infected after 5 days.

(b) Use a graphing utility to graph the model. Find the number of days it takes for 2000 students to become infected with the flu.

(c) According to this model, will all the students on campus become infected with the flu? Explain your reasoning.

Differentiating Exponential Functions   In Exercises 35–40, find the derivative of the function.

35. $y = 3e^{x^6}$

36. $y = 5e^{\sqrt[3]{x}}$

37. $y = \dfrac{x}{e^{2x}}$

38. $y = (2x - 1)e^x$

39. $y = \dfrac{5}{1 + e^{2x}}$

40. $y = \dfrac{10}{1 - 2e^x}$

Finding an Equation of a Tangent Line   In Exercises 41–44, find an equation of the tangent line to the graph of the function at the given point.

41. $y = e^{2-x}$, $(2, 1)$

42. $y = e^{2x^2}$, $(1, e^2)$

43. $y = x^3 e^{-x}$, $\left(3, \dfrac{27}{e^3}\right)$

44. $y = xe^x - e^x$, $(1, 0)$

Analyzing a Graph   In Exercises 45–48, analyze and sketch the graph of the function. Label any relative extrema, points of inflection, and asymptotes.

45. $f(x) = x^3 e^x$

46. $f(x) = \dfrac{e^x}{x^2}$

47. $f(x) = \dfrac{1}{xe^x}$

48. $f(x) = \dfrac{x^2}{e^x}$

Logarithmic and Exponential Forms of Equations In Exercises 49–52, write the logarithmic equation as an exponential equation, or vice versa.

49. $\ln 12 = 2.4849 \ldots$

50. $\ln 0.6 = -0.5108 \ldots$

51. $e^{1.5} = 4.4816 \ldots$

52. $e^{-4} = 0.0183 \ldots$

Graphing Logarithmic Functions In Exercises 53–56, sketch the graph of the function.

53. $y = \ln(4 - x)$

54. $y = \ln x - 3$

55. $y = \ln \dfrac{x}{3}$

56. $y = -2 \ln x$

Expanding Logarithmic Expressions In Exercises 57–62, use the properties of logarithms to rewrite the expression as a sum, difference, or multiple of logarithms. (Assume all variables are positive.)

57. $\ln \sqrt{x^2(x - 1)}$

58. $\ln y^4 \sqrt[3]{x^2 + 6}$

59. $\ln \dfrac{x^2}{(x + 1)^3}$

60. $\ln \dfrac{(x - 9)^4}{x^3}$

61. $\ln \left(\dfrac{1 - x}{5xz} \right)^3$

62. $\ln \left(\dfrac{2yz^2}{x + 7} \right)^2$

Condensing Logarithmic Expressions In Exercises 63–66, use the properties of logarithms to rewrite the expression as the logarithm of a single quantity. (Assume all variables are positive.)

63. $\ln(2x + 5) + \ln(x - 3)$

64. $\frac{1}{3} \ln(y^2 - 6) - 2 \ln(3z + 2)$

65. $4[\ln(x^3 - 1) + 2 \ln x - \ln(x - 5)]$

66. $\frac{1}{2}[\ln y + 3 \ln(x + 1) - \ln(x - 2)]$

Solving Exponential and Logarithmic Equations In Exercises 67–80, solve for x.

67. $e^{\ln x} = 12$

68. $e^{\ln(x + 2)} = 5$

69. $\ln x = 3$

70. $\ln 5x = 2$

71. $e^{-1.386x} = 0.25$

72. $e^{-0.01x} = 5.25$

73. $e^{2x - 1} - 6 = 2$

74. $4e^{2x - 3} - 5 = 7$

75. $\ln 2x - \ln(3x - 1) = 5$

76. $\ln x - \ln(2 - x) = 2$

77. $\ln x + \ln(x - 3) = 0$

78. $\ln x + \ln(x - 4) = 0$

79. $100(1.21)^x = 110$

80. $500(1.075)^{120x} = 100{,}000$

81. Compound Interest A deposit of \$400 is made in an account that earns interest at an annual rate of 2.5%. How long will it take for the balance to double when the interest is compounded (a) annually, (b) monthly, (c) daily, and (d) continuously?

82. Hourly Earnings The average hourly wages w (in dollars) for private industry employees in the United States from 2006 through 2014 can be modeled by

$$w = 17.95e^{0.0225t}$$

where $t = 6$ corresponds to 2006. *(Source: U.S. Bureau of Labor Statistics)*

(a) What was the average hourly wage in 2013?

(b) In what year will the average hourly wage be \$30?

83. Learning Theory Students in a psychology experiment were given an exam and then retested monthly with equivalent exams. The average scores S (on a 100-point scale) for the students can be modeled by

$$S = 75 - 6 \ln(t + 1), \quad 0 \le t \le 12$$

where t is the time in months.

(a) What was the average score on the original exam?

(b) What was the average score after 4 months?

(c) After how many months was the average score 60?

84. Demand The demand function for a product is given by

$$p = 9000 \left(1 - \dfrac{4}{4 + e^{-0.003x}} \right)$$

where p is the price per unit (in dollars) and x is the number of units sold. Find the number of units sold when (a) $p = \$200$ and (b) $p = \$1000$.

Differentiating Logarithmic Functions In Exercises 85–98, find the derivative of the function.

85. $f(x) = \ln 3x^2$

86. $y = \ln \sqrt{x}$

87. $y = \ln(x^2 - 2)^{2/3}$

88. $y = \ln \sqrt[3]{x^3 + 1}$

89. $f(x) = \ln(x^3 - 5)$

90. $f(x) = 6x^2 \ln x$

91. $y = \dfrac{\ln x}{x^4}$

92. $y = \dfrac{x^2}{\ln x}$

93. $y = \ln \dfrac{x^2}{x + 1}$

94. $f(x) = \ln \dfrac{x}{\sqrt{x + 1}}$

95. $f(x) = \ln(x^2 \sqrt{x + 1})$

96. $y = \ln \dfrac{x(4x - 1)}{x^3 - 2}$

97. $y = \ln \dfrac{e^x}{1 + e^x}$

98. $y = \ln(e^{2x} \sqrt{e^{2x} - 1})$

Evaluating Logarithms   In Exercises 99–102, evaluate the logarithm without using a calculator.

99. $\log_6 36$

100. $\log_2 32$

101. $\log_{10} 1$

102. $\log_4 \frac{1}{64}$

Changing Bases to Evaluate Logarithms   In Exercises 103–106, use the change-of-base formula and a calculator to evaluate the logarithm.

103. $\log_5 13$

104. $\log_4 18$

105. $\log_{16} 64$

106. $\log_4 \frac{2}{3}$

Differentiating Functions of Other Bases   In Exercises 107–112, find the derivative of the function.

107. $y = 5^{2x+1}$

108. $y = 8^{x^3}$

109. $y = \log_3(2x - 1)$

110. $y = \log_{16}(x^2 - 3x)$

111. $y = \log_{10} \dfrac{3}{x}$

112. $y = \log_2 \dfrac{1}{x^2}$

Finding Relative Extrema   In Exercises 113–116, find all relative extrema of the function. Use a graphing utility to verify your result.

113. $y = 3 \ln x - x$

114. $y = x - \ln 5x$

115. $y = \ln(x^2 + 2x + 7)$

116. $y = \ln(4 + 6x^2 - x^3)$

117. Music   The numbers of music albums downloaded D (in millions) from 2005 through 2013 can be modeled by

$$D = -179.8 + 116.84 \ln t$$

where $t = 5$ corresponds to 2005. Find the rates of change of the number of music albums downloaded in 2010 and 2012. *(Source: Recording Industry Association of America)*

118. Minimum Average Cost   The cost of producing x units of a product is modeled by

$$C = 200 + 75x - 300 \ln x, \quad x \geq 1.$$

(a) Find the average cost function $\bar{C}$.

(b) Find the minimum average cost analytically. Use a graphing utility to confirm your result.

Modeling Exponential Growth and Decay   In Exercises 119 and 120, find the exponential function $y = Ce^{kt}$ that passes through the two given points.

119. $(0, 3), (4, 1)$

120. $(1, 1), (5, 5)$

Modeling Radioactive Decay   In Exercises 121–126, complete the table for each radioactive isotope.

	Isotope	Half-life (in years)	Initial quantity	Amount after 1000 years	Amount after 10,000 years
121.	^{226}Ra	1599	8 grams		
122.	^{14}C	5715	5 grams		
123.	^{14}C	5715			6 grams
124.	^{226}Ra	1599		0.7 gram	
125.	^{239}Pu	24,100		2.4 grams	
126.	^{239}Pu	24,100			7.1 grams

Modeling Compound Interest   In Exercises 127–130, complete the table for an account in which interest is compounded continuously.

	Initial investment	Annual rate	Time to double	Amount after 10 years	Amount after 25 years
127.	$600	8%			
128.	$2000		7 years		
129.	$15,000			$18,321.04	
130.		4%			$9000.00

131. Medical Science   Soon after an injection, the concentration D (in milligrams per milliliter) of a drug in a patient's bloodstream is 500 milligrams per milliliter. After 6 hours, 50 milligrams per milliliter of the drug remains in the bloodstream.

(a) Find an exponential model for the concentration D after t hours.

(b) What is the concentration of the drug after 4 hours?

(c) Use a graphing utility to graph the function found in part (a).

132. Population Growth   The number of a certain type of bacteria increases continuously at a rate proportional to the number present. After 2 hours, there are 200 bacteria, and after 4 hours, there are 300 bacteria.

(a) Find an exponential model for the population P after t hours.

(b) How many bacteria will there be after 7 hours?

(c) How long will it take for the population to double?

TEST YOURSELF

See *CalcChat.com* for tutorial help and worked-out solutions to odd-numbered exercises.

Take this test as you would take a test in class. When you are done, check your work against the answers given in the back of the book.

In Exercises 1–4, use the properties of exponents to simplify the expression.

1. $3^2(3^{-2})$

2. $\left(\dfrac{2^3}{2^5}\right)^{-1}$

3. $(e^{1/2})(e^4)$

4. $(e^4)^5$

In Exercises 5–10, sketch the graph of the function.

5. $f(x) = 5^{x-2}$

6. $f(x) = 7^{-x} - 3$

7. $f(x) = e^{x-3}$

8. $f(x) = 8 + \ln x^2$

9. $f(x) = \ln(x + 6)$

10. $f(x) = 0.5 \ln x$

In Exercises 11–13, use the properties of logarithms to rewrite the expression as a sum, difference, or multiple of logarithms. (Assume all variables are positive.)

11. $\ln \dfrac{3}{2}$

12. $\ln z\sqrt{x + y}$

13. $\ln\left(\dfrac{x + 1}{yz}\right)^3$

In Exercises 14–16, use the properties of logarithms to rewrite the expression as the logarithm of a single quantity. (Assume all variables are positive.)

14. $\ln y + \ln(x + 1)$

15. $3 \ln x - 2 \ln(x - 1)$

16. $\ln x + 4 \ln y - \frac{1}{2}\ln(z + 4)$

In Exercises 17–19, solve for x.

17. $\ln 3x = 3.5$

18. $10e^{2x+1} = 900$

19. $50(1.06)^x = 1500$

20. A deposit of $500 is made in an account that earns interest at an annual rate of 4%. How long will it take for the balance to double when the interest is compounded (a) annually, (b) monthly, (c) daily, and (d) continuously?

In Exercises 21–26, find the derivative of the function.

21. $y = e^{-3x} + 5$

22. $y = 7e^{x+2} + 2x$

23. $y = 6^{3x+1}$

24. $y = \ln(7 - x^3)$

25. $y = \ln \dfrac{5x}{x + 2}$

26. $y = e^{x^4} \ln 4x^2$

27. The sales S (in billions of dollars) of organic food in the United States from 2007 through 2012 can be modeled by

$$S = 11.9e^{0.0815t}$$

where $t = 7$ corresponds to 2007.   *(Source: Statista)*

(a) Find the sales in 2010.

(b) At what rate were the sales changing in 2010?

28. What percent of a present amount of radioactive plutonium (^{239}Pu) will remain after 15,000 years? (The half-life of ^{239}Pu is 24,100 years.)

29. A population is growing continuously at the rate of 1.75% per year. Find the time necessary for the population to double in size.

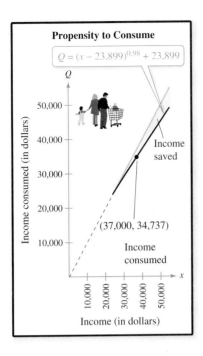

Propensity to Consume

$$Q = (x - 23{,}899)^{0.98} + 23{,}899$$

Income consumed (in dollars)

Income saved

(37,000, 34,737)

Income consumed

Income (in dollars)

Example 8 on page 328 shows how integration can be used to analyze the marginal propensity to consume.

5 Integration and Its Applications

iStockphoto.com/pixdeluxe

5.1 Antiderivatives and Indefinite Integrals

Horry County, South Carolina

In Exercise 71 on page 321, you will use integration to find a model for the population of a county.

■ Understand the definition of antiderivative and use indefinite integral notation for antiderivatives.
■ Use basic integration rules to find antiderivatives.
■ Use initial conditions to find particular solutions of indefinite integrals.
■ Use antiderivatives to solve real-life problems.

Antiderivatives

In Chapter 2, you were concerned primarily with the problem: *given a function, find its derivative.* Some important applications of calculus involve the inverse problem: *given a derivative, find the function.* For instance, consider the derivative $f'(x) = 3x^2$. To determine the function f, you might come up with

$$f(x) = x^3 \quad \text{because} \quad \frac{d}{dx}[x^3] = 3x^2.$$

This operation of determining the original function from its derivative is the inverse operation of differentiation. It is called **antidifferentiation.**

Definition of Antiderivative

A function F is an **antiderivative** of a function f when for every x in the domain of f, it follows that $F'(x) = f(x)$.

If $F(x)$ is an antiderivative of $f(x)$, then $F(x) + C$, where C is any constant, is also an antiderivative of $f(x)$. For example,

$$F(x) = x^3, \quad G(x) = x^3 - 5, \quad \text{and} \quad H(x) = x^3 + 0.3$$

are all antiderivatives of $3x^2$ because the derivative of each is $3x^2$. As it turns out, *all* antiderivatives of $3x^2$ are of the form $x^3 + C$. So, the process of antidifferentiation does not determine a single function, but rather a *family* of functions, each differing from the others by a constant.

The antidifferentiation process is also called **integration** and is denoted by

$$\int \qquad \text{Integral sign}$$

which is called an **integral sign.** The symbol

$$\int f(x)\, dx \qquad \text{Indefinite integral}$$

is the **indefinite integral** of $f(x)$, and it denotes the family of antiderivatives of $f(x)$. That is, if $F'(x) = f(x)$ for all x, then you can write

$$\int f(x)\, dx = F(x) + C$$

where $f(x)$ is the **integrand** and C is the **constant of integration.** The differential dx in the indefinite integral identifies the variable of integration. That is, the symbol $\int f(x)\, dx$ denotes the "antiderivative of f *with respect to x*" just as the symbol dy/dx denotes the "derivative of y *with respect to x.*"

STUDY TIP

In this text, the phrase "$F(x)$ is an antiderivative of $f(x)$" is used synonymously with "F is an antiderivative of f."

Edhar/Shutterstock.com

Finding Antiderivatives

The inverse relationship between the operations of integration and differentiation can be shown symbolically, as follows.

$$\frac{d}{dx}\left[\int f(x)\,dx\right] = f(x) \qquad \text{Differentiation is the inverse of integration.}$$

$$\int f'(x)\,dx = f(x) + C \qquad \text{Integration is the inverse of differentiation.}$$

This inverse relationship between integration and differentiation allows you to obtain integration formulas directly from differentiation formulas. The following summary lists the integration formulas that correspond to some of the differentiation formulas you have studied.

Basic Integration Rules

1. $\displaystyle\int k\,dx = kx + C, \quad k$ is a constant.   Constant Rule

2. $\displaystyle\int kf(x)\,dx = k\int f(x)\,dx$ Constant Multiple Rule

3. $\displaystyle\int [f(x) + g(x)]\,dx = \int f(x)\,dx + \int g(x)\,dx$ Sum Rule

4. $\displaystyle\int [f(x) - g(x)]\,dx = \int f(x)\,dx - \int g(x)\,dx$ Difference Rule

5. $\displaystyle\int x^n\,dx = \frac{x^{n+1}}{n+1} + C, \quad n \neq -1$ Simple Power Rule

Be sure you see that the Simple Power Rule has the restriction that n cannot be -1. So, you *cannot* use the Simple Power Rule to evaluate the integral

$$\int \frac{1}{x}\,dx.$$

To evaluate this integral, you need the Log Rule, which is described in Section 5.3.

EXAMPLE 1   Finding Indefinite Integrals

Find each indefinite integral.

a. $\displaystyle\int \frac{1}{2}\,dx$ **b.** $\displaystyle\int 1\,dx$ **c.** $\displaystyle\int -5\,dt$

SOLUTION

a. $\displaystyle\int \frac{1}{2}\,dx = \frac{1}{2}x + C$ **b.** $\displaystyle\int 1\,dx = x + C$ **c.** $\displaystyle\int -5\,dt = -5t + C$

✓ **Checkpoint 1**   *Worked-out solution available at LarsonAppliedCalculus.com*

Find each indefinite integral.

a. $\displaystyle\int 5\,dx$ **b.** $\displaystyle\int -1\,dr$ **c.** $\displaystyle\int 2\,dt$

EXAMPLE 2   Finding an Indefinite Integral

$$\int 3x \, dx = 3 \int x \, dx \qquad \text{Constant Multiple Rule}$$

$$= 3 \int x^1 \, dx \qquad \text{Rewrite } x \text{ as } x^1.$$

$$= 3 \left(\frac{x^2}{2} \right) + C \qquad \text{Simple Power Rule with } n = 1$$

$$= \frac{3}{2} x^2 + C \qquad \text{Simplify.}$$

✓ **Checkpoint 2**   Worked-out solution available at LarsonAppliedCalculus.com

Find $\int 5x \, dx$. ■

In finding indefinite integrals, a strict application of the basic integration rules tends to produce cumbersome constants of integration. For instance, in Example 2, you could have written

$$\int 3x \, dx = 3 \int x \, dx = 3 \left(\frac{x^2}{2} + C \right) = \frac{3}{2} x^2 + 3C.$$

However, because C represents *any* constant, it is unnecessary to write $3C$ as the constant of integration. You can simply write $\frac{3}{2} x^2 + C$.

In Example 2, note that the general pattern of integration is similar to that of differentiation.

Original Integral: $\int 3x \, dx$ ⟹ Rewrite: $3 \int x^1 \, dx$ ⟹ Integrate: $3 \left(\frac{x^2}{2} \right) + C$ ⟹ Simplify: $\frac{3}{2} x^2 + C$

EXAMPLE 3   Rewriting Before Integrating

	Original Integral	Rewrite	Integrate	Simplify
a.	$\int \dfrac{1}{x^3} \, dx$	$\int x^{-3} \, dx$	$\dfrac{x^{-2}}{-2} + C$	$-\dfrac{1}{2x^2} + C$
b.	$\int \sqrt{x} \, dx$	$\int x^{1/2} \, dx$	$\dfrac{x^{3/2}}{3/2} + C$	$\dfrac{2}{3} x^{3/2} + C$

✓ **Checkpoint 3**   Worked-out solution available at LarsonAppliedCalculus.com

Find each indefinite integral.

a. $\int \dfrac{1}{x^2} \, dx$ **b.** $\int \sqrt[3]{x} \, dx$ ■

Remember that you can check your answer to an antidifferentiation problem by differentiating. For instance, in Example 3(b), you can confirm that $\frac{2}{3} x^{3/2} + C$ is the correct antiderivative by differentiating to obtain

$$\frac{d}{dx} \left[\frac{2}{3} x^{3/2} + C \right] = \left(\frac{2}{3} \right) \left(\frac{3}{2} \right) x^{1/2} = \sqrt{x}.$$

With the five basic integration rules, you can integrate *any* polynomial function, as demonstrated in the next example.

EXAMPLE 4   **Integrating Polynomial Functions**

Find (a) $\int (x + 2)\, dx$ and (b) $\int (3x^4 - 5x^2 + x)\, dx$.

SOLUTION

a. $\int (x + 2)\, dx = \int x\, dx + \int 2\, dx$ Apply Sum Rule.

$= \dfrac{x^2}{2} + C_1 + 2x + C_2$ Apply Simple Power and Constant Rules.

$= \dfrac{x^2}{2} + 2x + C$ $C = C_1 + C_2$

The second line in this solution is usually omitted.

b. $\int (3x^4 - 5x^2 + x)\, dx = 3\left(\dfrac{x^5}{5}\right) - 5\left(\dfrac{x^3}{3}\right) + \dfrac{x^2}{2} + C$

$= \dfrac{3}{5}x^5 - \dfrac{5}{3}x^3 + \dfrac{1}{2}x^2 + C$

✓ **Checkpoint 4**   *Worked-out solution available at LarsonAppliedCalculus.com*

Find (a) $\int (x + 4)\, dx$ and (b) $\int (4x^3 - 5x + 2)\, dx$.

EXAMPLE 5   **Rewriting Before Integrating**

Find $\int \dfrac{x - 1}{\sqrt{x}}\, dx$.

SOLUTION   Begin by rewriting the quotient in the integrand as a difference. Then rewrite each term using rational exponents.

$\int \dfrac{x - 1}{\sqrt{x}}\, dx = \int \left(\dfrac{x}{\sqrt{x}} - \dfrac{1}{\sqrt{x}}\right) dx$ Rewrite as a difference.

$= \int (x^{1/2} - x^{-1/2})\, dx$ Rewrite using rational exponents.

$= \int x^{1/2}\, dx - \int x^{-1/2}\, dx$ Apply Difference Rule.

$= \dfrac{x^{3/2}}{3/2} - \dfrac{x^{1/2}}{1/2} + C$ Apply Simple Power Rule.

$= \dfrac{2}{3}x^{3/2} - 2x^{1/2} + C$ Simplify.

$= \dfrac{2}{3}\sqrt{x}(x - 3) + C$ Factor.

✓ **Checkpoint 5**   *Worked-out solution available at LarsonAppliedCalculus.com*

Find $\int \dfrac{x + 2}{\sqrt{x}}\, dx$.

STUDY TIP

When integrating quotients, *do not* integrate the numerator and denominator separately. For instance, in Example 5, be sure you understand that

$\int \dfrac{x - 1}{\sqrt{x}}\, dx = \dfrac{2}{3}\sqrt{x}(x - 3) + C$

is not the same as

$\dfrac{\int (x - 1)\, dx}{\int \sqrt{x}\, dx} = \dfrac{\frac{1}{2}x^2 - x + C_1}{\frac{2}{3}x\sqrt{x} + C_2}.$

ALGEBRA TUTOR

For help on the algebra in Example 5, see Example 1(a) in the *Chapter 5 Algebra Tutor,* on page 366.

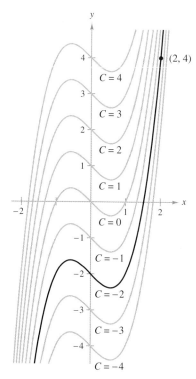

$$F(x) = x^3 - x + C$$

FIGURE 5.1

Particular Solutions

You have already seen that the equation $y = \int f(x)\,dx$ has many solutions, each differing from the others by a constant. This means that the graphs of any two antiderivatives of f are vertical translations of each other. For example, Figure 5.1 shows the graphs of several antiderivatives of the form

$$y = F(x) = \int (3x^2 - 1)\,dx = x^3 - x + C$$

for various integer values of C. Each of these antiderivatives is a solution of the *differential equation*

$$\frac{dy}{dx} = 3x^2 - 1.$$

A **differential equation** in x and y is an equation that involves x, y, and derivatives of y. The **general solution** of $dy/dx = 3x^2 - 1$ is $F(x) = x^3 - x + C$.

In many applications of integration, you are given enough information to determine a **particular solution.** To do this, you need to know the value of $y = F(x)$ for only one value of x. This information is called an **initial condition.** For example, in Figure 5.1, only one curve passes through the point $(2, 4)$. To find this curve, use the information below.

$$F(x) = x^3 - x + C \qquad \text{General solution}$$
$$F(2) = 4 \qquad \text{Initial condition}$$

By using the initial condition in the general solution, you can determine that $F(2) = 2^3 - 2 + C = 4$, which implies that $C = -2$. So, the particular solution is

$$F(x) = x^3 - x - 2. \qquad \text{Particular solution}$$

EXAMPLE 6   Finding a Particular Solution

Find the general solution of

$$F'(x) = 2x - 2$$

and find the particular solution that satisfies the initial condition $F(1) = 2$.

SOLUTION   Begin by integrating to find the general solution.

$$F(x) = \int (2x - 2)\,dx \qquad \text{Integrate } F'(x) \text{ to obtain } F(x).$$

$$= x^2 - 2x + C \qquad \text{General solution}$$

Using the initial condition $F(1) = 2$, you can write

$$F(1) = 1^2 - 2(1) + C = 2$$

which implies that $C = 3$. So, the particular solution is

$$F(x) = x^2 - 2x + 3. \qquad \text{Particular solution}$$

This solution is shown graphically in Figure 5.2. Note that each of the gray curves represents a solution of the equation $F'(x) = 2x - 2$. The black curve, however, is the only solution that passes through the point $(1, 2)$, which means that $F(x) = x^2 - 2x + 3$ is the only solution that satisfies the initial condition.

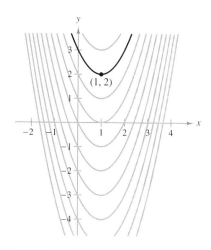

FIGURE 5.2

✓ **Checkpoint 6**   *Worked-out solution available at LarsonAppliedCalculus.com*

Find the general solution of $F'(x) = 4x + 2$, and find the particular solution that satisfies the initial condition $F(1) = 8$. ■

Applications

In Chapter 2, you used the general position function (neglecting air resistance) for a falling object

$$s(t) = -16t^2 + v_0 t + s_0$$

where $s(t)$ is the height (in feet) and t is the time (in seconds). In the next example, integration is used to *derive* this function.

EXAMPLE 7   Deriving a Position Function

A ball is thrown upward with an initial velocity of 64 feet per second from an initial height of 80 feet, as shown in Figure 5.3. Derive the position function giving the height s (in feet) as a function of the time t (in seconds). When does the ball hit the ground?

SOLUTION   Let $t = 0$ represent the initial time. Then the two given conditions can be written as

$s(0) = 80$	Initial height is 80 feet.
$s'(0) = 64.$	Initial velocity is 64 feet per second.

Because the acceleration due to gravity is -32 feet per second per second, you can integrate the acceleration function to find the velocity function, as shown.

$s''(t) = -32$	Acceleration function
$s'(t) = \int -32\, dt$	Integrate $s''(t)$ to obtain $s'(t)$.
$\quad = -32t + C_1$	Velocity function

Using the initial velocity, you obtain $s'(0) = -32(0) + C_1 = 64$, which implies that $C_1 = 64$. Next, integrate the velocity function to find the position function.

$s'(t) = -32t + 64$	Velocity function
$s(t) = \int (-32t + 64)\, dt$	Integrate $s'(t)$ to obtain $s(t)$.
$\quad = -16t^2 + 64t + C_2$	Position function

Using the initial height, you obtain $s(0) = -16(0)^2 + 64(0) + C_2 = 80$, which implies that $C_2 = 80$. So, the position function is

$$s(t) = -16t^2 + 64t + 80.$$ Position function

To find the time when the ball hits the ground, set the position function equal to 0 and solve for t.

$-16t^2 + 64t + 80 = 0$	Set $s(t)$ equal to zero.
$-16(t + 1)(t - 5) = 0$	Factor.
$t = -1, \quad t = 5$	Solve for t.

Because the time must be positive, you can conclude that the ball hits the ground 5 seconds after it is thrown.

✓ **Checkpoint 7**   *Worked-out solution available at LarsonAppliedCalculus.com*

Derive the position function when a ball is thrown upward with an initial velocity of 32 feet per second from an initial height of 48 feet. When does the ball hit the ground? With what velocity does the ball hit the ground? ▪

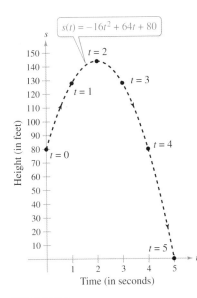

$s(t) = -16t^2 + 64t + 80$

FIGURE 5.3

EXAMPLE 8   Finding a Cost Function

The marginal cost of producing x units of a product is modeled by

$$\frac{dC}{dx} = 32 - 0.04x. \qquad \text{Marginal cost}$$

It costs $50 to produce one unit. Find the cost of producing 200 units.

SOLUTION   To find the cost function, integrate the marginal cost function.

$$C = \int (32 - 0.04x)\, dx \qquad \text{Integrate } \frac{dC}{dx} \text{ to obtain } C.$$

$$= 32x - 0.04\left(\frac{x^2}{2}\right) + K$$

$$= 32x - 0.02x^2 + K \qquad \text{Cost function}$$

To solve for K, use the initial condition $C = 50$ when $x = 1$.

$$50 = 32(1) - 0.02(1)^2 + K \qquad \text{Substitute 50 for } C \text{ and 1 for } x.$$

$$18.02 = K \qquad \text{Solve for } K.$$

So, the cost function is given by

$$C = 32x - 0.02x^2 + 18.02 \qquad \text{Cost function}$$

which implies that the cost of producing 200 units is

$$C = 32(200) - 0.02(200)^2 + 18.02$$

$$= \$5618.02.$$

✓ **Checkpoint 8**   *Worked-out solution available at LarsonAppliedCalculus.com*

The marginal cost of producing x units of a product is modeled by

$$\frac{dC}{dx} = 28 - 0.02x.$$

It costs $40 to produce one unit. Find the cost of producing 200 units.

> **STUDY TIP**
>
> In Example 8, note that K is used to represent the constant of integration rather than C. This is done to avoid confusion between the constant C and the cost function
>
> $$C = 32x - 0.02x^2 + 18.02.$$

SUMMARIZE   (Section 5.1)

1. State the definition of antiderivative *(page 312)*. For examples of antiderivatives, see Examples 1, 2, 3, 4, and 5.

2. State the Constant Rule *(page 313)*. For an example of the Constant Rule, see Example 1.

3. State the Constant Multiple Rule *(page 313)*. For an example of the Constant Multiple Rule, see Example 2.

4. State the Sum Rule *(page 313)*. For an example of the Sum Rule, see Example 4.

5. State the Difference Rule *(page 313)*. For an example of the Difference Rule, see Example 5.

6. State the Simple Power Rule *(page 313)*. For examples of the Simple Power Rule, see Examples 2, 3, 4, and 5.

7. Describe a real-life example of how antidifferentiation can be used to find a cost function *(page 318, Example 8)*.

SKILLS WARM UP 5.1

The following warm-up exercises involve skills that were covered in a previous course or in earlier sections. You will use these skills in the exercise set for this section. For additional help, review Appendix A.3 and Section 1.2.

In Exercises 1–6, rewrite the expression using rational exponents.

1. $\dfrac{\sqrt{x}}{x}$

2. $\sqrt[3]{2x}\,(2x)$

3. $\sqrt{5x^3} + \sqrt{x^5}$

4. $\dfrac{1}{\sqrt{x}} + \dfrac{1}{\sqrt[3]{x^2}}$

5. $\dfrac{(x+1)^3}{\sqrt{x+1}}$

6. $\dfrac{\sqrt{x^3}}{\sqrt[3]{x}}$

In Exercises 7–10, let $(x, y) = (2, 2)$, and solve the equation for C.

7. $y = x^2 + 5x + C$ **8.** $y = 3x^3 - 6x + C$ **9.** $y = -16x^2 + 26x + C$ **10.** $y = -\tfrac{1}{2}x^4 + 5x + C$

Exercises 5.1   See CalcChat.com for tutorial help and worked-out solutions to odd-numbered exercises.

Integration and Differentiation In Exercises 1–6, verify the statement by showing that the derivative of the right side is equal to the integrand on the left side.

1. $\displaystyle\int 4x\,dx = 2x^2 + C$

2. $\displaystyle\int 5x^4\,dx = x^5 + C$

3. $\displaystyle\int\left(-\dfrac{9}{x^4}\right)dx = \dfrac{3}{x^3} + C$

4. $\displaystyle\int\dfrac{4}{\sqrt{x}}\,dx = 8\sqrt{x} + C$

5. $\displaystyle\int\left(4x^3 - \dfrac{1}{x^2}\right)dx = x^4 + \dfrac{1}{x} + C$

6. $\displaystyle\int\left(1 - \dfrac{1}{\sqrt[3]{x^2}}\right)dx = x - 3\sqrt[3]{x} + C$

 Finding Indefinite Integrals In Exercises 7–18, find the indefinite integral. Check your result by differentiating. See Examples 1 and 2.

7. $\displaystyle\int du$

8. $\displaystyle\int dr$

9. $\displaystyle\int 6\,dx$

10. $\displaystyle\int -4\,dx$

11. $\displaystyle\int 7x\,dx$

12. $\displaystyle\int 2x\,dx$

13. $\displaystyle\int -9t^2\,dt$

14. $\displaystyle\int 3t^4\,dt$

15. $\displaystyle\int 5x^{-3}\,dx$

16. $\displaystyle\int 8y^{-5}\,dy$

17. $\displaystyle\int y^{3/2}\,dy$

18. $\displaystyle\int v^{-1/2}\,dv$

 Rewriting Before Integrating In Exercises 19–24, find the indefinite integral. See Example 3.

Original Integral	Rewrite	Integrate	Simplify

19. $\displaystyle\int \sqrt[3]{x^2}\,dx$

20. $\displaystyle\int \dfrac{1}{x^4}\,dx$

21. $\displaystyle\int \dfrac{1}{x\sqrt{x}}\,dx$

22. $\displaystyle\int x\sqrt[4]{x^3}\,dx$

23. $\displaystyle\int \dfrac{1}{2x^3}\,dx$

24. $\displaystyle\int \dfrac{1}{(3x)^2}\,dx$

 Finding Indefinite Integrals In Exercises 25–36, find the indefinite integral. Check your result by differentiating. See Examples 4 and 5.

25. $\displaystyle\int (x+3)\,dx$

26. $\displaystyle\int (5-x)\,dx$

27. $\displaystyle\int (x^5 - 8)\,dx$

28. $\displaystyle\int (x^2 + 7)\,dx$

29. $\displaystyle\int (x^2 + 5x + 1)\,dx$

30. $\displaystyle\int (x^3 - 4x + 2)\,dx$

31. $\displaystyle\int (3x^3 - 6x^2 + 2)\,dx$

32. $\displaystyle\int (2x^4 - 9x^2 + 8)\,dx$

33. $\displaystyle\int \dfrac{2x^3 - 1}{x^3}\,dx$

34. $\displaystyle\int \dfrac{t^2 + 2}{t^2}\,dt$

35. $\displaystyle\int \dfrac{5x + 4}{\sqrt[3]{x}}\,dx$

36. $\displaystyle\int \dfrac{14x - 3}{\sqrt[4]{x}}\,dx$

Using Technology In Exercises 37 and 38, use a symbolic integration utility to find the indefinite integral. Verify your results analytically.

37. $\int (x + 1)(3x - 2)\, dx$ **38.** $\int (2t - 1)^2\, dt$

Interpreting a Graph In Exercises 39–42, the graph of the derivative of a function is given. Sketch the graphs of *two* functions that have the given derivative. (There is more than one correct answer.)

39.

40.

41.

42.

 Finding a Particular Solution In Exercises 43–50, find the particular solution that satisfies the differential equation and the initial condition. *See Example 6.*

43. $f'(x) = 6x;\ f(0) = 8$

44. $f'(x) = 9x^2;\ f(0) = -1$

45. $f'(x) = 3x^2 + 4;\ f(-1) = -6$

46. $f'(x) = \frac{1}{5}x - 2;\ f(10) = -10$

47. $f'(x) = 20x^3 - 4x;\ f(1) = 3$

48. $f'(x) = \sqrt{x} + 2x;\ f(4) = 25$

49. $f'(x) = \dfrac{2 - x}{x^3},\ x > 0;\ f(2) = \dfrac{3}{4}$

50. $f'(x) = \dfrac{x^2 - 5}{x^2},\ x > 0;\ f(1) = 2$

 Finding a Particular Solution In Exercises 51–54, find a function f that satisfies the differential equation and the initial conditions.

51. $f''(x) = 2,\ f'(2) = 5,\ f(2) = 10$

52. $f''(x) = x^2,\ f'(0) = 6,\ f(0) = 3$

53. $f''(x) = x^{-2/3},\ f'(8) = 6,\ f(0) = 0$

54. $f''(x) = x^{-3/2},\ f'(1) = 2,\ f(9) = -4$

 Finding a Cost Function In Exercises 55–58, find the cost function for the given marginal cost and fixed cost. *See Example 8.*

Marginal Cost	Fixed Cost $(x = 0)$
55. $\dfrac{dC}{dx} = 85$	\$5500
56. $\dfrac{dC}{dx} = \dfrac{1}{50}x + 10$	\$1000
57. $\dfrac{dC}{dx} = \dfrac{1}{20\sqrt{x}} + 4$	\$750
58. $\dfrac{dC}{dx} = \dfrac{\sqrt[4]{x}}{10} + 10$	\$2300

Revenue and Demand In Exercises 59 and 60, find the revenue and demand functions for the given marginal revenue. (Use the fact that $R = 0$ when $x = 0$.)

59. $\dfrac{dR}{dx} = 225 - 3x$ **60.** $\dfrac{dR}{dx} = 310 - 4x$

Profit In Exercises 61–64, find the profit function for the given marginal profit and initial condition.

Marginal Profit	Initial Condition
61. $\dfrac{dP}{dx} = -18x + 1650$	$P(15) = \$22{,}725$
62. $\dfrac{dP}{dx} = -40x + 250$	$P(5) = \$650$
63. $\dfrac{dP}{dx} = -24x + 805$	$P(12) = \$8000$
64. $\dfrac{dP}{dx} = -30x + 920$	$P(8) = \$6500$

Vertical Motion In Exercises 65–68, use $s''(t) = -32$ feet per second per second as the acceleration due to gravity. (Neglect air resistance.) *See Example 7.*

65. The Grand Canyon is 6000 feet deep at the deepest point. A rock is dropped from the rim above this point. Express the height s (in feet) of the rock as a function of the time t (in seconds). How long will it take the rock to hit the canyon floor?

66. A ball is thrown upward with an initial velocity of 60 feet per second from an initial height of 16 feet. Express the height s (in feet) of the ball as a function of the time t (in seconds). How long will the ball be in the air?

67. With what initial velocity must an object be thrown upward (from ground level) to reach the top of the Washington Monument (555 feet)?

68. With what initial velocity must an object be thrown upward from a height of 5 feet to reach a maximum height of 230 feet?

69. Cost A company produces a product for which the marginal cost of producing x units is modeled by $dC/dx = 2x - 12$ and the fixed costs are $125.

(a) Find the total cost function and the average cost function.

(b) Find the total cost of producing 50 units.

(c) In part (b), how much of the total cost is fixed? How much is variable? Give examples of fixed costs associated with the manufacturing of a product. Give examples of variable costs.

70. Gardening An evergreen nursery usually sells a certain shrub after 6 years of growth and shaping. The growth rate during those 6 years is approximated by $dh/dt = 1.5t + 5$, where t is the time (in years) and h is the height (in centimeters). The seedlings are 12 centimeters tall when planted ($t = 0$).

(a) Find the height function.

(b) How tall are the shrubs when they are sold?

71. Population Growth The growth rate of the population P of Horry County in South Carolina from 1970 through 2013 can be modeled by

$$\frac{dP}{dt} = 145.52t + 2002.3$$

where t is the time in years, with $t = 0$ corresponding to 1970. The county's population was 289,650 in 2013. *(Source: U.S. Census Bureau)*

(a) Find the model for Horry County's population.

(b) Use the model to predict the population in 2021. Does your answer seem reasonable? Explain your reasoning.

 72. HOW DO YOU SEE IT? The graph shows the rate of change of the revenue of a company from 2000 through 2015.

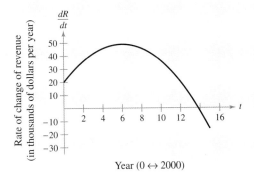

Year (0 ↔ 2000)

(a) Approximate the rate of change of the revenue in 2012. Explain your reasoning.

(b) Approximate the year when the revenue is maximum. Explain your reasoning.

73. Vital Statistics The rate of increase of the number of working married couples M (in millions) in the United States from 2008 through 2014 can be modeled by

$$\frac{dM}{dt} = 0.0918t^2 - 1.988t + 10.28$$

where t is the time in years, with $t = 8$ corresponding to 2008. The number of working married couples in 2014 was 31.7 million. *(Source: U.S. Census Bureau)*

(a) Find the model for the number of working married couples in the United States.

(b) Use the model to predict the number of working married couples in the United States in 2019. Does your answer seem reasonable? Explain your reasoning.

74. Internet Users The rate of growth of the number of Internet users I (in billions) in the world from 2000 through 2014 can be modeled by

$$\frac{dI}{dt} = -0.000176t^3 + 0.00243t^2 + 0.0070t + 0.092$$

where t is the time in years, with $t = 0$ corresponding to 2000. The number of Internet users in 2014 was 2.9 billion. *(Source: International Telecommunication Union)*

(a) Find the model for the number of Internet users in the world.

(b) Use the model to predict the number of Internet users in the world in 2018. Does your answer seem reasonable? Explain your reasoning.

75. Economics: Marginal Benefits and Costs The table gives the marginal benefit and marginal cost of producing x units of a product for a given company. (a) Plot the points in each column and use the *regression* feature of a graphing utility to find a linear model for marginal benefit and a quadratic model for marginal cost as functions of the units produced x. (b) Use integration to find the benefit B and cost C equations. Assume $B(0) = 0$ and $C(0) = 425$. (c) Find the intervals in which the benefit exceeds the cost of producing x units. Make a recommendation for how many units the company should produce based on your findings. *(Source: Adapted from Taylor, Economics, Fifth Edition)*

Number of units	1	2	3	4	5
Marginal benefit	330	320	290	270	250
Marginal cost	150	120	100	110	120

Number of units	6	7	8	9	10
Marginal benefit	230	210	190	170	160
Marginal cost	140	160	190	250	320

Spreadsheet at LarsonAppliedCalculus.com

5.2 Integration by Substitution and the General Power Rule

■ Use the General Power Rule to find indefinite integrals.
■ Use substitution to find indefinite integrals.
■ Use the General Power Rule to solve real-life problems.

In Exercise 51 on page 330, you will use integration to find a model for the cost of producing a product.

The General Power Rule

In Section 5.1, you used the Simple Power Rule

$$\int x^n\, dx = \frac{x^{n+1}}{n+1} + C, \quad n \neq -1$$

to find antiderivatives of functions expressed as powers of x alone. In this section, you will study a technique for finding antiderivatives of more complicated functions.

To begin, consider how you might find the antiderivative of

$$2x(x^2 + 1)^3.$$

Because you are hunting for a function whose derivative is $2x(x^2 + 1)^3$, you might discover the antiderivative as shown.

$$\frac{d}{dx}[(x^2 + 1)^4] = 4(x^2 + 1)^3(2x) \qquad \text{Use Chain Rule.}$$

$$\frac{d}{dx}\left[\frac{(x^2 + 1)^4}{4}\right] = (x^2 + 1)^3(2x) \qquad \text{Divide each side by 4.}$$

$$\frac{(x^2 + 1)^4}{4} + C = \int 2x(x^2 + 1)^3\, dx \qquad \text{Write in integral form.}$$

The key to this solution is the presence of the factor $2x$ in the integrand. In other words, this solution works because $2x$ is precisely the derivative of $(x^2 + 1)$. Letting $u = x^2 + 1$, you can write

$$\int \overbrace{(x^2 + 1)^3}^{u^3}\, \underbrace{2x\, dx}_{du} = \int u^3\, du$$

$$= \frac{u^4}{4} + C.$$

This is an example of the **General Power Rule** for integration.

General Power Rule for Integration

If u is a differentiable function of x, then

$$\int u^n \frac{du}{dx}\, dx = \int u^n\, du$$

$$= \frac{u^{n+1}}{n+1} + C, \quad n \neq -1.$$

When using the General Power Rule, you must first identify a factor u of the integrand that is raised to a power. Then, you must show that its derivative du/dx is also a factor of the integrand. This is demonstrated in Example 1.

EXAMPLE 1   Applying the General Power Rule

Find each indefinite integral.

a. $\displaystyle\int 3(3x - 1)^4 \, dx$ **b.** $\displaystyle\int (2x + 1)(x^2 + x) \, dx$

c. $\displaystyle\int 3x^2 \sqrt{x^3 - 2} \, dx$ **d.** $\displaystyle\int \frac{-4x}{(1 - 2x^2)^2} \, dx$

SOLUTION

a. $\displaystyle\int 3(3x - 1)^4 \, dx = \int \overbrace{(3x - 1)^4}^{u^n} \overbrace{(3)}^{\frac{du}{dx}} \, dx$ Let $u = 3x - 1$.

$\displaystyle \qquad\qquad\qquad = \frac{(3x - 1)^5}{5} + C$ General Power Rule

b. $\displaystyle\int (2x + 1)(x^2 + x) \, dx = \int \overbrace{(x^2 + x)}^{u^n} \overbrace{(2x + 1)}^{\frac{du}{dx}} \, dx$ Let $u = x^2 + x$.

$\displaystyle \qquad\qquad\qquad = \frac{(x^2 + x)^2}{2} + C$ General Power Rule

c. $\displaystyle\int 3x^2 \sqrt{x^3 - 2} \, dx = \int \overbrace{(x^3 - 2)^{1/2}}^{u^n} \overbrace{(3x^2)}^{\frac{du}{dx}} \, dx$ Let $u = x^3 - 2$.

$\displaystyle \qquad\qquad\qquad = \frac{(x^3 - 2)^{3/2}}{3/2} + C$ General Power Rule

$\displaystyle \qquad\qquad\qquad = \frac{2}{3}(x^3 - 2)^{3/2} + C$ Simplify.

d. $\displaystyle\int \frac{-4x}{(1 - 2x^2)^2} \, dx = \int \overbrace{(1 - 2x^2)^{-2}}^{u^n} \overbrace{(-4x)}^{\frac{du}{dx}} \, dx$ Let $u = 1 - 2x^2$.

$\displaystyle \qquad\qquad\qquad = \frac{(1 - 2x^2)^{-1}}{-1} + C$ General Power Rule

$\displaystyle \qquad\qquad\qquad = -\frac{1}{1 - 2x^2} + C$ Simplify.

> **STUDY TIP**
>
> Example 1(b) illustrates a case of the General Power Rule that is sometimes overlooked—when the power is $n = 1$. In this case, the rule takes the form
>
> $$\int u \frac{du}{dx} \, dx = \frac{u^2}{2} + C.$$

✓ **Checkpoint 1**   *Worked-out solution available at LarsonAppliedCalculus.com*

Find each indefinite integral.

a. $\displaystyle\int (3x^2 + 6)(x^3 + 6x)^2 \, dx$ **b.** $\displaystyle\int 2x \sqrt{x^2 - 2} \, dx$ ■

Remember that you can verify the result of an indefinite integral by differentiating the function. For instance, you can check the answer to Example 1(a) as follows.

$\displaystyle \frac{d}{dx}\left[\frac{(3x - 1)^5}{5} + C \right] = \left(\frac{1}{5} \right)(5)(3x - 1)^4(3)$ Apply Chain Rule.

$\displaystyle \qquad\qquad\qquad\qquad = 3(3x - 1)^4$ Simplify.

Many times, part of the derivative du/dx is missing from the integrand, and in *some* cases you can make the necessary adjustments to apply the General Power Rule.

ALGEBRA TUTOR

For help on the algebra in Example 2, see Example 1(b) in the *Chapter 5 Algebra Tutor*, on page 366.

EXAMPLE 2   Multiplying and Dividing by a Constant

Find $\displaystyle\int x(3 - 4x^2)^2\, dx$.

SOLUTION   Let $u = 3 - 4x^2$. To apply the General Power Rule, you need to create $du/dx = -8x$ as a factor of the integrand. You can accomplish this by multiplying and dividing by the constant -8.

$$\int x(3 - 4x^2)^2\, dx = \int \left(-\frac{1}{8}\right)\overbrace{(3 - 4x^2)^2}^{u^n}\,\overbrace{(-8x)}^{\frac{du}{dx}}\, dx \qquad \text{Multiply and divide by } -8.$$

$$= -\frac{1}{8}\int (3 - 4x^2)^2(-8x)\,dx \qquad \text{Factor } -\tfrac{1}{8} \text{ out of integrand.}$$

$$= \left(-\frac{1}{8}\right)\left[\frac{(3 - 4x^2)^3}{3}\right] + C \qquad \text{General Power Rule}$$

$$= -\frac{(3 - 4x^2)^3}{24} + C \qquad \text{Simplify.}$$

STUDY TIP

Try using the Chain Rule to check the result of Example 2. After differentiating

$$-\tfrac{1}{24}(3 - 4x^2)^3 + C$$

and simplifying, you should obtain the original integrand.

✓ **Checkpoint 2**   *Worked-out solution available at LarsonAppliedCalculus.com*

Find $\displaystyle\int x^3(3x^4 + 1)^2\, dx$.

EXAMPLE 3   Multiplying and Dividing by a Constant

Find $\displaystyle\int (x^2 + 2x)^3(x + 1)\, dx$.

SOLUTION   Let $u = x^2 + 2x$. To apply the General Power Rule, you need to create $du/dx = 2x + 2$ as a factor of the integrand. You can accomplish this by multiplying and dividing by the constant 2.

$$\int (x^2 + 2x)^3(x + 1)\, dx = \int \left(\frac{1}{2}\right)\overbrace{(x^2 + 2x)^3}^{u^n}\,\overbrace{(2)(x + 1)}^{\frac{du}{dx}}\, dx \qquad \text{Multiply and divide by 2.}$$

$$= \frac{1}{2}\int (x^2 + 2x)^3(2x + 2)\, dx \qquad \text{Rewrite integrand.}$$

$$= \frac{1}{2}\left[\frac{(x^2 + 2x)^4}{4}\right] + C \qquad \text{General Power Rule}$$

$$= \frac{1}{8}(x^2 + 2x)^4 + C \qquad \text{Simplify.}$$

✓ **Checkpoint 3**   *Worked-out solution available at LarsonAppliedCalculus.com*

Find $\displaystyle\int (x^3 - 3x)^2(x^2 - 1)\, dx$.   ▪

EXAMPLE 4   A Failure of the General Power Rule

Find $\int -8(3 - 4x^2)^2 \, dx$.

SOLUTION   Let $u = 3 - 4x^2$. To apply the General Power Rule, you must create $du/dx = -8x$ as a factor of the integrand. In Examples 2 and 3, this was done by multiplying and dividing by a constant, and then factoring that constant out of the integrand. This strategy *does not* work with variables. That is,

$$\int -8(3 - 4x^2)^2 \, dx \neq \frac{1}{x}\int (3 - 4x^2)^2(-8x) \, dx.$$

To find this indefinite integral, you can expand the integrand and use the Simple Power Rule.

$$\int -8(3 - 4x^2)^2 \, dx = \int (-72 + 192x^2 - 128x^4) \, dx$$

$$= -72x + 64x^3 - \frac{128}{5}x^5 + C$$

✓ **Checkpoint 4**   *Worked-out solution available at LarsonAppliedCalculus.com*

Find $\int 2(3x^4 + 1)^2 \, dx$.   ■

When an integrand contains an extra constant factor that is not needed as part of du/dx, you can simply move the factor outside the integral sign, as shown in the next example.

EXAMPLE 5   Applying the General Power Rule

Find $\int 7x^2\sqrt{x^3 + 1} \, dx$.

SOLUTION   Let $u = x^3 + 1$. Then you need to create $du/dx = 3x^2$ by multiplying and dividing by 3. The constant factor $\frac{7}{3}$ is not needed as part of du/dx, and can be moved outside the integral sign.

$$\int 7x^2\sqrt{x^3 + 1} \, dx = \int 7x^2(x^3 + 1)^{1/2} \, dx \qquad \text{Rewrite with rational exponent.}$$

$$= \int \frac{7}{3}(x^3 + 1)^{1/2}(3x^2) \, dx \qquad \text{Multiply and divide by 3.}$$

$$= \frac{7}{3}\int (x^3 + 1)^{1/2}(3x^2) \, dx \qquad \text{Factor } \frac{7}{3} \text{ outside integral.}$$

$$= \frac{7}{3}\left[\frac{(x^3 + 1)^{3/2}}{3/2}\right] + C \qquad \text{General Power Rule}$$

$$= \frac{14}{9}(x^3 + 1)^{3/2} + C \qquad \text{Simplify.}$$

✓ **Checkpoint 5**   *Worked-out solution available at LarsonAppliedCalculus.com*

Find $\int 5x\sqrt{x^2 - 1} \, dx$.   ■

Substitution

The integration technique used in Examples 1, 2, 3, and 5 depends on your ability to recognize or create an integrand of the form

$$u^n \frac{du}{dx}.$$

With more complicated integrands, it is difficult to recognize the steps needed to fit the integrand to a basic integration formula. When this occurs, an alternative procedure called **substitution** or **change of variables** can be helpful. With this procedure, you completely rewrite the integral in terms of u and du. That is, if $u = f(x)$, then $du = f'(x)\, dx$, and the General Power Rule takes the form

$$\int u^n \frac{du}{dx}\, dx = \int u^n\, du, \quad n \neq -1. \qquad \text{General Power Rule}$$

EXAMPLE 6   Integration by Substitution

Find $\int \sqrt{1 - 3x}\, dx$.

SOLUTION   Begin by letting $u = 1 - 3x$. Then, $du/dx = -3$ and $du = -3\, dx$. This implies that

$$dx = -\frac{1}{3}\, du$$

and you can find the indefinite integral as shown.

$$\int \sqrt{1 - 3x}\, dx = \int (1 - 3x)^{1/2}\, dx \qquad \text{Rewrite with rational exponent.}$$

$$= \int u^{1/2}\left(-\frac{1}{3}\, du\right) \qquad \text{Substitute for } x \text{ and } dx.$$

$$= -\frac{1}{3}\int u^{1/2}\, du \qquad \text{Factor } -\tfrac{1}{3} \text{ out of integrand.}$$

$$= \left(-\frac{1}{3}\right)\left(\frac{u^{3/2}}{3/2}\right) + C \qquad \text{Apply Power Rule.}$$

$$= -\frac{2}{9}u^{3/2} + C \qquad \text{Simplify.}$$

$$= -\frac{2}{9}(1 - 3x)^{3/2} + C \qquad \text{Substitute } 1 - 3x \text{ for } u.$$

You can check this result by differentiating.

$$\frac{d}{dx}\left[-\frac{2}{9}(1 - 3x)^{3/2} + C\right] = \left(-\frac{2}{9}\right)\left(\frac{3}{2}\right)(1 - 3x)^{1/2}(-3)$$

$$= \left(-\frac{1}{3}\right)(-3)(1 - 3x)^{1/2}$$

$$= \sqrt{1 - 3x}$$

✓ *Checkpoint 6*   Worked-out solution available at LarsonAppliedCalculus.com

Find $\int \sqrt{1 - 2x}\, dx$ by the method of substitution.   ■

The basic steps for integration by substitution are outlined in the guidelines below.

Guidelines for Integration by Substitution

1. Let u be a function of x (usually part of the integrand).

2. Rewrite the integral in terms of the variable u.

3. Find the resulting integral in terms of u.

4. Rewrite the antiderivative as a function of x.

5. Check your answer by differentiating.

EXAMPLE 7   Integration by Substitution

Find $\int x\sqrt{x^2 - 1}\, dx$.

SOLUTION   Consider the substitution $u = x^2 - 1$, which produces

$$du = 2x\, dx.$$

To create $2x\, dx$ as part of the integral, multiply and divide by 2.

$$\int x\sqrt{x^2 - 1}\, dx = \frac{1}{2}\int (x^2 - 1)^{1/2}\, 2x\, dx \qquad \text{Multiply and divide by 2.}$$

$$= \frac{1}{2}\int u^{1/2}\, du \qquad \text{Substitute for } x \text{ and } dx.$$

$$= \frac{1}{2}\left(\frac{u^{3/2}}{3/2}\right) + C \qquad \text{Apply Power Rule.}$$

$$= \frac{1}{3}u^{3/2} + C \qquad \text{Simplify.}$$

$$= \frac{1}{3}(x^2 - 1)^{3/2} + C \qquad \text{Substitute for } u.$$

You can check this result by differentiating.

$$\frac{d}{dx}\left[\frac{1}{3}(x^2 - 1)^{3/2} + C\right] = \frac{1}{3}\left(\frac{3}{2}\right)(x^2 - 1)^{1/2}(2x)$$

$$= \frac{1}{2}(2x)(x^2 - 1)^{1/2}$$

$$= x\sqrt{x^2 - 1}$$

✓ **Checkpoint 7**   *Worked-out solution available at LarsonAppliedCalculus.com*

Find $\int x\sqrt{x^2 + 4}\, dx$ by the method of substitution.   ■

To become efficient at integration, you should learn to use *both* techniques discussed in this section. For simpler integrals, you should use pattern recognition and create du/dx by multiplying and dividing by an appropriate constant. For more complicated integrals, you should use a formal change of variables, as shown in Examples 6 and 7. For the integrals in this section's exercise set, try working several of the problems twice—once with pattern recognition and once using formal substitution.

Extended Application: Propensity to Consume

In 2013, the U.S. poverty level for a family of four was about \$23,900. Families at or below the poverty level tend to consume 100% of their income—that is, they use all their income to purchase necessities such as food, clothing, and shelter. As income level increases, the average consumption tends to drop below 100%. For instance, a family earning \$30,000 may be able to save \$600 and so consume only \$29,400, or 98% of their income. As the income increases, the ratio of consumption to savings tends to decrease. The rate of change of consumption with respect to income is called the **marginal propensity to consume.**   *(Source: U.S. Census Bureau)*

Propensity to Consume

$Q = (x - 23{,}899)^{0.98} + 23{,}899$

FIGURE 5.4

EXAMPLE 8   Analyzing Consumption

For a family of four in 2013, the marginal propensity to consume income x (in dollars) can be modeled by

$$\frac{dQ}{dx} = \frac{0.98}{(x - 23{,}899)^{0.02}}, \quad x \geq 23{,}900$$

where Q represents the income consumed (in dollars). Use the model to estimate the amount consumed by a family of four whose 2013 income was \$37,000.

SOLUTION   Begin by integrating dQ/dx to find a model for the consumption Q.

$$Q = \int \frac{0.98}{(x - 23{,}899)^{0.02}} \, dx \qquad \text{Integrate } \frac{dQ}{dx} \text{ to obtain } Q.$$

$$= \int 0.98(x - 23{,}899)^{-0.02} \, dx \qquad \text{Rewrite.}$$

$$= (x - 23{,}899)^{0.98} + C \qquad \text{General Power Rule}$$

To solve for C, use the initial condition that $Q = 23{,}900$ when $x = 23{,}900$.

$$23{,}900 = (23{,}900 - 23{,}899)^{0.98} + C$$

$$23{,}900 = 1 + C$$

$$23{,}899 = C$$

So, the model is $Q = (x - 23{,}899)^{0.98} + 23{,}899$. Using this model, you can estimate that a family of four with an income of $x = 37{,}000$ consumed about

$$Q = (37{,}000 - 23{,}899)^{0.98} + 23{,}899 \approx \$34{,}737.$$

The graph of Q is shown in Figure 5.4.

✓ **Checkpoint 8**   *Worked-out solution available at LarsonAppliedCalculus.com*

According to the model in Example 8, at what income level would a family of four consume \$32,000?

SUMMARIZE   (Section 5.2)

1. State the General Power Rule for integration *(page 322)*. For examples of the General Power Rule, see Examples 1, 2, 3, and 5.

2. List the guidelines for integration by substitution *(page 327)*. For examples of integration by substitution, see Examples 6 and 7.

3. Describe a real-life example of how the General Power Rule can be used to analyze the marginal propensity to consume *(page 328, Example 8)*.

SKILLS WARM UP 5.2

The following warm-up exercises involve skills that were covered in earlier sections. You will use these skills in the exercise set for this section. For additional help, review Section 5.1.

In Exercises 1–9, find the indefinite integral.

1. $\displaystyle\int (2x^3 + 1)\, dx$

2. $\displaystyle\int (x^{1/2} + 3x - 4)\, dx$

3. $\displaystyle\int \frac{1}{x^5}\, dx$

4. $\displaystyle\int \frac{1}{8x^2}\, dx$

5. $\displaystyle\int (1 + 2t)t^{3/2}\, dt$

6. $\displaystyle\int \sqrt{x}(2x - 1)\, dx$

7. $\displaystyle\int \frac{5x^3 + 2}{x^2}\, dx$

8. $\displaystyle\int \frac{2x^2 - 5}{x^4}\, dx$

9. $\displaystyle\int \frac{8x^2 + 3}{\sqrt{x}}\, dx$

Exercises 5.2

See *CalcChat.com* for tutorial help and worked-out solutions to odd-numbered exercises.

Finding *u* and *du/dx* In Exercises 1–8, identify u and du/dx for the integral $\int u^n (du/dx)\, dx$.

1. $\displaystyle\int (5x^2 + 1)^2(10x)\, dx$

2. $\displaystyle\int (3 - 4x^2)^3(-8x)\, dx$

3. $\displaystyle\int \sqrt{1 - x^2}(-2x)\, dx$

4. $\displaystyle\int 3x^2\sqrt{x^3 + 1}\, dx$

5. $\displaystyle\int \left(4 + \frac{1}{x^2}\right)^5\left(\frac{-2}{x^3}\right) dx$

6. $\displaystyle\int \frac{1}{(1 + 2x)^2}(2)\, dx$

7. $\displaystyle\int (1 + \sqrt{x})^3\left(\frac{1}{2\sqrt{x}}\right) dx$

8. $\displaystyle\int (4 - \sqrt{x})^2\left(\frac{-1}{2\sqrt{x}}\right) dx$

 Applying the General Power Rule In Exercises 9–34, find the indefinite integral. Check your result by differentiating. *See Examples 1, 2, 3, and 5.*

9. $\displaystyle\int (x - 1)^4\, dx$

10. $\displaystyle\int (x - 3)^{5/2}\, dx$

11. $\displaystyle\int (7 - 2x)^2(-2)\, dx$

12. $\displaystyle\int (x^2 - 1)^3(2x)\, dx$

13. $\displaystyle\int (x^2 + 3x)(2x + 3)\, dx$

14. $\displaystyle\int (x^4 + 3x^2)(4x^3 + 6x)\, dx$

15. $\displaystyle\int \sqrt{4x^2 - 5}(8x)\, dx$

16. $\displaystyle\int \sqrt[3]{1 - 2x^2}(-4x)\, dx$

17. $\displaystyle\int \frac{6x}{(3x^2 - 5)^4}\, dx$

18. $\displaystyle\int \frac{-12x^2}{(1 - 4x^3)^2}\, dx$

19. $\displaystyle\int x^2(2x^3 - 1)^4\, dx$

20. $\displaystyle\int x(7 - 6x^2)^5\, dx$

21. $\displaystyle\int t\sqrt{t^2 + 6}\, dt$

22. $\displaystyle\int t^4\sqrt[3]{t^5 - 9}\, dt$

23. $\displaystyle\int \frac{x^4}{(3 - 2x^5)^3}\, dx$

24. $\displaystyle\int \frac{x^2}{(3x^3 + 8)^2}\, dx$

25. $\displaystyle\int (x^2 - 6x)^4(x - 3)\, dx$

26. $\displaystyle\int (4x^3 + 8x)^3(3x^2 + 2)\, dx$

27. $\displaystyle\int \frac{x^2 + 3}{(x^3 + 9x - 4)^2}\, dx$

28. $\displaystyle\int \frac{x - 2}{\sqrt{x^2 - 4x + 3}}\, dx$

29. $\displaystyle\int 5x\sqrt[3]{1 - x^2}\, dx$

30. $\displaystyle\int 9x^3\sqrt{x^4 + 2}\, dx$

31. $\displaystyle\int \frac{6x}{(1 + x^2)^3}\, dx$

32. $\displaystyle\int \frac{4x + 6}{(x^2 + 3x + 7)^3}\, dx$

33. $\displaystyle\int \frac{-3}{\sqrt{2t + 3}}\, dt$

34. $\displaystyle\int \frac{3x^2}{\sqrt{1 - x^3}}\, dx$

 Integration by Substitution In Exercises 35–42, use the method of substitution to find the indefinite integral. Check your result by differentiating. *See Examples 6 and 7.*

35. $\displaystyle\int \sqrt[3]{4x + 3}\, dx$

36. $\displaystyle\int (5x - 3)^{2/3}\, dx$

37. $\displaystyle\int x(6x^2 - 7)^3\, dx$

38. $\displaystyle\int t\sqrt{t^2 + 1}\, dt$

39. $\displaystyle\int \frac{2}{\sqrt{7x - 1}}\, dx$

40. $\displaystyle\int \frac{3}{\sqrt{2x + 1}}\, dx$

41. $\displaystyle\int \frac{x^2 + 1}{\sqrt{x^3 + 3x + 4}}\, dx$

42. $\displaystyle\int \frac{x^2 + 3}{\sqrt[3]{x^3 + 9x}}\, dx$

Comparing Methods In Exercises 43–46, (a) perform the integration in two ways: once using the Simple Power Rule and once using the General Power Rule. (b) Explain the difference in the results. (c) Which method do you prefer? Explain your reasoning.

43. $\displaystyle\int (x - 1)^2\, dx$

44. $\displaystyle\int (3 - x)^2\, dx$

45. $\displaystyle\int x(x^2 + 2)^2\, dx$

46. $\displaystyle\int x(2x^2 + 1)^2\, dx$

Finding an Equation of a Function In Exercises 47–50, find an equation of the function f that has the given derivative and whose graph passes through the given point.

47. $f'(x) = -3x^2(2 - x^3)^4$; $(0, 7)$

48. $f'(x) = 12x^3(3x^4 - 2)^3$; $(1, -3)$

49. $f'(x) = 2x(4x^2 - 10)^2$; $(2, 10)$

50. $f'(x) = x\sqrt{1 - x^2}$; $\left(0, \frac{7}{3}\right)$

51. Cost   The marginal cost of a product is modeled by

$$\frac{dC}{dx} = \frac{4}{\sqrt{x + 1}}$$

where x is the number of units. When $x = 15, C = 50$.

(a) Find the cost function.

(b) Find the cost of producing 50 units.

52. Cost   The marginal cost of a product is modeled by

$$\frac{dC}{dx} = \frac{12}{\sqrt[3]{12x + 1}}$$

where x is the number of units. When $x = 13, C = 100$.

(a) Find the cost function.

(b) Find the cost of producing 30 units.

Supply   In Exercises 53 and 54, find the supply function $x = f(p)$ that satisfies the initial condition.

53. $\dfrac{dx}{dp} = p\sqrt{p^2 - 25}$

   $x = 600$ when $p = \$13$

54. $\dfrac{dx}{dp} = \dfrac{10}{\sqrt{p - 3}}$

   $x = 100$ when $p = \$3$

Demand   In Exercises 55 and 56, find the demand function $x = f(p)$ that satisfies the initial condition.

55. $\dfrac{dx}{dp} = -\dfrac{6000p}{(p^2 - 16)^{3/2}}$

   $x = 5000$ when $p = \$5$

56. $\dfrac{dx}{dp} = -\dfrac{400}{(0.02p - 1)^3}$

   $x = 10,000$ when $p = \$100$

57. Gardening   An evergreen nursery usually sells a type of shrub after 5 years of growth and shaping. The growth rate during those 5 years is approximated by

$$\frac{dh}{dt} = \frac{17.6t}{\sqrt{17.6t^2 + 1}}$$

where t is the time (in years) and h is the height (in inches). The seedlings are 6 inches tall when planted $(t = 0)$.

(a) Find the height function.

(b) How tall are the shrubs when they are sold?

58. HOW DO YOU SEE IT?   The graph shows the rate of change of the revenue of a company from 2000 through 2015.

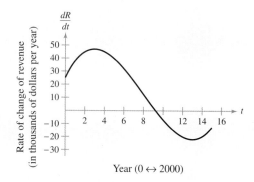

(a) Approximate the rate of change of the revenue in 2004. Explain your reasoning.

(b) Is $R(7) - R(6) > 0$? Explain your reasoning.

(c) Approximate the years in which the graph of the revenue is concave upward and the years in which it is concave downward. Approximate the years of any points of inflection.

Marginal Propensity to Consume In Exercises 59 and 60, (a) use the marginal propensity to consume, dQ/dx, to write Q as a function of x, where x is the income (in dollars) and Q is the income consumed (in dollars). Assume that families who have annual incomes of $25,000 or less consume 100% of their income. (b) Use the result of part (a) and a spreadsheet to complete the table showing the income consumed and the income saved, $x - Q$, for various incomes. (c) Use a graphing utility to represent graphically the income consumed and saved. See Example 8.

x	25,000	50,000	100,000	150,000
Q				
$x - Q$				

59. $\dfrac{dQ}{dx} = \dfrac{0.95}{(x - 24{,}999)^{0.05}}$, $x \geq 25{,}000$

60. $\dfrac{dQ}{dx} = \dfrac{0.93}{(x - 24{,}999)^{0.07}}$, $x \geq 25{,}000$

Integration Using Technology   In Exercises 61 and 62, use a symbolic integration utility to find the indefinite integral. Verify the result by differentiating.

61. $\displaystyle \int \frac{1}{\sqrt{x} + \sqrt{x + 1}}\, dx$

62. $\displaystyle \int \frac{x}{\sqrt{3x + 2}}\, dx$

5.3 Exponential and Logarithmic Integrals

■ Use the Exponential Rule to find indefinite integrals.
■ Use the Log Rule to find indefinite integrals.

Using the Exponential Rule

Each of the differentiation rules for exponential functions has a corresponding integration rule.

Population of Bacteria

In Exercise 51 on page 337, you will use integration to find a model for a population of bacteria.

Integrals of Exponential Functions

Let u be a differentiable function of x.

$$\int e^x \, dx = e^x + C \qquad \text{Simple Exponential Rule}$$

$$\int e^u \frac{du}{dx} \, dx = \int e^u \, du = e^u + C \qquad \text{General Exponential Rule}$$

EXAMPLE 1   Integrating Exponential Functions

Find each indefinite integral.

a. $\int 2e^x \, dx$ **b.** $\int 2e^{2x} \, dx$ **c.** $\int (e^x + x) \, dx$

SOLUTION

a. $\displaystyle \int 2e^x \, dx = 2 \int e^x \, dx$ Constant Multiple Rule

$\qquad\qquad = 2e^x + C$ Simple Exponential Rule

b. $\displaystyle \int 2e^{2x} \, dx = \int e^{2x}(2) \, dx$ Let $u = 2x$, then $\frac{du}{dx} = 2$.

$\qquad\qquad = \int e^u \frac{du}{dx} \, dx$ Substitute u and $\frac{du}{dx}$.

$\qquad\qquad = e^u + C$ General Exponential Rule

$\qquad\qquad = e^{2x} + C$ Substitute for u.

c. $\displaystyle \int (e^x + x) \, dx = \int e^x \, dx + \int x \, dx$ Sum Rule

$\qquad\qquad = e^x + \frac{x^2}{2} + C$ Simple Exponential and Power Rules

You can check each of these results by differentiating. For instance, in part (a),

$$\frac{d}{dx}[2e^x + C] = 2e^x.$$

✔ **Checkpoint 1**   *Worked-out solution available at LarsonAppliedCalculus.com*

Find each indefinite integral.

a. $\int 3e^x \, dx$ **b.** $\int 5e^{5x} \, dx$ **c.** $\int (e^x - x) \, dx$

■

EXAMPLE 2   Integrating an Exponential Function

Find $\int e^{3x+1}\, dx$.

SOLUTION   Let $u = 3x + 1$; then $du/dx = 3$. You can introduce the missing factor of 3 in the integrand by multiplying and dividing by 3.

$$\int e^{3x+1}\, dx = \frac{1}{3}\int e^{3x+1}(3)\, dx \qquad \text{Multiply and divide by 3.}$$

$$= \frac{1}{3}\int e^{u}\frac{du}{dx}\, dx \qquad \text{Substitute } u \text{ and } \frac{du}{dx}.$$

$$= \frac{1}{3}e^{u} + C \qquad \text{General Exponential Rule}$$

$$= \frac{1}{3}e^{3x+1} + C \qquad \text{Substitute for } u.$$

✓ **Checkpoint 2**   *Worked-out solution available at LarsonAppliedCalculus.com*

Find $\int e^{2x+3}\, dx$.

EXAMPLE 3   Integrating an Exponential Function

Find $\int 5xe^{-x^2}\, dx$.

SOLUTION   Let $u = -x^2$; then $du/dx = -2x$. You can create the factor $-2x$ in the integrand by multiplying and dividing by -2.

$$\int 5xe^{-x^2}\, dx = \int\left(-\frac{5}{2}\right)e^{-x^2}(-2x)\, dx \qquad \text{Multiply and divide by } -2.$$

$$= -\frac{5}{2}\int e^{-x^2}(-2x)\, dx \qquad \text{Factor } -\tfrac{5}{2} \text{ out of the integrand.}$$

$$= -\frac{5}{2}\int e^{u}\frac{du}{dx}\, dx \qquad \text{Substitute } u \text{ and } \frac{du}{dx}.$$

$$= -\frac{5}{2}e^{u} + C \qquad \text{General Exponential Rule}$$

$$= -\frac{5}{2}e^{-x^2} + C \qquad \text{Substitute for } u.$$

✓ **Checkpoint 3**   *Worked-out solution available at LarsonAppliedCalculus.com*

Find $\int 4xe^{x^2}\, dx$. ■

Remember that you cannot introduce a missing *variable* in the integrand. For instance, you cannot find

$$\int e^{x^2}\, dx$$

by multiplying and dividing by $2x$ and then factoring $1/(2x)$ out of the integrand. That is,

$$\int e^{x^2}\, dx \neq \frac{1}{2x}\int e^{x^2}(2x)\, dx.$$

Using the Log Rule

When the Power Rules for integration were introduced in Sections 5.1 and 5.2, you saw that they work for powers other than $n = -1$.

$$\int x^n \, dx = \frac{x^{n+1}}{n+1} + C, \quad n \neq -1 \qquad \text{Simple Power Rule}$$

$$\int u^n \frac{du}{dx} \, dx = \int u^n \, du = \frac{u^{n+1}}{n+1} + C, \quad n \neq -1 \qquad \text{General Power Rule}$$

The Log Rule for integration allows you to integrate functions of the form $\int x^{-1} \, dx$ and $\int u^{-1} \, du$.

Log Rule for Integration

Let u be a differentiable function of x.

$$\int \frac{1}{x} \, dx = \ln|x| + C \qquad \text{Simple Log Rule}$$

$$\int \frac{du/dx}{u} \, dx = \int \frac{1}{u} \, du = \ln|u| + C \qquad \text{General Log Rule}$$

You can verify each of these rules by differentiating. For instance, to verify the Simple Log Rule, notice that

$$\frac{d}{dx}[\ln x] = \frac{1}{x} \quad \text{and} \quad \frac{d}{dx}[\ln(-x)] = \frac{-1}{-x} = \frac{1}{x}.$$

STUDY TIP

Notice the absolute values in the Log Rule. For those special cases in which u or x cannot be negative, you can omit the absolute value. For instance, in Example 4(b), it is not necessary to write the antiderivative as $\ln|x^2| + C$ because x^2 cannot be negative.

EXAMPLE 4   Using the Log Rule for Integration

Find each indefinite integral.

a. $\displaystyle\int \frac{4}{x} \, dx$ **b.** $\displaystyle\int \frac{2x}{x^2} \, dx$ **c.** $\displaystyle\int \frac{3}{3x+1} \, dx$

SOLUTION

a. $\displaystyle\int \frac{4}{x} \, dx = 4 \int \frac{1}{x} \, dx$ Constant Multiple Rule

      $= 4 \ln|x| + C$ Simple Log Rule

b. $\displaystyle\int \frac{2x}{x^2} \, dx = \int \frac{du/dx}{u} \, dx$ Let $u = x^2$; then $\frac{du}{dx} = 2x$.

      $= \ln|u| + C$ General Log Rule

      $= \ln x^2 + C$ Substitute for u.

c. $\displaystyle\int \frac{3}{3x+1} \, dx = \int \frac{du/dx}{u} \, dx$ Let $u = 3x + 1$; then $\frac{du}{dx} = 3$.

      $= \ln|u| + C$ General Log Rule

      $= \ln|3x+1| + C$ Substitute for u.

✓ **Checkpoint 4**   *Worked-out solution available at LarsonAppliedCalculus.com*

Find each indefinite integral.

a. $\displaystyle\int \frac{2}{x} \, dx$ **b.** $\displaystyle\int \frac{3x^2}{x^3} \, dx$ **c.** $\displaystyle\int \frac{2}{2x+1} \, dx$

EXAMPLE 5   Using the Log Rule for Integration

Find $\int \dfrac{1}{2x - 1}\, dx$.

SOLUTION Let $u = 2x - 1$; then $du/dx = 2$. You can create the necessary factor of 2 in the integrand by multiplying and dividing by 2.

$$
\begin{aligned}
\int \frac{1}{2x - 1}\, dx &= \frac{1}{2} \int \frac{2}{2x - 1}\, dx && \text{Multiply and divide by 2.}\\[2mm]
&= \frac{1}{2} \int \frac{du/dx}{u}\, dx && \text{Substitute } u \text{ and } \frac{du}{dx}.\\[2mm]
&= \frac{1}{2} \ln|u| + C && \text{General Log Rule}\\[2mm]
&= \frac{1}{2} \ln|2x - 1| + C && \text{Substitute for } u.
\end{aligned}
$$

✓ **Checkpoint 5**   *Worked-out solution available at LarsonAppliedCalculus.com*

Find $\int \dfrac{1}{4x + 1}\, dx$.

EXAMPLE 6   Using the Log Rule for Integration

Find $\int \dfrac{6x}{x^2 + 1}\, dx$.

SOLUTION Let $u = x^2 + 1$; then

$$\frac{du}{dx} = 2x.$$

You can create the necessary factor of $2x$ in the integrand by factoring a 3 out of the integrand.

$$
\begin{aligned}
\int \frac{6x}{x^2 + 1}\, dx &= 3 \int \frac{2x}{x^2 + 1}\, dx && \text{Factor 3 out of integrand.}\\[2mm]
&= 3 \int \frac{du/dx}{u}\, dx && \text{Substitute } u \text{ and } \frac{du}{dx}.\\[2mm]
&= 3 \ln|u| + C && \text{General Log Rule}\\[2mm]
&= 3 \ln(x^2 + 1) + C && \text{Substitute for } u.
\end{aligned}
$$

✓ **Checkpoint 6**   *Worked-out solution available at LarsonAppliedCalculus.com*

Find $\int \dfrac{3x}{x^2 + 4}\, dx$.   ■

ALGEBRA TUTOR

For help on the algebra at the right, see Example 2(d) in the *Chapter 5 Algebra Tutor*, on page 367.

Integrals to which the Log Rule can be applied are often given in disguised form. For instance, when a rational function has a numerator of degree greater than or equal to that of the denominator, you should use long division to rewrite the integrand. Here is an example.

$$
\begin{aligned}
\int \frac{x^2 + 6x + 1}{x^2 + 1}\, dx &= \int \left(1 + \frac{6x}{x^2 + 1}\right) dx\\[2mm]
&= x + 3 \ln(x^2 + 1) + C
\end{aligned}
$$

The next example summarizes some additional situations in which it is helpful to rewrite the integrand in order to recognize the antiderivative.

ALGEBRA TUTOR

For help on the algebra in Example 7, see Example 2(a)–(c) in the *Chapter 5 Algebra Tutor*, on page 367.

EXAMPLE 7   Rewriting Before Integrating

Find each indefinite integral.

a. $\displaystyle\int \frac{3x^2 + 2x - 1}{x^2}\, dx$ **b.** $\displaystyle\int \frac{1}{1 + e^{-x}}\, dx$ **c.** $\displaystyle\int \frac{x^2 + x + 1}{x - 1}\, dx$

SOLUTION

a. Begin by rewriting the integrand as the sum of three fractions.

$$\int \frac{3x^2 + 2x - 1}{x^2}\, dx = \int \left(\frac{3x^2}{x^2} + \frac{2x}{x^2} - \frac{1}{x^2} \right) dx$$

$$= \int \left(3 + \frac{2}{x} - \frac{1}{x^2} \right) dx$$

$$= 3x + 2\ln|x| + \frac{1}{x} + C$$

b. Begin by rewriting the integrand by multiplying and dividing by e^x.

$$\int \frac{1}{1 + e^{-x}}\, dx = \int \left(\frac{e^x}{e^x} \right) \frac{1}{1 + e^{-x}}\, dx$$

$$= \int \frac{e^x}{e^x + 1}\, dx$$

$$= \ln(e^x + 1) + C$$

c. Begin by dividing the numerator by the denominator.

$$\int \frac{x^2 + x + 1}{x - 1}\, dx = \int \left(x + 2 + \frac{3}{x - 1} \right) dx$$

$$= \frac{x^2}{2} + 2x + 3\ln|x - 1| + C$$

✓ *Checkpoint 7*   *Worked-out solution available at LarsonAppliedCalculus.com*

Find each indefinite integral.

a. $\displaystyle\int \frac{4x^2 - 3x + 2}{x^2}\, dx$ **b.** $\displaystyle\int \frac{2}{e^{-x} + 1}\, dx$ **c.** $\displaystyle\int \frac{x^2 + 2x + 4}{x + 1}\, dx$

SUMMARIZE   (Section 5.3)

1. State the Simple Exponential Rule *(page 331)*. For an example of the Simple Exponential Rule, see Example 1.

2. State the General Exponential Rule *(page 331)*. For examples of the General Exponential Rule, see Examples 2 and 3.

3. State the Simple Log Rule *(page 333)*. For an example of the Simple Log Rule, see Example 4.

4. State the General Log Rule *(page 333)*. For examples of the General Log Rule, see Examples 5 and 6.

SKILLS WARM UP 5.3

The following warm-up exercises involve skills that were covered in earlier sections. You will use these skills in the exercise set for this section. For additional help, review Section 5.1.

In Exercises 1–4, use long division to rewrite the quotient.

1. $\dfrac{x^2 + 4x + 2}{x + 2}$

2. $\dfrac{x^2 - 8x + 16}{x - 3}$

3. $\dfrac{x^3 + 4x^2 - 30x - 4}{x^2 - 4x}$

4. $\dfrac{x^4 - x^3 + x^2 + 15x + 2}{x^2 + 5}$

In Exercises 5–8, find the indefinite integral.

5. $\displaystyle\int \left(x^3 + \dfrac{1}{x^2}\right) dx$

6. $\displaystyle\int \dfrac{x^2 + 2x}{x} \, dx$

7. $\displaystyle\int \dfrac{x^3 + 4}{x^2} \, dx$

8. $\displaystyle\int \dfrac{4x - 5}{x^4} \, dx$

Exercises 5.3

See *CalcChat.com* for tutorial help and worked-out solutions to odd-numbered exercises.

 Integrating an Exponential Function In Exercises 1–12, find the indefinite integral. *See Examples 1, 2, and 3.*

1. $\displaystyle\int 7e^{7x} \, dx$

2. $\displaystyle\int -3e^{-3x} \, dx$

3. $\displaystyle\int (-4e^x + x^2) \, dx$

4. $\displaystyle\int (6e^x - 4x^3) \, dx$

5. $\displaystyle\int e^{5x-3} \, dx$

6. $\displaystyle\int e^{-6x+5} \, dx$

7. $\displaystyle\int 9xe^{-2x^2} \, dx$

8. $\displaystyle\int 3xe^{5x^2} \, dx$

9. $\displaystyle\int 5x^2 e^{x^3} \, dx$

10. $\displaystyle\int 7x^3 e^{2x^4} \, dx$

11. $\displaystyle\int (2x + 1)e^{x^2+x} \, dx$

12. $\displaystyle\int (x - 4)e^{x^2-8x} \, dx$

 Using the Log Rule for Integration In Exercises 13–30, find the indefinite integral. *See Examples 4, 5, and 6.*

13. $\displaystyle\int \dfrac{4x^3}{x^4} \, dx$

14. $\displaystyle\int \dfrac{6x^5}{x^6} \, dx$

15. $\displaystyle\int \dfrac{5}{5x + 2} \, dx$

16. $\displaystyle\int \dfrac{4}{4x - 7} \, dx$

17. $\displaystyle\int \dfrac{1}{3 - 2x} \, dx$

18. $\displaystyle\int \dfrac{1}{6x - 5} \, dx$

19. $\displaystyle\int \dfrac{2}{3x + 5} \, dx$

20. $\displaystyle\int \dfrac{5}{7x + 2} \, dx$

21. $\displaystyle\int \dfrac{x}{x^2 + 1} \, dx$

22. $\displaystyle\int \dfrac{x^2}{3 - x^3} \, dx$

23. $\displaystyle\int \dfrac{x^2}{x^3 + 1} \, dx$

24. $\displaystyle\int \dfrac{x}{x^2 + 4} \, dx$

25. $\displaystyle\int \dfrac{x + 3}{x^2 + 6x + 7} \, dx$

26. $\displaystyle\int \dfrac{x^2 + 2x + 3}{x^3 + 3x^2 + 9x + 1} \, dx$

27. $\displaystyle\int \dfrac{1}{x \ln x} \, dx$

28. $\displaystyle\int \dfrac{1}{x(\ln x)^2} \, dx$

29. $\displaystyle\int \dfrac{e^{-x}}{1 - e^{-x}} \, dx$

30. $\displaystyle\int \dfrac{e^x}{1 + e^x} \, dx$

Finding Indefinite Integrals In Exercises 31–46, use any basic integration formula or formulas to find the indefinite integral. State which integration formula(s) you used to find the integral.

31. $\displaystyle\int \dfrac{x^3 - 8x}{2x^2} \, dx$

32. $\displaystyle\int \dfrac{x - 1}{4x} \, dx$

33. $\displaystyle\int \dfrac{8x^3 + 3x^2 + 6}{x^3} \, dx$

34. $\displaystyle\int \dfrac{2x^3 - 6x^2 - 5x}{x^2} \, dx$

35. $\displaystyle\int \dfrac{e^{2x} + 2e^x + 1}{e^x} \, dx$

36. $\displaystyle\int \dfrac{e^{5x} - 3e^{3x} + e^x}{e^{3x}} \, dx$

37. $\displaystyle\int e^x \sqrt{1 - e^x} \, dx$

38. $\displaystyle\int (6x + e^x)\sqrt{3x^2 + e^x} \, dx$

39. $\displaystyle\int \dfrac{1 + e^{-x}}{1 + xe^{-x}} \, dx$

40. $\displaystyle\int \dfrac{2(e^x - e^{-x})}{(e^x + e^{-x})^2} \, dx$

41. $\displaystyle\int \dfrac{5}{e^{-5x} + 7} \, dx$

42. $\displaystyle\int \dfrac{3}{1 + e^{-3x}} \, dx$

43. $\int \dfrac{x^2 + 2x + 5}{x - 1}\, dx$ **44.** $\int \dfrac{x^3 - 36x + 3}{x + 6}\, dx$

45. $\int \dfrac{x - 3}{x + 3}\, dx$ **46.** $\int \dfrac{x^2 + x + 1}{x^2 + 1}\, dx$

 Finding an Equation of a Function In Exercises 47–50, find an equation of the function f that has the given derivative and whose graph passes through the given point.

47. $f'(x) = \dfrac{1}{x^2} e^{2/x};\ (4, 6)$ **48.** $f'(x) = \dfrac{2}{1 + e^{-x}};\ (0, 3)$

49. $f'(x) = \dfrac{x^2 + 4x + 3}{x - 1};\ (2, 4)$

50. $f'(x) = \dfrac{x^3 - 4x^2 + 3}{x - 3};\ (4, -1)$

51. Biology A population P of bacteria is growing at the rate of

$$\frac{dP}{dt} = \frac{3000}{1 + 0.25t}$$

where t is the time (in days). When $t = 0$, the population is 1000.

(a) Find a model for the population.

(b) What is the population after 3 days?

(c) After how many days will the population be 12,000?

52. Biology Because of an insufficient oxygen supply, the trout population P in a lake is dying. The population's rate of change can be modeled by

$$\frac{dP}{dt} = -125e^{-t/20}$$

where t is the time (in days). When $t = 0$, the population is 2500.

(a) Find a model for the population.

(b) What is the population after 15 days?

(c) How long will it take for the entire trout population to die?

53. Demand The marginal price for the demand of a product can be modeled by

$$\frac{dp}{dx} = 0.1e^{-x/500}$$

where x is the quantity demanded. When the demand is 600 units, the price p is \$30.

(a) Find the demand function.

(b) Use a graphing utility to graph the demand function. Does price increase or decrease as demand increases?

(c) Use the *zoom* and *trace* features of the graphing utility to find the quantity demanded when the price is \$22.

 54. HOW DO YOU SEE IT? The graph shows the rate of change of the revenue of a company from 2000 through 2015.

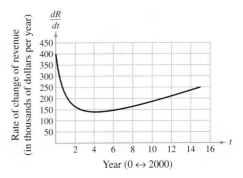

(a) Approximate the rate of change of the revenue in 2010. Explain your reasoning.

(b) Approximate the year when the rate of change of the revenue is the greatest. Explain your reasoning.

(c) Approximate the year when the revenue is maximum. Explain your reasoning.

55. Revenue The rate of change in revenue for the Priceline Group from 2004 through 2013 can be modeled by

$$\frac{dR}{dt} = 0.07e^{0.2387t}$$

where R is the revenue (in billions of dollars) and t is the time (in years), with $t = 4$ corresponding to 2004. In 2013, the revenue for the Priceline Group was \$6.8 billion. *(Source: The Priceline Group)*

(a) Find a model for the revenue of the Priceline Group.

(b) Find the revenue of the Priceline Group in 2012.

56. Revenue The rate of change in revenue for DIRECTV from 2004 through 2013 can be modeled by

$$\frac{dR}{dt} = 3.17 - \frac{6.671}{t}$$

where R is the revenue (in billions of dollars) and t is the time (in years), with $t = 4$ corresponding to 2004. In 2010, the revenue for DIRECTV was \$24.1 billion. *(Source: DIRECTV)*

(a) Find a model for the revenue of DIRECTV.

(b) Find the revenue of DIRECTV in 2013.

True or False In Exercises 57 and 58, determine whether the statement is true or false. If it is false, explain why or give an example that shows it is false.

57. $(\ln x)^{1/2} = \tfrac{1}{2}(\ln x)$

58. $\displaystyle\int \frac{1}{x}\, dx = \ln|ax| + C, \quad a \neq 0$

QUIZ YOURSELF   See *CalcChat.com* for tutorial help and worked-out solutions to odd-numbered exercises.

Take this quiz as you would take a quiz in class. When you are done, check your work against the answers given in the back of the book.

In Exercises 1–9, find the indefinite integral. Check your result by differentiating.

1. $\displaystyle\int 3\,dx$

2. $\displaystyle\int 10x\,dx$

3. $\displaystyle\int \frac{1}{x^6}\,dx$

4. $\displaystyle\int (x^2 - 2x + 15)\,dx$

5. $\displaystyle\int (6x + 1)^3(6)\,dx$

6. $\displaystyle\int x(5x^2 - 2)^4\,dx$

7. $\displaystyle\int (x^2 - 5x)(2x - 5)\,dx$

8. $\displaystyle\int \frac{4x^2}{(x^3 + 3)^3}\,dx$

9. $\displaystyle\int 2\sqrt{7x + 4}\,dx$

In Exercises 10 and 11, find the particular solution that satisfies the differential equation and the initial condition.

10. $f'(x) = 16x;\ \ f(0) = 1$

11. $f'(x) = 6x^2 + 1;\ \ f(1) = 9$

12. The marginal cost of producing x units of a product is modeled by

$$\frac{dC}{dx} = 16 - 0.06x.$$

It costs \$25 to produce one unit. Find (a) the cost function C (in dollars), (b) the fixed cost (when $x = 0$), and (c) the total cost of producing 500 units.

13. Find an equation of the function f whose graph passes through the point $(0, 1)$ and whose derivative is

$$f'(x) = 5x\sqrt[4]{x^2 + 16}.$$

14. The number of bolts B produced by a foundry changes according to the model

$$\frac{dB}{dt} = \frac{250t}{\sqrt{t^2 + 36}},\ \ 0 \le t \le 40$$

where t is the time (in hours). Find the number of bolts produced in (a) 8 hours and (b) 40 hours.

In Exercises 15–20, find the indefinite integral.

15. $\displaystyle\int 10e^{10x+7}\,dx$

16. $\displaystyle\int 9x^2 e^{x^3}\,dx$

17. $\displaystyle\int (x - 3)e^{x^2 - 6x}\,dx$

18. $\displaystyle\int \frac{9}{9x - 4}\,dx$

19. $\displaystyle\int \frac{1}{3 - 8x}\,dx$

20. $\displaystyle\int \frac{x}{3x^2 + 4}\,dx$

21. The rate of change in revenue for Under Armour from 2004 through 2013 can be modeled by

$$\frac{dR}{dt} = 35.38t - \frac{612.804}{t}$$

where R is the revenue (in millions of dollars) and t is the time (in years), with $t = 4$ corresponding to 2004. In 2004, the revenue for Under Armour was \$205.2 million. *(Source: Under Armour, Inc.)*

(a) Find a model for the revenue of Under Armour.

(b) Find the revenue of Under Armour in 2010.

5.4 Area and the Fundamental Theorem of Calculus

■ Understand the relationship between area and definite integrals.
■ Evaluate definite integrals using the Fundamental Theorem of Calculus.
■ Use definite integrals to solve marginal analysis problems.
■ Find the average values of functions over closed intervals.
■ Use properties of even and odd functions to help evaluate definite integrals.
■ Find the amounts of annuities.

Area and Definite Integrals

From your study of geometry, you know that area is a number that defines the size of a bounded region. For simple regions, such as rectangles, triangles, and circles, area can be found using geometric formulas.

In this section, you will learn how to use calculus to find the areas of nonstandard regions, such as the region R shown in the figure at the right.

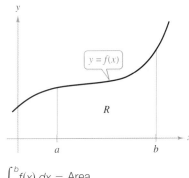

$y = f(x)$

R

$$\int_a^b f(x)\,dx = \text{Area}$$

One- to Four-Family Homes

Mortgage debt outstanding (in trillions of dollars)

Year (0 ↔ 2000)

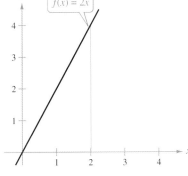

In Exercise 79 on page 350, you will use integration to find a model for the mortgage debt outstanding for one- to four-family homes.

Definition of a Definite Integral

Let f be nonnegative and continuous on the closed interval $[a, b]$. The area of the region bounded by the graph of f, the x-axis, and the lines $x = a$ and $x = b$ is denoted by

$$\text{Area} = \int_a^b f(x)\,dx.$$

The expression $\int_a^b f(x)\,dx$ is called the **definite integral** from a to b, where a is the **lower limit of integration** and b is the **upper limit of integration.**

$f(x) = 2x$

FIGURE 5.5

EXAMPLE 1   Evaluating a Definite Integral Using a Geometric Formula

The definite integral

$$\int_0^2 2x\,dx$$

represents the area of the region bounded by the graph of $f(x) = 2x$, the x-axis, and the line $x = 2$, as shown in Figure 5.5. The region is triangular, with a height of 4 units and a base of 2 units. Using the formula for the area of a triangle, you have

$$\int_0^2 2x\,dx = \frac{1}{2}(\text{base})(\text{height}) = \frac{1}{2}(2)(4) = 4.$$

✓ **Checkpoint 1**   Worked-out solution available at LarsonAppliedCalculus.com

Evaluate the definite integral using a geometric formula. Illustrate your answer with an appropriate sketch.

$$\int_0^3 4x\,dx$$

The Fundamental Theorem of Calculus

Consider the function A, which denotes the area of the region shown in Figure 5.6.

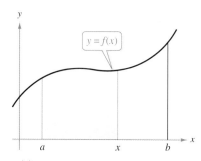

$A(x) =$ Area from a to x

FIGURE 5.6

To discover the relationship between A and f, let x increase by an amount Δx. This increases the area by ΔA. Let $f(m)$ and $f(M)$ denote the minimum and maximum values of f on the interval $[x, x + \Delta x]$.

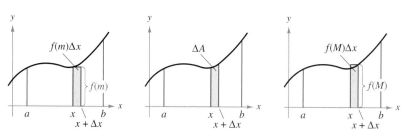

FIGURE 5.7

As indicated in Figure 5.7, you can write the inequality below.

$$f(m)\,\Delta x \le \quad \Delta A \quad \le f(M)\,\Delta x \qquad \text{See Figure 5.7.}$$

$$f(m) \le \quad \frac{\Delta A}{\Delta x} \quad \le f(M) \qquad \text{Divide each term by } \Delta x.$$

$$\lim_{\Delta x \to 0} f(m) \le \lim_{\Delta x \to 0} \frac{\Delta A}{\Delta x} \le \lim_{\Delta x \to 0} f(M) \qquad \text{Take limit of each term.}$$

$$f(x) \le \quad A'(x) \quad \le f(x) \qquad \text{Definition of derivative of } A(x)$$

So, $f(x) = A'(x)$, and $A(x) = F(x) + C$, where $F'(x) = f(x)$. Because $A(a) = 0$, it follows that $C = -F(a)$. So, $A(x) = F(x) - F(a)$, which implies that

$$A(b) = \int_a^b f(x)\,dx = F(b) - F(a).$$

This equation tells you that *if you can find an antiderivative for f,* then you can use the antiderivative to evaluate the definite integral $\int_a^b f(x)\,dx$. This result is called the **Fundamental Theorem of Calculus.**

The Fundamental Theorem of Calculus

If f is nonnegative and continuous on the closed interval $[a, b]$, then

$$\int_a^b f(x)\,dx = F(b) - F(a)$$

where F is any function such that $F'(x) = f(x)$ for all x in $[a, b]$.

Guidelines for Using the Fundamental Theorem of Calculus

1. The Fundamental Theorem of Calculus describes a way of *evaluating* a definite integral, not a procedure for finding antiderivatives.

2. In applying the Fundamental Theorem, it is helpful to use the notation

$$\int_a^b f(x)\, dx = F(x)\Big]_a^b = F(b) - F(a).$$

For instance, to evaluate $\int_1^3 x^3\, dx$, you can write

$$\int_1^3 x^3\, dx = \frac{x^4}{4}\Big]_1^3$$

$$= \frac{3^4}{4} - \frac{1^4}{4}$$

$$= 20.$$

3. The constant of integration C can be dropped because

$$\int_a^b f(x)\, dx = \Big[F(x) + C\Big]_a^b$$

$$= [F(b) + C] - [F(a) + C]$$

$$= F(b) - F(a) + C - C$$

$$= F(b) - F(a).$$

In the development of the Fundamental Theorem of Calculus, f was assumed to be nonnegative on the closed interval $[a, b]$. As such, the definite integral was defined as an area. Now, with the Fundamental Theorem, the definition of a definite integral can be extended to include functions that are negative on all or part of the closed interval $[a, b]$. Specifically, if f is *any* function that is continuous on a closed interval $[a, b]$, then the **definite integral** of $f(x)$ from a to b is defined to be

$$\int_a^b f(x)\, dx = F(b) - F(a)$$

where F is an antiderivative of f. Remember that definite integrals do not necessarily represent areas and can be negative, zero, or positive.

STUDY TIP

Be sure you see the distinction between indefinite and definite integrals. The *indefinite integral*

$$\int f(x)\, dx$$

denotes a *family of functions*, each of which is an antiderivative of f, whereas the *definite integral*

$$\int_a^b f(x)\, dx$$

is a *number*.

Properties of Definite Integrals

Let f and g be continuous on the closed interval $[a, b]$.

1. $\displaystyle\int_a^b k f(x)\, dx = k \int_a^b f(x)\, dx,$ k is a constant.

2. $\displaystyle\int_a^b [f(x) \pm g(x)]\, dx = \int_a^b f(x)\, dx \pm \int_a^b g(x)\, dx$

3. $\displaystyle\int_a^b f(x)\, dx = \int_a^c f(x)\, dx + \int_c^b f(x)\, dx,$ $a < c < b$

4. $\displaystyle\int_a^a f(x)\, dx = 0$

5. $\displaystyle\int_a^b f(x)\, dx = -\int_b^a f(x)\, dx$

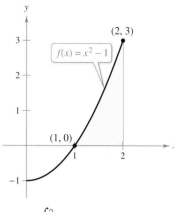

Area $= \int_1^2 (x^2 - 1)\, dx$

FIGURE 5.8

EXAMPLE 2   Finding Area by the Fundamental Theorem

Find the area of the region bounded by the x-axis and the graph of

$$f(x) = x^2 - 1, \quad 1 \le x \le 2.$$

SOLUTION   Note that $f(x) \ge 0$ on the interval $1 \le x \le 2$, as shown in Figure 5.8. So, you can represent the area of the region by a definite integral. To find the area, use the Fundamental Theorem of Calculus.

$$\text{Area} = \int_1^2 (x^2 - 1)\, dx \qquad \text{Integrate from } x = 1 \text{ to } x = 2.$$

$$= \left[\frac{x^3}{3} - x \right]_1^2 \qquad \text{Find antiderivative.}$$

$$= \left(\frac{2^3}{3} - 2 \right) - \left(\frac{1^3}{3} - 1 \right) \qquad \text{Apply Fundamental Theorem.}$$

$$= \frac{2}{3} - \left(-\frac{2}{3} \right)$$

$$= \frac{4}{3} \qquad \text{Simplify.}$$

So, the area of the region is $\frac{4}{3}$ square units.

✓ **Checkpoint 2**   *Worked-out solution available at LarsonAppliedCalculus.com*

Find the area of the region bounded by the x-axis and the graph of

$$f(x) = x^2 + 1, \quad 2 \le x \le 3.$$

EXAMPLE 3   Evaluating a Definite Integral

Evaluate the definite integral

$$\int_0^1 (4t + 1)^2\, dt$$

and sketch the region whose area is represented by the integral.

SOLUTION

$$\int_0^1 (4t + 1)^2\, dt = \frac{1}{4} \int_0^1 (4t + 1)^2 (4)\, dt \qquad \text{Multiply and divide by 4.}$$

$$= \frac{1}{4} \left[\frac{(4t + 1)^3}{3} \right]_0^1 \qquad \text{Find antiderivative.}$$

$$= \frac{1}{4} \left[\left(\frac{5^3}{3} \right) - \left(\frac{1}{3} \right) \right] \qquad \text{Apply Fundamental Theorem.}$$

$$= \frac{1}{4} \left(\frac{124}{3} \right)$$

$$= \frac{31}{3} \qquad \text{Simplify.}$$

The region is shown in Figure 5.9.

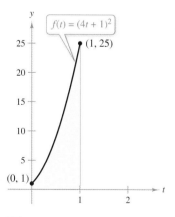

FIGURE 5.9

✓ **Checkpoint 3**   *Worked-out solution available at LarsonAppliedCalculus.com*

Evaluate $\displaystyle\int_0^1 (2t + 3)^3\, dt.$

EXAMPLE 4   **Evaluating Definite Integrals**

Evaluate each definite integral.

a. $\displaystyle\int_0^3 e^{2x}\, dx$ **b.** $\displaystyle\int_1^2 \frac{1}{x}\, dx$ **c.** $\displaystyle\int_1^4 -3\sqrt{x}\, dx$

SOLUTION

a. $\displaystyle\int_0^3 e^{2x}\, dx = \frac{1}{2}e^{2x}\Big]_0^3 = \frac{1}{2}(e^6 - e^0) \approx 201.21$

b. $\displaystyle\int_1^2 \frac{1}{x}\, dx = \ln x\Big]_1^2 = \ln 2 - \ln 1 = \ln 2 \approx 0.69$

> **STUDY TIP**
>
> In Example 4(c), note that the value of a definite integral can be negative.

c. $\displaystyle\int_1^4 -3\sqrt{x}\, dx = -3\int_1^4 x^{1/2}\, dx$ Rewrite with rational exponent.

$$= -3\left[\frac{x^{3/2}}{3/2}\right]_1^4$$ Find antiderivative.

$$= -2x^{3/2}\Big]_1^4$$

$$= -2(4^{3/2} - 1^{3/2})$$ Apply Fundamental Theorem.

$$= -2(8 - 1)$$

$$= -14$$ Simplify.

✓ **Checkpoint 4**   *Worked-out solution available at LarsonAppliedCalculus.com*

Evaluate each definite integral.

a. $\displaystyle\int_0^1 e^{4x}\, dx$ **b.** $\displaystyle\int_2^5 -\frac{1}{x}\, dx$

EXAMPLE 5   **A Definite Integral Involving Absolute Value**

Evaluate $\displaystyle\int_0^2 |2x - 1|\, dx$.

SOLUTION   The region represented by the definite integral is shown in Figure 5.10. From the definition of absolute value, you can write

$$|2x - 1| = \begin{cases} -(2x - 1), & x < \frac{1}{2} \\ 2x - 1, & x \geq \frac{1}{2} \end{cases}.$$

Using Property 3 of definite integrals, rewrite the integral as two definite integrals.

$$\int_0^2 |2x - 1|\, dx = \int_0^{1/2} -(2x - 1)\, dx + \int_{1/2}^2 (2x - 1)\, dx$$

$$= \left[-x^2 + x\right]_0^{1/2} + \left[x^2 - x\right]_{1/2}^2$$

$$= \left(-\frac{1}{4} + \frac{1}{2}\right) - (0 + 0) + (4 - 2) - \left(\frac{1}{4} - \frac{1}{2}\right)$$

$$= \frac{5}{2}$$

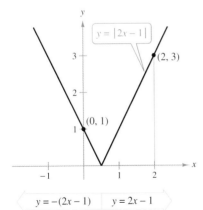

$y = |2x - 1|$

$(2, 3)$

$(0, 1)$

$y = -(2x - 1)$ $y = 2x - 1$

FIGURE 5.10

✓ **Checkpoint 5**   *Worked-out solution available at LarsonAppliedCalculus.com*

Evaluate $\displaystyle\int_0^5 |x - 2|\, dx$.

TECH TUTOR

Symbolic integration utilities can be used to evaluate definite integrals as well as to find indefinite integrals. If you have access to such a program, try using it to evaluate several of the definite integrals in this section.

Marginal Analysis

You have already studied *marginal analysis* in the context of derivatives and differentials (Sections 2.3 and 3.8). There, you were given a cost, revenue, or profit function, and you used the derivative to approximate the additional cost, revenue, or profit obtained by selling one additional unit. In this section, you will examine the reverse process. That is, you will be given the marginal cost, marginal revenue, or marginal profit and you will use a definite integral to find the exact increase or decrease in cost, revenue, or profit obtained by selling one or several additional units.

For instance, you are asked to find the additional revenue obtained by increasing sales from x_1 to x_2 units. When you know the revenue function R, you can find the additional revenue by subtracting $R(x_1)$ from $R(x_2)$. When you do not know R, you can use the marginal revenue function dR/dx to find the additional revenue by using a definite integral.

$$\int_{x_1}^{x_2} \frac{dR}{dx}\, dx = R(x_2) - R(x_1)$$

EXAMPLE 6   **Using Marginal Analysis**

The marginal profit for a product is modeled by

$$\frac{dP}{dx} = -0.0005x + 12.2.$$

a. Find the change in profit when sales increase from 100 to 101 units.

b. Find the change in profit when sales increase from 100 to 110 units.

SOLUTION

a. The change in profit obtained by increasing sales from 100 to 101 units is

$$\int_{100}^{101} \frac{dP}{dx}\, dx = \int_{100}^{101} (-0.0005x + 12.2)\, dx$$
$$= \left[-0.00025x^2 + 12.2x \right]_{100}^{101}$$
$$= \$12.15.$$

b. The change in profit obtained by increasing sales from 100 to 110 units is

$$\int_{100}^{110} \frac{dP}{dx}\, dx = \int_{100}^{110} (-0.0005x + 12.2)\, dx$$
$$= \left[-0.00025x^2 + 12.2x \right]_{100}^{110}$$
$$= \$121.48.$$

✓ *Checkpoint 6*   *Worked-out solution available at LarsonAppliedCalculus.com*

The marginal profit for a product is modeled by

$$\frac{dP}{dx} = -0.0002x + 14.2.$$

a. Find the change in profit when sales increase from 100 to 101 units.

b. Find the change in profit when sales increase from 100 to 110 units.

Average Value

The *average value* of a function on a closed interval is defined below.

Definition of the Average Value of a Function

If f is continuous on $[a, b]$, then the **average value** of f on $[a, b]$ is

$$\text{Average value of } f \text{ on } [a, b] = \frac{1}{b - a} \int_a^b f(x) \, dx.$$

	A	B
1	t	$P = 0.0004t^2 + 0.053t + 2.49$
2	1	2.5434
3	2	2.5976
4	3	2.6526
5	4	2.7084
6	5	2.765
7	6	2.8224
8	7	2.8806
9	8	2.9396
10	9	2.9994
11	10	3.06
12	11	3.1214
13	12	3.1836
14	13	3.2466
15	14	3.3104
16	15	3.375
17	16	3.4404
18	17	3.5066
19	18	3.5736
20	19	3.6414
21	20	3.71
22	21	3.7794
23	22	3.8496
24	23	3.9206
25	24	3.9924
26	25	4.065
27		
28	**Sum**	81.685
29	**Average**	3.2674

EXAMPLE 7   Finding the Average Population

The world population P (in billions of people) from 1951 through 2000 can be modeled by

$$P = 0.0004t^2 + 0.053t + 2.49, \quad 1 \le t \le 50$$

where t is the year, with $t = 1$ corresponding to 1951. Find the average population of the world from 1951 through 1975.   *(Source: U.S. Census Bureau)*

SOLUTION   To find the average population of the world from 1951 through 1975, integrate P over the interval $[1, 25]$.

$$\begin{aligned}
\text{Average population} &= \frac{1}{25 - 1} \int_1^{25} (0.0004t^2 + 0.053t + 2.49) \, dt \\
&= \frac{1}{24} \left[\frac{0.0004t^3}{3} + \frac{0.053t^2}{2} + 2.49t \right]_1^{25} \\
&= \frac{1}{24}(78.3792) \\
&= 3.2658 \text{ billion people}
\end{aligned}$$

So, the average population of the world from 1951 through 1975 was about 3.3 billion people. (See Figure 5.11.)

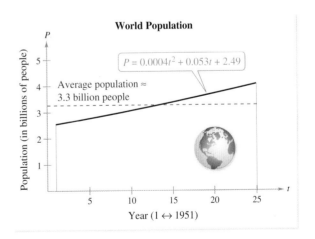

World Population

$P = 0.0004t^2 + 0.053t + 2.49$

Average population ≈ 3.3 billion people

Population (in billions of people)

Year (1 ↔ 1951)

FIGURE 5.11

✓ *Checkpoint 7*   *Worked-out solution available at LarsonAppliedCalculus.com*

Use the model in Example 7 to find the average population of the world from 1976 through 2000. Compare your result to the result of Example 7.   ■

Even and Odd Functions

Several common functions have graphs that are symmetric with respect to the y-axis or the origin, as shown in Figure 5.12. If the graph of f is symmetric with respect to the y-axis, as in Figure 5.12(a), then

$$f(-x) = f(x) \qquad \text{Even function}$$

and f is called an **even function.** If the graph of f is symmetric with respect to the origin, as in Figure 5.12(b), then

$$f(-x) = -f(x) \qquad \text{Odd function}$$

and f is called an **odd function.**

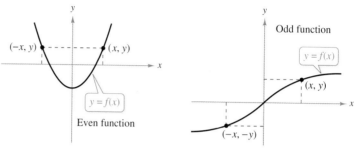

(a) y-axis symmetry   (b) Origin symmetry

FIGURE 5.12

Integration of Even and Odd Functions

1. If f is an *even* function, then $\displaystyle\int_{-a}^{a} f(x)\,dx = 2\int_{0}^{a} f(x)\,dx.$

2. If f is an *odd* function, then $\displaystyle\int_{-a}^{a} f(x)\,dx = 0.$

EXAMPLE 8   Integrating Even and Odd Functions

Evaluate each definite integral.

a. $\displaystyle\int_{-2}^{2} x^2\,dx$ **b.** $\displaystyle\int_{-2}^{2} x^3\,dx$

SOLUTION

a. Because $f(x) = x^2$ is an even function,

$$\int_{-2}^{2} x^2\,dx = 2\int_{0}^{2} x^2\,dx = 2\left[\frac{x^3}{3}\right]_{0}^{2} = 2\left(\frac{8}{3} - 0\right) = \frac{16}{3}.$$

b. Because $f(x) = x^3$ is an odd function,

$$\int_{-2}^{2} x^3\,dx = 0.$$

✓ **Checkpoint 8**   *Worked-out solution available at LarsonAppliedCalculus.com*

Evaluate each definite integral.

a. $\displaystyle\int_{-1}^{1} x^4\,dx$ **b.** $\displaystyle\int_{-1}^{1} x^5\,dx$

Annuity

A sequence of equal payments made at regular time intervals over a period of time is called an **annuity.** Some examples of annuities are payroll savings plans, monthly home mortgage payments, and individual retirement accounts. The **amount of an annuity** is the sum of the payments plus the interest earned.

Amount of an Annuity

If c represents a continuous income function in dollars per year (where t is the time in years), r represents the interest rate (in decimal form) compounded continuously, and T represents the term of the annuity in years, then the **amount of an annuity** is

$$\text{Amount of an annuity} = e^{rT} \int_0^T c(t)e^{-rt}\, dt.$$

EXAMPLE 9   Finding the Amount of an Annuity

You deposit $2000 each year for 15 years in an individual retirement account (IRA) paying 5% interest. How much will you have in your IRA after 15 years?

SOLUTION   The income function for your deposit is

$$c(t) = 2000.$$

So, the amount of the annuity after 15 years will be

$$
\begin{aligned}
\text{Amount of an annuity} &= e^{rT} \int_0^T c(t)e^{-rt}\, dt \\
&= e^{(0.05)(15)} \int_0^{15} 2000e^{-0.05t}\, dt \\
&= 2000e^{0.75} \left[-\frac{e^{-0.05t}}{0.05} \right]_0^{15} \\
&= \$44,680.00.
\end{aligned}
$$

 Checkpoint 9   *Worked-out solution available at LarsonAppliedCalculus.com*

You deposit $1000 each year in a savings account paying 4% interest. How much will be in the account after 10 years?

SUMMARIZE   (Section 5.4)

1. State the definition of a definite integral *(page 339).* For an example of a definite integral, see Example 1.

2. State the Fundamental Theorem of Calculus *(page 340).* For examples of the Fundamental Theorem of Calculus, see Examples 2 and 3.

3. State the properties of definite integrals *(page 341).* For examples of the properties of definite integrals, see Examples 4 and 5.

4. State the definition of the average value of a function *(page 345).* For an example of finding the average value of a function, see Example 7.

5. State the rules for integrating even and odd functions *(page 346).* For an example of integrating even and odd functions, see Example 8.

In Exercises 1–4, find the indefinite integral.

1. $\displaystyle\int (3x + 7)\, dx$

2. $\displaystyle\int \left(x^{3/2} + 2\sqrt{x}\right) dx$

3. $\displaystyle\int \frac{1}{5x}\, dx$

4. $\displaystyle\int e^{-4x}\, dx$

In Exercises 5–7, integrate the marginal function.

5. $\dfrac{dC}{dx} = 0.02x^{3/2} + 29{,}500$

6. $\dfrac{dR}{dx} = 9000 + 2x$

7. $\dfrac{dP}{dx} = 0.03x^2 + 4600$

Exercises 5.4

See *CalcChat.com* for tutorial help and worked-out solutions to odd-numbered exercises.

 Evaluating a Definite Integral Using a Geometric Formula In Exercises 1–6, sketch the region whose area is represented by the definite integral. Then use a geometric formula to evaluate the integral. *See Example 1.*

1. $\displaystyle\int_1^3 5\, dx$

2. $\displaystyle\int_2^5 4\, dx$

3. $\displaystyle\int_0^4 x\, dx$

4. $\displaystyle\int_0^3 \frac{x}{3}\, dx$

5. $\displaystyle\int_{-3}^3 \sqrt{9 - x^2}\, dx$

6. $\displaystyle\int_{-4}^4 \sqrt{16 - x^2}\, dx$

Using Properties of Definite Integrals In Exercises 7 and 8, use the values $\int_0^5 f(x)\, dx = 6$ and $\int_0^5 g(x)\, dx = 2$ to evaluate each definite integral.

7. (a) $\displaystyle\int_0^5 [f(x) + g(x)]\, dx$ (b) $\displaystyle\int_0^5 [f(x) - g(x)]\, dx$

(c) $\displaystyle\int_0^5 -4f(x)\, dx$ (d) $\displaystyle\int_0^5 [f(x) - 3g(x)]\, dx$

8. (a) $\displaystyle\int_0^5 2g(x)\, dx$ (b) $\displaystyle\int_5^0 f(x)\, dx$

(c) $\displaystyle\int_5^5 f(x)\, dx$ (d) $\displaystyle\int_0^5 [f(x) - f(x)]\, dx$

 Finding Area by the Fundamental Theorem In Exercises 9–16, find the area of the region. *See Example 2.*

9. $y = x - x^2$

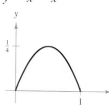

10. $y = 1 - x^4$

11. $y = \dfrac{1}{x^2}$

12. $y = \dfrac{2}{\sqrt{x}}$

13. $y = 3e^{-x/2}$

14. $y = 2e^{x/4}$

15. $y = \dfrac{x^2 + 4}{x}$

16. $y = \dfrac{x - 2}{x}$

 Evaluating a Definite Integral In Exercises 17–38, evaluate the definite integral. *See Examples 3 and 4.*

17. $\displaystyle\int_0^2 5x\, dx$

18. $\displaystyle\int_2^7 3v\, dv$

19. $\displaystyle\int_{-1}^0 (x - 2)\, dx$

20. $\displaystyle\int_{-2}^0 (4x + 9)\, dx$

21. $\displaystyle\int_{-1}^1 (4 - 3t)^2\, dt$

22. $\displaystyle\int_{-1}^1 (1 - 2x)^2\, dx$

23. $\int_{2}^{5} (x - 4)^3 \, dx$

24. $\int_{1}^{3} (x - 3)^4 \, dx$

25. $\int_{-1}^{1} \left(\sqrt[3]{t} - 2 \right) dt$

26. $\int_{1}^{4} \left(6\sqrt{x} + x \right) dx$

27. $\int_{-1}^{0} (t^{1/3} - t^{2/3}) \, dt$

28. $\int_{0}^{4} (x^{1/2} + x^{1/4}) \, dx$

29. $\int_{2}^{8} \frac{3}{x} \, dx$

30. $\int_{2}^{7} -\frac{8}{x} \, dx$

31. $\int_{0}^{4} \frac{1}{\sqrt{2x + 1}} \, dx$

32. $\int_{0}^{2} \frac{x}{\sqrt{1 + 2x^2}} \, dx$

33. $\int_{1}^{2} e^{1 - x} \, dx$

34. $\int_{-1}^{1} (e^x - e^{-x}) \, dx$

35. $\int_{0}^{1} e^{2x} \sqrt{e^{2x} + 1} \, dx$

36. $\int_{0}^{3} \frac{e^{-x}}{\sqrt{e^{-x} + 1}} \, dx$

37. $\int_{0}^{2} \frac{x}{1 + 4x^2} \, dx$

38. $\int_{0}^{1} \frac{e^{2x}}{e^{2x} + 1} \, dx$

 Definite Integral Involving Absolute Value In Exercises 39–42, evaluate the definite integral. *See Example 5.*

39. $\int_{-2}^{1} |4x| \, dx$

40. $\int_{-1}^{3} \left| \frac{x}{3} \right| dx$

41. $\int_{2}^{8} |3x - 9| \, dx$

42. $\int_{0}^{3} |2x - 3| \, dx$

 Finding the Area of a Region In Exercises 43–46, find the area of the region bounded by the graphs of the equations. Use a graphing utility to verify your results.

43. $y = 3x^2 + 1$, $y = 0$, $x = 0$, and $x = 2$

44. $y = 1 + \sqrt{x}$, $y = 0$, $x = 0$, and $x = 4$

45. $y = \frac{4}{x}$, $y = 0$, $x = 1$, and $x = e$

46. $y = e^x$, $y = 0$, $x = 1$, and $x = 4$

 Marginal Analysis In Exercises 47–52, find the change in cost C, revenue R, or profit P for the given marginal function. In each case, assume that the number of units x increases by 3 from the specified value of x. *See Example 6.*

Marginal Function	Number of Units, x
47. $\frac{dC}{dx} = 2.25$	$x = 100$
48. $\frac{dC}{dx} = \frac{20,000}{x^2}$	$x = 10$
49. $\frac{dR}{dx} = 48 - 3x$	$x = 12$
50. $\frac{dR}{dx} = 75\left(20 + \frac{900}{x} \right)$	$x = 500$
51. $\frac{dP}{dx} = \frac{400 - x}{150}$	$x = 200$
52. $\frac{dP}{dx} = 12.5\left(40 - 3\sqrt{x} \right)$	$x = 125$

 Average Value of a Function In Exercises 53–60, find the average value of the function on the interval. Then find all x-values in the interval for which the function is equal to its average value.

53. $f(x) = 6x$; $[0, 3]$ **54.** $f(x) = x^3$; $[1, 3]$

55. $f(x) = 4 - x^2$; $[-2, 2]$

56. $f(x) = x - 2\sqrt{x}$; $[0, 4]$

57. $f(x) = 2e^x$; $[-1, 1]$ **58.** $f(x) = e^{x/5}$; $[-2, 5]$

59. $f(x) = \frac{3}{x + 2}$; $[1, 5]$ **60.** $f(x) = \frac{1}{(x - 3)^2}$; $[0, 2]$

 Integrating Even and Odd Functions In Exercises 61–64, evaluate the definite integral using the properties of even and odd functions. *See Example 8.*

61. $\int_{-1}^{1} 3x^4 \, dx$

62. $\int_{-2}^{2} (x^3 - 4x) \, dx$

63. $\int_{-1}^{1} (2t^5 - 2t) \, dt$

64. $\int_{-2}^{2} \left(\frac{1}{2}t^6 + 1 \right) dt$

65. Using Properties of Definite Integrals Use the value $\int_{0}^{1} x^2 \, dx = \frac{1}{3}$ to evaluate each definite integral. Explain your reasoning.

(a) $\int_{-1}^{0} x^2 \, dx$ (b) $\int_{-1}^{1} x^2 \, dx$ (c) $\int_{0}^{1} -x^2 \, dx$

66. Using Properties of Definite Integrals Use the value $\int_{0}^{4} x^3 \, dx = 64$ to evaluate each definite integral. Explain your reasoning.

(a) $\int_{-4}^{0} x^3 \, dx$ (b) $\int_{-4}^{4} x^3 \, dx$ (c) $\int_{0}^{4} 2x^3 \, dx$

 Finding the Amount of an Annuity In Exercises 67–70, find the amount of an annuity with income function $c(t)$, interest rate r, and term T. *See Example 9.*

67. $c(t) = \$250$, $r = 8\%$, $T = 6$ years

68. $c(t) = \$500$, $r = 7\%$, $T = 4$ years

69. $c(t) = \$1500$, $r = 2\%$, $T = 10$ years

70. $c(t) = \$2000$, $r = 3\%$, $T = 15$ years

Capital Accumulation In Exercises 71–74, you are given the rate of investment dI/dt. Find the capital accumulation over a five-year period by evaluating the definite integral

$$\text{Capital accumulation} = \int_0^5 \frac{dI}{dt}\, dt$$

where t is the time (in years).

71. $\dfrac{dI}{dt} = 500$ **72.** $\dfrac{dI}{dt} = 100t$

73. $\dfrac{dI}{dt} = 500\sqrt{t+1}$

74. $\dfrac{dI}{dt} = \dfrac{12{,}000t}{\left(t^2 + 2\right)^2}$

75. Cost The total cost of purchasing a piece of equipment and maintaining it for x years can be modeled by

$$C = 5000\left(25 + 3\int_0^x t^{1/4}\, dt\right).$$

Find the total cost after (a) 1 year, (b) 5 years, and (c) 10 years.

76. Depreciation A company purchases a new machine for which the rate of depreciation can be modeled by

$$\frac{dV}{dt} = 10{,}000(t - 6), \quad 0 \le t \le 5$$

where V is the value of the machine after t years. Set up and evaluate the definite integral that yields the total loss of value of the machine over the first 3 years.

77. Compound Interest A deposit of $2250 is made in a savings account at an annual interest rate of 6%, compounded continuously. Find the average balance in the account during the first 5 years.

78. **HOW DO YOU SEE IT?** A college graduate has two job offers. The starting salary for each is $32,000, and after 8 years of service each will pay $54,000. The salary increase for each offer is shown in the figure. From a strictly monetary viewpoint, which is the better offer? Explain your reasoning.

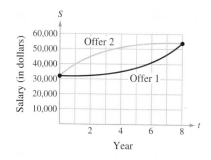

79. Mortgage Debt The rate of change of mortgage debt outstanding for one- to four-family homes in the United States from 2000 through 2013 can be modeled by

$$\frac{dM}{dt} = -1.079 + 2.5550t^{0.5} - 0.75428t + 0.0000061e^t$$

where M is the mortgage debt outstanding (in trillions of dollars) and t is the year, with $t = 0$ corresponding to 2000. In 2000, the mortgage debt outstanding in the United States was $5.1 trillion. *(Source: Board of Governors of the Federal Reserve System)*

(a) Write a model for the debt as a function of t.

(b) What was the average mortgage debt outstanding for 2000 through 2013?

Business Capsule

Former U.S. Senate staffers Heath Hall and Brett Thompson wanted to bring award-winning barbecue to Washington, DC. In 2008, with an initial investment of $10,000, they created Pork Barrel BBQ. The company now earns over $4 million in annual sales, with their sauces and rubs found in over 6000 stores in the United States and around the world. In 2011, they opened their first Pork Barrel BBQ Restaurant in Alexandria, Virginia. The pair continues to win barbecue competitions each year.

80. Research Project Use your school's library, the Internet, or some other reference source to research a small company similar to the one described above. Describe the impact of different factors, such as start-up capital and market conditions, on a company's revenue.

5.5 The Area of a Region Bounded by Two Graphs

Airplane Engines

In Exercise 51 on page 358, you will use integration to find the amount saved on fuel costs by switching to more efficient airplane engines.

■ Find the areas of regions bounded by two graphs.
■ Find consumer and producer surpluses.
■ Use the areas of regions bounded by two graphs to solve real-life problems.

Area of a Region Bounded by Two Graphs

With a few modifications, you can extend the use of definite integrals from finding the area of a region *under a graph* to finding the area of a region *bounded by two graphs*. To see how this is done, consider the region bounded by the graphs of

$$f, \quad g, \quad x = a, \quad \text{and} \quad x = b$$

as shown in Figure 5.13. If the graphs of both f and g lie above the x-axis, then you can interpret the area of the region between the graphs as the area of the region under the graph of g subtracted from the area of the region under the graph of f, as shown in Figure 5.13.

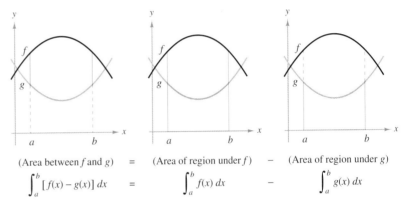

(Area between f and g)	=	(Area of region under f)	−	(Area of region under g)
$\int_a^b [f(x) - g(x)]\, dx$	=	$\int_a^b f(x)\, dx$	−	$\int_a^b g(x)\, dx$

FIGURE 5.13

Although Figure 5.13 depicts the graphs of f and g lying above the x-axis, this is not necessary, and the same integrand

$$[f(x) - g(x)]$$

can be used as long as both functions are continuous and $g(x) \le f(x)$ on the interval $[a, b]$.

Area of a Region Bounded by Two Graphs

If f and g are continuous on $[a, b]$ and $g(x) \le f(x)$ for all x in $[a, b]$, then the area of the region bounded by the graphs of

$$f, \quad g, \quad x = a, \quad \text{and} \quad x = b$$

(see Figure 5.14) is given by

$$A = \int_a^b [f(x) - g(x)]\, dx.$$

FIGURE 5.14

EXAMPLE 1 Finding the Area Bounded by Two Graphs

Find the area of the region bounded by the graphs of $y = x^2 + 2$ and $y = x$ for $0 \le x \le 1$.

SOLUTION Begin by sketching the graphs of both functions, as shown in Figure 5.15. From the figure, you can see that $x \le x^2 + 2$ for all x in $[0, 1]$. So, you can let $f(x) = x^2 + 2$ and $g(x) = x$. Then find the area as shown.

$$\text{Area} = \int_a^b [f(x) - g(x)] \, dx \qquad \text{Area between } f \text{ and } g$$

$$= \int_0^1 [(x^2 + 2) - (x)] \, dx \qquad \text{Substitute for } f \text{ and } g.$$

$$= \int_0^1 (x^2 - x + 2) \, dx$$

$$= \left[\frac{x^3}{3} - \frac{x^2}{2} + 2x \right]_0^1 \qquad \text{Find antiderivative.}$$

$$= \frac{11}{6} \text{ square units} \qquad \text{Apply Fundamental Theorem.}$$

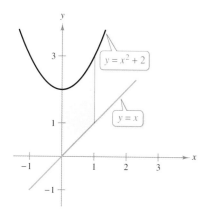

FIGURE 5.15

✓ **Checkpoint 1** *Worked-out solution available at LarsonAppliedCalculus.com*

Find the area of the region bounded by the graphs of $y = x^2 + 1$ and $y = x$ for $0 \le x \le 2$. Sketch the region bounded by the graphs.

EXAMPLE 2 Finding the Area Between Intersecting Graphs

Find the area of the region bounded by the graphs of $y = 2 - x^2$ and $y = x$.

SOLUTION Because the values of a and b are not given, you must determine them by finding the x-coordinates of the points of intersection of the two graphs. To do this, equate the two functions and solve for x.

$$2 - x^2 = x \qquad \text{Equate functions.}$$

$$-x^2 - x + 2 = 0 \qquad \text{Write in general form.}$$

$$-(x + 2)(x - 1) = 0 \qquad \text{Factor.}$$

$$x = -2, x = 1 \qquad \text{Solve for } x.$$

So, $a = -2$ and $b = 1$. In Figure 5.16, you can see that the graph of $f(x) = 2 - x^2$ lies above the graph of $g(x) = x$ for all x in the interval $[-2, 1]$.

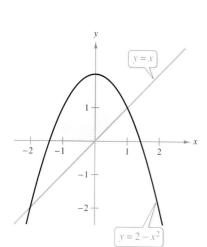

FIGURE 5.16

$$\text{Area} = \int_a^b [f(x) - g(x)] \, dx \qquad \text{Area between } f \text{ and } g$$

$$= \int_{-2}^1 [(2 - x^2) - (x)] \, dx \qquad \text{Substitute for } f \text{ and } g.$$

$$= \int_{-2}^1 (-x^2 - x + 2) \, dx$$

$$= \left[-\frac{x^3}{3} - \frac{x^2}{2} + 2x \right]_{-2}^1 \qquad \text{Find antiderivative.}$$

$$= \frac{9}{2} \text{ square units} \qquad \text{Apply Fundamental Theorem.}$$

✓ **Checkpoint 2** *Worked-out solution available at LarsonAppliedCalculus.com*

Find the area of the region bounded by the graphs of $y = 3 - x^2$ and $y = 2x$. ■

EXAMPLE 3 Finding an Area Below the x-Axis

Find the area of the region bounded by the graph of

$$y = x^2 - 3x - 4$$

and the x-axis.

SOLUTION Begin by finding the x-intercepts of the graph. To do this, set the function equal to zero and solve for x.

$x^2 - 3x - 4 = 0$	Set function equal to 0.
$(x - 4)(x + 1) = 0$	Factor.
$x = 4, x = -1$	Solve for x.

From Figure 5.17, you can see that $x^2 - 3x - 4 \leq 0$ for all x in the interval $[-1, 4]$.

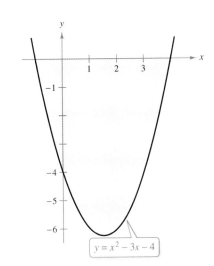

$y = x^2 - 3x - 4$

FIGURE 5.17

So, you can let

$$f(x) = 0 \quad \text{and} \quad g(x) = x^2 - 3x - 4$$

and find the area as shown.

$$\text{Area} = \int_a^b [f(x) - g(x)]\, dx \qquad \text{Area between } f \text{ and } g$$

$$= \int_{-1}^4 [(0) - (x^2 - 3x - 4)]\, dx \qquad \text{Substitute for } f \text{ and } g.$$

$$= \int_{-1}^4 (-x^2 + 3x + 4)\, dx$$

$$= \left[-\frac{x^3}{3} + \frac{3x^2}{2} + 4x \right]_{-1}^4 \qquad \text{Find antiderivative.}$$

$$= \frac{125}{6} \text{ square units} \qquad \text{Apply Fundamental Theorem.}$$

✓ **Checkpoint 3** *Worked-out solution available at LarsonAppliedCalculus.com*

Find the area of the region bounded by the graph of

$$y = x^2 - x - 2$$

and the x-axis.

TECH TUTOR

Most graphing utilities can display regions that are bounded by two graphs. For instance, to graph the region in Example 3, set the viewing window to $-1 \leq x \leq 4$ and $-7 \leq y \leq 1$. Consult your user's manual for specific keystrokes on how to shade the graph. You should obtain the graph below.

$y = 0$

$y = x^2 - 3x - 4$

Sometimes two graphs intersect at more than two points. To determine the area of the region bounded by two such graphs, you must find *all* points of intersection and check to see which graph is above the other in each interval determined by the points. This process is demonstrated in the next example.

EXAMPLE 4   Using Multiple Points of Intersection

Find the area of the region bounded by the graphs of

$$f(x) = 3x^3 - x^2 - 10x \quad \text{and} \quad g(x) = -x^2 + 2x.$$

SOLUTION   To find the points of intersection of the two graphs, set the functions equal to each other and solve for x.

$f(x) = g(x)$	Set $f(x)$ equal to $g(x)$.
$3x^3 - x^2 - 10x = -x^2 + 2x$	Substitute for $f(x)$ and $g(x)$.
$3x^3 - 12x = 0$	Write in general form.
$3x(x^2 - 4) = 0$	
$3x(x - 2)(x + 2) = 0$	Factor.
$x = 0, x = 2, x = -2$	Solve for x.

So, the graphs of f and g intersect at the points $(-2, -8)$, $(0, 0)$, and $(2, 0)$. These three points of intersection determine two intervals of integration:

$$[-2, 0] \quad \text{and} \quad [0, 2].$$

In the figure at the right, you can see that

$$g(x) \le f(x)$$

for all x in the interval $[-2, 0]$, and that

$$f(x) \le g(x)$$

for all x in the interval $[0, 2]$. So, you must use two integrals to determine the area of the region bounded by the graphs of f and g: one for the interval $[-2, 0]$ and one for the interval $[0, 2]$.

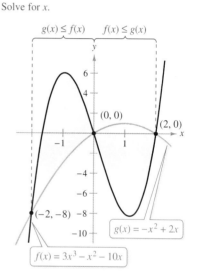

$$
\begin{aligned}
\text{Area} &= \int_{-2}^{0} [f(x) - g(x)]\, dx + \int_{0}^{2} [g(x) - f(x)]\, dx \\
&= \int_{-2}^{0} (3x^3 - 12x)\, dx + \int_{0}^{2} (-3x^3 + 12x)\, dx \\
&= \left[\frac{3x^4}{4} - 6x^2 \right]_{-2}^{0} + \left[-\frac{3x^4}{4} + 6x^2 \right]_{0}^{2} \\
&= (0 - 0) - (12 - 24) + (-12 + 24) - (0 + 0) \\
&= 24
\end{aligned}
$$

So, the region has an area of 24 square units.

✓ *Checkpoint 4*   *Worked-out solution available at LarsonAppliedCalculus.com*

Find the area of the region bounded by the graphs of

$$f(x) = x^3 + 2x^2 - 3x \quad \text{and} \quad g(x) = x^2 + 3x.$$

Sketch a graph of the region.

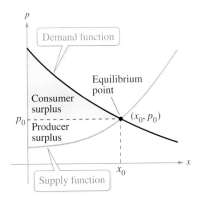

FIGURE 5.18

Consumer Surplus and Producer Surplus

In Section 1.2, you learned that a demand function relates the price of a product to the consumer demand. You also learned that a supply function relates the price of a product to producers' willingness to supply the product. The point (x_0, p_0) at which a demand function $p = D(x)$ and a supply function $p = S(x)$ intersect is the equilibrium point.

Economists call the area of the region bounded by the graph of the demand function, the horizontal line $p = p_0$, and the vertical line $x = 0$ the **consumer surplus,** as shown in Figure 5.18. Consumer surplus is the difference between the amount consumers would be willing to pay and the actual amount paid for a product. The area of the region bounded by the graph of the supply function, the horizontal line $p = p_0$, and the vertical line $x = 0$ is called the **producer surplus,** as shown in Figure 5.18. Producer surplus is the difference between the amount a producer receives for selling a product and the minimum price needed to get the producer to supply the product.

EXAMPLE 5   Finding Surpluses

The demand and supply functions for a product are modeled by

$$Demand: \quad p = -0.36x + 9 \quad and \quad Supply: \quad p = 0.14x + 2$$

where p is the price (in dollars) and x is the number of units (in millions). Find the consumer and producer surpluses for this product.

SOLUTION   By equating the demand and supply functions, you can determine that the equilibrium point occurs when $x = 14$ (million) and the price is \$3.96 per unit.

$$\text{Consumer surplus} = \int_0^{14} (\text{demand function} - \text{price})\, dx$$

$$= \int_0^{14} [(-0.36x + 9) - 3.96]\, dx$$

$$= \left[-0.18x^2 + 5.04x \right]_0^{14}$$

$$= 35.28$$

The consumer surplus is \$35.28.

$$\text{Producer surplus} = \int_0^{14} (\text{price} - \text{supply function})\, dx$$

$$= \int_0^{14} [3.96 - (0.14x + 2)]\, dx$$

$$= \left[-0.07x^2 + 1.96x \right]_0^{14}$$

$$= 13.72$$

The producer surplus is \$13.72. The consumer surplus and producer surplus are shown in Figure 5.19.

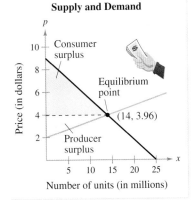

Supply and Demand

Price (in dollars)

Number of units (in millions)

FIGURE 5.19

✓ *Checkpoint 5*   *Worked-out solution available at LarsonAppliedCalculus.com*

The demand and supply functions for a product are modeled by

$$Demand: \quad p = -0.2x + 8 \quad and \quad Supply: \quad p = 0.1x + 2$$

where p is the price (in dollars) and x is the number of units (in millions). Find the consumer and producer surpluses for this product.

Application

In addition to consumer and producer surpluses, there are many other types of applications involving the area of a region bounded by two graphs.

EXAMPLE 6   Energy Consumption

The projected total energy consumption C (in quadrillion Btu per year) in the United States from 2020 through 2040 is given by the model

$$C_1 = 0.628t + 92.42, \quad 20 \le t \le 40$$

where $t = 20$ corresponds to 2020. Determine how much energy will be saved when the actual consumption follows the model

$$C_2 = 0.00135t^3 - 0.1186t^2 + 3.620t + 64.91, \quad 20 \le t \le 40.$$

(Source: U.S. Energy Information Administration)

U.S. Energy Consumption

SOLUTION   The energy saved can be represented as the area of the region between the graphs of C_1 and C_2, as shown in Figure 5.20.

$$\text{Energy saved} = \int_{20}^{40} (C_1 - C_2) \, dt$$

$$= \int_{20}^{40} (-0.00135t^3 + 0.1186t^2 - 2.992t + 27.51) \, dt$$

$$= \left[-\frac{0.00135}{4}t^4 + \frac{0.1186}{3}t^3 - \frac{2.992}{2}t^2 + 27.51t \right]_{20}^{40}$$

$$\approx 158.87$$

FIGURE 5.20

So, the energy saved during the 20-year period would be about 158.87 quadrillion Btu.

✓ **Checkpoint 6**   *Worked-out solution available at LarsonAppliedCalculus.com*

The projected fuel cost C (in millions of dollars per year) for a trucking company from 2020 to 2032 is

$$C_1 = 2.21t + 5.6, \quad 20 \le t \le 32$$

where $t = 20$ corresponds to 2020. After purchasing more efficient truck engines, the company expects fuel costs to follow the model

$$C_2 = 2.04t + 4.7, \quad 20 \le t \le 32.$$

How much money will the company save with the more efficient engines?

SUMMARIZE   (Section 5.5)

1. State the definition of the area of a region bounded by two graphs *(page 351)*. For examples of finding the area of a region bounded by two graphs, see Examples 1, 2, 3, and 4.

2. Describe a real-life example of how finding the area of a region bounded by two graphs can be used to find the consumer and producer surpluses for a product *(page 355, Example 5)*.

3. Describe a real-life example of how finding the area of a region bounded by two graphs can be used to analyze energy consumption *(page 356, Example 6)*.

SKILLS WARM UP 5.5

The following warm-up exercises involve skills that were covered in earlier sections. You will use these skills in the exercise set for this section. For additional help, review Section 1.2.

In Exercises 1–4, simplify the expression.

1. $(-x^2 + 4x + 3) - (x + 1)$

2. $(-2x^2 + 3x + 9) - (-x + 5)$

3. $(-x^3 + 3x^2 - 1) - (x^2 - 4x + 4)$

4. $(3x + 1) - (-x^3 + 9x + 2)$

In Exercises 5–8, find the points of intersection of the graphs of the functions.

5. $f(x) = x^2 - 4x + 4$, $g(x) = 4$

6. $f(x) = -3x^2$, $g(x) = 6 - 9x$

7. $f(x) = x^2$, $g(x) = -x + 6$

8. $f(x) = \frac{1}{2}x^3$, $g(x) = 2x$

Exercises 5.5

See *CalcChat.com* for tutorial help and worked-out solutions to odd-numbered exercises.

 Finding the Area Bounded by Two Graphs In Exercises 1–8, find the area of the region. *See Examples 1, 2, 3, and 4.*

1. $f(x) = x^2 - 6x$
   $g(x) = 0$

2. $f(x) = x^2 + 2x + 1$
   $g(x) = 2x + 5$

3. $f(x) = x^2 - 4x + 3$
   $g(x) = -x^2 + 2x + 3$

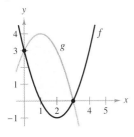

4. $f(x) = x^2$
   $g(x) = x^3$

5. $f(x) = e^x - 1$
   $g(x) = 0$

6. $f(x) = -x + 3$
   $g(x) = \dfrac{2}{x}$

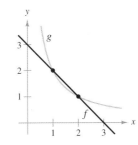

7. $f(x) = 3(x^3 - x)$
   $g(x) = 0$

8. $f(x) = (x - 1)^3$
   $g(x) = x - 1$

 Finding the Region In Exercises 9–12, the integrand of the definite integral is a difference of two functions. Sketch the graph of each function and shade the region whose area is represented by the integral.

9. $\displaystyle\int_0^4 \left[(x + 1) - \tfrac{1}{2}x\right] dx$

10. $\displaystyle\int_{-4}^0 \left[(x - 6) - (x^2 + 5x - 6)\right] dx$

11. $\displaystyle\int_{-2}^2 \left[2x^2 - (x^4 - 2x^2)\right] dx$

12. $\displaystyle\int_{-2}^0 \left[\left(\tfrac{1}{3}x^3 - x\right) - \tfrac{1}{3}x\right] dx$

Think About It In Exercises 13 and 14, determine which value best approximates the area of the region bounded by the graphs of *f* and *g*. Make your selection based on a sketch of the region and not by performing any calculations.

13. $f(x) = x + 1$, $g(x) = (x - 1)^2$
   (a) -2 (b) 2 (c) 10 (d) 4 (e) 8

14. $f(x) = 2 - \tfrac{1}{2}x$, $g(x) = 2 - \sqrt{x}$
   (a) 1 (b) 6 (c) -3 (d) 3 (e) 4

Finding the Area Bounded by Two Graphs In Exercises 15–30, sketch the region bounded by the graphs of the functions and find the area of the region. *See Examples 1, 2, 3, and 4.*

15. $y = x^2 - 1, y = -x + 2, x = 0, x = 1$

16. $y = -x^3 + 3, y = x, x = -1, x = 1$

17. $y = x^2 - 4x + 3, y = 3 + 4x - x^2$

18. $y = 8 - x^2, y = x^2$

19. $y = \dfrac{1}{x^2}, y = 0, x = 1, x = 5$

20. $y = x^2 - x - 6, y = 0, x = -2, x = 1$

21. $f(x) = 1 - x, g(x) = x^4 - x$

22. $f(x) = \sqrt{3x} + 1, g(x) = x + 1$

23. $f(x) = x^3 + 4x^2, g(x) = x + 4$

24. $f(x) = \sqrt[3]{x}, g(x) = x$

25. $y = xe^{-x^2}, y = -1, x = 0, x = 1$

26. $y = \dfrac{e^{1/x}}{x^2}, y = 3, x = 1, x = 3$

27. $f(x) = e^{0.5x}, g(x) = -\dfrac{1}{x}, x = 1, x = 2$

28. $f(x) = \dfrac{1}{x}, g(x) = -e^x, x = \dfrac{1}{2}, x = 1$

29. $y = \dfrac{8}{x}, y = x^2, y = 0, x = 1, x = 4$

30. $y = x^2 - 2x + 1, y = x^2 - 10x + 25, y = 0$

Writing Integrals In Exercises 31–34, use a graphing utility to graph the region bounded by the graphs of the functions. Write the definite integral that represents the area of the region. (*Hint:* More than one integral may be necessary.)

31. $f(x) = 2x, \ g(x) = 4 - 2x, \ h(x) = 0$

32. $f(x) = x(x^2 - 3x + 3), g(x) = x^2$

33. $y = \dfrac{4}{x}, y = x, x = 1, x = 4$

34. $y = x^3 - 4x^2 + 1, y = x - 3$

Finding Area In Exercises 35–40, use a graphing utility to graph the region bounded by the graphs of the functions. Find the area of the region by hand.

35. $f(x) = x^2 - 4x, g(x) = 0$

36. $f(x) = 3 - 2x - x^2, g(x) = 0$

37. $f(x) = x^2 + 2x + 1, g(x) = x + 1$

38. $f(x) = -x^2 + 4x + 2, g(x) = x + 2$

39. $f(x) = x^4 - 2x^2; g(x) = 2x^2$

40. $f(x) = x^3 - 2x^2; g(x) = x^2 - 2x$

Area of a Region In Exercises 41 and 42, use integration to find the area of the triangular region having the given vertices.

41. $(0, 0), (4, 0), (4, 4)$ 42. $(0, 0), (6, 0), (4, 3)$

Consumer and Producer Surpluses In Exercises 43–48, find the consumer and producer surpluses by using the demand and supply functions, where p is the price (in dollars) and x is the number of units (in millions). *See Example 5.*

Demand Function	Supply Function
43. $p = 50 - 0.5x$	$p = 0.125x$
44. $p = 300 - x$	$p = 100 + x$
45. $p = 200 - 0.4x$	$p = 100 + 1.6x$
46. $p = 975 - 23x$	$p = 42x$
47. $p = 42 - 0.015x^2$	$p = 0.01x^2 + 2$
48. $p = 62 - 0.3x$	$p = 0.002x^2 + 12$

Revenue In Exercises 49 and 50, two models, R_1 and R_2, are given for revenue (in billions of dollars) for a large corporation. Both models are estimates of revenues for 2020 through 2025, where $t = 20$ corresponds to 2020. Which model projects the greater revenue? How much more total revenue does that model project over the six-year period?

49. $R_1 = 7.21 + 0.58t, R_2 = 7.21 + 0.45t$

50. $R_1 = 7.21 + 0.26t + 0.02t^2, R_2 = 7.21 + 0.1t + 0.01t^2$

51. **Fuel Cost** The projected fuel cost C (in millions of dollars) for an airline from 2020 through 2030 is

$$C_1 = 568.5 + 7.15t$$

where $t = 20$ corresponds to 2020. After purchasing more efficient airplane engines, the fuel cost is expected to decrease and to follow the model

$$C_2 = 525.6 + 6.43t.$$

How much money will the airline save with the more efficient engines? Explain your reasoning.

52. **Health** An epidemic was spreading such that t weeks after its outbreak it had infected

$$N_1(t) = 0.1t^2 + 0.5t + 150, \quad 0 \le t \le 50$$

people. Twenty-five weeks after the outbreak, a vaccine was developed and administered to the public. At that point, the number of people infected was governed by the model

$$N_2(t) = -0.2t^2 + 6t + 200.$$

Approximate the number of people that the vaccine prevented from becoming ill during the epidemic.

53. Consumer Trends For the years 2003 through 2012, the per capita consumption C of all fruit (in pounds) in the United States can be modeled by

$$C(t) = \begin{cases} -0.059t^2 - 3.69t + 293.6, & 3 \le t \le 9 \\ -3.245t^2 + 65.475t - 73.67, & 9 < t \le 12 \end{cases}$$

where t is the year, with $t = 3$ corresponding to 2003. *(Source: U.S. Department of Agriculture)*

 (a) Use a graphing utility to graph this model.

(b) Suppose the fruit consumption from 2010 through 2012 had continued to follow the model for 2003 through 2009. How many more or fewer pounds of fruit would have been consumed from 2010 through 2012?

54. HOW DO YOU SEE IT? A state legislature is debating two proposals for eliminating the annual budget deficits after 10 years. The rate of decrease of the deficits for each proposal is shown in the figure.

Year

(a) What does the area between the two graphs represent?

(b) From the viewpoint of minimizing the cumulative state deficit, which is the better proposal? Explain your reasoning.

55. Cost, Revenue, and Profit The revenue (in millions of dollars) from a manufacturing process is projected to follow the model $R = 100$ for 10 years. Over the same period of time, the cost (in millions of dollars) is projected to follow the model $C = 60 + 0.2t^2$, where t is the time (in years). Approximate the profit over the 10-year period.

56. Cost, Revenue, and Profit Repeat Exercise 55 for revenue and cost models given by $R = 100 + 0.08t$ and $C = 60 + 0.2t^2$. Did the profit increase or decrease? Explain why.

57. Consumer and Producer Surpluses Factory orders for an air conditioner are about 6000 units per week when the price is $331 and about 8000 units per week when the price is $303. The supply function is given by $p = 0.0275x$. Find the consumer and producer surpluses. (Assume the demand function is linear.)

58. Consumer and Producer Surpluses Repeat Exercise 57 with a demand of about 6000 units per week when the price is $325 and about 8000 units per week when the price is $300. Find the consumer and producer surpluses. (Assume the demand function is linear.)

59. Lorenz Curve Economists use *Lorenz curves* to illustrate the distribution of income in a country. Letting x represent the percent of families in a country and y the percent of total income, the model $y = x$ would represent a country in which each family had the same income. The Lorenz curve, $y = f(x)$, represents the actual income distribution. The area between these two models, for $0 \le x \le 100$, indicates the "income inequality" of a country. In 2013, the Lorenz curve for the United States could be modeled by

$$y = (0.00063x^2 + 0.0190x + 1.723)^2, \quad 0 \le x \le 100$$

where x is measured from the poorest to the wealthiest families. *(Source: U.S. Census Bureau)*

(a) Find the income inequality for the United States in 2013.

(b) Use the Lorenz curve to complete the table, which lists the percent of total income earned by each quintile in the United States in 2013.

Quintile	Lowest	2nd	3rd	4th	Highest
Percent					

60. Project: Social Security For a project analyzing the receipts and expenditures for the Old-Age and Survivors Insurance Trust Fund (Social Security Trust Fund) from 1995 through 2014, visit this text's website at *LarsonAppliedCalculus.com*. *(Source: U.S. Social Security Administration)*

5.6 The Definite Integral as the Limit of a Sum

In Exercise 30 on page 365, you will use the Midpoint Rule to estimate the surface area of a golf green.

■ Use the Midpoint Rule to approximate definite integrals.
■ Understand the definite integral as the limit of a sum.

The Midpoint Rule

In Section 5.4, you learned that you cannot use the Fundamental Theorem of Calculus to evaluate a definite integral unless you can find an antiderivative of the integrand. When you cannot find an antiderivative of an integrand, you can use an approximation technique. One such technique, the **Midpoint Rule,** is demonstrated in Example 1.

EXAMPLE 1   Approximating the Area of a Plane Region

Use the five rectangles in Figure 5.21 to approximate the area of the region bounded by the graph of $f(x) = -x^2 + 5$, the x-axis, and the lines $x = 0$ and $x = 2$.

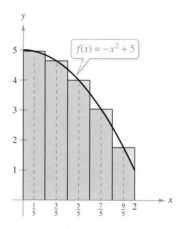

FIGURE 5.21

SOLUTION   You can find the heights of the five rectangles by evaluating f at the midpoint of each of the following intervals.

$$\left[0, \frac{2}{5}\right], \quad \left[\frac{2}{5}, \frac{4}{5}\right], \quad \left[\frac{4}{5}, \frac{6}{5}\right], \quad \left[\frac{6}{5}, \frac{8}{5}\right], \quad \left[\frac{8}{5}, \frac{10}{5}\right]$$

The width of each rectangle is $\frac{2}{5}$. So, the sum of the five areas is

$$
\begin{aligned}
\text{Area} &\approx \frac{2}{5}f\left(\frac{1}{5}\right) + \frac{2}{5}f\left(\frac{3}{5}\right) + \frac{2}{5}f\left(\frac{5}{5}\right) + \frac{2}{5}f\left(\frac{7}{5}\right) + \frac{2}{5}f\left(\frac{9}{5}\right) \\
&= \frac{2}{5}\left[f\left(\frac{1}{5}\right) + f\left(\frac{3}{5}\right) + f\left(\frac{5}{5}\right) + f\left(\frac{7}{5}\right) + f\left(\frac{9}{5}\right)\right] \\
&= \frac{2}{5}\left(\frac{124}{25} + \frac{116}{25} + \frac{100}{25} + \frac{76}{25} + \frac{44}{25}\right) \\
&= \frac{920}{125} \\
&= 7.36.
\end{aligned}
$$

✓ **Checkpoint 1**   *Worked-out solution available at LarsonAppliedCalculus.com*

Use four rectangles to approximate the area of the region bounded by the graph of $f(x) = x^2 + 1$, the x-axis, $x = 0$, and $x = 2$. ■

For the region in Example 1, you can find the exact area with a definite integral. That is,

$$\text{Area} = \int_0^2 (-x^2 + 5)\, dx = \frac{22}{3} \approx 7.33.$$

The approximation procedure used in Example 1 is the **Midpoint Rule.** You can use the Midpoint Rule to approximate *any* definite integral—not just those representing areas. The basic steps are summarized below.

Guidelines for Using the Midpoint Rule

To approximate the definite integral $\int_a^b f(x)\, dx$ with the Midpoint Rule, use the steps below.

1. Divide the interval $[a, b]$ into n subintervals, each of width

$$\Delta x = \frac{b - a}{n}.$$

2. Find the midpoint of each subinterval.

$$\text{Midpoints} = \{x_1, x_2, x_3, \dots, x_n\}$$

3. Evaluate f at each midpoint and form the sum as shown.

$$\int_a^b f(x)\, dx \approx \frac{b - a}{n}[f(x_1) + f(x_2) + f(x_3) + \cdots + f(x_n)]$$

An important characteristic of the Midpoint Rule is that the approximation tends to improve as n increases. The table below shows the approximations for the area of the region described in Example 1 for various values of n. For example, when $n = 10$, the Midpoint Rule yields

$$\int_0^2 (-x^2 + 5)\, dx \approx \frac{2}{10}\left[f\left(\frac{1}{10}\right) + f\left(\frac{3}{10}\right) + \cdots + f\left(\frac{19}{10}\right)\right]$$
$$= 7.34.$$

n	5	10	15	20	25	30
Approximation	7.3600	7.3400	7.3363	7.3350	7.3344	7.3341

Note that as n increases, the approximation gets closer and closer to the exact value of the integral, which was found to be

$$\frac{22}{3} \approx 7.3333.$$

STUDY TIP

In Example 1, the Midpoint Rule is used to approximate an integral whose exact value can be found with the Fundamental Theorem of Calculus. This was done to illustrate the accuracy of the rule. In practice, of course, you would use the Midpoint Rule to approximate the values of definite integrals for which you cannot find an antiderivative. Examples 2 and 3 illustrate such integrals.

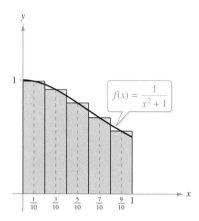

FIGURE 5.22

EXAMPLE 2 Using the Midpoint Rule

Use the Midpoint Rule with $n = 5$ to approximate the area of the region bounded by the graph of

$$f(x) = \frac{1}{x^2 + 1}$$

the x-axis, and the lines $x = 0$ and $x = 1$.

SOLUTION The region is shown in Figure 5.22. With $n = 5$, the interval $[0, 1]$ is divided into five subintervals.

$$\left[0, \frac{1}{5}\right], \quad \left[\frac{1}{5}, \frac{2}{5}\right], \quad \left[\frac{2}{5}, \frac{3}{5}\right], \quad \left[\frac{3}{5}, \frac{4}{5}\right], \quad \left[\frac{4}{5}, 1\right]$$

The midpoints of these intervals are $\frac{1}{10}, \frac{3}{10}, \frac{5}{10}, \frac{7}{10},$ and $\frac{9}{10}$. Because each subinterval has a width of $\Delta x = (1 - 0)/5 = \frac{1}{5}$, you can approximate the value of the definite integral as shown.

$$\int_0^1 \frac{1}{x^2 + 1} \, dx \approx \frac{1}{5}\left(\frac{1}{1.01} + \frac{1}{1.09} + \frac{1}{1.25} + \frac{1}{1.49} + \frac{1}{1.81}\right)$$
$$\approx 0.786$$

The actual area of this region is $\pi/4 \approx 0.785$. So, the approximation is off by about 0.001.

✓ Checkpoint 2   Worked-out solution available at LarsonAppliedCalculus.com

Use the Midpoint Rule with $n = 4$ to approximate the area of the region bounded by the graph of $f(x) = 1/(x^2 + 2)$, the x-axis, and the lines $x = 0$ and $x = 1$.

EXAMPLE 3 Using the Midpoint Rule

Use the Midpoint Rule with $n = 10$ to approximate the area of the region bounded by the graph of $f(x) = \sqrt{x^2 + 1}$, the x-axis, and the lines $x = 1$ and $x = 3$.

SOLUTION The region is shown in Figure 5.23. After dividing the interval $[1, 3]$ into 10 subintervals, you can determine that the midpoints of these intervals are

$$\frac{11}{10}, \quad \frac{13}{10}, \quad \frac{3}{2}, \quad \frac{17}{10}, \quad \frac{19}{10}, \quad \frac{21}{10}, \quad \frac{23}{10}, \quad \frac{5}{2}, \quad \frac{27}{10}, \quad \text{and} \quad \frac{29}{10}.$$

Because each subinterval has a width of $\Delta x = (3 - 1)/10 = \frac{1}{5}$, you can approximate the value of the definite integral as shown.

$$\int_1^3 \sqrt{x^2 + 1} \, dx \approx \frac{1}{5}\left[\sqrt{(1.1)^2 + 1} + \sqrt{(1.3)^2 + 1} + \cdots + \sqrt{(2.9)^2 + 1}\right]$$
$$\approx 4.504$$

It can be shown that the actual area is

$$\frac{1}{2}\left[3\sqrt{10} + \ln\left(3 + \sqrt{10}\right) - \sqrt{2} - \ln\left(1 + \sqrt{2}\right)\right] \approx 4.505.$$

So, the approximation is off by about 0.001.

✓ Checkpoint 3   Worked-out solution available at LarsonAppliedCalculus.com

Use the Midpoint Rule with $n = 4$ to approximate the area of the region bounded by the graph of $f(x) = \sqrt{x^2 - 1}$, the x-axis, and the lines $x = 2$ and $x = 4$. ■

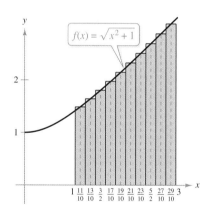

FIGURE 5.23

STUDY TIP

The Midpoint Rule is necessary for solving certain real-life problems, such as measuring irregular areas like bodies of water (see Exercise 29).

The Definite Integral as the Limit of a Sum

Consider the closed interval $[a, b]$, divided into n subintervals whose midpoints are x_i and whose widths are $\Delta x = (b - a)/n$. In this section, you have seen that the midpoint approximation

$$\int_a^b f(x)\, dx \approx f(x_1)\, \Delta x + f(x_2)\, \Delta x + f(x_3)\, \Delta x + \cdots + f(x_n)\, \Delta x$$

$$= [f(x_1) + f(x_2) + f(x_3) + \cdots + f(x_n)]\, \Delta x$$

becomes better and better as n increases. In fact, the limit of this sum as n approaches infinity is exactly equal to the definite integral. That is,

$$\int_a^b f(x)\, dx = \lim_{n \to \infty} [f(x_1) + f(x_2) + f(x_3) + \cdots + f(x_n)]\, \Delta x.$$

It can be shown that this limit is valid as long as x_i is *any* point in the *i*th interval.

EXAMPLE 4   **Approximating a Definite Integral**

Use the Midpoint Rule program in Appendix E or a symbolic integration utility to approximate the definite integral

$$\int_0^1 e^{-x^2}\, dx.$$

SOLUTION Using the Midpoint Rule program (see Figure 5.24), you can complete the following table.

n	10	20	30	40	50
Approximation	0.7471	0.7469	0.7469	0.7468	0.7468

From the table, it appears that

$$\int_0^1 e^{-x^2}\, dx \approx 0.7468.$$

Using a symbolic integration utility, the value of the integral is approximately 0.7468241328.

✓ **Checkpoint 4**   *Worked-out solution available at LarsonAppliedCalculus.com*

Use the Midpoint Rule program in Appendix E or a symbolic integration utility to approximate the definite integral

$$\int_0^1 e^{x^2}\, dx.$$ ■

```
UPPER LIMIT
?1
N DIVISIONS
?50
APPROXIMATION
          .7468363957
                  Done
```

FIGURE 5.24

SUMMARIZE (Section 5.6)

1. Describe how to approximate the area of a region using rectangles *(page 360, Example 1)*.

2. State the guidelines for using the Midpoint Rule *(page 361)*. For examples of using the Midpoint Rule, see Examples 2 and 3.

3. State the definite integral as the limit of a sum *(page 363)*.

SKILLS WARM UP 5.6   The following warm-up exercises involve skills that were covered in a previous course or in earlier sections. You will use these skills in the exercise set for this section. For additional help, review Appendix A.2 and Section 3.6.

In Exercises 1–6, find the midpoint of the interval.

1. $\left[0, \frac{1}{3}\right]$

2. $\left[\frac{1}{10}, \frac{2}{10}\right]$

3. $\left[\frac{3}{20}, \frac{4}{20}\right]$

4. $\left[1, \frac{7}{6}\right]$

5. $\left[-2, -\frac{29}{15}\right]$

6. $\left[\frac{26}{9}, 3\right]$

In Exercises 7–10, find the limit.

7. $\lim\limits_{x \to \infty} \dfrac{2x^2 + 4x - 1}{3x^2 - 2x}$

8. $\lim\limits_{x \to \infty} \dfrac{4x + 5}{7x - 5}$

9. $\lim\limits_{x \to \infty} \dfrac{x - 7}{x^2 + 1}$

10. $\lim\limits_{x \to \infty} \dfrac{5x^3 + 1}{x^3 + x^2 + 4}$

Exercises 5.6

See *CalcChat.com* for tutorial help and worked-out solutions to odd-numbered exercises.

Approximating the Area of a Plane Region In Exercises 1–6, use the rectangles to approximate the area of the region. Compare your result with the exact area obtained using a definite integral. *See Example 1.*

1. $f(x) = -x + 3, [1, 3]$

2. $f(x) = \dfrac{1}{x}, [1, 5]$

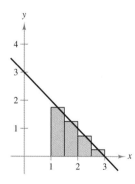

3. $f(x) = \sqrt{x}, [0, 1]$

4. $f(x) = 1 - x^2, [-1, 1]$

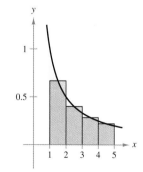

5. $f(x) = x^3 + 1, [0, 1]$

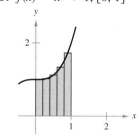

6. $f(x) = e^{-x/2}, [0, 3]$

Using the Midpoint Rule In Exercises 7–14, use the Midpoint Rule with $n = 5$ to approximate the area of the region bounded by the graph of f and the x-axis over the interval. Sketch the region. *See Examples 2 and 3.*

	Function	Interval
7.	$f(x) = x^2$	$[1, 6]$
8.	$f(x) = 4 - x^2$	$[-1, 0]$
9.	$f(x) = x\sqrt{x + 4}$	$[0, 1]$
10.	$f(x) = \sqrt{9 - x^2}$	$[1, 3]$
11.	$f(x) = (x^2 + 1)^{2/3}$	$[0, 5]$
12.	$f(x) = (x^3 + 4)^{3/4}$	$[-1, 1]$
13.	$f(x) = \dfrac{8}{x^2 + 1}$	$[-5, 5]$
14.	$f(x) = \dfrac{6}{1 + e^{-x}}$	$[1, 4]$

Using the Midpoint Rule In Exercises 15–20, use the Midpoint Rule with $n = 4$ to approximate the area of the region bounded by the graph of f and the x-axis over the interval. Sketch the region. *See Examples 2 and 3.*

	Function	*Interval*
15.	$f(x) = \frac{1}{2}x^3$	$[1, 3]$
16.	$f(x) = x^2 - x^3$	$[0, 1]$
17.	$f(x) = \sqrt{x^3 + 6}$	$[1, 5]$
18.	$f(x) = \sqrt{x^2 + 3}$	$[-1, 1]$
19.	$f(x) = \dfrac{1}{(x^2 + 1)^2}$	$[-1, 3]$
20.	$f(x) = \dfrac{10}{x^2 + 5}$	$[0, 2]$

Approximating a Definite Integral In Exercises 21–24, use the Midpoint Rule program in Appendix E or a symbolic integration utility to approximate the definite integral. If you use the Midpoint Rule program, complete the table. *See Example 4.*

n	10	20	30	40	50
Approximation					

21. $\displaystyle\int_0^4 \sqrt{2 + 3x^2}\, dx$ **22.** $\displaystyle\int_0^2 \frac{5}{x^3 + 1}\, dx$

23. $\displaystyle\int_1^3 x\sqrt[3]{x + 1}\, dx$ **24.** $\displaystyle\int_1^6 \frac{4}{\sqrt{1 + x^2}}\, dx$

Making a Closer Approximation In Exercises 25–28, use the Midpoint Rule program in Appendix E to approximate the definite integral. How large must n be to obtain an approximation that is correct to within 0.01?

25. $\displaystyle\int_0^4 (2x^2 + 3)\, dx$ **26.** $\displaystyle\int_1^2 (x^3 - 1)\, dx$

27. $\displaystyle\int_1^4 \frac{1}{x + 1}\, dx$ **28.** $\displaystyle\int_2^5 \sqrt{x + 3}\, dx$

29. Surface Area Use the Midpoint Rule to estimate the surface area of the pond shown in the figure.

30. Surface Area Use the Midpoint Rule to estimate the surface area of the golf green shown in the figure.

31. Surface Area Use the Midpoint Rule to estimate the surface area of the oil spill shown in the figure.

32. HOW DO YOU SEE IT? The graph shows three areas representing awake time, REM (rapid eye movement) sleep time, and non-REM sleep time over a typical individual's lifetime. *(Source: Adapted from Bernstein/Clarke-Stewart/ Roy/Wickens, Psychology, Seventh Edition)*

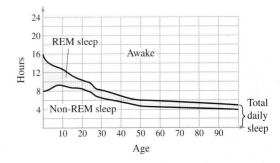

(a) Make generalizations about the amount of total sleep time (REM and non-REM) an individual gets as he or she gets older.

(b) How would you use the Midpoint Rule to estimate the amount of REM sleep time an individual gets between birth and age 10?

33. Numerical Approximation Use the Midpoint Rule with $n = 4$ to approximate π, where

$$\pi = \int_0^1 \frac{4}{1 + x^2}\, dx.$$

Then use a graphing utility to evaluate the definite integral. Compare your results.

ALGEBRA TUTOR

"Unsimplifying" an Algebraic Expression

In algebra it is often helpful to write an expression in its simplest form. In this chapter, you have seen that the reverse is often true in integration. That is, to fit an integrand to an integration formula, it often helps to "unsimplify" the expression. To do this, you use the same algebraic rules, but your goal is different. Here are some examples.

EXAMPLE 1   Rewriting Algebraic Expressions

Rewrite each algebraic expression as indicated in the example.

a. $\dfrac{x - 1}{\sqrt{x}}$

b. $x(3 - 4x^2)^2$

c. $7x^2\sqrt{x^3 + 1}$

d. $5xe^{-x^2}$

SOLUTION

a. $\dfrac{x - 1}{\sqrt{x}} = \dfrac{x}{\sqrt{x}} - \dfrac{1}{\sqrt{x}}$ Example 5, page 315
Rewrite as two fractions.

$\quad\quad = \dfrac{x^1}{x^{1/2}} - \dfrac{1}{x^{1/2}}$ Rewrite with rational exponents.

$\quad\quad = x^{1 - (1/2)} - x^{-1/2}$ Properties of exponents

$\quad\quad = x^{1/2} - x^{-1/2}$ Simplify exponent.

b. $x(3 - 4x^2)^2 = \dfrac{-8}{-8}x(3 - 4x^2)^2$ Example 2, page 324
Multiply and divide by -8.

$\quad\quad = \left(-\dfrac{1}{8}\right)(-8)x(3 - 4x^2)^2$ Regroup.

$\quad\quad = \left(-\dfrac{1}{8}\right)(3 - 4x^2)^2(-8x)$ Regroup.

c. $7x^2\sqrt{x^3 + 1} = 7x^2(x^3 + 1)^{1/2}$ Example 5, page 325
Rewrite with rational exponent.

$\quad\quad = \dfrac{3}{3}(7x^2)(x^3 + 1)^{1/2}$ Multiply and divide by 3.

$\quad\quad = \dfrac{7}{3}(3x^2)(x^3 + 1)^{1/2}$ Regroup.

$\quad\quad = \dfrac{7}{3}(x^3 + 1)^{1/2}(3x^2)$ Regroup.

d. $5xe^{-x^2} = \dfrac{-2}{-2}(5x)e^{-x^2}$ Example 3, page 332
Multiply and divide by -2.

$\quad\quad = \left(-\dfrac{5}{2}\right)(-2x)e^{-x^2}$ Regroup.

$\quad\quad = \left(-\dfrac{5}{2}\right)e^{-x^2}(-2x)$ Regroup.

EXAMPLE 2   Rewriting Algebraic Expressions

Rewrite each algebraic expression.

a. $\dfrac{3x^2 + 2x - 1}{x^2}$

b. $\dfrac{1}{1 + e^{-x}}$

c. $\dfrac{x^2 + x + 1}{x - 1}$

d. $\dfrac{x^2 + 6x + 1}{x^2 + 1}$

SOLUTION

a. $\dfrac{3x^2 + 2x - 1}{x^2} = \dfrac{3x^2}{x^2} + \dfrac{2x}{x^2} - \dfrac{1}{x^2}$

Example 7(a), page 335
Rewrite as separate fractions.

$\qquad = 3 + \dfrac{2}{x} - x^{-2}$

Properties of exponents

$\qquad = 3 + 2\left(\dfrac{1}{x}\right) - x^{-2}$

Regroup.

b. $\dfrac{1}{1 + e^{-x}} = \left(\dfrac{e^x}{e^x}\right)\dfrac{1}{1 + e^{-x}}$

Example 7(b), page 335
Multiply and divide by e^x.

$\qquad = \dfrac{e^x}{e^x + e^x(e^{-x})}$

Multiply.

$\qquad = \dfrac{e^x}{e^x + e^{x-x}}$

Property of exponents

$\qquad = \dfrac{e^x}{e^x + e^0}$

Simplify exponent.

$\qquad = \dfrac{e^x}{e^x + 1}$

$e^0 = 1$

c. $\dfrac{x^2 + x + 1}{x - 1} = x + 2 + \dfrac{3}{x - 1}$

Example 7(c), page 335
Use long division as shown below.

$$
\begin{array}{r}
x + 2 \\
x - 1 \overline{)\, x^2 + x + 1} \\
\underline{x^2 - x} \\
2x + 1 \\
\underline{2x - 2} \\
3
\end{array}
$$

d. $\dfrac{x^2 + 6x + 1}{x^2 + 1} = 1 + \dfrac{6x}{x^2 + 1}$

Bottom of page 334
Use long division as shown below.

$$
\begin{array}{r}
1 \\
x^2 + 1 \overline{)\, x^2 + 6x + 1} \\
\underline{x^2 \qquad\;\; + 1} \\
6x
\end{array}
$$

SUMMARY AND STUDY STRATEGIES

After studying this chapter, you should have acquired the following skills.
The exercise numbers are keyed to the Review Exercises that begin on page 370.
Answers to odd-numbered Review Exercises are given in the back of the text.*

Section 5.1

Review Exercises

■ Use basic integration rules to find indefinite integrals.   *1–14*

$$\int k\, dx = kx + C,\ k \text{ is a constant}$$

$$\int kf(x)\, dx = k \int f(x)\, dx$$

$$\int [f(x) + g(x)]\, dx = \int f(x)\, dx + \int g(x)\, dx$$

$$\int [f(x) - g(x)]\, dx = \int f(x)\, dx - \int g(x)\, dx$$

$$\int x^n\, dx = \frac{x^{n+1}}{n+1} + C,\quad n \neq -1$$

■ Use initial conditions to find particular solutions of indefinite integrals.   *15–18*
■ Use antiderivatives to solve real-life problems.   *19, 20*

Section 5.2

■ Use the General Power Rule or integration by substitution to find indefinite integrals.   *21–32*

$$\int u^n \frac{du}{dx}\, dx = \int u^n\, du = \frac{u^{n+1}}{n+1} + C,\quad n \neq -1$$

■ Use the General Power Rule or integration by substitution to solve real-life problems.   *33, 34*

Section 5.3

■ Use the Exponential and Log Rules to find indefinite integrals.   *35–46*

$$\int e^x\, dx = e^x + C \qquad\qquad \int \frac{1}{x}\, dx = \ln|x| + C$$

$$\int e^u \frac{du}{dx}\, dx = \int e^u\, du = e^u + C \qquad \int \frac{du/dx}{u}\, dx = \int \frac{1}{u}\, du = \ln|u| + C$$

Section 5.4

■ Find the areas of regions using a geometric formula.   *47–50*
■ Use properties of definite integrals.   *51, 52*
■ Find the areas of regions bounded by the graph of a function and the *x*-axis.   *53–58*

---

* Several study aids are available to help you master the material in this chapter. The *Student Solutions Manual* and *CalcChat.com* have solutions to all odd-numbered exercises, and *CalcView.com* has video solutions for selected exercises. *LarsonAppliedCalculus.com* offers algebra help, data spreadsheets, and much more. A *Graphing Technology Guide* with step-by-step commands for a variety of graphing calculators is at *CengageBrain.com*.

Section 5.4 (continued)

Review Exercises

■ Use the Fundamental Theorem of Calculus to evaluate definite integrals.   *59–70*

$$\int_a^b f(x)\,dx = F(x)\Big]_a^b = F(b) - F(a), \quad \text{where} \quad F'(x) = f(x)$$

■ Find the average values of functions over closed intervals.   *71–76*

$$\text{Average value} = \frac{1}{b-a}\int_a^b f(x)\,dx$$

■ Use properties of even and odd functions to help evaluate definite integrals.   *77–80*

Even function: $f(-x) = f(x)$

If f is an *even* function, then $\int_{-a}^a f(x)\,dx = 2\int_0^a f(x)\,dx.$

Odd function: $f(-x) = -f(x)$

If f is an *odd* function, then $\int_{-a}^a f(x)\,dx = 0.$

■ Find the amounts of annuities.   *81, 82*

■ Use definite integrals to solve marginal analysis problems.   *83, 84*

■ Use average values to solve real-life problems.   *85, 86*

Section 5.5

■ Find the areas of regions bounded by two graphs.   *87–94*

$$A = \int_a^b [f(x) - g(x)]\,dx$$

■ Find consumer and producer surpluses.   *95–98*

■ Use the areas of regions bounded by two graphs to solve real-life problems.   *99–102*

Section 5.6

■ Use the Midpoint Rule to approximate values of definite integrals.   *103–114*

$$\int_a^b f(x)\,dx \approx \frac{b-a}{n}[f(x_1) + f(x_2) + f(x_3) + \cdots + f(x_n)]$$

■ Use the Midpoint Rule to solve real-life problems.   *115*

Study Strategies

■ **Indefinite and Definite Integrals**   When integrating, remember that an indefinite integral denotes a *family of antiderivatives*, each differing by a constant C, whereas a definite integral is a *number*.

■ **Checking Antiderivatives by Differentiating**   When finding an antiderivative, remember that you can check your result by differentiating. For example, you can confirm that the antiderivative

$$\int (3x^3 - 4x)\,dx = \frac{3}{4}x^4 - 2x^2 + C \quad \text{is correct by differentiating to obtain} \quad \frac{d}{dx}\left[\frac{3}{4}x^4 - 2x^2 + C\right] = 3x^3 - 4x.$$

Because the derivative is equal to the original integrand, you know that the antiderivative is correct.

■ **Grouping Symbols and the Fundamental Theorem**   When using the Fundamental Theorem of Calculus to evaluate a definite integral, you can avoid sign errors by using grouping symbols. Here is an example.

$$\int_1^3 (x^3 - 9x)\,dx = \left[\frac{x^4}{4} - \frac{9x^2}{2}\right]_1^3 = \left[\frac{3^4}{4} - \frac{9(3^2)}{2}\right] - \left[\frac{1^4}{4} - \frac{9(1^2)}{2}\right] = \frac{81}{4} - \frac{81}{2} - \frac{1}{4} + \frac{9}{2} = -16$$

Finding Indefinite Integrals   In Exercises 1–14, find the indefinite integral. Check your result by differentiating.

1. $\displaystyle\int 16\,dx$

2. $\displaystyle\int -9\,dx$

3. $\displaystyle\int \frac{3}{5}x\,dx$

4. $\displaystyle\int 20x\,dx$

5. $\displaystyle\int 3x^2\,dx$

6. $\displaystyle\int 8x^3\,dx$

7. $\displaystyle\int \frac{2}{3\sqrt[3]{x}}\,dx$

8. $\displaystyle\int \frac{14}{x^{5/2}}\,dx$

9. $\displaystyle\int (2x^2 + 5x)\,dx$

10. $\displaystyle\int (4 - 7x - 6x^2)\,dx$

11. $\displaystyle\int \left(\sqrt[3]{x^4} + 3x\right)dx$

12. $\displaystyle\int \left(\frac{4}{\sqrt{x}} + \sqrt{x}\right)dx$

13. $\displaystyle\int \frac{2x^4 - 1}{\sqrt{x}}\,dx$

14. $\displaystyle\int \frac{1 - 3x}{x^3}\,dx$

Finding a Particular Solution   In Exercises 15–18, find the particular solution that satisfies the differential equation and the initial condition.

15. $f'(x) = 12x;\ f(0) = -3$

16. $f'(x) = 3x^2 - 8x;\ f(1) = 12$

17. $f'(x) = \sqrt{x};\ f(9) = 4$

18. $f'(x) = \dfrac{x^3 - 7}{x^2};\ f(2) = 7$

19. Vertical Motion   An object is projected upward from the ground with an initial velocity of 80 feet per second. Express the height s (in feet) of the object as a function of the time t (in seconds). How long will the object be in the air? (Use $s''(t) = -32$ feet per second per second as the acceleration due to gravity and neglect air resistance.)

20. Revenue   A company produces a new product for which the rate of change of the revenue can be modeled by

$$\frac{dR}{dt} = 0.675t^{3/2},\quad 0 \le t \le 225$$

where t is the time (in weeks). When $t = 0$, $R = 0$.

(a) Find a model for the revenue function.

(b) What is the revenue after 20 weeks?

(c) When will the weekly revenue be $27,000?

Applying the General Power Rule   In Exercises 21–32, find the indefinite integral. Check your result by differentiating.

21. $\displaystyle\int (x + 4)^3\,dx$

22. $\displaystyle\int (x - 6)^{4/3}\,dx$

23. $\displaystyle\int (5x + 1)^4(5)\,dx$

24. $\displaystyle\int (4x^3 + 2)\sqrt{x^4 + 2x}\,dx$

25. $\displaystyle\int (x^2 + 8x)^2(x + 4)\,dx$

26. $\displaystyle\int (6x - 2)^4\,dx$

27. $\displaystyle\int 4x^2(3x^3 + 1)^2\,dx$

28. $\displaystyle\int (2x^4 - 16x)^3(x^3 - 2)\,dx$

29. $\displaystyle\int \frac{x^2}{(2x^3 - 5)^3}\,dx$

30. $\displaystyle\int \frac{x^2}{(x^3 - 4)^2}\,dx$

31. $\displaystyle\int \frac{2}{\sqrt{7x - 1}}\,dx$

32. $\displaystyle\int \frac{4x}{\sqrt{1 - 3x^2}}\,dx$

33. Production   The rate of change of the output of a small sawmill is modeled by

$$\frac{dP}{dt} = 2t(0.001t^2 + 0.5)^{1/4},\quad 0 \le t \le 40$$

where t is the time (in hours) and P is the output (in board-feet). Find the output of the sawmill after (a) 6 hours and (b) 12 hours.

34. Cost The marginal cost for a catering service to cater to x people can be modeled by

$$\frac{dC}{dx} = \frac{20x}{\sqrt{x^2 + 1000}}.$$

When $x = 200$, the cost C (in dollars) is \$4049.69.

(a) Find the cost function.

(b) Find the costs of catering to 500 people and 1000 people.

Using the Exponential and Log Rules In Exercises 35–46, find the indefinite integral.

35. $\displaystyle\int 4e^{4x}\, dx$

36. $\displaystyle\int -8e^{-8x}\, dx$

37. $\displaystyle\int e^{-5x}\, dx$

38. $\displaystyle\int e^{9x+4}\, dx$

39. $\displaystyle\int 7xe^{3x^2}\, dx$

40. $\displaystyle\int (3t - 3)e^{t^2 - 2t}\, dt$

41. $\displaystyle\int \frac{1}{x - 6}\, dx$

42. $\displaystyle\int \frac{1}{1 - 4x}\, dx$

43. $\displaystyle\int \frac{4}{6x - 1}\, dx$

44. $\displaystyle\int \frac{5}{2x + 3}\, dx$

45. $\displaystyle\int \frac{x^2}{1 - x^3}\, dx$

46. $\displaystyle\int \frac{2x^2 + 5x - 8}{x^3}\, dx$

Evaluating a Definite Integral Using a Geometric Formula In Exercises 47–50, sketch the region whose area is represented by the definite integral. Then use a geometric formula to evaluate the integral.

47. $\displaystyle\int_2^6 2\, dx$

48. $\displaystyle\int_0^6 \frac{x}{2}\, dx$

49. $\displaystyle\int_0^4 (4 - x)\, dx$

50. $\displaystyle\int_{-2}^2 \sqrt{4 - x^2}\, dx$

51. Using Properties of Definite Integrals Given

$$\int_2^6 f(x)\, dx = 10 \quad \text{and} \quad \int_2^6 g(x)\, dx = 3$$

evaluate each definite integral.

(a) $\displaystyle\int_2^6 [f(x) + g(x)]\, dx$ (b) $\displaystyle\int_2^6 [f(x) - g(x)]\, dx$

(c) $\displaystyle\int_2^6 [2f(x) - 3g(x)]\, dx$ (d) $\displaystyle\int_2^6 5f(x)\, dx$

52. Using Properties of Definite Integrals Given

$$\int_0^3 f(x)\, dx = 4 \quad \text{and} \quad \int_3^6 f(x)\, dx = -1$$

evaluate each definite integral.

(a) $\displaystyle\int_0^6 f(x)\, dx$ (b) $\displaystyle\int_6^3 f(x)\, dx$

(c) $\displaystyle\int_3^3 f(x)\, dx$ (d) $\displaystyle\int_3^6 -10f(x)\, dx$

Finding Area by the Fundamental Theorem In Exercises 53–58, find the area of the region.

53. $f(x) = 4 - x^2$

54. $f(x) = 9 - x^2$

55. $f(x) = \dfrac{4}{\sqrt{x}}$

56. $f(x) = 2e^{x/2}$

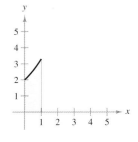

57. $f(x) = \dfrac{2}{x + 1}$

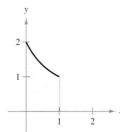

58. $f(x) = \dfrac{x - 1}{x}$

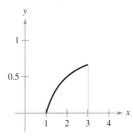

Evaluating a Definite Integral In Exercises 59–70, evaluate the definite integral.

59. $\displaystyle\int_0^4 (2 + x)\,dx$

60. $\displaystyle\int_{-3}^0 (t^2 + 3)\,dt$

61. $\displaystyle\int_{-1}^1 \left(4t^3 - 2\sqrt[3]{t}\right)dt$

62. $\displaystyle\int_{-2}^2 (x^4 + 2x^2 - 5)\,dx$

63. $\displaystyle\int_{-2}^0 (4x + 5)^3\,dx$

64. $\displaystyle\int_0^3 (2x - 3)^2\,dx$

65. $\displaystyle\int_3^9 \frac{5}{x}\,dx$

66. $\displaystyle\int_1^2 \left(\frac{1}{x^2} - \frac{1}{x^3}\right)dx$

67. $\displaystyle\int_1^5 \frac{1}{\sqrt{4 + x}}\,dx$

68. $\displaystyle\int_3^6 \frac{x}{3\sqrt{x^2 - 8}}\,dx$

69. $\displaystyle\int_0^{\ln 5} e^{x/5}\,dx$

70. $\displaystyle\int_{-1}^1 3xe^{x^2 - 1}\,dx$

Average Value of a Function In Exercises 71–76, find the average value of the function on the interval. Then find all x-values in the interval for which the function is equal to its average value.

71. $f(x) = 3x;\ [0, 2]$

72. $f(x) = x^2 + 2;\ [-3, 3]$

73. $f(x) = -2e^x;\ [0, 3]$

74. $f(x) = e^{5-x};\ [2, 5]$

75. $f(x) = \dfrac{1}{\sqrt{x}};\ [4, 9]$

76. $f(x) = \dfrac{1}{(x + 5)^2};\ [-1, 6]$

Integrating Even and Odd Functions In Exercises 77–80, evaluate the definite integral using the properties of even and odd functions.

77. $\displaystyle\int_{-2}^2 6x^5\,dx$

78. $\displaystyle\int_{-4}^4 5x^6\,dx$

79. $\displaystyle\int_{-3}^3 (x^4 + x^2)\,dx$

80. $\displaystyle\int_{-1}^1 (x^3 - x)\,dx$

Finding the Amount of an Annuity In Exercises 81 and 82, find the amount of an annuity with income function $c(t)$, interest rate r, and term T.

81. $c(t) = \$3000,\ r = 6\%,\ T = 5$ years

82. $c(t) = \$1200,\ r = 7\%,\ T = 8$ years

83. Cost The marginal cost of serving an additional typical client at a law firm can be modeled by

$$\frac{dC}{dx} = 675 + 0.5x$$

where x is the number of clients. Find the change in cost C (in dollars) when x increases from 50 to 51 clients.

84. Profit The marginal profit obtained by selling x dollars of automobile insurance can be modeled by

$$\frac{dP}{dx} = 0.4\left(1 - \frac{5000}{x}\right),\ x \geq 5000.$$

Find the change in the profit P (in dollars) when x increases from \$75,000 to \$100,000.

85. Compound Interest A deposit of \$500 is made in a savings account at an annual interest rate of 4%, compounded continuously. Find the average balance in the account during the first 6 years.

86. Sales The rate of change in sales for Garmin from 2008 through 2013 can be modeled by

$$\frac{dS}{dt} = -0.0972t^2 + 2.146t - 11.79$$

where S is the sales (in billions of dollars) and t is the time in years, with $t = 8$ corresponding to 2008. In 2009, the sales for Garmin were \$2.9 billion. *(Source: Garmin, Ltd.)*

(a) Find the model for the sales of Garmin.

(b) What were the average sales of Garmin from 2008 through 2013?

Finding the Area Bounded by Two Graphs In Exercises 87–94, sketch the region bounded by the graphs of the functions and find the area of the region.

87. $y = -\dfrac{1}{x^3},\ y = 0,\ x = 1,\ x = 3$

88. $y = x^2 + 4x - 5,\ y = 4x - 1$

89. $y = (x - 3)^2,\ y = 8 - (x - 3)^2$

90. $y = 4 - x,\ y = x^2 - 5x + 8,\ x = 0$

91. $y = \dfrac{5}{\sqrt{x + 1}},\ y = 0,\ x = 0,\ x = 8$

92. $y = \dfrac{2}{x^2},\ y = \dfrac{7}{2} - \dfrac{3}{2}x,\ x = 1,\ x = 2$

93. $y = x,\ y = x^3$

94. $y = x^3 - 4x,\ y = -x^2 - 2x$

Consumer and Producer Surpluses In Exercises 95–98, find the consumer and producer surpluses by using the demand and supply functions, where p is the price (in dollars) and x is the number of units (in millions).

Demand Function	Supply Function
95. $p = 36 - 0.35x$	$p = 0.05x$
96. $p = 200 - 0.2x$	$p = 50 + 1.3x$
97. $p = 250 - x^2$	$p = 150 + x$
98. $p = 500 - x$	$p = 0.008x^2 + 170$

99. Revenue For the years 2020 through 2025, two models, R_1 and R_2, used to project the revenue (in millions of dollars) for a company are

$$R_1 = 24.3 + 8.24t$$

and

$$R_2 = 21.6 + 9.36t$$

where $t = 20$ corresponds to 2020. Which model projects the greater revenue? How much more total revenue does that model project over the six-year period?

100. Revenue For the years 2004 through 2013, the revenue (in billions of dollars) for Southwest Airlines can be modeled by

$$R = \begin{cases} -0.050t^2 + 1.73t + 0.3, & 4 \le t \le 8 \\ -0.2250t^3 + 7.111t^2 - 72.04t + 246.7, & 8 < t \le 13 \end{cases}$$

where t is the year, with $t = 4$ corresponding to 2004. *(Source: Southwest Airlines)*

(a) Use a graphing utility to graph this model.

(b) Suppose the revenue from 2009 through 2013 had continued to follow the model for 2004 through 2008. How much more or less would the revenue have been for Southwest Airlines?

101. Cost, Revenue, and Profit The revenue (in millions of dollars) from a manufacturing process is projected to follow the model

$$R = 70$$

for 10 years. Over the same period of time, the cost (in millions of dollars) is projected to follow the model

$$C = 30 + 0.3t^2$$

where t is the time (in years). Approximate the profit over the 10-year period.

102. Cost, Revenue, and Profit Repeat Exercise 101 for revenue and cost models given by

$$R = 70 + 0.1t$$

and

$$C = 30 + 0.3t^2.$$

Did the profit increase or decrease? Explain why.

Approximating the Area of a Plane Region In Exercises 103 and 104, use the rectangles to approximate the area of the region. Compare your result with the exact area obtained using a definite integral.

103. $f(x) = \dfrac{x}{3}$, $[0, 3]$ 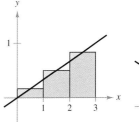   **104.** $f(x) = x^2 + 1$, $[0, 1]$

Using the Midpoint Rule In Exercises 105–110, use the Midpoint Rule with $n = 4$ to approximate the area of the region bounded by the graph of f and the x-axis over the interval. Sketch the region.

Function	Interval
105. $f(x) = x^3$	$[0, 2]$
106. $f(x) = 2x - x^3$	$[0, 1]$
107. $f(x) = (x^2 - 1)^{1/3}$	$[-1, 1]$
108. $f(x) = \sqrt{x^3 + 8}$	$[-2, 2]$
109. $f(x) = \dfrac{9}{x^2 + 3}$	$[2, 3]$
110. $f(x) = \dfrac{3x}{x^3 + 2}$	$[0, 4]$

Using the Midpoint Rule In Exercises 111–114, use the Midpoint Rule with $n = 6$ to approximate the area of the region bounded by the graph of f and the x-axis over the interval. Sketch the region.

Function	Interval
111. $f(x) = 9 - x^2$	$[-3, 3]$
112. $f(x) = x\sqrt{x + 1}$	$[0, 2]$
113. $f(x) = \dfrac{3}{x^2 + 1}$	$[-6, 6]$
114. $f(x) = \dfrac{5}{(x + 2)^2}$	$[0, 1]$

115. Surface Area Use the Midpoint Rule to estimate the surface area of the swamp shown in the figure.

TEST YOURSELF   See *CalcChat.com* for tutorial help and worked-out solutions to odd-numbered exercises.

Take this test as you would take a test in class. When you are done, check your work against the answers given in the back of the book.

In Exercises 1–6, find the indefinite integral.

1. $\displaystyle\int (9x^2 - 4x + 13)\, dx$

2. $\displaystyle\int (x + 1)^2\, dx$

3. $\displaystyle\int x^3\sqrt{x^4 - 7}\, dx$

4. $\displaystyle\int \frac{5x - 6}{\sqrt{x}}\, dx$

5. $\displaystyle\int 15e^{3x}\, dx$

6. $\displaystyle\int \frac{3}{4x - 1}\, dx$

In Exercises 7 and 8, find the particular solution that satisfies the differential equation and the initial condition.

7. $f'(x) = 5x + 2; f(2) = -4$

8. $f'(x) = e^x + 1; f(0) = 1$

In Exercises 9–14, evaluate the definite integral.

9. $\displaystyle\int_0^1 16x\, dx$

10. $\displaystyle\int_{-2}^2 (3 - 2x)^3\, dx$

11. $\displaystyle\int_{-1}^1 (x^3 + x^2)\, dx$

12. $\displaystyle\int_{-1}^2 \frac{2x}{\sqrt{x^2 + 1}}\, dx$

13. $\displaystyle\int_2^4 e^{x/8}\, dx$

14. $\displaystyle\int_{-2}^3 \frac{1}{x + 3}\, dx$

15. The rate of change in sales of Ross Stores from 2004 through 2013 can be modeled by

$$\frac{dS}{dt} = 0.2895e^{0.096t}$$

where S is the sales (in billions of dollars) and t is the time (in years), with $t = 4$ corresponding to 2004. In 2008, the sales of Ross Stores were \$6.5 billion. *(Source: Ross Stores, Inc.)*

(a) Find a model for the sales of Ross Stores.

(b) What were the average sales of Ross Stores from 2004 through 2013?

In Exercises 16–19, sketch the region bounded by the graphs of the functions and find the area of the region.

16. $f(x) = 6, g(x) = x^2 - x - 6$

17. $f(x) = 7 - x, g(x) = -\frac{1}{2}x^2 + x + 7$

18. $f(x) = \sqrt[3]{x}, g(x) = x^2$

19. $f(x) = x^3 + 3x^2 + 1, g(x) = x + 4$

20. The demand and supply functions for a product are modeled by

Demand: $p = -0.625x + 10$ and   *Supply:* $p = 0.25x + 3$

where p is the price (in dollars) and x is the number of units (in millions). Find the consumer and producer surpluses for this product.

In Exercises 21 and 22, use the Midpoint Rule with $n = 4$ to approximate the area of the region bounded by the graph of f and the x-axis over the interval. Compare your result with the exact area obtained using a definite integral. Sketch the region.

21. $f(x) = 3x^2, [0, 1]$

22. $f(x) = x^2 + 2, [-1, 1]$

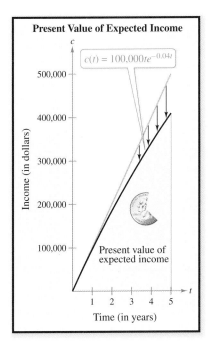

Present Value of Expected Income

$c(t) = 100,000te^{-0.04t}$

Income (in dollars)

500,000

400,000

300,000

200,000

100,000

Present value of
expected income

Time (in years)

1 2 3 4 5

Example 7 on page 382 shows how integration by parts can be used to find the present value of a company's future income.

6 Techniques of Integration

iStockphoto.com/PeopleImages

6.1 Integration by Parts and Present Value

■ Use integration by parts to find indefinite and definite integrals.
■ Find the present value of future income.

Integration by Parts

In this section, you will study an integration technique called **integration by parts.** This technique can be applied to a wide variety of functions and is particularly useful for integrands involving the products of algebraic and exponential or logarithmic functions. For instance, integration by parts works well with integrals such as

$$\int x^2 e^x \, dx \quad \text{and} \quad \int x \ln x \, dx.$$

Integration by parts is based on the Product Rule for differentiation.

$$\frac{d}{dx}[uv] = u\frac{dv}{dx} + v\frac{du}{dx} \qquad \text{Product Rule}$$

$$uv = \int u\frac{dv}{dx}\,dx + \int v\frac{du}{dx}\,dx \qquad \text{Integrate each side.}$$

$$uv = \int u\,dv + \int v\,du \qquad \text{Write in differential form.}$$

$$\int u\,dv = uv - \int v\,du \qquad \text{Rewrite.}$$

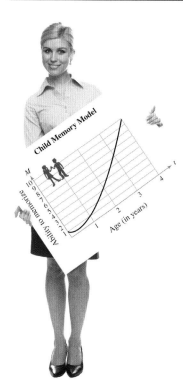

In Exercise 65 on page 384, you will use integration by parts to find the average value of a memory model for children.

Integration by Parts

Let u and v be differentiable functions of x.

$$\int u\,dv = uv - \int v\,du$$

Note that the formula for integration by parts expresses the original integral in terms of another integral. Depending on the choices of u and dv, it may be easier to evaluate the second integral than the original one. Because the choices of u and dv are critical in the integration by parts process, the following guidelines are provided.

Guidelines for Integration by Parts

1. Try letting dv be the most complicated portion of the integrand that fits a basic integration rule. Then u will be the remaining factor(s) of the integrand.

2. Try letting u be the portion of the integrand whose derivative is a function simpler than u. Then dv will be the remaining factor(s) of the integrand.

Note that dv always includes the dx of the original integrand.

When using integration by parts, note that you can first choose dv or first choose u. After you choose, however, the choice of the other factor is determined—it must be the remaining portion of the integrand. Also note that dv must contain the differential dx of the original integral.

EXAMPLE 1   **Integration by Parts**

Find $\int xe^x\, dx$.

SOLUTION   To apply integration by parts, you must rewrite the original integral in the form $\int u\, dv$. That is, you must break $xe^x\, dx$ into two factors—one "part" representing u and the other "part" representing dv. There are several ways to do this.

$$\int \underbrace{(x)}_{u}\underbrace{(e^x\, dx)}_{dv} \qquad \int \underbrace{(e^x)}_{u}\underbrace{(x\, dx)}_{dv} \qquad \int \underbrace{(1)}_{u}\underbrace{(xe^x\, dx)}_{dv} \qquad \int \underbrace{(xe^x)}_{u}\underbrace{(dx)}_{dv}$$

The guidelines on the preceding page suggest the first option because $dv = e^x\, dx$ is the most complicated portion of the integrand that fits a basic integration formula *and* because the derivative of $u = x$ is simpler than x.

$$dv = e^x\, dx \quad \Longrightarrow \quad v = \int dv = \int e^x\, dx = e^x$$

$$u = x \quad \Longrightarrow \quad du = dx$$

Next, you can apply the integration by parts formula as shown.

$$\int u\, dv = uv - \int v\, du \qquad \text{Integration by parts formula}$$

$$\int xe^x\, dx = xe^x - \int e^x\, dx \qquad \text{Substitute.}$$

$$= xe^x - e^x + C \qquad \text{Integrate } \int e^x\, dx.$$

You can check this result by differentiating.

$$\frac{d}{dx}[xe^x - e^x + C] = xe^x + e^x(1) - e^x = xe^x$$

✓ **Checkpoint 1**   *Worked-out solution available at LarsonAppliedCalculus.com*

Find $\int xe^{2x}\, dx$.   ■

In Example 1, notice that you do not need to include a constant of integration when solving $v = \int e^x\, dx = e^x$. To see why this is true, try replacing e^x by $e^x + C_1$ in the solution.

$$\int xe^x\, dx = x(e^x + C_1) - \int (e^x + C_1)\, dx$$

$$= xe^x + C_1 x - e^x - C_1 x + C$$

$$= xe^x - e^x + C$$

After integrating, you can see that the terms involving C_1 subtract out.

TECH TUTOR

If you have access to a symbolic integration utility, try using it to solve several of the exercises in this section. Note that the form of the integral may be slightly different from what you obtain when solving the exercise by hand.

EXAMPLE 2 Integration by Parts

Find $\displaystyle\int x^2 \ln x\, dx$.

SOLUTION In this case, x^2 is more easily integrated than $\ln x$. Furthermore, the derivative of $\ln x$ is simpler than $\ln x$. So, you should choose $dv = x^2\, dx$.

$$dv = x^2\, dx \quad\Longrightarrow\quad v = \int dv = \int x^2\, dx = \frac{x^3}{3}$$

$$u = \ln x \quad\Longrightarrow\quad du = \frac{1}{x}\, dx$$

Next, apply the integration by parts formula.

$$\int u\, dv = uv - \int v\, du \qquad\qquad \text{Integration by parts formula}$$

$$\int x^2 \ln x\, dx = \frac{x^3}{3}\ln x - \int \left(\frac{x^3}{3}\right)\left(\frac{1}{x}\right) dx \qquad \text{Substitute.}$$

$$= \frac{x^3}{3}\ln x - \frac{1}{3}\int x^2\, dx \qquad\qquad \text{Simplify.}$$

$$= \frac{x^3}{3}\ln x - \frac{x^3}{9} + C \qquad\qquad \text{Integrate.}$$

✓ **Checkpoint 2** *Worked-out solution available at LarsonAppliedCalculus.com*

Find $\displaystyle\int x \ln x\, dx$.

EXAMPLE 3 Integrating by Parts with a Single Term

Find $\displaystyle\int \ln x\, dx$.

SOLUTION This integrand is unusual because it has only one term. In such cases, you should choose $dv = dx$ and choose u to be the single term.

$$dv = dx \quad\Longrightarrow\quad v = \int dv = \int dx = x$$

$$u = \ln x \quad\Longrightarrow\quad du = \frac{1}{x}\, dx$$

Next, apply the integration by parts formula.

$$\int u\, dv = uv - \int v\, du \qquad\qquad \text{Integration by parts formula}$$

$$\int \ln x\, dx = x \ln x - \int (x)\left(\frac{1}{x}\right) dx \qquad \text{Substitute.}$$

$$= x \ln x - \int dx \qquad\qquad \text{Simplify.}$$

$$= x \ln x - x + C \qquad\qquad \text{Integrate.}$$

✓ **Checkpoint 3** *Worked-out solution available at LarsonAppliedCalculus.com*

Find $\displaystyle\int \ln 2x\, dx$.

EXAMPLE 4   **Using Integration by Parts Repeatedly**

Find $\int x^2 e^x \, dx$.

SOLUTION   The factors x^2 and e^x are both easy to integrate. Notice, however, that the derivative of x^2 becomes simpler, whereas the derivative of e^x does not. So, you should let $u = x^2$ and let $dv = e^x \, dx$.

$$dv = e^x \, dx \quad \Longrightarrow \quad v = \int dv = \int e^x \, dx = e^x$$

$$u = x^2 \quad \Longrightarrow \quad du = 2x \, dx$$

Next, apply the integration by parts formula.

$$\int x^2 e^x \, dx = x^2 e^x - \int 2x e^x \, dx \qquad \text{First application of integration by parts}$$

This first use of integration by parts has succeeded in simplifying the original integral, but the integral on the right still does not fit a basic integration rule. To evaluate that integral, you can apply integration by parts again. This time, let $u = 2x$ and $dv = e^x \, dx$.

$$dv = e^x \, dx \quad \Longrightarrow \quad v = \int dv = \int e^x \, dx = e^x$$

$$u = 2x \quad \Longrightarrow \quad du = 2 \, dx$$

Next, apply the integration by parts formula.

$$\int x^2 e^x \, dx = x^2 e^x - \int 2x e^x \, dx \qquad \text{First application of integration by parts}$$

$$= x^2 e^x - \left(2x e^x - \int 2e^x \, dx \right) \qquad \text{Second application of integration by parts}$$

$$= x^2 e^x - 2x e^x + 2e^x + C \qquad \text{Integrate.}$$

$$= e^x(x^2 - 2x + 2) + C \qquad \text{Simplify.}$$

You can confirm this result by differentiating.

$$\frac{d}{dx}[e^x(x^2 - 2x + 2) + C] = e^x(2x - 2) + (x^2 - 2x + 2)(e^x)$$

$$= 2x e^x - 2e^x + x^2 e^x - 2x e^x + 2e^x$$

$$= x^2 e^x$$

✓ **Checkpoint 4**   *Worked-out solution available at LarsonAppliedCalculus.com*

Find $\int x^3 e^x \, dx$.   ▪

When making repeated applications of integration by parts, you need to be careful not to interchange the substitutions in successive applications. For instance, in Example 4, the first substitution was $dv = e^x \, dx$ and $u = x^2$. If, in the second application, you had switched the substitution to $dv = 2x \, dx$ and $u = e^x$, you would have obtained

$$\int x^2 e^x \, dx = x^2 e^x - \int 2x e^x \, dx$$

$$= x^2 e^x - \left(x^2 e^x - \int x^2 e^x \, dx \right)$$

$$= \int x^2 e^x \, dx$$

thereby undoing the previous integration and returning to the *original* integral.

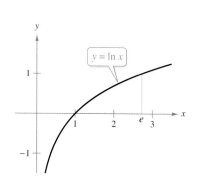

FIGURE 6.1

EXAMPLE 5   Evaluating a Definite Integral

Evaluate $\displaystyle\int_1^e \ln x \, dx$.

SOLUTION   Integration by parts was used to find the antiderivative of $\ln x$ in Example 3. Using this result, you can evaluate the definite integral as shown.

$$\int_1^e \ln x \, dx = \left[x \ln x - x \right]_1^e \qquad \text{Use result of Example 3.}$$
$$= (e \ln e - e) - (1 \ln 1 - 1) \qquad \text{Apply Fundamental Theorem.}$$
$$= (e - e) - (0 - 1)$$
$$= 1 \qquad \text{Simplify.}$$

The area represented by this definite integral is shown in Figure 6.1.

✓ *Checkpoint 5*   *Worked-out solution available at LarsonAppliedCalculus.com*

Use the result of Example 4 to evaluate $\displaystyle\int_0^1 x^2 e^x \, dx$. ■

ALGEBRA TUTOR

For help on the algebra in Example 5, see Example 1 in the *Chapter 6 Algebra Tutor*, on page 412.

Before starting the exercises in this section, remember that it is not enough to know *how* to use the various integration techniques. You also must know *when* to use them. Integration is first and foremost a problem of recognition—recognizing which formula or technique to apply to obtain an antiderivative. Often, a slight alteration of an integrand will necessitate the use of a different integration technique. Here are some examples.

Integral	*Technique*	*Antiderivative*
$\displaystyle\int x \ln x \, dx$	Integration by parts	$\dfrac{x^2}{2} \ln x - \dfrac{x^2}{4} + C$
$\displaystyle\int \dfrac{\ln x}{x} \, dx$	Power Rule: $\displaystyle\int u^n \dfrac{du}{dx} \, dx$	$\dfrac{(\ln x)^2}{2} + C$
$\displaystyle\int \dfrac{1}{x \ln x} \, dx$	Log Rule: $\displaystyle\int \dfrac{1}{u} \dfrac{du}{dx} \, dx$	$\ln|\ln x| + C$

As you gain experience in using integration by parts, your skill in determining u and dv will improve. The next summary lists several common integrals with suggestions for the choices of u and dv.

Summary of Common Integrals Using Integration by Parts

1. For integrals of the form

$$\int x^n e^{ax} \, dx$$

let $u = x^n$ and $dv = e^{ax} \, dx$. (See Examples 1 and 4.)

2. For integrals of the form

$$\int x^n \ln x \, dx$$

let $u = \ln x$ and $dv = x^n \, dx$. (See Examples 2 and 3.)

Present Value

Recall from Section 4.2 that the present value of a future payment is the amount that would have to be deposited today to produce the future payment. What is the present value of a future payment of $1000 one year from now? Because of inflation, $1000 today buys more than $1000 will buy a year from now. The definition below considers only the effect of inflation.

Present Value

If c represents a continuous income function in dollars per year and the annual rate of inflation is r (in decimal form), then the actual total income over t_1 years is

$$\text{Actual income over } t_1 \text{ years} = \int_0^{t_1} c(t)\, dt$$

and its **present value** is

$$\text{Present value} = \int_0^{t_1} c(t)e^{-rt}\, dt.$$

Ignoring inflation, the equation for present value also applies to an interest-bearing account, where the annual interest rate r is compounded continuously and c is an income function in dollars per year.

EXAMPLE 6 Finding Present Value

You have just won $1,000,000 in a state lottery. You will be paid an annuity of $50,000 a year for 20 years. When the annual rate of inflation is 3%, what is the present value of this income?

SOLUTION The income function for your winnings is given by $c(t) = 50,000$. So,

$$\begin{aligned}
\text{Actual income} &= \int_0^{20} 50,000\, dt \\
&= \left[50,000t \right]_0^{20} \\
&= \$1,000,000.
\end{aligned}$$

Because you do not receive this entire amount now, its present value is

$$\begin{aligned}
\text{Present value} &= \int_0^{20} 50,000e^{-0.03t}\, dt \\
&= \left[\frac{50,000}{-0.03}e^{-0.03t} \right]_0^{20} \\
&= \$751,980.61.
\end{aligned}$$

This present value represents the amount that the state must deposit now to cover your payments over the next 20 years. This shows why state lotteries are so profitable—for the states!

✓ **Checkpoint 6** *Worked-out solution available at LarsonAppliedCalculus.com*

Find the present value of the income from the lottery ticket in Example 6 when the annual rate of inflation is 7%.

Expected Income

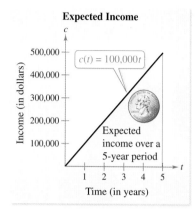

(a)

Present Value of Expected Income

(b)

FIGURE 6.2

| EXAMPLE 7 | Finding Present Value |

A company expects its income during the next 5 years to be given by

$$c(t) = 100,000t, \quad 0 \le t \le 5. \qquad \text{See Figure 6.2(a).}$$

Assuming an annual inflation rate of 4%, can the company claim that the present value of this income is at least $1 million?

SOLUTION   The present value is

$$\text{Present value} = \int_0^5 100,000te^{-0.04t}\, dt = 100,000\int_0^5 te^{-0.04t}\, dt.$$

Using integration by parts, let $dv = e^{-0.04t}\, dt$.

$$dv = e^{-0.04t}\, dt \qquad \Longrightarrow \qquad v = \int dv = \int e^{-0.04t}\, dt = -25e^{-0.04t}$$

$$u = t \qquad \Longrightarrow \qquad du = dt$$

This implies that

$$\int te^{-0.04t}\, dt = -25te^{-0.04t} + 25\int e^{-0.04t}\, dt$$

$$= -25te^{-0.04t} - 625e^{-0.04t}$$

$$= -25e^{-0.04t}(t + 25).$$

So, the present value is

$$\text{Present value} = 100,000\int_0^5 te^{-0.04t}\, dt \qquad \text{See Figure 6.2(b).}$$

$$= 100,000\left[-25e^{-0.04t}(t + 25) \right]_0^5$$

$$= \$1,095,193.52.$$

So, the company can claim that the present value of its expected income during the next 5 years is at least $1 million.

✓ **Checkpoint 7**   *Worked-out solution available at LarsonAppliedCalculus.com*

A company expects its income during the next 10 years to be given by $c(t) = 20,000t$, for $0 \le t \le 10$. Assuming an annual inflation rate of 5%, what is the present value of this income? ■

SUMMARIZE   (Section 6.1)

1. State the integration by parts formula *(page 376)*. For examples of using this formula, see Examples 1, 2, 3, 4, and 7.

2. State the guidelines for integration by parts *(page 376)*. For an example of using these guidelines, see Example 1.

3. Give a summary of the common integrals using integration by parts *(page 380)*. For examples of these common integrals, see Examples 1, 2, 3, and 4.

4. Describe a real-life example of how integration by parts can be used to find the present value of an annuity *(page 381, Example 6)*.

SKILLS WARM UP 6.1

The following warm-up exercises involve skills that were covered in earlier sections. You will use these skills in the exercise set for this section. For additional help, review Sections 4.3, 4.5, and 5.5.

In Exercises 1–6, find $f'(x)$.

1. $f(x) = \ln(x + 1)$

2. $f(x) = \ln(x^2 - 1)$

3. $f(x) = e^{x^3}$

4. $f(x) = e^{-x^2}$

5. $f(x) = x^4 e^x$

6. $f(x) = xe^{-4x}$

In Exercises 7–10, find the area of the region bounded by the graphs of f and g.

7. $f(x) = -x^2 + 4$, $g(x) = x^2 - 4$

8. $f(x) = -x^2 + 2$, $g(x) = 1$

9. $f(x) = 4x$, $g(x) = x^2 - 5$

10. $f(x) = x^3 - 3x^2 + 2$, $g(x) = x - 1$

Exercises 6.1

See *CalcChat.com* for tutorial help and worked-out solutions to odd-numbered exercises.

 Setting Up Integration by Parts In Exercises 1–4, identify u and dv for finding the integral using integration by parts. (Do not evaluate the integral.)

1. $\int xe^{9x}\,dx$

2. $\int x^2 e^{3x}\,dx$

3. $\int x \ln 2x\,dx$

4. $\int \ln 4x\,dx$

 Integration by Parts In Exercises 5–16, use integration by parts to find the indefinite integral. *See Examples 1, 2, 3, and 4.*

5. $\int xe^{3x}\,dx$

6. $\int xe^{-6x}\,dx$

7. $\int x^3 \ln x\,dx$

8. $\int \sqrt{x}\,\ln x\,dx$

9. $\int \ln 5x\,dx$

10. $\int \ln(3x)^2\,dx$

11. $\int x^2 e^{-x}\,dx$

12. $\int x^2 e^{7x}\,dx$

13. $\int x^2 \sqrt{x + 4}\,dx$

14. $\int x^2 \sqrt{x - 3}\,dx$

15. $\int x^2 (\ln x)^3\,dx$

16. $\int \frac{3x^3}{e^x}\,dx$

Finding an Indefinite Integral In Exercises 17–38, find the indefinite integral. (*Hint:* Integration by parts is not required for all the integrals.)

17. $\int e^{4x}\,dx$

18. $\int e^{-2x}\,dx$

19. $\int xe^{4x}\,dx$

20. $\int xe^{-2x}\,dx$

21. $\int \frac{x^2}{e^{x/4}}\,dx$

22. $\int \frac{3}{2}x^2 e^{x/2}\,dx$

23. $\int t \ln(t + 1)\,dt$

24. $\int (2x - 1)e^{7x}\,dx$

25. $\int \frac{e^{1/t}}{t^2}\,dt$

26. $\int \frac{1}{x(\ln x)^3}\,dx$

27. $\int x(\ln x)^2\,dx$

28. $\int \ln 6x\,dx$

29. $\int \frac{(\ln x)^2}{x}\,dx$

30. $\int \frac{1}{x \ln 3x}\,dx$

31. $\int \frac{\ln x}{x^2}\,dx$

32. $\int \frac{\ln 2x}{x^2}\,dx$

33. $\int x\sqrt{x - 3}\,dx$

34. $\int x\sqrt[3]{x + 2}\,dx$

35. $\int \frac{x}{\sqrt{x - 1}}\,dx$

36. $\int \frac{x}{\sqrt{2 + 3x}}\,dx$

37. $\int \frac{xe^{2x}}{(2x + 1)^2}\,dx$

38. $\int \frac{x^3 e^{x^2}}{(x^2 + 1)^2}\,dx$

Evaluating a Definite Integral In Exercises 39–46, use integration by parts to evaluate the definite integral. *See Example 5.*

39. $\displaystyle\int_1^e x^5 \ln x \, dx$

40. $\displaystyle\int_1^e 2x \ln x \, dx$

41. $\displaystyle\int_0^1 \ln(1 + 2x) \, dx$

42. $\displaystyle\int_0^4 \frac{x}{e^{x/2}} \, dx$

43. $\displaystyle\int_0^8 x \sqrt{x + 1} \, dx$

44. $\displaystyle\int_0^{12} \frac{x}{\sqrt{x + 4}} \, dx$

45. $\displaystyle\int_1^2 x^2 e^{2x} \, dx$

46. $\displaystyle\int_0^2 \frac{x^2}{e^{3x}} \, dx$

Area of a Region In Exercises 47–52, find the area of the region bounded by the graphs of the equations. Use a graphing utility to verify your results.

47. $y = (x + 4)e^x$, $y = 0$, $x = -2$, $x = 1$

48. $y = (x^2 - 1)e^x$, $y = 0$, $x = -1$, $x = 1$

49. $y = \frac{1}{9}xe^{-x/3}$, $y = 0$, $x = 0$, $x = 3$

50. $y = x^{-3} \ln x$, $y = 0$, $x = e$

51. $y = x^4 \ln x$, $y = 0$, $x = 1$, $x = e$

52. $y = \dfrac{\ln x}{x^2}$, $y = 0$, $x = 1$, $x = e$

Verifying Formulas In Exercises 53 and 54, use integration by parts to verify the formula.

53. $\displaystyle\int x^n \ln x \, dx = \frac{x^{n+1}}{(n + 1)^2}[-1 + (n + 1) \ln x] + C$,

$n \neq -1$

54. $\displaystyle\int x^n e^{ax} \, dx = \frac{x^n e^{ax}}{a} - \frac{n}{a}\int x^{n-1} e^{ax} \, dx$, $n > 0$

Using Formulas In Exercises 55–58, find the integral by using the appropriate formula from Exercises 53 and 54.

55. $\displaystyle\int x^2 e^{5x} \, dx$

56. $\displaystyle\int xe^{-3x} \, dx$

57. $\displaystyle\int x^{-4} \ln x \, dx$

58. $\displaystyle\int x^{3/2} \ln x \, dx$

Integration Using Technology In Exercises 59–62, use a symbolic integration utility to evaluate the integral.

59. $\displaystyle\int_0^2 t^3 e^{-4t} \, dt$

60. $\displaystyle\int_1^4 (x^2 + 4) \ln x \, dx$

61. $\displaystyle\int_0^5 x^4(25 - x^2)^{3/2} \, dx$

62. $\displaystyle\int_1^e x^9 \ln x \, dx$

63. Demand A manufacturing company forecasts that the demand x (in units) for its product over the next 10 years can be modeled by

$$x = 500(20 + te^{-0.1t}), \quad 0 \le t \le 10$$

where t is the time in years.

(a) Use a graphing utility to decide whether the company is forecasting an increase or a decrease in demand over the decade.

(b) Find the total demand over the next 10 years.

(c) Find the average annual demand during the 10-year period.

64. Capital Campaign The board of trustees of a college is planning a five-year capital gifts campaign to raise money for the college. The goal is to have an annual gift income I that is modeled by

$$I = 2000(375 + 68te^{-0.2t}), \quad 0 \le t \le 5$$

where t is the time in years.

(a) Use a graphing utility to decide whether the board of trustees expects the gift income to increase or decrease over the five-year period.

(b) Find the expected total gift income over the five-year period.

(c) Determine the average annual gift income over the five-year period.

65. Memory Model A model for the ability M of a child to memorize, measured on a scale from 0 to 10, is

$$M = 1 + 1.6t \ln t, \quad 0 < t \le 4$$

where t is the child's age in years.

(a) Find the average value of this model between the child's first and second birthdays.

(b) Find the average value of this model between the child's third and fourth birthdays.

66. Revenue A company sells a seasonal product. The revenue R (in dollars) generated by sales of the product can be modeled by

$$R = 410.5t^2 e^{-t/30} + 25,000, \quad 0 \le t \le 365$$

where t is the time in days.

(a) Find the average daily revenue during the first quarter, which is given by $0 \le t \le 90$.

(b) Find the average daily revenue during the fourth quarter, which is given by $274 \le t \le 365$.

(c) Find the total daily revenue during the year.

Finding Present Value In Exercises 67–72, find the present value of the income c (in dollars) over t_1 years at the given annual inflation rate r. *See Examples 6 and 7.*

67. $c = 5000$, $r = 4\%$, $t_1 = 4$ years

68. $c = 450$, $r = 5\%$, $t_1 = 10$ years

69. $c = 100{,}000 + 4000t$, $r = 5\%$, $t_1 = 10$ years

70. $c = 30{,}000 + 500t$, $r = 7\%$, $t_1 = 6$ years

71. $c = 1000 + 50te^{t/2}$, $r = 6\%$, $t_1 = 4$ years

72. $c = 5000 + 25te^{t/10}$, $r = 6\%$, $t_1 = 10$ years

73. Present Value   You have just won \$2,000,000 in a state lottery. You will be paid an annuity of \$80,000 a year for 25 years. When the annual rate of inflation is 5%, what is the present value of this income?

74. Present Value   You have just won \$72,500,000 in a lottery. You will be paid an annuity of \$2,500,000 a year for 29 years. When the annual rate of inflation is 3%, what is the present value of this income?

75. Present Value   A company expects its income c during the next 4 years to be modeled by

$$c = 150{,}000 + 75{,}000t, \quad 0 \le t \le 4.$$

(a) Find the actual income for the business over the 4 years.

(b) Assuming an annual inflation rate of 4%, what is the present value of this income?

76. Present Value   A professional athlete signs a three-year contract in which the earnings c can be modeled by $2{,}400{,}000 + 600{,}000t$, where t represents the year.

(a) Find the actual value of the athlete's contract.

(b) Assuming an annual inflation rate of 3%, what is the present value of the contract?

77. Present Value   A professional athlete signs a four-year contract in which the earnings c can be modeled by $c = 3{,}000{,}000 + 750{,}000t$, where t represents the year.

(a) Find the actual value of the athlete's contract.

(b) Assuming an annual inflation rate of 5%, what is the present value of the contract?

78. HOW DO YOU SEE IT?   The graphs of two equations show the expected income and the present value of the expected income for a company. Which graph represents the expected income and which graph represents the present value of the expected income? Explain your reasoning.

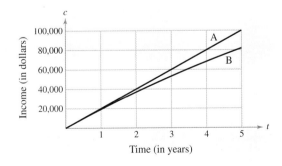

Time (in years)

Future Value   In Exercises 79 and 80, find the future value of the income (in dollars) given by $f(t)$ over t_1 years at annual interest rate r. If the function f represents a continuous investment over a period of t_1 years at an annual interest rate r (in decimal form) compounded continuously, then the future value of the investment is given by

$$\text{Future value} = e^{rt_1} \int_0^{t_1} f(t)e^{-rt}\, dt.$$

79. $f(t) = 3000$, $r = 8\%$, $t_1 = 10$ years

80. $f(t) = 3000e^{0.05t}$, $r = 10\%$, $t_1 = 5$ years

81. Finance: Future Value   Use the equation from Exercises 79 and 80 to calculate the following. *(Source: Adapted from Garman/Forgue, Personal Finance, Eighth Edition)*

(a) The future value of \$1200 saved each year for 10 years earning 7% interest

(b) A person who wishes to invest \$1200 each year finds one investment choice that is expected to pay 9% interest per year and another, riskier choice that may pay 10% interest per year. What is the difference in return (future value) if the investment is made for 15 years?

82. College Tuition Fund   Assume your grandparents had continuously invested in a college fund according to the model $f(t) = 450t$ for 18 years, at an annual interest rate of 7%. Use the equation from Exercises 79 and 80 to answer the following. Assume the tuition cost is the same for the four years.

(a) In 2014, the tuition cost of attending The Pennsylvania State University for 1 year was estimated to be \$30,452. Will the fund have grown enough to allow you to cover 4 years of tuition at The Pennsylvania State University? *(Source: The Pennsylvania State University)*

(b) In 2014, the tuition cost of attending The Ohio State University for 1 year was estimated to be \$26,537. Will the fund have grown enough to allow you to cover 4 years of tuition at The Ohio State University? *(Source: The Ohio State University)*

83. Midpoint Rule   Use a program similar to the Midpoint Rule program in Appendix E with $n = 10$ to approximate

$$\int_1^4 \frac{4}{\sqrt{x} + \sqrt[3]{x}}\, dx.$$

84. Midpoint Rule   Use a program similar to the Midpoint Rule program in Appendix E with $n = 12$ to approximate the area of the region bounded by the graphs of

$$y = \frac{10}{\sqrt{xe^x}}, \quad y = 0, \quad x = 1, \quad \text{and} \quad x = 4.$$

6.2 Integration Tables

■ Use integration tables to find indefinite and definite integrals.
■ Use reduction formulas to find indefinite integrals.
■ Use integration tables to solve real-life problems.

Integration Tables

You have studied several integration techniques that can be used with the basic integration formulas. Certainly these techniques and formulas do not cover every possible method for finding an antiderivative, but they do cover most of the important ones.

In this section, you will expand the list of integration formulas to form a table of integrals. As you add new integration formulas to the basic list, two effects occur. On one hand, it becomes increasingly difficult to memorize, or even become familiar with, the entire list of formulas. On the other hand, with a longer list you need fewer techniques for fitting an integral to one of the formulas on the list. The procedure of integrating by means of a long list of formulas is called **integration by tables.** (The table in Appendix C constitutes only a partial listing of integration formulas. Much longer lists exist, some of which contain several hundred formulas.)

Integration by tables should not be considered a trivial task. It requires considerable thought and insight, and it often requires substitution. Many people find a table of integrals to be a valuable supplement to the integration techniques discussed in this text. As you gain competence in the use of integration tables, you will improve in the use of the various integration techniques. In doing so, you should find that a combination of techniques and tables is the most versatile approach to integration.

Each integration formula in Appendix C can be developed using one or more of the techniques you have studied. You should try to verify several of the formulas. For instance, Formula 17

$$\int \frac{\sqrt{a + bu}}{u}\, du = 2\sqrt{a + bu} + a \int \frac{1}{u\sqrt{a + bu}}\, du \qquad \text{Formula 17}$$

can be verified using integration by parts, Formula 39

$$\int \frac{1}{1 + e^u}\, du = u - \ln(1 + e^u) + C \qquad \text{Formula 39}$$

can be verified using substitution, and Formula 44

$$\int (\ln u)^2\, du = u[2 - 2\ln u + (\ln u)^2] + C \qquad \text{Formula 44}$$

can be verified using integration by parts twice.

In the table of integrals in Appendix C, the formulas have been classified according to the form of the integrand. Several of the forms are listed below.

- Forms involving u^n
- Forms involving $a + bu$
- Forms involving $\sqrt{a + bu}$
- Forms involving $u^2 - a^2$
- Forms involving $\sqrt{u^2 \pm a^2}$
- Forms involving $\sqrt{a^2 - u^2}$
- Forms involving e^u
- Forms involving $\ln u$

Revenue of a New Product

In Exercise 59 on page 392, you will use a formula from the integration table in Appendix C to find the total revenue of a new product during its first 2 years.

TECH TUTOR

Throughout this section, remember that a symbolic integration utility can be used instead of integration tables. If you have access to such a utility, try using it to find the indefinite integrals in Examples 1 and 2.

EXAMPLE 1   Using Integration Tables

Find $\displaystyle\int \frac{x}{\sqrt{x-1}}\,dx$.

SOLUTION   Because the expression inside the radical is linear, you should consider forms involving $\sqrt{a+bu}$, as in Formula 19.

$$\int \frac{u}{\sqrt{a+bu}}\,du = -\frac{2(2a-bu)}{3b^2}\sqrt{a+bu} + C \qquad \text{Formula 19}$$

Using this formula, let $a = -1$, $b = 1$, and $u = x$. Then $du = dx$, and you obtain

$$\int \frac{x}{\sqrt{x-1}}\,dx = -\frac{2(-2-x)}{3}\sqrt{x-1} + C \qquad \begin{array}{l}\text{Substitute values}\\ \text{of } a, b, \text{ and } u.\end{array}$$

$$= \frac{2}{3}(2+x)\sqrt{x-1} + C. \qquad \text{Simplify.}$$

✔ **Checkpoint 1**   *Worked-out solution available at LarsonAppliedCalculus.com*

Use the integration table in Appendix C to find

$$\int \frac{x}{\sqrt{2+x}}\,dx.$$

EXAMPLE 2   Using Integration Tables

Find $\displaystyle\int x\sqrt{x^4-9}\,dx$.

SOLUTION   Because it is not clear which formula to use, you can begin by letting $u = x^2$ and $du = 2x\,dx$. With these substitutions, you can write the integral as shown.

$$\int x\sqrt{x^4-9}\,dx = \frac{1}{2}\int \sqrt{(x^2)^2-9}\,(2x)\,dx \qquad \text{Multiply and divide by 2.}$$

$$= \frac{1}{2}\int \sqrt{u^2-9}\,du \qquad \text{Substitute } u \text{ and } du.$$

Now, it appears that you can use Formula 23.

$$\int \sqrt{u^2-a^2}\,du = \frac{1}{2}\left(u\sqrt{u^2-a^2} - a^2\ln\left|u+\sqrt{u^2-a^2}\right|\right) + C$$

Letting $a = 3$, you obtain

$$\int x\sqrt{x^4-9}\,dx = \frac{1}{2}\int \sqrt{u^2-a^2}\,du$$

$$= \frac{1}{2}\left[\frac{1}{2}\left(u\sqrt{u^2-a^2} - a^2\ln\left|u+\sqrt{u^2-a^2}\right|\right)\right] + C$$

$$= \frac{1}{4}\left(x^2\sqrt{x^4-9} - 9\ln\left|x^2+\sqrt{x^4-9}\right|\right) + C.$$

✔ **Checkpoint 2**   *Worked-out solution available at LarsonAppliedCalculus.com*

Use the integration table in Appendix C to find

$$\int \frac{\sqrt{x^2+16}}{x}\,dx.$$

EXAMPLE 3   Using Integration Tables

Find $\displaystyle\int \frac{1}{x\sqrt{x+1}}\,dx.$

SOLUTION   Considering forms involving $\sqrt{a+bu}$, where $a=1$, $b=1$, and $u=x$, you can use Formula 15.

$$\int \frac{1}{u\sqrt{a+bu}}\,du = \frac{1}{\sqrt{a}}\ln\left|\frac{\sqrt{a+bu}-\sqrt{a}}{\sqrt{a+bu}+\sqrt{a}}\right| + C, \quad a>0$$

So,

$$\int \frac{1}{x\sqrt{x+1}}\,dx = \int \frac{1}{u\sqrt{a+bu}}\,du$$

$$= \frac{1}{\sqrt{a}}\ln\left|\frac{\sqrt{a+bu}-\sqrt{a}}{\sqrt{a+bu}+\sqrt{a}}\right| + C$$

$$= \ln\left|\frac{\sqrt{x+1}-1}{\sqrt{x+1}+1}\right| + C.$$

✓ **Checkpoint 3**   Worked-out solution available at LarsonAppliedCalculus.com

Use the integration table in Appendix C to find $\displaystyle\int \frac{1}{x^2-4}\,dx.$

EXAMPLE 4   Using Integration Tables

Evaluate $\displaystyle\int_0^2 \frac{x}{1+e^{-x^2}}\,dx.$

SOLUTION   Of the forms involving e^u, Formula 39

$$\int \frac{1}{1+e^u}\,du = u - \ln(1+e^u) + C$$

seems most appropriate. To use this formula, let $u = -x^2$ and $du = -2x\,dx$.

$$\int \frac{x}{1+e^{-x^2}}\,dx = -\frac{1}{2}\int \frac{1}{1+e^{-x^2}}(-2x)\,dx$$

$$= -\frac{1}{2}\int \frac{1}{1+e^u}\,du$$

$$= -\frac{1}{2}[u - \ln(1+e^u)] + C$$

$$= -\frac{1}{2}[-x^2 - \ln(1+e^{-x^2})] + C$$

$$= \frac{1}{2}[x^2 + \ln(1+e^{-x^2})] + C$$

So, the value of the definite integral is

$$\int_0^2 \frac{x}{1+e^{-x^2}}\,dx = \frac{1}{2}\left[x^2 + \ln(1+e^{-x^2})\right]_0^2 \approx 1.66. \qquad \text{See Figure 6.3.}$$

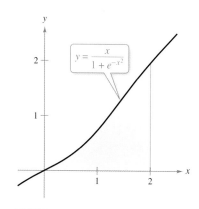

$y = \dfrac{x}{1+e^{-x^2}}$

FIGURE 6.3

✓ **Checkpoint 4**   Worked-out solution available at LarsonAppliedCalculus.com

Use the integration table in Appendix C to evaluate $\displaystyle\int_0^1 \frac{x^2}{1+e^{x^3}}\,dx.$

Reduction Formulas

Several of the formulas in the integration table have the form

$$\int f(x)\, dx = g(x) + \int h(x)\, dx$$

where the right side contains an integral. Such integration formulas are called **reduction formulas** because they reduce the original integral to the sum of a function and a simpler integral.

ALGEBRA TUTOR

For help on the algebra in Example 5, see Example 3 in the *Chapter 6 Algebra Tutor,* on page 413.

EXAMPLE 5 Using a Reduction Formula

Find $\int x^2 e^x\, dx$.

SOLUTION Using Formula 38

$$\int u^n e^u\, du = u^n e^u - n \int u^{n-1} e^u\, du$$

you can let $u = x$ and $n = 2$. Then $du = dx$, and you can write

$$\int x^2 e^x\, dx = x^2 e^x - 2 \int x e^x\, dx.$$

Then, using Formula 37

$$\int u e^u\, du = (u - 1)e^u + C$$

you can write

$$\int x^2 e^x\, dx = x^2 e^x - 2 \int x e^x\, dx$$
$$= x^2 e^x - 2(x - 1)e^x + C$$
$$= x^2 e^x - 2xe^x + 2e^x + C$$
$$= e^x(x^2 - 2x + 2) + C.$$

You can check this result by differentiating.

$$\frac{d}{dx}[e^x(x^2 - 2x + 2) + C] = e^x(2x - 2) + (x^2 - 2x + 2)(e^x)$$
$$= 2xe^x - 2e^x + x^2 e^x - 2xe^x + 2e^x$$
$$= x^2 e^x$$

✓ *Checkpoint 5*   *Worked-out solution available at LarsonAppliedCalculus.com*

Use the integration table in Appendix C to find $\int (\ln x)^3\, dx$. ▪

TECH TUTOR

You have now studied two ways to find the indefinite integral in Example 5. Example 5 uses an integration table, and Example 4 in Section 6.1 uses integration by parts. A third way would be to use a symbolic integration utility.

Application

Integration can be used to find the probability that an event will occur. In such an application, the real-life situation is modeled by a *probability density function f*, and the probability that x will lie between a and b is represented by

$$P(a \le x \le b) = \int_a^b f(x) \, dx.$$

The probability $P(a \le x \le b)$ must be a number between 0 and 1.

EXAMPLE 6   Finding a Probability

A psychologist finds that the probability that a participant in a memory experiment will recall between a and b percent (in decimal form) of the material is

$$P(a \le x \le b) = \int_a^b \frac{1}{e-2} x^2 e^x \, dx, \quad 0 \le a \le b \le 1.$$

Find the probability that a randomly chosen participant will recall between 0% and 87.5% of the material.

SOLUTION   You can use the Constant Multiple Rule to rewrite the integral as

$$\frac{1}{e-2} \int_a^b x^2 e^x \, dx.$$

Note that the integrand is the same as the one in Example 5. Use the result of Example 5 to find the probability with $a = 0$ and $b = 0.875$.

$$
\begin{aligned}
P(0 \le x \le 0.875) &= \frac{1}{e-2} \int_0^{0.875} x^2 e^x \, dx \\
&= \frac{1}{e-2} \left[e^x (x^2 - 2x + 2) \right]_0^{0.875} \\
&\approx 0.608
\end{aligned}
$$

So, the probability is about 60.8%, as indicated in Figure 6.4.

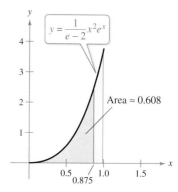

Area ≈ 0.608

FIGURE 6.4

 Checkpoint 6   *Worked-out solution available at LarsonAppliedCalculus.com*

Use Example 6 to find the probability that a participant will recall between 0% and 62.5% of the material. ■

SUMMARIZE   (Section 6.2)

1. Describe what is meant by integration by tables *(page 386)*. For examples of integration by tables, see Examples 1, 2, 3, and 4.

2. Describe what is meant by a reduction formula *(page 389)*. For an example of a reduction formula, see Example 5.

3. Describe a real-life example of how integration by tables can be used to analyze the results of a memory experiment *(page 390, Example 6)*.

The following warm-up exercises involve skills that were covered in a previous course or in earlier sections. You will use these skills in the exercise set for this section. For additional help, review Appendix A.4 and Section 6.1.

In Exercises 1–4, expand the expression.

1. $(x + 4)^2$

2. $(x - 1)^2$

3. $\left(x + \tfrac{1}{2}\right)^2$

4. $\left(x - \tfrac{1}{3}\right)^2$

In Exercises 5 and 6, use integration by parts to find the indefinite integral.

5. $\displaystyle\int 2xe^x \, dx$

6. $\displaystyle\int 3x^2 \ln x \, dx$

Exercises 6.2

See *CalcChat.com* for tutorial help and worked-out solutions to odd-numbered exercises.

 Using Integration Tables In Exercises 1–8, use the indicated formula from the integration table in Appendix C to find the indefinite integral. *See Examples 1, 2, and 3.*

1. $\displaystyle\int \frac{x}{(2 + 3x)^2} \, dx$, Formula 4

2. $\displaystyle\int \frac{1}{x(3 + 7x)^2} \, dx$, Formula 11

3. $\displaystyle\int 8x^2\sqrt{x^2 + 16} \, dx$, Formula 24

4. $\displaystyle\int \frac{4}{x^2 - 9} \, dx$, Formula 21

5. $\displaystyle\int \frac{2x}{\sqrt{x^4 - 9}} \, dx$, Formula 27

6. $\displaystyle\int x\sqrt{4x^4 + 25} \, dx$, Formula 23

7. $\displaystyle\int x^3 e^{x^2} \, dx$, Formula 37

8. $\displaystyle\int \frac{x}{1 + e^{x^2}} \, dx$, Formula 39

 Using Integration Tables In Exercises 9–36, use the integration table in Appendix C to find the indefinite integral. *See Examples 1, 2, 3, and 5.*

9. $\displaystyle\int \frac{1}{x(1 + x)} \, dx$

10. $\displaystyle\int \frac{1}{x^2(5 + 2x)} \, dx$

11. $\displaystyle\int \frac{1}{x\sqrt{x^2 + 49}} \, dx$

12. $\displaystyle\int \frac{1}{\sqrt{x^2 - 1}} \, dx$

13. $\displaystyle\int \frac{1}{x\sqrt{4 - x^2}} \, dx$

14. $\displaystyle\int \frac{\sqrt{x^2 - 9}}{x^2} \, dx$

15. $\displaystyle\int 3x \ln 3x \, dx$

16. $\displaystyle\int (\ln 5x)^2 \, dx$

17. $\displaystyle\int \frac{6x}{1 + e^{3x^2}} \, dx$

18. $\displaystyle\int 7x^3 e^{7x^2} \, dx$

19. $\displaystyle\int \frac{x}{x^4 - 36} \, dx$

20. $\displaystyle\int t^2\sqrt{2 + 9t^2} \, dt$

21. $\displaystyle\int x^2\sqrt{3 + x} \, dx$

22. $\displaystyle\int \frac{\sqrt{3 + 4t}}{t} \, dt$

23. $\displaystyle\int \sqrt{3 + x^2} \, dx$

24. $\displaystyle\int \frac{\sqrt{x^2 + 10}}{x} \, dx$

25. $\displaystyle\int \frac{x^2}{1 + x} \, dx$

26. $\displaystyle\int \frac{1}{1 + e^{2x}} \, dx$

27. $\displaystyle\int \frac{1}{x^2\sqrt{1 - x^2}} \, dx$

28. $\displaystyle\int \frac{1}{x^2\sqrt{x^2 - 4}} \, dx$

29. $\displaystyle\int \frac{8x^2}{(9 - 2x)^5} \, dx$

30. $\displaystyle\int \frac{2x}{(1 - 3x)^2} \, dx$

31. $\displaystyle\int 4x^2 \ln 2x \, dx$

32. $\displaystyle\int xe^{x^2} \, dx$

33. $\displaystyle\int \frac{x^2}{(3x - 5)^2} \, dx$

34. $\displaystyle\int \frac{1}{2x^2(2x - 1)^2} \, dx$

35. $\displaystyle\int \frac{\ln x}{x(4 + 3\ln x)} \, dx$

36. $\displaystyle\int (\ln x)^3 \, dx$

 Using Integration Tables In Exercises 37–44, use the integration table in Appendix C to evaluate the definite integral. *See Example 4.*

37. $\displaystyle\int_0^1 \frac{x}{\sqrt{1 + x}} \, dx$

38. $\displaystyle\int_0^5 \frac{x}{\sqrt{5 + 2x}} \, dx$

39. $\displaystyle\int_3^6 \frac{x}{4x - 7} \, dx$

40. $\displaystyle\int_2^4 \frac{x^2}{3x - 5} \, dx$

41. $\displaystyle\int_0^4 \frac{6}{1 + e^{0.5x}} \, dx$

42. $\displaystyle\int_1^2 x^3 \ln x^2 \, dx$

43. $\displaystyle\int_1^3 \frac{1}{x^2\sqrt{x^2 + 5}} \, dx$

44. $\displaystyle\int_0^3 \frac{x}{(1 + 3x)^4} \, dx$

Area of a Region In Exercises 45–50, use the integration table in Appendix C to find the exact area of the region bounded by the graphs of the equations. Use a graphing utility to verify your results.

45. $y = \dfrac{1}{(16 - x^2)^{3/2}}$, $y = 0$, $x = -2$, $x = 2$

46. $y = \dfrac{1}{(x^2 + 0.25)^{3/2}}$, $y = 0$, $x = 0$, $x = 1$

47. $y = \dfrac{1}{9x^2(2 + 3x)}$, $y = 0$, $x = 1$, $x = 2$

48. $y = \dfrac{10}{x(2 + 5x)}$, $y = 0$, $x = 1$, $x = 2$

49. $y = x^2\sqrt{x^2 + 4}$, $y = 0$, $x = \sqrt{5}$

50. $y = x \ln x^2$, $y = 0$, $x = 4$

 Finding Indefinite Integrals Using Two Methods In Exercises 51–54, find the indefinite integral (a) using the integration table in Appendix C and (b) using integration by parts.

51. $\displaystyle \int \ln \frac{x}{3}\, dx$

52. $\displaystyle \int 4xe^{4x}\, dx$

53. $\displaystyle \int \frac{x}{\sqrt{7x - 3}}\, dx$

54. $\displaystyle \int 7x \ln 7x\, dx$

55. Probability The probability of recalling between a and b percent (in decimal form) of the material learned in a memory experiment is modeled by

$$P(a \le x \le b) = \int_a^b \frac{75}{14}\left(\frac{x}{\sqrt{4 + 5x}}\right) dx$$

where $0 \le a \le b \le 1$. What is the probability of recalling (a) between 40% and 80% and (b) between 0% and 50% of the material?

56. Probability The probability of finding between a and b percent iron (in decimal form) in ore samples is modeled by

$$P(a \le x \le b) = \int_a^b 2x^3 e^{x^2}\, dx, \quad 0 \le a \le b \le 1.$$

What is the probability of finding (a) between 0% and 25% and (b) between 50% and 100% iron in a sample?

Population Growth In Exercises 57 and 58, use a graphing utility to graph the growth function. Use the integration table in Appendix C to find the average value of the growth function over the interval, where N is the size of a population and t is the time in days.

57. $N = \dfrac{5000}{1 + e^{4.8 - 1.9t}}$, $[0, 2]$

58. $N = \dfrac{375}{1 + e^{4.20 - 0.25t}}$, $[21, 28]$

59. Revenue The revenue (in dollars) for a new product is modeled by $R = 10{,}000[1 - 1/(1 + 0.1t^2)^{1/2}]$, where t is the time in years. Estimate the total revenue of the product over its first 2 years on the market.

 60. **HOW DO YOU SEE IT?** The graph shows the rate of change of the sales of a new product.

(a) Approximate the rate of change of the sales after 16 weeks. Explain your reasoning.

(b) Approximate the weeks for which the sales are increasing. Explain your reasoning.

61. Consumer and Producer Surpluses The demand and supply functions for a product are modeled by

Demand: $p = 60/\sqrt{x^2 + 81}$,   *Supply:* $p = x/3$

where p is the price (in dollars) and x is the number of units (in millions). Find the consumer and producer surpluses for this product.

62. Project: Purchasing Power of the Dollar For a project analyzing the purchasing power of the dollar from 1987 through 2013, visit this text's website at *LarsonAppliedCalculus.com*. (*Source: U.S. Bureau of Labor Statistics*)

QUIZ YOURSELF   See *CalcChat.com* for tutorial help and worked-out solutions to odd-numbered exercises.

Take this quiz as you would take a quiz in class. When you are done, check your work against the answers given in the back of the book.

In Exercises 1–6, use integration by parts to find the indefinite integral.

1. $\displaystyle\int xe^{5x}\,dx$

2. $\displaystyle\int \ln x^3\,dx$

3. $\displaystyle\int (x+1)\ln x\,dx$

4. $\displaystyle\int x\sqrt[3]{x+6}\,dx$

5. $\displaystyle\int x\ln\sqrt{x}\,dx$

6. $\displaystyle\int x^2 e^{-2x}\,dx$

7. A manufacturing company forecasts that the demand x (in units) for its product over the next 5 years can be modeled by

$$x = 1000(45 + 20te^{-0.5t}),\ 0 \le t \le 5$$

where t is the time in years.

(a) Find the total demand over the next 5 years.

(b) Find the average annual demand during the 5-year period.

8. A small business expects its income c during the next 7 years to be given by

$$c(t) = 32{,}000t,\ 0 \le t \le 7.$$

(a) Find the actual income for the business over the 7 years.

(b) Assuming an annual inflation rate of 3.3%, what is the present value of this income?

In Exercises 9–14, use the integration table in Appendix C to find the indefinite integral.

9. $\displaystyle\int \frac{x}{1+2x}\,dx$

10. $\displaystyle\int \frac{1}{x(0.1+0.2x)}\,dx$

11. $\displaystyle\int \frac{\sqrt{x^2-81}}{x^2}\,dx$

12. $\displaystyle\int \frac{1}{x\sqrt{4+9x}}\,dx$

13. $\displaystyle\int \frac{2x}{1+e^{4x^2}}\,dx$

14. $\displaystyle\int 2x(x^2+1)e^{x^2+1}\,dx$

15. The revenue (in millions of dollars) for a new product is modeled by

$$R = \sqrt{144t^2 + 400}$$

where t is the time in years.

(a) Estimate the total revenue of the product over its first 3 years on the market.

(b) Estimate the total revenue of the product over its first 6 years on the market.

In Exercises 16–21, use integration by parts or the integration table in Appendix C to evaluate the definite integral.

16. $\displaystyle\int_{-2}^{0} xe^{x/2}\,dx$

17. $\displaystyle\int_{1}^{2} 5x\ln x\,dx$

18. $\displaystyle\int_{0}^{8} \frac{x}{\sqrt{x+8}}\,dx$

19. $\displaystyle\int_{1}^{e} (\ln 6x)^2\,dx$

20. $\displaystyle\int_{2}^{3} \frac{1}{x^2\sqrt{9-x^2}}\,dx$

21. $\displaystyle\int_{4}^{6} \frac{2x}{x^4-4}\,dx$

6.3 Numerical Integration

■ Use the Trapezoidal Rule to approximate definite integrals.
■ Use Simpson's Rule to approximate definite integrals.
■ Analyze the approximate errors in the Trapezoidal Rule and Simpson's Rule.

The Trapezoidal Rule

In Section 5.6, you studied one technique for approximating the value of a *definite* integral— the Midpoint Rule. In this section, you will study two other approximation techniques: the **Trapezoidal Rule** and **Simpson's Rule.**

To develop the Trapezoidal Rule, consider a function f that is nonnegative and continuous on the closed interval $[a, b]$. To approximate the area represented by

$$\int_a^b f(x)\, dx$$

partition the interval into n subintervals, each of width

$$\Delta x = \frac{b - a}{n}. \qquad \text{Width of each subinterval}$$

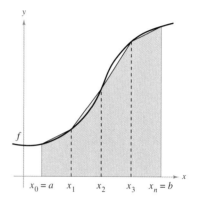

The area of the region can be approximated using four trapezoids.
FIGURE 6.5

U.S. Resident Population

Year (5 ↔ 2005)

In Exercise 43 on page 401, you will use Simpson's Rule to find the average median age of the U.S. resident population from 2005 through 2013.

Next, form n trapezoids, as shown in Figure 6.5. As you can see in Figure 6.6, the area of the first trapezoid is

$$\text{Area of first trapezoid} = \left(\frac{b - a}{n}\right)\left[\frac{f(x_0) + f(x_1)}{2}\right].$$

The areas of the other trapezoids follow a similar pattern, and the sum of the n areas is

$$\text{Area} = \left(\frac{b - a}{n}\right)\left[\frac{f(x_0) + f(x_1)}{2} + \frac{f(x_1) + f(x_2)}{2} + \cdots + \frac{f(x_{n-1}) + f(x_n)}{2}\right]$$

$$= \left(\frac{b - a}{2n}\right)[f(x_0) + f(x_1) + f(x_1) + f(x_2) + \cdots + f(x_{n-1}) + f(x_n)]$$

$$= \left(\frac{b - a}{2n}\right)[f(x_0) + 2f(x_1) + 2f(x_2) + \cdots + 2f(x_{n-1}) + f(x_n)].$$

Although this development assumes f to be continuous *and* nonnegative on $[a, b]$, the resulting formula is valid as long as f is continuous on $[a, b]$.

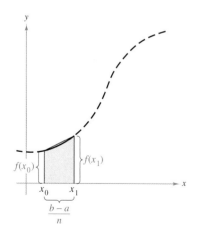

FIGURE 6.6

The Trapezoidal Rule

If f is continuous on $[a, b]$, then

$$\int_a^b f(x)\, dx \approx \left(\frac{b - a}{2n}\right)[f(x_0) + 2f(x_1) + \cdots + 2f(x_{n-1}) + f(x_n)].$$

Note that the coefficients in the Trapezoidal Rule have the following pattern.

$$1 \quad 2 \quad 2 \quad 2 \ldots 2 \quad 2 \quad 1$$

Four Subintervals
FIGURE 6.7

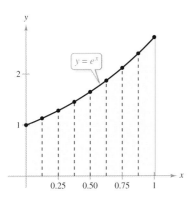

Eight Subintervals
FIGURE 6.8

EXAMPLE 1 Using the Trapezoidal Rule

Use the Trapezoidal Rule to approximate $\int_0^1 e^x\, dx$. Compare the results for $n = 4$ and $n = 8$.

SOLUTION When $n = 4$, the width of each subinterval is

$$\frac{1-0}{4} = \frac{1}{4}$$

and the endpoints of the subintervals are

$$x_0 = 0, \quad x_1 = \frac{1}{4}, \quad x_2 = \frac{1}{2}, \quad x_3 = \frac{3}{4}, \quad \text{and} \quad x_4 = 1$$

as indicated in Figure 6.7. So, by the Trapezoidal Rule,

$$\int_0^1 e^x\, dx = \frac{1}{8}(e^0 + 2e^{0.25} + 2e^{0.5} + 2e^{0.75} + e^1)$$

$$\approx 1.7272. \qquad \text{Approximation using } n = 4$$

When $n = 8$, the width of each subinterval is

$$\frac{1-0}{8} = \frac{1}{8}$$

and the endpoints of the subintervals are

$$x_0 = 0, \quad x_1 = \frac{1}{8}, \quad x_2 = \frac{1}{4}, \quad x_3 = \frac{3}{8}, \quad x_4 = \frac{1}{2},$$

$$x_5 = \frac{5}{8}, \quad x_6 = \frac{3}{4}, \quad x_7 = \frac{7}{8}, \quad \text{and} \quad x_8 = 1$$

as indicated in Figure 6.8. So, by the Trapezoidal Rule,

$$\int_0^1 e^x\, dx = \frac{1}{16}(e^0 + 2e^{0.125} + 2e^{0.25} + \cdots + 2e^{0.875} + e^1)$$

$$\approx 1.7205. \qquad \text{Approximation using } n = 8$$

Of course, for *this particular* integral, you could have found an antiderivative and used the Fundamental Theorem of Calculus to find the exact value of the definite integral. The exact value is

$$\int_0^1 e^x\, dx = e - 1 \qquad \text{Exact value}$$

which is approximately 1.718282.

✓ *Checkpoint 1*   *Worked-out solution available at LarsonAppliedCalculus.com*

Use the Trapezoidal Rule with $n = 4$ to approximate $\int_0^1 e^{2x}\, dx$. ■

There are two important points that should be made concerning the Trapezoidal Rule. First, the approximation tends to become more accurate as n increases. For instance, in Example 1, when $n = 16$, the Trapezoidal Rule yields an approximation of 1.7188. Second, although you could have used the Fundamental Theorem of Calculus to evaluate the integral in Example 1, this theorem cannot be used to evaluate an integral as simple as $\int_0^1 e^{x^2}\, dx$. Yet the Trapezoidal Rule can be easily applied to estimate this integral.

TECH TUTOR

A symbolic integration utility can be used to evaluate a definite integral. Use a symbolic integration utility to approximate the integral

$$\int_0^1 e^{x^2}\, dx.$$

Simpson's Rule

One way to view the Trapezoidal Rule is to say that f is approximated by a first-degree polynomial on each subinterval. In Simpson's Rule, f is approximated by a second-degree polynomial on each subinterval.

To develop Simpson's Rule, partition the interval $[a, b]$ into an *even number n* of subintervals, each of width

$$\Delta x = \frac{b - a}{n}.$$

On the subinterval $[x_0, x_2]$, approximate the function f by the second-degree polynomial $p(x)$ that passes through the points

$$(x_0, f(x_0)), \quad (x_1, f(x_1)), \quad \text{and} \quad (x_2, f(x_2))$$

as shown in Figure 6.9. The Fundamental Theorem of Calculus can be used to show that

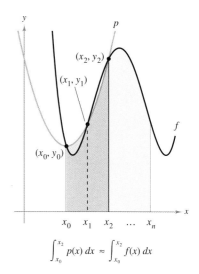

$$\int_{x_0}^{x_2} p(x)\, dx \approx \int_{x_0}^{x_2} f(x)\, dx$$

FIGURE 6.9

$$\int_{x_0}^{x_2} f(x)\, dx \approx \int_{x_0}^{x_2} p(x)\, dx$$

$$= \left(\frac{x_2 - x_0}{6}\right)\left[p(x_0) + 4p\left(\frac{x_0 + x_2}{2}\right) + p(x_2)\right]$$

$$= \frac{2[(b - a)/n]}{6}[p(x_0) + 4p(x_1) + p(x_2)]$$

$$= \left(\frac{b - a}{3n}\right)[f(x_0) + 4f(x_1) + f(x_2)].$$

Repeating this process on the subintervals $[x_{i-2}, x_i]$ produces

$$\int_a^b f(x)\, dx \approx \left(\frac{b - a}{3n}\right)[f(x_0) + 4f(x_1) + f(x_2) + f(x_2) + 4f(x_3) +$$

$$f(x_4) + \cdots + f(x_{n-2}) + 4f(x_{n-1}) + f(x_n)].$$

By grouping like terms, you can obtain the approximation shown below, which is known as Simpson's Rule. This rule is named after the English mathematician Thomas Simpson (1710–1761).

Simpson's Rule (*n* Is Even)

If f is continuous on $[a, b]$ and n is an even integer, then

$$\int_a^b f(x)\, dx \approx \left(\frac{b - a}{3n}\right)[f(x_0) + 4f(x_1) + 2f(x_2) + 4f(x_3) +$$

$$\cdots + 4f(x_{n-1}) + f(x_n)].$$

Note that the coefficients in Simpson's Rule have the following pattern.

1   4   2   4   2   4 ... 4   2   4   1

The Trapezoidal Rule and Simpson's Rule are necessary for solving certain real-life problems, such as approximating the present value of an income. You will see such problems in the exercise set for this section.

In Example 1, the Trapezoidal Rule was used to estimate the value of

$$\int_0^1 e^x \, dx.$$

The next example uses Simpson's Rule to approximate the same integral.

EXAMPLE 2   Using Simpson's Rule

Use Simpson's Rule to approximate

$$\int_0^1 e^x \, dx.$$

Compare the results for $n = 4$ and $n = 8$.

SOLUTION   When $n = 4$, the width of each subinterval is

$$\frac{1 - 0}{4} = \frac{1}{4}$$

and the endpoints of the subintervals are

$$x_0 = 0, \quad x_1 = \frac{1}{4}, \quad x_2 = \frac{1}{2}, \quad x_3 = \frac{3}{4}, \quad \text{and} \quad x_4 = 1$$

as indicated in Figure 6.10. So, by Simpson's Rule

$$\int_0^1 e^x \, dx = \frac{1}{12}(e^0 + 4e^{0.25} + 2e^{0.5} + 4e^{0.75} + e^1)$$

$$\approx 1.718319. \qquad \text{Approximation using } n = 4$$

When $n = 8$, the width of each subinterval is $(1 - 0)/8 = \frac{1}{8}$ and the endpoints of the subintervals are

$$x_0 = 0, \quad x_1 = \frac{1}{8}, \quad x_2 = \frac{1}{4}, \quad x_3 = \frac{3}{8}, \quad x_4 = \frac{1}{2},$$

$$x_5 = \frac{5}{8}, \quad x_6 = \frac{3}{4}, \quad x_7 = \frac{7}{8}, \quad \text{and} \quad x_8 = 1$$

as indicated in Figure 6.11. So, by Simpson's Rule

$$\int_0^1 e^x \, dx = \frac{1}{24}(e^0 + 4e^{0.125} + 2e^{0.25} + \cdots + 4e^{0.875} + e^1)$$

$$\approx 1.718284. \qquad \text{Approximation using } n = 8$$

Recall that the exact value of this integral is

$$\int_0^1 e^x \, dx = e - 1 \qquad \text{Exact value}$$

which is approximately

$$1.718282. \qquad \text{Approximate value}$$

So, with only eight subintervals, you obtained an approximation that is correct to within 0.000002—an impressive result.

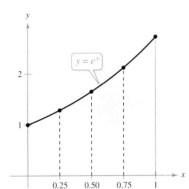

y

$y = e^x$

2

1

0.25   0.50   0.75   1   x

Four Subintervals

FIGURE 6.10

y

$y = e^x$

2

1

0.25   0.50   0.75   1   x

Eight Subintervals

FIGURE 6.11

STUDY TIP

Comparing the results of Examples 1 and 2, you can see that for a given value of n, Simpson's Rule tends to be more accurate than the Trapezoidal Rule.

✓ **Checkpoint 2**   *Worked-out solution available at LarsonAppliedCalculus.com*

Use Simpson's Rule with $n = 4$ to approximate

$$\int_0^1 e^{2x} \, dx.$$

Error Analysis

In Examples 1 and 2, you were able to calculate the exact value of the integral and compare that value with the approximations to see how good they were. In practice, you need to have a different way of telling how good an approximation is. Such a way is provided in the next result.

TECH TUTOR

A program for several models of graphing utilities that uses Simpson's Rule to approximate the definite integral $\int_a^b f(x)\, dx$ can be found in Appendix E.

Errors in the Trapezoidal Rule and Simpson's Rule

The errors E in approximating

$$\int_a^b f(x)\, dx$$

are as shown.

Trapezoidal Rule: $\quad |E| \le \dfrac{(b-a)^3}{12n^2}\Big[\max|f''(x)|\Big], \quad a \le x \le b$

Simpson's Rule: $\quad |E| \le \dfrac{(b-a)^5}{180n^4}\Big[\max|f^{(4)}(x)|\Big], \quad a \le x \le b$

This result indicates that the errors generated by the Trapezoidal Rule and Simpson's Rule have upper bounds that are dependent on the extreme values of

$$f''(x) \quad \text{and} \quad f^{(4)}(x)$$

in the interval $[a, b]$. Furthermore, the bounds for the errors can be made arbitrarily small by *increasing n*. To determine what value of n to choose, consider the steps below.

Trapezoidal Rule

1. Find $f''(x)$.

2. Find the maximum of $|f''(x)|$ on the interval $[a, b]$.

3. Set up the inequality

$$|E| \le \frac{(b-a)^3}{12n^2}\Big[\max|f''(x)|\Big].$$

4. For an error less than ε (the lowercase Greek letter epsilon), solve for n in the inequality

$$\frac{(b-a)^3}{12n^2}\Big[\max|f''(x)|\Big] < \varepsilon.$$

5. Partition $[a, b]$ into n subintervals and apply the Trapezoidal Rule.

Simpson's Rule

1. Find $f^{(4)}(x)$.

2. Find the maximum of $|f^{(4)}(x)|$ on the interval $[a, b]$.

3. Set up the inequality

$$|E| \le \frac{(b-a)^5}{180n^4}\Big[\max|f^{(4)}(x)|\Big].$$

4. For an error less than ε, solve for n in the inequality

$$\frac{(b-a)^5}{180n^4}\Big[\max|f^{(4)}(x)|\Big] < \varepsilon.$$

5. Partition $[a, b]$ into n subintervals and apply Simpson's Rule.

ALGEBRA TUTOR

For help on the algebra in Example 3, see Example 4 in the Chapter 6 *Algebra Tutor*, on page 413.

EXAMPLE 3 The Approximate Error in the Trapezoidal Rule

Use the Trapezoidal Rule to estimate the value of $\int_0^1 e^{-x^2}\,dx$ such that the error in the approximation of the integral is less than 0.01.

SOLUTION

1. Begin by finding the second derivative of $f(x) = e^{-x^2}$.

$$f(x) = e^{-x^2}$$
$$f'(x) = -2xe^{-x^2}$$
$$f''(x) = 4x^2 e^{-x^2} - 2e^{-x^2}$$
$$= 2e^{-x^2}(2x^2 - 1)$$

2. f'' has only one critical number in the interval $[0, 1]$, and the maximum value of $|f''(x)|$ on this interval is $|f''(0)| = 2$.

3. The error E using the Trapezoidal Rule is bounded by

$$|E| \le \frac{(b-a)^3}{12n^2}(2) = \frac{1}{12n^2}(2) = \frac{1}{6n^2}.$$

4. To ensure that the approximation has an error of less than 0.01, you should choose n such that

$$\frac{1}{6n^2} < 0.01.$$

Solving for n, you can determine that n must be 5 or more.

5. Partition $[0, 1]$ into five subintervals, as shown in Figure 6.12. Then apply the Trapezoidal Rule to obtain

$$\int_0^1 e^{-x^2}\,dx = \frac{1}{10}\left(\frac{1}{e^0} + \frac{2}{e^{0.04}} + \frac{2}{e^{0.16}} + \frac{2}{e^{0.36}} + \frac{2}{e^{0.64}} + \frac{1}{e^1}\right) \approx 0.744.$$

So, with an error less than 0.01, you know that

$$0.734 \le \int_0^1 e^{-x^2}\,dx \le 0.754.$$

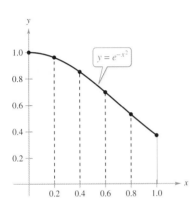

$y = e^{-x^2}$

FIGURE 6.12

✓ *Checkpoint 3*   *Worked-out solution available at LarsonAppliedCalculus.com*

Use the Trapezoidal Rule to estimate the value of

$$\int_0^1 \sqrt{1 + x^2}\,dx$$

such that the error in the approximation of the integral is less than 0.01.

SUMMARIZE (Section 6.3)

1. State the Trapezoidal Rule *(page 394)*. For an example of the Trapezoidal Rule, see Example 1.

2. State Simpson's Rule *(page 396)*. For an example of Simpson's Rule, see Example 2.

3. State the approximate errors in the Trapezoidal Rule and Simpson's Rule *(page 398)*. For an example of using the approximate error in the Trapezoidal Rule, see Example 3.

SKILLS WARM UP 6.3

The following warm-up exercises involve skills that were covered in a previous course or in earlier sections. You will use these skills in the exercise set for this section. For additional help, review Appendix A.1 and Sections 2.2, 2.6, 3.2, 4.3, and 4.5.

In Exercises 1–6, find the indicated derivative.

1. $f(x) = \dfrac{1}{x^2}, f''(x)$

2. $f(x) = \ln(2x + 1), f^{(4)}(x)$

3. $f(x) = 2 \ln x, f^{(4)}(x)$

4. $f(x) = x^3 - 2x^2 + 7x - 12, f''(x)$

5. $f(x) = e^{3x}, f^{(4)}(x)$

6. $f(x) = e^{x^2}, f''(x)$

In Exercises 7 and 8, find the absolute maximum of f on the interval.

7. $f(x) = -x^2 + 6x + 9, [0, 4]$

8. $f(x) = \dfrac{8}{x^3}, [1, 2]$

In Exercises 9 and 10, solve for n. (Assume n is positive.)

9. $\dfrac{1}{4n^2} < 0.001$

10. $\dfrac{1}{16n^4} < 0.0001$

Exercises 6.3

See *CalcChat.com* for tutorial help and worked-out solutions to odd-numbered exercises.

Using the Trapezoidal Rule and Simpson's Rule In Exercises 1–10, use the Trapezoidal Rule and Simpson's Rule to approximate the value of the definite integral for the indicated value of n. Compare these results with the exact value of the definite integral. Round your answers to four decimal places. *See Examples 1 and 2.*

1. $\displaystyle\int_0^2 x^2 \, dx, n = 4$

2. $\displaystyle\int_0^1 \left(\dfrac{x^3}{2} + 1\right) dx, n = 4$

3. $\displaystyle\int_1^3 (4 - x^2) \, dx, n = 4$

4. $\displaystyle\int_1^2 \dfrac{2}{x} \, dx, n = 8$

5. $\displaystyle\int_0^4 \sqrt{x} \, dx, n = 8$

6. $\displaystyle\int_0^8 \sqrt[3]{x} \, dx, n = 8$

7. $\displaystyle\int_3^4 \dfrac{1}{x - 2} \, dx, n = 4$

8. $\displaystyle\int_0^2 e^{-4x} \, dx, n = 8$

9. $\displaystyle\int_0^1 xe^{3x^2} \, dx, n = 4$

10. $\displaystyle\int_0^2 x\sqrt{x^2 + 1} \, dx, n = 4$

Using the Trapezoidal Rule and Simpson's Rule In Exercises 11–20, approximate the value of the definite integral using (a) the Trapezoidal Rule and (b) Simpson's Rule for the indicated value of n. Round your answers to three decimal places. *See Examples 1 and 2.*

11. $\displaystyle\int_0^1 \dfrac{1}{1 + x^2} \, dx, n = 4$

12. $\displaystyle\int_0^4 \dfrac{8}{x^2 + 3} \, dx, n = 4$

13. $\displaystyle\int_0^2 \sqrt{9 - x^3} \, dx, n = 4$

14. $\displaystyle\int_0^1 \sqrt[3]{4 - x^2} \, dx, n = 8$

15. $\displaystyle\int_0^1 e^{x^2} \, dx, n = 8$

16. $\displaystyle\int_0^2 e^{-x^2} \, dx, n = 4$

17. $\displaystyle\int_0^4 \dfrac{1}{\sqrt[3]{x^2 + 1}} \, dx, n = 8$

18. $\displaystyle\int_0^2 \dfrac{1}{\sqrt{1 + x^3}} \, dx, n = 4$

19. $\displaystyle\int_0^3 \dfrac{1}{2 - 2x + x^2} \, dx, n = 6$

20. $\displaystyle\int_0^3 \dfrac{x}{2 + x + x^2} \, dx, n = 6$

Using Simpson's Rule In Exercises 21–24, use the Simpson's Rule program in Appendix E with $n = 100$ to approximate the definite integral.

21. $\displaystyle\int_1^4 x\sqrt{x + 4} \, dx$

22. $\displaystyle\int_1^5 x^2\sqrt{x + 6} \, dx$

23. $\displaystyle\int_2^5 10xe^{-x} \, dx$

24. $\displaystyle\int_3^6 9x^2 e^{-x+1} \, dx$

Marginal Analysis In Exercises 25 and 26, use the Simpson's Rule program in Appendix E with $n = 4$ to approximate the change in revenue from the marginal revenue function dR/dx. Assume that the number of units sold x increases from 14 to 16.

25. $\dfrac{dR}{dx} = 5\sqrt{8000 - x^3}$

26. $\dfrac{dR}{dx} = 50\sqrt{x}\sqrt{20 - x}$

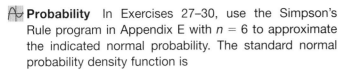

Probability   In Exercises 27–30, use the Simpson's Rule program in Appendix E with $n = 6$ to approximate the indicated normal probability. The standard normal probability density function is

$$f(x) = \frac{1}{\sqrt{2\pi}} e^{-x^2/2}.$$

If x is chosen at random from a population with this density, then the probability that x lies in the interval $[a, b]$ is

$$P(a \le x \le b) = \int_a^b f(x)\, dx.$$

27. $P(0 \le x \le 1)$ **28.** $P(0 \le x \le 2)$

29. $P(0 \le x \le 3.5)$ **30.** $P(0 \le x \le 1.5)$

Surveying   In Exercises 31 and 32, use the Trapezoidal Rule to estimate the number of square feet of land in the lot, where x and y are measured in feet, as shown in the figure. The land is bounded by a stream and two straight roads. (The number of subintervals is shown in the figure.)

31.

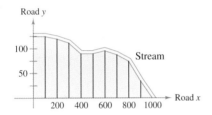

x	0	100	200	300	400	500
y	125	125	120	112	90	90

x	600	700	800	900	1000
y	95	88	75	35	0

Spreadsheet at LarsonAppliedCalculus.com

32.

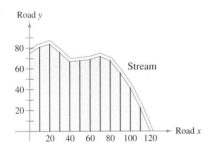

x	0	10	20	30	40	50	60
y	75	81	84	76	67	68	69

x	70	80	90	100	110	120
y	72	68	56	42	23	0

Spreadsheet at LarsonAppliedCalculus.com

Error Analysis   In Exercises 33–36, use the error formulas to find bounds for the error in approximating the definite integral using (a) the Trapezoidal Rule and (b) Simpson's Rule. Let $n = 4$.

33. $\displaystyle\int_0^2 (x^2 + 2x)\, dx$ **34.** $\displaystyle\int_0^1 \frac{1}{x+1}\, dx$

35. $\displaystyle\int_0^1 e^{x^3}\, dx$ **36.** $\displaystyle\int_0^1 e^{2x^2}\, dx$

Error Analysis   In Exercises 37–40, use the error formulas to find n such that the error in the approximation of the definite integral is less than 0.0001 using (a) the Trapezoidal Rule and (b) Simpson's Rule. *See Example 3.*

37. $\displaystyle\int_0^2 x^4\, dx$ **38.** $\displaystyle\int_1^3 \frac{1}{x}\, dx$

39. $\displaystyle\int_1^3 e^{2x}\, dx$ **40.** $\displaystyle\int_3^5 \ln x\, dx$

Present Value   In Exercises 41 and 42, use the Simpson's Rule program in Appendix E with $n = 8$ to approximate the present value of the income $c(t)$ over t_1 years at the given annual interest rate r. Then use the integration capabilities of a graphing utility to approximate the present value. Compare the results. (Present value is defined in Section 6.1.)

41. $c(t) = 6000 + 200\sqrt{t}$, $r = 7\%$, $t_1 = 4$

42. $c(t) = 200{,}000 + 15{,}000\sqrt[3]{t}$, $r = 10\%$, $t_1 = 8$

43. Median Age   The table shows the median ages of the U.S. resident population for the years 2005 through 2013. *(Source: U.S. Census Bureau)*

Year	2005	2006	2007	2008	2009
Median age	36.2	36.3	36.5	36.7	36.8

Year	2010	2011	2012	2013
Median age	37.2	37.3	37.5	37.6

Spreadsheet at LarsonAppliedCalculus.com

(a) Use Simpson's Rule to estimate the average median age over the time period.

(b) A model for the data is

$$A = 37.2 - 0.54t + 0.084t^2 - 0.0031t^3$$

for $5 \le t \le 13$, where A is the median age and t is the year, with $t = 5$ corresponding to 2005. Use integration to find the average median age over the time period.

(c) Compare the results of parts (a) and (b).

44. Electricity   The table shows the residential prices of electricity (in cents per kilowatt-hour) for the years 2006 through 2014. *(Source: U.S. Energy Information Administration)*

Year	2006	2007	2008	2009	2010
Price	10.40	10.65	11.26	11.51	11.54

Year	2011	2012	2013	2014
Price	11.72	11.88	12.12	12.54

Spreadsheet at LarsonAppliedCalculus.com

(a) Use Simpson's Rule to estimate the average residential price of electricity over the time period.

(b) A model for the data is

$$E = 516.23254e^{-t} + 0.0182t^3 - 0.595t^2 + 6.58t - 12.9$$

for $6 \leq t \leq 14$, where E is the residential price of electricity (in cents per kilowatt-hour) and t is the year, with $t = 6$ corresponding to 2006. Use integration to find the average residential price of electricity over the time period.

(c) Compare the results of parts (a) and (b).

45. Medicine   A body assimilates a 12-hour cold tablet at a rate modeled by $dC/dt = 8 - \ln(t^2 - 2t + 4)$, $0 \leq t \leq 12$, where dC/dt is measured in milligrams per hour and t is the time in hours. Use Simpson's Rule with $n = 8$ to estimate the total amount of the drug absorbed into the body during the 12 hours.

46. HOW DO YOU SEE IT?   The graph shows the weekly revenue (in thousands of dollars) for a company.

(a) Which gives a more accurate approximation of the total weekly revenue for the first 4 weeks using the Trapezoidal Rule, $n = 8$ or $n = 16$?

(b) Which gives a more accurate approximation of the total weekly revenue for the first 4 weeks, the Trapezoidal Rule with $n = 8$ or Simpson's Rule with $n = 8$?

47. Consumer Trends   The rate of change in the number of subscribers S to a newly introduced magazine is modeled by

$$\frac{dS}{dt} = 1000t^2e^{-t}, \ 0 \leq t \leq 6$$

where t is the time in years. Use Simpson's Rule with $n = 12$ to estimate the total increase in the number of subscribers during the first 6 years.

48. Using Simpson's Rule   Prove that Simpson's Rule is exact when approximating the integral of a cubic polynomial function, and demonstrate the result with $n = 2$ for

$$\int_0^1 x^3 \, dx.$$

Business Capsule

Susie Wang and Ric Kostick graduated in 2002 from the University of California at Berkeley with degrees in mathematics. Together they launched a cosmetics brand called 100% Pure, which uses fruit and vegetable pigments to color cosmetics and uses only organic ingredients for the purest skin care. The company grew quickly and now has annual sales of over $40 million. Wang and Kostick attribute their success to applying what they learned from their studies. "Mathematics teaches you logic, discipline, and accuracy, which help you with all aspects of daily life," says Ric Kostick.

49. Research Project   Use your school's library, the Internet, or some other reference source to research the opportunity cost of attending graduate school for 2 years to receive a Masters of Business Administration (MBA) degree rather than working for 2 years with a bachelor's degree. Write a short paper describing these costs.

6.4 Improper Integrals

Height of 30- to 39-Year-Old American Women

In Exercise 27 on page 411, you will evaluate an improper integral to determine the probability that a 30- to 39-year-old woman is 6 feet or taller.

■ Recognize improper integrals.
■ Evaluate improper integrals with infinite limits of integration.
■ Use improper integrals to solve real-life problems.
■ Find the present value of a perpetuity.

Improper Integrals

The definition of a definite integral

$$\int_a^b f(x)\,dx$$

requires that the interval $[a, b]$ be finite. Furthermore, the Fundamental Theorem of Calculus, by which you have been evaluating definite integrals, requires that f be continuous on $[a, b]$. Some integrals do not satisfy these requirements because of one of the conditions below.

1. One or both of the limits of integration are infinite.

2. The function f has an *infinite discontinuity* in the interval $[a, b]$.

Integrals having either of these characteristics are called **improper integrals.** In this section, you will study integrals where one or both limits of integration are infinite. For instance, the integral

$$\int_0^\infty e^{-x}\,dx$$

is improper because one limit of integration is infinite, as shown in Figure 6.13.

FIGURE 6.13

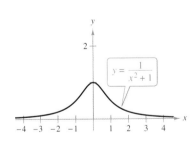

FIGURE 6.14

Similarly, the integral

$$\int_{-\infty}^\infty \frac{1}{x^2 + 1}\,dx$$

is improper because both limits of integration are infinite, as shown in Figure 6.14.
    The integrals

$$\int_1^5 \frac{1}{\sqrt{x - 1}}\,dx \quad \text{and} \quad \int_{-2}^2 \frac{1}{(x + 1)^2}\,dx$$

are improper because their integrands have an **infinite discontinuity**—that is, they approach infinity somewhere in the interval of integration. Evaluating an integral whose integrand has an infinite discontinuity is beyond the scope of this text.

Integrals with Infinite Limits of Integration

To see how to evaluate an improper integral, consider the integral shown in Figure 6.15.

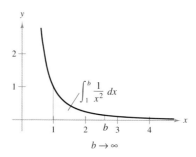

FIGURE 6.15

As long as b is a real number that is greater than 1 (no matter how large), this is a definite integral whose value is

$$\int_1^b \frac{1}{x^2}\,dx = \left[-\frac{1}{x}\right]_1^b = -\frac{1}{b} + 1 = 1 - \frac{1}{b}.$$

The table shows the values of this integral for several values of b.

b	2	5	10	100	1000	10,000
$\int_1^b \frac{1}{x^2}\,dx = 1 - \frac{1}{b}$	0.5000	0.8000	0.9000	0.9900	0.9990	0.9999

From this table, it appears that the value of the integral is approaching a limit as b increases without bound. This limit is denoted by the *improper integral* shown below.

$$\int_1^\infty \frac{1}{x^2}\,dx = \lim_{b\to\infty}\int_1^b \frac{1}{x^2}\,dx = \lim_{b\to\infty}\left(1 - \frac{1}{b}\right) = 1$$

This improper integral can be interpreted as the area of the *unbounded* region between the graph of $f(x) = 1/x^2$ and the x-axis (to the right of $x = 1$).

Definition of Improper Integrals (Infinite Limits of Integration)

1. If f is continuous on the interval $[a, \infty)$, then

$$\int_a^\infty f(x)\,dx = \lim_{b\to\infty}\int_a^b f(x)\,dx.$$

2. If f is continuous on the interval $(-\infty, b]$, then

$$\int_{-\infty}^b f(x)\,dx = \lim_{a\to-\infty}\int_a^b f(x)\,dx.$$

3. If f is continuous on the interval $(-\infty, \infty)$, then

$$\int_{-\infty}^\infty f(x)\,dx = \int_{-\infty}^c f(x)\,dx + \int_c^\infty f(x)\,dx$$

where c is any real number.

In the first two cases, if the limit exists, then the improper integral **converges;** otherwise, the improper integral **diverges.** In the third case, the integral on the left diverges when either one of the integrals on the right diverges.

EXAMPLE 1   Evaluating an Improper Integral

Determine the convergence or divergence of $\displaystyle\int_{1}^{\infty} \frac{1}{x}\,dx$.

SOLUTION   Begin by applying the definition of an improper integral.

$$\int_{1}^{\infty} \frac{1}{x}\,dx = \lim_{b \to \infty} \int_{1}^{b} \frac{1}{x}\,dx \qquad \text{Definition of improper integral}$$

$$= \lim_{b \to \infty} \Big[\ln x\Big]_{1}^{b} \qquad \text{Apply Log Rule.}$$

$$= \lim_{b \to \infty} (\ln b - 0) \qquad \text{Apply Fundamental Theorem.}$$

$$= \infty \qquad \text{Evaluate limit.}$$

Because the limit does not exist, the improper integral diverges.

✓ **Checkpoint 1**   *Worked-out solution available at LarsonAppliedCalculus.com*

Determine the convergence or divergence of each improper integral.

a. $\displaystyle\int_{1}^{\infty} \frac{1}{x^3}\,dx$ **b.** $\displaystyle\int_{1}^{\infty} \frac{1}{\sqrt{x}}\,dx$

As you begin to work with improper integrals, you will find that integrals that appear to be similar can have very different values. For instance, consider the two improper integrals

$$\int_{1}^{\infty} \frac{1}{x}\,dx = \infty \qquad \text{Divergent integral}$$

and

$$\int_{1}^{\infty} \frac{1}{x^2}\,dx = 1. \qquad \text{Convergent integral}$$

The first integral diverges and the second converges to 1. Graphically, this means that the areas shown in Figure 6.16 are very different. The region lying between the graph of

$$y = \frac{1}{x}$$

and the *x*-axis (for $x \geq 1$) has an *infinite* area, and the region lying between the graph of

$$y = \frac{1}{x^2}$$

and the *x*-axis (for $x \geq 1$) has a *finite* area.

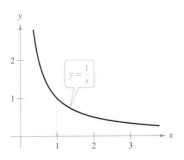

Diverges (infinite area)   Converges (finite area)

FIGURE 6.16

ALGEBRA TUTOR

For help on the algebra in Example 2, see Example 2(a) in the *Chapter 6 Algebra Tutor*, on page 412.

EXAMPLE 2   Evaluating an Improper Integral

Evaluate the improper integral.

$$\int_{-\infty}^{0} \frac{1}{(1-2x)^{3/2}}\, dx$$

SOLUTION   Begin by applying the definition of an improper integral.

$$\int_{-\infty}^{0} \frac{1}{(1-2x)^{3/2}}\, dx = \lim_{a \to -\infty} \int_{a}^{0} \frac{1}{(1-2x)^{3/2}}\, dx \qquad \text{Definition of improper integral}$$

$$= \lim_{a \to -\infty} \left[\frac{1}{\sqrt{1-2x}} \right]_{a}^{0} \qquad \text{Find antiderivative.}$$

$$= \lim_{a \to -\infty} \left(1 - \frac{1}{\sqrt{1-2a}} \right) \qquad \text{Apply Fundamental Theorem.}$$

$$= 1 - 0 \qquad \text{Evaluate limit.}$$

$$= 1 \qquad \text{Simplify.}$$

So, the improper integral converges to 1. As shown in Figure 6.17, this implies that the region lying between the graph of $y = 1/(1-2x)^{3/2}$ and the x-axis (for $x \le 0$) has an area of 1 square unit.

FIGURE 6.17

✓ *Checkpoint 2*   Worked-out solution available at LarsonAppliedCalculus.com

Evaluate the improper integral, if possible.

$$\int_{-\infty}^{0} \frac{1}{(x-1)^{?}}\, dx$$

ALGEBRA TUTOR

For help on the algebra in Example 3, see Example 2(b) in the *Chapter 6 Algebra Tutor*, on page 412.

EXAMPLE 3   Evaluating an Improper Integral

Evaluate the improper integral.

$$\int_{0}^{\infty} 2xe^{-x^2}\, dx$$

SOLUTION   Begin by applying the definition of an improper integral.

$$\int_{0}^{\infty} 2xe^{-x^2}\, dx = \lim_{b \to \infty} \int_{0}^{b} 2xe^{-x^2}\, dx \qquad \text{Definition of improper integral}$$

$$= \lim_{b \to \infty} \left[-e^{-x^2} \right]_{0}^{b} \qquad \text{Apply Exponential Rule.}$$

$$= \lim_{b \to \infty} \left(-e^{-b^2} + 1 \right) \qquad \text{Apply Fundamental Theorem.}$$

$$= 0 + 1 \qquad \text{Evaluate limit.}$$

$$= 1 \qquad \text{Simplify.}$$

So, the improper integral converges to 1. As shown in Figure 6.18, this implies that the region lying between the graph of $y = 2xe^{-x^2}$ and the x-axis (for $x \ge 0$) has an area of 1 square unit.

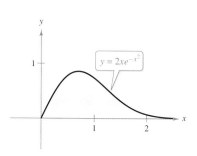

FIGURE 6.18

✓ *Checkpoint 3*   Worked-out solution available at LarsonAppliedCalculus.com

Evaluate the improper integral, if possible.

$$\int_{-\infty}^{0} e^{2x}\, dx$$

■

Application

In Section 4.3, you studied the graph of the *normal probability density function*

$$f(x) = \frac{1}{\sigma\sqrt{2\pi}} e^{-(x-\mu)^2/(2\sigma^2)}.$$

This function is used in statistics to represent a population that is normally distributed with a mean of μ and a standard deviation of σ. Specifically, when an outcome x is chosen at random from the population, the probability that x will have a value between a and b is

$$P(a \leq x \leq b) = \int_a^b \frac{1}{\sigma\sqrt{2\pi}} e^{-(x-\mu)^2/(2\sigma^2)}\, dx.$$

As shown in Figure 6.19, the probability $P(-\infty < x < \infty)$ is

$$P(-\infty < x < \infty) = \int_{-\infty}^{\infty} \frac{1}{\sigma\sqrt{2\pi}} e^{-(x-\mu)^2/(2\sigma^2)}\, dx = 1.$$

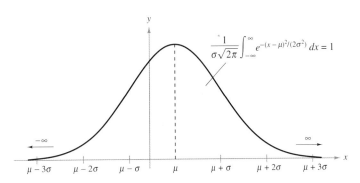

FIGURE 6.19

EXAMPLE 4   Finding a Probability

The mean height of American men (from 20 to 29 years old) is 69 inches, and the standard deviation is 3 inches. A 20- to 29-year-old man is chosen at random from the population. What is the probability that he is 6 feet tall or taller? *(Source: U.S. National Center for Health Statistics)*

SOLUTION   Note that the mean and standard deviation are given in inches and the height of the man chosen at random is given in feet. To calculate the probability, you need to use the same units for these quantities. Because it is easier to convert feet to inches, use 72 inches (1 foot = 12 inches) for the man's height. So, the probability can be written as $P(72 \leq x < \infty)$. Using a mean of $\mu = 69$ and a standard deviation of $\sigma = 3$, the probability $P(72 \leq x < \infty)$ is given by the improper integral

$$P(72 \leq x < \infty) = \int_{72}^{\infty} \frac{1}{3\sqrt{2\pi}} e^{-(x-69)^2/18}\, dx.$$

Using a symbolic integration utility, you can approximate the value of this integral to be about 0.159. So, the probability that the man is 6 feet tall or taller is about 15.9%.

✔ *Checkpoint 4*   *Worked-out solution available at LarsonAppliedCalculus.com*

Use Example 4 to find the probability that a 20- to 29-year-old man chosen at random from the population is 6 feet 6 inches tall or taller. ■

Many professional basketball players are over $6\frac{1}{2}$ feet tall. When a man is chosen at random from the population, the probability that he is $6\frac{1}{2}$ feet tall or taller is less than half of one percent.

Present Value of a Perpetuity

Recall from Section 6.1 that for an interest-bearing account, the present value over t_1 years is

$$\text{Present value} = \int_0^{t_1} c(t)e^{-rt}\,dt$$

where c represents a continuous income function (in dollars per year) and the annual interest rate r (in decimal form) is compounded continuously. If the size of an annuity's payment is a constant number of dollars P, then $c(t)$ is equal to P and the present value is

$$\text{Present value} = \int_0^{t_1} Pe^{-rt}\,dt = P\int_0^{t_1} e^{-rt}\,dt.$$

Present value of an annuity
with payment P

Consider an annuity, such as a scholarship fund, that pays the same amount each year *forever*. Because the annuity continues indefinitely, the number of years t_1 approaches infinity. Such an annuity is called a **perpetual annuity** or a **perpetuity.** This situation can be represented by the following improper integral.

$$\text{Present value} = P\int_0^{\infty} e^{-rt}\,dt$$

Present value of an perpetuity
with payment P

This integral is simplified as follows.

$$P\int_0^{\infty} e^{-rt}\,dt = P\lim_{b\to\infty}\int_0^{b} e^{-rt}\,dt \qquad \text{Definition of improper integral}$$

$$= P\lim_{b\to\infty}\left[-\frac{e^{-rt}}{r}\right]_0^b \qquad \text{Apply Exponential Rule.}$$

$$= P\lim_{b\to\infty}\left(-\frac{e^{-rb}}{r} + \frac{1}{r}\right) \qquad \text{Apply Fundamental Theorem.}$$

$$= P\left(0 + \frac{1}{r}\right) \qquad \text{Evaluate limit.}$$

$$= \frac{P}{r} \qquad \text{Simplify.}$$

So, the improper integral converges to P/r. As shown in Figure 6.20, this implies that the region lying between the graph of

$$y = Pe^{-rt}$$

and the t-axis (for $t \geq 0$) has an area equal to the annual payment P divided by the annual interest rate r.

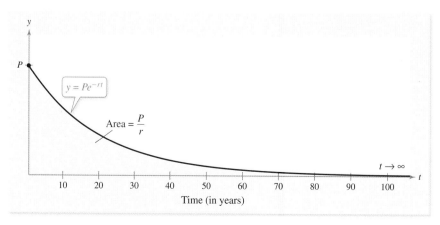

FIGURE 6.20

The present value of a perpetuity is defined as follows.

Present Value of a Perpetuity

If P represents the size of each annual payment in dollars and r represents the annual interest rate (in decimal form) compounded continuously, then the present value of a perpetuity is

$$\text{Present value} = P \int_0^\infty e^{-rt}\, dt = \frac{P}{r}.$$

This definition is useful in determining the amount of money needed to start an endowment, such as a scholarship fund, as shown in Example 5.

EXAMPLE 5   Finding Present Value

You want to start a scholarship fund at your alma mater. You plan to give one $9000 scholarship annually beginning one year from now, and you have at most $120,000 to start the fund. You also want the scholarship to be given out indefinitely. Assuming an annual interest rate of 8% (compounded continuously), do you have enough money to start the scholarship fund?

SOLUTION To answer this question, you must find the present value of the scholarship fund. Because the scholarship is to be given out each year indefinitely, the time period is infinite. The fund is a perpetuity with $P = 9000$ and $r = 0.08$. The present value is

$$\begin{aligned}
\text{Present value} &= \frac{P}{r} \\
&= \frac{9000}{0.08} \\
&= 112{,}500.
\end{aligned}$$

The amount you need to start the scholarship fund is $112,500. So, you have enough money to start the scholarship fund.

✓ **Checkpoint 5**   *Worked-out solution available at LarsonAppliedCalculus.com*

In Example 5, do you have enough money to start a scholarship fund that pays $10,000 annually? Explain why or why not. ■

SUMMARIZE   (Section 6.4)

1. Describe the different types of improper integrals *(page 403)*. For examples of evaluating improper integrals, see Examples 1, 2, and 3.

2. Define the term *converges* as it applies to improper integrals *(page 404)*. For examples of improper integrals that converge, see Examples 2 and 3.

3. Define the term *diverges* as it applies to improper integrals *(page 404)*. For an example of an improper integral that diverges, see Example 1.

4. Describe a real-life example of how an improper integral can be used to find a probability *(page 407, Example 4)*.

5. Describe a real-life example of how an improper integral can be used to find the present value of a perpetuity *(page 409, Example 5)*.

In Exercises 1–6, find the limit (if it exists).

1. $\lim\limits_{x \to 2}(2x + 5)$

2. $\lim\limits_{x \to 3}\left(\dfrac{3}{x} - x + 2\right)$

3. $\lim\limits_{x \to -4}\dfrac{x + 4}{x^2 - 16}$

4. $\lim\limits_{x \to 0}\dfrac{x^2 - 2x}{x^3 + 3x^2}$

5. $\lim\limits_{x \to 1}\dfrac{1}{\sqrt{x - 1}}$

6. $\lim\limits_{x \to -3}\dfrac{x^2 + 2x - 3}{x + 3}$

In Exercises 7–10, evaluate the expression (a) when $x = b$ and (b) when $x = 0$.

7. $\dfrac{4}{3}(2x - 1)^3$

8. $\dfrac{1}{x - 5} + \dfrac{3}{(x - 2)^2}$

9. $\ln(5 - 3x^2) - \ln(x + 1)$

10. $e^{3x^2} + e^{-3x^2}$

Exercises 6.4

See *CalcChat.com* for tutorial help and worked-out solutions to odd-numbered exercises.

Determining Whether an Integral Is Improper In Exercises 1–6, decide whether the integral is improper. Explain your reasoning.

1. $\displaystyle\int_0^1 \dfrac{1}{3x - 2}\,dx$

2. $\displaystyle\int_1^3 \dfrac{1}{x^2}\,dx$

3. $\displaystyle\int_0^1 \dfrac{2x - 5}{x^2 - 5x + 6}\,dx$

4. $\displaystyle\int_{-2}^2 \ln(x^2)\,dx$

5. $\displaystyle\int_3^\infty e^{x^2}\,dx$

6. $\displaystyle\int_{-\infty}^\infty \dfrac{1}{x^2 + 3}\,dx$

 Evaluating an Improper Integral In Exercises 7–20, determine whether the improper integral diverges or converges. Evaluate the integral if it converges. *See Examples 1, 2, and 3.*

7. $\displaystyle\int_1^\infty \dfrac{1}{x^4}\,dx$

8. $\displaystyle\int_1^\infty \dfrac{1}{\sqrt[3]{x}}\,dx$

9. $\displaystyle\int_0^\infty e^{x/3}\,dx$

10. $\displaystyle\int_0^\infty \dfrac{5}{e^{2x}}\,dx$

11. $\displaystyle\int_5^\infty \dfrac{1}{\sqrt{2x - 1}}\,dx$

12. $\displaystyle\int_5^\infty \dfrac{x}{\sqrt{x^2 - 16}}\,dx$

13. $\displaystyle\int_{-\infty}^{-1} e^x\,dx$

14. $\displaystyle\int_{-\infty}^{-1} \dfrac{1}{x^2}\,dx$

15. $\displaystyle\int_1^\infty \dfrac{e^{\sqrt{x}}}{\sqrt{x}}\,dx$

16. $\displaystyle\int_{-\infty}^0 \dfrac{x}{x^2 + 1}\,dx$

17. $\displaystyle\int_4^\infty \dfrac{1}{x(\ln x)^3}\,dx$

18. $\displaystyle\int_0^\infty \dfrac{7x}{(3x^2 + 5)^2}\,dx$

19. $\displaystyle\int_{-\infty}^\infty 2xe^{-3x^2}\,dx$

20. $\displaystyle\int_{-\infty}^\infty x^2 e^{-x^3}\,dx$

 Area of a Region In Exercises 21–26, find the area of the unbounded shaded region.

21. $y = e^{-x}$

22. $y = e^{x/4}$

23. $y = -\dfrac{7}{(x - 1)^3}$

24. $y = \dfrac{5}{\sqrt{4 - x}}$

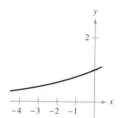

25. $y = \dfrac{6x}{x^2 + 1}$

26. $y = \dfrac{16x}{x^2 + 4}$

27. Women's Height The mean height of American women (from 30 to 39 years old) is 64 inches, and the standard deviation is 3 inches. Use a symbolic integration utility or a graphing utility to find the probability that a 30- to 39-year-old woman chosen at random is

(a) between 5 and 6 feet tall.

(b) 5 feet 8 inches or taller.

(c) 6 feet or taller.

(Source: U.S. National Center for Health Statistics)

28. **HOW DO YOU SEE IT?** The graph shows the probability density function for a car brand that has a mean fuel efficiency of 26 miles per gallon and a standard deviation of 2.4 miles per gallon.

(a) Which is greater, the probability of choosing a car at random that gets between 26 and 28 miles per gallon or the probability of choosing a car at random that gets between 22 and 24 miles per gallon?

(b) Which is greater, the probability of choosing a car at random that gets between 20 and 22 miles per gallon or the probability of choosing a car at random that gets at least 30 miles per gallon?

29. Quality Control A company manufactures wooden yardsticks. The lengths of the yardsticks are normally distributed with a mean of 36 inches and a standard deviation of 0.2 inch. Use a symbolic integration utility or a graphing utility to find the probability that a yardstick chosen at random is

(a) longer than 35.5 inches.

(b) longer than 35.9 inches.

30. Quality Control A company manufactures compact fluorescent light bulbs. The life spans of the light bulbs are normally distributed with a mean of 9000 hours and a standard deviation of 500 hours. Use a symbolic integration utility or a graphing utility to find the probability that a light bulb chosen at random has a life span that is

(a) 8000 hours or longer.

(b) 11,000 hours or longer.

 Endowment In Exercises 31–34, determine the amount of money required to set up a charitable endowment that pays the amount P each year indefinitely for the annual interest rate r compounded continuously. *See Example 5.*

31. $P = \$12,000,\ r = 6\%$

32. $P = \$10,000,\ r = 7\%$

33. $P = \$5000,\ r = 7.5\%$

34. $P = \$3500,\ r = 3.5\%$

35. Scholarship Fund You want to start a scholarship fund at your alma mater. You plan to give one $18,000 scholarship annually beginning one year from now, and you have at most $400,000 to start the fund. You also want the scholarship to be given out indefinitely. Assuming an annual interest rate of 5% compounded continuously, do you have enough money to start the scholarship fund?

36. Charitable Foundation A charitable foundation wants to help schools buy tablet computers. The foundation plans to donate $35,000 each year to one school beginning one year from now, and the foundation has at most $500,000 to start the fund. The foundation wants the donation to be given out indefinitely. Assuming an annual interest rate of 8% compounded continuously, does the foundation have enough money to start the school fund?

37. Present Value A business is expected to yield a continuous flow of profit at the rate of $500,000 per year. Assuming an annual interest rate of 9% compounded continuously, what is the present value of the business

(a) for 20 years?

(b) forever?

38. Present Value A farm is expected to yield a continuous flow of profit at the rate of $75,000 per year. Assuming an annual interest rate of 8% compounded continuously, what is the present value of the farm

(a) for 20 years?

(b) forever?

Capitalized Cost In Exercises 39–42, the capitalized cost C of an asset is given by

$$C = C_0 + \int_0^n c(t)e^{-rt}\, dt$$

where C_0 is the original investment, t is the time in years, r is the annual interest rate (in decimal form) compounded continuously, and $c(t)$ is the annual cost of maintenance (in dollars). Find the capitalized cost of an asset (a) for 5 years, (b) for 10 years, and (c) forever.

39. $C_0 = \$650,000,\ c(t) = 25,000,\ r = 10\%$

40. $C_0 = \$800,000,\ c(t) = 30,000,\ r = 4\%$

41. $C_0 = \$300,000,\ c(t) = 15,000t,\ r = 6\%$

42. $C_0 = \$650,000,\ c(t) = 25,000(1 + 0.08t),\ r = 12\%$

ALGEBRA TUTOR

Algebra and Integration Techniques

Integration techniques involve many different algebraic skills. For a definite integral, you need a variety of algebraic skills to apply the Fundamental Theorem of Calculus and evaluate the resulting expression. Study the examples in this Algebra Tutor. Be sure that you understand the algebra used in each step.

EXAMPLE 1   Evaluating an Expression

Evaluate the expression

$$(e \ln e - e) - (1 \ln 1 - 1).$$

SOLUTION   Recall that

$$\ln e = 1 \quad \text{because} \quad e^1 = e$$

and

$$\ln 1 = 0 \quad \text{because} \quad e^0 = 1.$$

$$
\begin{aligned}
(e \ln e - e) - (1 \ln 1 - 1) && \text{Example 5, page 380} \\
= [e(1) - e] - [1(0) - 1] && \text{Logarithmic properties} \\
= (e - e) - (0 - 1) && \text{Multiply.} \\
= 0 - (-1) && \text{Simplify.} \\
= 1 && \text{Simplify.}
\end{aligned}
$$

EXAMPLE 2   Finding Limits

Find the limit.

a. $\displaystyle\lim_{a \to -\infty} \left(1 - \frac{1}{\sqrt{1 - 2a}}\right)$

b. $\displaystyle\lim_{b \to \infty} \left(-e^{-b^2} + 1\right)$

SOLUTION

a. $\displaystyle\lim_{a \to -\infty} \left(1 - \frac{1}{\sqrt{1 - 2a}}\right)$ Example 2, page 406

$$
\begin{aligned}
&= \lim_{a \to -\infty} 1 - \lim_{a \to -\infty} \left(\frac{1}{\sqrt{1 - 2a}}\right) && \lim_{x \to -\infty} [f(x) - g(x)] = \lim_{x \to -\infty} f(x) - \lim_{x \to -\infty} g(x) \\
&= 1 - 0 && \text{Evaluate limits.} \\
&= 1 && \text{Simplify.}
\end{aligned}
$$

b. $\displaystyle\lim_{b \to \infty} \left(-e^{-b^2} + 1\right)$ Example 3, page 406

$$
\begin{aligned}
&= \lim_{b \to \infty} \left(-e^{-b^2}\right) + \lim_{b \to \infty} 1 && \lim_{x \to \infty} [f(x) + g(x)] = \lim_{x \to \infty} f(x) + \lim_{x \to \infty} g(x) \\
&= \lim_{b \to \infty} \left(\frac{1}{-e^{b^2}}\right) + \lim_{b \to \infty} 1 && \text{Rewrite with positive exponent.} \\
&= 0 + 1 && \text{Evaluate limits.} \\
&= 1 && \text{Simplify.}
\end{aligned}
$$

EXAMPLE 3   Algebra and Integration Techniques

Simplify the expression

$$x^2e^x - 2(x - 1)e^x.$$

SOLUTION

$x^2e^x - 2(x - 1)e^x$	Example 5, page 389
$= x^2e^x - 2(xe^x - e^x)$	Multiply factors.
$= x^2e^x - 2xe^x + 2e^x$	Multiply factors.
$= e^x(x^2 - 2x + 2)$	Factor.

EXAMPLE 4   Solving a Rational Inequality

Solve the rational inequality

$$\frac{1}{6n^2} < 0.01$$

for n, where n is a positive integer.

SOLUTION

$\dfrac{1}{6n^2} < 0.01$	Example 3, page 399
$\dfrac{1}{6n^2} < \dfrac{1}{100}$	Rewrite decimal as a fraction.
$\dfrac{100}{6n^2} < 1$	Multiply each side by 100.
$100 < 6n^2$	Multiply each side by $6n^2$ $(n > 0)$.
$\dfrac{100}{6} < n^2$	Divide each side by 6.
$\dfrac{50}{3} < n^2$	Simplify.
$\sqrt{\dfrac{50}{3}} < n$	Take positive square root of each side $(n > 0)$.

Because n is a positive integer and

$$\sqrt{\frac{50}{3}} \approx 4.08$$

n must be 5 or more. You can check this result using a graphing utility. Let $y_1 = 1/(6x^2)$ and $y_2 = 0.01$. Then use the *intersect* feature (see figure) to determine that $x \approx 4.08$. So, the solution found algebraically is correct.

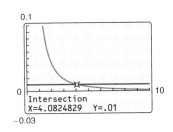

SUMMARY AND STUDY STRATEGIES

After studying this chapter, you should have acquired the following skills.
The exercise numbers are keyed to the Review Exercises that begin on page 416.
Answers to odd-numbered Review Exercises are given in the back of the text.*

Section 6.1

Review Exercises

■ Use integration by parts to find indefinite and definite integrals.

1–12

$$\int u \, dv = uv - \int v \, du$$

For integrals of the form

$$\int x^n e^{ax} \, dx$$

let $u = x^n$ and $dv = e^{ax} \, dx$.

For integrals of the form

$$\int x^n \ln x \, dx$$

let $u = \ln x$ and $dv = x^n \, dx$.

■ Find the present value of future income.

13–18

$$\text{Actual income over } t_1 \text{ years} = \int_0^{t_1} c(t) \, dt$$

$$\text{Present value} = \int_0^{t_1} c(t) e^{-rt} \, dt$$

Section 6.2

■ Use integration tables to find indefinite and definite integrals.   19–32
■ Use integration tables to solve real-life problems.   33, 34

Section 6.3

■ Use the Trapezoidal Rule and Simpson's Rule to approximate definite integrals.   35–46

Trapezoidal Rule:

$$\int_a^b f(x) \, dx \approx \left(\frac{b - a}{2n}\right)[f(x_0) + 2f(x_1) + \cdots + 2f(x_{n-1}) + f(x_n)]$$

Simpson's Rule:

$$\int_a^b f(x) \, dx \approx \left(\frac{b - a}{3n}\right)[f(x_0) + 4f(x_1) + 2f(x_2) + 4f(x_3) + \cdots + 4f(x_{n-1}) + f(x_n)]$$

* Several study aids are available to help you master the material in this chapter. The *Student Solutions Manual* and *CalcChat.com* have solutions to all odd-numbered exercises, and *CalcView.com* has video solutions for selected exercises. *LarsonAppliedCalculus.com* offers algebra help, data spreadsheets, and much more. A *Graphing Technology Guide* with step-by-step commands for a variety of graphing calculators is at *CengageBrain.com*.

Section 6.3   (continued)   **Review Exercises**

■ Analyze the approximate errors in the Trapezoidal Rule and Simpson's Rule.   *47–50*

Errors in the Trapezoidal Rule:

$$|E| \le \frac{(b-a)^3}{12n^2}[\max|f''(x)|], \quad a \le x \le b$$

Errors in Simpson's Rule:

$$|E| \le \frac{(b-a)^5}{180n^4}[\max|f^{(4)}(x)|], \quad a \le x \le b$$

Section 6.4

■ Evaluate improper integrals with infinite limits of integration.   *51–56*

$$\int_a^\infty f(x)\,dx = \lim_{b \to \infty} \int_a^b f(x)\,dx$$

$$\int_{-\infty}^b f(x)\,dx = \lim_{a \to -\infty} \int_a^b f(x)\,dx$$

$$\int_{-\infty}^\infty f(x)\,dx = \int_{-\infty}^c f(x)\,dx + \int_c^\infty f(x)\,dx$$

■ Find the area of an unbounded region.   *57–60*

■ Find the present value of a perpetuity.   *61–64*

$$\text{Present value} = P\int_0^\infty e^{-rt}\,dt = \frac{P}{r}$$

Study Strategies

■ **Use a Variety of Approaches**   To be efficient at finding antiderivatives, you need to use a variety of approaches.

1. Check to see whether the integral fits one of the basic integration formulas—you should have these formulas memorized.

2. Try an integration technique such as substitution or integration by parts to rewrite the integral in a form that fits one of the basic integration formulas.

3. Use a table of integrals.

4. Use a symbolic integration utility.

■ **Use Numerical Integration**   When solving a definite integral, remember that you cannot apply the Fundamental Theorem of Calculus unless you can find an antiderivative of the integrand. This is not always possible—even with a symbolic integration utility. In such cases, you can use a numerical technique such as the Midpoint Rule, the Trapezoidal Rule, or Simpson's Rule to approximate the value of the integral.

Review Exercises

See *CalcChat.com* for tutorial help and worked-out solutions to odd-numbered exercises.

Integration by Parts   In Exercises 1–8, use integration by parts to find the indefinite integral.

1. $\displaystyle\int (x + 1)e^x \, dx$

2. $\displaystyle\int xe^{-3x} \, dx$

3. $\displaystyle\int \frac{\ln x}{\sqrt{x}} \, dx$

4. $\displaystyle\int x \ln 4x \, dx$

5. $\displaystyle\int x\sqrt{x - 5} \, dx$

6. $\displaystyle\int \frac{x}{\sqrt{2x + 7}} \, dx$

7. $\displaystyle\int 2x^2 e^{2x} \, dx$

8. $\displaystyle\int (\ln x)^3 \, dx$

Evaluating a Definite Integral   In Exercises 9–12, use integration by parts to evaluate the definite integral.

9. $\displaystyle\int_1^e 6x \ln x \, dx$

10. $\displaystyle\int_0^4 \ln(1 + 3x) \, dx$

11. $\displaystyle\int_0^1 \frac{x}{e^{x/4}} \, dx$

12. $\displaystyle\int_{-1}^1 x^2 e^{3x} \, dx$

Finding Present Value   In Exercises 13–16, find the present value of the income c (in dollars) over t_1 years at the given annual inflation rate r.

13. $c = 20{,}000$, $r = 4\%$, $t_1 = 5$ years

14. $c = 1500t$, $r = 7\%$, $t_1 = 10$ years

15. $c = 24{,}000t$, $r = 5\%$, $t_1 = 10$ years

16. $c = 20{,}000 + 100te^{t/2}$, $r = 5\%$, $t_1 = 5$ years

17. Present Value   A company expects its income c during the next 4 years to be modeled by

$$c = 200{,}000 + 50{,}000t, \quad 0 \le t \le 4.$$

(a) Find the actual income for the business over the 4 years.

(b) Assuming an annual inflation rate of 6%, what is the present value of this income?

18. Present Value   A company expects its income c during the next 7 years to be modeled by

$$c = 400{,}000 + 175{,}000t, \quad 0 \le t \le 7.$$

(a) Find the actual income for the business over the 7 years.

(b) Assuming an annual inflation rate of 4%, what is the present value of this income?

Using Integration Tables   In Exercises 19–22, use the indicated formula from the integration table in Appendix C to find the indefinite integral.

19. $\displaystyle\int \frac{x^2}{2 + 3x} \, dx$, Formula 6

20. $\displaystyle\int \frac{1}{1 + e^{6x}} \, dx$, Formula 40

21. $\displaystyle\int \frac{1}{x\sqrt{x^2 + 64}} \, dx$, Formula 28

22. $\displaystyle\int x^5 \ln x \, dx$, Formula 43

Using Integration Tables   In Exercises 23–30, use the integration table in Appendix C to find the indefinite integral.

23. $\displaystyle\int \frac{1}{x(2 + 3x)^2} \, dx$

24. $\displaystyle\int \frac{x^2}{(7 + 2x)^3} \, dx$

25. $\displaystyle\int \frac{\sqrt{x^2 + 25}}{x} \, dx$

26. $\displaystyle\int \frac{x^2}{\sqrt{x^2 - 6}} \, dx$

27. $\displaystyle\int \frac{1}{4x^2 - 49} \, dx$

28. $\displaystyle\int (\ln 3x)^2 \, dx$

29. $\displaystyle\int \frac{\sqrt{1 + x}}{x} \, dx$

30. $\displaystyle\int \frac{1}{(x^2 - 16)^2} \, dx$

Using Integration Tables   In Exercises 31 and 32, use the integration table in Appendix C to evaluate the definite integral.

31. $\displaystyle\int_1^2 \frac{1}{x(4 + 3x)} \, dx$

32. $\displaystyle\int_{-1}^3 \frac{1}{(x^2 + 2)^{3/2}} \, dx$

33. Probability   The probability of recalling between a and b percent (in decimal form) of the material learned in a memory experiment is modeled by

$$P(a \le x \le b) = \int_a^b \frac{96}{11}\left(\frac{x}{\sqrt{9 + 16x}}\right) dx,$$

where $0 \le a \le b \le 1$. What is the probability of recalling (a) between 0% and 80% and (b) between 0% and 50% of the material?

34. Probability   The probability of locating between a and b percent of the oil and gas deposits (in decimal form) in a region is modeled by

$$P(a \le x \le b) = \int_a^b 1.5x^2 e^{x^{1.5}} \, dx, \quad 0 \le a \le b \le 1.$$

What is the probability of locating (a) between 40% and 60% and (b) between 0% and 50% of the deposits?

Using the Trapezoidal Rule and Simpson's Rule
In Exercises 35–40, use the Trapezoidal Rule and Simpson's Rule to approximate the value of the definite integral for the indicated value of n. Compare these results with the exact value of the definite integral. Round your answers to four decimal places.

35. $\int_{1}^{3} \frac{1}{x^2} dx$, $n = 4$

36. $\int_{1}^{2} \frac{1}{x^3} dx$, $n = 8$

37. $\int_{0}^{2} (x^2 + 1) dx$, $n = 8$

38. $\int_{0}^{1} (2 - x^3) dx$, $n = 4$

39. $\int_{0}^{4} e^{-x/2} dx$, $n = 4$

40. $\int_{0}^{8} \sqrt{x + 3} dx$, $n = 8$

Using the Trapezoidal Rule and Simpson's Rule In Exercises 41–46, approximate the value of the definite integral using (a) the Trapezoidal Rule and (b) Simpson's Rule for the indicated value of n. Round your answers to three decimal places.

41. $\int_{1}^{2} \frac{1}{1 + \ln x} dx$, $n = 4$

42. $\int_{0}^{2} \frac{1}{\sqrt{x^2 + 1}} dx$, $n = 8$

43. $\int_{0}^{1} \frac{x^{3/2}}{2 - x^2} dx$, $n = 4$

44. $\int_{-1}^{1} e^{x^4} dx$, $n = 4$

45. $\int_{0}^{8} \frac{3}{x^2 + 2} dx$, $n = 8$

46. $\int_{0}^{1} \sqrt{1 - x} dx$, $n = 4$

Error Analysis In Exercises 47 and 48, use the error formulas to find bounds for the error in approximating the definite integral using (a) the Trapezoidal Rule and (b) Simpson's Rule for the indicated value of n.

47. $\int_{0}^{1} e^{3x} dx$, $n = 4$ **48.** $\int_{2}^{4} \frac{1}{x - 1} dx$, $n = 8$

Error Analysis In Exercises 49 and 50, use the error formulas to find n such that the error in the approximation of the definite integral is less than 0.0001 using (a) the Trapezoidal Rule and (b) Simpson's Rule.

49. $\int_{0}^{3} x^5 dx$ **50.** $\int_{0}^{5} e^{x/5} dx$

Evaluating an Improper Integral In Exercises 51–56, determine whether the improper integral diverges or converges. Evaluate the integral if it converges.

51. $\int_{-\infty}^{-1} \frac{1}{x^5} dx$ **52.** $\int_{1}^{\infty} \frac{1}{\sqrt[4]{x}} dx$

53. $\int_{-\infty}^{0} \frac{1}{\sqrt[3]{8 - x}} dx$ **54.** $\int_{0}^{\infty} e^{-2x} dx$

55. $\int_{1}^{\infty} \frac{\ln x}{x} dx$ **56.** $\int_{0}^{\infty} \frac{e^x}{1 + e^x} dx$

Area of a Region In Exercises 57–60, find the area of the unbounded shaded region.

57. $y = e^{-x/4}$ **58.** $y = \dfrac{2x}{x^2 + 2}$

59. $y = \dfrac{3}{(1 - 3x)^{2/3}}$ **60.** $y = 4xe^{-2x^2}$

     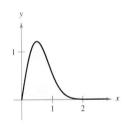

Endowment In Exercises 61 and 62, determine the amount of money required to set up a charitable endowment that pays the amount P each year indefinitely for the annual interest rate r compounded continuously.

61. $P = \$8000$, $r = 3\%$

62. $P = \$15,000$, $r = 5\%$

63. Scholarship Fund You want to start a scholarship fund at your alma mater. You plan to give one $21,000 scholarship annually beginning one year from now, and you have at most $325,000 to start the fund. You also want the scholarship to be given out indefinitely. Assuming an annual interest rate of 7% compounded continuously, do you have enough money to start the scholarship fund?

64. Present Value You are considering buying a franchise that yields a continuous flow of profit at the rate of $100,000 per year. Assuming an annual interest rate of 6% compounded continuously, what is the present value of the franchise (a) for 15 years and (b) forever?

Take this test as you would take a test in class. When you are done, check your work against the answers given in the back of the book.

In Exercises 1–3, use integration by parts to find the indefinite integral.

1. $\displaystyle\int xe^{x+1}\,dx$ **2.** $\displaystyle\int x^2 \ln 9x\,dx$ **3.** $\displaystyle\int x^2 e^{-x/3}\,dx$

4. The sales S (in billions of dollars) for Dollar General from 2008 through 2013 can be modeled by

$$S = 2.1\sqrt{t}\,\ln t - 2$$

where t is the year, with $t = 8$ corresponding to 2008.   *(Source: Dollar General Corporation)*

(a) Find the total sales for the years 2008 through 2013.

(b) Find the average sales for the years 2008 through 2013.

In Exercises 5–7, use the integration table in Appendix C to find the indefinite integral.

5. $\displaystyle\int \frac{x}{(7 + 2x)^2}\,dx$ **6.** $\displaystyle\int \frac{3x^2}{1 + e^{x^3}}\,dx$ **7.** $\displaystyle\int 2x^5\sqrt{x^4 - 9}\,dx$

In Exercises 8–10, use integration by parts or the integration table in Appendix C to evaluate the definite integral.

8. $\displaystyle\int_0^1 \ln(3 - 2x)\,dx$ **9.** $\displaystyle\int_3^6 \frac{x}{\sqrt{x - 2}}\,dx$ **10.** $\displaystyle\int_{-3}^{-1} \frac{\sqrt{x^2 + 49}}{x}\,dx$

11. Use the Trapezoidal Rule with $n = 4$ to approximate

$$\int_2^5 (x^2 - 2x)\,dx.$$

Compare your result with the exact value of the definite integral.

12. Use Simpson's Rule with $n = 4$ to approximate

$$\int_0^1 9xe^{3x}\,dx.$$

Compare your result with the exact value of the definite integral.

13. Use the error formulas to find n such that the error in the approximations of

$$\int_0^1 (2x^6 + 1)\,dx$$

is less than 0.01 using (a) the Trapezoidal Rule and (b) Simpson's Rule.

In Exercises 14–17, determine whether the improper integral diverges or converges. Evaluate the integral if it converges.

14. $\displaystyle\int_0^\infty e^{-3x}\,dx$ **15.** $\displaystyle\int_1^\infty \frac{2}{\sqrt{x}}\,dx$

16. $\displaystyle\int_{-\infty}^0 \frac{1}{(4x - 1)^{2/3}}\,dx$ **17.** $\displaystyle\int_0^\infty 2x^3 e^{-x^4}\,dx$

18. A business is expected to yield a continuous flow of profit at the rate of $600,000 per year. Assuming an annual interest rate of 5% compounded continuously, what is the present value of the business (a) for 20 years and (b) forever?

7 Functions of Several Variables

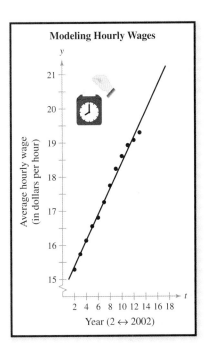

Example 3 on page 477 shows how least squares regression analysis can be used to find the best-fitting line that models hourly wages for production workers in manufacturing industries.

iStockphoto.com/mediaphotos

7.1 The Three-Dimensional Coordinate System

■ Plot points in space.
■ Find distances between points in space and find midpoints of line segments in space.
■ Write the standard forms of the equations of spheres and find the centers and radii of spheres.
■ Sketch the coordinate plane traces of surfaces.

The Three-Dimensional Coordinate System

Recall from Section 1.1 that the Cartesian plane is determined by two perpendicular real number lines called the *x*-axis and the *y*-axis. These axes together with their point of intersection (the origin) allow you to develop a two-dimensional coordinate system for identifying points in a plane. To identify a point in space, you must introduce a third dimension to the model. The geometry of this three-dimensional model is called **solid analytic geometry.**

You can construct a **three-dimensional coordinate system** by passing a *z*-axis perpendicular to both the *x*- and *y*-axes at the origin. Figure 7.1 shows the positive portion of each coordinate axis. Taken as pairs, the axes determine three **coordinate planes:** the **xy-plane,** the **xz-plane,** and the **yz-plane.** These three coordinate planes separate the three-dimensional coordinate system into eight **octants.** The first octant is the one for which all three coordinates are positive. In this three-dimensional system, a point *P* in space is determined by an ordered triple (x, y, z), where *x*, *y*, and *z* are as follows.

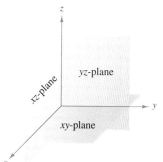

FIGURE 7.1

x = directed distance from *yz*-plane to *P*
y = directed distance from *xz*-plane to *P*
z = directed distance from *xy*-plane to *P*

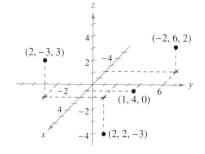

In Exercise 57 on page 426, you will model the shape of a spherical building using the standard equation of a sphere.

EXAMPLE 1 Plotting Points in Space

Plot the points in the same three-dimensional coordinate system.

a. $(2, -3, 3)$ **b.** $(-2, 6, 2)$
c. $(1, 4, 0)$ **d.** $(2, 2, -3)$

SOLUTION To plot the point $(2, -3, 3)$, notice that

$$x = 2, \quad y = -3, \quad \text{and} \quad z = 3.$$

To help visualize the point, locate the point $(2, -3)$ in the *xy*-plane (denoted by a cross in Figure 7.2). The point lies three units above the cross. You can plot the other points in a similar manner, as shown in Figure 7.2.

✓ **Checkpoint 1** *Worked-out solution available at LarsonAppliedCalculus.com*

Plot the points in the same three-dimensional coordinate system.

a. $(2, 5, 1)$ **b.** $(-2, -4, 3)$ **c.** $(4, 0, -5)$

FIGURE 7.2

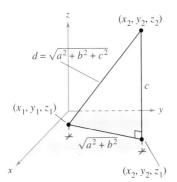

FIGURE 7.3

The Distance and Midpoint Formulas

Many of the formulas established for the two-dimensional coordinate system can be extended to three dimensions. For example, to find the distance between two points in space, you can use the Pythagorean Theorem twice with $a = |x_2 - x_1|$, $b = |y_2 - y_1|$, and $c = |z_2 - z_1|$, as shown in Figure 7.3. By doing this, you will obtain the formula for the distance between two points in space.

Distance Formula in Space

The distance d between the points (x_1, y_1, z_1) and (x_2, y_2, z_2) is

$$d = \sqrt{(x_2 - x_1)^2 + (y_2 - y_1)^2 + (z_2 - z_1)^2}.$$

EXAMPLE 2 Finding the Distance Between Two Points in Space

Find the distance between $(1, 0, 2)$ and $(2, 4, -3)$.

SOLUTION

$$
\begin{aligned}
d &= \sqrt{(x_2 - x_1)^2 + (y_2 - y_1)^2 + (z_2 - z_1)^2} &&\text{Distance Formula in Space} \\
&= \sqrt{(2 - 1)^2 + (4 - 0)^2 + (-3 - 2)^2} &&\text{Substitute.} \\
&= \sqrt{1 + 16 + 25} &&\text{Simplify.} \\
&= \sqrt{42} &&\text{Simplify.}
\end{aligned}
$$

✓ **Checkpoint 2** Worked-out solution available at LarsonAppliedCalculus.com

Find the distance between $(2, 3, -1)$ and $(0, 5, 3)$.

Notice the similarity between the Distance Formulas in the plane and in space. The Midpoint Formulas in the plane and in space are also similar.

Midpoint Formula in Space

The midpoint of the line segment joining the points (x_1, y_1, z_1) and (x_2, y_2, z_2) is

$$\text{Midpoint} = \left(\frac{x_1 + x_2}{2}, \frac{y_1 + y_2}{2}, \frac{z_1 + z_2}{2} \right).$$

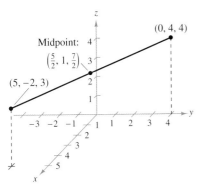

FIGURE 7.4

EXAMPLE 3 Using the Midpoint Formula in Space

Find the midpoint of the line segment joining

$$(5, -2, 3) \quad \text{and} \quad (0, 4, 4).$$

SOLUTION Using the Midpoint Formula in space, the midpoint is

$$\left(\frac{5 + 0}{2}, \frac{-2 + 4}{2}, \frac{3 + 4}{2} \right) = \left(\frac{5}{2}, 1, \frac{7}{2} \right)$$

as shown in Figure 7.4.

✓ **Checkpoint 3** Worked-out solution available at LarsonAppliedCalculus.com

Find the midpoint of the line segment joining

$$(3, -2, 0) \quad \text{and} \quad (-8, 6, -4).$$

The Equation of a Sphere

A **sphere** with center at (h, k, j) and radius r is defined to be the set of all points (x, y, z) such that the distance between (x, y, z) and (h, k, j) is r, as shown in Figure 7.5. Using the Distance Formula, this condition can be written as

$$\sqrt{(x - h)^2 + (y - k)^2 + (z - j)^2} = r.$$

By squaring each side of this equation, you obtain the standard equation of a sphere.

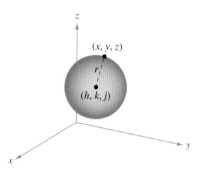

Sphere: Radius r, Center (h, k, j)
FIGURE 7.5

Standard Equation of a Sphere

The **standard equation of a sphere** with center at (h, k, j) and radius r is

$$(x - h)^2 + (y - k)^2 + (z - j)^2 = r^2.$$

EXAMPLE 4   Finding the Equation of a Sphere

Find the standard equation of the sphere with center at $(2, 4, 3)$ and radius 3. Does this sphere intersect the xy-plane?

SOLUTION

$$(x - h)^2 + (y - k)^2 + (z - j)^2 = r^2 \qquad \text{Write standard equation.}$$
$$(x - 2)^2 + (y - 4)^2 + (z - 3)^2 = 3^2 \qquad \text{Substitute.}$$
$$(x - 2)^2 + (y - 4)^2 + (z - 3)^2 = 9 \qquad \text{Simplify.}$$

In Figure 7.6, note that the center of the sphere lies three units above the xy-plane. The sphere has a radius of 3, so it must intersect the xy-plane—at the point $(2, 4, 0)$.

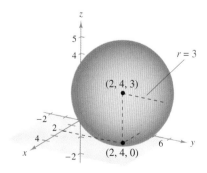

FIGURE 7.6

✓ *Checkpoint 4*   *Worked-out solution available at LarsonAppliedCalculus.com*

Find the standard equation of the sphere with center at $(4, 3, 2)$ and radius 5.   ■

EXAMPLE 5   Finding the Equation of a Sphere

Find the equation of the sphere that has the points

$$(3, -2, 6) \quad \text{and} \quad (-1, 4, 2)$$

as endpoints of a diameter.

SOLUTION   By the Midpoint Formula, the center of the sphere is

$$(h, k, j) = \left(\frac{3 + (-1)}{2}, \frac{-2 + 4}{2}, \frac{6 + 2}{2} \right) \qquad \text{Apply Midpoint Formula.}$$

$$= (1, 1, 4). \qquad \text{Simplify.}$$

By the Distance Formula, the radius is

$$r = \sqrt{(3 - 1)^2 + (-2 - 1)^2 + (6 - 4)^2} \qquad \text{Apply Distance Formula.}$$

$$= \sqrt{4 + 9 + 4} \qquad \text{Simplify.}$$

$$= \sqrt{17}. \qquad \text{Simplify.}$$

So, the standard equation of the sphere is

$$(x - h)^2 + (y - k)^2 + (z - j)^2 = r^2 \qquad \text{Write formula for a sphere.}$$

$$(x - 1)^2 + (y - 1)^2 + (z - 4)^2 = 17. \qquad \text{Substitute.}$$

✓ **Checkpoint 5**   *Worked-out solution available at LarsonAppliedCalculus.com*

Find the equation of the sphere that has the points $(-2, 5, 7)$ and $(4, 1, -3)$ as endpoints of a diameter.

EXAMPLE 6   Finding the Center and Radius of a Sphere

Find the center and radius of the sphere whose equation is

$$x^2 + y^2 + z^2 - 2x + 4y - 6z + 8 = 0.$$

SOLUTION   You can obtain the standard equation of the sphere by completing the square. To do this, begin by grouping terms with the same variable. Then add "the square of half the coefficient of each linear term" to each side of the equation. So, to complete the square of $(x^2 - 2x)$, add $\left[\frac{1}{2}(-2) \right]^2 = 1$ to each side. To complete the square of $(y^2 + 4y)$, add $\left[\frac{1}{2}(4) \right]^2 = 4$ to each side. To complete the square of $(z^2 - 6z)$, add $\left[\frac{1}{2}(-6) \right]^2 = 9$ to each side.

$$x^2 + y^2 + z^2 - 2x + 4y - 6z + 8 = 0$$

$$(x^2 - 2x + \quad) + (y^2 + 4y + \quad) + (z^2 - 6z + \quad) = -8$$

$$(x^2 - 2x + 1) + (y^2 + 4y + 4) + (z^2 - 6z + 9) = -8 + 1 + 4 + 9$$

$$(x - 1)^2 + (y + 2)^2 + (z - 3)^2 = 6$$

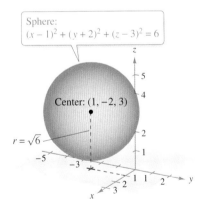

Sphere:
$(x - 1)^2 + (y + 2)^2 + (z - 3)^2 = 6$

Center: $(1, -2, 3)$

$r = \sqrt{6}$

FIGURE 7.7

So, the center of the sphere is $(1, -2, 3)$, and its radius is $\sqrt{6}$, as shown in Figure 7.7.

✓ **Checkpoint 6**   *Worked-out solution available at LarsonAppliedCalculus.com*

Find the center and radius of the sphere whose equation is

$$x^2 + y^2 + z^2 + 6x - 8y + 2z - 10 = 0. \qquad ■$$

Note in Example 6 that the points satisfying the equation of the sphere are "surface points," not "interior points." In general, the collection of points satisfying an equation involving x, y, and z is called a **surface in space.**

Traces of Surfaces

Finding the intersection of a surface with one of the three coordinate planes (or with a plane parallel to one of the three coordinate planes) helps visualize the surface. Such an intersection is called a **trace** of the surface. For example, the *xy*-trace of a surface consists of all points that are common to both the surface *and* the *xy*-plane. Similarly, the *xz*-trace of a surface consists of all points that are common to both the surface *and* the *xz*-plane.

EXAMPLE 7   Finding a Trace of a Surface

Sketch the *xy*-trace of the sphere given by $(x - 3)^2 + (y - 2)^2 + (z + 4)^2 = 5^2$.

SOLUTION   To find the *xy*-trace of this surface, use the fact that every point in the *xy*-plane has a *z*-coordinate of zero. By substituting $z = 0$ into the original equation, the resulting equation will represent the intersection of the surface with the *xy*-plane.

$$(x - 3)^2 + (y - 2)^2 + (z + 4)^2 = 5^2 \qquad \text{Write original equation.}$$
$$(x - 3)^2 + (y - 2)^2 + (0 + 4)^2 = 25 \qquad \text{Let } z = 0 \text{ to find } xy\text{-trace.}$$
$$(x - 3)^2 + (y - 2)^2 + 16 = 25 \qquad \text{Simplify.}$$
$$(x - 3)^2 + (y - 2)^2 = 9 \qquad \text{Subtract 16 from each side.}$$
$$(x - 3)^2 + (y - 2)^2 = 3^2 \qquad \text{Equation of circle}$$

From this equation, you can see that the *xy*-trace is a circle of radius 3, as shown in Figure 7.8.

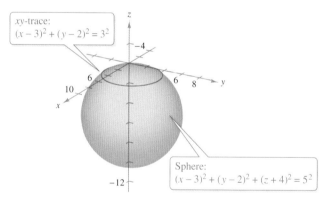

FIGURE 7.8

✓ *Checkpoint 7*   *Worked-out solution available at LarsonAppliedCalculus.com*

Find the equation of the *xy*-trace of the sphere given by

$$(x + 1)^2 + (y - 2)^2 + (z + 3)^2 = 5^2.$$

SUMMARIZE   (Section 7.1)

1. State the Distance Formula in space *(page 421)*. For an example of the Distance Formula in space, see Example 2.

2. State the Midpoint Formula in space *(page 421)*. For an example of the Midpoint Formula in space, see Example 3.

3. State the standard equation of a sphere *(page 422)*. For examples of finding equations of spheres, see Examples 4 and 5.

4. Explain what is meant by the trace of a surface *(page 424)*. For an example of finding the trace of a surface, see Example 7.

SKILLS WARM UP 7.1

The following warm-up exercises involve skills that were covered in earlier sections. You will use these skills in the exercise set for this section. For additional help, review Sections 1.1 and 1.2.

In Exercises 1–6, find the distance between the points and find the midpoint of the line segment joining the points.

1. $(5, 1), (3, 5)$ **2.** $(2, 3), (-1, -1)$ **3.** $(-1, -2), (3, 2)$

4. $(-4, 3), (2, -1)$ **5.** $(-5, 4), (-5, -4)$ **6.** $(-3, 6), (-3, -2)$

In Exercises 7 and 8, find the standard form of the equation of the circle with the given characteristics.

7. Center: $(2, 3)$; radius: 2 **8.** Endpoints of a diameter: $(4, 0), (-2, 8)$

Exercises 7.1

See *CalcChat.com* for tutorial help and worked-out solutions to odd-numbered exercises.

 Plotting Points in Space In Exercises 1–4, plot the points in the same three-dimensional coordinate system. *See Example 1.*

1. $(-3, 0, -1), (0, -1, -4), (-2, -3, 0), (1, 3, 4)$

2. $(0, 1, -5), (4, 0, 5), \left(-1, \frac{1}{2}, 0\right), (-3, -1, 3)$

3. $(2, 1, 3), (-1, 2, 1), (3, -2, 5), \left(\frac{3}{2}, 4, -2\right)$

4. $(-5, -2, 2), (5, -2, -2), (1, 3, 1), (-2, 4, -3)$

 Finding the Coordinates of a Point in Space In Exercises 5–8, find the coordinates of the point.

5. The point is located three units behind the yz-plane, four units to the right of the xz-plane, and five units above the xy-plane.

6. The point is located seven units in front of the yz-plane, two units to the left of the xz-plane, and one unit below the xy-plane.

7. The point is located on the x-axis, 10 units in front of the yz-plane.

8. The point is located in the yz-plane, three units to the right of the xz-plane, and two units above the xy-plane.

9. Think About It What is the z-coordinate of any point in the xy-plane?

10. Think About It What is the y-coordinate of any point in the xz-plane?

 Finding the Distance Between Two Points in Space In Exercises 11–14, find the distance between the two points. *See Example 2.*

11. $(4, 1, 5), (8, 2, 6)$

12. $(8, -2, 2), (8, -2, 4)$

13. $(6, -9, 1), (-2, -1, 5)$

14. $(-4, -1, 1), (2, -1, 5)$

 Using the Midpoint Formula in Space In Exercises 15–18, find the midpoint of the line segment joining the two points. *See Example 3.*

15. $(4, 0, -6), (8, 8, 20)$ **16.** $(0, -2, 5), (4, 2, 7)$

17. $(6, -4, 2), (-2, 1, 3)$ **18.** $(-9, -2, 5), (2, 3, -7)$

Using the Midpoint Formula in Space In Exercises 19–22, find (x, y, z).

19.

20.

21.

22.

 Identifying Triangles In Exercises 23–26, find the lengths of the sides of the triangle with the given vertices, and determine whether the triangle is a right triangle, an isosceles triangle, or neither.

23. $(0, 0, 0), (2, 2, 1), (2, -4, 4)$

24. $(5, 3, 4), (7, 1, 3), (3, 5, 3)$

25. $(-1, 0, -2), (-1, 5, 2), (-3, -1, 1)$

26. $(5, 0, 0), (0, 2, 0), (0, 0, -3)$

27. Think About It The triangle in Exercise 23 is translated five units downward in the negative z-direction. Determine the coordinates of the translated triangle.

28. Think About It The triangle in Exercise 24 is translated three units to the right in the positive y-direction. Determine the coordinates of the translated triangle.

 Finding the Equation of a Sphere In Exercises 29–38, find the standard equation of the sphere with the given characteristics. *See Examples 4 and 5.*

29.   30.

31. 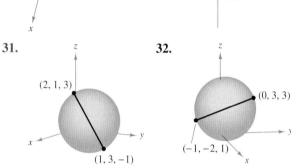   32.

33. Center: $(3, -2, -3)$; radius: 4

34. Center: $(4, -1, 1)$; radius: 5

35. Endpoints of a diameter: $(-3, 6, 1), (1, -5, 2)$

36. Endpoints of a diameter: $(2, 9, 11), (2, -3, -6)$

37. Center: $(-4, 3, 6)$; tangent to the xy-plane

38. Center: $(1, 2, 0)$; tangent to the yz-plane

 Finding the Center and Radius of a Sphere In Exercises 39–44, find the center and radius of the sphere. *See Example 6.*

39. $x^2 + y^2 + z^2 - 12x = 0$

40. $x^2 + y^2 + z^2 - 8y = 0$

41. $x^2 + y^2 + z^2 + 4x - 2y + 8z - 4 = 0$

42. $x^2 + y^2 + z^2 - 4y + 6z + 4 = 0$

43. $2x^2 + 2y^2 + 2z^2 - 4x - 12y - 8z + 3 = 0$

44. $4x^2 + 4y^2 + 4z^2 - 8x + 16y + 11 = 0$

 Finding a Trace of a Sphere In Exercises 45–48, find the equation of the xy-trace of the sphere. Then sketch the xy-trace of the sphere. *See Example 7.*

45. $(x - 1)^2 + (y + 3)^2 + (z - 2)^2 = 29$

46. $(x + 1)^2 + (y + 2)^2 + (z - 2)^2 = 20$

47. $x^2 + y^2 + z^2 - 6x - 10y + 6z + 30 = 0$

48. $x^2 + y^2 + z^2 - 4y + 2z - 60 = 0$

Finding a Trace of a Sphere In Exercises 49–52, find the equation of the yz-trace of the sphere. Then sketch the yz-trace of the sphere. *See Example 7*

49. $x^2 + (y - 4)^2 + z^2 = 16$

50. $(x + 2)^2 + (y - 3)^2 + z^2 = 13$

51. $x^2 + y^2 + z^2 - 4x - 4y - 6z - 12 = 0$

52. $x^2 + y^2 + z^2 - 6x - 10y + 6z + 30 = 0$

Finding a Trace of a Surface In Exercises 53–56, sketch the trace of the intersection of each plane with the given sphere.

53. $x^2 + y^2 + z^2 = 25$

   (a) $z = 3$ (b) $x = 4$

54. $x^2 + y^2 + z^2 = 169$

   (a) $x = 5$ (b) $y = 12$

55. $x^2 + y^2 + z^2 - 4x - 6y + 9 = 0$

   (a) $x = 2$ (b) $y = 3$

56. $x^2 + y^2 + z^2 - 8x - 6z + 16 = 0$

   (a) $x = 4$ (b) $z = 3$

57. Architecture A spherical building has a diameter of 165 feet. The center of the building is placed at the origin of a three-dimensional coordinate system. What is the equation of the sphere that models the shape of the building?

58. HOW DO YOU SEE IT? Crystals are classified according to their symmetry.

(a) Crystals shaped like cubes are classified as isometric. The vertices of an isometric crystal mapped onto a three-dimensional coordinate system are shown in the figure. Determine (x, y, z).

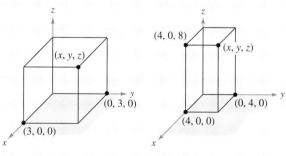

Figure for (a)   Figure for (b)

(b) Crystals shaped like rectangular prisms are classified as tetragonal. The vertices of a tetragonal crystal mapped onto a three-dimensional coordinate system are shown in the figure. Determine (x, y, z).

7.2 Surfaces in Space

- Sketch planes in space.
- Draw planes in space with different numbers of intercepts.
- Classify quadric surfaces in space.

Equatorial
radius = 3963 mi

Polar
radius = 3950 mi

In Exercise 49 on page 435,
you will write an equation of an ellipsoid
that models the shape of the Earth.

Equations of Planes in Space

In Section 7.1, you studied one type of surface in space—a sphere. In this section, you will study a second type—a plane in space. The graph of the equation

$$ax + by + cz = d \qquad \text{Equation of a plane}$$

is a **plane in space** with a, b, and c not all zero. Note the similarity of this equation to an equation of a line, $ax + by = c$, in the two-dimensional coordinate system. In fact, when you intersect the plane represented by this equation with each of the three coordinate planes, you will obtain traces that are lines, as shown in Figure 7.9.

In Figure 7.9, the points where the plane intersects the three coordinate axes are the x-, y-, and z-intercepts of the plane. By connecting these three points, you can form a triangular region, which helps you visualize the plane in space.

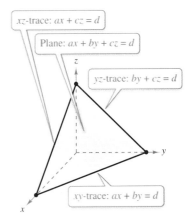

xz-trace: $ax + cz = d$

Plane: $ax + by + cz = d$

yz-trace: $by + cz = d$

xy-trace: $ax + by = d$

FIGURE 7.9

EXAMPLE 1 Sketching a Plane in Space

Find the x-, y-, and z-intercepts of the plane

$$3x + 2y + 4z = 12.$$

Then sketch the plane.

SOLUTION To find the x-intercept, let both y and z be zero.

$$3x + 2(0) + 4(0) = 12 \qquad \text{Substitute 0 for } y \text{ and } z.$$
$$3x = 12 \qquad \text{Simplify.}$$
$$x = 4 \qquad \text{Solve for } x.$$

So, the x-intercept is $(4, 0, 0)$. To find the y-intercept, let x and z be zero and conclude that $y = 6$. So, the y-intercept is $(0, 6, 0)$. Similarly, by letting x and y be zero, you can determine that $z = 3$ and that the z-intercept is $(0, 0, 3)$. Figure 7.10 shows the triangular portion of the plane formed by connecting the three intercepts

$$(4, 0, 0), \quad (0, 6, 0), \quad \text{and} \quad (0, 0, 3).$$

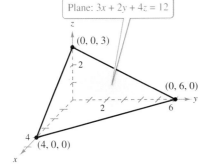

Plane: $3x + 2y + 4z = 12$

$(0, 0, 3)$

$(0, 6, 0)$

$(4, 0, 0)$

Sketch Made by Connecting
Intercepts: $(4, 0, 0)$, $(0, 6, 0)$, $(0, 0, 3)$
FIGURE 7.10

✓ **Checkpoint 1** *Worked-out solution available at LarsonAppliedCalculus.com*

Find the x-, y-, and z-intercepts of the plane

$$2x + 4y + z = 8.$$

Then sketch the plane. ■

iStockphoto.com/4x6

Drawing Planes in Space

The planes shown in Figures 7.9 and 7.10 have three intercepts. When this occurs, you can draw the plane by sketching the triangular region formed by connecting the three intercepts.

It is possible for a plane in space to have fewer than three intercepts. This occurs when one or two of the coefficients in the equation $ax + by + cz = d$ is zero. Figure 7.11 shows some planes in space that have only one intercept, and Figure 7.12 shows some that have only two intercepts. In each figure, note the use of dashed lines and shading to give the illusion of three dimensions.

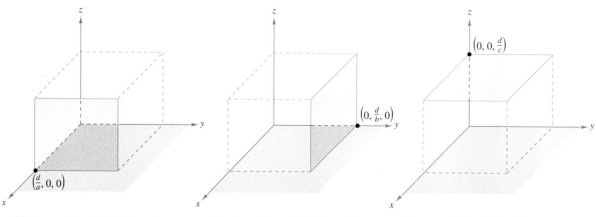

Plane $ax = d$ is parallel to the yz-plane.   Plane $by = d$ is parallel to the xz-plane.   Plane $cz = d$ is parallel to the xy-plane.

Planes Parallel to Coordinate Planes
FIGURE 7.11

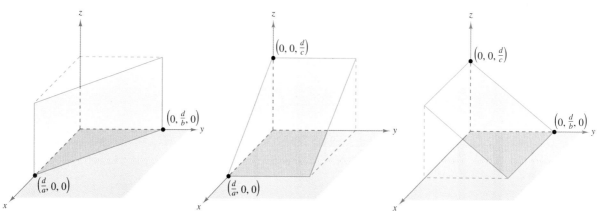

Plane $ax + by = d$ is parallel to the z-axis.   Plane $ax + cz = d$ is parallel to the y-axis.   Plane $by + cz = d$ is parallel to the x-axis.

Planes Parallel to Coordinate Axes
FIGURE 7.12

When an equation of a plane has a missing variable, such as

$$2x + z = 1 \qquad \text{See Figure 7.13.}$$

the plane must be *parallel to the axis* represented by the missing variable, as shown in Figure 7.12. When two variables are missing from an equation of a plane, the plane is *parallel to the coordinate plane* represented by the missing variables, as shown in Figure 7.11.

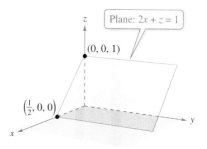

Plane $2x + z = 1$ is parallel to the y-axis.
FIGURE 7.13

Quadric Surfaces

A third common type of surface in space is a **quadric surface.** Quadric surfaces are the three-dimensional analogs of conic sections. The equation of a quadric surface in space is a second-degree equation in three variables, such as

$$Ax^2 + By^2 + Cz^2 + Dx + Ey + Fz + G = 0.$$
Second-degree equation

There are six basic types of quadric surfaces.

1. Elliptic cone   2. Elliptic paraboloid

3. Hyperbolic paraboloid   4. Ellipsoid

5. Hyperboloid of one sheet   6. Hyperboloid of two sheets

The six types are summarized on the next two pages. Notice that each surface is pictured with two types of three-dimensional sketches. The computer-generated sketches use traces with hidden lines to give the illusion of three dimensions. The artist-rendered sketches use shading to create the same illusion.

All of the quadric surfaces on the next two pages are centered at the origin and have axes along the coordinate axes. Moreover, only one of several possible orientations of each surface is shown. When the surface has a different center or is oriented along a different axis, its standard equation will change accordingly. For instance, the ellipsoid

$$\frac{x^2}{1^2} + \frac{y^2}{3^2} + \frac{z^2}{2^2} = 1$$

has $(0, 0, 0)$ as its center, but the ellipsoid

$$\frac{(x-2)^2}{1^2} + \frac{(y+1)^2}{3^2} + \frac{(z-4)^2}{2^2} = 1$$

has $(2, -1, 4)$ as its center. A computer-generated graph of the first ellipsoid is shown in Figure 7.14.

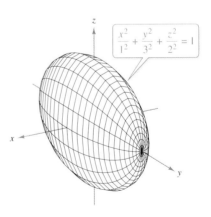

FIGURE 7.14

TECH TUTOR

If you have access to a three-dimensional graphing utility, try using it to graph the surface in Figure 7.14. When you do this, you will discover that sketching surfaces in space is not a simple task—even with a graphing utility.

Elliptic Cone

$$\frac{x^2}{a^2} + \frac{y^2}{b^2} - \frac{z^2}{c^2} = 0$$

Trace	*Plane*
Ellipse	Parallel to xy-plane
Hyperbola	Parallel to xz-plane
Hyperbola	Parallel to yz-plane

The axis of the cone corresponds to the variable whose coefficient is negative. The traces in the coordinate planes parallel to this axis are intersecting lines.

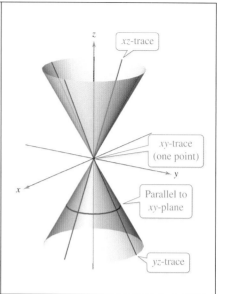

Elliptic Paraboloid

$$z = \frac{x^2}{a^2} + \frac{y^2}{b^2}$$

Trace	*Plane*
Ellipse	Parallel to xy-plane
Parabola	Parallel to xz-plane
Parabola	Parallel to yz-plane

The axis of the paraboloid corresponds to the variable raised to the first power.

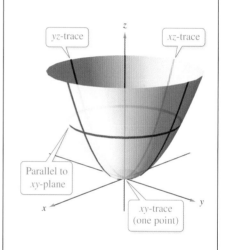

Hyperbolic Paraboloid

$$z = \frac{y^2}{b^2} - \frac{x^2}{a^2}$$

Trace	*Plane*
Hyperbola	Parallel to xy-plane
Parabola	Parallel to xz-plane
Parabola	Parallel to yz-plane

The axis of the paraboloid corresponds to the variable raised to the first power.

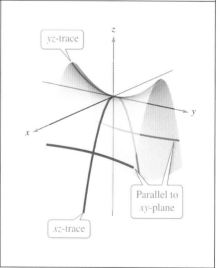

Ellipsoid

$$\frac{x^2}{a^2} + \frac{y^2}{b^2} + \frac{z^2}{c^2} = 1$$

Trace	Plane
Ellipse	Parallel to xy-plane
Ellipse	Parallel to xz-plane
Ellipse	Parallel to yz-plane

The surface is a sphere when the coefficients a, b, and c are equal and nonzero.

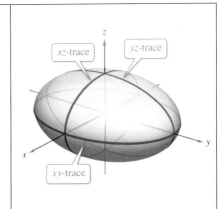

Hyperboloid of One Sheet

$$\frac{x^2}{a^2} + \frac{y^2}{b^2} - \frac{z^2}{c^2} = 1$$

Trace	Plane
Ellipse	Parallel to xy-plane
Hyperbola	Parallel to xz-plane
Hyperbola	Parallel to yz-plane

The axis of the hyperboloid corresponds to the variable whose coefficient is negative.

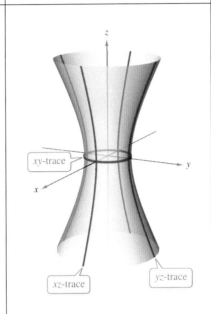

Hyperboloid of Two Sheets

$$\frac{z^2}{c^2} - \frac{x^2}{a^2} - \frac{y^2}{b^2} = 1$$

Trace	Plane
Ellipse	Parallel to xy-plane
Hyperbola	Parallel to xz-plane
Hyperbola	Parallel to yz-plane

The axis of the hyperboloid corresponds to the variable whose coefficient is positive. There is no trace in the coordinate plane perpendicular to this axis.

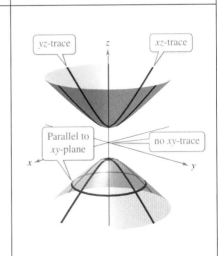

When classifying quadric surfaces, note that the two types of paraboloids have one variable raised to the first power. The other four types of quadric surfaces have equations that are of second degree in *all* three variables.

EXAMPLE 2 Classifying a Quadric Surface

Describe the traces of the surface given by $x - y^2 - z^2 = 0$ in the xy-plane, the xz-plane, and the plane given by $x = 1$. Then classify the surface.

SOLUTION Because x is raised only to the first power, the surface is a paraboloid whose axis is the x-axis. In standard form, the equation is $x = y^2 + z^2$. The traces in the xy-plane, the xz-plane, and the plane given by $x = 1$ are as shown.

Trace in xy-plane $(z = 0)$: $x = y^2$ Parabola

Trace in xz-plane $(y = 0)$: $x = z^2$ Parabola

Trace in plane $x = 1$: $y^2 + z^2 = 1$ Circle

These three traces are shown in Figure 7.15.

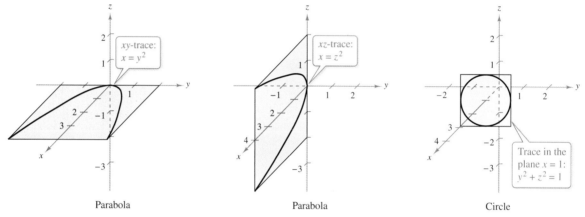

FIGURE 7.15

From the traces, you can see that the surface is an elliptic (or circular) paraboloid, as shown in Figure 7.16.

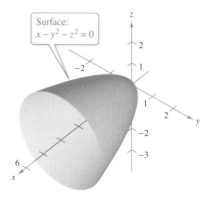

Elliptic Paraboloid
FIGURE 7.16

✓ *Checkpoint 2*   *Worked-out solution available at LarsonAppliedCalculus.com*

Describe the traces of the surface given by $x^2 + y^2 - z^2 = 1$ in the xy-plane, the yz-plane, the xz-plane, and the plane given by $z = 3$. Then classify the surface. ■

EXAMPLE 3 **Classifying Quadric Surfaces**

Classify the surface given by each equation.

a. $x^2 - 4y^2 - 4z^2 - 4 = 0$

b. $x^2 + 4y^2 + z^2 - 4 = 0$

SOLUTION

a. The equation $x^2 - 4y^2 - 4z^2 - 4 = 0$ can be written in standard form as

$$\frac{x^2}{4} - y^2 - z^2 = 1.$$ Standard form

From the standard form, you can see that the graph is a hyperboloid of two sheets, with the x-axis as its axis, as shown in Figure 7.17(a).

b. The equation $x^2 + 4y^2 + z^2 - 4 = 0$ can be written in standard form as

$$\frac{x^2}{4} + y^2 + \frac{z^2}{4} = 1.$$ Standard form

From the standard form, you can see that the graph is an ellipsoid, as shown in Figure 7.17(b).

(a)

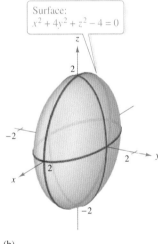

(b)

FIGURE 7.17

✓ **Checkpoint 3**   *Worked-out solution available at LarsonAppliedCalculus.com*

Classify the surface given by each equation.

a. $4x^2 + 9y^2 - 36z = 0$

b. $36x^2 + 16y^2 - 144z^2 = 0$

SUMMARIZE   (Section 7.2)

1. Describe how to sketch a plane in space *(pages 427 and 428)*. For an example of sketching a plane in space, see Example 1.

2. List the six basic types of quadric surfaces *(page 429)*. For examples of classifying quadric surfaces, see Examples 2 and 3.

In Exercises 1–4, find the x- and y-intercepts of the graph of the equation.

1. $3x + 4y = 12$

2. $6x + y = -8$

3. $-2x + y = -2$

4. $3x - 2y = 9$

In Exercises 5 and 6, write the equation of the sphere in standard form.

5. $18x^2 + 18y^2 + 18z^2 = 18$

6. $9x^2 + 9y^2 + 9z^2 = 36$

Exercises 7.2   See *CalcChat.com* for tutorial help and worked-out solutions to odd-numbered exercises.

Sketching a Plane in Space In Exercises 1–12, find the x-, y-, and z-intercepts of the plane. Then sketch the plane. *See Example 1.*

1. $x + y + z = 3$

2. $x + 2y + z = 4$

3. $4x + 2y + 6z = 12$

4. $3x + 6y + 2z = 6$

5. $2x - y + 3z = 4$

6. $5x - 2y + 4z = 10$

7. $z = 8$

8. $x = 5$

9. $x + y = 6$

10. $x + 2y = 4$

11. $x - 3z = 3$

12. $5y - 3z = 15$

Comparing Planes In Exercises 13–22, determine whether the planes

$$a_1x + b_1y + c_1z = d_1$$

and

$$a_2x + b_2y + c_2z = d_2$$

are parallel, perpendicular, or neither. The planes are parallel when there exists a nonzero constant k such that $a_1 = ka_2$, $b_1 = kb_2$, and $c_1 = kc_2$, and are perpendicular when $a_1a_2 + b_1b_2 + c_1c_2 = 0$.

13. $5x - 3y + z = 4,\ x + 4y + 7z = 1$

14. $3x + y - 4z = 3,\ -9x - 3y + 12z = 4$

15. $x - 5y - z = 1,\ 5x - 25y - 5z = -3$

16. $x + 3y - 2z = 6,\ -2x - 6y + 4z = -12$

17. $x + 3y + z = 7,\ x - 5z = 0$

18. $2x - z = 1,\ 4x + y + 8z = 10$

19. $x + 2y = 3,\ 4x + 8y = 5$

20. $2x + y = 3,\ 3x - 5z = 0$

21. $x = 3,\ z = -1$

22. $x = -2,\ y = 4$

Matching In Exercises 23–28, match the equation with its graph. [The graphs are labeled (a)–(f).]

23. $\dfrac{x^2}{9} + \dfrac{y^2}{16} + \dfrac{z^2}{9} = 1$

24. $15x^2 - 4y^2 + 15z^2 = -4$

25. $4x^2 - y^2 + 4z^2 = 4$

26. $y^2 = 4x^2 + 9z^2$

27. $4x^2 - 4y + z^2 = 0$

28. $4x^2 - y^2 + 4z = 0$

Classifying a Quadric Surface In Exercises 29–34, describe the traces of the surface in the given planes. Then classify the surface. *See Example 2.*

29. $z = x^2 - y^2$

   (a) xy-plane   (b) $x = 3$ (c) xz-plane

30. $y = x^2 + z^2$

   (a) xy-plane   (b) $y = 1$ (c) yz-plane

31. $\dfrac{x^2}{4} + y^2 + z^2 = 1$

   (a) xy-plane   (b) xz-plane   (c) yz-plane

32. $y^2 + z^2 - x^2 = 1$

   (a) xy-plane   (b) xz-plane   (c) yz-plane

33. $z^2 - \dfrac{x^2}{3} - \dfrac{y^2}{8} = 1$

   (a) xz-plane   (b) $x = 2$ (c) $z = -4$

34. $y^2 + \dfrac{z^2}{4} - x^2 = 0$

   (a) $y = -1$ (b) $z = 4$ (c) yz-plane

Classifying a Quadric Surface In Exercises 35–48, classify the quadric surface. *See Example 3.*

35. $x^2 + \dfrac{y^2}{4} + z^2 = 1$

36. $z = y^2 - \dfrac{x^2}{9}$

37. $z = \dfrac{x^2}{36} + \dfrac{y^2}{49}$

38. $\dfrac{x^2}{9} + \dfrac{y^2}{16} + \dfrac{z^2}{16} = 1$

39. $z^2 = x^2 + \dfrac{y^2}{64} + 1$

40. $4y = x^2 + z^2$

41. $x^2 - y^2 + z = 0$

42. $9x^2 + 4y^2 - 8z^2 = 72$

43. $2x^2 - y^2 + 2z^2 = -4$

44. $3z = -y^2 + x^2$

45. $z^2 = 9x^2 + y^2$

46. $4z = 16x^2 + 8y^2$

47. $25x^2 + 25y^2 - z^2 = 5$

48. $z^2 = 2x^2 + 2y^2$

49. Physical Science Because of the forces caused by its rotation, Earth is an oblate ellipsoid rather than a sphere. The equatorial radius is 3963 miles and the polar radius is 3950 miles. Find an equation of the ellipsoid. (Assume that the center of Earth is at the origin and that the trace formed by the plane $z = 0$ corresponds to the equator.)

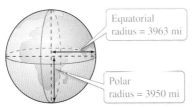

Equatorial radius = 3963 mi

Polar radius = 3950 mi

50. **HOW DO YOU SEE IT?** The four figures are graphs of the quadric surface $z = x^2 + y^2$. Match each of the four graphs with the point in space from which the paraboloid is viewed. The four points are $(0, 0, 20)$, $(0, 20, 0)$, $(20, 0, 0)$, and $(10, 10, 20)$.

(a)   (b)

(c)   (d)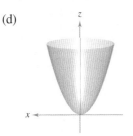

51. Modeling Data Personal consumption expenditures (in billions of dollars) for several types of recreation from 2008 through 2013 are shown in the table, where x is the expenditures on amusement parks and campgrounds, y is the expenditures on live entertainment (excluding sports), and z is the expenditures on spectator sports.

Year	2008	2009	2010	2011	2012	2013
x	38.3	37.2	38.8	41.3	44.4	46.5
y	24.3	25.2	26.3	28.3	28.4	28.7
z	19.4	18.8	19.2	20.4	21.2	22.3

Spreadsheet at LarsonAppliedCalculus.com

A model for the data in the table is given by

$$-0.391x + 0.069y + z = 6.$$

(Source: U.S. Bureau of Economic Analysis)

(a) Add a fifth row to the table using the model to approximate z for the given values of x and y. Compare the approximations with the actual values of z.

(b) According to this model, increases in expenditures of recreation types y and z would correspond to what kind of change in expenditures of recreation type x?

7.3 Functions of Several Variables

■ Evaluate functions of several variables.
■ Find the domains and ranges of functions of two variables.
■ Read contour maps and sketch level curves of functions of two variables.
■ Use functions of several variables to answer questions about real-life situations.

In Exercise 51 on page 443, you will read a weather map and identify areas of high and low pressure.

Functions of Several Variables

So far in this text, you have studied functions of a single (independent) variable. Many quantities in business, science, and technology, however, are functions of two or more variables. For instance, the demand function for a product is often dependent on the price *and* the advertising, rather than on the price alone. The notation for a function of two or more variables is similar to that for a function of a single variable. Here are two examples.

$$z = f(\underbrace{x, y}_{\text{2 variables}}) = x^2 + xy \qquad \text{Function of two variables}$$

and

$$w = f(\underbrace{x, y, z}_{\text{3 variables}}) = x + 2y - 3z \qquad \text{Function of three variables}$$

Definition of a Function of Two Variables

Let D be a set of ordered pairs of real numbers. If to each ordered pair (x, y) in D there corresponds a unique real number $f(x, y)$, then f is a **function of x and y.** The set D is the **domain** of f, and the corresponding set of values for $f(x, y)$ is the **range** of f. Functions of three, four, or more variables are defined similarly.

For the function $z = f(x, y)$, x and y are the **independent variables** and z is the **dependent variable.**

EXAMPLE 1   Evaluating Functions of Several Variables

a. For $f(x, y) = 2x^2 - y^2$, find $f(2, 3)$.

b. For $f(x, y, z) = e^x(y + z)$, find $f(0, -1, 4)$.

SOLUTION

a. Replace x with 2 and y with 3 in $f(x, y) = 2x^2 - y^2$ and simplify.

$$f(2, 3) = 2(2)^2 - (3)^2 = 8 - 9 = -1$$

b. Replace x with 0, y with -1, and z with 4 in $f(x, y, z) = e^x(y + z)$ and simplify.

$$f(0, -1, 4) = e^0(-1 + 4) = (1)(3) = 3$$

✓ **Checkpoint 1**   *Worked-out solution available at LarsonAppliedCalculus.com*

a. For $f(x, y) = x^2 + 2xy$, find $f(2, -1)$.

b. For $f(x, y, z) = \dfrac{2x^2z}{y^3}$, find $f(-3, 2, 1)$.

The Domain and Range of a Function of Two Variables

A function of two variables can be represented graphically as a surface in space by letting

$$z = f(x, y). \qquad \text{Function of two variables}$$

When sketching the graph of a function of x and y, remember that even though the graph is three-dimensional, the domain of the function is two-dimensional—it consists of the points in the xy-plane for which the function is defined. As with functions of a single variable, unless specifically restricted, the domain of a function of two variables is assumed to be the set of all points (x, y) for which the defining equation has meaning. In other words, to each point (x, y) in the domain of f there corresponds a point (x, y, z) on the surface, and conversely, to each point (x, y, z) on the surface there corresponds a point (x, y) in the domain of f.

EXAMPLE 2   Finding the Domain and Range of a Function

Find the domain and range of the function

$$f(x, y) = \sqrt{64 - x^2 - y^2}.$$

SOLUTION   Because no restrictions are given, the domain is assumed to be the set of all points for which the defining equation makes sense.

$$64 - x^2 - y^2 \geq 0 \qquad \text{Quantity inside radical must be nonnegative.}$$
$$-x^2 - y^2 \geq -64 \qquad \text{Subtract 64 from each side.}$$
$$x^2 + y^2 \leq 64 \qquad \text{Multiply each side by } -1 \text{ and reverse the inequality symbol.}$$

So, the domain of f is the set of all points that lie on or inside the circle

$$x^2 + y^2 = 8^2$$

as shown in Figure 7.18. The range of f is

$$0 \leq z \leq 8. \qquad \text{Range of the function}$$

As shown in Figure 7.19, the graph of the function is a hemisphere.

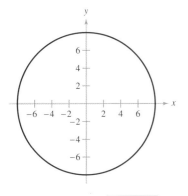

Domain of $f(x, y) = \sqrt{64 - x^2 - y^2}$

FIGURE 7.18

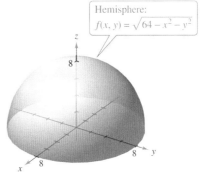

Hemisphere:
$f(x, y) = \sqrt{64 - x^2 - y^2}$

Domain: $x^2 + y^2 \leq 64$
Range: $0 \leq z \leq 8$

FIGURE 7.19

✓ **Checkpoint 2**   *Worked-out solution available at LarsonAppliedCalculus.com*

Consider the function

$$f(x, y) = \sqrt{9 - x^2 - y^2}.$$

a. Find the domain of f.

b. Find the range of f.

Contour Maps and Level Curves

A **contour map** of a surface is created by *projecting* traces, taken in evenly spaced planes that are parallel to the xy-plane, onto the xy-plane. Each projection is a **level curve** of the surface.

Contour maps are used to create weather, topographical, and population density maps. For instance, Figure 7.20(a) shows a graph of a "mountain and valley" surface given by $z = f(x, y)$. Each of the level curves in Figure 7.20(b) represents the intersection of the surface $z = f(x, y)$ with a plane $z = c$, where $c = 828, 830, \ldots, 854$.

(a) Surface

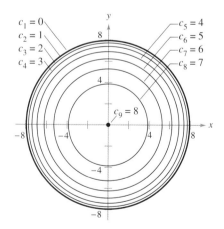

(b) Contour map

FIGURE 7.20

EXAMPLE 3 Sketching a Contour Map

The hemisphere $f(x, y) = \sqrt{64 - x^2 - y^2}$ is shown in Figure 7.21. Sketch a contour map of this surface using level curves corresponding to $c = 0, 1, 2, \ldots, 8$.

SOLUTION For each value of c, the equation $f(x, y) = c$ is a circle (or point) in the xy-plane. For instance, when $c_1 = 0$, the level curve is

$$x^2 + y^2 = 8^2 \qquad \text{Circle of radius 8}$$

which is a circle of radius 8. Figure 7.22 shows the nine level curves for the hemisphere.

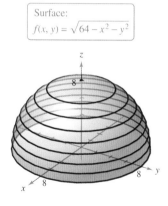

Hemisphere
FIGURE 7.21

Contour map
FIGURE 7.22

✓ **Checkpoint 3**   *Worked-out solution available at LarsonAppliedCalculus.com*

Describe the level curves of $f(x, y) = \sqrt{9 - x^2 - y^2}$. Sketch the level curves for $c = 0, 1, 2,$ and 3.

Applications

The **Cobb-Douglas production function** is used in economics to represent the numbers of units produced by varying amounts of labor and capital. Let x represent the number of units of labor and let y represent the number of units of capital. Then, the number of units produced is modeled by

$$f(x, y) = Cx^a y^{1-a}$$

where C and a are constants, with $0 < a < 1$.

EXAMPLE 4   Using a Production Function

A manufacturer estimates that its production (measured in units of a product) can be modeled by $f(x, y) = 100x^{0.6}y^{0.4}$, where the labor x is measured in person-hours and the capital y is measured in thousands of dollars.

a. What is the production level when $x = 1000$ and $y = 500$?

b. What is the production level when $x = 2000$ and $y = 1000$?

c. How does doubling the amounts of labor and capital from part (a) to part (b) affect the production?

SOLUTION

a. When $x = 1000$ and $y = 500$, the production level is

$$f(1000, 500) = 100(1000)^{0.6}(500)^{0.4} \approx 75{,}786 \text{ units.}$$

b. When $x = 2000$ and $y = 1000$, the production level is

$$f(2000, 1000) = 100(2000)^{0.6}(1000)^{0.4} \approx 151{,}572 \text{ units.}$$

c. When the amounts of labor and capital are doubled, the production level also doubles. In Exercise 45, you are asked to show that this is characteristic of the Cobb-Douglas production function.

A contour map of this function is shown in Figure 7.23. Note that the level curves occur at increments of 10,000.

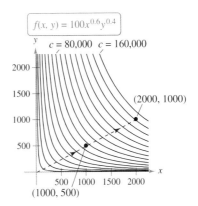

Level Curves (at Increments of 10,000)

FIGURE 7.23

✓ *Checkpoint 4*   Worked-out solution available at LarsonAppliedCalculus.com

Use the Cobb-Douglas production function in Example 4 to find the production levels when $x = 1500$ and $y = 1000$ and when $x = 1000$ and $y = 1500$. Use your results to determine which variable has a greater influence on production.   ■

EXAMPLE 5   Finding Monthly Payments

The monthly payment M for an installment loan of P dollars taken out over t years at an annual interest rate of r (in decimal form) is

$$M = f(P, r, t) = \frac{\dfrac{Pr}{12}}{1 - \left[\dfrac{1}{1 + (r/12)}\right]^{12t}}.$$

a. Find the monthly payment for a home mortgage of $100,000 taken out for 30 years at an annual interest rate of 4%. How much is the total amount paid?

b. Find the monthly payment for a car loan of $22,000 taken out for 5 years at an annual interest rate of 2.5%. How much is the total amount paid?

SOLUTION

a. When $P = \$100{,}000$, $r = 0.04$, and $t = 30$, the monthly payment is

$$M = f(100{,}000, 0.04, 30)$$

$$= \frac{\dfrac{(100{,}000)(0.04)}{12}}{1 - \left[\dfrac{1}{1 + (0.04/12)}\right]^{12(30)}}$$

$$= \$477.42.$$

Because the mortgage is for 30 years, the total number of payments is $12(30) = 360$. So, the total amount paid is $360(477.42) = \$171{,}871.20$.

b. When $P = \$22{,}000$, $r = 0.025$, and $t = 5$, the monthly payment is

$$M = f(22{,}000, 0.025, 5)$$

$$= \frac{\dfrac{(22{,}000)(0.025)}{12}}{1 - \left[\dfrac{1}{1 + (0.025/12)}\right]^{12(5)}}$$

$$= \$390.44.$$

Because the loan is for 5 years, the total number of payments is $12(5) = 60$. So, the total amount paid is $60(390.44) = \$23{,}426.40$.

✓ *Checkpoint 5*   *Worked-out solution available at LarsonAppliedCalculus.com*

Find the monthly payment M for a home mortgage of $100,000 taken out for 30 years at an annual interest rate of 3%. How much is the total amount paid?   ■

SUMMARIZE   (Section 7.3)

1. State the definition of a function of two variables *(page 436)*. For an example of evaluating a function of two variables, see Example 1(a).

2. Describe how a contour map of a surface is created *(page 438)*. For an example of sketching a contour map, see Example 3.

3. State the Cobb-Douglas production function *(page 439)*. For an example of using the Cobb-Douglas production function, see Example 4.

4. Describe a real-life example of how a function of several variables can be used to find the monthly payment for a loan *(page 440, Example 5)*.

SKILLS WARM UP 7.3

The following warm-up exercises involve skills that were covered in a previous course or in earlier sections. You will use these skills in the exercise set for this section. For additional help, review Section 1.4.

In Exercises 1–4, evaluate the function when $x = -3$.

1. $f(x) = 5 - 2x$ **2.** $f(x) = -x^2 + 4x + 5$ **3.** $y = \sqrt{4x^2 - 3x + 4}$ **4.** $y = \sqrt[3]{34 - 4x + 2x^2}$

In Exercises 5–8, find the domain of the function.

5. $f(x) = 5x^2 + 3x - 2$ **6.** $g(x) = \dfrac{1}{2x} - \dfrac{2}{x + 3}$ **7.** $h(y) = \sqrt{y - 5}$ **8.** $f(y) = \sqrt{y^2 - 1}$

In Exercises 9 and 10, evaluate the expression.

9. $(476)^{0.65}$ **10.** $(251)^{0.35}$

Exercises 7.3

See *CalcChat.com* for tutorial help and worked-out solutions to odd-numbered exercises.

Evaluating Functions of Several Variables In Exercises 1–14, find and simplify the function values. *See Example 1.*

1. $f(x, y) = 2x - y + 3$

  (a) $f(0, 2)$ (b) $f(-1, 0)$ (c) $f(5, 30)$
  (d) $f(3, y)$ (e) $f(x, 4)$ (f) $f(5, t)$

2. $f(x, y) = 4 - x^2 - 4y^2$

  (a) $f(0, 0)$ (b) $f(0, 1)$ (c) $f(2, 3)$
  (d) $f(1, y)$ (e) $f(x, 0)$ (f) $f(t, 1)$

3. $f(x, y) = xe^y$

  (a) $f(5, 0)$ (b) $f(3, 2)$ (c) $f(2, -1)$
  (d) $f(4, y)$ (e) $f(x, \ln 6)$ (f) $f(t, t)$

4. $g(x, y) = \ln|x + y|$

  (a) $g(2, 3)$ (b) $g(5, 6)$ (c) $g(e, 0)$
  (d) $g(0, 1)$ (e) $g(2, -3)$ (f) $g(e, e)$

5. $h(x, y, z) = \dfrac{xy}{z}$

  (a) $h(2, 3, 9)$ (b) $h(1, 0, 1)$

6. $f(x, y, z) = \dfrac{\sqrt{y + z}}{x}$

  (a) $f(1, 5, 4)$ (b) $f(6, 8, -3)$

7. $V(r, h) = \pi r^2 h$

  (a) $V(3, 10)$ (b) $V(5, 2)$

8. $F(r, N) = 500\left(1 + \dfrac{r}{12}\right)^N$

  (a) $F(0.09, 60)$ (b) $F(0.14, 240)$

9. $A(P, r, t) = P\left[\left(1 + \dfrac{r}{12}\right)^{12t} - 1\right]\left(1 + \dfrac{12}{r}\right)$

  (a) $A(100, 0.10, 10)$ (b) $A(275, 0.0925, 40)$

10. $A(P, r, t) = Pe^{rt}$

  (a) $A(500, 0.10, 5)$ (b) $A(1500, 0.12, 20)$

11. $f(x, y) = \displaystyle\int_x^y (2t - 3)\, dt$

  (a) $f(1, 2)$ (b) $f(1, 4)$

12. $g(x, y) = \displaystyle\int_x^y \dfrac{1}{t}\, dt$

  (a) $g(4, 1)$ (b) $g(6, 3)$

13. $f(x, y) = x^2 - 2y$

  (a) $f(x + \Delta x, y)$ (b) $\dfrac{f(x, y + \Delta y) - f(x, y)}{\Delta y}$

14. $f(x, y) = 3xy + y^2$

  (a) $f(x + \Delta x, y)$ (b) $\dfrac{f(x, y + \Delta y) - f(x, y)}{\Delta y}$

Finding the Domain and Range of a Function In Exercises 15–30, find the domain and range of the function. *See Example 2.*

15. $f(x, y) = \sqrt{16 - x^2 - y^2}$

16. $z = \sqrt{4 - x^2 - y^2}$

17. $f(x, y) = x^2 + y^2$

18. $f(x, y) = x^2 + y^2 - 1$

19. $f(x, y) = e^{x/y}$ **20.** $f(x, y) = e^{xy}$

21. $z = \ln(5 - x - y)$ **22.** $f(x, y) = \ln(2x + 3y)$

23. $z = \sqrt{9 - 3x^2 - y^2}$ **24.** $z = \sqrt{4 - x^2 - 4y^2}$

25. $z = \dfrac{y}{x}$ **26.** $f(x, y) = \dfrac{x}{y}$

27. $f(x, y) = \dfrac{1}{xy}$ **28.** $g(x, y) = \dfrac{1}{x - y}$

29. $h(x, y) = x\sqrt{y}$ **30.** $f(x, y) = \sqrt{xy}$

Matching In Exercises 31–34, match the graph of the surface with one of the contour maps. [The contour maps are labeled (a)–(d).]

(a)

(b)

(c)

(d)

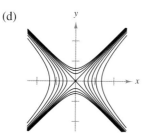

31. $f(x, y) = x^2 + \dfrac{y^2}{4}$ **32.** $f(x, y) = e^{1 - x^2 + y^2}$

33. $f(x, y) = e^{1 - x^2 - y^2}$ **34.** $f(x, y) = \ln|y - x^2|$

Sketching a Contour Map In Exercises 35–42, describe the level curves of the function. Sketch a contour map of the surface using level curves for the given *c*-values. *See Example 3.*

Function	c-Values
35. $z = x + y$	$c = -1, 0, 2, 4$
36. $z = 6 - 2x - 3y$	$c = 0, 2, 4, 6, 8, 10$
37. $z = \sqrt{25 - x^2 - y^2}$	$c = 0, 1, 2, 3, 4, 5$
38. $f(x, y) = x^2 + 4y^2$	$c = 0, 2, 4, 6, 8$
39. $f(x, y) = xy$	$c = \pm 1, \pm 2, \ldots, \pm 6$

Function	c-Values
40. $z = e^{-xy}$	$c = 1, 2, 3, 4, \frac{1}{2}, \frac{1}{3}, \frac{1}{4}$
41. $f(x, y) = \dfrac{x}{x^2 + y^2}$	$c = \pm \frac{1}{2}, \pm 1, \pm \frac{3}{2}, \pm 2$
42. $f(x, y) = \ln(x - y)$	$c = 0, \pm \frac{1}{2}, \pm 1, \pm \frac{3}{2}, \pm 2$

43. Cobb-Douglas Production Function A manufacturer estimates that its production can be modeled by $f(x, y) = 100x^{0.8}y^{0.2}$. Estimate the production level when $x = 1000$ and $y = 2000$.

44. Cobb-Douglas Production Function A manufacturer estimates that its production can be modeled by $f(x, y) = 100x^{0.75}y^{0.25}$. Estimate the production level when $x = 1500$ and $y = 1000$.

45. Cobb-Douglas Production Function Use the Cobb-Douglas production function $f(x, y) = Cx^a y^{1-a}$ to show that when both the number of units of labor and the number of units of capital are doubled, the production level is also doubled.

46. Cobb-Douglas Production Function Show that the Cobb-Douglas production function $z = Cx^a y^{1-a}$ can be rewritten as

$$\ln \frac{z}{y} = \ln C + a \ln \frac{x}{y}.$$

47. Profit A sporting goods manufacturer produces regulation soccer balls at two plants. The costs of producing x_1 units at Location 1 and x_2 units at Location 2 are given by

$$C_1(x_1) = 0.02x_1^2 + 4x_1 + 500$$

and

$$C_2(x_2) = 0.05x_2^2 + 4x_2 + 275$$

respectively. If the product sells for \$50 per unit, then the profit function for the product is given by

$$P(x_1, x_2) = 50(x_1 + x_2) - C_1(x_1) - C_2(x_2).$$

Evaluate each of the following.

(a) $P(250, 150)$

(b) $P(300, 200)$

(c) $P(600, 400)$

48. Queuing Model The average length of time that a customer waits in line for service is given by

$$W(x, y) = \frac{1}{x - y}, \quad y < x$$

where y is the average arrival rate, written as the number of customers per unit of time, and x is the average service rate, written in the same units. Evaluate each of the following.

(a) $W(15, 10)$ (b) $W(12, 9)$

(c) $W(12, 6)$ (d) $W(4, 2)$

49. Investment In 2015, an investment of $2000 was made in a bond earning 10% compounded annually. The investor pays tax at rate R (in decimal form), and the annual rate of inflation is I (in decimal form). In the year 2025, the value V of the bond in constant 2015 dollars is given by

$$V(I, R) = 2000\left[\frac{1 + 0.10(1 - R)}{1 + I}\right]^{10}.$$

Use this function of two variables and a spreadsheet to complete the table.

	Inflation Rate		
Tax Rate	0	0.03	0.05
0			
0.28			
0.35			

50. Investment A principal of $5000 is deposited in a savings account that earns an interest rate of r (in decimal form), compounded continuously. The amount $A(r, t)$ after t years is

$$A(r, t) = 5000e^{rt}.$$

Use this function of two variables and a spreadsheet to complete the table.

	Number of Years				
Rate	5	10	15	20	25
0.02					
0.03					
0.04					
0.05					
0.06					

51. Meteorology Meteorologists measure the atmospheric pressure in millibars. From these observations they create weather maps on which the curves of equal atmospheric pressure (isobars) are drawn (see figure). On the map, the closer the isobars, the higher the wind speed. Match points A, B, and C with (a) highest pressure, (b) lowest pressure, and (c) highest wind speed.

52. **HOW DO YOU SEE IT?** The contour map of the Southern Hemisphere shown in the figure was computer generated using data collected by satellite instrumentation. Color is used to show the "ozone hole" in Earth's atmosphere. The purple and blue areas represent the lowest levels of ozone, and the yellow and red areas represent the highest levels. *(Source: National Aeronautics and Space Administration)*

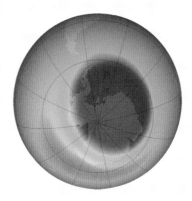

(a) Do the level curves correspond to equally spaced ozone levels? Explain.

(b) Describe how to obtain a more detailed contour map.

53. Shareholder's Equity The shareholder's equity z (in millions of dollars) for Skechers from 2009 through 2013 can be modeled by $z = 0.105x + 0.339y + 257$, where x is the sales (in millions of dollars) and y is the total assets (in millions of dollars). *(Source: Skechers U.S.A., Inc.)*

(a) Find the shareholder's equity when $x = 1000$ and $y = 500$.

(b) Which of the two variables in this model has the greater influence on shareholder's equity? Explain.

54. Monthly Payments You are taking out a car loan for $16,000, and you are given the options below. Find the monthly payment and the total amount of money you will pay for each loan. Which option would you choose? Explain your reasoning.

(a) A fixed annual rate of 5%, over a term of 3 years.

(b) A fixed annual rate of 3.5%, over a term of 4 years.

55. Monthly Payments You are taking out a home mortgage for $120,000, and you are given the options below. Find the monthly payment and the total amount of money you will pay for each mortgage. Which option would you choose? Explain your reasoning.

(a) A fixed annual rate of 8%, over a term of 20 years.

(b) A fixed annual rate of 7%, over a term of 30 years.

(c) A fixed annual rate of 7%, over a term of 15 years.

7.4 Partial Derivatives

■ Find the first partial derivatives of functions of two variables.

■ Find the slopes of surfaces in the x- and y-directions and use partial derivatives to answer questions about real-life situations.

■ Find the partial derivatives of functions of three variables.

■ Find higher-order partial derivatives.

Partial Derivatives of a Function of Two Variables

Real-life applications of functions of several variables are often concerned with how changes in one of the variables will affect the values of the functions. For instance, an economist who wants to determine the effect of a tax increase on the economy might make calculations using different tax rates while holding all other variables, such as unemployment, constant.

You can follow a similar procedure to find the rate of change of a function f with respect to one of its independent variables. That is, you find the derivative of f with respect to one independent variable while holding the other variable(s) constant. This process is called **partial differentiation,** and each derivative is called a **partial derivative.** A function of several variables has as many partial derivatives as it has independent variables.

In Exercise 62 on page 453, you will use partial derivatives to find the marginal revenues of a pharmaceutical corporation at two locations that produce the same medicine.

Partial Derivatives of a Function of Two Variables

If $z = f(x, y)$, then the **first partial derivatives of f with respect to x and y** are the functions $\partial z/\partial x$ and $\partial z/\partial y$, defined as shown.

$$\frac{\partial z}{\partial x} = \lim_{\Delta x \to 0} \frac{f(x + \Delta x, y) - f(x, y)}{\Delta x} \qquad y \text{ is held constant.}$$

$$\frac{\partial z}{\partial y} = \lim_{\Delta y \to 0} \frac{f(x, y + \Delta y) - f(x, y)}{\Delta y} \qquad x \text{ is held constant.}$$

This definition indicates that if $z = f(x, y)$, then to find $\partial z/\partial x$, you *consider y to be constant* and differentiate with respect to x. Similarly, to find $\partial z/\partial y$, you *consider x to be constant* and differentiate with respect to y.

STUDY TIP

The notation $\partial z/\partial x$ is read as "the partial derivative of z with respect to x," and $\partial z/\partial y$ is read as "the partial derivative of z with respect to y."

EXAMPLE 1 Finding Partial Derivatives

Find $\partial z/\partial x$ and $\partial z/\partial y$ for the function

$$z = 3x - x^2 y^2 + 2x^3 y.$$

SOLUTION

$$\frac{\partial z}{\partial x} = 3 - 2xy^2 + 6x^2 y \qquad \text{Hold } y \text{ constant and differentiate with respect to } x.$$

$$\frac{\partial z}{\partial y} = -2x^2 y + 2x^3 \qquad \text{Hold } x \text{ constant and differentiate with respect to } y.$$

✓ **Checkpoint 1**   Worked-out solution available at LarsonAppliedCalculus.com

Find $\partial z/\partial x$ and $\partial z/\partial y$ for $z = 2x^2 - 4x^2 y^3 + y^4$.

Notation for First Partial Derivatives

The first partial derivatives of $z = f(x, y)$ are denoted by

$$\frac{\partial z}{\partial x} = f_x(x, y) = z_x = \frac{\partial}{\partial x}[f(x, y)] \qquad \text{Partial derivative with respect to } x$$

and

$$\frac{\partial z}{\partial y} = f_y(x, y) = z_y = \frac{\partial}{\partial y}[f(x, y)]. \qquad \text{Partial derivative with respect to } y$$

The values of the first partial derivatives at the point (a, b) are denoted by

$$\frac{\partial z}{\partial x}\bigg|_{(a, b)} = f_x(a, b)$$

and

$$\frac{\partial z}{\partial y}\bigg|_{(a, b)} = f_y(a, b).$$

TECH TUTOR

Symbolic differentiation utilities can be used to find partial derivatives of a function of two variables. Try using a symbolic differentiation utility to find the first partial derivatives of the function in Example 2.

EXAMPLE 2   Finding and Evaluating Partial Derivatives

Find the first partial derivatives of

$$f(x, y) = xe^{x^2 y}$$

and evaluate each at the point $(1, \ln 2)$.

SOLUTION   To find the first partial derivative with respect to x, hold y constant and differentiate using the Product Rule.

$$f_x(x, y) = x\frac{\partial}{\partial x}[e^{x^2 y}] + e^{x^2 y}\frac{\partial}{\partial x}[x] \qquad \text{Apply Product Rule.}$$
$$= xe^{x^2 y}(2xy) + e^{x^2 y} \qquad y \text{ is held constant.}$$
$$= e^{x^2 y}(2x^2 y + 1) \qquad \text{Simplify.}$$

At the point $(1, \ln 2)$, the value of this derivative is

$$f_x(1, \ln 2) = e^{(1)^2(\ln 2)}[2(1)^2(\ln 2) + 1] \qquad \text{Substitute for } x \text{ and } y.$$
$$= 2(2 \ln 2 + 1) \qquad \text{Simplify.}$$
$$\approx 4.773. \qquad \text{Use a calculator.}$$

To find the first partial derivative with respect to y, hold x constant and differentiate to obtain

$$f_y(x, y) = xe^{x^2 y}(x^2) \qquad \text{Apply Constant Multiple Rule.}$$
$$= x^3 e^{x^2 y}. \qquad \text{Simplify.}$$

At the point $(1, \ln 2)$, the value of this derivative is

$$f_y(1, \ln 2) = (1)^3 e^{(1)^2(\ln 2)} \qquad \text{Substitute for } x \text{ and } y.$$
$$= 2. \qquad \text{Simplify.}$$

✓ **Checkpoint 2**   Worked-out solution available at LarsonAppliedCalculus.com

Find the first partial derivatives of

$$f(x, y) = x^2 y^3$$

and evaluate each at the point $(1, 2)$.

Graphical Interpretation of Partial Derivatives

Earlier in the text, you studied graphical interpretations of the derivative of a function of a single variable. There, you found that $f'(x_0)$ represents the slope of the tangent line to the graph of $y = f(x)$ at the point (x_0, y_0). The partial derivatives of a function of two variables also have useful graphical interpretations. Consider the function

$z = f(x, y).$ Function of two variables

As shown in Figure 7.24(a), the graph of this function is a surface in space. If $y = y_0$, then

$z = f(x, y_0)$ Function of one variable

is a function of one variable. The graph of this function is the curve that is the intersection of the plane $y = y_0$ and the surface $z = f(x, y)$. On this curve, the partial derivative

$f_x(x, y_0)$ Slope in x-direction

represents the slope in the plane $y = y_0$, as shown in Figure 7.24(a). Similarly, if $x = x_0$, then

$z = f(x_0, y)$ Function of one variable

is a function of one variable. Its graph is the intersection of the plane $x = x_0$ and the surface $z = f(x, y)$. On this curve, the partial derivative

$f_y(x_0, y)$ Slope in y-direction

represents the slope in the plane $x = x_0$, as shown in Figure 7.24(b).

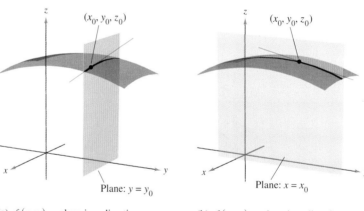

(a) $f_x(x, y_0)$ = slope in x-direction   (b) $f_y(x_0, y)$ = slope in y-direction

FIGURE 7.24

Informally, $f_x(x_0, y_0)$ and $f_y(x_0, y_0)$ at the point (x_0, y_0, z_0) denote the **slopes of the surface in the x- and y-directions,** respectively.

Guidelines for Finding the Slopes of a Surface at a Point

Let (x_0, y_0, z_0) be a point on the surface of

$z = f(x, y).$

1. Find the partial derivatives of f with respect to x and y.

2. The slope in the x-direction at (x_0, y_0, z_0) is $f_x(x_0, y_0)$.

3. The slope in the y-direction at (x_0, y_0, z_0) is $f_y(x_0, y_0)$.

EXAMPLE 3   **Finding Slopes in the x- and y-Directions**

Find the slopes of the surface

$$f(x, y) = -\frac{x^2}{2} - y^2 + \frac{25}{8}$$

at the point $\left(\frac{1}{2}, 1, 2\right)$ in

a. the x-direction.

b. the y-direction.

SOLUTION

a. To find the slope in the x-direction, hold y constant and differentiate with respect to x to obtain

$$f_x(x, y) = -x. \qquad \text{Partial derivative with respect to } x$$

At the point $\left(\frac{1}{2}, 1, 2\right)$, the slope in the x-direction is

$$f_x\left(\frac{1}{2}, 1\right) = -\frac{1}{2} \qquad \text{Slope in x-direction}$$

as shown in Figure 7.25(a).

b. To find the slope in the y-direction, hold x constant and differentiate with respect to y to obtain

$$f_y(x, y) = -2y. \qquad \text{Partial derivative with respect to } y$$

At the point $\left(\frac{1}{2}, 1, 2\right)$, the slope in the y-direction is

$$f_y\left(\frac{1}{2}, 1\right) = -2 \qquad \text{Slope in y-direction}$$

as shown in Figure 7.25(b).

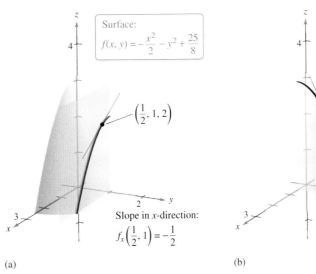

(a)

(b)

FIGURE 7.25

✓ *Checkpoint 3*   *Worked-out solution available at LarsonAppliedCalculus.com*

Find the slopes of the surface

$$f(x, y) = 4x^2 + 9y^2 + 36$$

at the point $(1, -1, 49)$ in

a. the x-direction.

b. the y-direction.

For 2015, Hampton Hotels was chosen as the number one franchise by *Entrepreneur* magazine. In early 2015, Hampton Hotels had more than 1900 franchises worldwide. What type of product would be complementary to a hotel? What type of product would be a substitute?

Consumer products in the same market or in related markets can be classified as **complementary products** or **substitute products.** Two products are complementary when an *increase* in demand of one product results in an *increase* in demand of the other. For instance, when mobile phones are sold, consumers are more likely to buy applications for use on the phone. Two products are substitutes when an *increase* in demand of one product results in a *decrease* in demand of the other. So, substitute products are competitors, such as cable television providers and satellite television providers.

Here is a test to determine whether two products in the same market are complementary, substitutes, or neither. In the test, note that (1) the demand functions for each product are dependent on the prices for *both* products and (2) the partial derivative of each demand function is taken with respect to the price of the other product.

Test for Complementary and Substitute Products

The demand functions for two products in the same market are

$$x_1 = f(p_1, p_2) \quad \text{and} \quad x_2 = g(p_1, p_2)$$

where p_1 and p_2 are the prices per unit for the two products, and x_1 and x_2 are the numbers of units sold. The two products are complementary when

$$\frac{\partial x_1}{\partial p_2} < 0 \quad \text{and} \quad \frac{\partial x_2}{\partial p_1} < 0. \qquad \text{Complementary products}$$

The products are substitutes when

$$\frac{\partial x_1}{\partial p_2} > 0 \quad \text{and} \quad \frac{\partial x_2}{\partial p_1} > 0. \qquad \text{Substitute products}$$

If the test fails any one of the conditions, then the products are neither complementary nor substitutes.

EXAMPLE 4   Complementary and Substitute Products

The demand functions for two products in the same market are

$$x_1 = 200(p_2 - p_1) \qquad \text{Demand for Product 1}$$

and

$$x_2 = 500 + 100p_1 - 180p_2 \qquad \text{Demand for Product 2}$$

where p_1 and p_2 are the prices per unit (in dollars), and x_1 and x_2 are the numbers of units sold. Determine whether the products are complementary, substitutes, or neither.

SOLUTION The partial derivatives are

$$\frac{\partial x_1}{\partial p_2} = 200 \quad \text{and} \quad \frac{\partial x_2}{\partial p_1} = 100.$$

Because both partial derivatives are positive, the products are substitutes.

✓ *Checkpoint 4*   *Worked-out solution available at LarsonAppliedCalculus.com*

The demand functions for two products in the same market are

$$x_1 = 100 - 2p_1 + 1.5p_2 \quad \text{and} \quad x_2 = 145 + 0.5p_1 - 0.75p_2$$

where p_1 and p_2 are the prices per unit (in dollars), and x_1 and x_2 are the numbers of units sold. Determine whether the products are complementary, substitutes, or neither. ■

Partial Derivatives of a Function of Three Variables

The concept of a partial derivative can be extended naturally to functions of three or more variables. For instance, the function

$$w = f(x, y, z) \qquad \text{Function of three variables}$$

has three partial derivatives, each of which is formed by holding two of the variables constant. That is, to define the partial derivative of w with respect to x, hold y *and* z constant and write

$$\frac{\partial w}{\partial x} = f_x(x, y, z) = \lim_{\Delta x \to 0} \frac{f(x + \Delta x, y, z) - f(x, y, z)}{\Delta x}.$$

To define the partial derivative of w with respect to y, hold x *and* z constant and write

$$\frac{\partial w}{\partial y} = f_y(x, y, z) = \lim_{\Delta y \to 0} \frac{f(x, y + \Delta y, z) - f(x, y, z)}{\Delta y}.$$

To define the partial derivative of w with respect to z, hold x *and* y constant and write

$$\frac{\partial w}{\partial z} = f_z(x, y, z) = \lim_{\Delta z \to 0} \frac{f(x, y, z + \Delta z) - f(x, y, z)}{\Delta z}.$$

EXAMPLE 5 Finding Partial Derivatives of a Function

Find the first partial derivatives of $w = xe^{xy + 2z}$ with respect to x, y, and z.

SOLUTION Holding y and z constant, you obtain

$$\begin{aligned}
\frac{\partial w}{\partial x} &= x\frac{\partial}{\partial x}[e^{xy + 2z}] + e^{xy + 2z}\frac{\partial}{\partial x}[x] && \text{Apply Product Rule.}\\
&= x(e^{xy + 2z})(y) + e^{xy + 2z}(1) && \text{Hold } y \text{ and } z \text{ constant.}\\
&= (xy + 1)e^{xy + 2z}. && \text{Simplify.}
\end{aligned}$$

Holding x and z constant, you obtain

$$\begin{aligned}
\frac{\partial w}{\partial y} &= x\frac{\partial}{\partial y}[e^{xy + 2z}] && \text{Apply Constant Multiple Rule.}\\
&= x(e^{xy + 2z})(x) && \text{Hold } x \text{ and } z \text{ constant.}\\
&= x^2 e^{xy + 2z}. && \text{Simplify.}
\end{aligned}$$

Holding x and y constant, you obtain

$$\begin{aligned}
\frac{\partial w}{\partial z} &= x\frac{\partial}{\partial z}[e^{xy + 2z}] && \text{Apply Constant Multiple Rule.}\\
&= x(e^{xy + 2z})(2) && \text{Hold } x \text{ and } y \text{ constant.}\\
&= 2xe^{xy + 2z}. && \text{Simplify.}
\end{aligned}$$

✓ **Checkpoint 5** *Worked-out solution available at LarsonAppliedCalculus.com*

Find the first partial derivatives of $w = x^2 y \ln(xz)$ with respect to x, y, and z. ■

In Example 5, the Product Rule is used only when finding the partial derivative with respect to x. For

$$\frac{\partial w}{\partial y} \quad \text{and} \quad \frac{\partial w}{\partial z}$$

x is considered to be constant, so the Constant Multiple Rule is used.

TECH TUTOR

A symbolic differentiation utility can be used to find the partial derivatives of a function of three or more variables. Try using a symbolic differentiation utility to find the partial derivative $f_y(x, y, z)$ for the function in Example 5.

Higher-Order Partial Derivatives

As with ordinary derivatives, it is possible to take second-, third-, and higher-order partial derivatives of a function of several variables, provided such derivatives exist. Higher-order derivatives are denoted by the order in which the differentiation occurs. For instance, there are four different ways to find a second partial derivative of $z = f(x, y)$.

1. $\dfrac{\partial}{\partial x}\left(\dfrac{\partial f}{\partial x}\right) = \dfrac{\partial^2 f}{\partial x^2} = f_{xx}$ Differentiate twice with respect to x.

2. $\dfrac{\partial}{\partial y}\left(\dfrac{\partial f}{\partial y}\right) = \dfrac{\partial^2 f}{\partial y^2} = f_{yy}$ Differentiate twice with respect to y.

3. $\dfrac{\partial}{\partial y}\left(\dfrac{\partial f}{\partial x}\right) = \dfrac{\partial^2 f}{\partial y \partial x} = f_{xy}$ Differentiate first with respect to x and then with respect to y.

4. $\dfrac{\partial}{\partial x}\left(\dfrac{\partial f}{\partial y}\right) = \dfrac{\partial^2 f}{\partial x \partial y} = f_{yx}$ Differentiate first with respect to y and then with respect to x.

The third and fourth cases are **mixed partial derivatives.** Notice that with the two types of notation for mixed partials, different conventions are used for indicating the order of differentiation. For instance, the partial derivative

$$\frac{\partial}{\partial y}\left(\frac{\partial f}{\partial x}\right) = \frac{\partial^2 f}{\partial y \partial x} \qquad \text{Right-to-left order}$$

indicates differentiation with respect to x first, but the partial derivative

$$(f_y)_x = f_{yx} \qquad \text{Left-to-right order}$$

indicates differentiation with respect to y first. To remember this, note that in each case you differentiate first with respect to the variable "nearest" f.

EXAMPLE 6   Finding Second Partial Derivatives

Find the second partial derivatives of

$$f(x, y) = 3xy^2 - 2y + 5x^2y^2$$

and determine the value of $f_{xy}(-1, 2)$.

SOLUTION   Begin by finding the first partial derivatives with respect to x and y.

$$f_x(x, y) = 3y^2 + 10xy^2 \qquad\qquad f_y(x, y) = 6xy - 2 + 10x^2y$$

Then, differentiating with respect to x and y produces

$$f_{xx}(x, y) = 10y^2, \qquad\qquad f_{yy}(x, y) = 6x + 10x^2,$$

$$f_{xy}(x, y) = 6y + 20xy, \qquad\qquad f_{yx}(x, y) = 6y + 20xy.$$

The value of f_{xy} at the point $(-1, 2)$ is

$$f_{xy}(-1, 2) = 6(2) + 20(-1)(2) = 12 - 40 = -28.$$

✓ **Checkpoint 6**   Worked-out solution available at LarsonAppliedCalculus.com

Find the second partial derivatives of

$$f(x, y) = 4x^2y^2 + 2x + 4y^2. \qquad\qquad ■$$

Notice in Example 6 that the two mixed partials are equal. It can be shown that when a function has continuous second partial derivatives, then the order in which the partial derivatives are taken is irrelevant.

A function of two variables has two first partial derivatives and four second partial derivatives. For a function of three variables, there are three first partials

$$f_x, \quad f_y, \quad \text{and} \quad f_z$$

and nine second partials

$$f_{xx}, \quad f_{xy}, \quad f_{xz}, \quad f_{yx}, \quad f_{yy}, \quad f_{yz}, \quad f_{zx}, \quad f_{zy}, \quad \text{and} \quad f_{zz}$$

of which six are mixed partials. To find partial derivatives of order three and higher, follow the same pattern used to find second partial derivatives. For instance, if $z = f(x, y)$, then

$$z_{xxx} = \frac{\partial}{\partial x}\left(\frac{\partial^2 f}{\partial x^2}\right) = \frac{\partial^3 f}{\partial x^3} \quad \text{and} \quad z_{xxy} = \frac{\partial}{\partial y}\left(\frac{\partial^2 f}{\partial x^2}\right) = \frac{\partial^3 f}{\partial y \partial x^2}.$$

EXAMPLE 7   Finding Second Partial Derivatives

Find the second partial derivatives of

$$f(x, y, z) = ye^x + x \ln z.$$

SOLUTION   Begin by finding the first partial derivatives.

$$f_x(x, y, z) = ye^x + \ln z, \quad f_y(x, y, z) = e^x, \quad f_z(x, y, z) = \frac{x}{z}$$

Then, differentiate with respect to x, y, and z to find the nine second partial derivatives.

$$f_{xx}(x, y, z) = ye^x, \quad f_{xy}(x, y, z) = e^x, \quad f_{xz}(x, y, z) = \frac{1}{z}$$

$$f_{yx}(x, y, z) = e^x, \quad f_{yy}(x, y, z) = 0, \quad f_{yz}(x, y, z) = 0$$

$$f_{zx}(x, y, z) = \frac{1}{z}, \quad f_{zy}(x, y, z) = 0, \quad f_{zz}(x, y, z) = -\frac{x}{z^2}$$

✓ **Checkpoint 7**   Worked-out solution available at LarsonAppliedCalculus.com

Find the second partial derivatives of $f(x, y, z) = xe^y + 2xz + y^2$.   ■

SUMMARIZE   (Section 7.4)

1. State the definition of partial derivatives of a function of two variables *(page 444)*. For an example of finding the partial derivatives of a function of two variables, see Example 1.

2. Explain the notation used for first partial derivatives *(page 445)*. For examples of using the notation and evaluating partial derivatives, see Example 2.

3. State the guidelines for finding the slopes of a surface at a point *(page 446)*. For an example of finding slopes, see Example 3.

4. State the test for complementary and substitute products *(page 448)*. For an example of using the test, see Example 4.

5. Explain how to find the partial derivatives of a function of three variables *(page 449)*. For an example of finding the partial derivatives of a function of three variables, see Example 5.

6. List the different ways to find the second partial derivatives of a function of two variables *(page 450)*. For an example of finding the second partial derivatives of a function of two variables, see Example 6.

SKILLS WARM UP 7.4

The following warm-up exercises involve skills that were covered in earlier sections. You will use these skills in the exercise set for this section. For additional help, review Sections 2.2, 2.4, 2.5, 4.3, and 4.5.

In Exercises 1–8, find the derivative of the function.

1. $f(x) = 6x^2 - 9x + 4$

2. $g(x) = (3 - x^2)^3$

3. $g(t) = te^{t^3+1}$

4. $f(x) = e^{2x}\sqrt{1 - e^{2x}}$

5. $f(x) = \ln(3 - 2x)$

6. $u(t) = \ln\sqrt{t^3 - 6t}$

7. $g(x) = \dfrac{5x^2}{(4x - 1)^2}$

8. $f(x) = \dfrac{(x + 2)^3}{(x^2 - 9)^2}$

In Exercises 9 and 10, find the slope of the graph of the function at the point (2, 4).

9. $f(x) = x^2 e^{x-2}$

10. $g(x) = x\sqrt{x^2 - x + 2}$

Exercises 7.4

See *CalcChat.com* for tutorial help and worked-out solutions to odd-numbered exercises.

 Finding Partial Derivatives In Exercises 1–14, find the first partial derivatives. *See Example 1.*

1. $z = 3x + 5y - 1$

2. $z = x^2 - 2y$

3. $z = 6x - x^2y + 8y^2$

4. $z = 4xy^2 - x^2y^3 + 7$

5. $h(x, y) = e^{-(x^2+y^2)}$

6. $g(x, y) = e^{x/y}$

7. $f(x, y) = \sqrt{x^2 + y^2}$

8. $z = \sqrt[3]{2x^3 + 5y}$

9. $z = y^2 e^{2xy}$

10. $z = xe^{x+y}$

11. $g(x, y) = \ln(x^2 + y^2)$

12. $z = \ln\dfrac{x + y}{x - y}$

13. $f(x, y) = \dfrac{x^4y}{3y + 2}$

14. $f(x, y) = \dfrac{xy}{x^2 + y^2}$

 Finding Partial Derivatives In Exercises 15 and 16, use the limit definition of partial derivatives to find $\partial z/\partial x$ and $\partial z/\partial y$.

15. $z = 3x + 2y$

16. $z = x^2 - 2xy + y^2$

 Finding and Evaluating Partial Derivatives In Exercises 17–24, find the first partial derivatives and evaluate each at the given point. *See Example 2.*

17. $f(x, y) = 3x^2 + xy - y^2$; (2, 1)

18. $f(x, y) = x^2 - 3xy + y^2$; (1, −1)

19. $f(x, y) = e^xy^2$; (ln 3, 2)

20. $f(x, y) = 3e^{xy}$; (1, ln 5)

21. $f(x, y) = \ln(3x + 5y)$; (1, 0)

22. $f(x, y) = \ln\sqrt{xy}$; (−1, −1)

23. $f(x, y) = \dfrac{xy}{x - y}$; (2, −2)

24. $f(x, y) = \dfrac{4xy}{\sqrt{x^2 + y^2}}$; (1, 0)

Finding Slopes in the x- and y-Directions In Exercises 25–28, find the slopes of the surface at the given point in (a) the x-direction and (b) the y-direction. *See Example 3.*

25. $z = xy$

(1, 2, 2)

26. $z = \sqrt{25 - x^2 - y^2}$

(3, 0, 4)

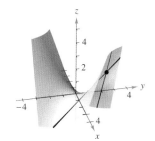

27. $z = 4 - x^2 - y^2$

(1, 1, 2)

28. $z = x^2 - y^2$

(−2, 1, 3)

 Finding Partial Derivatives In Exercises 29–32, find the first partial derivatives with respect to x, y, and z. *See Example 5.*

29. $w = xy^2z^4 + 9xy - z$

30. $w = x^3yz^2 + xy - 4yz$

31. $w = \dfrac{2z}{x + y}$

32. $w = \dfrac{xy}{x + y + z}$

Finding and Evaluating Partial Derivatives In Exercises 33–40, find the first partial derivatives with respect to *x*, *y*, and *z*, and evaluate each at the given point.

33. $w = 2xz^2 + 3xyz - 6y^2z$; $(1, -1, 2)$

34. $w = 3x^2y - 5xyz + 10yz^2$; $(3, 4, -2)$

35. $w = \sqrt{x^2 + y^2 + z^2}$; $(2, -1, 2)$

36. $w = \sqrt{3x^2 + y^2 - 2z^2}$; $(1, -2, 1)$

37. $w = y^3z^2e^{2x^2}$; $\left(\frac{1}{2}, -1, 2\right)$

38. $w = xye^{9z^3}$; $\left(2, 3, -\frac{1}{3}\right)$

39. $w = \ln(5x + 2y^3 - 3z)$; $(4, 1, -1)$

40. $w = \ln\sqrt{x^2 + y^2 + z^2}$; $(3, 0, 4)$

Using First Partial Derivatives In Exercises 41–44, find all values of *x* and *y* such that $f_x(x, y) = 0$ and $f_y(x, y) = 0$ simultaneously.

41. $f(x, y) = x^2 + 4xy + y^2 - 4x + 16y + 3$

42. $f(x, y) = 3x^3 - 12xy + y^3$

43. $f(x, y) = \dfrac{1}{x} + \dfrac{1}{y} + xy$

44. $f(x, y) = \ln(x^2 + y^2 + 1)$

 Finding Second Partial Derivatives In Exercises 45–52, find the four second partial derivatives. *See Example 6.*

45. $z = x^3 - 4y^2$ **46.** $z = 2x^2 + y^5$

47. $z = x^4 - 2xy + 3y^3$ **48.** $z = y^3 - 6x^2y^2 - 1$

49. $z = (3x^4 - 2y^3)^3$ **50.** $z = \sqrt{9 - x^2 - y^2}$

51. $z = \dfrac{x^2 - y^2}{2xy}$ **52.** $z = \dfrac{x}{x + y}$

Finding and Evaluating Second Partial Derivatives In Exercises 53–56, find the four second partial derivatives and evaluate each at the given point.

53. $f(x, y) = x^4 - 3x^2y^2 + y^2$; $(1, 0)$

54. $f(x, y) = x^3 + 2xy^3 - 3y$; $(3, 2)$

55. $f(x, y) = y^3e^{x^2}$; $(1, -1)$

56. $f(x, y) = \ln(x - y)$; $(0, -2)$

 Finding Second Partial Derivatives In Exercises 57–60, find the nine second partial derivatives. *See Example 7.*

57. $w = x^2 - 3xy + 4yz + z^3$

58. $w = x^2y^3 + 2xyz - 3yz$

59. $w = \dfrac{4xz}{x + y}$ **60.** $w = \dfrac{xy}{x + y + z}$

61. Marginal Cost A company manufactures mountain bikes and racing bikes. The cost function for producing *x* mountain bikes and *y* racing bikes is given by

$$C = 10\sqrt{xy} + 149x + 189y + 675.$$

(a) Find the marginal costs ($\partial C/\partial x$ and $\partial C/\partial y$) when $x = 120$ and $y = 160$.

(b) When additional production is required, which model of bicycle results in the cost increasing at a higher rate? How can this be determined from the cost model?

62. Marginal Revenue A pharmaceutical corporation has two locations that produce the same over-the-counter medicine. If x_1 and x_2 are the numbers of units produced at Location 1 and Location 2, respectively, then the total revenue for the product is given by

$$R = 200x_1 + 200x_2 - 4x_1^2 - 8x_1x_2 - 4x_2^2.$$

When $x_1 = 4$ and $x_2 = 12$, find

(a) the marginal revenue for Location 1, $\partial R/\partial x_1$.

(b) the marginal revenue for Location 2, $\partial R/\partial x_2$.

63. Marginal Productivity Consider the Cobb-Douglas production function $f(x, y) = 200x^{0.7}y^{0.3}$. When $x = 1000$ and $y = 500$, find

(a) the marginal productivity of labor, $\partial f/\partial x$.

(b) the marginal productivity of capital, $\partial f/\partial y$.

64. Marginal Productivity Repeat Exercise 63 for the production function given by $f(x, y) = 100x^{0.75}y^{0.25}$.

 Complementary and Substitute Products In Exercises 65 and 66, determine whether the demand functions describe products in the same market that are complementary, substitutes, or neither. Using the notation of Example 4, let x_1 and x_2 be the demands for two products whose prices per unit (in dollars) are p_1 and p_2, respectively. *See Example 4.*

65. $x_1 = 150 - 2p_1 - \frac{5}{2}p_2$, $x_2 = 350 - \frac{3}{2}p_1 - 3p_2$

66. $x_1 = 150 - 2p_1 + 1.8p_2$, $x_2 = 350 + \frac{3}{4}p_1 - 1.9p_2$

67. Expenditures The expenditures *z* (in billions of dollars) for spectator sports from 2008 through 2013 can be modeled by

$$z = 0.391x - 0.069y + 6$$

where *x* is the expenditures (in billions of dollars) on amusement parks and campgrounds, and *y* is the expenditures (in billions of dollars) on live entertainment (excluding sports). *(Source: U.S. Bureau of Economic Analysis)*

(a) Find $\partial z/\partial x$ and $\partial z/\partial y$.

(b) Interpret the partial derivatives in the context of the problem.

68. Shareholder's Equity   The shareholder's equity z (in millions of dollars) for Skechers from 2009 through 2013 can be modeled by

$$z = 0.105x + 0.339y + 257$$

where x is the sales (in millions of dollars) and y is the total assets (in millions of dollars).   *(Source: Skechers U.S.A., Inc.)*

(a) Find $\partial z/\partial x$ and $\partial z/\partial y$.

(b) Interpret the partial derivatives in the context of the problem.

69. Psychology   Early in the twentieth century, an intelligence test called the *Stanford-Binet Test* (more commonly known as the *IQ test*) was developed. In this test, an individual's mental age M is divided by the individual's chronological age C and the quotient is multiplied by 100. The result is the individual's IQ.

$$IQ(M, C) = \frac{M}{C} \times 100$$

Find the partial derivatives of IQ with respect to M and with respect to C. Evaluate the partial derivatives at the point $(12, 10)$ and interpret the result.   *(Source: Adapted from Bernstein/Clark-Stewart/Roy/Wickens, Psychology, Fourth Edition)*

70.   HOW DO YOU SEE IT?   Use the graph of the surface to determine the sign of each partial derivative. Explain your reasoning.

(a) $f_x(4, 1)$ (b) $f_y(4, 1)$

(c) $f_x(-1, -2)$ (d) $f_y(-1, -2)$

71. Investment   The value of an investment of $1000 earning 10% interest compounded annually is

$$V(I, R) = 1000\left[\frac{1 + 0.10(1 - R)}{1 + I}\right]^{10}$$

where I is the annual rate of inflation (in decimal form) and R is the tax rate (in decimal form) for the person making the investment. Calculate $V_I(0.03, 0.28)$ and $V_R(0.03, 0.28)$. Determine whether the tax rate or the rate of inflation is the greater "negative" influence on the growth of the investment.

72. Think About It   Let N be the number of applicants to a university, p the charge for food and housing at the university, and t the tuition. Suppose that N is a function of p and t such that $\partial N/\partial p < 0$ and $\partial N/\partial t < 0$. What information is gained by noticing that both partials are negative?

73. Marginal Utility   The utility function $U = f(x, y)$ is a measure of the utility (or satisfaction) derived by a person from the consumption of two products x and y. The utility function for two products is $U = -5x^2 + xy - 3y^2$.

(a) Determine the marginal utility of product x.

(b) Determine the marginal utility of product y.

(c) When $x = 2$ and $y = 3$, should a person consume one more unit of product x or one more unit of product y? Explain your reasoning.

(d) Use a three-dimensional graphing utility to graph the function. Interpret the marginal utilities of products x and y graphically.

Business Capsule

Daniel and Stephanie Rensing founded The Smart Baker in 2009, which produces a line of aprons, towels, and other kitchen and baking tools. They appeared on the television show *Shark Tank* in March 2012. After the show aired, sales increased by 5000 percent compared to the previous year. Reruns of the episode continue to produce a bump in sales.

74. Research Project   Use your school's library, the Internet, or some other reference source to research a company that increased demand for its product by media or social media exposure. Use graphs to show how a change in demand is related to a change in the marginal utility of a product or service.

7.5 Extrema of Functions of Two Variables

■ Understand the relative extrema of functions of two variables.
■ Use the First-Partials Test to find the relative extrema of functions of two variables.
■ Use the Second-Partials Test to find the relative extrema of functions of two variables.
■ Use relative extrema to answer questions about real-life situations.

In Exercise 45 on page 462, you will find the dimensions of a rectangular package of maximum volume that can be sent by a shipping company.

Relative Extrema

Earlier in the text, you learned how to use derivatives to find the relative minimum and relative maximum values of a function of a single variable. In this section, you will learn how to use partial derivatives to find the relative minimum and relative maximum values of a function of two variables.

Relative Extrema of a Function of Two Variables

Let f be a function defined on a region containing (x_0, y_0). The function f has a **relative maximum** at (x_0, y_0) when there is a circular region R centered at (x_0, y_0) such that

$$f(x, y) \le f(x_0, y_0) \qquad \text{f has a relative maximum at (x_0, y_0).}$$

for all (x, y) in R. The function f has a **relative minimum** at (x_0, y_0) when there is a circular region R centered at (x_0, y_0) such that

$$f(x, y) \ge f(x_0, y_0) \qquad \text{f has a relative minimum at (x_0, y_0).}$$

for all (x, y) in R.

To say that f has a relative maximum at (x_0, y_0) means that the point (x_0, y_0, z_0) is at least as high as all nearby points on the graph of $z = f(x, y)$. Similarly, f has a relative minimum at (x_0, y_0) when (x_0, y_0, z_0) is at least as low as all nearby points on the graph. (See Figure 7.26.)

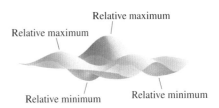

Relative maximum

Relative maximum

Relative minimum

Relative minimum

Relative Extrema

FIGURE 7.26

As in single-variable calculus, you need to distinguish between relative extrema and absolute extrema of a function of two variables. The number $f(x_0, y_0)$ is an absolute maximum of f in the region R when it is greater than or equal to all other function values in the region. (An absolute minimum of f in a region is defined similarly.) For instance, the function

$$f(x, y) = -(x^2 + y^2)$$

is a paraboloid, opening downward, with vertex at $(0, 0, 0)$. (See Figure 7.27.) The number

$$f(0, 0) = 0$$

is an absolute maximum of the function over the entire xy-plane.

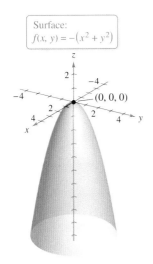

Surface:
$f(x, y) = -(x^2 + y^2)$

$(0, 0, 0)$

f has an absolute maximum at $(0, 0, 0)$.

FIGURE 7.27

The First-Partials Test for Relative Extrema

To locate the relative extrema of a function of two variables, you can use a procedure that is similar to the First-Derivative Test used for functions of a single variable. In the test described below, note the use of the term *open region*. An open region in the xy-plane is similar to an open interval on the real number line. For instance, the region R consisting of the interior of the circle $x^2 + y^2 = 1$ is an open region. If the region R consists of the interior of the circle *and* the points on the circle, then R is a *closed region*.

First-Partials Test for Relative Extrema

If f has a relative extremum at (x_0, y_0) on an open region R in the xy-plane, and the first partial derivatives of f exist in R, then

$$f_x(x_0, y_0) = 0$$

and

$$f_y(x_0, y_0) = 0$$

as shown in Figure 7.28.

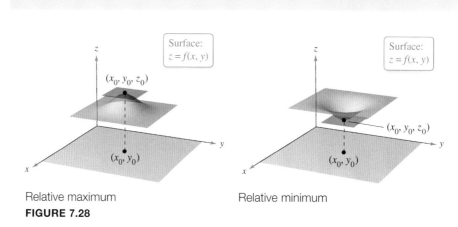

Relative maximum   Relative minimum

FIGURE 7.28

Let f be defined on an open region R containing (x_0, y_0). The point (x_0, y_0) is a **critical point** of f when one of the following is true.

1. Both $f_x(x_0, y_0)$ and $f_y(x_0, y_0)$ are zero.

2. Either $f_x(x_0, y_0)$ or $f_y(x_0, y_0)$ is undefined.

The First-Partials Test states that if the first partial derivatives exist, then you need only examine values of $f(x, y)$ at critical points to find the relative extrema. As is true for a function of a single variable, however, the critical points of a function of two variables do not always yield relative extrema. For instance, the point $(0, 0)$ is a critical point of the surface shown in Figure 7.29, but $f(0, 0)$ is not a relative extremum of the function. Such points are called **saddle points** of the function.

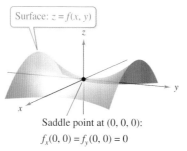

Saddle point at $(0, 0, 0)$:
$$f_x(0, 0) = f_y(0, 0) = 0$$

FIGURE 7.29

Surface:
$f(x, y) = 2x^2 + y^2 + 8x - 6y + 20$

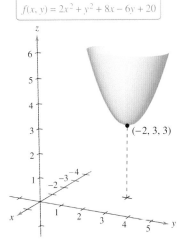

$(-2, 3, 3)$

FIGURE 7.30

EXAMPLE 1 Finding Relative Extrema

Find the relative extrema of

$$f(x, y) = 2x^2 + y^2 + 8x - 6y + 20.$$

SOLUTION Begin by finding the first partial derivatives of f.

$$f_x(x, y) = 4x + 8 \quad \text{and} \quad f_y(x, y) = 2y - 6$$

Because these partial derivatives are defined for all points in the xy-plane, the only critical points are those for which both first partial derivatives are zero. To locate these points, set $f_x(x, y)$ and $f_y(x, y)$ equal to 0, and solve the resulting system of equations.

$$4x + 8 = 0 \qquad \text{Set } f_x(x, y) \text{ equal to 0.}$$
$$2y - 6 = 0 \qquad \text{Set } f_y(x, y) \text{ equal to 0.}$$

The solution of this system is $x = -2$ and $y = 3$. So, the point $(-2, 3)$ is the only critical number of f. In Figure 7.30, you can see that this critical point yields a relative minimum of the function. So, the function has only one relative extremum, which is

$$f(-2, 3) = 3. \qquad \text{Relative minimum}$$

✓ **Checkpoint 1** Worked-out solution available at LarsonAppliedCalculus.com

Find the relative extrema of

$$f(x, y) = x^2 + 2y^2 + 16x - 8y + 8.$$

Example 1 shows a relative minimum occurring at one type of critical point—the type for which both $f_x(x, y)$ and $f_y(x, y)$ are zero. The next example shows a relative maximum that occurs at the other type of critical point—the type for which either $f_x(x, y)$ or $f_y(x, y)$ is undefined.

Surface:
$f(x, y) = 1 - (x^2 + y^2)^{1/3}$

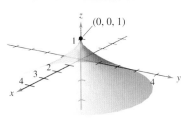

$(0, 0, 1)$

$f_x(x, y)$ and $f_y(x, y)$ are undefined at $(0, 0)$.

FIGURE 7.31

EXAMPLE 2 Finding Relative Extrema

Find the relative extrema of

$$f(x, y) = 1 - (x^2 + y^2)^{1/3}.$$

SOLUTION Begin by finding the first partial derivatives of f.

$$f_x(x, y) = -\frac{2x}{3(x^2 + y^2)^{2/3}} \quad \text{and} \quad f_y(x, y) = -\frac{2y}{3(x^2 + y^2)^{2/3}}$$

These partial derivatives are defined for all points in the xy-plane *except* the point $(0, 0)$. So, $(0, 0)$ is a critical point of f. Moreover, this is the only critical point, because there are no other values of x and y for which either partial derivative is undefined or for which both partial derivatives are zero. In Figure 7.31, you can see that this critical point yields a relative maximum of the function. So, the function has only one relative extremum, which is

$$f(0, 0) = 1. \qquad \text{Relative maximum}$$

✓ **Checkpoint 2** Worked-out solution available at LarsonAppliedCalculus.com

Find the relative extrema of

$$f(x, y) = \sqrt{1 - \frac{x^2}{16} - \frac{y^2}{4}}.$$

The Second-Partials Test for Relative Extrema

For functions such as those in Examples 1 and 2, you can determine the *types* of extrema at the critical points by sketching the graph of the function. For more complicated functions, a graphical approach is not so easy to use. The **Second-Partials Test** is an analytical test that can be used to determine whether a critical number yields a relative minimum, a relative maximum, or neither.

Second-Partials Test for Relative Extrema

Let f have continuous second partial derivatives on an open region containing (a, b) for which

$$f_x(a, b) = 0 \quad \text{and} \quad f_y(a, b) = 0.$$

To test for relative extrema of f, consider the quantity

$$d = f_{xx}(a, b) f_{yy}(a, b) - [f_{xy}(a, b)]^2.$$

1. If $d > 0$ and $f_{xx}(a, b) > 0$, then f has a **relative minimum** at (a, b).
2. If $d > 0$ and $f_{xx}(a, b) < 0$, then f has a **relative maximum** at (a, b).
3. If $d < 0$, then $(a, b, f(a, b))$ is a **saddle point**.
4. The test gives no information when $d = 0$.

Note that if $d > 0$, then $f_{xx}(a, b)$ and $f_{yy}(a, b)$ must have the same sign. So, you can replace $f_{xx}(a, b)$ with $f_{yy}(a, b)$ in the first two parts of the test.

ALGEBRA TUTOR

For help in solving the system of equations

$$y - x^3 = 0$$
$$x - y^3 = 0$$

in Example 3, see Example 1(a) in the *Chapter 7 Algebra Tutor*, on page 496.

$\left(-1, -1, \frac{1}{2}\right)$

$(0, 0, 0)$ $\left(1, 1, \frac{1}{2}\right)$

$f(x, y) = xy - \frac{1}{4}x^4 - \frac{1}{4}y^4$

FIGURE 7.32

EXAMPLE 3   Applying the Second-Partials Test

Find the relative extrema and saddle points of $f(x, y) = xy - \frac{1}{4}x^4 - \frac{1}{4}y^4$.

SOLUTION   Begin by finding the critical points of f. Because

$$f_x(x, y) = y - x^3 \quad \text{and} \quad f_y(x, y) = x - y^3$$

are defined for all points in the xy-plane, the only critical points are those for which both first partial derivatives are zero. By solving the equations

$$y - x^3 = 0 \quad \text{and} \quad x - y^3 = 0$$

simultaneously, you can determine that the critical points are $(1, 1)$, $(-1, -1)$, and $(0, 0)$. Furthermore, because

$$f_{xx}(x, y) = -3x^2, \quad f_{yy}(x, y) = -3y^2, \quad \text{and} \quad f_{xy}(x, y) = 1$$

you can use the quantity $d = f_{xx}(a, b) f_{yy}(a, b) - [f_{xy}(a, b)]^2$ to classify the critical points, as shown.

Critical Point	d	$f_{xx}(x, y)$	Conclusion
$(1, 1)$	$(-3)(-3) - 1 = 8$	-3	Relative maximum
$(-1, -1)$	$(-3)(-3) - 1 = 8$	-3	Relative maximum
$(0, 0)$	$(0)(0) - 1 = -1$	0	Saddle point

The graph of f is shown in Figure 7.32.

✓ Checkpoint 3   Worked-out solution available at LarsonAppliedCalculus.com

Find the relative extrema and saddle points of $f(x, y) = \dfrac{y^2}{16} - \dfrac{x^2}{4}$.   ■

Applications

EXAMPLE 4   Finding a Maximum Profit

A company makes two substitute products whose demand functions are given by

$$x_1 = 200(p_2 - p_1)$$ Demand for Product 1

and

$$x_2 = 500 + 100p_1 - 180p_2$$ Demand for Product 2

where p_1 and p_2 are the prices per unit (in dollars) and x_1 and x_2 are the numbers of units sold. The costs of producing the two products are \$0.50 and \$0.75 per unit, respectively. Find the prices that will yield a maximum profit.

SOLUTION   The cost function is

$$C = 0.5x_1 + 0.75x_2$$ Write cost function.
$$= 0.5(200)(p_2 - p_1) + 0.75(500 + 100p_1 - 180p_2)$$ Substitute.
$$= 375 - 25p_1 - 35p_2.$$ Simplify.

The revenue function is

$$R = p_1x_1 + p_2x_2$$ Write revenue function.
$$= p_1(200)(p_2 - p_1) + p_2(500 + 100p_1 - 180p_2)$$ Substitute.
$$= -200p_1^2 - 180p_2^2 + 300p_1p_2 + 500p_2.$$ Simplify.

This implies that the profit function is

$$P = R - C$$ Write profit function.
$$= -200p_1^2 - 180p_2^2 + 300p_1p_2 + 500p_2 - (375 - 25p_1 - 35p_2)$$
$$= -200p_1^2 - 180p_2^2 + 300p_1p_2 + 25p_1 + 535p_2 - 375.$$

Next, find the first partial derivatives of P.

$$\frac{\partial P}{\partial p_1} = -400p_1 + 300p_2 + 25 \qquad \frac{\partial P}{\partial p_2} = 300p_1 - 360p_2 + 535$$

By setting the first partial derivatives equal to zero and solving the equations

$$-400p_1 + 300p_2 + 25 = 0$$
$$300p_1 - 360p_2 + 535 = 0$$

simultaneously, you can conclude that the solution is $p_1 = \$3.14$ and $p_2 = \$4.10$. From the graph of P shown at the right, you can see that this critical number yields a maximum. So, the maximum profit is

$$P(3.14, 4.10) = \$761.48.$$

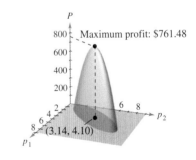

Maximum profit: \$761.48

✓ **Checkpoint 4**   *Worked-out solution available at LarsonAppliedCalculus.com*

Find the prices that will yield a maximum profit for the products in Example 4 when the costs of producing the two products are \$0.75 and \$0.50 per unit, respectively. ■

In Example 4, to convince yourself that the maximum profit is \$761.48, try substituting other prices into the profit function. For each pair of prices, you will obtain a profit that is less than \$761.48. For instance, when $p_1 = \$2$ and $p_2 = \$3$, the profit is $P(2, 3) = \$660$.

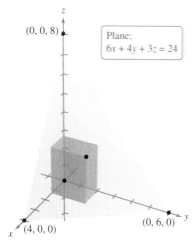

(0, 0, 8)

Plane:
$6x + 4y + 3z = 24$

(4, 0, 0) (0, 6, 0)

FIGURE 7.33

EXAMPLE 5 Finding a Maximum Volume

A rectangular box is resting on the xy-plane with one vertex at the origin. The opposite vertex lies in the plane

$$6x + 4y + 3z = 24$$

as shown in Figure 7.33. Find the maximum volume of the box.

SOLUTION Let x, y, and z represent the length, width, and height of the box. Because one vertex of the box lies in the plane given by $6x + 4y + 3z = 24$, you know that

$$z = \tfrac{1}{3}(24 - 6x - 4y). \qquad \text{Solve for } z.$$

Now, you can write the volume of the box as a function of two variables.

$$
\begin{aligned}
V &= xyz & \text{Volume} = (\text{length})(\text{width})(\text{height})\\
&= xy\left(\tfrac{1}{3}\right)(24 - 6x - 4y) & \text{Substitute for } z.\\
&= \tfrac{1}{3}(24xy - 6x^2y - 4xy^2) & \text{Simplify.}
\end{aligned}
$$

Next, find the first partial derivatives of V.

$$
\begin{aligned}
V_x &= \tfrac{1}{3}(24y - 12xy - 4y^2) & \text{Partial with respect to } x\\
&= \tfrac{1}{3}y(24 - 12x - 4y) & \text{Factor.}\\
V_y &= \tfrac{1}{3}(24x - 6x^2 - 8xy) & \text{Partial with respect to } y\\
&= \tfrac{1}{3}x(24 - 6x - 8y) & \text{Factor.}
\end{aligned}
$$

By solving the equations

$$
\begin{aligned}
\tfrac{1}{3}y(24 - 12x - 4y) &= 0 & \text{Set } V_x \text{ equal to 0.}\\
\tfrac{1}{3}x(24 - 6x - 8y) &= 0 & \text{Set } V_y \text{ equal to 0.}
\end{aligned}
$$

simultaneously, you can conclude that the solutions are $(0, 0)$, $(0, 6)$, $(4, 0)$, and $\left(\tfrac{4}{3}, 2\right)$. Using the Second-Partials Test, you can determine that the maximum volume occurs when the length is $x = \tfrac{4}{3}$ and the width is $y = 2$. For these values, the height of the box is

$$z = \tfrac{1}{3}\left[24 - 6\left(\tfrac{4}{3}\right) - 4(2)\right] = \tfrac{8}{3}.$$

So, the maximum volume is

$$V = xyz = \left(\tfrac{4}{3}\right)(2)\left(\tfrac{8}{3}\right) = \tfrac{64}{9} \text{ cubic units.}$$

✓ **Checkpoint 5** *Worked-out solution available at LarsonAppliedCalculus.com*

Find the maximum volume of a box that is resting on the xy-plane with one vertex at the origin and the opposite vertex in the plane $2x + 4y + z = 8$. ∎

ALGEBRA TUTOR

For help in solving the system of equations

$$y(24 - 12x - 4y) = 0$$
$$x(24 - 6x - 8y) = 0$$

in Example 5, see Example 2(a) in the *Chapter 7 Algebra Tutor*, on page 497.

SUMMARIZE (Section 7.5)

1. State the definition of relative extrema of a function of two variables *(page 455)*. For examples of relative extrema, see Figure 7.26.

2. State the First-Partials Test for relative extrema *(page 456)*. For examples of using the First-Partials Test to find relative extrema, see Examples 1 and 2.

3. State the Second-Partials Test for relative extrema *(page 458)*. For examples of using the Second-Partials Test to find relative extrema, see Examples 3 and 5.

4. Describe a real-life example of how relative extrema can be used to find a company's maximum profit *(page 459, Example 4)*.

SKILLS WARM UP 7.5
The following warm-up exercises involve skills that were covered in earlier sections. You will use these skills in the exercise set for this section. For additional help, review Section 7.4.

In Exercises 1–8, solve the system of equations.

1. $\begin{cases} 5x = 15 \\ 3x - 2y = 5 \end{cases}$

2. $\begin{cases} \frac{1}{2}y = 3 \\ -x + 5y = 19 \end{cases}$

3. $\begin{cases} x + y = 5 \\ x - y = -3 \end{cases}$

4. $\begin{cases} x + y = 8 \\ 2x - y = -2 \end{cases}$

5. $\begin{cases} x - 2y = 4 \\ 5x - 3y = 13 \end{cases}$

6. $\begin{cases} 2x - 4y = 14 \\ 3x + y = 7 \end{cases}$

7. $\begin{cases} x^2 + x = 0 \\ 2yx + y = 0 \end{cases}$

8. $\begin{cases} 3y^2 + 6y = 0 \\ xy + x + 2 = 0 \end{cases}$

In Exercises 9–14, find all first and second partial derivatives.

9. $z = 4x^3 - 3y^2$

10. $z = 2x^5 - y^3$

11. $z = x^4 - \sqrt{xy} + 2y$

12. $z = 2x^2 - 3xy + y^2$

13. $z = ye^{xy^2}$

14. $z = xe^{xy}$

Exercises 7.5
See *CalcChat.com* for tutorial help and worked-out solutions to odd-numbered exercises.

Applying the Second-Partials Test In Exercises 1–18, find the relative extrema and saddle points of the function. *See Examples 1, 2, and 3.*

1. $f(x, y) = x^2 + y^2 + 8x - 12y - 3$
2. $f(x, y) = x^2 + y^2 + 2x - 6y + 6$
3. $f(x, y) = -3x^2 - 4y^2 - 6x - 2y + 9$
4. $f(x, y) = 7x^2 + 2y^2 - 7x + 16y - 13$
5. $f(x, y) = \sqrt{x^2 + y^2 + 1}$
6. $f(x, y) = -4(x^2 + y^2 + 81)^{1/4}$
7. $f(x, y) = (3x^2 + 6y^2 + 8)^{1/3}$
8. $f(x, y) = \sqrt{25 - (x - 2)^2 - y^2}$
9. $f(x, y) = (x - 1)^2 + (y - 3)^2$
10. $f(x, y) = 9 - (x - 3)^2 - (y + 2)^2$
11. $f(x, y) = -5x^2 + 4xy - y^2 + 16x + 10$
12. $f(x, y) = x^2 + 6xy + 10y^2 - 4y + 4$
13. $f(x, y) = -x^3 + 4xy - 2y^2 + 1$
14. $f(x, y) = 2xy - \frac{1}{2}x^4 - \frac{1}{2}y^4 + 1$
15. $f(x, y) = \frac{1}{2}xy$

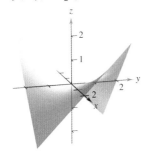

16. $f(x, y) = x + y + 2xy - x^2 - y^2$

17. $f(x, y) = (x + y)e^{1 - x^2 - y^2}$

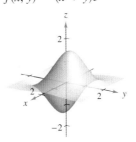

18. $f(x, y) = 3e^{-(x^2 + y^2)}$

Think About It In Exercises 19–24, determine whether there is a relative maximum, a relative minimum, a saddle point, or insufficient information to determine the nature of the function $f(x, y)$ at the critical point (x_0, y_0).

19. $f_{xx}(x_0, y_0) = 9$, $f_{yy}(x_0, y_0) = 4$, $f_{xy}(x_0, y_0) = 6$

20. $f_{xx}(x_0, y_0) = -3$, $f_{yy}(x_0, y_0) = -8$, $f_{xy}(x_0, y_0) = 2$

21. $f_{xx}(x_0, y_0) = -9$, $f_{yy}(x_0, y_0) = 6$, $f_{xy}(x_0, y_0) = 10$

22. $f_{xx}(x_0, y_0) = 25$, $f_{yy}(x_0, y_0) = 5$, $f_{xy}(x_0, y_0) = -11$

23. $f_{xx}(x_0, y_0) = 4$, $f_{yy}(x_0, y_0) = 8$, $f_{xy}(x_0, y_0) = 3$

24. $f_{xx}(x_0, y_0) = 8$, $f_{yy}(x_0, y_0) = 7$, $f_{xy}(x_0, y_0) = 9$

Analyzing a Function In Exercises 25–30, find the critical points, relative extrema, and saddle points of the function. List the critical points for which the Second-Partials Test fails.

25. $f(x, y) = x^3 + y^3$

26. $f(x, y) = x^3 + y^3 - 3x^2 + 6y^2 + 3x + 12y + 7$

27. $f(x, y) = (xy)^2$

28. $f(x, y) = \sqrt{x^2 + y^2}$

29. $f(x, y) = x^{2/3} + y^{2/3}$

30. $f(x, y) = (x^2 + y^2)^{2/3}$

 Analyzing a Function of Three Variables In Exercises 31 and 32, find the critical points of the function and, from the form of the function, determine whether a relative maximum or a relative minimum occurs at each point.

31. $f(x, y, z) = (x - 1)^2 + (y + 3)^2 + z^2$

32. $f(x, y, z) = 6 - [x(y + 2)(z - 1)]^2$

Finding Positive Numbers In Exercises 33–36, find three positive numbers x, y, and z that satisfy the given conditions.

33. The sum is 45 and the product is a maximum.

34. The sum is 32 and $P = xy^2z$ is a maximum.

35. The sum is 60 and the sum of the squares is a minimum.

36. The product is 1 and the sum of the squares is a minimum.

37. Revenue A company manufactures running shoes and basketball shoes. The total revenue from x_1 units of running shoes and x_2 units of basketball shoes is

$$R = -5x_1^2 - 8x_2^2 - 2x_1x_2 + 42x_1 + 102x_2$$

where x_1 and x_2 are in thousands of units. Find x_1 and x_2 so as to maximize the revenue.

38. Revenue A retail outlet sells two types of riding lawn mowers, the prices of which are p_1 and p_2. Find p_1 and p_2 so as to maximize total revenue, where

$$R = 870p_1 + 950p_2 + 1.5p_1p_2 - 1.5p_1^2 - p_2^2.$$

 Revenue In Exercises 39 and 40, find p_1 and p_2, the prices per unit (in dollars), so as to maximize the total revenue $R = x_1p_1 + x_2p_2$, where x_1 and x_2 are the numbers of units sold, for a retail outlet that sells two substitute products with the given demand functions. *See Example 4.*

39. $x_1 = 1000 - 2p_1 + p_2$, $x_2 = 1500 + 2p_1 - 1.5p_2$

40. $x_1 = 600 - 4p_1 + 2p_2$, $x_2 = 870 + 4p_1 - 3p_2$

41. Profit A corporation manufactures a product for a high-performance automobile engine at two locations. The cost of producing x_1 units at Location 1 is

$$C_1 = 0.05x_1^2 + 15x_1 + 5400$$

and the cost of producing x_2 units at Location 2 is

$$C_2 = 0.03x_2^2 + 15x_2 + 6100.$$

The demand function for the product is

$$p = 225 - 0.4(x_1 + x_2)$$

and the total revenue function is

$$R = [225 - 0.4(x_1 + x_2)](x_1 + x_2).$$

Find the production levels at the two locations that will maximize the profit $P = R - C_1 - C_2$.

42. Profit A corporation manufactures candles at two locations. The cost of producing x_1 units at Location 1 is $C_1 = 0.02x_1^2 + 4x_1 + 500$ and the cost of producing x_2 units at Location 2 is $C_2 = 0.05x_2^2 + 4x_2 + 275$. The candles sell for \$15 per unit. Find the quantity that should be produced at each location to maximize the profit $P = 15(x_1 + x_2) - C_1 - C_2$.

 Finding a Maximum Volume In Exercises 43 and 44, find the maximum volume of a box that is resting on the xy-plane with one vertex at the origin and the opposite vertex in the given plane. *See Example 5.*

43. $5x + 2y + z = 10$ **44.** $3x + 4y + 2z = 12$

45. Volume Find the dimensions of a rectangular package of maximum volume that may be sent by a shipping company, assuming that the sum of the length and the girth (perimeter of a cross section) cannot exceed 96 inches.

46. Volume Repeat Exercise 45, assuming that the sum of the length and the girth cannot exceed 144 inches.

47. Cost A manufacturer makes an open-top wooden crate having a volume of 18 cubic feet. Material costs are $0.20 per square foot for the base and $0.15 per square foot for the sides. Find the dimensions that minimize the cost of each crate. What is the minimum cost?

48. Cost A home improvement contractor is painting the walls and ceiling of a rectangular room. The volume of the room is 1584 cubic feet. The cost of wall paint is $0.06 per square foot and the cost of ceiling paint is $0.11 per square foot. Find the room dimensions that minimize the cost of the paint. What is the minimum cost?

49. Cost An automobile manufacturer has determined that its annual labor and equipment cost (in millions of dollars) can be modeled by

$$C(x, y) = 2x^2 + 3y^2 - 15x - 20y + 4xy + 39$$

where x is the amount spent per year (in millions of dollars) on labor and y is the amount spent per year (in millions of dollars) on equipment. Find the values of x and y that minimize the annual labor and equipment cost. What is this minimum cost?

50. Medicine In order to treat a certain bacterial infection, a combination of two drugs is being tested. Studies have shown that the duration of the infection in laboratory tests can be modeled by

$$D(x, y) = x^2 + 2y^2 - 18x - 24y + 2xy + 120$$

where x is the dosage (in hundreds of milligrams) of the first drug and y is the dosage (in hundreds of milligrams) of the second drug. Find the amount of each drug necessary to minimize the duration of the infection.

51. Biology A lake is to be stocked with smallmouth and largemouth bass. Let x represent the number of smallmouth bass and let y represent the number of largemouth bass. The weight of each fish is dependent on the population densities. After a six-month period, the weight of a single smallmouth bass is given by

$$W_1 = 3 - 0.002x - 0.001y$$

and the weight of a single largemouth bass is given by

$$W_2 = 4.5 - 0.004x - 0.005y.$$

Assuming that no fish die during the six-month period, how many smallmouth and largemouth bass should be stocked in the lake so that the *total* weight T of bass in the lake is a maximum?

52. **HOW DO YOU SEE IT?** The figure shows the level curves for a function $f(x, y)$. What, if any, information can be said about f at the points A, B, C, and D? Explain your reasoning.

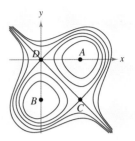

53. Hardy-Weinberg Law Common blood types are determined genetically by the three alleles A, B, and O. (An allele is any of a group of possible mutational forms of a gene.) A person whose blood type is AA, BB, or OO is homozygous. A person whose blood type is AB, AO, or BO is heterozygous. The Hardy-Weinberg Law states that the proportion P of heterozygous individuals in any given population is

$$P(p, q, r) = 2pq + 2pr + 2qr$$

where p represents the percent of allele A in the population, q represents the percent of allele B in the population, and r represents the percent of allele O in the population. Use the fact that $p + q + r = 1$ (the sum of the three must equal 100%) to show that the maximum proportion of heterozygous individuals in any population is $\frac{2}{3}$.

54. Shannon Diversity Index One way to measure species diversity is to use the Shannon diversity index H. A habitat consists of three species A, B, and C, and its Shannon diversity index is

$$H = -x \ln x - y \ln y - z \ln z$$

where x is the percent of species A in the habitat, y is the percent of species B in the habitat, and z is the percent of species C in the habitat. Use the fact that $x + y + z = 1$ (the sum of the three must equal 100%) to show that the maximum value of H occurs when

$$x = y = z = \frac{1}{3}.$$

What is the maximum value of H?

True or False? In Exercises 55 and 56, determine whether the statement is true or false. If it is false, explain why or give an example that shows it is false.

55. A saddle point always occurs at a critical point.

56. If $f(x, y)$ has a relative maximum at (x_0, y_0, z_0), then

$$f_x(x_0, y_0) = f_y(x_0, y_0) = 0.$$

QUIZ YOURSELF
See *CalcChat.com* for tutorial help and worked-out solutions to odd-numbered exercises.

Take this quiz as you would take a quiz in class. When you are done, check your work against the answers given in the back of the book.

In Exercises 1–3, (a) plot the points in a three-dimensional coordinate system, (b) find the distance between the two points, and (c) find the midpoint of the line segment joining the two points.

1. $(1, 3, 2), (-1, 2, 0)$ **2.** $(2, 1, 4), (0, -5, -6)$ **3.** $(-2, 0, 3), (5, -1, 2)$

In Exercises 4 and 5, find the standard equation of the sphere with the given characteristics.

4. Center: $(-1, 7, 4)$; radius: 6

5. Endpoints of a diameter: $(0, 3, 1), (2, 5, -5)$

6. Find the center and radius of the sphere whose equation is

$$x^2 + y^2 + z^2 - 8x - 2y - 6z - 23 = 0.$$

In Exercises 7–9, find the *x*-, *y*-, and *z*-intercepts of the plane. Then sketch the plane.

7. $2x + 3y + z = 6$ **8.** $x - 2z = 4$ **9.** $y = 3$

In Exercises 10–12, classify the quadric surface.

10. $\dfrac{x^2}{4} + \dfrac{y^2}{9} + \dfrac{z^2}{16} = 1$ **11.** $z^2 - x^2 - y^2 = 25$ **12.** $64y - 8x^2 - z^2 = 0$

In Exercises 13–15, find $f(1, 0)$ and $f(4, -1)$.

13. $f(x, y) = x - 9y^2$ **14.** $f(x, y) = \sqrt{4x^2 + y}$ **15.** $f(x, y) = \ln(x - 2y)$

In Exercises 16 and 17, find the domain and range of the function.

16. $f(x, y) = \sqrt{81 - x^2 - y^2}$ **17.** $f(x, y) = e^{x+y}$

18. Describe the level curves of $f(x, y) = x^2 + y$. Sketch the level curves for $c = 0, 1, 2, 3, 4,$ and 5.

19. Find the monthly payment for a boat loan of \$29,000 taken out for 6 years at an annual interest rate of 5.5%. How much is the total amount paid?

In Exercises 20–23, find the first partial derivatives and evaluate each at the point $(-2, 3)$.

20. $f(x, y) = x^2 + 2y^2 - 3x - y + 1$ **21.** $f(x, y) = \dfrac{3x - y^2}{x + y}$

22. $f(x, y) = x^3 e^{2y}$ **23.** $f(x, y) = \ln(2x + 7y)$

In Exercises 24 and 25, find the relative extrema and saddle points of the function.

24. $f(x, y) = 3x^2 + y^2 - 2xy - 6x + 2y$

25. $f(x, y) = -x^3 + 4xy - 2y^2 + 1$

26. A company manufactures two types of wood-burning stoves: a freestanding model and a fireplace-insert model. The total cost (in thousands of dollars) for producing *x* freestanding stoves and *y* fireplace-insert stoves can be modeled by

$$C(x, y) = \tfrac{1}{16}x^2 + y^2 - 10x - 40y + 820.$$

Find the values of *x* and *y* that minimize the total cost. What is the minimum cost?

7.6 Lagrange Multipliers

■ Understand the Method of Lagrange Multipliers.
■ Use Lagrange multipliers to solve constrained optimization problems.

Lagrange Multipliers with One Constraint

In Example 5 in Section 7.5, you were asked to find the dimensions of the rectangular box of maximum volume that would fit in the first octant beneath the plane

$$6x + 4y + 3z = 24$$

as shown again in the figure at the right. Another way of stating this problem is to say that you are asked to find the maximum of

$$V = xyz \qquad \text{Objective function}$$

subject to the constraint

$$6x + 4y + 3z - 24 = 0. \qquad \text{Constraint}$$

This type of problem is called a **constrained optimization** problem. In Section 7.5, you answered this question by solving for z in the constraint equation and then rewriting V as a function of two variables.

In this section, you will study a different (and often better) way to solve constrained optimization problems. This method involves the use of variables called **Lagrange multipliers,** named after the French mathematician Joseph Louis Lagrange (1736–1813).

(0, 0, 8)

Plane:
$6x + 4y + 3z = 24$

$\left(\frac{4}{3}, 2, \frac{8}{3}\right)$

(0, 6, 0)

(4, 0, 0)

Method of Lagrange Multipliers

If $f(x, y)$ has a maximum or minimum subject to the constraint $g(x, y) = 0$, then it will occur at one of the critical numbers of the function F defined by

$$F(x, y, \lambda) = f(x, y) - \lambda g(x, y).$$

The variable λ (the lowercase Greek letter lambda) is called a **Lagrange multiplier.** To find the minimum or maximum of f, use the following steps.

1. Solve the following system of equations.

$$F_x(x, y, \lambda) = 0$$
$$F_y(x, y, \lambda) = 0$$
$$F_\lambda(x, y, \lambda) = 0$$

2. Evaluate f at each solution point obtained in the first step. The greatest value yields the maximum of f subject to the constraint $g(x, y) = 0$, and the least value yields the minimum of f subject to the constraint $g(x, y) = 0$.

When using the Method of Lagrange Multipliers for functions of three variables, F has the form $F(x, y, z, \lambda) = f(x, y, z) - \lambda g(x, y, z)$. The system of equations used in Step 1 is

$$F_x(x, y, z, \lambda) = 0$$
$$F_y(x, y, z, \lambda) = 0$$
$$F_z(x, y, z, \lambda) = 0$$
$$F_\lambda(x, y, z, \lambda) = 0.$$

In Exercise 39 on page 472, you will use Lagrange multipliers to find the dimensions that will minimize the cost of fencing in two enclosures.

Constrained Optimization Problems

EXAMPLE 1   Using Lagrange Multipliers

Find the maximum of

$$V = xyz$$ Objective function

subject to the constraint

$$6x + 4y + 3z - 24 = 0.$$ Constraint

SOLUTION First, let $f(x, y, z) = xyz$ and $g(x, y, z) = 6x + 4y + 3z - 24$. Then, define a new function F as

$$F(x, y, z, \lambda) = f(x, y, z) - \lambda g(x, y, z)$$
$$= xyz - \lambda(6x + 4y + 3z - 24).$$

To find the critical numbers of F, begin by finding the partial derivatives of F with respect to x, y, z, and λ. Then, set the partial derivatives equal to zero.

$$F_x(x, y, z, \lambda) = yz - 6\lambda \qquad\Longrightarrow\qquad yz - 6\lambda = 0$$
$$F_y(x, y, z, \lambda) = xz - 4\lambda \qquad\Longrightarrow\qquad xz - 4\lambda = 0$$
$$F_z(x, y, z, \lambda) = xy - 3\lambda \qquad\Longrightarrow\qquad xy - 3\lambda = 0$$
$$F_\lambda(x, y, z, \lambda) = -6x - 4y - 3z + 24 \qquad\Longrightarrow\qquad -6x - 4y - 3z + 24 = 0$$

Solving for λ in the first equation produces

$$yz - 6\lambda = 0 \qquad\Longrightarrow\qquad \lambda = \frac{yz}{6}.$$

Substituting for λ in the second and third equations produces the following.

$$xz - 4\left(\frac{yz}{6}\right) = 0 \qquad\Longrightarrow\qquad y = \frac{3}{2}x$$

$$xy - 3\left(\frac{yz}{6}\right) = 0 \qquad\Longrightarrow\qquad z = 2x$$

Next, substitute for y and z in the equation $F_\lambda(x, y, z, \lambda) = 0$ and solve for x.

$$F_\lambda(x, y, z, \lambda) = 0$$
$$-6x - 4y - 3z + 24 = 0$$
$$-6x - 4\left(\tfrac{3}{2}x\right) - 3(2x) + 24 = 0$$
$$-18x = -24$$
$$x = \tfrac{4}{3}$$

Using this x-value, you can conclude that the critical values are $x = \frac{4}{3}$, $y = 2$, and $z = \frac{8}{3}$, which implies that the maximum is

$$V = xyz$$ Write objective function.

$$= \left(\frac{4}{3}\right)(2)\left(\frac{8}{3}\right)$$ Substitute values of x, y, and z.

$$= \frac{64}{9} \text{ cubic units.}$$ Maximum volume

✓ **Checkpoint 1**   *Worked-out solution available at LarsonAppliedCalculus.com*

Find the maximum volume of $V = xyz$ subject to the constraint

$$2x + 4y + z - 8 = 0.$$

For some industrial applications, a simple robot can cost more than a year's wages and benefits for one employee. So, manufacturers must carefully balance the amount of money spent on labor and capital.

TECH TUTOR

You can use a spreadsheet to solve constrained optimization problems. Try using a spreadsheet to solve the problem in Example 2. (Consult the user's manual of a spreadsheet software program for specific instructions on how to solve a constrained optimization problem.)

EXAMPLE 2   Finding a Maximum Production Level

A manufacturer's production is modeled by the Cobb-Douglas function

$$f(x, y) = 100x^{3/4}y^{1/4} \qquad \text{Objective function}$$

where x represents the units of labor (at \$150 per unit) and y represents the units of capital (at \$250 per unit). The total costs for labor and capital cannot exceed \$50,000. Will the manufacturer's maximum production level exceed 16,000 units?

SOLUTION   Because total labor and capital expenses cannot exceed \$50,000, the constraint is

$$150x + 250y = 50,000 \qquad \text{Constraint}$$
$$150x + 250y - 50,000 = 0. \qquad \text{Write in general form.}$$

To find the maximum production level, begin by writing the function

$$F(x, y, \lambda) = 100x^{3/4}y^{1/4} - \lambda(150x + 250y - 50,000).$$

Then find the partial derivatives of F with respect to x, y, and λ.

$$F_x(x, y, \lambda) = 75x^{-1/4}y^{1/4} - 150\lambda$$
$$F_y(x, y, \lambda) = 25x^{3/4}y^{-3/4} - 250\lambda$$
$$F_\lambda(x, y, \lambda) = -150x - 250y + 50,000$$

Next, set the partial derivatives equal to zero to obtain the following system of equations.

$$75x^{-1/4}y^{1/4} - 150\lambda = 0 \qquad \text{Equation 1}$$
$$25x^{3/4}y^{-3/4} - 250\lambda = 0 \qquad \text{Equation 2}$$
$$-150x - 250y + 50,000 = 0 \qquad \text{Equation 3}$$

By solving for λ in the first equation

$$75x^{-1/4}y^{1/4} - 150\lambda = 0 \qquad \text{Equation 1}$$
$$\lambda = \tfrac{1}{2}x^{-1/4}y^{1/4} \qquad \text{Solve for } \lambda.$$

and substituting for λ in Equation 2, you obtain

$$25x^{3/4}y^{-3/4} - 250\left(\tfrac{1}{2}\right)x^{-1/4}y^{1/4} = 0 \qquad \text{Substitute in Equation 2.}$$
$$25x - 125y = 0 \qquad \text{Multiply by } x^{1/4}y^{3/4}.$$
$$x = 5y. \qquad \text{Solve for } x.$$

So, $x = 5y$. By substituting for x in Equation 3, you obtain

$$-150(5y) - 250y + 50,000 = 0 \qquad \text{Substitute in Equation 3.}$$
$$-1000y = -50,000 \qquad \text{Simplify.}$$
$$y = 50. \qquad \text{Solve for } y.$$

When $y = 50$ units of capital, it follows that $x = 5(50) = 250$ units of labor. So, the maximum production level is

$$f(250, 50) = 100(250)^{3/4}(50)^{1/4} \qquad \text{Substitute for } x \text{ and } y.$$
$$\approx 16,719 \text{ units.} \qquad \text{Maximum production level}$$

You can conclude that the maximum production level will exceed 16,000 units.

✓ **Checkpoint 2**   Worked-out solution available at LarsonAppliedCalculus.com

In Example 2, suppose that each labor unit costs \$200 and each capital unit costs \$250. Find the maximum production level when labor and capital cannot exceed \$50,000.

Economists call the Lagrange multiplier obtained in a production function the **marginal productivity of money.** For instance, in Example 2, the marginal productivity of money when $x = 250$ and $y = 50$ is

$$\lambda = \tfrac{1}{2}x^{-1/4}y^{1/4} = \tfrac{1}{2}(250)^{-1/4}(50)^{1/4} \approx 0.334.$$

This means that for each additional dollar spent on production, approximately 0.334 additional unit of the product can be produced.

EXAMPLE 3   Finding a Maximum Production Level

The manufacturer in Example 2 now has $70,000 available for labor and capital. What is the maximum number of units that can be produced?

SOLUTION   You could rework the entire problem, as demonstrated in Example 2. However, because the only change in the problem is the availability of additional money to spend on labor and capital, you can use the fact that the marginal productivity of money is

$$\lambda \approx 0.334.$$

Because an additional $20,000 is available and the maximum production level in Example 2 was 16,719 units, you can conclude that the maximum production level is now

$$16{,}719 + (0.334)(20{,}000) \approx 23{,}400 \text{ units.}$$

Try using the procedure demonstrated in Example 2 to confirm this result.

✓ **Checkpoint 3**   *Worked-out solution available at LarsonAppliedCalculus.com*

The manufacturer in Example 2 now has $80,000 available for labor and capital. What is the maximum number of units that can be produced?

TECH TUTOR

You can use a three-dimensional graphing utility to confirm graphically the results of Examples 2 and 3. Begin by graphing the surface $f(x, y) = 100x^{3/4}y^{1/4}$. Then graph the vertical plane given by $150x + 250y = 50,000$. As shown below, the maximum production level corresponds to the highest point on the intersection of the surface and the plane.

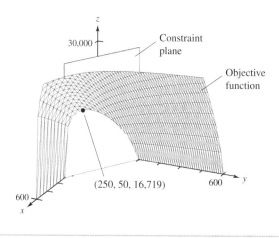

The constrained optimization problem in Example 4 is represented graphically in Figure 7.34. The graph of the objective function is a paraboloid and the graph of the constraint is a vertical plane. In the "unconstrained" optimization problem on page 459, the maximum profit occurred at the vertex of the paraboloid. In this "constrained" problem, however, the maximum profit corresponds to the highest point on the curve that is the intersection of the paraboloid and the vertical "constraint" plane.

FIGURE 7.34

In Example 4 in Section 7.5, you found the maximum profit for two substitute products whose demand functions are given by

$$x_1 = 200(p_2 - p_1)$$ Demand for Product 1
$$x_2 = 500 + 100p_1 - 180p_2.$$ Demand for Product 2

With this model, the total demand, $x_1 + x_2$, is completely determined by the prices p_1 and p_2. In many real-life situations, this assumption is too simplistic; regardless of the prices of the substitute brands, the annual total demands for some products, such as toothpaste, are relatively constant. In such situations, the total demand is **limited,** and variations in price do not affect the total demand as much as they affect the market share of the substitute brands.

EXAMPLE 4   **Finding a Maximum Profit**

A company makes two substitute products whose demand functions are given by

$$x_1 = 200(p_2 - p_1)$$ Demand for Product 1
$$x_2 = 500 + 100p_1 - 180p_2$$ Demand for Product 2

where p_1 and p_2 are the prices per unit (in dollars) and x_1 and x_2 are the numbers of units sold. The costs of producing the two products are $0.50 and $0.75 per unit, respectively. The total demand is limited to 200 units per year. Find the prices that will yield a maximum profit.

SOLUTION   From Example 4 in Section 7.5, the profit function is modeled by

$$P = -200p_1^2 - 180p_2^2 + 300p_1p_2 + 25p_1 + 535p_2 - 375.$$

The total demand for the two products is

$$x_1 + x_2 = 200(p_2 - p_1) + 500 + 100p_1 - 180p_2$$
$$= 200p_2 - 200p_1 + 500 + 100p_1 - 180p_2$$
$$= -100p_1 + 20p_2 + 500.$$

Because the total demand is limited to 200 units, the constraint is

$$-100p_1 + 20p_2 + 500 = 200.$$ Constraint

Using Lagrange multipliers, you can determine that the maximum profit occurs when $p_1 = \$3.94$ and $p_2 = \$4.69$. This corresponds to an annual profit of $712.21.

✔ *Checkpoint 4*   *Worked-out solution available at LarsonAppliedCalculus.com*

In Example 4, find the prices that will yield a maximum profit when the total demand is limited to 250 units per year.   ■

SUMMARIZE   (Section 7.6)

1. Explain the Method of Lagrange Multipliers *(page 465)*. For an example of how to use Lagrange multipliers to solve a constrained optimization problem, see Example 1.

2. Describe a real-life example of using Lagrange multipliers to find a manufacturer's maximum production level *(page 467, Example 2)*.

3. Describe a real-life example of using Lagrange multipliers to find a company's maximum profit *(page 469, Example 4)*.

SKILLS WARM UP 7.6

The following warm-up exercises involve skills that were covered in earlier sections. You will use these skills in the exercise set for this section. For additional help, review Section 7.4.

In Exercises 1–6, solve the system of linear equations.

1. $\begin{cases} 4x - 6y = 3 \\ 2x + 3y = 2 \end{cases}$

2. $\begin{cases} 6x - 6y = 5 \\ -3x - y = 1 \end{cases}$

3. $\begin{cases} 5x - y = 25 \\ x - 5y = 15 \end{cases}$

4. $\begin{cases} 4x - 9y = 5 \\ -x + 8y = -2 \end{cases}$

5. $\begin{cases} 2x - y + z = 3 \\ 2x + 2y + z = 4 \\ -x + 2y + 3z = -1 \end{cases}$

6. $\begin{cases} 3x + 4y + z = 2 \\ x - 2y - 2z = -7 \\ 4x - 6y + 4z = 0 \end{cases}$

In Exercises 7–10, find the first partial derivatives.

7. $f(x, y) = x^2y + xy^2$

8. $f(x, y) = 25(xy + y^2)^2$

9. $f(x, y, z) = x(x^2 + 3y^2 - z^2)$

10. $f(x, y, z) = z(xy + xz + yz)$

Exercises 7.6

See *CalcChat.com* for tutorial help and worked-out solutions to odd-numbered exercises.

 Using Lagrange Multipliers In Exercises 1–12, use Lagrange multipliers to find the indicated extremum. Assume that x and y are positive. *See Example 1.*

1. Maximize $f(x, y) = xy$
   Constraint: $x + y - 14 = 0$

2. Maximize $f(x, y) = xy$
   Constraint: $x + 3y - 6 = 0$

3. Minimize $f(x, y) = x^2 + y^2$
   Constraint: $3x + y = 10$

4. Minimize $f(x, y) = x^2 + y^2$
   Constraint: $-2x - 4y = -5$

5. Minimize $f(x, y) = x^2 - y^2$
   Constraint: $x - 2y + 6 = 0$

6. Maximize $f(x, y) = x^2 - y^2$
   Constraint: $2y - x^2 = 0$

7. Maximize $f(x, y) = 2x + 2xy + y$
   Constraint: $2x + y = 100$

8. Minimize $f(x, y) = 3x + y + 10$
   Constraint: $x^2y = 6$

9. Maximize $f(x, y) = \sqrt{6 - x^2 - y^2}$
   Constraint: $x + y - 2 = 0$

10. Minimize $f(x, y) = \sqrt{x^2 + y^2}$
    Constraint: $2x + 4y - 15 = 0$

11. Maximize $f(x, y) = e^{xy}$
    Constraint: $x^2 + y^2 - 8 = 0$

12. Maximize $f(x, y) = e^{2xy}$
    Constraint: $2x^2 + \frac{1}{2}y^2 - 1 = 0$

 Using Lagrange Multipliers In Exercises 13–18, use Lagrange multipliers to find the indicated extremum. Assume that x, y, and z are positive. *See Example 1.*

13. Maximize $f(x, y, z) = xyz$
    Constraint: $x + y + z - 6 = 0$

14. Maximize $f(x, y, z) = x^2y^2z^2$
    Constraint: $x^2 + y^2 + z^2 - 9 = 0$

15. Minimize $f(x, y, z) = x^2 + y^2 + z^2$
    Constraint: $x + y + z = 1$

16. Minimize $f(x, y, z) = 2x^2 + 3y^2 + 2z^2$
    Constraint: $x + y + z - 24 = 0$

17. Maximize $f(x, y, z) = x + y + z$
    Constraint: $x^2 + y^2 + z^2 = 1$

18. Maximize $f(x, y, z) = 3x + 2y + z$
    Constraint: $3x^2 + y^2 + z^2 = 6$

Finding Positive Numbers In Exercises 19–22, find three positive numbers x, y, and z that satisfy the given conditions.

19. The sum is 90 and the product is a maximum.

20. The sum is 80 and $P = x^2yz$ is a maximum.

21. The sum is 120 and the sum of the squares is a minimum.

22. The sum is 36 and the sum of the cubes is a minimum.

Finding Distance In Exercises 23–28, find the minimum distance from the curve or surface to the given point. (*Hint:* Start by minimizing the square of the distance.)

23. Line: $x + y = 6$, $(0, 0)$

Minimize $d^2 = x^2 + y^2$

24. Line: $2x + 3y = -1$, $(0, 0)$

Minimize $d^2 = x^2 + y^2$

25. Parabola: $y = x^2$, $(-3, 0)$

Minimize $d^2 = (x + 3)^2 + y^2$

26. Circle: $(x - 4)^2 + y^2 = 4$, $(0, 10)$

Minimize $d^2 = x^2 + (y - 10)^2$

27. Plane: $x + y + z = 1$, $(2, 1, 1)$

Minimize $d^2 = (x - 2)^2 + (y - 1)^2 + (z - 1)^2$

28. Cone: $z = \sqrt{x^2 + y^2}$, $(4, 0, 0)$

Minimize $d^2 = (x - 4)^2 + y^2 + z^2$

29. Volume A rectangular box is resting on the xy-plane with one vertex at the origin. The opposite vertex lies in the plane

$$2x + 3y + 5z = 90.$$

Find the dimensions that maximize the volume. (*Hint:* Maximize $V = xyz$ subject to the constraint $2x + 3y + 5z - 90 = 0$.)

30. Volume Find the dimensions of the rectangular package of maximum volume subject to the constraint that the sum of the length and the girth cannot exceed 108 inches (see figure). (*Hint:* Maximize $V = xyz$ subject to the constraint $x + 2y + 2z = 108$.)

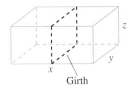

Figure for 30   Figure for 31

31. Cost In redecorating an office, the cost for new carpeting is $4 per square foot and the cost of wallpapering a wall is $3 per square foot. Find the dimensions of the largest office that can be redecorated for $1728 (see figure). (*Hint:* Maximize $V = xyz$ subject to $4xy + 6xz + 6yz = 1728$.)

32. Cost A cargo container (in the shape of a rectangular solid) must have a volume of 480 cubic feet. The bottom will cost $5 per square foot to construct and the sides and the top will cost $3 per square foot to construct. Find the dimensions of the container that has a minimum cost.

33. Cost A manufacturer has an order for 1000 units of fine paper that can be produced at two locations. Let x_1 and x_2 be the numbers of units produced at the two locations. The cost function is modeled by

$$C = 0.25x_1^2 + 25x_1 + 0.05x_2^2 + 12x_2.$$

Find the number of units that should be produced at each location to minimize the cost.

34. Cost A manufacturer has an order for 2000 units of all-terrain vehicle tires that can be produced at two locations. Let x_1 and x_2 be the numbers of units produced at the two locations. The cost function is modeled by

$$C = 0.25x_1^2 + 10x_1 + 0.15x_2^2 + 12x_2.$$

Find the number of units that should be produced at each location to minimize the cost.

35. Production The production function for a manufacturer is given by

$$f(x, y) = 100x^{0.25}y^{0.75}$$

where x is the number of units of labor (at $48 per unit) and y is the number of units of capital (at $36 per unit). The total cost for labor and capital cannot exceed $100,000.

(a) Find the maximum production level for this manufacturer.

(b) Find the marginal productivity of money.

(c) Use the marginal productivity of money to find the maximum number of units that can be produced when $125,000 is available for labor and capital.

(d) Use the marginal productivity of money to find the maximum number of units that can be produced when $350,000 is available for labor and capital.

36. Production Repeat Exercise 35 for the production function given by

$$f(x, y) = 100x^{0.6}y^{0.4}.$$

37. Least-Cost Rule The production function for a company is given by

$$f(x, y) = 100x^{0.7}y^{0.3}$$

where x is the number of units of labor (at $50 per unit) and y is the number of units of capital (at $100 per unit). Management sets a production goal of 20,000 units.

(a) Find the numbers of units of labor and capital needed to meet the production goal while minimizing the cost.

(b) Show that the conditions of part (a) are met when

$$\frac{\text{Marginal productivity of labor}}{\text{Marginal productivity of capital}} = \frac{\text{Unit price of labor}}{\text{Unit price of capital}}.$$

This proportion is called the *Least-Cost Rule* (or *Equimarginal Rule*).

38. Least-Cost Rule Repeat Exercise 37 for the production function given by

$$f(x, y) = 100^{0.4}y^{0.6}.$$

39. Construction An animal shelter plans to use the side of a barn as a boundary for two adjacent rectangular enclosures (see figure). Fencing for the perimeter costs $10 per foot. To separate the enclosures, a fence that costs $4 per foot will divide the region. The total area of the two enclosures is to be 6000 square feet.

(a) Find the dimensions that will minimize the cost of the fencing.

(b) What is the minimum cost?

40. Office Space Partitions will be used in an office to form four equal work areas with a total area of 360 square feet (see figure). The partitions that are x feet long cost $100 per foot and the partitions that are y feet long cost $120 per foot.

(a) Find the dimensions x and y that will minimize the cost of the partitions.

(b) What is the minimum cost?

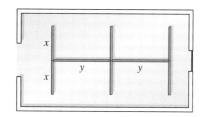

41. Biology A microbiologist must prepare a culture medium in which to grow a certain type of bacteria. The proportion of salt in this medium is given by

$$S = 12xyz$$

where x, y, and z are the amounts (in liters) of the three nutrient solutions to be mixed in the medium. For the bacteria to grow, the medium must be 13% salt. Nutrient solutions x, y, and z cost $1, $2, and $3 per liter, respectively. How much of each nutrient solution should be used to minimize the cost of the culture medium?

42. Biology Repeat Exercise 41 for a salt-content model given by

$$S = 0.01x^2y^2z^2.$$

43. Nutrition The number of grams of your favorite ice cream can be modeled by

$$G = (x, y, z) = 0.05x^2 + 0.16xy + 0.25z^2$$

where x is the number of fat grams, y is the number of carbohydrate grams, and z is the number of protein grams. Find the maximum number of grams of ice cream you can eat without consuming more than 400 calories. Assume that there are 9 calories per fat gram, 4 calories per carbohydrate gram, and 4 calories per protein gram.

44. HOW DO YOU SEE IT? The graphs show the constraint and several level curves of the objective function. Use the graph to approximate the indicated extrema.

(a) Maximize $z = xy$

Constraint: $2x + y = 4$

(b) Minimize $z = x^2 + y^2$

Constraint: $x + y - 4 = 0$

45. Advertising A private golf club is determining how to spend its $8100 advertising budget. The club knows from prior experience that the number of responses A is given by

$$A = 0.0001t^2pr^{1.5}$$

where t is the number of cable television ads, p is the number of newspaper ads, and r is the number of radio ads. A cable television ad costs $90, a newspaper ad costs $36, and a radio ad costs $45.

(a) How much should be spent on each type of advertising to obtain the maximum number of responses? (Assume the golf club uses each type of advertising.)

(b) What is the maximum number of responses expected?

7.7 Least Squares Regression Analysis

■ Find the sum of the squared errors for mathematical models.
■ Find the least squares regression lines for data.

Measuring the Accuracy of a Mathematical Model

When finding a mathematical model to represent a collection of data, the goals are simplicity and accuracy. Of course, these goals often conflict. For instance, a simple linear model for the points shown in Figure 7.35(a) is

$$f(x) = 1.9x - 5. \qquad \text{Linear model}$$

Figure 7.35(b), however, shows that by choosing the slightly more complicated quadratic model

$$g(x) = 0.20x^2 - 0.7x + 1 \qquad \text{Quadratic model}$$

you can obtain significantly greater accuracy.*

In Exercise 16 on page 478, you will find the least squares regression line that models the demand of a tool at a hardware store in terms of the price.

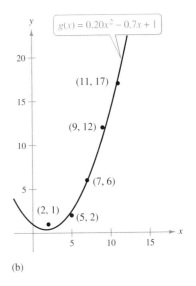

(a)   (b)

FIGURE 7.35

To measure how well the model $y = f(x)$ fits a collection of points, find the sum of the squares of the differences between the actual y-values and the values given by the model. This sum is called the **sum of the squared errors** and is denoted by S. Graphically, S can be interpreted as the sum of the squares of the vertical distances between the graph of f and the given points in the plane, as shown in the figure at the right. If the model is a perfect fit, then

$$S = 0. \qquad \text{Sum of squares for a perfect model}$$

However, when a perfect fit is not feasible, you should use a model that minimizes S.

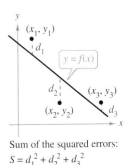

Sum of the squared errors:
$S = d_1^2 + d_2^2 + d_3^2$

---

* An analytic method for finding a quadratic model for a collection of data is not given in this text. You can perform this task using a graphing utility or a spreadsheet software program that has a built-in program for finding the least squares regression quadratic.

Definition of the Sum of the Squared Errors

The **sum of the squared errors** for the model $y = f(x)$ with respect to the points

$$(x_1, y_1), (x_2, y_2), \ldots, (x_n, y_n)$$

is given by

$$S = [f(x_1) - y_1]^2 + [f(x_2) - y_2]^2 + \cdots + [f(x_n) - y_n]^2.$$

EXAMPLE 1   Finding the Sum of the Squared Errors

Find the sum of the squared errors for the linear model

$$f(x) = 1.9x - 5 \qquad \text{Linear model [see Figure 7.35(a)]}$$

and the quadratic model

$$g(x) = 0.20x^2 - 0.7x + 1 \qquad \text{Quadratic model [see Figure 7.35(b)]}$$

with respect to the points

$$(2, 1), (5, 2), (7, 6), (9, 12), (11, 17).$$

Then decide which model is a better fit.

SOLUTION   Begin by evaluating each model at the given x-values, as shown in the table.

x	2	5	7	9	11
Actual y-values	1	2	6	12	17
Linear model, $f(x)$	-1.2	4.5	8.3	12.1	15.9
Quadratic model, $g(x)$	0.4	2.5	5.9	10.9	17.5

For the linear model f, the sum of the squared errors is

$$S = (-1.2 - 1)^2 + (4.5 - 2)^2 + (8.3 - 6)^2 + (12.1 - 12)^2 + (15.9 - 17)^2$$
$$= 17.6.$$

Similarly, the sum of the squared errors for the quadratic model g is

$$S = (0.4 - 1)^2 + (2.5 - 2)^2 + (5.9 - 6)^2 + (10.9 - 12)^2 + (17.5 - 17)^2$$
$$= 2.08.$$

Because the sum of the squared errors for the quadratic model is less than the sum of the squared errors for the linear model, the quadratic model is a better fit.

✓ **Checkpoint 1**   Worked-out solution available at LarsonAppliedCalculus.com

Find the sum of the squared errors for the linear model

$$f(x) = 2.9x - 6 \qquad \text{Linear model}$$

and the quadratic model

$$g(x) = 0.20x^2 + 0.5x - 1 \qquad \text{Quadratic model}$$

with respect to the points

$$(2, 1), (4, 5), (6, 9), (8, 16), (10, 24).$$

Then decide which model is a better fit.

Least Squares Regression Line

The sum of the squared errors can be used to determine which of several models is the best fit for a collection of data. In general, if the sum of the squared errors of f is less than the sum of the squared errors of g, then f is said to be a better fit for the data than g. In regression analysis, you consider all possible models of a certain type. The one that is defined to be the best-fitting model is the one with the least sum of the squared errors. Example 2 shows how to use the optimization techniques described in Section 7.5 to find the best-fitting linear model for a collection of data.

EXAMPLE 2   **Finding the Best Linear Model**

Find the values of a and b such that the linear model

$$f(x) = ax + b$$

has a minimum sum of the squared errors for the points

$$(-3, 0), (-1, 1), (0, 2), (2, 3).$$

SOLUTION   The sum of the squared errors is

$$S = [f(x_1) - y_1]^2 + [f(x_2) - y_2]^2 + [f(x_3) - y_3]^2 + [f(x_4) - y_4]^2$$
$$= (-3a + b - 0)^2 + (-a + b - 1)^2 + (b - 2)^2 + (2a + b - 3)^2$$
$$= 14a^2 - 4ab + 4b^2 - 10a - 12b + 14.$$

To find the values of a and b for which S is a minimum, you can use the techniques described in Section 7.5. That is, find the partial derivatives of S.

$$\frac{\partial S}{\partial a} = 28a - 4b - 10 \qquad \text{Differentiate with respect to } a.$$

$$\frac{\partial S}{\partial b} = -4a + 8b - 12 \qquad \text{Differentiate with respect to } b.$$

ALGEBRA TUTOR

For help in solving the system of equations in Example 2, see Example 2(b) in the *Chapter 7 Algebra Tutor*, on page 497.

Next, set each partial derivative equal to zero.

$$28a - 4b - 10 = 0 \qquad \text{Set } \partial S/\partial a \text{ equal to 0.}$$
$$-4a + 8b - 12 = 0 \qquad \text{Set } \partial S/\partial b \text{ equal to 0.}$$

The solution of this system of linear equations is

$$a = \frac{8}{13} \quad \text{and} \quad b = \frac{47}{26}.$$

So, the best-fitting linear model for the given points is

$$f(x) = \frac{8}{13}x + \frac{47}{26}.$$

The graph of this model is shown in the figure at the right.

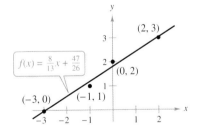

✓ **Checkpoint 2**   Worked-out solution available at LarsonAppliedCalculus.com

Find the values of a and b such that the linear model

$$f(x) = ax + b$$

has a minimum sum of the squared errors for the points

$$(-2, 0), (0, 2), (2, 5), (4, 7).$$

The line in Example 2 is called the **least squares regression line** for the given data. The solution shown in Example 2 can be generalized to find a formula for the least squares regression line. Consider the linear model

$$f(x) = ax + b$$

and the points

$$(x_1, y_1), (x_2, y_2), \ldots, (x_n, y_n).$$

The sum of the squared errors is

$$S = [f(x_1) - y_1]^2 + [f(x_2) - y_2]^2 + \cdots + [f(x_n) - y_n]^2$$
$$= (ax_1 + b - y_1)^2 + (ax_2 + b - y_2)^2 + \cdots + (ax_n + b - y_n)^2.$$

To minimize S, set the partial derivatives $\partial S/\partial a$ and $\partial S/\partial b$ equal to zero and solve for a and b. The results are summarized below.

The Least Squares Regression Line

The **least squares regression line** for the points

$$(x_1, y_1), (x_2, y_2), \ldots, (x_n, y_n)$$

is $f(x) = ax + b$, where

$$a = \frac{n \sum\limits_{i=1}^{n} x_i y_i - \sum\limits_{i=1}^{n} x_i \sum\limits_{i=1}^{n} y_i}{n \sum\limits_{i=1}^{n} x_i^2 - \left(\sum\limits_{i=1}^{n} x_i \right)^2} \quad \text{and} \quad b = \frac{1}{n}\left(\sum\limits_{i=1}^{n} y_i - a \sum\limits_{i=1}^{n} x_i \right).$$

The summation notation

$$\sum_{i=1}^{n} x_i$$

where Σ is the Greek letter sigma, is used to indicate the sum of the numbers

$$x_1 + x_2 + \cdots + x_n.$$

Similarly,

$$\sum_{i=1}^{n} x_i y_i = x_1 y_1 + x_2 y_2 + \cdots + x_n y_n, \quad \sum_{i=1}^{n} x_i^2 = x_1^2 + x_2^2 + \cdots + x_n^2,$$

and so on.

In the formula for the least squares regression line, note that if the x-values are symmetrically spaced about zero, then

$$\sum_{i=1}^{n} x_i = 0$$

and the formulas for a and b simplify to

$$a = \frac{n \sum\limits_{i=1}^{n} x_i y_i}{n \sum\limits_{i=1}^{n} x_i^2} \quad \text{and} \quad b = \frac{1}{n} \sum_{i=1}^{n} y_i.$$

Note also that only the *development* of the least squares regression line involves partial derivatives. The *application* of this formula is a matter of computing the values of a and b. This task is performed much more simply on a calculator or a computer than by hand.

EXAMPLE 3   Modeling Hourly Wages

The average hourly wages y (in dollars per hour) for production workers in manufacturing industries from 2002 through 2013 are shown in the table. Find the least squares regression line for the data and use the result to estimate the average hourly wage in 2017.   *(Source: U.S. Bureau of Labor Statistics)*

Year	2002	2003	2004	2005	2006	2007
y	15.29	15.74	16.14	16.56	16.81	17.26

Year	2008	2009	2010	2011	2012	2013
y	17.75	18.24	18.61	18.93	19.08	19.30

Spreadsheet at LarsonAppliedCalculus.com

SOLUTION   Let t represent the year, with $t = 2$ corresponding to 2002. Then, you need to find the linear model that best fits the points

$(2, 15.29), (3, 15.74), (4, 16.14), (5, 16.56), (6, 16.81), (7, 17.26), (8, 17.75),$
$(9, 18.24), (10, 18.61), (11, 18.93), (12, 19.08), (13, 19.30).$

Using a calculator with a built-in least squares regression program, you can determine that the best-fitting line is $y = 0.380t + 14.62$. To estimate the average hourly wage in 2017, let $t = 17$ in the model.

$$y = 0.380(17) + 14.62 = 6.46 + 14.62 = 21.08$$

So, you can estimate that the average hourly wage in 2017 will be $21.08, as shown in Figure 7.36.

Modeling Hourly Wages

FIGURE 7.36

✓ *Checkpoint 3*   *Worked-out solution available at LarsonAppliedCalculus.com*

The numbers of cellular phone subscribers y (in millions) for the years 2004 through 2013 are shown in the table. Find the least squares regression line for the data and use the result to estimate the number of subscribers in 2017. Let t represent the year, with $t = 4$ corresponding to 2004.   *(Source: CTIA-The Wireless Association)*

Year	2004	2005	2006	2007	2008
y	182.1	207.9	233.0	255.4	270.3

Year	2009	2010	2011	2012	2013
y	285.6	296.3	316.0	326.5	335.7

Spreadsheet at LarsonAppliedCalculus.com

SUMMARIZE   (Section 7.7)

1. State the definition of the sum of the squared errors *(page 474)*. For an example of finding the sum of the squared errors, see Example 1.

2. State the definition of the least squares regression line *(page 476)*. For an example of finding the least squares regression line, see Example 2.

3. Describe a real-life example of modeling hourly wages using the least squares regression line *(page 477, Example 3)*.

SKILLS WARM UP 7.7

The following warm-up exercises involve skills that were covered in a previous course or in earlier sections. You will use these skills in the exercise set for this section. For additional help, review Appendix A.3 and Sections 1.3 and 7.4.

In Exercises 1 and 2, evaluate the expression.

1. $(2.5 - 1)^2 + (3.25 - 2)^2 + (4.1 - 3)^2$

2. $(1.1 - 1)^2 + (2.08 - 2)^2 + (2.95 - 3)^2$

In Exercises 3–6, find an equation of the line that passes through the points. Then sketch the line.

3. $(0, 1), (3, 7)$

4. $(1, 3), (4, 0)$

5. $(-2, -3), (8, 2)$

6. $(-4, -2), (2, -3.5)$

In Exercises 7 and 8, find the partial derivatives of S.

7. $S = a^2 + 6b^2 - 4a - 8b - 4ab + 6$

8. $S = 4a^2 + 9b^2 - 6a - 4b - 2ab + 8$

Exercises 7.7

See *CalcChat.com* for tutorial help and worked-out solutions to odd-numbered exercises.

 Finding the Sum of the Squared Errors In Exercises 1–4, find the sum of the squared errors for the linear model $f(x)$ and the quadratic model $g(x)$ using the given points. Then decide which model is a better fit. *See Example 1.*

1. $f(x) = 1.6x + 6$, $g(x) = 0.29x^2 + 2.2x + 6$
$(-3, 2), (-2, 2), (-1, 4), (0, 6), (1, 8)$

2. $f(x) = -0.7x + 2$, $g(x) = 0.06x^2 - 0.7x + 1$
$(-3, 4), (-1, 2), (1, 1), (3, 0)$

3. $f(x) = -1.4x + 8$, $g(x) = 0.14x^2 - 1.6x + 7$
$(-2, 11), (-1, 10), (0, 7), (3, 4)$

4. $f(x) = 2.0x - 3$, $g(x) = 0.14x^2 + 1.3x - 3$
$(-1, -4), (1, -3), (2, 0), (4, 5), (6, 9)$

 Finding the Least Squares Regression Line In Exercises 5–8, find the least squares regression line for the given points. Then plot the points and sketch the regression line. *See Example 2.*

5. $(-2, -1), (0, 0), (2, 3)$

6. $(-3, 0), (-1, 1), (1, 1), (3, 2)$

7. $(-2, 4), (-1, 1), (0, -1), (1, -3)$

8. $(-5, -3), (-4, -2), (-2, -1), (-1, 1)$

Finding the Least Squares Regression Line In Exercises 9–14, use the regression capabilities of a graphing utility or a spreadsheet to find the least squares regression line for the given points.

9. $(-4, -1), (-2, 0), (2, 4), (4, 5)$

10. $(-5, 1), (1, 3), (2, 3), (2, 5)$

11. $(0, 6), (4, 3), (5, 0), (8, -4), (10, -5)$

12. $(-10, 10), (-5, 8), (3, 6), (7, 4), (5, 0)$

13. $(6, 4), (1, -2), (3, 1), (8, 6), (11, 8), (10, 8)$

14. $(0, -2), (5, -15), (-4, 9), (-1, 1), (2, -9), (1, -5)$

15. Revenue The table gives the revenues y (in billions of dollars) for Costco Wholesale from 2009 through 2014. *(Source: Costco Wholesale Corporation)*

DATA

Year	2009	2010	2011
Revenue, y	71.4	77.9	88.9

Year	2012	2013	2014
Revenue, y	99.1	105.2	112.6

Spreadsheet at LarsonAppliedCalculus.com

(a) Use the regression capabilities of a graphing utility or a spreadsheet to find the least squares regression line for the data. Let $t = 9$ represent 2009.

(b) Estimate the revenue in 2017.

(c) In what year will the revenue reach $180 billion?

16. Demand A hardware retailer wants to know the demand y for a tool as a function of price x. The monthly sales for four different prices of the tool are listed in the table.

Price, x	$25	$30	$35	$40
Demand, y	82	75	67	55

(a) Use the regression capabilities of a graphing utility or a spreadsheet to find the least squares regression line for the data.

(b) Estimate the demand when the price is $32.95.

(c) What price will create a demand of 100 tools?

17. Agriculture An agronomist used four test plots to determine the relationship between the wheat yield y (in bushels per acre) and the amount of fertilizer x (in pounds per acre). The results are shown in the table.

Fertilizer, x	100	150	200	250
Yield, y	35	44	50	56

 (a) Use the regression capabilities of a graphing utility or a spreadsheet to find the least squares regression line for the data.

(b) Estimate the yield for a fertilizer application of 160 pounds per acre.

18. HOW DO YOU SEE IT? Match the regression equation with the appropriate graph. Explain your reasoning. (Note that the x- and y-axes are broken.)

(a) $y = 0.22x - 7.5$ (b) $y = -0.35x + 11.5$

(c) $y = 0.09x + 19.8$ (d) $y = -1.29x + 89.8$

(i)   (ii)

(iii)   (iv)

Determining Correlation In Exercises 19–24, plot the points and determine whether the data have positive, negative, or no linear correlation (see figures below). Then use a graphing utility to find the value of r and confirm your result. The number r is called the *correlation coefficient*. It is a measure of how well the model fits the data. Correlation coefficients vary between -1 and 1, and the closer $|r|$ is to 1, the better the model.

$r = 0.981$

Positive correlation

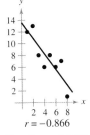

$r = -0.866$

Negative correlation

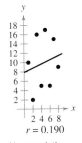

$r = 0.190$

No correlation

19. $(1, 4), (2, 6), (3, 8), (4, 11), (5, 13), (6, 15)$

20. $(1, 7.5), (2, 7), (3, 7), (4, 6), (5, 5), (6, 4.9)$

21. $(1, 3), (2, 6), (3, 2), (4, 3), (5, 9), (6, 1)$

22. $(1, 36), (2, 10), (3, 0), (4, 4), (5, 16), (6, 36)$

23. $(0.5, 9), (1, 8.5), (1.5, 7), (2, 5.5), (2.5, 5), (3, 3.5)$

24. $(0.5, 2), (0.75, 1.75), (1, 3), (1.5, 3.2), (2, 3.7), (2.6, 4)$

True or False? In Exercises 25–30, determine whether the statement is true or false. If it is false, explain why or give an example that shows it is false.

25. Data that are modeled by

$$y = 3.29x - 4.17$$

have a negative correlation.

26. Data that are modeled by

$$y = -0.238x + 25$$

have a negative correlation.

27. When the correlation coefficient is $r \approx -0.98781$, the model is a good fit.

28. A correlation coefficient of $r \approx 0.201$ implies that the data have no correlation.

29. A linear regression model with a positive correlation coefficient will have a slope that is greater than 0.

30. When the correlation coefficient for a linear regression model is close to -1, the regression line cannot be used to describe the data.

31. Project: Financial Data For a project analyzing the revenues per share, revenues, and shareholder's equity of Amazon.com from 2008 through 2014, visit the text's website at *LarsonAppliedCalculus.com*. *(Source: Amazon.com)*

7.8   Double Integrals and Area in the Plane

■ Evaluate double integrals.
■ Use double integrals to find the areas of regions.

Double Integrals

In Section 7.4, you learned that it is meaningful to differentiate functions of several variables with respect to one variable while holding the other variable(s) constant. You can *integrate* functions of several variables by a similar procedure. For instance, consider the partial derivative

$$f_x(x, y) = 2xy.$$ Partial with respect to x

By holding y constant, you can integrate with respect to x to obtain

$$f(x, y) = \int f_x(x, y)\, dx$$ Integrate with respect to x.

$$= \int 2xy\, dx$$ Hold y constant.

$$= y \int 2x\, dx$$ Factor out constant y.

$$= y(x^2) + C(y)$$ Antiderivative of $2x$ is x^2.

$$= x^2 y + C(y).$$ $C(y)$ is a function of y.

This procedure is called **partial integration with respect to x.** Note that the "constant of integration" $C(y)$ is a function of y, because y is fixed during integration with respect to x. Similarly, for the partial derivative

$$f_y(x, y) = x^2 + 2$$ Partial with respect to y

you can hold x constant and integrate with respect to y to obtain

$$f(x, y) = \int f_y(x, y)\, dy$$ Integrate with respect to y.

$$= \int (x^2 + 2)\, dy$$ Hold x constant.

$$= (x^2 + 2) \int dy$$ Factor out constant $x^2 + 2$.

$$= (x^2 + 2)(y) + C(x)$$ Antiderivative of 1 is y.

$$= x^2 y + 2y + C(x).$$ $C(x)$ is a function of x.

In this case, the "constant of integration" $C(x)$ is a function of x, because x is fixed during integration with respect to y.

To evaluate a definite integral of a function of several variables, you can apply the Fundamental Theorem of Calculus to one variable while holding the other variable(s) constant, as shown.

$$\int_1^{2y} 2xy\, dx = x^2 y \Big]_1^{2y} = (2y)^2 y - (1)^2 y = 4y^3 - y$$

x is the variable of integration and y is fixed.   Replace x by the limits of integration.   The result is a function of y.

Note that you omit the constant of integration, just as you do for a definite integral of a function of one variable.

In Exercise 29 on page 487, you will use a double integral to find the area of a region.

$y = 4 - x^2$

EXAMPLE 1   Finding Partial Integrals

a. $\displaystyle\int_1^x (2x^2y^{-2} + 2y)\, dy = \left[\dfrac{-2x^2}{y} + y^2 \right]_1^x$ Hold x constant.

$\qquad\qquad\qquad\qquad = \left(\dfrac{-2x^2}{x} + x^2 \right) - \left(\dfrac{-2x^2}{1} + 1 \right)$

$\qquad\qquad\qquad\qquad = 3x^2 - 2x - 1$

b. $\displaystyle\int_y^{5y} \sqrt{x - y}\, dx = \left[\dfrac{2}{3}(x - y)^{3/2} \right]_y^{5y}$ Hold y constant.

$\qquad\qquad\qquad\qquad = \dfrac{2}{3}\left[(5y - y)^{3/2} - (y - y)^{3/2} \right]$

$\qquad\qquad\qquad\qquad = \dfrac{16}{3}y^{3/2}$

✓ **Checkpoint 1**   *Worked-out solution available at LarsonAppliedCalculus.com*

Find each partial integral.

a. $\displaystyle\int_1^x (4xy + y^3)\, dy$ **b.** $\displaystyle\int_y^{y^2} \dfrac{1}{x + y}\, dx$ ■

In Example 1(a), note that the definite integral defines a function of x and can *itself* be integrated. An "integral of an integral" is called a **double integral.** With a function of two variables, there are two types of double integrals.

$$\int_a^b \int_{g_1(x)}^{g_2(x)} f(x, y)\, dy\, dx = \int_a^b \left[\int_{g_1(x)}^{g_2(x)} f(x, y)\, dy \right] dx$$

$$\int_a^b \int_{g_1(y)}^{g_2(y)} f(x, y)\, dx\, dy = \int_a^b \left[\int_{g_1(y)}^{g_2(y)} f(x, y)\, dx \right] dy$$

EXAMPLE 2   Evaluating a Double Integral

$$\int_1^2 \int_0^x (2xy + 3)\, dy\, dx = \int_1^2 \left[\int_0^x (2xy + 3)\, dy \right] dx$$

$$= \int_1^2 \left[xy^2 + 3y \right]_0^x dx$$

$$= \int_1^2 (x^3 + 3x)\, dx$$

$$= \left[\dfrac{x^4}{4} + \dfrac{3x^2}{2} \right]_1^2$$

$$= \left(\dfrac{2^4}{4} + \dfrac{3(2^2)}{2} \right) - \left(\dfrac{1^4}{4} + \dfrac{3(1^2)}{2} \right)$$

$$= \dfrac{33}{4}$$

✓ **Checkpoint 2**   *Worked-out solution available at LarsonAppliedCalculus.com*

Evaluate $\displaystyle\int_1^2 \int_0^x (5x^2y - 2)\, dy\, dx$.   ■

Finding Area with a Double Integral

One application of double integrals is finding the area of a plane region. For instance, consider the region R that is bounded by

$$a \le x \le b \quad \text{and} \quad g_1(x) \le y \le g_2(x)$$

as shown in Figure 7.37. Using the techniques described in Section 5.5, you know that the area of R is

$$\int_a^b [g_2(x) - g_1(x)]\, dx. \qquad \text{Area of } R$$

This same area is also given by the double integral

$$\int_a^b \int_{g_1(x)}^{g_2(x)} dy\, dx \qquad \text{Area of } R$$

because

$$\int_a^b \int_{g_1(x)}^{g_2(x)} dy\, dx = \int_a^b \left[y \right]_{g_1(x)}^{g_2(x)} dx = \int_a^b [g_2(x) - g_1(x)]\, dx.$$

Figure 7.38 shows the two basic types of plane regions whose areas can be determined by a double integral.

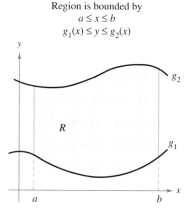

Region is bounded by
$$a \le x \le b$$
$$g_1(x) \le y \le g_2(x)$$

FIGURE 7.37

STUDY TIP

To designate a double integral or an area of a region without specifying a particular order of integration, you can use the symbol

$$\int_R \int dA$$

where $dA = dx\, dy$ or $dA = dy\, dx$.

Determining Area in the Plane by Double Integrals

Region is bounded by
$$a \le x \le b$$
$$g_1(x) \le y \le g_2(x)$$

$$\text{Area} = \int_a^b \int_{g_1(x)}^{g_2(x)} dy\, dx$$

Region is bounded by
$$c \le y \le d$$
$$h_1(y) \le x \le h_2(y)$$

$$\text{Area} = \int_c^d \int_{h_1(y)}^{h_2(y)} dx\, dy$$

FIGURE 7.38

STUDY TIP

In Figure 7.38, note that the horizontal or vertical orientation of the narrow rectangle indicates the order of integration. The "outer" variable of integration always corresponds to the width of the rectangle. Notice also that the outer limits of integration for a double integral are constant, whereas the inner limits may be functions of the outer variable.

EXAMPLE 3   **Finding Area with a Double Integral**

Use a double integral to find the area of the rectangular region shown in Figure 7.39.

SOLUTION   The bounds for x are $1 \leq x \leq 5$ and the bounds for y are $2 \leq y \leq 4$. So, the area of the region is

$$\int_1^5 \int_2^4 dy\, dx = \int_1^5 \left[y \right]_2^4 dx \qquad \text{Integrate with respect to } y.$$

$$= \int_1^5 (4 - 2)\, dx \qquad \text{Apply Fundamental Theorem of Calculus.}$$

$$= \int_1^5 2\, dx \qquad \text{Simplify.}$$

$$= \left[2x \right]_1^5 \qquad \text{Integrate with respect to } x.$$

$$= 10 - 2 \qquad \text{Apply Fundamental Theorem of Calculus.}$$

$$= 8 \text{ square units.} \qquad \text{Simplify.}$$

You can confirm this by noting that the rectangle measures 2 units by 4 units.

✓ **Checkpoint 3**   *Worked-out solution available at LarsonAppliedCalculus.com*

Use a double integral to find the area of the rectangular region shown in Example 3 by integrating with respect to x and then with respect to y.

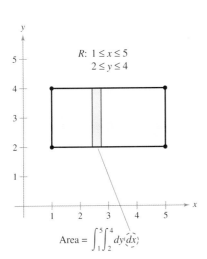

$R: 1 \leq x \leq 5$
$\quad 2 \leq y \leq 4$

Area $= \int_1^5 \int_2^4 dy\, dx$

FIGURE 7.39

EXAMPLE 4   **Finding Area with a Double Integral**

Use a double integral to find the area of the region bounded by the graphs of

$$y = x^2 \quad \text{and} \quad y = x^3.$$

SOLUTION   As shown in Figure 7.40, the two graphs intersect when $x = 0$ and $x = 1$. Choosing x to be the outer variable, the bounds for x are $0 \leq x \leq 1$. On the interval $0 \leq x \leq 1$, the region is bounded above by $y = x^2$ and below by $y = x^3$. So, the bounds for y are

$$x^3 \leq y \leq x^2.$$

This implies that the area of the region is

$$\int_0^1 \int_{x^3}^{x^2} dy\, dx = \int_0^1 \left[y \right]_{x^3}^{x^2} dx \qquad \text{Integrate with respect to } y.$$

$$= \int_0^1 (x^2 - x^3)\, dx \qquad \text{Apply Fundamental Theorem of Calculus.}$$

$$= \left[\frac{x^3}{3} - \frac{x^4}{4} \right]_0^1 \qquad \text{Integrate with respect to } x.$$

$$= \frac{1}{3} - \frac{1}{4} \qquad \text{Apply Fundamental Theorem of Calculus.}$$

$$= \frac{1}{12} \text{ square unit.} \qquad \text{Simplify.}$$

$R: 0 \leq x \leq 1$
$\quad x^3 \leq y \leq x^2$

$(1, 1)$

$y = x^2$

$y = x^3$

Area $= \int_0^1 \int_{x^3}^{x^2} dy\, dx$

FIGURE 7.40

✓ **Checkpoint 4**   *Worked-out solution available at LarsonAppliedCalculus.com*

Use a double integral to find the area of the region bounded by the graphs of

$$y = 2x \quad \text{and} \quad y = x^2.$$

In setting up double integrals, the most difficult task is likely to be determining the correct limits of integration. This can be simplified by making a sketch of the region R and identifying the appropriate bounds for x and y.

EXAMPLE 5   Changing the Order of Integration

For the double integral

$$\int_0^2 \int_{y^2}^4 dx\, dy$$

a. sketch the region R whose area is represented by the integral,

b. rewrite the integral so that x is the outer variable, and

c. show that both orders of integration yield the same value.

SOLUTION

a. From the limits of integration, you know that

$$y^2 \le x \le 4 \qquad \text{Inner limits of integration}$$

which means that the region R is bounded on the left by the parabola $x = y^2$ and on the right by the line $x = 4$. Furthermore, because

$$0 \le y \le 2 \qquad \text{Outer limits of integration}$$

you know that the region lies above the x-axis, as shown in Figure 7.41.

b. If you interchange the order of integration so that x is the outer variable, then x will have constant bounds of integration given by

$$0 \le x \le 4. \qquad \text{Outer limits of integration}$$

By solving for y in the equation $x = y^2$, you can conclude that the bounds for y are

$$0 \le y \le \sqrt{x} \qquad \text{Inner limits of integration}$$

as shown in Figure 7.42. So, with x as the outer variable, the integral can be written as

$$\int_0^4 \int_0^{\sqrt{x}} dy\, dx.$$

c. Evaluating the double integral with the order $dx\, dy$, you have

$$\int_0^2 \int_{y^2}^4 dx\, dy = \int_0^2 \left[x \right]_{y^2}^4 dy = \int_0^2 (4 - y^2)\, dy = \left[4y - \frac{y^3}{3} \right]_0^2 = \frac{16}{3}.$$

Evaluating the double integral with the order $dy\, dx$, you have

$$\int_0^4 \int_0^{\sqrt{x}} dy\, dx = \int_0^4 \left[y \right]_0^{\sqrt{x}} dx = \int_0^4 \sqrt{x}\, dx = \left[\frac{2}{3} x^{3/2} \right]_0^4 = \frac{16}{3}.$$

So, both orders of integration yield the same value.

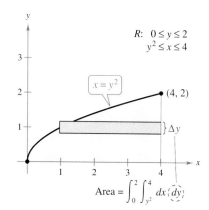

$R:\ 0 \le y \le 2$
$y^2 \le x \le 4$

$x = y^2$

$(4, 2)$

$\}\Delta y$

Area $= \int_0^2 \int_{y^2}^4 dx\, \widehat{(dy)}$

FIGURE 7.41

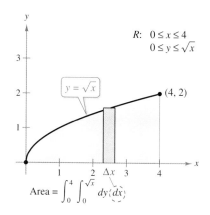

$R:\ 0 \le x \le 4$
$0 \le y \le \sqrt{x}$

$y = \sqrt{x}$

$(4, 2)$

Δx

Area $= \int_0^4 \int_0^{\sqrt{x}} dy\, \widehat{(dx)}$

FIGURE 7.42

✔ *Checkpoint 5*   *Worked-out solution available at LarsonAppliedCalculus.com*

For the double integral $\displaystyle \int_0^2 \int_{2y}^4 dx\, dy,$

a. sketch the region R whose area is represented by the integral,

b. rewrite the integral so that x is the outer variable, and

c. show that both orders of integration yield the same value.

EXAMPLE 6   Finding Area with a Double Integral

Use a double integral to calculate the area denoted by

$$\int_R \int dA$$

where R is the region bounded by $y = x$ and $y = x^2 - x$.

SOLUTION   Begin by sketching the region R, as shown in Figure 7.43. From the sketch, you can see that vertical rectangles of width dx are more convenient than horizontal ones. So, x is the outer variable of integration and its constant bounds are $0 \le x \le 2$. This implies that the bounds for y are $x^2 - x \le y \le x$, and the area is

$$\int_R \int dA = \int_0^2 \int_{x^2-x}^x dy\, dx \qquad \text{Substitute bounds for region.}$$

$$= \int_0^2 \Big[y \Big]_{x^2-x}^x dx \qquad \text{Integrate with respect to } y.$$

$$= \int_0^2 \big[x - (x^2 - x) \big]\, dx \qquad \text{Apply Fundamental Theorem of Calculus.}$$

$$= \int_0^2 (2x - x^2)\, dx \qquad \text{Simplify.}$$

$$= \Big[x^2 - \frac{x^3}{3} \Big]_0^2 \qquad \text{Integrate with respect to } x.$$

$$= 4 - \frac{8}{3} \qquad \text{Apply Fundamental Theorem of Calculus.}$$

$$= \frac{4}{3} \text{ square units.} \qquad \text{Simplify.}$$

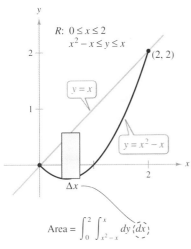

$R:\ 0 \le x \le 2$
$x^2 - x \le y \le x$

$(2, 2)$

$y = x$

$y = x^2 - x$

Δx

$\text{Area} = \int_0^2 \int_{x^2-x}^x dy\, dx$

FIGURE 7.43

✓ *Checkpoint 6*   *Worded-out solution available at LarsonAppliedCalculus.com*

Use a double integral to calculate the area denoted by

$$\int_R \int dA$$

where R is the region bounded by $y = 2x + 3$ and $y = x^2$.   ■

As you are working the exercises for this section, you should be aware that the primary uses of double integrals will be discussed in Section 7.9. Double integrals have been introduced by way of areas in the plane so that you can gain practice in finding the limits of integration. When setting up a double integral, remember that your first step should be to sketch the region R. After doing this, you have two choices of integration orders: $dx\, dy$ or $dy\, dx$.

SUMMARIZE   (Section 7.8)

1. Describe a procedure for finding a partial integral with respect to one variable *(page 480)*. For an example of finding a partial integral with respect to x or to y, see Example 1.

2. Explain how to determine the area of a region in the plane using a double integral *(page 482)*. For examples of finding area using a double integral, see Examples 3, 4, 5, and 6.

SKILLS WARM UP 7.8   The following warm-up exercises involve skills that were covered in earlier sections. You will use these skills in the exercise set for this section. For additional help, review Sections 5.4 and 5.5.

In Exercises 1–12, evaluate the definite integral.

1. $\int_0^1 dx$

2. $\int_0^2 3 \, dy$

3. $\int_{-1}^4 6x^2 \, dx$

4. $\int_0^1 2x^3 \, dx$

5. $\int_1^2 (x^3 - 2x + 4) \, dx$

6. $\int_0^2 (4 - y^2) \, dy$

7. $\int_1^3 \frac{4}{5x^3} \, dx$

8. $\int_1^4 \frac{2}{\sqrt{x}} \, dx$

9. $\int_0^2 \frac{2x}{x^2 + 1} \, dx$

10. $\int_2^e \frac{1}{y - 1} \, dy$

11. $\int_0^2 xe^{x^2 + 1} \, dx$

12. $\int_0^1 e^{-2y} \, dy$

In Exercises 13–16, sketch the region bounded by the graphs of the functions.

13. $y = x, \, y = 0, \, x = 3$

14. $y = x, \, y = 3, \, x = 0$

15. $y = 16 - x^2, \, y = 0, \, x = 0$

16. $y = x^2, \, y = 4x$

Exercises 7.8   See *CalcChat.com* for tutorial help and worked-out solutions to odd-numbered exercises.

 Finding Partial Integrals   In Exercises 1–10, find the partial integral. *See Example 1.*

1. $\int_0^x (2x - y) \, dy$

2. $\int_0^y (5x + 8y) \, dx$

3. $\int_x^{x^2} \frac{y}{x} \, dy$

4. $\int_1^{2y} \frac{y}{x} \, dx$

5. $\int_2^y (3x^2y + x) \, dx$

6. $\int_4^x (xy^3 + 4y^{-2}) \, dy$

7. $\int_{x^3}^{\sqrt{x}} (x^2 + 3y^2) \, dy$

8. $\int_{-\sqrt{1-y^2}}^{\sqrt{1-y^2}} (x^2 + y^2) \, dx$

9. $\int_1^{e^y} \frac{y \ln x}{x} \, dx$

10. $\int_y^3 \frac{xy}{\sqrt{x^2 + 1}} \, dx$

 Evaluating a Double Integral   In Exercises 11–24, evaluate the double integral. *See Example 2.*

11. $\int_0^1 \int_0^2 (x + y) \, dy \, dx$

12. $\int_0^2 \int_0^3 (6y - x^2) \, dy \, dx$

13. $\int_{-3}^2 \int_{-2}^4 xy \, dx \, dy$

14. $\int_{-1}^2 \int_{-3}^3 (x^2 - xy^2) \, dy \, dx$

15. $\int_0^2 \int_0^{6x^2} x^3 \, dy \, dx$

16. $\int_0^1 \int_0^y (6x + 5y^3) \, dx \, dy$

17. $\int_0^2 \int_0^{\sqrt{1-y^2}} -5xy \, dx \, dy$

18. $\int_0^2 \int_{3y^2 - 6y}^{2y - y^2} 3y \, dx \, dy$

19. $\int_0^1 \int_0^{3x} (3x^2 + 3y^2 + 1) \, dy \, dx$

20. $\int_0^1 \int_y^{2y} (1 + 2x^2 + 2y^2) \, dx \, dy$

21. $\int_0^1 \int_0^x \sqrt{1 - x^2} \, dy \, dx$

22. $\int_0^4 \int_0^x \frac{2}{x^2 + 1} \, dy \, dx$

23. $\int_0^\infty \int_0^\infty e^{-(x+y)/2} \, dy \, dx$

24. $\int_0^\infty \int_0^\infty xye^{-(x^2+y^2)} \, dx \, dy$

 Finding Area with a Double Integral   In Exercises 25–30, use a double integral to find the area of the specified region. *See Example 3.*

25.

26.

27.

28.

29.

30.

 Finding Area with a Double Integral In Exercises 31–36, use a double integral to find the area of the region bounded by the graphs of the equations. *See Example 4.*

31. $y = 9 - x^2$, $y = 0$ **32.** $y = x^{3/2}$, $y = x$

33. $2x - 3y = 0$, $x + y = 5$, $y = 0$

34. $\sqrt{x} + \sqrt{y} = 2$, $x = 0$, $y = 0$

35. $y = x$, $y = 2x$, $x = 2$ **36.** $y = 4 - x^2$, $y = x + 2$

 Changing the Order of Integration In Exercises 37–44, sketch the region R whose area is given by the double integral. Then change the order of integration and show that both orders yield the same value. *See Example 5.*

37. $\displaystyle\int_0^1 \int_0^2 dy\, dx$ **38.** $\displaystyle\int_1^3 \int_2^4 dx\, dy$

39. $\displaystyle\int_0^1 \int_{2y}^2 dx\, dy$ **40.** $\displaystyle\int_0^4 \int_0^{3x} dy\, dx$

41. $\displaystyle\int_0^2 \int_{x/2}^1 dy\, dx$ **42.** $\displaystyle\int_0^4 \int_{\sqrt{x}}^2 dy\, dx$

43. $\displaystyle\int_0^1 \int_{y^2}^{\sqrt[3]{y}} dx\, dy$ **44.** $\displaystyle\int_{-2}^2 \int_0^{4-y^2} dx\, dy$

Finding Area with a Double Integral In Exercises 45–48, use a double integral to calculate the area denoted by

$$\int_R \int dA$$

where R is the region bounded by the given equations. *See Example 6.*

45. $y = x + 6$, $y = x^2$ **46.** $y = x$, $y = x^2 - 2x$

47. $x - y = 1$, $\sqrt{5x} - y = 1$

48. $2x - y + 4 = 0$, $2x^2 - y = 0$

49. Think About It Explain why you need to change the order of integration to evaluate the double integral. Then evaluate the double integral.

(a) $\displaystyle\int_0^3 \int_y^3 e^{x^2}\, dx\, dy$ (b) $\displaystyle\int_0^2 \int_x^2 e^{-y^2}\, dy\, dx$

50. HOW DO YOU SEE IT? Complete the double integrals so that each one represents the area of the region R (see figure).

(a) Area $= \displaystyle\int\int dx\, dy$ (b) Area $= \displaystyle\int\int dy\, dx$

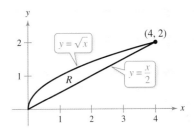

Evaluating a Double Integral In Exercises 51–58, use a symbolic integration utility to evaluate the double integral.

51. $\displaystyle\int_0^1 \int_0^2 e^{-x^2 - y^2}\, dx\, dy$

52. $\displaystyle\int_0^2 \int_{x^2}^{2x} (x^3 + 3y^2)\, dy\, dx$

53. $\displaystyle\int_1^2 \int_0^x e^{xy}\, dy\, dx$

54. $\displaystyle\int_1^2 \int_y^{2y} \ln(x + y)\, dx\, dy$

55. $\displaystyle\int_0^1 \int_x^1 \sqrt{1 - x^2}\, dy\, dx$

56. $\displaystyle\int_0^3 \int_0^{x^2} \sqrt{x}\sqrt{1 + x}\, dy\, dx$

57. $\displaystyle\int_0^2 \int_{\sqrt{4-x^2}}^{4-x^2/4} \frac{xy}{x^2 + y^2 + 1}\, dy\, dx$

58. $\displaystyle\int_0^4 \int_0^y \frac{2}{(x + 1)(y + 1)}\, dx\, dy$

True or False? In Exercises 59 and 60, determine whether the statement is true or false. If it is false, explain why or give an example that shows it is false.

59. $\displaystyle\int_{-1}^1 \int_{-2}^2 y\, dy\, dx = \int_{-1}^1 \int_{-2}^2 y\, dx\, dy$

60. $\displaystyle\int_0^1 \int_0^x f(x, y)\, dy\, dx = \int_0^1 \int_0^y f(x, y)\, dx\, dy$

7.9 Applications of Double Integrals

■ Use double integrals to find the volumes of solid regions.
■ Use double integrals to find the average values of real-life models.

In Exercise 31 on page 495, you will use a double integral to find the average weekly profit of a company.

Volume of a Solid Region

In Section 7.8, you used double integrals as an alternative way to find the area of a plane region. In this section, you will study the primary uses of double integrals: to find the volume of a solid region and to find the average value of a function.

Consider a function $z = f(x, y)$ that is continuous and nonnegative over a region R. Let S be the solid region that lies between the xy-plane and the surface

$$z = f(x, y)$$

directly above the region R, as shown in the figure at the right. You can find the volume of S by integrating

$$f(x, y)$$

over the region R.

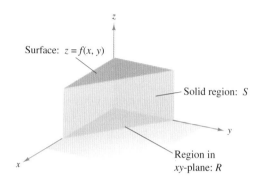

Determining Volume with Double Integrals

If R is a bounded region in the xy-plane and f is continuous and nonnegative over R, then the **volume of the solid region** between the surface

$$z = f(x, y)$$

and R is given by the double integral

$$\iint_R f(x, y)\, dA$$

where $dA = dx\, dy$ or $dA = dy\, dx$.

You can use the following guidelines when finding the volume of a solid region.

Guidelines for Finding the Volume of a Solid Region

1. Write the equation of the surface in the form

$$z = f(x, y)$$

   and sketch the solid region.

2. Sketch the region R in the xy-plane and determine the order and limits of integration.

3. Evaluate the double integral

$$\iint_R f(x, y)\, dA$$

   using the order and limits determined in the second step.

iStockphoto.com/4x6

EXAMPLE 1   Finding the Volume of a Solid Region

Find the volume of the solid region bounded in the first octant by the plane

$$z = 2 - x - 2y.$$

SOLUTION

1. The equation of the surface is already in the form $z = f(x, y)$. A graph of the solid region is shown in Figure 7.44.

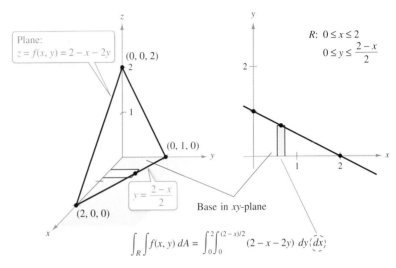

$$\int_R \int f(x, y) \, dA = \int_0^2 \int_0^{(2-x)/2} (2 - x - 2y) \, dy \, dx$$

FIGURE 7.44

2. Sketch the region R in the xy-plane. In Figure 7.44, you can see that the region R is bounded by the lines $x = 0$, $y = 0$, and $y = \frac{1}{2}(2 - x)$. One way to set up the double integral is to choose x as the outer variable. With that choice, the bounds for x are $0 \le x \le 2$ and the bounds for y are $0 \le y \le \frac{1}{2}(2 - x)$.

3. The volume of the solid region is

$$V = \int_0^2 \int_0^{(2-x)/2} (2 - x - 2y) \, dy \, dx$$

$$= \int_0^2 \left[(2 - x)y - y^2 \right]_0^{(2-x)/2} dx$$

$$= \int_0^2 \left\{ (2 - x)\left(\frac{1}{2}\right)(2 - x) - \left[\frac{1}{2}(2 - x)\right]^2 \right\} dx$$

$$= \frac{1}{4} \int_0^2 (2 - x)^2 \, dx$$

$$= \frac{1}{4} \left[-\frac{1}{3}(2 - x)^3 \right]_0^2$$

$$= \frac{2}{3} \text{ cubic unit.}$$

✓ **Checkpoint 1**   *Worked-out solution available at LarsonAppliedCalculus.com*

Find the volume of the solid region bounded in the first octant by the plane $z = 4 - 2x - y$. ■

Example 1 uses $dy \, dx$ as the order of integration. The other order, $dx \, dy$, as indicated in Figure 7.45, produces the same result. Try verifying this.

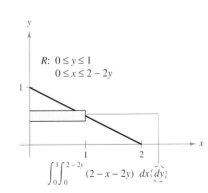

FIGURE 7.45

In Example 1, the problem could be solved with either order of integration. Moreover, had you used the order $dx\,dy$, you would have obtained a double integral of comparable difficulty. There are, however, some occasions in which one order of integration is much more convenient than the other. Example 2 shows such a case.

EXAMPLE 2   Comparing Different Orders of Integration

Find the volume of the solid region bounded by the surface

$$f(x, y) = e^{-x^2} \qquad \text{Surface}$$

and the planes $z = 0$, $y = 0$, $y = x$, and $x = 1$, as shown in Figure 7.46.

SOLUTION   In the xy-plane, the bounds of region R are the lines

$$y = 0, \quad x = 1, \quad \text{and} \quad y = x.$$

The two possible orders of integration are indicated in Figure 7.47.

FIGURE 7.46

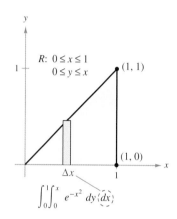

FIGURE 7.47

By setting up the corresponding integrals, you can see that the order $dy\,dx$ produces an integral that is easier to evaluate than the order $dx\,dy$.

$$V = \int_0^1 \int_0^x e^{-x^2}\, dy\, dx$$

$$= \int_0^1 \left[e^{-x^2} y \right]_0^x dx$$

$$= \int_0^1 x e^{-x^2}\, dx$$

$$= \left[-\frac{1}{2} e^{-x^2} \right]_0^1$$

$$= -\frac{1}{2}\left(\frac{1}{e} - 1 \right)$$

$$\approx 0.316 \text{ cubic unit}$$

TECH TUTOR

Use a symbolic integration utility to evaluate the double integral in Example 2.

✓ *Checkpoint 2*   Worked-out solution available at LarsonAppliedCalculus.com

Find the volume bounded by the surface

$$f(x, y) = e^{x^2}$$

and the planes $z = 0$, $y = 0$, $y = 2x$, and $x = 1$.

In the guidelines for finding the volume of a solid region given at the beginning of this section, the first step suggests that you sketch the three-dimensional solid region. This is a good suggestion, but it is not always feasible and is not as important as making a sketch of the two-dimensional region R.

EXAMPLE 3   Finding the Volume of a Solid Region

Find the volume of the solid region bounded above by the surface

$$f(x, y) = 6x^2 - 2xy$$

and below by the plane region R shown in Figure 7.48.

SOLUTION   Because the region R is bounded by the parabola

$$y = 3x - x^2$$

and the line

$$y = x$$

the limits for y are $x \leq y \leq 3x - x^2$. The limits for x are $0 \leq x \leq 2$, and the volume of the solid region is

$$
\begin{aligned}
V &= \int_0^2 \int_x^{3x-x^2} (6x^2 - 2xy)\, dy\, dx \\
&= \int_0^2 \left[6x^2y - xy^2 \right]_x^{3x-x^2} dx \\
&= \int_0^2 \left[(18x^3 - 6x^4 - 9x^3 + 6x^4 - x^5) - (6x^3 - x^3) \right] dx \\
&= \int_0^2 (4x^3 - x^5)\, dx \\
&= \left[x^4 - \frac{x^6}{6} \right]_0^2 \\
&= \frac{16}{3} \text{ cubic units.}
\end{aligned}
$$

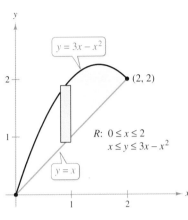

$y = 3x - x^2$

$(2, 2)$

$R: \ 0 \leq x \leq 2$
$x \leq y \leq 3x - x^2$

$y = x$

FIGURE 7.48

✓ **Checkpoint 3**   *Worked-out solution available at LarsonAppliedCalculus.com*

Find the volume of the solid region bounded above by the surface

$$f(x, y) = 4x^2 + 2xy$$

and below by the plane region R shown in the figure.

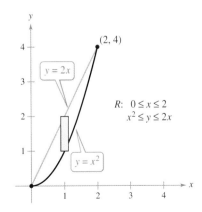

$(2, 4)$

$y = 2x$

$R: \ 0 \leq x \leq 2$
$x^2 \leq y \leq 2x$

$y = x^2$

A *population density function*

$$p = f(x, y)$$

is a model that describes the density (in people per square unit) of a region. To find the population of a region R, evaluate the double integral

$$\int_R \int f(x, y) \, dA.$$

EXAMPLE 4   Finding the Population of a City

The population density (in people per square mile) of the city shown in Figure 7.49 can be modeled by

$$f(x, y) = \frac{50,000}{x + |y| + 1}$$

where x and y are measured in miles. Approximate the population of the city. Is the average population density of the city more than 10,000 people per square mile?

SOLUTION   Because the model involves the absolute value of y, it follows that the population density is symmetrical about the x-axis. So, the population in the first quadrant is equal to the population in the fourth quadrant. This means that you can find the total population by doubling the population in the first quadrant.

$$\begin{aligned}
\text{Population} &= 2 \int_0^4 \int_0^5 \frac{50,000}{x + y + 1} \, dy \, dx \\
&= 100,000 \int_0^4 \int_0^5 \frac{1}{x + y + 1} \, dy \, dx \\
&= 100,000 \int_0^4 \left[\ln(x + y + 1) \right]_0^5 dx \\
&= 100,000 \int_0^4 \left[\ln(x + 6) - \ln(x + 1) \right] dx \\
&= 100,000 \left[(x + 6) \ln(x + 6) - (x + 6) - (x + 1) \ln(x + 1) + (x + 1) \right]_0^4 \\
&= 100,000 \left[(x + 6) \ln(x + 6) - (x + 1) \ln(x + 1) - 5 \right]_0^4 \\
&= 100,000 [10 \ln(10) - 5 \ln(5) - 5 - 6 \ln(6) + 5] \\
&\approx 422,810 \text{ people}
\end{aligned}$$

So, the population of the city is about 422,810. Because the city covers a region 4 miles wide and 10 miles long, its area is 40 square miles. So, the average population density is

$$\text{Average population density} = \frac{422,810}{40}$$

$$\approx 10,570 \text{ people per square mile.}$$

So, you can conclude that the average population density of the city is more than 10,000 people per square mile.

✓ *Checkpoint 4*   *Worked-out solution available at LarsonAppliedCalculus.com*

In Example 4, what integration technique was used to integrate

$$\int \left[\ln(x + 6) - \ln(x + 1) \right] dx?$$

$R: \ 0 \le x \le 4$
$ -5 \le y \le 5$

Ocean   City

FIGURE 7.49

Average Value of a Function over a Region

Average Value of a Function over a Region

If f is integrable over the plane region R with area A, then its **average value** over R is

$$\text{Average value} = \frac{1}{A} \int_R \int f(x, y) \, dA.$$

EXAMPLE 5   Finding Average Profit

A manufacturer determines that the profit P (in dollars) for selling x units of one product and y units of a second product is modeled by $P = -(x - 200)^2 - (y - 100)^2 + 5000$. The weekly sales for the first product vary between 150 and 200 units, and the weekly sales for the second product vary between 80 and 100 units. Estimate the average weekly profit for the two products.

SOLUTION   Because $150 \le x \le 200$ and $80 \le y \le 100$, you can estimate the weekly profit to be the average of the profit function over the rectangular region shown in Figure 7.50. Because the area of this rectangular region is $(50)(20) = 1000$, it follows that the average profit is

$$
\begin{aligned}
\text{Average profit} &= \frac{1}{1000} \int_{150}^{200} \int_{80}^{100} \left[-(x - 200)^2 - (y - 100)^2 + 5000 \right] dy \, dx \\
&= \frac{1}{1000} \int_{150}^{200} \left[-(x - 200)^2 y - \frac{(y - 100)^3}{3} + 5000y \right]_{80}^{100} dx \\
&= \frac{1}{1000} \int_{150}^{200} \left[-20(x - 200)^2 + \frac{292{,}000}{3} \right] dx \\
&= \frac{1}{3000} \left[-20(x - 200)^3 + 292{,}000x \right]_{150}^{200} \\
&\approx \$4033.
\end{aligned}
$$

FIGURE 7.50

✓ **Checkpoint 5**   Worked-out solution available at LarsonAppliedCalculus.com

Find the average value of $f(x, y) = 4 - \frac{1}{2}x - \frac{1}{2}y$ over the region $0 \le x \le 2$ and $0 \le y \le 2$. ■

SUMMARIZE   (Section 7.9)

1. State the definition of the volume of a solid region using double integrals *(page 488)*. For examples of finding the volume of a solid region, see Examples 1, 2, and 3.

2. Give the guidelines for finding the volume of a solid region *(page 488)*. For examples of using these guidelines, see Examples 1, 2, and 3.

3. Describe a real-life example of how a double integral can be used to find the population of a city *(page 492, Example 4)*.

4. State the definition of the average value of a function over a region *(page 493)*. For an example of finding the average value of a function, see Example 5.

SKILLS WARM UP 7.9

The following warm-up exercises involve skills that were covered in earlier sections. You will use these skills in the exercise set for this section. For additional help, review Section 7.8.

In Exercises 1–4, sketch the region R whose area is given by the double integral.

1. $\int_0^2 \int_0^4 dy \, dx$

2. $\int_2^3 \int_1^3 dx \, dy$

3. $\int_0^4 \int_0^{2x+1} dy \, dx$

4. $\int_0^6 \int_{y/2}^3 dx \, dy$

In Exercises 5–10, evaluate the double integral.

5. $\int_0^1 \int_1^2 dy \, dx$

6. $\int_0^3 \int_1^3 dx \, dy$

7. $\int_0^1 \int_0^x x \, dy \, dx$

8. $\int_0^4 \int_1^y y \, dx \, dy$

9. $\int_1^3 \int_x^{x^2} 2 \, dy \, dx$

10. $\int_0^1 \int_x^{-x^2+2} dy \, dx$

Exercises 7.9

See *CalcChat.com* for tutorial help and worked-out solutions to odd-numbered exercises.

 Finding the Volume of a Solid Region In Exercises 1–6, find the volume of the solid region bounded in the first octant by the plane. *See Example 1.*

1. $z = 5 - x - y$

2. $z = 6 - x - 3y$

3. $z = 8 - 8x - 2y$

4. $z = 16 - 8x - 4y$

5. $z = 2 - 6x - 6y$

6. $z = 1 - 4x - 2y$

 Comparing Different Orders of Integration In Exercises 7–12, set up the integrals for both orders of integration and use the more convenient order to find the volume of the solid region bounded by the surface $f(x, y)$ and the planes. *See Example 2.*

7. $f(x, y) = xy$

Planes: $z = 0, y = 0, y = 5, x = 0, x = 3$

8. $f(x, y) = x$

Planes: $z = 0, y = \sqrt{25 - x^2}, y = 0$

9. $f(x, y) = \dfrac{y}{x^2 + y^2}$

Planes: $z = 0, y = x, y = 2x, x = 2$

10. $f(x, y) = \dfrac{y}{1 + x^2}$

Planes: $z = 0, y = 0, y = \sqrt{x}, x = 4$

11. $f(x, y) = \dfrac{1}{\sqrt{x^2 + 4}}$

Planes: $z = 0, y = 0, y = 3x, x = 1$

12. $f(x, y) = \dfrac{1}{\sqrt{2x^2 + 1}}$

Planes: $z = 0, y = 0, y = 3x, x = 4$

 Finding the Volume of a Solid Region In Exercises 13–20, use a double integral to find the volume of the solid region.

13.

$0 \le x \le 4$
$0 \le y \le 2$

14.

$0 \le x \le 4$
$0 \le y \le 2$

15.

16.

17. **18.**

19.

$z = 4 - x^2 - y^2$

$-1 \le x \le 1$
$-1 \le y \le 1$

20.

$x^2 + z^2 = 1$

$x = 1$ $y = x$

 Finding the Volume of a Solid Region In Exercises 21–24, use a double integral to find the volume of the solid region bounded by the graphs of the equations. *See Example 3.*

21. $z = x$, $z = 0$, $y = x$, $y = 0$, $x = 0$, $x = 4$

22. $z = xy$, $z = 0$, $y = 2x$, $y = 0$, $x = 0$, $x = 3$

23. $z = 9 - x^2$, $z = 0$, $y = x + 2$, $y = 0$, $x = 0$, $x = 2$

24. $z = x + y$, $x^2 + y^2 = 4$ (first octant)

25. Population Density The population density (in people per square mile) for a coastal town can be modeled by

$$f(x, y) = \frac{120{,}000}{(2 + x + y)^3}$$

where x and y are measured in miles. What is the population inside the rectangular area defined by the vertices $(0, 0)$, $(2, 0)$, $(0, 2)$, and $(2, 2)$? Is the average population density of the town more than 2000 people?

26. Population Density The population density (in people per square mile) for a coastal town on an island can be modeled by

$$f(x, y) = \frac{5000xe^y}{1 + 2x^2}$$

where x and y are measured in miles. What is the population inside the rectangular area defined by the vertices $(0, 0)$, $(4, 0)$, $(0, -2)$, and $(4, -2)$? Is the average population density of the town more than 500 people?

 Average Value of a Function over a Region In Exercises 27–30, find the average value of $f(x, y)$ over the region R.

27. $f(x, y) = y$
R: rectangle with vertices $(0, 0)$, $(5, 0)$, $(5, 3)$, $(0, 3)$

28. $f(x, y) = xy$
R: rectangle with vertices $(0, 0)$, $(4, 0)$, $(4, 2)$, $(0, 2)$

29. $f(x, y) = x^2 + y^2$
R: square with vertices $(0, 0)$, $(2, 0)$, $(2, 2)$, $(0, 2)$

30. $f(x, y) = e^{x+y}$
R: triangle with vertices $(0, 0)$, $(0, 1)$, $(1, 1)$

31. Average Weekly Profit A firm's weekly profit P (in dollars) in marketing two products is given by

$$P = 192x_1 + 576x_2 - x_1^2 - 5x_2^2 - 2x_1x_2 - 5000$$

where x_1 and x_2 represent the numbers of units of each product sold weekly. Estimate the average weekly profit when x_1 varies between 40 and 50 units and x_2 varies between 45 and 50 units.

32. Average Weekly Profit After a change in marketing, the weekly profit of the firm in Exercise 31 is given by

$$P = 200x_1 + 580x_2 - x_1^2 - 5x_2^2 - 2x_1x_2 - 7500.$$

Estimate the average weekly profit when x_1 varies between 55 and 65 units and x_2 varies between 50 and 60 units.

33. Average Revenue A company sells two products whose demand functions are given by

$$x_1 = 500 - 3p_1 \quad \text{and} \quad x_2 = 750 - 2.4p_2.$$

The total revenue is given by

$$R = x_1p_1 + x_2p_2.$$

Estimate the average revenue when price p_1 varies between \$50 and \$75 and price p_2 varies between \$100 and \$150.

34. **HOW DO YOU SEE IT?** The figure below shows Erie County, New York. Let $f(x, y)$ represent the total annual snowfall at the point (x, y) in the county, where R is the county. Interpret each of the following.

(a) $\displaystyle\int_R\int f(x, y)\, dA$

(b) $\displaystyle\frac{\int_R\int f(x, y)\, dA}{\int_R\int dA}$

Buffalo

ERIE

35. Average Production The Cobb-Douglas production function for an automobile manufacturer is

$$f(x, y) = 100x^{0.6}y^{0.4}$$

where x is the number of units of labor and y is the number of units of capital. Estimate the average production level when the number of units of labor x varies between 200 and 250 and the number of units of capital y varies between 300 and 325.

36. Average Production Repeat Exercise 35 for the production function given by

$$f(x, y) = x^{0.25}y^{0.75}.$$

ALGEBRA TUTOR

Solving Systems of Equations

Three of the sections in this chapter (7.5, 7.6, and 7.7) involve solutions of systems of equations. These systems can be linear or nonlinear, as shown below.

Nonlinear System in Two Variables

$$\begin{cases} 4x + 3y = 6 \\ x^2 - y = 4 \end{cases}$$

Linear System in Three Variables

$$\begin{cases} -x + 2y + 4z = 2 \\ 2x - y + z = 0 \\ 6x + 2z = 3 \end{cases}$$

There are many techniques for solving a system of linear equations. Two of the more common ones are listed here.

1. *Substitution:* Solve for one of the variables in one of the equations and substitute the value into another equation.

2. *Elimination:* Add multiples of one equation to a second equation to eliminate a variable in the second equation.

EXAMPLE 1   Solving Systems of Equations

Solve each system of equations.

a. $\begin{cases} y - x^3 = 0 \\ x - y^3 = 0 \end{cases}$ **b.** $\begin{cases} -400p_1 + 300p_2 = -25 \\ 300p_1 - 360p_2 = -535 \end{cases}$

SOLUTION

a. Example 3, page 458

$$\begin{cases} y - x^3 = 0 \\ x - y^3 = 0 \end{cases}$$ Equation 1 / Equation 2

$$y = x^3$$ Solve for y in Equation 1.

$$x - (x^3)^3 = 0$$ Substitute x^3 for y in Equation 2.

$$x - x^9 = 0$$ $(x^m)^n = x^{mn}$

$$x(x - 1)(x + 1)(x^2 + 1)(x^4 + 1) = 0$$ Factor.

Setting each factor equal to zero and solving yields the (real) solutions $x = 0$, $x = 1$, and $x = -1$. [Note that the factors $(x^2 + 1)$ and $(x^4 + 1)$ do not have real solutions.] So, the solutions of the system are $(0, 0)$, $(1, 1)$, and $(-1, -1)$.

b. Example 4, page 459

$$\begin{cases} -400p_1 + 300p_2 = -25 \\ 300p_1 - 360p_2 = -535 \end{cases}$$ Equation 1 / Equation 2

$$p_2 = \tfrac{1}{12}(16p_1 - 1)$$ Solve for p_2 in Equation 1.

$$300p_1 - 360\left(\tfrac{1}{12}\right)(16p_1 - 1) = -535$$ Substitute for p_2 in Equation 2.

$$300p_1 - 30(16p_1 - 1) = -535$$ Multiply factors.

$$-180p_1 = -565$$ Combine like terms.

$$p_1 = \tfrac{113}{36} \approx 3.14$$ Divide each side by -180.

$$p_2 = \tfrac{1}{12}\left[16\left(\tfrac{113}{36}\right) - 1\right]$$ Find p_2 by substituting p_1.

$$p_2 \approx 4.10$$ Solve for p_2.

So, to the nearest cent, $p_1 = \$3.14$ and $p_2 = \$4.10$.

EXAMPLE 2   Solving Systems of Equations

Solve each system of equations.

a. $\begin{cases} y(24 - 12x - 4y) = 0 \\ x(24 - 6x - 8y) = 0 \end{cases}$

b. $\begin{cases} 28a - 4b = 10 \\ -4a + 8b = 12 \end{cases}$

SOLUTION

a. Example 5, page 460

Before solving this system of equations, factor 4 out of the first equation and factor 2 out of the second equation.

$$\begin{cases} y(24 - 12x - 4y) = 0 & \text{Original Equation 1} \\ x(24 - 6x - 8y) = 0 & \text{Original Equation 2} \end{cases}$$

$$\begin{cases} y(4)(6 - 3x - y) = 0 & \text{Factor 4 out of Equation 1.} \\ x(2)(12 - 3x - 4y) = 0 & \text{Factor 2 out of Equation 2.} \end{cases}$$

$$\begin{cases} y(6 - 3x - y) = 0 & \text{Equation 1} \\ x(12 - 3x - 4y) = 0 & \text{Equation 2} \end{cases}$$

In each equation, either factor can be 0, so you obtain four different linear systems. For the first system, substitute $y = 0$ into the second equation to obtain $x = 4$.

$$\begin{cases} y = 0 \\ 12 - 3x - 4y = 0 \end{cases} \qquad (4, 0) \text{ is a solution.}$$

You can solve the second system by the method of elimination.

$$\begin{cases} 6 - 3x - y = 0 \\ 12 - 3x - 4y = 0 \end{cases} \qquad \left(\tfrac{4}{3}, 2\right) \text{ is a solution.}$$

The third system is already solved.

$$\begin{cases} y = 0 \\ x = 0 \end{cases} \qquad (0, 0) \text{ is a solution.}$$

You can solve the last system by substituting $x = 0$ into the first equation to obtain $y = 6$.

$$\begin{cases} 6 - 3x - y = 0 \\ x = 0 \end{cases} \qquad (0, 6) \text{ is a solution.}$$

b. Example 2, page 475

$$\begin{cases} 28a - 4b = 10 & \text{Equation 1} \\ -4a + 8b = 12 & \text{Equation 2} \end{cases}$$

$$-2a + 4b = 6 \qquad \text{Divide Equation 2 by 2.}$$

$$26a \qquad\quad = 16 \qquad \text{Add new equation to Equation 1.}$$

$$a \qquad\quad = \tfrac{8}{13} \qquad \text{Divide each side by 26.}$$

$$28\left(\tfrac{8}{13}\right) - 4b = 10 \qquad \text{Substitute for } a \text{ in Equation 1.}$$

$$b = \tfrac{47}{26} \qquad \text{Solve for } b.$$

SUMMARY AND STUDY STRATEGIES

After studying this chapter, you should have acquired the following skills.
The exercise numbers are keyed to the Review Exercises that begin on page 500.
Answers to odd-numbered Review Exercises are given in the back of the text.*

Section 7.1   **Review Exercises**

■ Plot points in space.   *1, 2*
■ Find the distance between two points in space.   *3, 4*
$$d = \sqrt{(x_2 - x_1)^2 + (y_2 - y_1)^2 + (z_2 - z_1)^2}$$
■ Find the midpoint of a line segment in space.   *5, 6*
$$\text{Midpoint} = \left(\frac{x_1 + x_2}{2}, \frac{y_1 + y_2}{2}, \frac{z_1 + z_2}{2}\right)$$
■ Write the standard forms of the equations of spheres.   *7–10*
$$(x - h)^2 + (y - k)^2 + (z - j)^2 = r^2$$
■ Find the centers and radii of spheres.   *11, 12*
■ Sketch the coordinate plane traces of spheres.   *13, 14*

Section 7.2

■ Sketch planes in space.   *15–18*
■ Classify quadric surfaces in space.   *19–26*

Section 7.3

■ Evaluate functions of several variables.   *27, 28*
■ Find the domains and ranges of functions of two variables.   *29–32*
■ Sketch level curves of functions of two variables.   *33–36*
■ Use functions of several variables to answer questions about real-life situations.   *37–40*

Section 7.4

■ Find the first partial derivatives of functions of several variables.   *41–50*
■ Find the slopes of surfaces in the *x*- and *y*-directions.   *51–54*
■ Find the second partial derivatives of functions of several variables. *55–60*
■ Use partial derivatives to answer questions about real-life situations. *61, 62*

Section 7.5

■ Find the relative extrema of functions of two variables.   *63–70*
■ Use relative extrema to answer questions about real-life situations.   *71, 72*

---

* Several study aids are available to help you master the material in this chapter. The *Student Solutions Manual* and
CalcChat.com have solutions to all odd-numbered exercises, and *CalcView.com* has video solutions for selected
exercises. *LarsonAppliedCalculus.com* offers algebra help, data spreadsheets, and much more. A *Graphing
Technology Guide* with step-by-step commands for a variety of graphing calculators is at *CengageBrain.com*.

Section 7.6

	Review Exercises
■ Use Lagrange multipliers to find extrema of functions of several variables.	*73–78*
■ Use Lagrange multipliers to answer questions about real-life situations.	*79, 80*

Section 7.7

■ Find the least squares regression line, $y = ax + b$, for data.   *81, 82*

$$a = \left[n \sum_{i=1}^{n} x_i y_i - \sum_{i=1}^{n} x_i \sum_{i=1}^{n} y_i \right] \bigg/ \left[n \sum_{i=1}^{n} x_i^2 - \left(\sum_{i=1}^{n} x_i \right)^2 \right], \quad b = \frac{1}{n} \left(\sum_{i=1}^{n} y_i - a \sum_{i=1}^{n} x_i \right)$$

■ Use least squares regression lines to model real-life data.   *83, 84*

Section 7.8

■ Evaluate double integrals.   *85–88*

■ Use double integrals to find the areas of regions.   *89–92*

Section 7.9

■ Use double integrals to find the volumes of solid regions.   *93–98*

$$\text{Volume} = \int_R \int f(x, y) \, dA$$

■ Use double integrals to find the average values of functions.   *99–104*

$$\text{Average value} = \frac{1}{A} \int_R \int f(x, y) \, dA$$

Study Strategies

■ **Comparing Two Dimensions with Three Dimensions** Many of the formulas and techniques in this chapter are generalizations of formulas and techniques used in earlier chapters of the text. Here are several examples.

Two-Dimensional Coordinate System	Three-Dimensional Coordinate System
Distance Formula	*Distance Formula*
$d = \sqrt{(x_2 - x_1)^2 + (y_2 - y_1)^2}$	$d = \sqrt{(x_2 - x_1)^2 + (y_2 - y_1)^2 + (z_2 - z_1)^2}$
Midpoint Formula	*Midpoint Formula*
$\text{Midpoint} = \left(\dfrac{x_1 + x_2}{2}, \dfrac{y_1 + y_2}{2} \right)$	$\text{Midpoint} = \left(\dfrac{x_1 + x_2}{2}, \dfrac{y_1 + y_2}{2}, \dfrac{z_1 + z_2}{2} \right)$
Equation of Circle	*Equation of Sphere*
$(x - h)^2 + (y - k)^2 = r^2$	$(x - h)^2 + (y - k)^2 + (z - j)^2 = r^2$
Equation of Line	*Equation of Plane*
$ax + by = c$	$ax + by + cz = d$
Derivative of $y = f(x)$	*Partial Derivative of $z = f(x, y)$*
$\dfrac{dy}{dx} = \lim_{\Delta x \to 0} \dfrac{f(x + \Delta x) - f(x)}{\Delta x}$	$\dfrac{\partial z}{\partial x} = \lim_{\Delta x \to 0} \dfrac{f(x + \Delta x, y) - f(x, y)}{\Delta x}$
Area of Region	*Volume of Solid Region*
$A = \displaystyle\int_a^b f(x) \, dx$	$V = \displaystyle\int_R \int f(x, y) \, dA$

Review Exercises

See *CalcChat.com* for tutorial help and worked-out solutions to odd-numbered exercises.

Plotting Points in Space   In Exercises 1 and 2, plot the points in the same three-dimensional coordinate system.

1. $(2, -1, 4), (-1, 3, -3), (-2, -2, 1), (3, 1, 2)$

2. $(1, -2, -3), (-4, -3, 5), \left(4, \frac{5}{2}, 1\right), (-2, 2, 2)$

Finding the Distance Between Two Points in Space   In Exercises 3 and 4, find the distance between the two points.

3. $(0, 2, 6), (3, 5, 8)$

4. $(-4, 1, 5), (1, 3, 7)$

Using the Midpoint Formula in Space   In Exercises 5 and 6, find the midpoint of the line segment joining the two points.

5. $(2, 6, 4), (-4, 2, 8)$

6. $(5, 0, 7), (-1, -2, 9)$

Finding the Equation of a Sphere   In Exercises 7–10, find the standard equation of the sphere with the given characteristics.

7. Center: $(0, 1, 9)$; radius: 7

8. Center: $(4, -5, 3)$; radius: 10

9. Endpoints of a diameter: $(3, -4, -1), (1, 0, -5)$

10. Endpoints of a diameter: $(3, 4, 0), (5, 8, 2)$

Finding the Center and Radius of a Sphere   In Exercises 11 and 12, find the center and radius of the sphere.

11. $x^2 + y^2 + z^2 + 8x + 2y - 14z - 15 = 0$

12. $x^2 + y^2 + z^2 + 4y - 10z - 7 = 0$

Finding a Trace of a Sphere   In Exercises 13 and 14, find the equation of the *xy*-trace of the sphere. Then sketch the *xy*-trace of the sphere.

13. $(x + 2)^2 + (y - 1)^2 + (z - 3)^2 = 25$

14. $(x - 1)^2 + (y + 3)^2 + (z - 6)^2 = 72$

Sketching a Plane in Space   In Exercises 15–18, find the *x*-, *y*-, and *z*-intercepts of the plane. Then sketch the plane.

15. $x + 2y + 3z = 6$ **16.** $4x - y + 2z = 8$

17. $2y + z = 4$ **18.** $10x - 2z = 10$

Classifying a Quadric Surface   In Exercises 19–26, classify the quadric surface.

19. $x^2 + \dfrac{y^2}{16} + \dfrac{z^2}{9} = 1$ **20.** $z = \dfrac{x^2}{9} + y^2$

21. $x^2 - \dfrac{y^2}{16} - \dfrac{z^2}{9} = 1$

22. $z = x^2 - \dfrac{y^2}{4}$

23. $3x^2 + 9y^2 - z^2 = 9$

24. $16x^2 + 16y^2 - 9z^2 = 0$

25. $z = \sqrt{x^2 + y^2}$

26. $-4x^2 + y^2 + z^2 = 4$

Evaluating Functions of Several Variables   In Exercises 27 and 28, find and simplify the function values.

27. $f(x, y) = xy^2$

(a) $f(2, 3)$ (b) $f(0, 1)$

(c) $f(-5, 7)$ (d) $f(-2, -4)$

28. $f(x, y) = \dfrac{x^2}{y}$

(a) $f(6, 9)$ (b) $f(8, 4)$

(c) $f(t, 2)$ (d) $f(r, r)$

Finding the Domain and Range of a Function   In Exercises 29–32, find the domain and range of the function.

29. $f(x, y) = \sqrt{1 - x^2 - y^2}$

30. $f(x, y) = x^2 + y^2 - 3$

31. $f(x, y) = ye^{1/x}$

32. $f(x, y) = \dfrac{1}{x + y}$

Sketching a Contour Map   In Exercises 33–36, describe the level curves of the function. Sketch a contour map of the surface using level curves for the given *c*-values.

Function	*c*-Values
33. $z = 10 - 2x - 5y$	$c = 0, 2, 4, 5, 10$
34. $z = \sqrt{36 - x^2 - y^2}$	$c = 0, 1, 2, 3$
35. $z = (xy)^2$	$c = 1, 4, 9, 16, 25$
36. $z = y - x^2$	$c = 0, \pm 1, \pm 2$

37. Cobb-Douglas Production Function   A manufacturer estimates that its production can be modeled by

$$f(x, y) = 100x^{0.73}y^{0.27}.$$

(a) Estimate the production level when $x = 1100$ and $y = 900$.

(b) Estimate the production level when $x = 1600$ and $y = 1000$.

38. Investment A principal of $2000 is deposited in a savings account that earns an interest rate of r (in decimal form), compounded continuously. The amount $A(r, t)$ after t years is

$$A(r, t) = 2000e^{rt}.$$

Use this function of two variables and a spreadsheet to complete the table.

	Number of Years			
Rate	5	10	15	20
0.02				
0.04				
0.06				
0.07				

39. Earnings per Share The earnings per share z (in dollars) for lululemon from 2007 through 2013 can be modeled by

$$z = -0.398 + 0.0029x - 0.0021y$$

where x is the sales (in millions of dollars) and y is the shareholder's equity (in millions of dollars). *(Source: lululemon athletica inc.)*

(a) Find the earnings per share when $x = 1000$ and $y = 400$.

(b) Which of the two variables in this model has the greater influence on the earnings per share? Explain.

40. Monthly Payments You are taking out a home mortgage for $200,000, and you are given the options below. Find the monthly payment and the total amount of money you will pay for each mortgage. Which option would you choose? Explain your reasoning.

(a) A fixed annual rate of 5%, over a term of 30 years.

(b) A fixed annual rate of 6%, over a term of 20 years.

(c) A fixed annual rate of 6.5%, over a term of 15 years.

Finding Partial Derivatives In Exercises 41–50, find the first partial derivatives.

41. $f(x, y) = x^2y + 3xy + 2x - 5y$

42. $f(x, y) = 4xy + xy^2 - 3x^2y - x + 7y$

43. $z = (xy + 2x + 4y)^2$

44. $z = \dfrac{x^3y^2}{2x - 9}$

45. $f(x, y) = \ln(5x + 4y)$

46. $f(x, y) = \ln\sqrt{2x + 3y}$

47. $f(x, y) = xe^y + ye^x$

48. $f(x, y) = x^2ye^{-2y}$

49. $w = xyz^2$

50. $w = 3xy - 5xz^2 + 2yz$

Finding Slopes in the x- and y-Directions In Exercises 51–54, find the slopes of the surface at the given point in (a) the x-direction and (b) the y-direction.

51. $z = 3xy$
$(-2, -3, 18)$

52. $z = y^2 - x^2$
$(1, 2, 3)$

53. $z = 8 - x^2 - y^2$
$(1, 1, 6)$

54. $z = \sqrt{100 - x^2 - y^2}$
$(0, 6, 8)$

Finding Second Partial Derivatives In Exercises 55–60, find the second partial derivatives.

55. $f(x, y) = 3x^2 - xy + 2y^3$

56. $f(x, y) = \dfrac{y}{x - y}$

57. $f(x, y) = \sqrt{1 + x + y}$

58. $f(x, y) = x^2e^{-y^2}$

59. $f(x, y, z) = xy + 5x^2yz^3 - 3y^3z$

60. $f(x, y, z) = \dfrac{3yz}{x + z}$

61. Marginal Cost A company manufactures cross-country skis and downhill skis. The cost function for producing x pairs of cross-country skis and y pairs of downhill skis is given by

$$C = 15(xy)^{1/3} + 99x + 139y + 2293.$$

(a) Find the marginal costs $(\partial C/\partial x$ and $\partial C/\partial y)$ when $x = 500$ and $y = 250$.

(b) When additional production is required, which model of skis results in the cost increasing at a higher rate? How can this be determined from the cost model?

62. Marginal Revenue At a baseball stadium, souvenir hats are sold at two locations. If x_1 and x_2 are the numbers of baseball hats sold at Location 1 and Location 2, respectively, then the total revenue for the hats is modeled by

$$R = 15x_1 + 16x_2 - \frac{1}{10}x_1^2 - \frac{1}{10}x_2^2 - \frac{1}{100}x_1 x_2.$$

When $x_1 = 50$ and $x_2 = 40$, find

(a) the marginal revenue for Location 1, $\partial R / \partial x_1$.

(b) the marginal revenue for Location 2, $\partial R / \partial x_2$.

Applying the Second-Partials Test In Exercises 63–70, find the relative extrema and saddle points of the function.

63. $f(x, y) = x^2 + 2y^2$

64. $f(x, y) = -x^2 - y^2 + 2x + 4y$

65. $f(x + y) = (x^2 + y^2 + 1)^{1/4}$

66. $f(x, y) = 1 - (x + 2)^2 + (y - 3)^2$

67. $f(x, y) = e^x - x + y^2$

68. $f(x, y) = y^2 + xy + 3y - 2x + 5$

69. $f(x, y) = x^3 - 3xy + y^2$

70. $f(x, y) = x^3 + y^3 - 3x - 3y + 2$

71. Revenue A company manufactures racing bikes and mountain bikes. The total revenue for x_1 units of racing bikes and x_2 units of mountain bikes is

$$R = -6x_1^2 - 10x_2^2 - 2x_1 x_2 + 32x_1 + 84x_2$$

where x_1 and x_2 are in thousands of units. Find x_1 and x_2 so as to maximize the revenue.

72. Profit A company manufactures a product at two locations. The cost of producing x_1 units at Location 1 is

$$C_1 = 0.03x_1^2 + 4x_1 + 300$$

and the cost of producing x_2 units at Location 2 is

$$C_2 = 0.05x_2^2 + 7x_2 + 175.$$

The product sells for $10 per unit. Find the quantity that should be produced at each location to maximize the profit

$$P = 10(x_1 + x_2) - C_1 - C_2.$$

Using Lagrange Multipliers In Exercises 73–78, use Lagrange multipliers to find the indicated extremum. Assume that the variables are positive.

73. Maximize $f(x, y) = 2xy$

Constraint: $2x + y - 12 = 0$

74. Maximize $f(x, y) = 4x + 3xy + y$

Constraint: $x + 4y - 29 = 0$

75. Minimize $f(x, y) = x^2 + y^2$

Constraint: $x + y = 4$

76. Minimize $f(x, y) = 3x^2 - y^2$

Constraint: $2x - 4y = -11$

77. Maximize $f(x, y, z) = 5xyz$

Constraint: $x + 3y + z - 6 = 0$

78. Maximize $f(x, y, z) = x^2 z + yz$

Constraint: $2x + y + z = 5$

79. Cost A manufacturer has an order for 1000 wooden benches that can be produced at two locations. Let x_1 and x_2 be the numbers of wooden benches produced at the two locations. The cost function is modeled by

$$C = 0.25x_1^2 + 10x_1 + 0.15x_2^2 + 12x_2.$$

Use Lagrange multipliers to find the number of wooden benches that should be produced at each location to minimize the cost.

80. Production The production function for a manufacturer is given by

$$f(x, y) = 100x^{0.8}y^{0.2}$$

where x is the number of units of labor (at $40 per unit) and y is the number of units of capital (at $35 per unit). The total cost for labor and capital cannot exceed $110,000.

(a) Use Lagrange multipliers to find the maximum production level for this manufacturer.

(b) Find the marginal productivity of money.

(c) Use the marginal productivity of money to find the maximum number of units that can be produced when $120,000 is available for labor and capital.

Finding the Least Squares Regression Line In Exercises 81 and 82, find the least squares regression line for the given points. Then plot the points and sketch the regression line.

81. $(-2, 4), (-1, 2), (1, -1), (2, -2)$

82. $(-3, -1), (-2, -1), (0, 0), (1, 1), (2, 1)$

83. Demand A store manager wants to know the demand y for a wireless Bluetooth speaker as a function of price x. The monthly sales for four different prices of the Bluetooth speaker are listed in the table.

Price, x	$190	$200	$210	$220
Demand, y	140	117	91	63

(a) Use the regression capabilities of a graphing utility or a spreadsheet to find the least squares regression line for the data.

(b) Estimate the demand when the price is $195.

(c) What price will create a demand of 160 speakers?

84. Master's Degrees The number of men x (in thousands) and the number of women y (in thousands) in the United States that attained a master's degree from 2007 through 2012 are shown in the table. *(Source: U.S. Department of Education)*

DATA	Year	2007	2008	2009
	Men, x	242.2	250.2	263.5
	Women, y	368.4	380.5	398.5

Year	2010	2011	2012
Men, x	275.2	291.6	302.2
Women, y	417.8	439.1	452.0

Spreadsheet at LarsonAppliedCalculus.com

(a) Use the regression capabilities of a graphing utility or a spreadsheet to find the least squares regression line for the data.

(b) Estimate the number of women that attain a master's degree when 400 thousand men attain a master's degree.

Evaluating a Double Integral In Exercises 85–88, evaluate the double integral.

85. $\displaystyle\int_0^1 \int_0^{1+x} (4x - 2y)\, dy\, dx$

86. $\displaystyle\int_{-3}^3 \int_0^4 (x - y^2)\, dx\, dy$

87. $\displaystyle\int_1^2 \int_1^{2y} \frac{x}{y^2}\, dx\, dy$ **88.** $\displaystyle\int_0^4 \int_0^{\sqrt{16-x^2}} 2x\, dy\, dx$

Finding Area with a Double Integral In Exercises 89–92, use a double integral to find the area of the region bounded by the graphs of the equations.

89. $y = 10 - x^2,\ y = 6$

90. $y = \dfrac{4}{x},\ y = 0,\ x = 1,\ x = 4$

91. $y = \sqrt{x + 3},\ y = \frac{1}{3}x + 1$

92. $y = x^2 - 2x - 2,\ y = -x$

Finding the Volume of a Solid Region In Exercises 93–96, use a double integral to find the volume of the solid region.

93.   **94.**

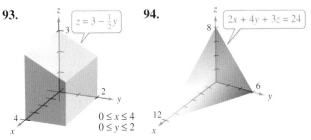

$0 \le x \le 4$
$0 \le y \le 2$

95.   **96.**

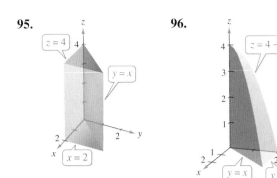

Finding the Volume of a Solid Region In Exercises 97 and 98, use a double integral to find the volume of the solid region bounded by the graphs of the equations.

97. $z = (xy)^2,\ z = 0,\ y = 0,\ y = 4,\ x = 0,\ x = 4$

98. $z = x + y,\ z = 0,\ x = 0,\ x = 3,\ y = x,\ y = 0$

Average Value of a Function over a Region In Exercises 99 and 100, find the average value of $f(x, y)$ over the region R.

99. $f(x, y) = 4xy$

R: rectangle with vertices $(0, 0),\ (2, 0),\ (2, 4),\ (0, 4)$

100. $f(x, y) = x^2 + 2xy + y^2$

R: rectangle with vertices $(0, 0),\ (2, 0),\ (2, 5),\ (0, 5)$

101. Average Weekly Profit A firm's weekly profit P (in dollars) in marketing two products is given by

$$P = 150x_1 + 400x_2 - x_1^2 - 5x_2^2 - 2x_1x_2 - 3000$$

where x_1 and x_2 represent the numbers of units of each product sold weekly. Estimate the average weekly profit when x_1 varies between 30 and 40 units and x_2 varies between 40 and 50 units.

102. Average Revenue A company sells two products whose demand functions are given by

$$x_1 = 500 - 2.5p_1 \quad \text{and} \quad x_2 = 750 - 3p_2.$$

The total revenue is given by

$$R = x_1p_1 + x_2p_2.$$

Estimate the average revenue when price p_1 varies between \$25 and \$50 and price p_2 varies between \$75 and \$125.

103. Average Production The Cobb-Douglas production function for a manufacturer is $f(x, y) = 100x^{0.65}y^{0.35}$, where x is the number of units of labor and y is the number of units of capital. Estimate the average production level when the number of units of labor x varies between 100 and 175 and the number of units of capital y varies between 200 and 250.

104. Average Production Repeat Exercise 103 for the production function given by $f(x, y) = 100x^{0.85}y^{0.15}$.

TEST YOURSELF

See *CalcChat.com* for tutorial help and worked-out solutions to odd-numbered exercises.

Take this test as you would take a test in class. When you are done, check your work against the answers given in the back of the book.

In Exercises 1–3, (a) plot the points in a three-dimensional coordinate system, (b) find the distance between the two points, and (c) find the midpoint of the line segment joining the two points.

1. $(1, -3, 0), (3, -1, 0)$ **2.** $(-2, 2, 3), (-4, 0, 2)$ **3.** $(3, -7, 2), (5, 11, -6)$

4. Find the center and radius of the sphere whose equation is

$$x^2 + y^2 + z^2 - 20x + 10y - 10z + 125 = 0.$$

In Exercise 5–7, classify the quadric surface.

5. $4x^2 + 2y^2 - z^2 = 16$ **6.** $36x^2 + 9y^2 - 4z^2 = 0$ **7.** $7x^2 - y^2 - 49z = 0$

In Exercises 8 and 9, find $f(1, 4)$, and $f(-6, -2)$.

8. $f(x, y) = x^2 + xy + 1$ **9.** $f(x, y) = \dfrac{x + 2y}{3x - y}$

In Exercises 10 and 11, find the first partial derivatives and evaluate each at the point $(10, -1)$.

10. $f(x, y) = 3x^2 + 9xy^2 - 2$ **11.** $f(x, y) = x\sqrt{x + y}$

In Exercises 12 and 13, find the second partial derivatives.

12. $z = 2x^3 + 5y^2 + 9$

13. $w = x^2y - 3y^3 - 2yz^2$

In Exercises 14 and 15, find the relative extrema and saddle points of the function.

14. $f(x, y) = 3x^2 + 4y^2 - 6x + 16y - 4$

15. $f(x, y) = 4xy - x^4 - y^4$

16. The production function for a manufacturer is given by

$$f(x, y) = 60x^{0.7}y^{0.3}$$

where x is the number of units of labor (at \$42 per unit) and y is the number of units of capital (at \$144 per unit). The total cost for labor and capital cannot exceed \$240,000. Find the maximum production level for this manufacturer.

17. Find the least squares regression line for the points $(1, 2)$, $(3, 3)$, $(6, 4)$, $(8, 6)$, and $(11, 7)$. Then plot the points and sketch the regression line.

In Exercises 18 and 19, evaluate the double integral.

18. $\displaystyle\int_0^1 \int_x^1 (30x^2y - 1)\, dy\, dx$ **19.** $\displaystyle\int_0^{\sqrt{e-1}} \int_0^{2y} \frac{1}{y^2 + 1}\, dx\, dy$

20. Use a double integral to find the area of the region bounded by the graphs of $y = 3$ and $y = x^2 - 2x + 3$.

21. Use a double integral to find the volume of the solid region bounded by the graphs of $z = 8 - 2x$, $z = 0$, $y = 0$, $y = 3$, $x = 0$, and $x = 4$.

22. Find the average value of $f(x, y) = x^2 + y$ over the region defined by a rectangle with vertices $(0, 0)$, $(1, 0)$, $(1, 3)$, and $(0, 3)$.

Appendices

A Precalculus Review

A.1 The Real Number Line and Order

■ Represent and classify real numbers.
■ Use inequalities to represent sets of real numbers.
■ Solve inequalities.
■ Use inequalities to model and solve real-life problems.

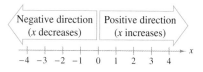

The Real Number Line

FIGURE A.1

Every point on the real number line corresponds to one and only one real number.

Every real number corresponds to one and only one point on the real number line.

FIGURE A.2

The Real Number Line

Real numbers can be represented with a coordinate system called the **real number line** (or x-axis), as shown in Figure A.1. The **positive direction** (to the right) is denoted by an arrowhead and indicates the direction of increasing values of x. The real number corresponding to a particular point on the real number line is called the **coordinate** of the point. As shown in Figure A.1, it is customary to label those points whose coordinates are integers.

The point on the real number line corresponding to zero is called the **origin.** Numbers to the right of the origin are **positive,** and numbers to the left of the origin are **negative.** The term **nonnegative** describes a number that is either positive or zero.

The importance of the real number line is that it provides you with a conceptually perfect picture of the real numbers. That is, each point on the real number line corresponds to one and only one real number, and each real number corresponds to one and only one point on the real number line. This type of relationship is called a **one-to-one correspondence** and is illustrated in Figure A.2.

Each of the four points in Figure A.2 corresponds to a real number that can be expressed as the ratio of two integers.

$$-2.6 = -\tfrac{13}{5} \qquad \tfrac{5}{4} \qquad -\tfrac{7}{3} \qquad 1.85 = \tfrac{37}{20}$$

Such numbers are called **rational.** Rational numbers have either terminating or infinitely repeating decimal representations.

Terminating Decimals

$$\frac{2}{5} = 0.4$$

$$\frac{7}{8} = 0.875$$

Infinitely Repeating Decimals

$$\frac{1}{3} = 0.333 \ldots = 0.\overline{3}*$$

$$\frac{12}{7} = 1.714285714285 \ldots = 1.\overline{714285}$$

Real numbers that are not rational are called **irrational,** and they cannot be represented as the ratio of two integers (or as terminating or infinitely repeating decimals). So, a decimal approximation is used to represent an irrational number. Some irrational numbers occur so frequently in applications that mathematicians have invented special symbols to represent them. For example, the symbols $\sqrt{2}$, π, and e represent irrational numbers whose decimal approximations are as shown. (See Figure A.3.)

$$\sqrt{2} \approx 1.4142135624$$
$$\pi \approx 3.1415926536$$
$$e \approx 2.7182818285$$

FIGURE A.3

* The bar indicates which digit or digits repeat infinitely.

Order and Intervals on the Real Number Line

One important property of the real numbers is that they are **ordered:** 0 is less than 1, -3 is less than -2.5, π is less than $\frac{22}{7}$, and so on. You can visualize this property on the real number line by observing that a is less than b if and only if a lies to the left of b on the real number line. Symbolically, "a is less than b" is denoted by the inequality $a < b$. For example, the inequality

$$\frac{3}{4} < 1$$

follows from the fact that $\frac{3}{4}$ lies to the left of 1 on the real number line, as shown in Figure A.4.

FIGURE A.4

When three real numbers a, x, and b are ordered such that $a < x$ and $x < b$, you can say that x is **between** a and b and write

$$a < x < b. \qquad \text{x is between a and b.}$$

The set of *all* real numbers between a and b is called the **open interval** between a and b and is denoted by (a, b). An interval of the form (a, b) does not contain the "endpoints" a and b. Intervals that include their endpoints are called **closed** and are denoted by $[a, b]$. Intervals of the form $[a, b)$ and $(a, b]$ are neither open nor closed. Figure A.5 shows the nine types of intervals on the real number line.

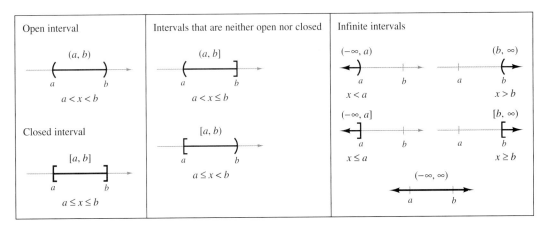

Intervals on the Real Number Line
FIGURE A.5

Note that a square bracket is used to denote "less than or equal to" ($\leq$) or "greater than or equal to" ($\geq$). Furthermore, the symbols

$$\infty \qquad \text{Positive infinity}$$

and

$$-\infty \qquad \text{Negative infinity}$$

denote **positive** and **negative infinity,** respectively. These symbols do not denote real numbers; they merely let you describe unbounded conditions more concisely. For instance, the interval $[b, \infty)$ is unbounded to the right because it includes *all* real numbers that are greater than or equal to b.

Solving Inequalities

In calculus, you are frequently required to "solve inequalities" involving variable expressions such as $3x - 4 < 5$. The number a is a **solution** of an inequality if the inequality is true when a is substituted for x. The set of all values of x that satisfy an inequality is called the **solution set** of the inequality. The following properties are useful for solving inequalities. (Similar properties are obtained when $<$ is replaced by $\leq$ and $>$ is replaced by $\geq$.)

Properties of Inequalities

Let a, b, c, and d be real numbers.

1. Transitive property: $a < b$ and $b < c$ $\implies$ $a < c$

2. Adding inequalities: $a < b$ and $c < d$ $\implies$ $a + c < b + d$

3. Multiplying by a (positive) constant: $a < b$ $\implies$ $ac < bc$, $c > 0$

4. Multiplying by a (negative) constant: $a < b$ $\implies$ $ac > bc$, $c < 0$

5. Adding a constant: $a < b$ $\implies$ $a + c < b + c$

6. Subtracting a constant: $a < b$ $\implies$ $a - c < b - c$

Note that you *reverse the inequality* when you multiply by a negative number. For example, if $x < 3$, then $-4x > -12$. This principle also applies to division by a negative number. So, if $-2x > 4$, then $x < -2$.

EXAMPLE 1   Solving an Inequality

Find the solution set of the inequality

$$3x - 4 < 5.$$

SOLUTION

$3x - 4 < 5$	Write original inequality.
$3x - 4 + 4 < 5 + 4$	Add 4 to each side.
$3x < 9$	Simplify.
$\frac{1}{3}(3x) < \frac{1}{3}(9)$	Multiply each side by $\frac{1}{3}$.
$x < 3$	Simplify.

So, the solution set is the interval $(-\infty, 3)$, as shown in Figure A.6. Once you have solved an inequality, it is a good idea to check some x-values in your solution set to see whether they satisfy the original inequality. You should also check some values outside your solution set to verify that they do *not* satisfy the inequality. For instance, Figure A.6 shows that when $x = 0$ or $x = 2$, the inequality is satisfied, but when $x = 4$, the inequality is not satisfied.

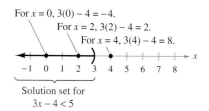

For $x = 0$, $3(0) - 4 = -4$.
For $x = 2$, $3(2) - 4 = 2$.
For $x = 4$, $3(4) - 4 = 8$.

Solution set for
$3x - 4 < 5$

FIGURE A.6

✓ *Checkpoint 1*   *Worked-out solution available at LarsonAppliedCalculus.com*

Find the solution set of the inequality

$$2x - 3 < 7.$$ ■

In Example 1, all five inequalities listed as steps in the solution have the same solution set, and they are called **equivalent inequalities**.

The inequality in Example 1 involves a first-degree polynomial. To solve inequalities involving polynomials of higher degree, you can use the fact that a polynomial can change signs *only* at its real zeros (the real numbers that make the polynomial zero). Between two consecutive real zeros, a polynomial must be entirely positive or entirely negative. This means that when the real zeros of a polynomial are put in order, they divide the real number line into **test intervals** in which the polynomial has no sign changes. That is, if a polynomial has the factored form

$$(x - r_1)(x - r_2), \ldots, (x - r_n), \qquad r_1 < r_2 < r_3 < \cdots < r_{n-1} < r_n$$

then the test intervals are

$$(-\infty, r_1), \quad (r_1, r_2), \quad \ldots, \quad (r_{n-1}, r_n), \quad \text{and} \quad (r_n, \infty).$$

For example, the polynomial

$$x^2 - x - 6 = (x + 2)(x - 3)$$

can change signs only at $x = -2$ and $x = 3$. To determine the sign of the polynomial in the intervals $(-\infty, -2)$, $(-2, 3)$, and $(3, \infty)$, you need to test only *one value* in each interval.

EXAMPLE 2   Solving a Polynomial Inequality

$$x^2 < x + 6 \qquad \text{Original inequality}$$
$$x^2 - x - 6 < 0 \qquad \text{Polynomial form}$$
$$(x + 2)(x - 3) < 0 \qquad \text{Factor.}$$

So, the polynomial $x^2 - x - 6$ has $x = -2$ and $x = 3$ as its zeros. You can solve the inequality by testing the sign of the polynomial in each of the intervals $(-\infty, -2)$, $(-2, 3)$, and $(3, \infty)$. In each interval, choose a representative x-value and evaluate the polynomial.

Interval	x-Value	Polynomial Value	Conclusion
$(-\infty, -2)$	$x = -3$	$(-3)^2 - (-3) - 6 = 6$	Positive
$(-2, 3)$	$x = 0$	$(0)^2 - (0) - 6 = -6$	Negative
$(3, \infty)$	$x = 4$	$(4)^2 - (4) - 6 = 6$	Positive

From this you can conclude that the inequality is satisfied for all x-values in $(-2, 3)$. This implies that the solution of the inequality $x^2 < x + 6$ is the interval $(-2, 3)$, as shown in Figure A.7. Note that the original inequality contains a "less than" symbol. This means that the solution set does not contain the endpoints of the test interval $(-2, 3)$.

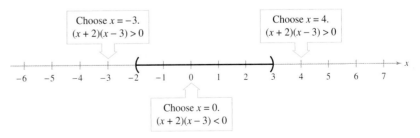

FIGURE A.7

✓ *Checkpoint 2*   Worked-out solution available at LarsonAppliedCalculus.com

Find the solution set of the inequality

$$x^2 > 3x + 10.$$

Application

Inequalities are frequently used to describe conditions that occur in business and science. For instance, the inequality

$$8.8 \le W \le 26.4$$

describes the typical weights W (in pounds) of adult rhesus monkeys. Example 3 shows how an inequality can be used to describe the production levels in a manufacturing plant.

EXAMPLE 3   Production Levels

In addition to fixed overhead costs of $500 per day, the cost of producing x units of an item is $2.50 per unit. During the month of August, the total cost of production varied from a high of $1325 to a low of $1200 per day. Find the high and low *production levels* during the month.

SOLUTION   Because it costs $2.50 to produce one unit, it costs $2.5x$ to produce x units. Furthermore, because the fixed cost per day is $500, the total daily cost C (in dollars) of producing x units is

$$C = 2.5x + 500.$$

Now, because the cost ranged from $1200 to $1325, you can write the following.

$1200 \le$	$2.5x + 500$	≤ 1325	Write original inequality.
$1200 - 500 \le$	$2.5x + 500 - 500$	$\le 1325 - 500$	Subtract 500 from each part.
$700 \le$	$2.5x$	≤ 825	Simplify.
$\dfrac{700}{2.5} \le$	$\dfrac{2.5x}{2.5}$	$\le \dfrac{825}{2.5}$	Divide each part by 2.5.
$280 \le$	x	≤ 330	Simplify.

So, the daily production levels during the month of August varied from a low of 280 units to a high of 330 units, as shown in Figure A.8.

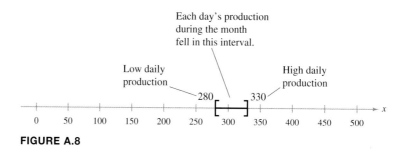

FIGURE A.8

✓ *Checkpoint 3*   *Worked-out solution available at LarsonAppliedCalculus.com*

Use the information in Example 3 to find the high and low production levels during the month of October, when the total cost of production varied from a high of $1500 to a low of $1000 per day. ■

Exercises A.1

See *CalcChat.com* for tutorial help and worked-out solutions to odd-numbered exercises.

Classifying Real Numbers In Exercises 1–8, determine whether the real number is rational or irrational.

1. 0.25

2. -3678

3. $4.3\overline{451}$

4. $2\sqrt{5} - 1$

5. $\dfrac{3\pi}{2}$

6. $\dfrac{22}{7}$

7. $\sqrt[3]{64}$

8. $2e$

Determining Solutions In Exercises 9–12, determine whether each given value of x is a solution of the inequality.

9. $5x - 12 > 0$

 (a) $x = 3$ (b) $x = -3$ (c) $x = \frac{5}{2}$

10. $6x + 4 < 2x$

 (a) $x = 2$ (b) $x = -3$ (c) $x = -1$

11. $0 < \dfrac{x - 2}{4} < 2$

 (a) $x = 4$ (b) $x = 8$ (c) $x = 0$

12. $-1 < \dfrac{3 - x}{2} \le 1$

 (a) $x = 0$ (b) $x = 1$ (c) $x = 5$

Solving an Inequality In Exercises 13–26, find the soution set of the inequality. Then graph the solution set on the real number line. *See Examples 1 and 2.*

13. $x - 5 \ge 7$

14. $2x + 6 > 9$

15. $4x + 1 \le 2x$

16. $7x + 4 < 5x$

17. $4 - 2x < 3x - 1$

18. $x - 4 \le 2x + 1$

19. $-4 < 2x - 3 < 4$

20. $0 \le x + 3 < 5$

21. $\dfrac{3}{4} > x + 1 > \dfrac{1}{4}$

22. $-1 < -\dfrac{x}{3} < 1$

23. $x^2 \ge 3x + 4$

24. $x^2 < 2x + 15$

25. $2x^2 - x \le 6$

26. $2x^2 + 1 > 9x - 3$

Writing Inequalities In Exercises 27–30, use inequality notation to describe the subset of real numbers.

27. Earnings per Share A company expects its earnings per share E for the next quarter to be no less than $4.10 and no more than $4.25.

28. Production The estimated daily oil production p at a refinery is greater than 2 million barrels but less than 2.4 million barrels.

29. Survey According to a survey, the percent p of Americans who now conduct most of their banking transactions online is no more than 51%.

30. Income The net income I of a company is expected to be no less than $239 million.

31. Physiology The maximum heart rate r (in beats per minute) of a person in normal health is related to the person's age by the equation

$$r = 220 - A$$

where A is the person's age (in years). Some physiologists recommend that during physical activity, a sedentary person should strive to increase his or her heart rate to at least 60% of the maximum heart rate, and a highly fit person should strive to increase his or her heart rate to at most 90% of the maximum heart rate. Use inequality notation to express the range of the target heart rate for physical activity for a 20-year-old.

32. Annual Operating Costs A utility company has a fleet of vans. The annual operating cost C (in dollars) of each van is estimated to be

$$C = 0.35m + 2500$$

where m is the number of miles driven. What number of miles will yield an annual operating cost for one van that is less than $13,000?

33. Profit The revenue for selling x units of a product is

$$R = 115.95x$$

and the cost of producing x units is

$$C = 95x + 750.$$

To obtain a profit, the revenue must be *greater than* the cost. For what values of x will this product return a profit?

34. Sales A doughnut shop sells a dozen doughnuts for $7.50. Beyond the fixed cost of $260 per day, it costs $3.25 for enough materials and labor to produce each dozen doughnuts. During the month of January, the daily profit varies between $250 and $675. Between what levels (in dozens) do the daily sales vary?

True or False? In Exercises 35 and 36, determine whether each statement is true or false, given $a < b$.

35. (a) $-2a < -2b$

 (b) $a + 2 < b + 2$

 (c) $6a < 6b$

 (d) $\dfrac{1}{a} < \dfrac{1}{b}$

36. (a) $a - 4 < b - 4$

 (b) $4 - a < 4 - b$

 (c) $-3b < -3a$

 (d) $\dfrac{a}{4} < \dfrac{b}{4}$

A.2 Absolute Value of a Real Number

■ Find the absolute values of real numbers.
■ Find the distance between two numbers on the real number line.
■ Write and solve absolute value inequalities.
■ Use intervals to model and solve real-life problems and find the midpoint of an interval.

Absolute Value of a Real Number

TECH TUTOR

Absolute value expressions can be evaluated on a graphing utility. When an expression such as $|3 - 8|$ is evaluated, parentheses should surround the expression, as in abs$(3 - 8)$.

Definition of Absolute Value

The **absolute value** of a real number a is

$$|a| = \begin{cases} a, & \text{if } a \geq 0 \\ -a, & \text{if } a < 0 \end{cases}.$$

The absolute value of a real number is either positive or zero. For example, consider $|-3|$. Because $-3 < 0$, you have $|-3| = -(-3) = 3$. Moreover, 0 is the only real number whose absolute value is 0. So, $|0| = 0$.

EXAMPLE 1   Evaluating Absolute Value Expressions

	Expression	*x-Value*	*Substitute*	*Value of Expression*
a.	$\|3x + 2\|$	$x = 7$	$\|3(7) + 2\|$	$\|21 + 2\| = \|23\| = 23$
b.	$\left\|1 - \dfrac{5}{0.03x}\right\|$	$x = 100$	$\left\|1 - \dfrac{5}{0.03(100)}\right\|$	$\left\|1 - \dfrac{5}{3}\right\| = \left\|-\dfrac{2}{3}\right\| = \dfrac{2}{3}$

✓ **Checkpoint 1**   *Worked-out solution available at LarsonAppliedCalculus.com*

Evaluate $\left|\dfrac{120}{x} + 6\right|$ for (a) $x = 4$ and (b) $x = -2$.   ■

Properties of Absolute Value

1. Multiplication: $|ab| = |a||b|$ **2.** Division: $\left|\dfrac{a}{b}\right| = \dfrac{|a|}{|b|}, \quad b \neq 0$

3. Power: $\quad\quad |a^n| = |a|^n$ **4.** Square root: $\sqrt{a^2} = |a|$

Be sure you understand the fourth property in this list. A common error in algebra is to assume that by squaring a number and then taking the square root, you obtain the original number. But this is true only if the original number is nonnegative. For instance, if $a = 2$, then $\sqrt{2^2} = \sqrt{4} = 2$. But, if $a = -2$, then $\sqrt{(-2)^2} = \sqrt{4} = 2$. The reason for this is that (by definition) the square root symbol $\sqrt{}$ denotes only the nonnegative root.

Distance on the Real Number Line

Consider two distinct points a and b on the real number line, where $a < b$, as shown in Figure A.9.

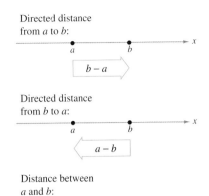

Directed distance from a to b:

Directed distance from b to a:

Distance between a and b:

$|a - b|$ or $|b - a|$

FIGURE A.9

1. The **directed distance from a to b** is

$b - a.$

2. The **directed distance from b to a** is

$a - b.$

3. The **distance between a and b** is

$$|a - b| \quad \text{or} \quad |b - a|.$$

In Figure A.9, note that because b is to the right of a, the directed distance from a to b (moving to the right) is positive. Moreover, because a is to the left of b, the directed distance from b to a (moving to the left) is negative. The distance *between* two points on the real number line can never be negative.

Distance Between Two Points on the Real Number Line

The distance d between points x_1 and x_2 on the real number line is given by

$$d = |x_2 - x_1|.$$

Note that the order of subtraction with x_1 and x_2 does not matter because

$$|x_2 - x_1| = |x_1 - x_2|.$$

EXAMPLE 2   Finding Distance on the Real Number Line

Determine the distance between -3 and 4 on the real number line. What is the directed distance from -3 to 4? What is the directed distance from 4 to -3?

SOLUTION   The distance between -3 and 4 is given by

$$|-3 - 4| = |-7| = 7 \qquad |a - b|$$

or

$$|4 - (-3)| = |7| = 7 \qquad |b - a|$$

as shown in Figure A.10.

Distance = 7

FIGURE A.10

The directed distance from -3 to 4 is

$$4 - (-3) = 7. \qquad b - a$$

The directed distance from 4 to -3 is

$$-3 - 4 = -7. \qquad a - b$$

✓ **Checkpoint 2**   *Worked-out solution available at LarsonAppliedCalculus.com*

Determine the distance between -2 and 6 on the real number line. What is the directed distance from -2 to 6? What is the directed distance from 6 to -2? ■

Absolute Value Inequalities

Two Basic Types of Inequalities Involving Absolute Value

Let a and d be real numbers, where $d > 0$.

$|x - a| \leq d$ if and only if $a - d \leq x \leq a + d$.

$|x - a| \geq d$ if and only if $x \leq a - d$ or $a + d \leq x$.

Inequality	Interpretation	Graph
$|x - a| \leq d$	All numbers x whose distance from a is less than or equal to d.	
$|x - a| \geq d$	All numbers x whose distance from a is greater than or equal to d.	

Be sure you see that inequalities of the form $|x - a| \geq d$ have solution sets consisting of two intervals. To describe the two intervals without using absolute values, you must use *two* separate inequalities, connected by an "or" to indicate union.

EXAMPLE 3   Writing and Solving an Absolute Value Inequality

Find the interval on the real number line that contains all numbers that lie no more than two units from 3.

SOLUTION   Let x be any point in this interval. You need to find all x such that the distance between x and 3 is less than or equal to 2. This distance can be represented by the absolute value inequality

$$|x - 3| \leq 2.$$

Requiring the absolute value of $x - 3$ to be less than or equal to 2 means that $x - 3$ must lie between -2 and 2. So, you can write

$$-2 \leq x - 3 \leq 2.$$

Solving this pair of inequalities, you have

$$-2 + 3 \leq x - 3 + 3 \leq 2 + 3$$
$$1 \leq \quad x \quad \leq 5. \qquad \text{Solution set}$$

So, the interval is $[1, 5]$, as shown in Figure A.11.

FIGURE A.11

✓ **Checkpoint 3**   *Worked-out solution available at LarsonAppliedCalculus.com*

Find the interval on the real number line that contains all numbers that lie no more than four units from 6.

Application

EXAMPLE 4   Quality Control

A large manufacturer hired a quality control firm to determine the reliability of a product. Using statistical methods, the firm determined that the manufacturer could expect $0.35\% \pm 0.17\%$ of the units to be defective. The manufacturer offers a money-back guarantee on this product. How much should be budgeted to cover the refunds on 100,000 units? (Assume that the retail price is $8.95.) Will the manufacturer have to establish a refund budget greater than $5000?

SOLUTION   Let r represent the percent of defective units (in decimal form). You know that r will differ from 0.0035 by at most 0.0017.

$$0.0035 - 0.0017 \le r \le 0.0035 + 0.0017$$
$$0.0018 \le r \le 0.0052 \qquad \text{Figure A.12(a)}$$

Now, letting x be the number of defective units out of 100,000, it follows that $x = 100,000r$ and you have

$$0.0018(100,000) \le 100,000r \le 0.0052(100,000)$$
$$180 \le \quad x \quad \le 520. \qquad \text{Figure A.12(b)}$$

Finally, letting C be the cost of refunds, you have $C = 8.95x$. So, the total cost of refunds for 100,000 units should fall within the interval given by

$$180(8.95) \le 8.95x \le 520(8.95)$$
$$\$1611 \le \quad C \quad \le \$4654. \qquad \text{Figure A.12(c)}$$

The manufacturer will *not* have to establish a refund budget greater than $5000.

✓ **Checkpoint 4**   *Worked-out solution available at LarsonAppliedCalculus.com*

Use the information in Example 4 to determine how much should be budgeted to cover refunds on 250,000 units.   ■

In Example 4, the manufacturer should expect to spend between $1611 and $4654 for refunds. Of course, the safer budget figure for refunds would be the higher of these estimates. From a statistical point of view, however, the most representative estimate would be the average of these two extremes. Graphically, the average of two numbers is the **midpoint** of the interval with the two numbers as endpoints, as shown in Figure A.13.

(a) Percent of defective units

(b) Number of defective units

(c) Cost of refunds

FIGURE A.12

$$\text{Midpoint} = \frac{1611 + 4654}{2} = 3132.5$$

FIGURE A.13

Midpoint of an Interval

The **midpoint** of the interval with endpoints a and b is found by taking the average of the endpoints.

$$\text{Midpoint} = \frac{a + b}{2}$$

Exercises A.2

See *CalcChat.com* for tutorial help and worked-out solutions to odd-numbered exercises.

Evaluating Absolute Value Expressions In Exercises 1–4, evaluate the expression for each value of *x*. *See Example 1.*

Expression	Values
1. $\lvert 2x + 5 \rvert$	(a) $x = 3$ (b) $x = -6$
2. $\lvert 4 - 6x \rvert$	(a) $x = 5$ (b) $x = -4$
3. $\left\lvert \dfrac{8}{x} - 9 \right\rvert$	(a) $x = 8$ (b) $x = -3$
4. $\left\lvert \dfrac{7}{0.2x} + 2 \right\rvert$	(a) $x = 10$ (b) $x = 35$

Finding Distance on the Real Number Line In Exercises 5–10, determine (a) the distance between *a* and *b*, (b) the directed distance from *a* to *b*, and (c) the directed distance from *b* to *a*. *See Example 2.*

5. $a = 126, b = 75$

6. $a = -126, b = -75$

7. $a = 9.34, b = -5.65$

8. $a = -2.05, b = 4.25$

9. $a = \frac{16}{5}, b = \frac{112}{75}$

10. $a = -\frac{18}{5}, b = \frac{61}{15}$

Writing an Absolute Value Inequality In Exercises 11–14, write an absolute value inequality that decribes the given interval on the real number line. *See Example 3.*

11. All numbers less than three units from 5

12. All numbers more than five units from 2

13. *y* is at least two units from *a*.

14. *y* is at most *h* units from *c*.

Solving an Absolute Value Inequality In Exercises 15–26, solve the inequality. Then graph the solution set on the real number line. *See Example 3.*

15. $\lvert x \rvert < 4$

16. $\lvert 2x \rvert < 6$

17. $\lvert x + 5 \rvert \leq 7$

18. $\lvert 2x + 1 \rvert < 5$

19. $\lvert 9 - 2x \rvert > 1$

20. $\lvert 11 - 4x \rvert \geq 9$

21. $\left\lvert \dfrac{x - 3}{2} \right\rvert \geq 5$

22. $\left\lvert 1 - \dfrac{2x}{3} \right\rvert \leq 1$

23. $\lvert x - a \rvert \leq b, \; b > 0$

24. $\lvert 2x - a \rvert \geq b, \; b > 0$

25. $\left\lvert a - \dfrac{5x}{2} \right\rvert > b, \; b > 0$

26. $\left\lvert \dfrac{3x - a}{4} \right\rvert < 2b, \; b > 0$

Finding a Midpoint In Exercises 27–32, find the midpoint of the given interval.

27. $[8, 24]$

28. $[7, 19]$

29. $[-6.85, 9.35]$

30. $[-4.6, -1.3]$

31. $\left[-\frac{1}{2}, \frac{3}{4} \right]$

32. $\left[\frac{5}{6}, \frac{5}{2} \right]$

33. Stock Price A stock market analyst predicts that over the next year, the price *p* of a stock will not change from its current price of $33.15 by more than $2. Use absolute values to write this prediction as an inequality.

34. Production The estimated daily production *x* at a refinery is given by

$$\lvert x - 200{,}000 \rvert \leq 25{,}000$$

where *x* is measured in barrels of oil. Determine the high and low production levels.

35. Manufacturing The acceptable weights for a 20-ounce cereal box are given by

$$\lvert x - 20 \rvert \leq 0.75$$

where *x* is measured in ounces. Determine the high and low weights for the cereal box.

36. Weight The American Kennel Club has developed guidelines for judging the features of various breeds of dogs. To not receive a penalty, the guidelines specify that the weights for male collies must satisfy the inequality

$$\left\lvert \frac{w - 67.5}{7.5} \right\rvert \leq 1$$

where *w* is the weight (in pounds). Determine the interval on the real number line in which these weights lie. *(Source: The American Kennel Club, Inc.)*

Budget In Exercises 37–40, (a) use absolute value notation to represent the two intervals in which expenses must lie if they are to be within $500 and within 5% of the specified budget amount and (b) using the more stringent constraint, determine whether the given expense is within the budget restriction.

Item	Budget	Expense
37. Utilities	$4750.00	$5116.37
38. Insurance	$15,000.00	$14,695.00
39. Maintenance	$20,000.00	$22,718.35
40. Taxes	$7500.00	$8691.00

41. Quality Control In determining the reliability of a product, a manufacturer determines that it should expect $0.05\% \pm 0.01\%$ of the units to be defective. The manufacturer offers a money-back guarantee on this product. How much should be budgeted to cover the refunds on 150,000 units? (Assume that the retail price is $195.99.)

A.3 Exponents and Radicals

■ Evaluate expressions involving exponents or radicals.
■ Simplify expressions with exponents.
■ Find the domains of algebraic expressions.

Expressions Involving Exponents or Radicals

Properties of Exponents

1. Whole-number exponents:
$$x^n = \underbrace{x \cdot x \cdot x \cdots x}_{n \text{ factors}}$$

2. Zero exponent: $x^0 = 1, \quad x \neq 0$

3. Negative exponents: $x^{-n} = \dfrac{1}{x^n}, \quad x \neq 0$

4. Radicals (principal nth root): $\sqrt[n]{x} = a \implies x = a^n$

5. Rational exponents $(1/n)$: $x^{1/n} = \sqrt[n]{x}$

6. Rational exponents (m/n): $x^{m/n} = (x^{1/n})^m = \left(\sqrt[n]{x}\right)^m$
$$x^{m/n} = (x^m)^{1/n} = \sqrt[n]{x^m}$$

7. Special convention (square root): $\sqrt[2]{x} = \sqrt{x}$

> **STUDY TIP**
>
> If n is even, then the principal nth root is positive. For example, $\sqrt{4} = +2$ and $\sqrt[4]{81} = +3$.

EXAMPLE 1 Evaluating Expressions

	Expression	*x-Value*	*Substitution*
a.	$y = -2x^2$	$x = 4$	$y = -2(4^2) = -2(16) = -32$
b.	$y = 3x^{-3}$	$x = -1$	$y = 3(-1)^{-3} = \dfrac{3}{(-1)^3} = \dfrac{3}{-1} = -3$
c.	$y = (-x)^2$	$x = \dfrac{1}{2}$	$y = \left(-\dfrac{1}{2}\right)^2 = \dfrac{1}{4}$
d.	$y = \dfrac{2}{x^{-2}}$	$x = 3$	$y = \dfrac{2}{3^{-2}} = 2(3^2) = 18$

✓ **Checkpoint 1**   *Worked-out solution available at LarsonAppliedCalculus.com*

Evaluate $y = 4x^{-2}$ for $x = 3$.

EXAMPLE 2 Evaluating Expressions

	Expression	*x-Value*	*Substitution*
a.	$y = 2x^{1/2}$	$x = 4$	$y = 2\sqrt{4} = 2(2) = 4$
b.	$y = \sqrt[3]{x^2}$	$x = 8$	$y = 8^{2/3} = (8^{1/3})^2 = 2^2 = 4$

✓ **Checkpoint 2**   *Worked-out solution available at LarsonAppliedCalculus.com*

Evaluate $y = 4x^{1/3}$ for $x = 8$.

Operations with Exponents

TECH TUTOR

Graphing utilities perform the established order of operations when evaluating an expression. To see this, try entering the expressions

$$1200\left(1 + \frac{0.09}{12}\right)^{12\cdot6}$$

and

$$1200 \times 1 + \left(\frac{0.09}{12}\right)^{12\cdot6}$$

into your graphing utility to see that the expressions result in different values.

Operations with Exponents

1. Multiplying like bases: $x^n x^m = x^{n+m}$ Add exponents.

2. Dividing like bases: $\dfrac{x^n}{x^m} = x^{n-m}$ Subtract exponents.

3. Removing parentheses: $(xy)^n = x^n y^n$

$$\left(\frac{x}{y}\right)^n = \frac{x^n}{y^n}$$

$$(x^n)^m = x^{nm}$$

4. Special conventions: $-x^n = -(x^n), \quad -x^n \neq (-x)^n$

$$cx^n = c(x^n), \quad cx^n \neq (cx)^n$$

$$x^{n^m} = x^{(n^m)}, \quad x^{n^m} \neq (x^n)^m$$

EXAMPLE 3   Simplifying Expressions with Exponents

Simplify each expression.

a. $2x^2(x^3)$ **b.** $(3x)^2\sqrt[3]{x}$ **c.** $\dfrac{3x^2}{(x^{1/2})^3}$

d. $\dfrac{5x^4}{(x^2)^3}$ **e.** $x^{-1}(2x^2)$ **f.** $\dfrac{-\sqrt{x}}{5x^{-1}}$

SOLUTION

a. $2x^2(x^3) = 2x^{2+3} = 2x^5$ $x^n x^m = x^{n+m}$

b. $(3x)^2\sqrt[3]{x} = 9x^2 x^{1/3} = 9x^{2+(1/3)} = 9x^{7/3}$ $x^n x^m = x^{n+m}$

c. $\dfrac{3x^2}{(x^{1/2})^3} = 3\left(\dfrac{x^2}{x^{3/2}}\right) = 3x^{2-(3/2)} = 3x^{1/2}$ $(x^n)^m = x^{nm}, \dfrac{x^n}{x^m} = x^{n-m}$

d. $\dfrac{5x^4}{(x^2)^3} = \dfrac{5x^4}{x^6} = 5x^{4-6} = 5x^{-2} = \dfrac{5}{x^2}$ $(x^n)^m = x^{nm}, \dfrac{x^n}{x^m} = x^{n-m}$

e. $x^{-1}(2x^2) = 2x^{-1}x^2 = 2x^{-1+2} = 2x$ $x^n x^m = x^{n+m}$

f. $\dfrac{-\sqrt{x}}{5x^{-1}} = -\dfrac{1}{5}\left(\dfrac{x^{1/2}}{x^{-1}}\right) = -\dfrac{1}{5}x^{(1/2)+1} = -\dfrac{1}{5}x^{3/2}$ $\dfrac{x^n}{x^m} = x^{n-m}$

✓ **Checkpoint 3**   Worked-out solution available at LarsonAppliedCalculus.com

Simplify each expression.

a. $3x^2(x^4)$

b. $(2x)^3\sqrt{x}$

c. $\dfrac{4x^2}{(x^{1/3})^2}$

Note in Example 3 that one characteristic of simplified expressions is the absence of negative exponents. Another characteristic of simplified expressions is that sums and differences are written in *factored form*. To do this, you can use the **Distributive Property.**

$$abx^n + acx^{n+m} = ax^n(b + cx^m)$$

Study the next example carefully to be sure that you understand the concepts involved in the factoring process.

EXAMPLE 4 Simplifying by Factoring

Simplify each expression by factoring.

a. $2x^2 - x^3$ **b.** $2x^3 + x^2$ **c.** $2x^{1/2} + 4x^{5/2}$ **d.** $2x^{-1/2} + 3x^{5/2}$

SOLUTION

a. $2x^2 - x^3 = x^2(2 - x)$

b. $2x^3 + x^2 = x^2(2x + 1)$

c. $2x^{1/2} + 4x^{5/2} = 2x^{1/2}(1 + 2x^2)$

d. $2x^{-1/2} + 3x^{5/2} = x^{-1/2}(2 + 3x^3) = \dfrac{2 + 3x^3}{\sqrt{x}}$

✓ *Checkpoint 4*   *Worked-out solution available at LarsonAppliedCalculus.com*

Simplify each expression by factoring.

a. $x^3 - 2x$

b. $2x^{1/2} + 8x^{3/2}$ ■

Many algebraic expressions obtained in calculus occur in unsimplified form. For instance, the two expressions shown in the following example are the result of an operation in calculus called *differentiation*. [The first is the derivative of $2(x + 1)^{3/2}(2x - 3)^{5/2}$, and the second is the derivative of $2(x + 1)^{1/2}(2x - 3)^{5/2}$.]

STUDY TIP

To check that a simplified expression is equivalent to the original expression, try substituting values for x into each expression.

EXAMPLE 5 Simplifying by Factoring

a. $3(x + 1)^{1/2}(2x - 3)^{5/2} + 10(x + 1)^{3/2}(2x - 3)^{3/2}$

$\qquad = (x + 1)^{1/2}(2x - 3)^{3/2}[3(2x - 3) + 10(x + 1)]$

$\qquad = (x + 1)^{1/2}(2x - 3)^{3/2}(6x - 9 + 10x + 10)$

$\qquad = (x + 1)^{1/2}(2x - 3)^{3/2}(16x + 1)$

b. $(x + 1)^{-1/2}(2x - 3)^{5/2} + 10(x + 1)^{1/2}(2x - 3)^{3/2}$

$\qquad = (x + 1)^{-1/2}(2x - 3)^{3/2}[(2x - 3) + 10(x + 1)]$

$\qquad = (x + 1)^{-1/2}(2x - 3)^{3/2}(2x - 3 + 10x + 10)$

$\qquad = (x + 1)^{-1/2}(2x - 3)^{3/2}(12x + 7)$

$\qquad = \dfrac{(2x - 3)^{3/2}(12x + 7)}{(x + 1)^{1/2}}$

✓ *Checkpoint 5*   *Worked-out solution available at LarsonAppliedCalculus.com*

Simplify the expression by factoring.

$$(x + 2)^{1/2}(3x - 1)^{3/2} + 4(x + 2)^{-1/2}(3x - 1)^{5/2}$$ ■

Example 6 shows some additional types of expressions that can occur in calculus. [The expression in Example 6(d) is an antiderivative of $(x + 1)^{2/3}(2x + 3)$, and the expression in Example 6(e) is the derivative of $(x + 2)^3/(x - 1)^3$.]

EXAMPLE 6   Factors Involving Quotients

Simplify each expression by factoring.

a. $\dfrac{3x^2 + x^4}{2x}$

b. $\dfrac{\sqrt{x} + x^{3/2}}{x}$

c. $(9x + 2)^{-1/3} + 18(9x + 2)$

d. $\dfrac{3}{5}(x + 1)^{5/3} + \dfrac{3}{4}(x + 1)^{8/3}$

e. $\dfrac{3(x + 2)^2(x - 1)^3 - 3(x + 2)^3(x - 1)^2}{[(x - 1)^3]^2}$

SOLUTION

a. $\dfrac{3x^2 + x^4}{2x} = \dfrac{x^2(3 + x^2)}{2x} = \dfrac{x^{2-1}(3 + x^2)}{2} = \dfrac{x(3 + x^2)}{2}$

b. $\dfrac{\sqrt{x} + x^{3/2}}{x} = \dfrac{x^{1/2}(1 + x)}{x} = \dfrac{1 + x}{x^{1-(1/2)}} = \dfrac{1 + x}{\sqrt{x}}$

c. $(9x + 2)^{-1/3} + 18(9x + 2) = (9x + 2)^{-1/3}[1 + 18(9x + 2)^{4/3}]$

$\qquad\qquad = \dfrac{1 + 18(9x + 2)^{4/3}}{\sqrt[3]{9x + 2}}$

d. $\dfrac{3}{5}(x + 1)^{5/3} + \dfrac{3}{4}(x + 1)^{8/3} = \dfrac{12}{20}(x + 1)^{5/3} + \dfrac{15}{20}(x + 1)^{8/3}$

$\qquad\qquad = \dfrac{3}{20}(x + 1)^{5/3}[4 + 5(x + 1)]$

$\qquad\qquad = \dfrac{3}{20}(x + 1)^{5/3}(4 + 5x + 5)$

$\qquad\qquad = \dfrac{3}{20}(x + 1)^{5/3}(5x + 9)$

e. $\dfrac{3(x + 2)^2(x - 1)^3 - 3(x + 2)^3(x - 1)^2}{[(x - 1)^3]^2}$

$\quad = \dfrac{3(x + 2)^2(x - 1)^2[(x - 1) - (x + 2)]}{(x - 1)^6}$

$\quad = \dfrac{3(x + 2)^2(x - 1 - x - 2)}{(x - 1)^{6-2}}$

$\quad = \dfrac{-9(x + 2)^2}{(x - 1)^4}$

✓ **Checkpoint 6**   *Worked-out solution available at LarsonAppliedCalculus.com*

Simplify the expression by factoring.

$$\dfrac{5x^3 + x^6}{3x}$$

Domain of an Algebraic Expression

When working with algebraic expressions involving x, you face the potential difficulty of substituting a value of x for which the expression is not defined (does not produce a real number). For example, the expression $\sqrt{2x + 3}$ is *not defined* when $x = -2$ because

$$\sqrt{2(-2) + 3} = \sqrt{-1}$$

is not a real number.

The set of all values for which an expression is defined is called its **domain.** So, the domain of $\sqrt{2x + 3}$ is the set of all values of x such that $\sqrt{2x + 3}$ is a real number. In order for $\sqrt{2x + 3}$ to represent a real number, it is necessary that

$2x + 3 \geq 0$.   Expression must be nonnegative.

In other words, $\sqrt{2x + 3}$ is defined only for those values of x that lie in the interval $\left[-\frac{3}{2}, \infty\right)$, as shown in Figure A.14.

FIGURE A.14

EXAMPLE 7   Finding the Domain of an Expression

Find the domain of each expression.

a. $\sqrt{3x - 2}$

b. $\dfrac{1}{\sqrt{3x - 2}}$

c. $\sqrt[3]{9x + 1}$

SOLUTION

a. The domain of $\sqrt{3x - 2}$ consists of all x such that

$3x - 2 \geq 0$ Expression must be nonnegative.

which implies that $x \geq \frac{2}{3}$. So, the domain is $\left[\frac{2}{3}, \infty\right)$.

b. The domain of $1/\sqrt{3x - 2}$ is the same as the domain of $\sqrt{3x - 2}$, except that $1/\sqrt{3x - 2}$ is not defined when $3x - 2 = 0$. Because this occurs when $x = \frac{2}{3}$, the domain is $\left(\frac{2}{3}, \infty\right)$.

c. Because $\sqrt[3]{9x + 1}$ is defined for all real numbers, its domain is $(-\infty, \infty)$.

✓ *Checkpoint 7*   *Worked-out solution available at LarsonAppliedCalculus.com*

Find the domain of each expression.

a. $\sqrt{x - 2}$

b. $\dfrac{1}{\sqrt{x - 2}}$

c. $\sqrt[3]{x - 2}$

Exercises A.3

Evaluating Expressions In Exercises 1–20, evaluate the expression for the given value of x. *See Examples 1 and 2.*

Expression	x-Value	Expression	x-Value
1. $-4x^3$	$x = 3$	**2.** $\dfrac{x^2}{3}$	$x = 6$
3. $5x^{-3}$	$x = 2$	**4.** $7x^{-2}$	$x = 5$
5. $\dfrac{1 + x^{-1}}{x^{-1}}$	$x = 3$	**6.** $x - 4x^{-2}$	$x = 3$
7. $3x^2 - 4x^3$	$x = -2$	**8.** $5(-x)^3$	$x = 3$
9. $6x^0 - (6x)^0$	$x = 10$	**10.** $\dfrac{1}{(-x)^{-3}}$	$x = 4$
11. $\sqrt[3]{x^4}$	$x = 64$	**12.** $\sqrt{x^3}$	$x = \frac{1}{9}$
13. $x^{-1/2}$	$x = 4$	**14.** $x^{-3/4}$	$x = 16$
15. $x^{-2/5}$	$x = -32$	**16.** $(x^{2/3})^3$	$x = 10$
17. $500x^{60}$	$x = 1.01$	**18.** $\dfrac{10,000}{x^{120}}$	$x = 1.1$
19. $\sqrt[3]{x}$	$x = -54$	**20.** $\sqrt[6]{x}$	$x = 325$

Simplifying Expressions with Exponents In Exercises 21–30, simplify the expression. *See Example 3.*

21. $10(x^2)^2$

22. $(4x^3)^2$

23. $6y^{-2}(2y^4)^{-3}$

24. $9z^{-3}(3x^4)^2$

25. $\dfrac{7x^2}{x^{-3}}$

26. $\dfrac{x^{-3}}{\sqrt{x}}$

27. $\dfrac{10(x + y)^3}{4(x + y)^{-2}}$

28. $\left(\dfrac{12s^2}{9s}\right)^3$

29. $\dfrac{3x\sqrt{x}}{x^{1/2}}$

30. $\left(\sqrt[3]{x^2}\right)^3$

Simplifying Radicals In Exercises 31–36, simplify the expression by removing all possible factors from the radical.

31. $\sqrt{8}$

32. $\sqrt[3]{\frac{16}{27}}$

33. $\sqrt[3]{54x^5}$

34. $\sqrt[5]{(3x^2y^3)^5}$

35. $\sqrt[3]{144x^9y^{-4}z^5}$

36. $\sqrt[4]{32xy^5z^{-8}}$

Simplifying by Factoring In Exercises 37–44, simplify the expression by factoring. *See Examples 4, 5, and 6.*

37. $3x^3 - 12x$

38. $8x^4 - 6x^2$

39. $7x^{7/2} + x^{-1/2}$

40. $5x^{3/2} - x^{-3/2}$

41. $(x - 2)^{1/3}(x + 1)^{3/2} - 6(x - 2)^{4/3}(x + 1)^{1/2}$

42. $2x(x - 1)^{9/2}(x + 4)^{2/3} - 4(x - 1)^{5/2}(x + 4)^{5/3}$

43. $\dfrac{8x^7 + x^4}{7x^2}$

44. $\dfrac{(x + 1)(x - 1)^2 - (x - 1)^3}{(x + 1)^2}$

Finding the Domain of an Expression In Exercises 45–52, find the domain of the expression. *See Example 7.*

45. $\sqrt{x - 4}$

46. $\sqrt{5 - 2x}$

47. $\sqrt[3]{x + 7}$

48. $\sqrt{4x^2 + 1}$

49. $\dfrac{1}{\sqrt{x - 8}}$

50. $\dfrac{1}{\sqrt[3]{x + 6}}$

51. $\dfrac{\sqrt{x + 2}}{1 - x}$

52. $\dfrac{1}{\sqrt{2x + 3}} + \sqrt{6 - 4x}$

Compound Interest In Exercises 53–56, a certificate of deposit has a principal of P dollars and an annual percentage rate of r (expressed as a decimal) compounded n times per year. The balance A in the account is given by

$$A = P\left(1 + \frac{r}{n}\right)^N$$

where N is the number of compoundings. Find the balance in the account.

53. $P = \$10,000$, $r = 6.5\%$, $n = 12$, $N = 120$

54. $P = \$7000$, $r = 5\%$, $n = 365$, $N = 1000$

55. $P = \$5000$, $r = 5.5\%$, $n = 4$, $N = 60$

56. $P = \$8000$, $r = 7\%$, $n = 6$, $N = 90$

57. Period of a Pendulum The period T (in seconds) of a pendulum is

$$T = 2\pi\sqrt{\frac{L}{32}}$$

where L is the length (in feet) of the pendulum. Find the period of a pendulum whose length is 4 feet.

58. Annuity After n annual payments of P dollars have been made into an annuity earning an annual percentage rate of r compounded annually, the balance A is

$$A = P(1 + r) + P(1 + r)^2 + \cdots + P(1 + r)^n.$$

Rewrite this formula by completing the following factorization.

$$A = P(1 + r)(\quad)$$

A.4 Factoring Polynomials

■ Use special products and factorization techniques to factor polynomials.
■ Use synthetic division to factor polynomials of degree three or more.
■ Use the Rational Zero Theorem to find the real zeros of polynomials.

Factorization Techniques

The **Fundamental Theorem of Algebra** states that every nth-degree polynomial

$$a_n x^n + a_{n-1} x^{n-1} + \cdots + a_1 x + a_0, \quad a_n \neq 0$$

has precisely n **zeros.** (The zeros may be repeated or imaginary.) The zeros of a polynomial in x are the values of x that make the polynomial zero. The problem of finding the zeros of a polynomial is equivalent to the problem of factoring the polynomial into linear factors.

Special Products and Factorization Techniques

Quadratic Formula

$$ax^2 + bx + c = 0 \implies x = \frac{-b \pm \sqrt{b^2 - 4ac}}{2a}$$

Example

$$x^2 + 3x - 1 = 0 \implies x = \frac{-3 \pm \sqrt{13}}{2}$$

Special Products

$x^2 - a^2 = (x - a)(x + a)$

$x^3 - a^3 = (x - a)(x^2 + ax + a^2)$

$x^3 + a^3 = (x + a)(x^2 - ax + a^2)$

$x^4 - a^4 = (x - a)(x + a)(x^2 + a^2)$

Examples

$x^2 - 9 = (x - 3)(x + 3)$

$x^3 - 8 = (x - 2)(x^2 + 2x + 4)$

$x^3 + 64 = (x + 4)(x^2 - 4x + 16)$

$x^4 - 16 = (x - 2)(x + 2)(x^2 + 4)$

*Binomial Theorem**

$(x + a)^2 = x^2 + 2ax + a^2$

$(x - a)^2 = x^2 - 2ax + a^2$

$(x + a)^3 = x^3 + 3ax^2 + 3a^2x + a^3$

$(x - a)^3 = x^3 - 3ax^2 + 3a^2x - a^3$

$(x + a)^4 = x^4 + 4ax^3 + 6a^2x^2 + 4a^3x + a^4$

$(x - a)^4 = x^4 - 4ax^3 + 6a^2x^2 - 4a^3x + a^4$

Examples

$(x + 3)^2 = x^2 + 6x + 9$

$(x^2 - 5)^2 = x^4 - 10x^2 + 25$

$(x + 2)^3 = x^3 + 6x^2 + 12x + 8$

$(x - 1)^3 = x^3 - 3x^2 + 3x - 1$

$(x + 2)^4 = x^4 + 8x^3 + 24x^2 + 32x + 16$

$(x - 4)^4 = x^4 - 16x^3 + 96x^2 - 256x + 256$

$$(x + a)^n = x^n + nax^{n-1} + \frac{n(n-1)}{2!}a^2x^{n-2} + \frac{n(n-1)(n-2)}{3!}a^3x^{n-3} + \cdots + na^{n-1}x + a^n$$

$$(x - a)^n = x^n - nax^{n-1} + \frac{n(n-1)}{2!}a^2x^{n-2} - \frac{n(n-1)(n-2)}{3!}a^3x^{n-3} + \cdots \pm na^{n-1}x \mp a^n$$

Factoring by Grouping

$$acx^3 + adx^2 + bcx + bd = ax^2(cx + d) + b(cx + d)$$
$$= (ax^2 + b)(cx + d)$$

Example

$$3x^3 - 2x^2 - 6x + 4 = x^2(3x - 2) - 2(3x - 2)$$
$$= (x^2 - 2)(3x - 2)$$

---

* In the expansions of $(x + a)^n$ and $(x - a)^n$, note the use of the factorial symbol !,
which is defined as follows: $0! = 1$, $1! = 1$, $2! = 2 \cdot 1 = 2$, $3! = 3 \cdot 2 \cdot 1 = 6$,
$4! = 4 \cdot 3 \cdot 2 \cdot 1 = 24$, and so on.

EXAMPLE 1   Applying the Quadratic Formula

Use the Quadratic Formula to find all real zeros of each polynomial.

a. $4x^2 + 6x + 1$

b. $x^2 + 6x + 9$

c. $2x^2 - 6x + 5$

SOLUTION

a. Using $a = 4$, $b = 6$, and $c = 1$, you can write

$$x = \frac{-b \pm \sqrt{b^2 - 4ac}}{2a}$$

$$= \frac{-6 \pm \sqrt{6^2 - 4(4)(1)}}{2(4)}$$

$$= \frac{-6 \pm \sqrt{36 - 16}}{8}$$

$$= \frac{-6 \pm \sqrt{20}}{8}$$

$$= \frac{-6 \pm 2\sqrt{5}}{8}$$

$$= \frac{2(-3 \pm \sqrt{5})}{2(4)}$$

$$= \frac{-3 \pm \sqrt{5}}{4}.$$

So, there are two real zeros:

$$x = \frac{-3 - \sqrt{5}}{4} \approx -1.309 \quad \text{and} \quad x = \frac{-3 + \sqrt{5}}{4} \approx -0.191.$$

b. In this case, $a = 1$, $b = 6$, and $c = 9$, and the Quadratic Formula yields

$$x = \frac{-b \pm \sqrt{b^2 - 4ac}}{2a} = \frac{-6 \pm \sqrt{36 - 36}}{2} = -\frac{6}{2} = -3.$$

So, there is one (repeated) real zero: $x = -3$.

c. For this quadratic equation, $a = 2$, $b = -6$, and $c = 5$. So,

$$x = \frac{-b \pm \sqrt{b^2 - 4ac}}{2a} = \frac{6 \pm \sqrt{36 - 40}}{4} = \frac{6 \pm \sqrt{-4}}{4}.$$

Because $\sqrt{-4}$ is imaginary, there are no real zeros.

STUDY TIP

Some zeros may be imaginary, such as the zeros of the polynomial in Example 1(c). You will only be asked to find real zeros in this text.

✓ **Checkpoint 1**   *Worked-out solution available at LarsonAppliedCalculus.com*

Use the Quadratic Formula to find all real zeros of each polynomial.

a. $2x^2 + 4x + 1$ **b.** $x^2 - 8x + 16$ **c.** $2x^2 - x + 5$ ■

The zeros in Example 1(a) are irrational, and the zeros in Example 1(c) are imaginary. In both of these cases the quadratic is said to be **irreducible** because it cannot be factored into linear factors with rational coefficients. The next example shows how to find the zeros associated with *reducible* quadratics. In this example, factoring is used to find the zeros of each quadratic. Try using the Quadratic Formula to obtain the same zeros.

Recall that the zeros of a polynomial in x are the values of x that make the polynomial zero. To find the zeros, factor the polynomial into linear factors and set each factor equal to zero. For instance, the zeros of $(x - 2)(x - 3)$ occur when $x - 2 = 0$ and $x - 3 = 0$.

EXAMPLE 2   **Finding Real Zeros by Factoring**

Find all the real zeros of each quadratic polynomial.

a. $x^2 - 5x + 6$ **b.** $x^2 - 6x + 9$ **c.** $2x^2 + 5x - 3$

SOLUTION

a. Because

$$x^2 - 5x + 6 = (x - 2)(x - 3)$$

the zeros are $x = 2$ and $x = 3$.

b. Because

$$x^2 - 6x + 9 = (x - 3)^2$$

the only zero is $x = 3$.

c. Because

$$2x^2 + 5x - 3 = (2x - 1)(x + 3)$$

the zeros are $x = \frac{1}{2}$ and $x = -3$.

✓ **Checkpoint 2**   *Worked-out solution available at LarsonAppliedCalculus.com*

Find all the real zeros of each quadratic polynomial.

a. $x^2 - 2x - 15$ **b.** $x^2 + 2x + 1$ **c.** $2x^2 - 7x + 6$

EXAMPLE 3   **Finding the Domain of a Radical Expression**

Find the domain of $\sqrt{x^2 - 3x + 2}$.

SOLUTION   This expression is defined only for x-values for which $x^2 - 3x + 2 \geq 0$. Because

$$x^2 - 3x + 2 = (x - 1)(x - 2)$$

you know that the zeros of the quadratic are $x = 1$ and $x = 2$. So, you need to test the sign of the quadratic in the three intervals $(-\infty, 1)$, $(1, 2)$, and $(2, \infty)$. After testing each of these intervals, you can see that the quadratic is negative in the center interval and positive in the outer two intervals. Moreover, because the quadratic is zero when $x = 1$ and $x = 2$, you can conclude that the domain of $\sqrt{x^2 - 3x + 2}$ is $(-\infty, 1] \cup [2, \infty)$, as shown in Figure A.15.

Values of $\sqrt{x^2 - 3x + 2}$

x	$\sqrt{x^2 - 3x + 2}$
0	$\sqrt{2}$
1	0
1.5	Undefined
2	0
3	$\sqrt{2}$

FIGURE A.15

✓ **Checkpoint 3**   *Worked-out solution available at LarsonAppliedCalculus.com*

Find the domain of

$$\sqrt{x^2 + x - 2}.$$

Factoring Polynomials of Degree Three or More

It can be difficult to find the zeros of polynomials of degree three or more. However, if one of the zeros of a polynomial is known, then you can use that zero to reduce the degree of the polynomial. For example, if you know that $x = 2$ is a zero of

$$x^3 - 4x^2 + 5x - 2$$

then you know that $(x - 2)$ is a factor, and you can use long division to factor the polynomial as shown.

$$x^3 - 4x^2 + 5x - 2 = (x - 2)(x^2 - 2x + 1)$$
$$= (x - 2)(x - 1)(x - 1)$$

As an alternative to long division, many people prefer to use **synthetic division** to reduce the degree of a polynomial.

Synthetic Division for a Cubic Polynomial

Given: $x = x_1$ is a zero of $ax^3 + bx^2 + cx + d$.

| | a | b | c | d |

x_1

a ⌣ 0

Coefficients for quadratic factor

Vertical pattern:
Add terms.

Diagonal pattern:
Multiply by x_1.

Performing synthetic division on the polynomial

$$x^3 - 4x^2 + 5x - 2$$

using the given zero, $x = 2$, produces the following.

```
2 | 1   -4    5   -2
  |      2   -4    2
  ---------------------
    1   -2    1    0
```

$$(x - 2)(x^2 - 2x + 1) = x^3 - 4x^2 + 5x - 2$$

When you use synthetic division, remember to take *all* coefficients into account—*even when some of them are zero.* For instance, when you know that $x = -2$ is a zero of $x^3 + 3x + 14$, you can apply synthetic division as shown.

```
-2 | 1    0    3    14
   |     -2    4   -14
   ----------------------
     1   -2    7     0
```

$$(x + 2)(x^2 - 2x + 7) = x^3 + 3x + 14$$

TECH TUTOR

You can use a graphing utility to find the zeros of a polynomial. For example, to find the zeros of $x^3 - 4x^2 + 5x - 2$, use the graphing utility's *equation editor* to enter $y_1 = x^3 - 4x^2 + 5x - 2$. Then use the *zero* or *root* feature to approximate any zeros. As shown below, one of the zeros of the polynomial is $x = 2$.

Using the same process, the other (repeated) zero is $x = 1$.

STUDY TIP

The algorithm for synthetic division given above works *only* for divisors of the form $x - x_1$. Remember that $x + x_1 = x - (-x_1)$.

The Rational Zero Theorem

There is a systematic way to find the *rational* zeros of a polynomial. You can use the **Rational Zero Theorem** (also called the Rational Root Theorem).

Rational Zero Theorem

If a polynomial

$$a_n x^n + a_{n-1} x^{n-1} + \cdots + a_1 x + a_0$$

has integer coefficients, then every *rational* zero is of the form

$$x = \frac{p}{q}$$

where p is a factor of a_0, and q is a factor of a_n.

EXAMPLE 4   Using the Rational Zero Theorem

Find all real zeros of the polynomial.

$$2x^3 + 3x^2 - 8x + 3$$

SOLUTION

$$(2)x^3 + 3x^2 - 8x + (3)$$

Factors of constant term: $\pm 1, \pm 3$

Factors of leading coefficient: $\pm 1, \pm 2$

The possible rational zeros are the factors of the constant term divided by the factors of the leading coefficient.

$$1, -1, 3, -3, \frac{1}{2}, -\frac{1}{2}, \frac{3}{2}, -\frac{3}{2}$$

By testing these possible zeros, you can see that $x = 1$ works.

$$2(1)^3 + 3(1)^2 - 8(1) + 3 = 2 + 3 - 8 + 3 = 0$$

Now, by synthetic division you have the following.

$$
\begin{array}{r|rrrr}
1 & 2 & 3 & -8 & 3 \\
  &   & 2 & 5 & -3 \\
\hline
  & 2 & 5 & -3 & 0 \\
\end{array}
$$

$$(x - 1)(2x^2 + 5x - 3) = 2x^3 + 3x^2 - 8x + 3$$

Finally, by factoring the quadratic

$$2x^2 + 5x - 3 = (2x - 1)(x + 3)$$

you have

$$2x^3 + 3x^2 - 8x + 3 = (x - 1)(2x - 1)(x + 3)$$

and you can conclude that the zeros are $x = 1$, $x = \frac{1}{2}$, and $x = -3$.

✓ **Checkpoint 4**   *Worked-out solution available at LarsonAppliedCalculus.com*

Find all real zeros of the polynomial.

$$2x^3 - 3x^2 - 3x + 2$$

STUDY TIP

In Example 4, you can check that the zeros are correct by substituting into the original polynomial.

Check that $x = 1$ is a zero.

$2(1)^3 + 3(1)^2 - 8(1) + 3$

$= 2 + 3 - 8 + 3$

$= 0$

Check that $x = \frac{1}{2}$ is a zero.

$2\left(\frac{1}{2}\right)^3 + 3\left(\frac{1}{2}\right)^2 - 8\left(\frac{1}{2}\right) + 3$

$= \frac{1}{4} + \frac{3}{4} - 4 + 3$

$= 0$

Check that $x = -3$ is a zero.

$2(-3)^3 + 3(-3)^2 - 8(-3) + 3$

$= -54 + 27 + 24 + 3$

$= 0$

Exercises A.4

See *CalcChat.com* for tutorial help and worked-out solutions to odd-numbered exercises.

Applying the Quadratic Formula In Exercises 1–8, use the Quadratic Formula to find all real zeros of the polynomial. *See Example 1.*

1. $6x^2 - 7x + 1$

2. $8x^2 - 2x - 1$

3. $4x^2 - 12x + 9$

4. $9x^2 + 12x + 4$

5. $y^2 + 4y + 1$

6. $y^2 + 5y + 7$

7. $2x^2 + 3x + 4$

8. $3x^2 - 8x - 4$

Factoring Polynomials In Exercises 9–18, write the second-degree polynomial as the product of two linear factors.

9. $x^2 - 4x + 4$

10. $x^2 + 10x + 25$

11. $x^2 - 3x - 18$

12. $x^2 - x - 20$

13. $4x^2 + 4x + 1$

14. $9x^2 - 12x + 4$

15. $3x^2 - 4x + 1$

16. $4x^2 + 19x + 12$

17. $x^2 - 4xy + 4y^2$

18. $x^2 - xy - 2y^2$

Factoring Polynomials In Exercises 19–34, completely factor the polynomial.

19. $81 - y^4$

20. $x^4 - 256$

21. $x^3 - 1$

22. $y^3 - 64$

23. $y^3 + 27$

24. $z^3 + 125$

25. $x^3 - y^3$

26. $(x - a)^3 + b^3$

27. $x^3 - 4x^2 - x + 4$

28. $x^3 - x^2 - x + 1$

29. $2x^3 - 3x^2 + 4x - 6$

30. $x^3 - 5x^2 - 5x + 25$

31. $2x^3 - 4x^2 - x + 2$

32. $4x^3 + 8x^2 - x - 2$

33. $x^4 - 15x^2 - 16$

34. $2x^4 - 49x^2 - 25$

Finding Real Zeros by Factoring In Exercises 35–54, find all real zeros of the polynomial. *See Example 2.*

35. $x^2 - 5x$

36. $2x^2 - 3x$

37. $x^2 - 81$

38. $x^2 - 25$

39. $x^2 - 3$

40. $x^2 - 8$

41. $(x - 3)^2 - 9$

42. $(x + 1)^2 - 36$

43. $x^2 + x - 2$

44. $x^2 + 5x + 6$

45. $x^2 - 5x - 6$

46. $x^2 + x - 20$

47. $3x^2 + 5x + 2$

48. $2x^2 - x - 1$

49. $x^3 + 343$

50. $x^3 - 216$

51. $x^4 - 1$

52. $x^4 - 625$

53. $x^3 - x^2 - 4x + 4$

54. $2x^3 + x^2 + 6x + 3$

Finding the Domain of a Radical Expression In Exercises 55–60, find the domain of the expression. *See Example 3.*

55. $\sqrt{x^2 - 49}$

56. $\sqrt{4 - x^2}$

57. $\sqrt{x^2 - 7x + 12}$

58. $\sqrt{x^2 - 8x + 15}$

59. $\sqrt{5x^2 + 6x + 1}$

60. $\sqrt{3x^2 - 10x + 3}$

Using Synthetic Division In Exercises 61–64, use synthetic division to complete the indicated factorization.

61. $x^3 - 3x^2 - 6x - 2 = (x + 1)(\quad)$

62. $x^3 - x^2 - 4x + 4 = (x + 2)(\quad)$

63. $2x^3 - 9x^2 + 10x - 3 = (x - 3)(\quad)$

64. $x^4 - 16x^3 + 96x^2 - 256x + 256 = (x - 4)(\quad)$

Using the Rational Zero Theorem In Exercises 65–74, use the Rational Zero Theorem to find all real zeros of the polynomial. *See Example 4.*

65. $x^3 - 28x - 48$

66. $x^3 - 7x - 6$

67. $x^3 - 6x^2 + 11x - 6$

68. $x^3 + 2x^2 - 5x - 6$

69. $x^3 - 3x^2 - 3x - 4$

70. $2x^3 - x^2 - 13x - 6$

71. $4x^3 + 11x^2 + 5x - 2$

72. $3x^3 + 4x^2 - 13x + 6$

73. $6x^3 - 11x^2 - 19x - 6$

74. $18x^3 - 9x^2 - 8x + 4$

75. Production Level The minimum average cost of producing x units of a product occurs when the production level is set at the (positive) solution of

$$0.0003x^2 - 1200 = 0.$$

How many solutions does this equation have? Find and interpret the solution(s) in the context of the problem. What production level will minimize the average cost?

76. Profit The profit P (in dollars) of a product is given by

$$P = -200x^2 + 2000x - 3800$$

where x is the number of units sold per day (in hundreds). Determine the interval for x such that the profit will be greater than $1000.

A.5 Fractions and Rationalization

■ Simplify rational expressions.
■ Add and subtract rational expressions.
■ Simplify rational expressions involving radicals.
■ Rationalize numerators and denominators of rational expressions.

Simplifying Rational Expressions

In this section, you will review operations involving fractional expressions such as

$$\frac{2}{x}, \quad \frac{x^2 + 2x - 4}{x + 6}, \quad \text{and} \quad \frac{1}{\sqrt{x^2 + 1}}.$$

The first two expressions have polynomials as both numerator and denominator and are called **rational expressions.** A rational expression is **proper** when the degree of the numerator is less than the degree of the denominator. For example,

$$\frac{x}{x^2 + 1}$$

is proper. If the degree of the numerator is greater than or equal to the degree of the denominator, then the rational expression is **improper.** For example,

$$\frac{x^2}{x^2 + 1} \quad \text{and} \quad \frac{x^3 + 2x + 1}{x + 1}$$

are both improper.

A fraction is in simplest form when its numerator and denominator have no factors in common aside from ± 1. To write a fraction in simplest form, divide out common factors.

$$\frac{a \cdot c}{b \cdot c} = \frac{a}{b}, \quad c \neq 0$$

The key to success in simplifying rational expressions lies in your ability to factor polynomials. When simplifying rational expressions, be sure to factor each polynomial completely before concluding that the numerator and denominator have no common factors.

EXAMPLE 1   Simplifying a Rational Expression

Write $\dfrac{12 + x - x^2}{2x^2 - 9x + 4}$ in simplest form.

SOLUTION

$$\frac{12 + x - x^2}{2x^2 - 9x + 4} = \frac{(4 - x)(3 + x)}{(2x - 1)(x - 4)} \qquad \text{Factor completely.}$$

$$= \frac{-(x - 4)(3 + x)}{(2x - 1)(x - 4)} \qquad (4 - x) = -(x - 4)$$

$$= -\frac{3 + x}{2x - 1}, \quad x \neq 4 \qquad \text{Divide out common factor.}$$

✓ **Checkpoint 1**   *Worked-out solution available at LarsonAppliedCalculus.com*

Write $\dfrac{x^2 + 8x - 20}{x^2 + 11x + 10}$ in simplest form.

STUDY TIP

To simplify a rational expression, it may be necessary to change the sign of a factor by factoring out (-1), as shown in Example 1.

Operations with Fractions

Operations with Fractions

1. Add fractions (find a common denominator):

$$\frac{a}{b} + \frac{c}{d} = \frac{a}{b}\left(\frac{d}{d}\right) + \frac{c}{d}\left(\frac{b}{b}\right) = \frac{ad}{bd} + \frac{bc}{bd} = \frac{ad + bc}{bd}, \quad b \neq 0, d \neq 0$$

2. Subtract fractions (find a common denominator):

$$\frac{a}{b} - \frac{c}{d} = \frac{a}{b}\left(\frac{d}{d}\right) - \frac{c}{d}\left(\frac{b}{b}\right) = \frac{ad}{bd} - \frac{bc}{bd} = \frac{ad - bc}{bd}, \quad b \neq 0, d \neq 0$$

3. Multiply fractions:

$$\left(\frac{a}{b}\right)\left(\frac{c}{d}\right) = \frac{ac}{bd}, \quad b \neq 0, d \neq 0$$

4. Divide fractions (invert and multiply):

$$\frac{a/b}{c/d} = \left(\frac{a}{b}\right)\left(\frac{d}{c}\right) = \frac{ad}{bc}, \quad b \neq 0, c \neq 0, d \neq 0$$

$$\frac{a/b}{c} = \frac{a/b}{c/1} = \left(\frac{a}{b}\right)\left(\frac{1}{c}\right) = \frac{a}{bc}, \quad b \neq 0, c \neq 0$$

5. Divide out common factors:

$$\frac{\cancel{a}b}{\cancel{a}c} = \frac{b}{c}, \quad a \neq 0, c \neq 0$$

$$\frac{ab + ac}{ad} = \frac{\cancel{a}(b + c)}{\cancel{a}d} = \frac{b + c}{d}, \quad a \neq 0, d \neq 0$$

EXAMPLE 2   **Adding and Subtracting Rational Expressions**

Perform each indicated operation and simplify.

a. $x + \dfrac{1}{x}$ **b.** $\dfrac{1}{x + 1} - \dfrac{2}{2x - 1}$

SOLUTION

a. $x + \dfrac{1}{x} = \dfrac{x^2}{x} + \dfrac{1}{x}$ Write with common denominator.

$\qquad\quad = \dfrac{x^2 + 1}{x}$ Add fractions.

b. $\dfrac{1}{x + 1} - \dfrac{2}{2x - 1} = \dfrac{(2x - 1)}{(x + 1)(2x - 1)} - \dfrac{2(x + 1)}{(x + 1)(2x - 1)}$

$\qquad\qquad\qquad\qquad = \dfrac{2x - 1 - 2x - 2}{2x^2 + x - 1}$

$\qquad\qquad\qquad\qquad = \dfrac{-3}{2x^2 + x - 1}$

✓ **Checkpoint 2**   *Worked-out solution available at LarsonAppliedCalculus.com*

Perform each indicated operation and simplify.

a. $x + \dfrac{2}{x}$ **b.** $\dfrac{2}{x + 1} - \dfrac{1}{2x + 1}$

In adding (or subtracting) fractions whose denominators have no common factors, it is convenient to use the following pattern.

$$\frac{a}{b} + \frac{c}{d} = \frac{a \quad c}{b \quad d} = \frac{ad + bc}{bd}$$

For instance, in Example 2(b), you could have used this pattern as shown.

$$\frac{1}{x + 1} - \frac{2}{2x - 1} = \frac{(2x - 1) - 2(x + 1)}{(x + 1)(2x - 1)}$$

$$= \frac{2x - 1 - 2x - 2}{(x + 1)(2x - 1)}$$

$$= \frac{-3}{2x^2 + x - 1}$$

In Example 2, the denominators of the rational expressions have no common factors. When the denominators do have common factors, it is best to find the least common denominator before adding or subtracting. For instance, when adding

$$\frac{1}{x} \quad \text{and} \quad \frac{2}{x^2}$$

you can recognize that the least common denominator is x^2 and write

$$\frac{1}{x} + \frac{2}{x^2} = \frac{x}{x^2} + \frac{2}{x^2} \qquad \text{Write with common denominator.}$$

$$= \frac{x + 2}{x^2}. \qquad \text{Add fractions.}$$

This is further demonstrated in Example 3.

EXAMPLE 3   Adding Rational Expressions

Add the rational expressions.

$$\frac{x}{x^2 - 1} + \frac{3}{x + 1}$$

SOLUTION Because $x^2 - 1 = (x - 1)(x + 1)$, the least common denominator is $x^2 - 1$.

$$\frac{x}{x^2 - 1} + \frac{3}{x + 1} = \frac{x}{(x - 1)(x + 1)} + \frac{3}{x + 1} \qquad \text{Factor.}$$

$$= \frac{x}{(x - 1)(x + 1)} + \frac{3(x - 1)}{(x - 1)(x + 1)} \qquad \text{Write with common denominator.}$$

$$= \frac{x + 3(x - 1)}{(x - 1)(x + 1)} \qquad \text{Add fractions.}$$

$$= \frac{x + 3x - 3}{(x - 1)(x + 1)} \qquad \text{Multiply.}$$

$$= \frac{4x - 3}{x^2 - 1} \qquad \text{Simplify.}$$

✓ *Checkpoint 3*   *Worked-out solution available at LarsonAppliedCalculus.com*

Add the rational expressions.

$$\frac{x}{x^2 - 4} + \frac{2}{x - 2}$$

■

EXAMPLE 4   Subtracting Rational Expressions

Subtract the rational expressions.

$$\frac{1}{2(x^2 + 2x)} - \frac{1}{4x}$$

SOLUTION   In this case, the least common denominator is $4x(x + 2)$.

$$\frac{1}{2(x^2 + 2x)} - \frac{1}{4x} = \frac{1}{2x(x + 2)} - \frac{1}{2(2x)} \qquad \text{Factor.}$$

$$= \frac{2}{2(2x)(x + 2)} - \frac{x + 2}{2(2x)(x + 2)} \qquad \begin{array}{l}\text{Write with common}\\ \text{denominator.}\end{array}$$

$$= \frac{2 - (x + 2)}{4x(x + 2)} \qquad \text{Subtract fractions.}$$

$$= \frac{2 - x - 2}{4x(x + 2)} \qquad \text{Remove parentheses.}$$

$$= \frac{-\cancel{x}}{4\cancel{x}(x + 2)} \qquad \begin{array}{l}\text{Divide out}\\ \text{common factor.}\end{array}$$

$$= \frac{-1}{4(x + 2)}, \quad x \neq 0 \qquad \text{Simplify.}$$

✓ **Checkpoint 4**   *Worked-out solution available at LarsonAppliedCalculus.com*

Subtract the rational expressions.

$$\frac{1}{3(x^2 + 2x)} - \frac{1}{3x}$$

EXAMPLE 5   Combining Three Rational Expressions

Perform the operations and simplify.

$$\frac{3}{x - 1} - \frac{2}{x} + \frac{x + 3}{x^2 - 1}$$

SOLUTION   Using the factored denominators $(x - 1)$, x, and $(x + 1)(x - 1)$, you can see that the least common denominator is $x(x + 1)(x - 1)$.

$$\frac{3}{x - 1} - \frac{2}{x} + \frac{x + 3}{x^2 - 1} = \frac{3(x)(x + 1)}{x(x + 1)(x - 1)} - \frac{2(x + 1)(x - 1)}{x(x + 1)(x - 1)} + \frac{(x + 3)(x)}{x(x + 1)(x - 1)}$$

$$= \frac{3(x)(x + 1) - 2(x + 1)(x - 1) + (x + 3)(x)}{x(x + 1)(x - 1)}$$

$$= \frac{3x^2 + 3x - 2x^2 + 2 + x^2 + 3x}{x(x + 1)(x - 1)}$$

$$= \frac{2x^2 + 6x + 2}{x(x + 1)(x - 1)}$$

$$= \frac{2(x^2 + 3x + 1)}{x(x + 1)(x - 1)}$$

✓ **Checkpoint 5**   *Worked-out solution available at LarsonAppliedCalculus.com*

Perform the operations and simplify.

$$\frac{4}{x} - \frac{2}{x^2} + \frac{4}{x + 3}$$

Expressions Involving Radicals

In calculus, the operation of differentiation tends to produce "messy" expressions when applied to fractional expressions. This is especially true when the fractional expressions involve radicals. When differentiation is used, it is important to be able to simplify these expressions in order to obtain more managcable forms. The expressions in Example 6 are the results of differentiation. In each case, note how much *simpler* the simplified form is than the original form.

EXAMPLE 6   Simplifying an Expression with Radicals

Simplify each expression.

a. $\dfrac{\sqrt{x+1} - \dfrac{x}{2\sqrt{x+1}}}{x+1}$

b. $\left(\dfrac{1}{x + \sqrt{x^2+1}}\right)\left(1 + \dfrac{2x}{2\sqrt{x^2+1}}\right)$

SOLUTION

a. $\dfrac{\sqrt{x+1} - \dfrac{x}{2\sqrt{x+1}}}{x+1} = \dfrac{\dfrac{2(x+1)}{2\sqrt{x+1}} - \dfrac{x}{2\sqrt{x+1}}}{x+1}$ Write with common denominator.

$= \dfrac{\dfrac{2x+2-x}{2\sqrt{x+1}}}{\dfrac{x+1}{1}}$ Subtract fractions.

$= \dfrac{x+2}{2\sqrt{x+1}}\left(\dfrac{1}{x+1}\right)$ To divide, invert and multiply.

$= \dfrac{x+2}{2(x+1)^{3/2}}$ Multiply.

b. $\left(\dfrac{1}{x + \sqrt{x^2+1}}\right)\left(1 + \dfrac{2x}{2\sqrt{x^2+1}}\right) = \left(\dfrac{1}{x + \sqrt{x^2+1}}\right)\left(1 + \dfrac{x}{\sqrt{x^2+1}}\right)$

$= \left(\dfrac{1}{x + \sqrt{x^2+1}}\right)\left(\dfrac{\sqrt{x^2+1}}{\sqrt{x^2+1}} + \dfrac{x}{\sqrt{x^2+1}}\right)$

$= \left(\dfrac{1}{x + \sqrt{x^2+1}}\right)\left(\dfrac{x + \sqrt{x^2+1}}{\sqrt{x^2+1}}\right)$

$= \dfrac{1}{\sqrt{x^2+1}}$

✓ **Checkpoint 6**   *Worked-out solution available at LarsonAppliedCalculus.com*

Simplify each expression.

a. $\dfrac{\sqrt{x+2} - \dfrac{x}{4\sqrt{x+2}}}{x+2}$

b. $\left(\dfrac{1}{x + \sqrt{x^2+4}}\right)\left(1 + \dfrac{x}{\sqrt{x^2+4}}\right)$

Rationalization Techniques

In working with quotients involving radicals, it is often convenient to move the radical expression from the denominator to the numerator, or vice versa. For example, you can move $\sqrt{2}$ from the denominator to the numerator in the following quotient by multiplying by $\sqrt{2}/\sqrt{2}$.

Radical in Denominator   *Rationalize*   *Radical in Numerator*

$$\frac{1}{\sqrt{2}} \quad\Longrightarrow\quad \frac{1}{\sqrt{2}}\left(\frac{\sqrt{2}}{\sqrt{2}}\right) \quad\Longrightarrow\quad \frac{\sqrt{2}}{2}$$

This process is called **rationalizing the denominator.** A similar process is used to **rationalize the numerator.**

STUDY TIP

The success of the second and third rationalizing techniques stems from the following.

$$\left(\sqrt{a} - \sqrt{b}\right)\left(\sqrt{a} + \sqrt{b}\right)$$
$$= a - b$$

Rationalizing Techniques

1. When the denominator is $\sqrt{a}$, multiply by $\dfrac{\sqrt{a}}{\sqrt{a}}$.

2. When the denominator is $\sqrt{a} - \sqrt{b}$, multiply by $\dfrac{\sqrt{a} + \sqrt{b}}{\sqrt{a} + \sqrt{b}}$.

3. When the denominator is $\sqrt{a} + \sqrt{b}$, multiply by $\dfrac{\sqrt{a} - \sqrt{b}}{\sqrt{a} - \sqrt{b}}$.

The same guidelines apply to rationalizing numerators.

EXAMPLE 7   Rationalizing Denominators and Numerators

Rationalize the denominator or numerator.

a. $\dfrac{3}{\sqrt{12}}$ **b.** $\dfrac{\sqrt{x+1}}{2}$ **c.** $\dfrac{1}{\sqrt{5} + \sqrt{2}}$ **d.** $\dfrac{1}{\sqrt{x} - \sqrt{x+1}}$

SOLUTION

a. $\dfrac{3}{\sqrt{12}} = \dfrac{3}{2\sqrt{3}} = \dfrac{3}{2\sqrt{3}}\left(\dfrac{\sqrt{3}}{\sqrt{3}}\right) = \dfrac{3\sqrt{3}}{2(3)} = \dfrac{\sqrt{3}}{2}$

b. $\dfrac{\sqrt{x+1}}{2} = \dfrac{\sqrt{x+1}}{2}\left(\dfrac{\sqrt{x+1}}{\sqrt{x+1}}\right) = \dfrac{x+1}{2\sqrt{x+1}}$

c. $\dfrac{1}{\sqrt{5} + \sqrt{2}} = \dfrac{1}{\sqrt{5} + \sqrt{2}}\left(\dfrac{\sqrt{5} - \sqrt{2}}{\sqrt{5} - \sqrt{2}}\right) = \dfrac{\sqrt{5} - \sqrt{2}}{5 - 2} = \dfrac{\sqrt{5} - \sqrt{2}}{3}$

d. $\dfrac{1}{\sqrt{x} - \sqrt{x+1}} = \dfrac{1}{\sqrt{x} - \sqrt{x+1}}\left(\dfrac{\sqrt{x} + \sqrt{x+1}}{\sqrt{x} + \sqrt{x+1}}\right)$

$$= \dfrac{\sqrt{x} + \sqrt{x+1}}{x - (x+1)}$$

$$= -\sqrt{x} - \sqrt{x+1}$$

✓ **Checkpoint 7**   *Worked-out solution available at LarsonAppliedCalculus.com*

Rationalize the denominator or numerator.

a. $\dfrac{5}{\sqrt{8}}$ **b.** $\dfrac{\sqrt{x+2}}{4}$ **c.** $\dfrac{1}{\sqrt{6} - \sqrt{3}}$ **d.** $\dfrac{1}{\sqrt{x} + \sqrt{x+2}}$

Simplifying a Rational Expression In Exercises 1–4, write the rational expression in simplest form. *See Example 1.*

1. $\dfrac{x^2 - 7x + 12}{x^2 + 3x - 18}$

2. $\dfrac{x^2 - 5x - 6}{x^2 + 11x + 10}$

3. $\dfrac{x^2 + 3x - 10}{2x^2 - x - 6}$

4. $\dfrac{3x^2 + 13x + 12}{x^2 - 4x - 21}$

Adding and Subtracting Rational Expressions In Exercises 5–16, perform the indicated operation(s) and simplify. *See Examples 2, 3, 4, and 5.*

5. $\dfrac{x}{x - 2} + \dfrac{3}{x - 2}$

6. $\dfrac{5x + 10}{2x - 1} - \dfrac{2x + 10}{2x - 1}$

7. $x^2 - \dfrac{4}{x^3}$

8. $3x + \dfrac{2}{x^2}$

9. $\dfrac{2}{x - 3} + \dfrac{5x}{3x + 4}$

10. $\dfrac{3}{3x - 1} - \dfrac{1}{x + 2}$

11. $\dfrac{6}{x^2 - 16} - \dfrac{1}{x - 4}$

12. $\dfrac{5}{x^2 - 9} + \dfrac{x}{x + 3}$

13. $\dfrac{x}{x^2 + x - 2} - \dfrac{1}{x + 2}$

14. $\dfrac{2}{x + 1} + \dfrac{3x - 2}{x^2 - 2x - 3}$

15. $\dfrac{2}{x^2 + 1} - \dfrac{1}{x} + \dfrac{1}{x^3 + x}$

16. $\dfrac{3}{x + 2} + \dfrac{3}{x - 2} + \dfrac{1}{x^2 - 4}$

Simplifying an Expression with Radicals In Exercises 17–28, simplify the expression. *See Example 6.*

17. $\dfrac{-x}{(x + 1)^{3/2}} + \dfrac{2}{(x + 1)^{1/2}}$

18. $2\sqrt{x}(x - 2) + \dfrac{(x - 2)^2}{2\sqrt{x}}$

19. $\dfrac{2 - t}{2\sqrt{1 + t}} - \sqrt{1 + t}$

20. $-\dfrac{\sqrt{x^2 + 1}}{x^2} + \dfrac{1}{\sqrt{x^2 + 1}}$

21. $\left(2x\sqrt{x^2 + 1} - \dfrac{x^3}{\sqrt{x^2 + 1}}\right) \div (x^2 + 1)$

22. $\left(\sqrt{x^3 + 1} - \dfrac{3x^3}{2\sqrt{x^3 + 1}}\right) \div (x^3 + 1)$

23. $\dfrac{(x^2 + 2)^{1/2} - x^2(x^2 + 2)^{-1/2}}{x^2}$

24. $\dfrac{x(x + 1)^{-1/2} - (x + 1)^{1/2}}{x^2}$

25. $\dfrac{\dfrac{\sqrt{x + 1}}{\sqrt{x}} - \dfrac{\sqrt{x}}{\sqrt{x + 1}}}{2(x + 1)}$

26. $\dfrac{\dfrac{2x^2}{3(x^2 - 1)^{2/3}} - (x^2 - 1)^{1/3}}{x^2}$

27. $\dfrac{-x^2}{(2x + 3)^{3/2}} + \dfrac{2x}{(2x + 3)^{1/2}}$

28. $\dfrac{-x}{2(3 + x^2)^{3/2}} + \dfrac{3}{(3 + x^2)^{1/2}}$

Rationalizing Denominators and Numerators In Exercises 29–40, rationalize the denominator or numerator and simplify. *See Example 7.*

29. $\dfrac{2}{\sqrt{10}}$

30. $\dfrac{3}{\sqrt{21}}$

31. $\dfrac{49(x - 3)}{\sqrt{x^2 - 9}}$

32. $\dfrac{10(x + 2)}{\sqrt{x^2 - x - 6}}$

33. $\dfrac{5}{\sqrt{14} - 2}$

34. $\dfrac{13}{6 + \sqrt{10}}$

35. $\dfrac{1}{\sqrt{6} + \sqrt{5}}$

36. $\dfrac{x}{\sqrt{2} - \sqrt{3}}$

37. $\dfrac{2}{\sqrt{x} + \sqrt{x - 2}}$

38. $\dfrac{10}{\sqrt{x} + \sqrt{x + 5}}$

39. $\dfrac{\sqrt{x + 2} - \sqrt{2}}{x}$

40. $\dfrac{\sqrt{x + 1} - 1}{x}$

41. **Installment Loan** The monthly payment M (in dollars) for an installment loan is modeled by

$$M = P\left[\dfrac{r/12}{1 - \left(\dfrac{1}{(r/12) + 1}\right)^N}\right]$$

where P is the amount of the loan (in dollars), r is the annual interest rate (in decimal form), and N is the number of monthly payments. Find the monthly payment for a loan of $10,000 at an annual interest rate of 7.5% ($r = 0.075$) for 5 years ($N = 60$ monthly payments).

42. **Inventory Cost** A retailer determines that the cost C (in dollars) of ordering and storing x units of a product is

$$C = 6x + \dfrac{900,000}{x}.$$

(a) Write the expression for cost as a single fraction.

(b) Which order size should the retailer place: 240 units, 390 units, or 480 units? Explain your reasoning.

B Alternative Introduction to the Fundamental Theorem of Calculus

In this appendix, a summation process is used to provide an alternative development of the definite integral. It is intended that this supplement follow Section 5.3 in the text. If used, this appendix should replace the material preceding Example 2 in Section 5.4. Example 1 below shows how the area of a region in the plane can be approximated by the use of rectangles.

EXAMPLE 1   Using Rectangles to Approximate the Area of a Region

Use the four rectangles shown in Figure B.1 to approximate the area of the region lying between the graph of

$$f(x) = \frac{x^2}{2}$$

and the x-axis, between $x = 0$ and $x = 4$.

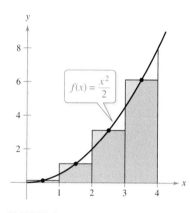

FIGURE B.1

SOLUTION   You can find the heights of the rectangles by evaluating the function f at each of the midpoints of the subintervals

$$[0, 1], \quad [1, 2], \quad [2, 3], \quad [3, 4].$$

Because the width of each rectangle is 1, the sum of the areas of the four rectangles is

$$S = \overbrace{(1)}^{\text{width}} \overbrace{f\left(\tfrac{1}{2}\right)}^{\text{height}} + \overbrace{(1)}^{\text{width}} \overbrace{f\left(\tfrac{3}{2}\right)}^{\text{height}} + \overbrace{(1)}^{\text{width}} \overbrace{f\left(\tfrac{5}{2}\right)}^{\text{height}} + \overbrace{(1)}^{\text{width}} \overbrace{f\left(\tfrac{7}{2}\right)}^{\text{height}}$$

$$= \frac{1}{8} + \frac{9}{8} + \frac{25}{8} + \frac{49}{8}$$

$$= \frac{84}{8}$$

$$= 10.5 \text{ square units.}$$

So, you can approximate the area of the region to be 10.5 square units.   ■

> **STUDY TIP**
>
> The approximation technique used in Example 1 is called the *Midpoint Rule*. The Midpoint Rule is discussed further in Section 5.6.

The procedure shown in Example 1 can be generalized. Let f be a continuous function defined on the closed interval $[a, b]$. To begin, partition the interval into n subintervals, each of width

$$\Delta x = \frac{b - a}{n}$$

as

$$a = x_0 < x_1 < x_2 < \cdots < x_{n-1} < x_n = b.$$

In each subinterval $[x_{i-1}, x_i]$, choose an arbitrary point c_i and form the sum

$$S = f(c_1)\,\Delta x + f(c_2)\,\Delta x + \cdots + f(c_{n-1})\,\Delta x + f(c_n)\,\Delta x.$$

This type of summation is called a **Riemann sum** and is often written using summation notation, as shown below.

$$S = \sum_{i=1}^{n} f(c_i)\,\Delta x, \quad x_{i-1} \le c_i \le x_i$$

For the Riemann sum in Example 1, the interval is $[a, b] = [0, 4]$, the number of subintervals is $n = 4$, the width of each subinterval is $\Delta x = 1$, and the point c_i in each subinterval is its midpoint. So, you can write the approximation in Example 1 as

$$S = \sum_{i=1}^{n} f(c_i)\,\Delta x$$

$$= \sum_{i=1}^{4} f(c_i)(1)$$

$$= \frac{1}{8} + \frac{9}{8} + \frac{25}{8} + \frac{49}{8}$$

$$= \frac{84}{8}.$$

EXAMPLE 2   Using a Riemann Sum to Approximate Area

Use a Riemann sum to approximate the area of the region bounded by the graph of

$$f(x) = -x^2 + 2x$$

and the x-axis, for $0 \le x \le 2$. In the Riemann sum, let $n = 6$ and choose c_i to be the left endpoint of each subinterval.

SOLUTION   Subdivide the interval $[0, 2]$ into six subintervals, each of width

$$\Delta x = \frac{2 - 0}{6}$$

$$= \frac{1}{3}$$

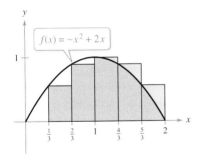

$f(x) = -x^2 + 2x$

FIGURE B.2

as shown in Figure B.2. Because c_i is the left endpoint of each subinterval, the Riemann sum is given by

$$S = \sum_{i=1}^{n} f(c_i)\,\Delta x$$

$$= \left[f(0) + f\left(\frac{1}{3}\right) + f\left(\frac{2}{3}\right) + f(1) + f\left(\frac{4}{3}\right) + f\left(\frac{5}{3}\right) \right]\left(\frac{1}{3}\right)$$

$$= \left[0 + \frac{5}{9} + \frac{8}{9} + 1 + \frac{8}{9} + \frac{5}{9} \right]\left(\frac{1}{3}\right)$$

$$= \frac{35}{27} \text{ square units}.$$

Example 2 illustrates an important point. If a function f is continuous and nonnegative over the interval $[a, b]$, then the Riemann sum

$$S = \sum_{i=1}^{n} f(c_i) \, \Delta x$$

can be used to approximate the area of the region bounded by the graph of f and the x-axis, between $x = a$ and $x = b$. Moreover, for a given interval, as the number of subintervals increases, the approximation to the actual area will improve. This is illustrated in the next two examples by using Riemann sums to approximate the area of a triangle.

EXAMPLE 3   Approximating the Area of a Triangle

Use a Riemann sum to approximate the area of the triangular region bounded by the graph of

$$f(x) = 2x$$

and the x-axis for $0 \le x \le 3$. Use a partition of six subintervals and choose c_i to be the left endpoint of each subinterval.

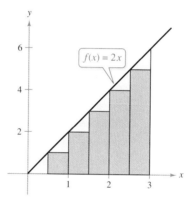

FIGURE B.3

SOLUTION   Subdivide the interval $[0, 3]$ into six subintervals, each of width

$$\Delta x = \frac{3 - 0}{6}$$

$$= \frac{1}{2}$$

as shown in Figure B.3. Because c_i is the left endpoint of each subinterval, the Riemann sum is given by

$$S = \sum_{i=1}^{n} f(c_i) \, \Delta x$$

$$= \left[f(0) + f\left(\frac{1}{2}\right) + f(1) + f\left(\frac{3}{2}\right) + f(2) + f\left(\frac{5}{2}\right) \right]\left(\frac{1}{2}\right)$$

$$= [0 + 1 + 2 + 3 + 4 + 5]\left(\frac{1}{2}\right)$$

$$= \frac{15}{2} \text{ square units.}$$

TECH TUTOR

Most graphing utilities are able to sum the first n terms of a sequence. Try using a graphing utility to verify the right Riemann sum in Example 3.

The approximations in Examples 2 and 3 are called **left Riemann sums,** because c_i was chosen to be the left endpoint of each subinterval. Using the right endpoints in Example 3, the **right Riemann sum** is $\frac{21}{2}$. Note that the exact area of the triangular region in Example 3 is

$$\text{Area} = \frac{1}{2}(\text{base})(\text{height})$$

$$= \frac{1}{2}(3)(6)$$

$$= 9 \text{ square units.}$$

So, the left Riemann sum gives an approximation that is less than the actual area, and the right Riemann sum gives an approximation that is greater than the actual area.

In Example 4, you will see that the approximation improves as the number of subintervals increases.

EXAMPLE 4   **Increasing the Number of Subintervals**

Let $f(x) = 2x$, $0 \le x \le 3$. Use a graphing utility to determine the left and right Riemann sums for $n = 10$, $n = 100$, and $n = 1000$ subintervals.

SOLUTION   A graphing utility program for this problem is shown in Figure B.4. [Note that the function $f(x) = 2x$ is entered as Y1.]

```
PROGRAM:RIEMANN
:Input ("ENTER VALU
E OF N",N)
:Input ("ENTER VALU
E OF A",A)
:Input ("ENTER VALU
E OF B",B)
:(B-A)/N→D
```

```
PROGRAM:RIEMANN
:0→R
:0→L
:A→X
:For(I,1,N)
:L+Y1→L
:A+ID→X
:R+Y1→R
```

```
PROGRAM:RIEMANN
:End
:LD→L
:RD→R
:Disp "LEFT SUM",L
:Disp "RIGHT SUM",
R
```

FIGURE B.4

Running this program for $n = 10$, $n = 100$, and $n = 1000$ gives the results shown in the table.

n	Left Riemann sum	Right Riemann sum
10	8.100	9.900
100	8.910	9.090
1000	8.991	9.009

From the results of Example 4, it appears that the Riemann sums are approaching the limit 9 as n approaches infinity. It is this observation that motivates the definition of a **definite integral.** In this definition, consider the partition of $[a, b]$ into n subintervals of equal width $\Delta x = (b - a)/n$, as shown.

$$a = x_0 < x_1 < x_2 < \cdots < x_{n-1} < x_n = b$$

Moreover, consider c_i to be an arbitrary point in the ith subinterval $[x_{i-1}, x_i]$. To say that the number of subintervals n approaches infinity is equivalent to saying that the width, Δx, of the subintervals approaches zero.

Definition of Definite Integral

If f is a continuous function defined on the closed interval $[a, b]$, then the **definite integral of f on $[a, b]$** is

$$\int_a^b f(x)\, dx = \lim_{\Delta x \to 0} \sum_{i=1}^n f(c_i)\, \Delta x$$

$$= \lim_{n \to \infty} \sum_{i=1}^n f(c_i)\, \Delta x.$$

If f is continuous and nonnegative on the interval $[a, b]$, then the definite integral of f on $[a, b]$ gives the area of the region bounded by the graph of f, the x-axis, and the vertical lines $x = a$ and $x = b$.

Evaluation of a definite integral by its limit definition can be difficult. However, there are times when a definite integral can be solved by recognizing that it represents the area of a common type of geometric figure.

EXAMPLE 5   The Areas of Common Geometric Figures

Sketch the region corresponding to each of the definite integrals. Then evaluate each definite integral using a geometric formula.

a. $\displaystyle\int_1^3 4\, dx$

b. $\displaystyle\int_0^3 (x + 2)\, dx$

c. $\displaystyle\int_{-2}^2 \sqrt{4 - x^2}\, dx$

SOLUTION   A sketch of each region is shown in Figure B.5.

a. The region associated with this definite integral is a rectangle of height 4 and width 2. Moreover, because the function $f(x) = 4$ is continuous and nonnegative on the interval $[1, 3]$, you can conclude that the area of the rectangle is given by the definite integral. So, the value of the definite integral is

$$\int_1^3 4\, dx = 4(2) = 8 \text{ square units.}$$

b. The region associated with this definite integral is a trapezoid with an altitude of 3 and parallel bases of lengths 2 and 5. The formula for the area of a trapezoid is $\frac{1}{2}h(b_1 + b_2)$, and so you have

$$\int_0^3 (x + 2)\, dx = \frac{1}{2}(3)(2 + 5)$$

$$= \frac{21}{2} \text{ square units.}$$

c. The region associated with this definite integral is a semicircle of radius 2. The formula for the area of a semicircle is $\frac{1}{2}\pi r^2$, and so you have

$$\int_{-2}^2 \sqrt{4 - x^2}\, dx = \frac{1}{2}\pi(2^2)$$

$$= 2\pi \text{ square units.}$$

(a) $\displaystyle\int_1^3 4\, dx$

Rectangle

(b) $\displaystyle\int_0^3 (x + 2)\, dx$

Trapezoid

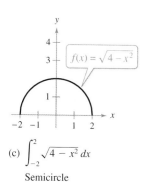

(c) $\displaystyle\int_{-2}^2 \sqrt{4 - x^2}\, dx$

Semicircle

FIGURE B.5

For some simple functions, it is possible to evaluate definite integrals by the Riemann sum definition. In the next example, you will use the fact that the sum of the first n integers is given by the formula

$$1 + 2 + \cdots + n = \sum_{i=1}^{n} i = \frac{n(n + 1)}{2}$$ See Exercise 29.

to compute the area of the triangular region in Examples 3 and 4.

EXAMPLE 6   Evaluating a Definite Integral by Its Definition

Evaluate $\displaystyle\int_0^3 2x \, dx$.

SOLUTION   Let

$$\Delta x = \frac{b - a}{n} = \frac{3}{n}$$

and choose c_i to be the right endpoint of each subinterval,

$$c_i = \frac{3i}{n}.$$

Then you have

$$\int_0^3 2x \, dx = \lim_{\Delta x \to 0} \sum_{i=1}^{n} f(c_i) \, \Delta x$$

$$= \lim_{n \to \infty} \sum_{i=1}^{n} 2\left(i \frac{3}{n}\right)\left(\frac{3}{n}\right)$$

$$= \lim_{n \to \infty} \frac{18}{n^2} \sum_{i=1}^{n} i$$

$$= \lim_{n \to \infty} \left(\frac{18}{n^2}\right)\left[\frac{n(n + 1)}{2}\right]$$

$$= \lim_{n \to \infty} \left(9 + \frac{9}{n}\right).$$

This limit can be evaluated in the same way that you calculated horizontal asymptotes in Section 3.6. In particular, as n approaches infinity, you see that $9/n$ approaches 0, and the limit above is 9. So, you can conclude that

$$\int_0^3 2x \, dx = 9.$$ ■

Example 6 shows that it can be difficult to evaluate the definite integral of even a simple function by using Riemann sums. A computer can help in calculating these sums for large values of n, but this procedure would give only an approximation of the definite integral. Fortunately, the **Fundamental Theorem of Calculus** provides a technique for evaluating definite integrals using antiderivatives, and for this reason it is often thought to be the most important theorem in calculus. In the remainder of this appendix, you will see how derivatives and integrals are related via the Fundamental Theorem of Calculus.

FIGURE B.6

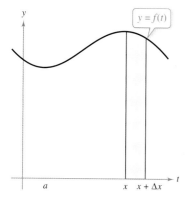

FIGURE B.7

To simplify the discussion, assume that f is a continuous nonnegative function defined on the interval $[a, b]$. Let $A(x)$ be the area of the region under the graph of f from a to x, as indicated in Figure B.6. The area under the shaded region in Figure B.7 is

$$A(x + \Delta x) - A(x).$$

If Δx is small, then this area is approximated by the area of the rectangle of height $f(x)$ and width Δx. So, you have

$$A(x + \Delta x) - A(x) \approx f(x)\, \Delta x.$$

Dividing by Δx produces

$$f(x) \approx \frac{A(x + \Delta x) - A(x)}{\Delta x}.$$

By taking the limit as Δx approaches 0, you can see that

$$f(x) = \lim_{\Delta x \to 0} \frac{A(x + \Delta x) - A(x)}{\Delta x} = A'(x)$$

and you can establish the fact that the area function $A(x)$ is an antiderivative of f. Although it was assumed that f is continuous and nonnegative, this development is valid when the function f is simply continuous on the closed interval $[a, b]$. This result is used in the proof of the Fundamental Theorem of Calculus.

Fundamental Theorem of Calculus

If f is a continuous function on the closed interval $[a, b]$, then

$$\int_a^b f(x)\, dx = F(b) - F(a)$$

where F is any function such that $F'(x) = f(x)$.

PROOF   From the discussion above, you know that

$$\int_a^x f(x)\, dx = A(x)$$

and in particular,

$$A(a) = \int_a^a f(x)\, dx = 0$$

and

$$A(b) = \int_a^b f(x)\, dx.$$

If F is *any* antiderivative of f, then you know that F differs from A by a constant. That is, $A(x) = F(x) + C$. So,

$$\int_a^b f(x)\, dx = A(b) - A(a)$$
$$= [F(b) + C] - [F(a) + C]$$
$$= F(b) + C - F(a) - C$$
$$= F(b) - F(a). \qquad \blacksquare$$

You are now ready to continue Section 5.4, on page 341, just after the statement of the Fundamental Theorem of Calculus.

Exercises B

See *CalcChat.com* for tutorial help and worked-out solutions to odd-numbered exercises.

Using Rectangles to Approximate the Area of a Region In Exercises 1 and 2, use the rectangles to approximate the area of the region. *See Example 1.*

1. $y = x + 1$

2. $y = 4 - x^2$

Using a Riemann Sum to Approximate Area In Exercises 3–8, use the left Riemann sum and the right Riemann sum to approximate the area of the region using the indicated number of subintervals. *See Examples 2 and 3.*

3. $y = \sqrt{x}$

4. $y = \sqrt{x} + 1$

5. $y = \dfrac{1}{x}$

6. $y = \dfrac{1}{x - 2}$

7. $y = \sqrt{1 - x^2}$

8. $y = \sqrt{x + 1}$

9. Comparing Riemann Sums Consider a triangle of area 2 square units bounded by the graphs of $y = x$, $y = 0$, and $x = 2$.

(a) Sketch the graph of the region.

(b) Divide the interval $[0, 2]$ into n equal subintervals and show that the endpoints of the subintervals are

$$0 < 1\left(\frac{2}{n}\right) < \cdots < (n - 1)\left(\frac{2}{n}\right) < n\left(\frac{2}{n}\right).$$

(c) Show that the left Riemann sum is

$$S_L = \sum_{i=1}^{n}\left[(i - 1)\left(\frac{2}{n}\right)\right]\left(\frac{2}{n}\right).$$

(d) Show that the right Riemann sum is

$$S_R = \sum_{i=1}^{n}\left[i\left(\frac{2}{n}\right)\right]\left(\frac{2}{n}\right).$$

(e) Use a graphing utility and the program in Example 4 to complete the table.

n	5	10	50	100
Left sum, S_L				
Right sum, S_R				

(f) Show that $\lim\limits_{n\to\infty} S_L = \lim\limits_{n\to\infty} S_R = 2$.

10. Comparing Riemann Sums Consider a trapezoid of area 4 square units bounded by the graphs of $y = x$, $y = 0$, $x = 1$, and $x = 3$.

(a) Sketch the graph of the region.

(b) Divide the interval $[1, 3]$ into n equal subintervals and show that the endpoints of the subintervals are

$$1 < 1 + 1\left(\frac{2}{n}\right) < \cdots < 1 + (n - 1)\left(\frac{2}{n}\right) < 1 + n\left(\frac{2}{n}\right).$$

(c) Show that the left Riemann sum is

$$S_L = \sum_{i=1}^{n}\left[1 + (i - 1)\left(\frac{2}{n}\right)\right]\left(\frac{2}{n}\right).$$

(d) Show that the right Riemann sum is

$$S_R = \sum_{i=1}^{n}\left[1 + i\left(\frac{2}{n}\right)\right]\left(\frac{2}{n}\right).$$

(e) Use a graphing utility and the program in Example 4 to complete the table.

n	5	10	50	100
Left sum, S_L				
Right sum, S_R				

(f) Show that $\lim\limits_{n\to\infty} S_L = \lim\limits_{n\to\infty} S_R = 4$.

Writing a Definite Integral In Exercises 11–18, set up a definite integral that yields the area of the region. (Do not evaluate the integral.)

11. $f(x) = 3$

12. $f(x) = 4 - 2x$

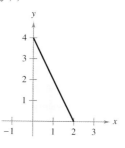

13. $f(x) = 4 - |x|$

14. $f(x) = x^2$

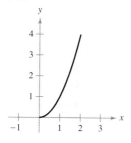

15. $f(x) = 4 - x^2$

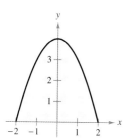

16. $f(x) = \dfrac{1}{x^2 + 1}$

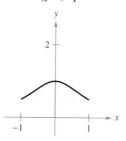

17. $f(x) = \sqrt{x + 1}$

18. $f(x) = (x^2 + 1)^2$

Finding Areas of Common Geometric Figures In Exercises 19–28, sketch the region whose area is given by the definite integral. Then use a geometric formula to evaluate the integral ($a > 0, r > 0$). *See Example 5.*

19. $\displaystyle\int_0^3 5 \, dx$

20. $\displaystyle\int_{-a}^a 2 \, dx$

21. $\displaystyle\int_0^2 3x \, dx$

22. $\displaystyle\int_0^4 \frac{x}{2} \, dx$

23. $\displaystyle\int_0^2 (2x + 5) \, dx$

24. $\displaystyle\int_0^5 (6 - x) \, dx$

25. $\displaystyle\int_{-1}^1 \left(1 - |x|\right) dx$

26. $\displaystyle\int_{-a}^a \left(a - |x|\right) dx$

27. $\displaystyle\int_{-3}^3 \sqrt{9 - x^2} \, dx$

28. $\displaystyle\int_{-r}^r \sqrt{r^2 - x^2} \, dx$

29. Proving a Sum Show that
$$\sum_{i=1}^n i = \frac{n(n + 1)}{2}.$$
(*Hint:* Add the two sums below.)
$$S = 1 + 2 + 3 + \cdots + (n - 2) + (n - 1) + n$$
$$S = n + (n - 1) + (n - 2) + \cdots + 3 + 2 + 1$$

30. Evaluating a Definite Integral by Its Definition Use the Riemann sum definition and the result of Exercise 29 to evaluate the definite integrals.

(a) $\displaystyle\int_1^2 x \, dx$ (b) $\displaystyle\int_0^4 3x \, dx$

Comparing a Sum with an Integral In Exercises 31 and 32, use the figure to fill in the blank with the symbol $<$, $>$, or $=$. Explain your reasoning.

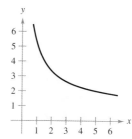

31. The interval $[1, 5]$ is partitioned into n subintervals of equal width Δx, and x_i is the left endpoint of the ith subinterval.

$$\sum_{i=1}^n f(x_i) \, \Delta x \qquad \int_1^5 f(x) \, dx$$

32. The interval $[1, 5]$ is partitioned into n subintervals of equal width Δx, and x_i is the right endpoint of the ith subinterval.

$$\sum_{i=1}^n f(x_i) \, \Delta x \qquad \int_1^5 f(x) \, dx$$

C Formulas

C.1 Differentiation and Integration Formulas

■ Use differentiation and integration tables to supplement differentiation and integration techniques.

Differentiation Formulas

1. $\dfrac{d}{dx}[cu] = cu'$

2. $\dfrac{d}{dx}[u \pm v] = u' \pm v'$

3. $\dfrac{d}{dx}[uv] = uv' + vu'$

4. $\dfrac{d}{dx}\left[\dfrac{u}{v}\right] = \dfrac{vu' - uv'}{v^2}$

5. $\dfrac{d}{dx}[c] = 0$

6. $\dfrac{d}{dx}[u^n] = nu^{n-1}u'$

7. $\dfrac{d}{dx}[x] = 1$

8. $\dfrac{d}{dx}[\ln u] = \dfrac{u'}{u}$

9. $\dfrac{d}{dx}[e^u] = e^u u'$

10. $\dfrac{d}{dx}[\sin u] = (\cos u)u'$

11. $\dfrac{d}{dx}[\cos u] = -(\sin u)u'$

12. $\dfrac{d}{dx}[\tan u] = (\sec^2 u)u'$

13. $\dfrac{d}{dx}[\cot u] = -(\csc^2 u)u'$

14. $\dfrac{d}{dx}[\sec u] = (\sec u \tan u)u'$

15. $\dfrac{d}{dx}[\csc u] = -(\csc u \cot u)u'$

Integration Formulas

Forms Involving u^n

1. $\displaystyle\int u^n \, du = \dfrac{u^{n+1}}{n+1} + C, \quad n \neq -1$

2. $\displaystyle\int \dfrac{1}{u} \, du = \ln|u| + C$

Forms Involving $a + bu$

3. $\displaystyle\int \dfrac{u}{a+bu} \, du = \dfrac{1}{b^2}(bu - a \ln|a + bu|) + C$

4. $\displaystyle\int \dfrac{u}{(a+bu)^2} \, du = \dfrac{1}{b^2}\left(\dfrac{a}{a+bu} + \ln|a + bu|\right) + C$

5. $\displaystyle\int \dfrac{u}{(a+bu)^n} \, du = \dfrac{1}{b^2}\left[\dfrac{-1}{(n-2)(a+bu)^{n-2}} + \dfrac{a}{(n-1)(a+bu)^{n-1}}\right] + C, \quad n \neq 1, 2$

6. $\displaystyle\int \dfrac{u^2}{a+bu} \, du = \dfrac{1}{b^3}\left[-\dfrac{bu}{2}(2a - bu) + a^2 \ln|a + bu|\right] + C$

7. $\displaystyle\int \dfrac{u^2}{(a+bu)^2} \, du = \dfrac{1}{b^3}\left(bu - \dfrac{a^2}{a+bu} - 2a \ln|a + bu|\right) + C$

8. $\displaystyle\int \dfrac{u^2}{(a+bu)^3} \, du = \dfrac{1}{b^3}\left[\dfrac{2a}{a+bu} - \dfrac{a^2}{2(a+bu)^2} + \ln|a + bu|\right] + C$

9. $\displaystyle\int \dfrac{u^2}{(a+bu)^n} \, du = \dfrac{1}{b^3}\left[\dfrac{-1}{(n-3)(a+bu)^{n-3}} + \dfrac{2a}{(n-2)(a+bu)^{n-2}} - \dfrac{a^2}{(n-1)(a+bu)^{n-1}}\right] + C, \quad n \neq 1, 2, 3$

10. $\displaystyle\int \dfrac{1}{u(a+bu)} \, du = \dfrac{1}{a}\ln\left|\dfrac{u}{a+bu}\right| + C$

Integration Formulas　(continued)

11. $\displaystyle\int \frac{1}{u(a + bu)^2}\,du = \frac{1}{a}\left(\frac{1}{a + bu} + \frac{1}{a}\ln\left|\frac{u}{a + bu}\right|\right) + C$

12. $\displaystyle\int \frac{1}{u^2(a + bu)}\,du = -\frac{1}{a}\left(\frac{1}{u} + \frac{b}{a}\ln\left|\frac{u}{a + bu}\right|\right) + C$

13. $\displaystyle\int \frac{1}{u^2(a + bu)^2}\,du = -\frac{1}{a^2}\left[\frac{a + 2bu}{u(a + bu)} + \frac{2b}{a}\ln\left|\frac{u}{a + bu}\right|\right] + C$

Forms Involving $\sqrt{a + bu}$

14. $\displaystyle\int u^n \sqrt{a + bu}\,du = \frac{2}{b(2n + 3)}\left[u^n(a + bu)^{3/2} - na\int u^{n-1}\sqrt{a + bu}\,du\right]$

15. $\displaystyle\int \frac{1}{u\sqrt{a + bu}}\,du = \frac{1}{\sqrt{a}}\ln\left|\frac{\sqrt{a + bu} - \sqrt{a}}{\sqrt{a + bu} + \sqrt{a}}\right| + C, \quad a > 0$

16. $\displaystyle\int \frac{1}{u^n\sqrt{a + bu}}\,du = \frac{-1}{a(n - 1)}\left[\frac{\sqrt{a + bu}}{u^{n-1}} + \frac{(2n - 3)b}{2}\int \frac{1}{u^{n-1}\sqrt{a + bu}}\,du\right], \quad n \ne 1$

17. $\displaystyle\int \frac{\sqrt{a + bu}}{u}\,du = 2\sqrt{a + bu} + a\int \frac{1}{u\sqrt{a + bu}}\,du$

18. $\displaystyle\int \frac{\sqrt{a + bu}}{u^n}\,du = \frac{-1}{a(n - 1)}\left[\frac{(a + bu)^{3/2}}{u^{n-1}} + \frac{(2n - 5)b}{2}\int \frac{\sqrt{a + bu}}{u^{n-1}}\,du\right], \quad n \ne 1$

19. $\displaystyle\int \frac{u}{\sqrt{a + bu}}\,du = -\frac{2(2a - bu)}{3b^2}\sqrt{a + bu} + C$

20. $\displaystyle\int \frac{u^n}{\sqrt{a + bu}}\,du = \frac{2}{(2n + 1)b}\left(u^n\sqrt{a + bu} - na\int \frac{u^{n-1}}{\sqrt{a + bu}}\,du\right)$

Forms Involving $u^2 - a^2, a > 0$

21. $\displaystyle\int \frac{1}{u^2 - a^2}\,du = -\int \frac{1}{a^2 - u^2}\,du = \frac{1}{2a}\ln\left|\frac{u - a}{u + a}\right| + C$

22. $\displaystyle\int \frac{1}{(u^2 - a^2)^n}\,du = \frac{-1}{2a^2(n - 1)}\left[\frac{u}{(u^2 - a^2)^{n-1}} + (2n - 3)\int \frac{1}{(u^2 - a^2)^{n-1}}\,du\right], \quad n \ne 1$

Forms Involving $\sqrt{u^2 \pm a^2}, a > 0$

23. $\displaystyle\int \sqrt{u^2 \pm a^2}\,du = \frac{1}{2}\left(u\sqrt{u^2 \pm a^2} \pm a^2 \ln\left|u + \sqrt{u^2 \pm a^2}\right|\right) + C$

24. $\displaystyle\int u^2\sqrt{u^2 \pm a^2}\,du = \frac{1}{8}\left[u(2u^2 \pm a^2)\sqrt{u^2 \pm a^2} - a^4 \ln\left|u + \sqrt{u^2 \pm a^2}\right|\right] + C$

25. $\displaystyle\int \frac{\sqrt{u^2 + a^2}}{u}\,du = \sqrt{u^2 + a^2} - a \ln\left|\frac{a + \sqrt{u^2 + a^2}}{u}\right| + C$

26. $\displaystyle\int \frac{\sqrt{u^2 \pm a^2}}{u^2}\,du = \frac{-\sqrt{u^2 \pm a^2}}{u} + \ln\left|u + \sqrt{u^2 \pm a^2}\right| + C$

27. $\displaystyle\int \frac{1}{\sqrt{u^2 \pm a^2}}\,du = \ln\left|u + \sqrt{u^2 \pm a^2}\right| + C$

28. $\displaystyle\int \frac{1}{u\sqrt{u^2 + a^2}}\,du = \frac{-1}{a}\ln\left|\frac{a + \sqrt{u^2 + a^2}}{u}\right| + C$

29. $\displaystyle\int \frac{u^2}{\sqrt{u^2 \pm a^2}}\,du = \frac{1}{2}\left(u\sqrt{u^2 \pm a^2} \mp a^2 \ln\left|u + \sqrt{u^2 \pm a^2}\right|\right) + C$

30. $\displaystyle\int \frac{1}{u^2\sqrt{u^2 \pm a^2}}\,du = \mp\frac{\sqrt{u^2 \pm a^2}}{a^2 u} + C$

31. $\displaystyle\int \frac{1}{(u^2 \pm a^2)^{3/2}}\,du = \frac{\pm u}{a^2\sqrt{u^2 \pm a^2}} + C$

Forms Involving $\sqrt{a^2 - u^2}, a > 0$

32. $\displaystyle\int \frac{\sqrt{a^2 - u^2}}{u}\,du = \sqrt{a^2 - u^2} - a\ln\left|\frac{a + \sqrt{a^2 - u^2}}{u}\right| + C$

33. $\displaystyle\int \frac{1}{u\sqrt{a^2 - u^2}}\,du = \frac{-1}{a}\ln\left|\frac{a + \sqrt{a^2 - u^2}}{u}\right| + C$

34. $\displaystyle\int \frac{1}{u^2\sqrt{a^2 - u^2}}\,du = \frac{-\sqrt{a^2 - u^2}}{a^2 u} + C$

35. $\displaystyle\int \frac{1}{(a^2 - u^2)^{3/2}}\,du = \frac{u}{a^2\sqrt{a^2 - u^2}} + C$

Forms Involving e^u

36. $\displaystyle\int e^u\,du = e^u + C$

37. $\displaystyle\int ue^u\,du = (u - 1)e^u + C$

38. $\displaystyle\int u^n e^u\,du = u^n e^u - n\int u^{n-1}e^u\,du$

39. $\displaystyle\int \frac{1}{1 + e^u}\,du = u - \ln(1 + e^u) + C$

40. $\displaystyle\int \frac{1}{1 + e^{nu}}\,du = u - \frac{1}{n}\ln(1 + e^{nu}) + C$

Forms Involving ln u

41. $\displaystyle\int \ln u\,du = u(-1 + \ln u) + C$

42. $\displaystyle\int u\ln u\,du = \frac{u^2}{4}(-1 + 2\ln u) + C$

43. $\displaystyle\int u^n\ln u\,du = \frac{u^{n+1}}{(n+1)^2}[-1 + (n+1)\ln u] + C, \quad n \neq -1$

44. $\displaystyle\int (\ln u)^2\,du = u[2 - 2\ln u + (\ln u)^2] + C$

45. $\displaystyle\int (\ln u)^n\,du = u(\ln u)^n - n\int (\ln u)^{n-1}\,du$

Forms Involving sin u or cos u

46. $\displaystyle\int \sin u\,du = -\cos u + C$

47. $\displaystyle\int \cos u\,du = \sin u + C$

48. $\displaystyle\int \sin^2 u\,du = \frac{1}{2}(u - \sin u\cos u) + C$

49. $\displaystyle\int \cos^2 u\,du = \frac{1}{2}(u + \sin u\cos u) + C$

50. $\displaystyle\int \sin^n u\,du = -\frac{\sin^{n-1}u\cos u}{n} + \frac{n-1}{n}\int \sin^{n-2}u\,du$

51. $\displaystyle\int \cos^n u\,du = \frac{\cos^{n-1}u\sin u}{n} + \frac{n-1}{n}\int \cos^{n-2}u\,du$

52. $\displaystyle\int u\sin u\,du = \sin u - u\cos u + C$

53. $\displaystyle\int u\cos u\,du = \cos u + u\sin u + C$

54. $\displaystyle\int u^n\sin u\,du = -u^n\cos u + n\int u^{n-1}\cos u\,du$

Integration Formulas   (continued)

55. $\displaystyle\int u^n \cos u \; du = u^n \sin u - n \int u^{n-1} \sin u \; du$

56. $\displaystyle\int \frac{1}{1 \pm \sin u} \; du = \tan u \mp \sec u + C$

57. $\displaystyle\int \frac{1}{1 \pm \cos u} \; du = -\cot u \pm \csc u + C$

58. $\displaystyle\int \frac{1}{\sin u \cos u} \; du = \ln|\tan u| + C$

Forms Involving tan u, cot u, sec u, or csc u

59. $\displaystyle\int \tan u \; du = -\ln|\cos u| + C$

60. $\displaystyle\int \cot u \; du = \ln|\sin u| + C$

61. $\displaystyle\int \sec u \; du = \ln|\sec u + \tan u| + C$

62. $\displaystyle\int \csc u \; du = -\ln|\csc u + \cot u| + C$

63. $\displaystyle\int \tan^2 u \; du = -u + \tan u + C$

64. $\displaystyle\int \cot^2 u \; du = -u - \cot u + C$

65. $\displaystyle\int \sec^2 u \; du = \tan u + C$

66. $\displaystyle\int \csc^2 u \; du = -\cot u + C$

67. $\displaystyle\int \tan^n u \; du = \frac{\tan^{n-1} u}{n-1} - \int \tan^{n-2} u \; du, \quad n \neq 1$

68. $\displaystyle\int \cot^n u \; du = -\frac{\cot^{n-1} u}{n-1} - \int \cot^{n-2} u \; du, \quad n \neq 1$

69. $\displaystyle\int \sec^n u \; du = \frac{\sec^{n-2} u \tan u}{n-1} + \frac{n-2}{n-1} \int \sec^{n-2} u \; du, \quad n \neq 1$

70. $\displaystyle\int \csc^n u \; du = -\frac{\csc^{n-2} u \cot u}{n-1} + \frac{n-2}{n-1} \int \csc^{n-2} u \; du, \quad n \neq 1$

71. $\displaystyle\int \frac{1}{1 \pm \tan u} \; du = \frac{1}{2}(u \pm \ln|\cos u \pm \sin u|) + C$

72. $\displaystyle\int \frac{1}{1 \pm \cot u} \; du = \frac{1}{2}(u \mp \ln|\sin u \pm \cos u|) + C$

73. $\displaystyle\int \frac{1}{1 \pm \sec u} \; du = u + \cot u \mp \csc u + C$

74. $\displaystyle\int \frac{1}{1 \pm \csc u} \; du = u - \tan u \pm \sec u + C$

C.2 Formulas from Business and Finance

■ Summary of business and finance formulas

Formulas from Business

Basic Terms

x = number of units produced (or sold)

p = price per unit

R = total revenue from selling x units

C = total cost of producing x units

$\overline{C}$ = average cost per unit

P = total profit from selling x units

Basic Equations

$$R = xp \qquad \overline{C} = \frac{C}{x} \qquad P = R - C$$

Typical Graphs of Supply and Demand Curves

Supply curves increase as price increases and demand curves decrease as price increases. The equilibrium point occurs when the supply and demand curves intersect.

Demand Function: $p = f(x)$ = price required to sell x units

$$\eta = \frac{p/x}{dp/dx} = \text{price elasticity of demand}$$

(When $|\eta| < 1$, the demand is inelastic. When $|\eta| > 1$, the demand is elastic.)

Typical Graphs of Revenue, Cost, and Profit Functions

Revenue Function

The low prices required to sell more units eventually result in a decreasing revenue.

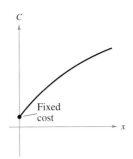

Cost Function

The total cost to produce x units includes the fixed cost.

Profit Function

The break-even point occurs when $R = C$.

Formulas from Business   (continued)

Marginals

$\dfrac{dR}{dx}$ = marginal revenue ≈ the *extra* revenue from selling one additional unit

$\dfrac{dC}{dx}$ = marginal cost ≈ the *extra* cost of producing one additional unit

$\dfrac{dP}{dx}$ = marginal profit ≈ the *extra* profit from selling one additional unit

Revenue Function

Formulas from Finance

Basic Terms

P = amount of deposit r = interest rate

n = number of times interest is compounded per year

t = number of years A = balance after t years

Compound Interest Formulas

1. Balance when interest is compounded n times per year: $A = P\left(1 + \dfrac{r}{n}\right)^{nt}$

2. Balance when interest is compounded continuously: $A = Pe^{rt}$

Effective Rate of Interest

$$r_{eff} = \left(1 + \frac{r}{n}\right)^{n} - 1$$

Present Value of a Future Investment

$$P = \frac{A}{\left(1 + \dfrac{r}{n}\right)^{nt}}$$

Balance of an Increasing Annuity After *n* Deposits of *P* per Year for *t* Years

$$A = P\left[\left(1 + \frac{r}{n}\right)^{nt} - 1\right]\left(1 + \frac{n}{r}\right)$$

Initial Deposit for a Decreasing Annuity with *n* Withdrawals of *W* per Year for *t* Years

$$P = W\left(\frac{n}{r}\right)\left\{1 - \left[\frac{1}{1 + (r/n)}\right]^{nt}\right\}$$

Monthly Installment *M* for a Loan of *P* Dollars over *t* Years at *r*% Interest

$$M = P\left\{\frac{r/12}{1 - \left[\dfrac{1}{1 + (r/12)}\right]^{12t}}\right\}$$

Amount of an Annuity

$$e^{rT}\int_{0}^{T} c(t)e^{-rt}\, dt$$

$c(t)$ is the continuous income function in dollars per year and T is the term of the annuity in years.

Answers to Selected Exercises

Chapter 1

Section 1.1   *(page 8)*

> **Skills Warm Up**   *(page 8)*
>
> **1.** $\frac{1}{2}$ **2.** -2 **3.** $3\sqrt{5}$ **4.** $2\sqrt{5}$ **5.** $5\sqrt{3}$
> **6.** $-\sqrt{2}$ **7.** $x = 19$ **8.** $y = 1$
> **9.** $x = -3, x = 9$ **10.** $y = -8, y = 4$

1.

3. (a)

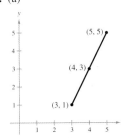

(b) $d = 2\sqrt{5}$
(c) Midpoint: $(4, 3)$

5. (a)

(b) $d = 4\sqrt{5}$
(c) Midpoint: $(-1, 3)$

7. (a)

(b) $d = 10$
(c) Midpoint: $(5, -8)$

9. (a)

(b) $d = 2\sqrt{10}$
(c) Midpoint: $\left(-\frac{1}{2}, -2\right)$

11. (a)

(b) $d = \sqrt{116.89}$ (c) Midpoint: $(0.25, 0.6)$

13. (a) $a = 4, b = 3, c = 5$
(b) $4^2 + 3^2 = 5^2$
15. (a) $a = 10, b = 3, c = \sqrt{109}$
(b) $10^2 + 3^2 = \left(\sqrt{109}\right)^2$

17. $d_1 = \sqrt{45}, d_2 = \sqrt{20},$
$d_3 = \sqrt{65}$
$d_1^{\,2} + d_2^{\,2} = d_3^{\,2}$

19. $d_1 = d_2 = d_3 = d_4$
$= \sqrt{5}$

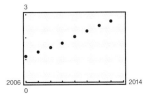

21. $x = 4, x = -2$ **23.** $y = \pm 2\sqrt{15}$ **25.** About 45 yd

27.

The number of Internet users increases each year.
29. (a) 14,500; 15,500; 16,500   (b) 5.3% decrease
31. (a) Revenue: \$1025.6 million
   Profit: \$61 million
(b) Actual 2012 revenue: \$1040.5 million
   Actual 2012 profit: \$57.3 million
(c) Yes, the revenue and profit increased in a linear pattern from 2011 to 2013.
(d) Expenses for 2011: \$734.1 million
   Expenses for 2012: \$983.2 million
   Expenses for 2013: \$1195.1 million
(e) Answers will vary.
33. (a)

(b) The larger the clinic, the more patients a doctor can treat.

35. $(-6, -6), (-4, -7), (-3, -5)$

37. $\left(\dfrac{3x_1 + x_2}{4}, \dfrac{3y_1 + y_2}{4}\right), \left(\dfrac{x_1 + x_2}{2}, \dfrac{y_1 + y_2}{2}\right),$

$\left(\dfrac{x_1 + 3x_2}{4}, \dfrac{y_1 + 3y_2}{4}\right)$

39. $x_1 + \left(\dfrac{x_2 - x_1}{3}\right) = \dfrac{3x_1 + x_2 - x_1}{3} = \dfrac{1}{3}(2x_1 + x_2)$

$y_1 + \left(\dfrac{y_2 - y_1}{3}\right) = \dfrac{3y_1 + y_2 - y_1}{3} = \dfrac{1}{3}(2y_1 + y_2)$

So, $\left(\dfrac{1}{3}[2x_1 + x_2], \dfrac{1}{3}[2y_1 + y_2]\right)$ is a point of trisection.

$\left(\dfrac{\frac{2}{3}x_1 + \frac{1}{3}x_2 + x_2}{2}, \dfrac{\frac{2}{3}y_1 + \frac{1}{3}y_2 + y_2}{2}\right)$

$= \left(\dfrac{\frac{2}{3}x_1 + \frac{4}{3}x_2}{2}, \dfrac{\frac{2}{3}y_1 + \frac{4}{3}y_2}{2}\right)$

$= \left(\dfrac{1}{3}x_1 + \dfrac{2}{3}x_2, \dfrac{1}{3}y_1 + \dfrac{2}{3}y_2\right)$

$= \left(\dfrac{1}{3}[x_1 + 2x_2], \dfrac{1}{3}[y_1 + 2y_2]\right)$

Section 1.2   *(page 19)*

Skills Warm Up   *(page 19)*

1. $y = \frac{1}{5}(x + 12)$ **2.** $y = x - 15$

3. $y = \dfrac{1}{x^3 + 2}$

4. $y = \pm\sqrt{x^2 + x - 6} = \pm\sqrt{(x + 3)(x - 2)}$

5. $y = -1 \pm \sqrt{9 - (x - 2)^2}$

6. $y = 5 \pm \sqrt{81 - (x + 6)^2}$ **7.** $y = -10$

8. $y = 5$ **9.** $y = -6$ **10.** $y = -1$

11. $(x - 2)(x - 1)$ **12.** $(x + 3)(x + 2)$

13. $\left(y - \frac{3}{2}\right)^2$ **14.** $\left(y - \frac{7}{2}\right)^2$

1. e   **2.** b   **3.** c   **4.** f   **5.** a   **6.** d

7.

9.

11.

13.

15.

17.

19.

21.
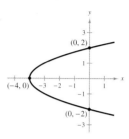

23. $(0, -3), \left(\frac{3}{2}, 0\right)$ **25.** $(0, -2), (-2, 0), (1, 0)$

27. $(-7, 0), (0, 0)$ **29.** $(-2, 0), (0, 2)$ **31.** $(0, 0)$

33. $x^2 + y^2 = 16$ **35.** $(x - 2)^2 + (y + 1)^2 = 9$

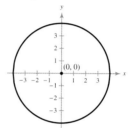

37. $(x + 1)^2 + (y - 1)^2 = 16$ **39.** $x^2 + y^2 = 100$

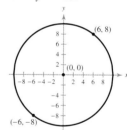

41. $(1, 1)$ **43.** $(1, 14), (-4, -1)$

45. $(0, 0), \left(\sqrt{2}, 2\sqrt{2}\right), \left(-\sqrt{2}, -2\sqrt{2}\right)$

47. $(-1, 0), (0, 1), (1, 0)$

49. 50,000 units   **51.** 193 units   **53.** 1250 units

55. (a) $C = 11.5x + 21,000;\ R = 19.9x$

(b) 2500 units

(c) 2619 units

57. $(10, 165)$

59. (a)

Year	2009	2010	2011	2012	2013
Revenue	0.82	1.52	2.31	3.35	4.52
Model	0.82	1.50	2.33	3.33	4.52

The model fits the data well.

(b) $14 billion

61. (a)

Year	2008	2009	2010
Degrees	747	793	854

Year	2011	2012	2016
Degrees	931	1024	1550

(b) Answers will vary.

(c) 2325; Yes, the number of associate's degrees should keep increasing over time.

63.

The greater the value of c, the steeper the line.

65.

$(-2.75, 0), (0, 1.815)$

67.

$(1.4780, 0), (12.8553, 0), (0, 2.3875)$

69.

$(0, 0.4167)$

71. Answers will vary.

Section 1.3   *(page 31)*

1. 2 **3.** 0 **5.** $m = 1, (0, 7)$ **7.** $m = -5, (0, 20)$

9. $m = -\frac{7}{6}, (0, 5)$ **11.** $m = 3, (0, -15)$

13. m is undefined; no y-intercept.   **15.** $m = 0, (0, 9)$

17.

19.

21.

23.

25.

27. $m = \frac{1}{4}$ **29.** $m = 3$ **31.** $m = -4$

33. m is undefined.   **35.** $m = 0$ **37.** $m = -\frac{24}{5}$

39. $m = 8$ **41.** $(0, 1), (1, 1), (3, 1)$

43. $(0, 10), (2, 4), (3, 1)$

45. $(3, -6), (9, -2), (12, 0)$ **47.** $(-8, 0), (-8, 2), (-8, 3)$

49. The points are not collinear. Explanations will vary.

51. The points are collinear. Explanations will vary.

53. $3x - 4y + 12 = 0$ **55.** $y - 7 = 0$

57. $4x + y + 6 = 0$ **59.** $3x - 12y - 8 = 0$

61. $y = 2x - 5$ **63.** $x - 2 = 0$

65. $y + 1 = 0$ **67.** $4x - y + 6 = 0$

69. $3x - 6y + 7 = 0$

71. $x - 3 = 0$ **73.** $y + 10 = 0$
75. (a) $x + y + 1 = 0$ (b) $x - y + 5 = 0$

77. (a) $6x + 8y - 3 = 0$ (b) $96x - 72y + 127 = 0$

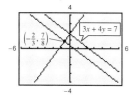

79. (a) $y = 0$ (b) $x + 1 = 0$

81. (a) $x - 1 = 0$ (b) $y - 1 = 0$

83. (a) The average salary increased the most from 2008 to 2009 and increased the least from 2010 to 2011.
 (b) $m = 1536$
 (c) The average salary increased $1536 per year over the 5 years between 2008 and 2013.

85. 12 ft

87. (a) $y = 22t + 5457$; The slope $m = 22$ indicates that the population increases by 22 thousand each year.
 (b) 5699 thousand (5,699,000)
 (c) 5709 thousand (5,709,000); The estimate was very close to the actual population.
 (d) The model could possibly be used to predict the population in 2018 if the population continues to grow at the same linear rate.

89. (a) $y = -205t + 1025$
 (b)
 (c) $410
 (d) 2.07 years

91. (a) Current job: $W = 0.07S + 2000$
  Job offer: $W = 0.05S + 2300$
 (b)

  (15,000, 3050); When your monthly sales are $15,000, your monthly wage is the same for both jobs.
 (c) No; Your monthly wage would be $3400 at your current job but only $3300 at the new job.

Quiz Yourself   *(page 34)*

1. (a)
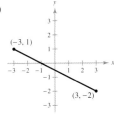
(-3, 1)
(3, -2)

(b) $d = 3\sqrt{5}$
(c) Midpoint: $(0, -0.5)$

2. (a)

$\left(\frac{1}{2}, 2\right)$
$\left(\frac{1}{4}, -\frac{3}{2}\right)$

(b) $d = \sqrt{12.3125}$
(c) Midpoint: $\left(\frac{3}{8}, \frac{1}{4}\right)$

3. (a)

(-12, 4)
(6, -2)

(b) $d = 6\sqrt{10}$
(c) Midpoint: $(-3, 1)$

4. $d_1 = \sqrt{5}$
$d_2 = \sqrt{45}$
$d_3 = \sqrt{50}$
$d_1{}^2 + d_2{}^2 = d_3{}^2$

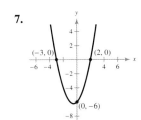
(2, 1) d_1 (4, 0)
d_2
d_3
(-1, -5)

5. 9901 thousand

6.

(0, 2)
(-0.4, 0)

7.

(-3, 0)   (2, 0)
(0, -6)

8.
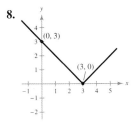
(0, 3)
(3, 0)

9. $x^2 + y^2 = 81$

(0, 0)

10. $(x + 1)^2 + y^2 = 36$
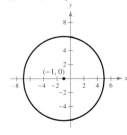
(-1, 0)

11. $(x - 2)^2 + (y + 2)^2 = 25$ **12.** 4735 units

(-1, 2)
(2, -2)

13. $y = -3$

14. $y = 2x - 1$
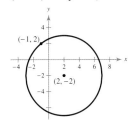

15. $y = -\frac{1}{3}x + 7$
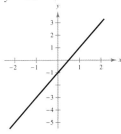

16. $y = -1.2x + 0.2$

17. $x = -2$

18. $y = 2$

19. (a) $y = -0.25x - 4.25$ (b) $y = 4x - 17$
20. 2019: \$2,270,000; 2022: \$2,622,500
21. $C = 0.56x + 218$
22. (a) $S = 1700t + 13{,}600$ (b) \$47,600

Section 1.4   *(page 44)*

Skills Warm Up   *(page 44)*

1. 20   **2.** -4 **3.** $x^2 + x - 6$

4. $x^3 + 9x^2 + 26x + 30$ **5.** $\dfrac{1}{x}$ **6.** $\dfrac{5x - 7}{x}$

7. $y = -2x + 17$ **8.** $y = \frac{6}{5}x^2 + \frac{1}{5}$

9. $y = 3 \pm \sqrt{5 + (x + 1)^2}$ **10.** $y = \pm\sqrt{4x^2 + 2}$

11. $y = 2x + \dfrac{1}{2}$ **12.** $y = \dfrac{x^3}{2} + \dfrac{1}{2}$

---

1. y is not a function of x.   **3.** y is a function of x.

5. y is a function of x.   **7.** y is a function of x.

9. y is not a function of x.   **11.** y is a function of x.

13. Domain: $(-\infty, \infty)$ **15.** Domain: $[-2, 2]$
Range: $(-\infty, \infty)$ Range: $[0, 2]$

17. Domain: $(-\infty, \infty)$ **19.** Domain: $(-\infty, 0) \cup (0, \infty)$
Range: $(-\infty, 4]$ Range: $y = -1$ or $y = 1$

21. Domain: $(-\infty, \infty)$ **23.** Domain:
Range: $(-\infty, \infty)$ $(-\infty, -4) \cup (-4, \infty)$
          Range: $(-\infty, 1) \cup (1, \infty)$

25. (a) -2 (b) 13 (c) $3x - 5$

27. (a) 5 (b) $-\dfrac{5}{3}$ (c) $\dfrac{1}{x + 4}$

29. $\Delta x + 2x - 5, \Delta x \neq 0$

31. $\dfrac{1}{\sqrt{5 + \Delta x} + \sqrt{5}},\ \Delta x \neq 0$

33. $-\dfrac{1}{(x + \Delta x - 2)(x - 2)},\ \Delta x \neq 0$

35. (a) $-x - 1$ (b) $5x - 9$ (c) $-6x^2 + 23x - 20$

(d) $\dfrac{2x - 5}{4 - 3x}$ (e) $3 - 6x$ (f) $19 - 6x$

37. (a) $x^2 + x$ (b) $x^2 - x + 2$

(c) $(x^2 + 1)(x - 1) = x^3 - x^2 + x - 1$

(d) $\dfrac{x^2 + 1}{x - 1}$ (e) $x^2 - 2x + 2$ (f) x^2

39. (a) 0 (b) 0 (c) $-\frac{1}{2}$ (d) 2

(e) $\sqrt{x^2 - 1}$ (f) $x - 1, x \geq 0$

41. $f^{-1}(x) = \frac{1}{4}x$
$f\big(f^{-1}(x)\big) = 4\big(\frac{1}{4}x\big) = x$
$f^{-1}(f(x)) = \frac{1}{4}(4x) = x$

43. $f^{-1}(x) = x - 12$
$f\big(f^{-1}(x)\big) = (x - 12) + 12 = x$
$f^{-1}(f(x)) = (x + 12) - 12 = x$

45. $f^{-1}(x) = \dfrac{x + 3}{2}$ **47.** $f^{-1}(x) = \dfrac{2}{3}(x - 1)$

49. $f^{-1}(x) = \sqrt[5]{x}$ **51.** $f^{-1}(x) = \dfrac{1}{x}$

53. $f^{-1}(x) = \sqrt{9 - x^2},\ \ 0 \leq x \leq 3$

55. $f^{-1}(x) = x^{3/2},\ \ x \geq 0$

57.

$f(x)$ is one-to-one. $f^{-1}(x) = \dfrac{3 - x}{7}$

59.   **61.**

$f(x)$ is not one-to-one. $f(x)$ is not one-to-one.

63. (a)   (b)

(c)   (d)

(e)   (f)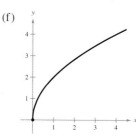

65. (a) $y = (x + 3)^2$ (b) $y = -(x + 6)^2 - 3$

67. (a) 2004: \$190 billion
2006: \$225 billion
2013: \$270 billion

(b) 2004: \$191.2 billion
2006: \$225.1 billion
2013: \$269.5 billion

The model fits the data well.

69. $R_T = R_1 + R_2 = -0.8t^2 - 7.22t + 1148, t = 9, 10, \ldots, 15$

71. (a) $C(x) = 2.89x + 8000$

(b) $\overline{C} = 2.89 + \dfrac{8000}{x}$

(c) More than 2000 units

73. (a) $C(x(t)) = 2800t + 500$

This function gives the production cost for t hours.

(b) \$11,700

(c) 6.25 hours

75. (a) $C(x) = 12.30x + 98,000$

(b) $R(x) = 17.98x$

(c) $P(x) = 5.68x - 98,000$

77. Answers will vary. Sample answer:

Domain of f: $[1, \infty)$

Range of f: $[0, \infty)$

$f^{-1}(x) = \sqrt{x} + 1$

Domain of f^{-1}: $[0, \infty)$

Range of f^{-1}: $[1, \infty)$

79. Answers will vary. Sample answer:

Domain of f: $[-4, \infty)$

Range of f: $[0, \infty)$

$f^{-1}(x) = x - 4$

Domain of f^{-1}: $[0, \infty)$

Range of f^{-1}: $[-4, \infty)$

81. Answers will vary. Sample answer:

Domain of f: $[0, \infty)$

Range of f: $(-\infty, 1]$

$f^{-1}(x) = \dfrac{\sqrt{-2(x - 1)}}{2}$

Domain of f^{-1}: $(-\infty, 1]$

Range of f^{-1}: $[0, \infty)$

83. Answers will vary. Sample answer:

Domain of f: $[-1, \infty)$

Range of f: $[-2, \infty)$

$f^{-1}(x) = x + 1$

Domain of f^{-1}: $[-2, \infty)$

Range of f^{-1}: $[-1, \infty)$

85. Answers will vary.

Section 1.5 *(page 57)*

Skills Warm Up *(page 57)*

1. $\frac{1}{3}x^2 + \frac{1}{6}x$ **2.** $x^2(x + 9)$ **3.** $x + 4$ **4.** $x + 6$

5. (a) 7   (b) $c^2 - 3c + 3$

(c) $x^2 + 2xh + h^2 - 3x - 3h + 3$

6. (a) -3 (b) 4   (c) $3t^2 + 4$ **7.** h **8.** 4

9. Domain: $(-\infty, 0) \cup (0, \infty)$

Range: $(-\infty, 0) \cup (0, \infty)$

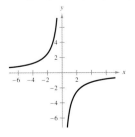

10. Domain: $[-5, 5]$

Range: $[0, 5]$

11. Domain: $(-\infty, \infty)$

Range: $[0, \infty)$

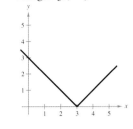

12. Domain:

$(-\infty, 0) \cup (0, \infty)$

Range: $-2, 2$

13. y is not a function of x.

14. y is a function of x.

1. (a) 4   (b) 1   **3.** (a) 1   (b) 3

5.

x	5.9	5.99	5.999	6
$f(x)$	2.96	2.996	2.9996	?

x	6.001	6.01	6.1
$f(x)$	3.0004	3.004	3.04

$\displaystyle\lim_{x \to 6} \dfrac{2x + 3}{5} = 3$

7.

x	3.9	3.99	3.999	4
$f(x)$	0.3448	0.3344	0.3334	?

x	4.001	4.01	4.1
$f(x)$	0.3332	0.3322	0.3226

$\displaystyle\lim_{x \to 4} \dfrac{x - 4}{x^2 - 5x + 4} = \dfrac{1}{3}$

9.

x	-0.1	-0.01	-0.001	0
$f(x)$	0.1252	0.1250	0.1250	?

x	0.001	0.01	0.1
$f(x)$	0.1250	0.1250	0.1248

$\displaystyle\lim_{x \to 0} \dfrac{\sqrt{x + 16} - 4}{x} = 0.125$

11.

x	-4.1	-4.01	-4.001	-4
$f(x)$	2.5	25	250	?

x	-3.999	-3.99	-3.9
$f(x)$	-250	-25	-2.5

The limit does not exist.

13. 6 **15.** -2 **17.** 49 **19.** 6

21. (a) 12 (b) 27 (c) $\frac{1}{3}$

23. (a) 4 (b) 48 (c) 256

25. -1 **27.** 0 **29.** 3 **31.** -2 **33.** $-\frac{3}{4}$ **35.** 2

37. -6 **39.** $\frac{7}{4}$ **41.** 12 **43.** 2 **45.** $2t - 5$

47. $\frac{1}{6}$ **49.** $\dfrac{1}{2\sqrt{5}}$ **51.** 2 **53.** Limit does not exist.

55. Limit does not exist. **57.** Limit does not exist.

59. (a) 1 (b) 1 (c) 1

61. (a) -1 (b) -1 (c) -1

63. (a) 6 (b) -6 (c) Limit does not exist.

65. $\lim\limits_{x\to -3^-} \dfrac{|x+3|}{x+3} = -1$, $\lim\limits_{x\to -3^+} \dfrac{|x+3|}{x+3} = 1$

67.

x	3	2.5	2.1	2.01
$f(x)$	0.6	1.33	7.32	74.81

x	2.001	2.0001	2
$f(x)$	749.81	7499.81	Undefined

The limit does not exist because f is unbounded as x approaches 2 from the right.

69.

x	-3	-2.5	-2.1	-2.01
$f(x)$	-1	-2	-10	-100

x	-2.001	-2.0001	-2
$f(x)$	-1000	$-10,000$	Undefined

The limit does not exist because f is unbounded as x approaches -2 from the left.

71.

Limit does not exist.

73.

$-\frac{17}{9} \approx -1.8889$

75. (a) $25,000 (b) 80%

(c) ∞; The cost function increases without bound as x approaches 100 from the left. Therefore, according to the model, it is not possible to remove 100% of the pollutants.

77. (a)

(b) For $x = 0.25$, $A \approx \$2685.06$.

For $x = \frac{1}{365}$, $A \approx \$2717.91$.

(c) $\lim\limits_{x\to 0^+} 1000(1 + 0.1x)^{10/x} = 1000e \approx \$2718.28$;

continuous compounding; The length x of the compounding period is approaching 0.

Section 1.6 *(page 67)*

Skills Warm Up *(page 67)*

1. $\dfrac{x+4}{x-8}$ **2.** $\dfrac{x+1}{x-3}$ **3.** $\dfrac{x+2}{2(x-3)}$ **4.** $\dfrac{x-4}{x-2}$

5. $x = 0, -7$ **6.** $x = -5, 1$ **7.** $x = -\frac{2}{3}, -2$

8. $x = 0, 3, -\frac{8}{3}$ **9.** 13 **10.** 3

---

1. Continuous; The function is a polynomial.

3. Not continuous $(x \neq \pm 4)$

5. Continuous; The rational function's domain is the set of real numbers.

7. Not continuous $(x \neq 3$ and $x \neq 5)$

9. Not continuous $(x \neq \pm 6)$

11. $(-\infty, 0)$ and $(0, \infty)$; Explanations will vary. There is a discontinuity at $x = 0$, because $f(0)$ is not defined.

13. $(-\infty, -1)$ and $(-1, \infty)$; Explanations will vary. There is a discontinuity at $x = -1$, because $f(-1)$ is not defined.

15. $(-\infty, \infty)$; Explanations will vary.

17. $(-\infty, -1)$, $(-1, 1)$, and $(1, \infty)$; Explanations will vary. There are discontinuities at $x = \pm 1$, because $f(\pm 1)$ are not defined.

19. $(-\infty, \infty)$; Explanations will vary.

21. $(-\infty, 4), (4, 5),$ and $(5, \infty)$; Explanations will vary. There are discontinuities at $x = 4$ and $x = 5$, because $f(4)$ and $f(5)$ are not defined.

23. $(-\infty, 4]$; Explanations will vary.

25. $[0, \infty)$; Explanations will vary.

27. $[-1, 3]$; Explanations will vary.

29. $(-\infty, 2), (2, \infty)$; Explanations will vary. There is a discontinuity at $x = 2$ because $\lim_{x \to 2} f(x)$ does not exist.

31. $(-\infty, -1)$ and $(-1, \infty)$; Explanations will vary. There is a discontinuity at $x = -1$, because $f(-1)$ is not defined.

33. $[-3, \infty)$; Explanations will vary.

35. Continuous on all intervals $\left(\dfrac{c}{2}, \dfrac{c}{2} + \dfrac{1}{2}\right)$, where c is an integer.

Explanations will vary. There are discontinuities at $x = \dfrac{c}{2}$, where c is an integer, because $\lim_{x \to c} f\left(\dfrac{c}{2}\right)$ does not exist.

37. Continuous on all intervals $(c, c + 1)$, where c is an integer. Explanations will vary. There are discontinuities at $x = c$, where c is an integer, because $\lim_{x \to c} f(c)$ does not exist.

39. $(1, \infty)$; Explanations will vary.

41.

Continuous on $(-\infty, 4)$ and $(4, \infty)$; removable discontinuity at $x = 4$

43.

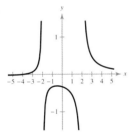

Continuous on $(-\infty, -2), (-2, 2),$ and $(2, \infty)$; nonremovable discontinuities at $x = -2$ and $x = 2$

45.

Continuous on $(-\infty, 0)$ and $(0, \infty)$; nonremovable discontinuity at $x = 0$

47. Continuous   **49.** Nonremovable discontinuity at $x = 2$

51.

Not continuous at $x = 2$ and $x = -1$, because $f(-1)$ and $f(2)$ are not defined.

53.

Not continuous at $x = 3$, because $\lim_{x \to 3} f(3)$ does not exist.

55.

Not continuous at all integers c, because $\lim_{x \to c} f(c)$ does not exist.

57. $a = 2$

59.

The graph of $f(x) = \dfrac{x^2 + x}{x}$ appears to be continuous on $[-4, 4]$, but f is not continuous at $x = 0$. Explanations will vary.

61. (a) $[0, 100)$; Explanations will vary.

(b)

Continuous; Explanations will vary.

(c) $6 million

63.

There are nonremovable discontinuities at $t = 1, 2, 3, 4, 5,$ and 6.

Explanations will vary.

65. (a)

> There are nonremovable discontinuities at $x = 1, 2,$ and 3. Explanations will vary.

(b) \$0.91

67. (a) No; The graph has nonremovable discontinuities at $t = \frac{1}{4}, \frac{1}{2}, \frac{3}{4}, 1, \frac{5}{4}, \ldots$

(b) \$8448.69

(c) \$11,379.17

69. (a) $C(x) = 3.50 - 1.90[\![1 - x]\!], x > 0$

(b)

> The graph has nonremovable discontinuities at $x = 1, 2, 3, \ldots$

Review Exercises for Chapter 1   *(page 74)*

1.

3. $\sqrt{29}$ **5.** $3\sqrt{2}$ **7.** $\dfrac{\sqrt{17}}{2}$ **9.** $(7, 4)$

11. $(-8, 6)$ **13.** $(1.15, -1.45)$

15. $P = R - C$; The difference in the heights of the bars that represent revenue and cost is equal to the height of the bar that represents profit.

17. $(3, 6), (4, 7), (6, 4)$

19.   **21.**

23.   **25.**

27.

29. $\left(-\frac{3}{4}, 0\right), (0, -3)$ **31.** $(-4, 0), (2, 0), (0, -8)$

33. $x^2 + y^2 = 64$ **35.** $(x - 2)^2 + y^2 = 9$

37. $(-2, 9)$ **39.** $(-1, -1), (0, 0), (1, 1)$

41. (a) $C = 10x + 200$ **43.** $p = \$46.40$
$R = 14x$ $x = 5000$ units

(b) 50 shirts

(c) 200 shirts

45. Slope: -1 **47.** Slope: 4
y-intercept: $(0, 12)$ y-intercept: $(0, -6)$

49. Slope: 0 (horizontal line)   **51.** Slope: $-\frac{2}{5}$
y-intercept: $\left(0, \frac{7}{4}\right)$ y-intercept: $(0, -1)$

    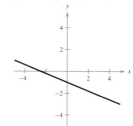

53. $\frac{6}{7}$ **55.** 0

57. $y = -2x + 5$ **59.** $y = -4$

61. $y = 2x - 9$

63. $x = 5$

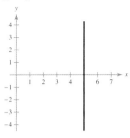

65. (a) $5x - 8y + 63 = 0$ (b) $x - 5y + 33 = 0$
   (c) $2x + y = 0$ (d) $2x + 3y - 12 = 0$

67. (a) $x = -10p + 1070$ (b) 725 units   (c) 650 units

69. y is a function of x.   **71.** y is not a function of x.

73. Domain: $(-\infty, \infty)$ **75.** Domain: $[-5, \infty)$
   Range: $(-\infty, \infty)$ Range: $[0, \infty)$

77. Domain: $(-\infty, -1) \cup (-1, 1) \cup (1, \infty)$
   Range: $(-\infty, 0) \cup \left(0, \frac{1}{2}\right) \cup \left(\frac{1}{2}, \infty\right)$

79. (a) 7   (b) -11 (c) $3x + 7$

81. (a) $x^2 + 3x + 1$ (b) $x^2 - 3x + 11$
   (c) $3x^3 - 5x^2 + 18x - 30$ (d) $\dfrac{6 + x^2}{3x - 5}$
   (e) $9x^2 - 30x + 31$ (f) $3x^2 + 13$

83.

f is one-to-one.
$f^{-1}(x) = \frac{1}{4}(x + 3)$

85.

f is one-to-one.
$f^{-1}(x) = x^2 + \frac{1}{2}$

87.

f does not have an inverse function.

89.

x	0.9	0.99	0.999	1	1.001	1.01	1.1
$f(x)$	0.6	0.96	0.996	?	1.004	1.04	1.4

$\lim\limits_{x \to 1} (4x - 3) = 1$

91.

x	-0.1	-0.01	-0.001	0
$f(x)$	0.205	0.204	0.204	?

x	0.001	0.01	0.1
$f(x)$	0.204	0.204	0.203

$\lim\limits_{x \to 0} \dfrac{\sqrt{x + 6} - \sqrt{6}}{x} \approx 0.204$

93. 8   **95.** 7   **97.** $-\frac{2}{5}$ **99.** Limit does not exist.

101. $-\frac{1}{4}$ **103.** $-\infty$ **105.** $-\dfrac{\sqrt{3}}{6}$

107. 5   **109.** $3x^2 - 1$

111. $(-\infty, \infty)$; For any c on the real number line, $F(c)$ is defined, $\lim\limits_{x \to c} f(x)$ exists, and $\lim\limits_{x \to c} f(x) = f(c)$.

113. $(-\infty, -4)$ and $(-4, \infty)$; $f(-4)$ is undefined.

115. $(-\infty, -1)$ and $(-1, \infty)$; $f(-1)$ is undefined.

117. $[8, \infty)$; For any $c > 8$ on the real number line, $F(c)$ is defined, $\lim\limits_{x \to c} f(x)$ exists, and $\lim\limits_{x \to c} f(x) = f(c)$.

119. Continuous on all intervals $(c, c + 1)$, where c is an integer; $\lim\limits_{x \to 0} f(c)$ does not exist.

121. $(-\infty, 0)$ and $(0, \infty)$; $\lim\limits_{x \to 0} f(x)$ does not exist.

123. $a = 2$

125. (a)

Explanations will vary. The function is defined for all values of x greater than zero. The function is discontinuous when $x = 5$, $x = 10$, and $x = 15$.

   (b) \$49.90

127. (a)

There are nonremovable discontinuities at $x = 24, 48, 72, \ldots$.

   (b) \$31

Chapter Test   *(page 78)*

1. (a) $d = 5\sqrt{2}$ (b) Midpoint: $(-1.5, 1.5)$
   (c) $m = -1$ (d) $y = -x$
   (e)

2. (a) $d = 2.5$ (b) Midpoint: $(1.25, 2)$
   (c) $m = 0$ (d) $y = 2$
   (e)

3. (a) $d = 2\sqrt{10}$ (b) Midpoint: $(-1, 2)$
(c) $m = \frac{1}{3}$ (d) $y = \frac{1}{3}x + \frac{7}{3}$
(e)

4. $(5.5, 53.45)$
5. $m = \frac{1}{5}$; $(0, -2)$

6. m is undefined; no y-intercept

7. $m = -2.5$; $(0, 6.25)$

8. $y = -\frac{1}{6}x - \frac{3}{2}$ **9.** $y = \frac{5}{2}x - 4$
10. (a)

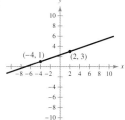

(b) Domain: $(-\infty, \infty)$
   Range: $(-\infty, \infty)$
(c) $f(-3) = -1$; $f(-2) = 1$; $f(3) = 11$
(d) The function is one-to-one.

11. (a)

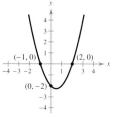

(b) Domain: $(-\infty, \infty)$
   Range: $(-2.25, \infty)$
(c) $f(-3) = 10$; $f(-2) = 4$; $f(3) = 4$
(d) The function is not one-to-one.

12. (a)

(b) Domain: $[-9, \infty)$
   Range: $[0, \infty)$
(c) $f(-3) = \sqrt{6}$; $f(-2) = \sqrt{7}$; $f(3) = 2\sqrt{3}$
(d) The function is one-to-one.

13. $f^{-1}(x) = \frac{1}{4}x - \frac{3}{2}$ **14.** $f^{-1}(x) = -\frac{1}{3}x^3 + \frac{8}{3}$
15. -1 **16.** Limit does not exist.   **17.** 2   **18.** $\frac{1}{6}$
19. $(-\infty, 6)$ and $(6, \infty)$; Explanations will vary. There is a discontinuity at $x = 6$, because $f(6)$ is not defined.
20. $(-\infty, 5]$; Explanations will vary.
21. $(-\infty, \infty)$; Explanations will vary.
22. (a)

Year	2008	2009	2010
Actual	8924	14,265	14,825
Model	9026	13,956	15,046

Year	2011	2012	2013
Actual	13,747	12,506	11,460
Model	13,924	12,219	11,558

The model fits the data well.
(b) 80,899 thousand (80,899,000); No, this would represent an increase of over 600% in a five-year period.

Chapter 2

Section 2.1   *(page 88)*

Skills Warm Up   *(page 88)*

1. $x = 3$ **2.** $y = 2$ **3.** $y = -2x + 7$

4. $y = 3x - 4$ **5.** $2x$ **6.** $3x^2$ **7.** $\dfrac{1}{x^2}$

8. $2x$ **9.** $(-\infty, \infty)$ **10.** $(-\infty, 1) \cup (1, \infty)$

11. $(-\infty, \infty)$ **12.** $(-\infty, 0) \cup (0, \infty)$

1.

3.

5.

7. $m = 1$ **9.** $m = 0$ **11.** $m = -\frac{1}{3}$

13. 2009: $m \approx 188$

2011: $m \approx 375$

The slope is the rate of change of revenue at the given point in time.

15. $t = 3$: $m \approx 8$

$t = 7$: $m \approx 1$

$t = 10$: $m \approx -10$

The slope is the rate of change of the average temperature at the given point in time.

17. $f'(x) = 0$

$f'(0) = 0$

19. $f'(x) = -4$

$f'(3) = -4$

21. $f'(x) = 4x$

$f'(2) = 8$

23. $f'(x) = 3x^2 - 4$

$f'(-1) = -1$

25. $f'(x) = \dfrac{1}{\sqrt{x}}$

$f'(4) = \dfrac{1}{2}$

27. $f(x) = 3$

$f(x + \Delta x) = 3$

$f(x + \Delta x) - f(x) = 0$

$\dfrac{f(x + \Delta x) - f(x)}{\Delta x} = 0$

$\displaystyle\lim_{\Delta x \to 0} \dfrac{f(x + \Delta x) - f(x)}{\Delta x} = 0$

29. $f(x) = -5x$

$f(x + \Delta x) = -5x - 5\Delta x$

$f(x + \Delta x) - f(x) = -5\Delta x$

$\dfrac{f(x + \Delta x) - f(x)}{\Delta x} = -5$

$\displaystyle\lim_{\Delta x \to 0} \dfrac{f(x + \Delta x) - f(x)}{\Delta x} = -5$

31. $g(s) = \dfrac{1}{3}s + 2$

$g(s + \Delta s) = \dfrac{1}{3}s + \dfrac{1}{3}\Delta s + 2$

$g(s + \Delta s) - g(s) = \dfrac{1}{3}\Delta s$

$\dfrac{g(s + \Delta s) - g(s)}{\Delta s} = \dfrac{1}{3}$

$\displaystyle\lim_{\Delta s \to 0} \dfrac{g(s + \Delta s) - g(s)}{\Delta s} = \dfrac{1}{3}$

33. $f(x) = 4x^2 - 5x$

$f(x + \Delta x) = 4x^2 + 8x\Delta x + 4(\Delta x)^2 - 5x - 5\Delta x$

$f(x + \Delta x) - f(x) = 8x\Delta x + 4(\Delta x)^2 - 5\Delta x$

$\dfrac{f(x + \Delta x) - f(x)}{\Delta x} = 8x + 4\Delta x - 5$

$\displaystyle\lim_{\Delta x \to 0} \dfrac{f(x + \Delta x) - f(x)}{\Delta x} = 8x - 5$

35. $h(t) = \sqrt{t - 3}$

$h(t + \Delta t) = \sqrt{t + \Delta t - 3}$

$h(t + \Delta t) - h(t) = \sqrt{t + \Delta t - 3} - \sqrt{t - 3}$

$\dfrac{h(t + \Delta t) - h(t)}{\Delta t} = \dfrac{1}{\sqrt{t + \Delta t - 3} + \sqrt{t - 3}}$

$\displaystyle\lim_{\Delta t \to 0} \dfrac{h(t + \Delta t) - h(t)}{\Delta t} = \dfrac{1}{2\sqrt{t - 3}}$

37. $f(t) = t^3 - 12t$

$f(t + \Delta t) = t^3 + 3t^2\Delta t + 3t(\Delta t)^2$

$\quad + (\Delta t)^3 - 12t - 12\Delta t$

$f(t + \Delta t) - f(t) = 3t^2\Delta t + 3t(\Delta t)^2 + (\Delta t)^3 - 12\Delta t$

$\dfrac{f(t + \Delta t) - f(t)}{\Delta t} = 3t^2 + 3t\Delta t + (\Delta t)^2 - 12$

$\displaystyle\lim_{\Delta t \to 0} \dfrac{f(t + \Delta t) - f(t)}{\Delta t} = 3t^2 - 12$

39. $f(x) = \dfrac{1}{x + 2}$

$f(x + \Delta x) = \dfrac{1}{x + \Delta x + 2}$

$f(x + \Delta x) - f(x) = \dfrac{-\Delta x}{(x + \Delta x + 2)(x + 2)}$

$\dfrac{f(x + \Delta x) - f(x)}{\Delta x} = \dfrac{-1}{(x + \Delta x + 2)(x + 2)}$

$\displaystyle\lim_{\Delta x \to 0} \dfrac{f(x + \Delta x) - f(x)}{\Delta x} = -\dfrac{1}{(x + 2)^2}$

41. $y = 2x - 2$ **43.** $y = -6x - 3$

45. $y = \dfrac{x}{4} + 2$

(4, 3)

47. $y = -5x - 2$

$\left(-\frac{1}{5}, -1\right)$

49. $y = -x + 1$ **51.** $y = -9x + 18, \ y = -9x - 18$

53. y is differentiable for all $x \neq -3$.
At $(-3, 0)$ the graph has a node.

55. y is differentiable for all $x \neq -\frac{1}{2}$.
At $\left(-\frac{1}{2}, 0\right)$ the graph has a vertical tangent.

57. y is differentiable for all $x = \pm 2$.
The function is not defined at $x = \pm 2$.

59. $f(x) = -3x + 2$

61.

$f'(x) = \frac{1}{2}x$

x	-2	$-\frac{3}{2}$	-1	$-\frac{1}{2}$
$f(x)$	1	0.5625	0.25	0.0625
$f'(x)$	-1	-0.75	-0.5	-0.25

x	0	$\frac{1}{2}$	1	$\frac{3}{2}$	2
$f(x)$	0	0.0625	0.25	0.5625	1
$f'(x)$	0	0.25	0.5	0.75	1

63.

$f'(x) = -\frac{3}{2}x^2$

x	-2	$-\frac{3}{2}$	-1	$-\frac{1}{2}$
$f(x)$	4	1.6875	0.5	0.0625
$f'(x)$	-6	-3.375	-1.5	-0.375

x	0	$\frac{1}{2}$	1	$\frac{3}{2}$	2
$f(x)$	0	-0.0625	-0.5	-1.6875	-4
$f'(x)$	0	-0.375	-1.5	-3.375	-6

65. $f'(x) = 2x - 4$

The x-intercept of the derivative indicates a point of horizontal tangency for f.

67. $f'(x) = 3x^2 - 3$

The x-intercepts of the derivative indicate points of horizontal tangency for f.

69. True   **71.** True

73.

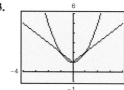

The graph of f is smooth at $(0, 1)$, but the graph of g has a sharp point at $(0, 1)$. The function f is differentiable at $x = 0$.

Section 2.2   *(page 100)*

Skills Warm Up   *(page 100)*

1. (a) 8   (b) 100   (c) $\frac{3}{2}$

2. (a) $\frac{1}{36}$ (b) $\frac{1}{32}$ (c) $\frac{1}{64}$

3. $4x(3x^2 + 1)$ **4.** $\frac{3}{2}x^{1/2}(x^{3/2} - 1)$ **5.** $\dfrac{1}{4x^{3/4}}$

6. $x^2 - \dfrac{1}{x^{1/2}} + \dfrac{1}{3x^{2/3}}$ **7.** $0, -\dfrac{2}{3}$

8. $0, \pm 1$ **9.** $-10, 2$ **10.** $2, \frac{4}{3}$

1. 0   **3.** $5x^4$ **5.** $9x^2$ **7.** $\frac{10}{3}x^3$ **9.** 4   **11.** $-3x^2$

13. $8x - 3$ **15.** $-6t + 2$ **17.** $16t^3 - 4t + 1$

19. $\dfrac{2}{3\sqrt[3]{x}}$ **21.** $\dfrac{16}{3}t^{1/3}$ **23.** $-\dfrac{8}{x^3} + 4x$

25. Function: $y = \dfrac{2}{7x^4}$

Rewrite: $y = \dfrac{2}{7}x^{-4}$

Differentiate: $y' = -\dfrac{8}{7}x^{-5}$

Simplify: $y' = -\dfrac{8}{7x^5}$

27. Function: $y = \dfrac{1}{(4x)^3}$

Rewrite: $y = \dfrac{1}{64}x^{-3}$

Differentiate: $y' = -\dfrac{3}{64}x^{-4}$

Simplify: $y' = -\dfrac{3}{64x^4}$

29. Function: $y = \dfrac{4}{(2x)^{-5}}$

Rewrite: $y = 128x^5$

Differentiate: $y' = 128(5)x^4$

Simplify: $y' = 640x^4$

31. Function: $y = 6\sqrt{x}$

Rewrite: $y = 6x^{1/2}$

Differentiate: $y' = 6\left(\dfrac{1}{2}\right)x^{-1/2}$

Simplify: $y' = \dfrac{3}{\sqrt{x}}$

33. Function: $y = \dfrac{1}{7\sqrt[6]{x}}$

Rewrite: $y = \dfrac{1}{7}x^{-1/6}$

Differentiate: $y' = \dfrac{1}{7}\left(-\dfrac{1}{6}\right)x^{-7/6}$

Simplify: $y' = -\dfrac{1}{42\sqrt[6]{x^7}}$

35. Function: $y = \sqrt[5]{8x}$

Rewrite: $y = 8^{1/5}x^{1/5}$

Differentiate: $y' = 8^{1/5}\left(\dfrac{1}{5}x^{-4/5}\right)$

Simplify: $y' = \dfrac{\sqrt[5]{8x}}{5x}$

37. $\dfrac{3}{2}$ **39.** -128 **41.** -11 **43.** -2

45. (a) $y = 2x - 2$ **47.** (a) $y = \dfrac{8}{15}x + \dfrac{22}{15}$

(b) and (c)   (b) and (c)

dy/dx=1.999992   dy/dx=.53333344

49. (a) $y = 36x - 54$

(b) and (c)

dy/dx=36.000003

51. $2x + \dfrac{4}{x^2} + \dfrac{6}{x^3}$ **53.** $2x - 2 + \dfrac{8}{x^5}$ **55.** $\dfrac{4}{5x^{1/5}} + 1$

57. $3x^2 + 1$ **59.** $\dfrac{2x^3 - 6}{x^3}$ **61.** $\dfrac{4x^3 - 2x - 10}{x^3}$

63. $(0, 3), (1, 2), (-1, 2)$ **65.** $(-5, -12.5)$ **67.** $(2, 7)$

69. (a)   (b) $f'(1) = g'(1) = 3$

(c)   (d) $f' = g' = 3x^2$ for every value of x.

71. $f'(x) = g'(x)$ **73.** $-5f'(x) = g'(x)$

75. (a) 2009: $m \approx 186.8$; 2011: $m \approx 386.8$

(b) These results are close to the estimates in Exercise 13 in Section 2.1.

(c) The slope of the graph at time t is the rate at which sales are increasing in millions of dollars per year.

77. (a) The men and women who seem to suffer most from migraines are those between 30 and 40 years old. More females than males suffer from migraines. Fewer people whose incomes are greater than or equal to $30,000 suffer from migraines than people whose incomes are less than $10,000.

(b) The derivatives are positive up to approximately 37 years old and negative after about 37 years of age. The percent of adults suffering from migraines increases up to about 37 years old, then decreases. The units of the derivative are percents of adults suffering from migraines per year.

79. $C = 7.75x + 500$

$C' = 7.75$, which equals the marginal cost.

81. $(0.11, 0.14), (1.84, -10.49)$

83. False. Let $f(x) = x$ and $g(x) = x + 1$.

Section 2.3   *(page 113)*

Skills Warm Up   *(page 113)*

1. 3   **2.** -6 **3.** -3 **4.** 2.4

5. $y' = 8x - 2$ **6.** $s' = -6t^2 + 16t - 7$

7. $s' = -32t + 24$ **8.** $y' = -32x + 54$

9. $A' = -\dfrac{3}{5}r^2 + \dfrac{3}{5}r + \dfrac{1}{2}$ **10.** $y' = 2x^2 - 4x + 7$

11. $y' = 12 - \dfrac{x}{2500}$ **12.** $y' = 74 - \dfrac{3x^2}{10,000}$

---

1. (a) $9.5 billion/yr   (b) $7.5 billion/yr

(c) $10.2 billion/yr   (d) $13.2 billion/yr

(e) $17.2 billion/yr   (f) $12.2 billion/yr

(g) $13.7 billion/yr   (h) $15.3 billion/yr

3. Average rate: 3   **5.** Average rate: -4

Instantaneous rates: Instantaneous rates:

$f'(1) = f'(2) = 3$ $h'(-2) = -8, h'(2) = 0$

7. Average rate: $\dfrac{45}{7}$ **9.** Average rate: $-\dfrac{1}{5}$

Instantaneous rates: Instantaneous rates:

$f'(1) = 4, f'(8) = 8$ $f'(1) = -1, f'(5) = -\dfrac{1}{25}$

CHAPTER 2

11. Average rate: -9

Instantaneous rates:

$f'(-2) = -24$, $f'(-1) = 0$

13. (a) -450

The number of visitors to the park is decreasing at an average rate of 450 thousand people per month from September to December.

(b) Answers will vary. Sample answer: $[4, 11]$; Both the instantaneous rate of change at $t = 8$ and the average rate of change over $[4, 11]$ are about 0.

15. (a) Average rate: 14 ft/sec

Instantaneous rates: $s'(0) = 30$ ft/sec

$s'(1) = -2$ ft/sec

(b) Average rate: -18 ft/sec

Instantaneous rates: $s'(1) = -2$ ft/sec

$s'(2) = -34$ ft/sec

(c) Average rate: -50 ft/sec

Instantaneous rates: $s'(2) = -34$ ft/sec

$s'(3) = -66$ ft/sec

(d) Average rate: -82 ft/sec

Instantaneous rates: $s'(3) = -66$ ft/sec

$s'(4) = -98$ ft/sec

17. (a) -80 ft/sec

(b) $s'(2) = -64$ ft/sec

$s'(3) = -96$ ft/sec

(c) $\dfrac{\sqrt{555}}{4} \approx 5.89$ sec

(d) $-8\sqrt{555} \approx -188.5$ ft/sec

19. 9800 dollars   **21.** $470 - 0.5x$ dollars, $0 \le x \le 940$

23. $50 - x$ dollars   **25.** $-18x^2 + 16x + 200$ dollars

27. $-4x + 72$ dollars   **29.** $0.0039x^2 + 12$ dollars

31. (a) \$0.60   (b) \$0.58

(c) The results are nearly the same.

33. (a) \$13.00   (b) \$12.96

(c) The results are nearly the same.

35. (a) $P(0) = 16{,}994{,}000$ people

$P(3) = 17{,}581{,}370$ people

$P(6) = 18{,}199{,}880$ people

$P(9) = 18{,}849{,}530$ people

$P(12) = 19{,}530{,}320$ people

$P(15) = 20{,}242{,}250$ people

$P(18) = 20{,}985{,}320$ people

$P(21) = 21{,}759{,}530$ people

The population is increasing from 1990 to 2011.

(b) $\dfrac{dP}{dt} = 3.46t + 190.6$

dP/dt represents the population growth rate.

(c) $P'(0) = 190{,}600$ people/yr

$P'(3) = 200{,}980$ people/yr

$P'(6) = 211{,}360$ people/yr

$P'(9) = 221{,}740$ people/yr

$P'(12) = 232{,}120$ people/yr

$P'(15) = 242{,}500$ people/yr

$P'(18) = 252{,}880$ people/yr

$P'(21) = 263{,}260$ people/yr

The rate of growth is increasing.

37. (a) $TR = -10Q^2 + 160Q$

(b) $(TR)' = MR = -20Q + 160$

(c)

Q	0	2	4	6	8	10
Model	160	120	80	40	0	-40
Table	—	130	90	50	10	-30

Answers will vary.

39. (a) $P = -0.0025x^2 + 2.65x - 25$

(b)

When $x = 300$, slope is positive.

When $x = 530$, slope is 0.

When $x = 700$, slope is negative.

(c) $P'(300) = 1.15$

$P'(530) = 0$

$P'(700) = -0.85$

41. (a) $C = \dfrac{39{,}000}{x}$

(b) $\dfrac{dC}{dx} = \dfrac{-39{,}000}{x^2}$

This is the rate of change of fuel cost.

(c)

x	10	15	20	25
C	3900	2600	1950	1560
dC/dx	-390	-173.33	-97.5	-62.4

x	30	35	40
C	1300	1114.29	975
dC/dx	-43.33	-31.84	-24.38

(d) The driver who gets 15 mi/gal; The value of dC/dx is a greater savings for $x = 15$ than for $x = 35$.

43. (a) \$445.37 per year

(b) \$702.73 per year

(c) \$840.07 per year

(d) The average rate of change from 2004 to 2006 is a better estimate because the data are closer to the year in question.

Section 2.4   *(page 124)*

Skills Warm Up   *(page 124)*

1. $2(3x^2 + 7x + 1)$ **2.** $4x^2(6 - 5x^2)$

3. $8x^2(x^2 + 2)^3 + (x^2 + 4)$

4. $(2x)(2x + 1)[2x + (2x + 1)^3]$

5. $\dfrac{23}{(2x + 7)^2}$ **6.** $-\dfrac{x^2 + 8x + 4}{(x^2 - 4)^2}$

7. $-\dfrac{2(x^2 + x - 1)}{(x^2 + 1)^2}$ **8.** $\dfrac{4(3x^4 - x^3 + 1)}{(1 - x^4)^2}$

9. $\dfrac{4x^3 - 3x^2 + 3}{x^2}$ **10.** $\dfrac{x^2 - 2x + 4}{(x-1)^2}$

11. 11 **12.** 0 **13.** $-\dfrac{1}{14}$ **14.** $\dfrac{17}{4}$

1. $f'(x) = (2x - 3)(-5) + (2)(1 - 5x) = -20x + 17$

3. $f'(x) = (6x - x^2)(3) + (6 - 2x)(4 + 3x)$
   $= -9x^2 + 28x + 24$

5. $f'(x) = x(2x) + 1(x^2 + 3) = 3x^2 + 3$

7. $h'(x) = \left(\dfrac{2}{x} - 3\right)(2x) + \left(-\dfrac{2}{x^2}\right)(x^2 + 7) = -\dfrac{14}{x^2} - 6x + 2$

9. $g'(x) = (x^2 - 4x + 3)(1) + (2x - 4)(x - 2)$
   $= 3x^2 - 12x + 11$

11. $h'(x) = \dfrac{(x - 5)(1) - (x)(1)}{(x - 5)^2} = -\dfrac{5}{(x - 5)^2}$

13. $f'(t) = \dfrac{(3t + 1)(4t) - (2t^2 - 3)3}{(3t + 1)^2} = \dfrac{6t^2 + 4t + 9}{(3t + 1)^2}$

15. $f'(t) = \dfrac{(t^2 - 8)(1) - (t + 6)(2t)}{(t^2 - 8)^2} = \dfrac{-t^2 - 12t - 8}{(t^2 - 8)^2}$

17. $f'(x) = \dfrac{(2x - 1)(2x + 6) - (x^2 + 6x + 5)(2)}{(2x - 1)^2}$
   $= \dfrac{2x^2 - 2x - 16}{(2x - 1)^2}$

19. $f'(x) = \dfrac{(3x - 1)\left(-\dfrac{2}{x^2}\right) - \left(6 + \dfrac{2}{x}\right)(3)}{(3x - 1)^2}$
   $= \dfrac{-18x^2 - 12x + 2}{x^2(3x - 1)^2}$

21. Function: $f(x) = \dfrac{x^3 + 6x}{3}$

Rewrite: $f(x) = \dfrac{x^3}{3} + 2x$

Differentiate: $f'(x) = x^2 + 2$
Simplify: $f'(x) = x^2 + 2$

23. Function: $y = \dfrac{7x^2}{5}$

Rewrite: $y = \dfrac{7}{5}x^2$

Differentiate: $y' = \dfrac{7}{5} \cdot 2x$

Simplify: $y' = \dfrac{14}{5}x$

25. Function: $y = \dfrac{7}{3x^3}$

Rewrite: $y = \dfrac{7}{3}x^{-3}$

Differentiate: $y' = -7x^{-4}$

Simplify: $y' = -\dfrac{7}{x^4}$

27. Function: $y = \dfrac{4x^2 - 3x}{8\sqrt{x}}$

Rewrite: $y = \dfrac{1}{2}x^{3/2} - \dfrac{3}{8}x^{1/2},\ x \neq 0$

Differentiate: $y' = \dfrac{3}{4}x^{1/2} - \dfrac{3}{16}x^{-1/2}$

Simplify: $y' = \dfrac{3}{4}\sqrt{x} - \dfrac{3}{16\sqrt{x}}$

29. Function: $y = \dfrac{x^2 - 4x + 3}{2(x - 1)}$

Rewrite: $y = \dfrac{1}{2}(x - 3),\ x \neq 1$

Differentiate: $y' = \dfrac{1}{2}(1),\ x \neq 1$

Simplify: $y' = \dfrac{1}{2},\ x \neq 1$

In Exercises 31–45, the differentiation rule(s) used may vary. A sample answer is provided.

31. $10x^4 + 12x^3 - 3x^2 - 18x - 15$; Product Rule

33. 2; Constant Multiple Rule

35. $\dfrac{x^4 - 6x^2 - 4x - 3}{(x^2 - 1)^2}$; Quotient Rule **37.** 1; Power Rule

39. $12t^2(2t^3 - 1)$; Product Rule

41. $\dfrac{3s^2 - 2s - 5}{2s^{3/2}}$; Quotient Rule

43. $\dfrac{12x^2 + 12x - 2}{(4x + 2)^2}$; Quotient Rule

45. $6x^2 + 22x - 15$; Product Rule

47. $y = 3x + 3$ **49.** $y = -7x + 4$

51. $y = \dfrac{1}{5}x + \dfrac{6}{5}$ **53.** $y = \dfrac{31}{5}x + \dfrac{26}{5}$

55. $(0, 0), (2, 4)$ **57.** $(0, 0), (\sqrt[3]{-4}, -2.117)$

59. **61.**

63. -1.87 units/dollar **65.** 31.55 bacteria/h

67. (a) -0.480/wk (b) 0.120/wk (c) 0.015/wk
   Each rate in parts (a), (b), and (c) is the rate at which the level of oxygen in the pond is changing at that particular time.

69. (a)

(b) At $x = 6.683$, $\dfrac{C}{x} = \dfrac{dC}{dx} \approx 20.50$.

So, the point of intersection is $(6.683, 20.50)$. At this point the average cost is at a minimum.

71. (a) -38.125 (b) -10.37 (c) -3.80

Increasing the order size reduces the cost per item; An order size of 2000 should be chosen because the cost per item is the least when $x = 20$.

73. $f'(2) = 0$ **75.** $f'(2) = 14$ **77.** Answers will vary.

Quiz Yourself *(page 127)*

1. $f(x) = 5x + 3$

$f(x + \Delta x) = 5x + 5\Delta x + 3$

$f(x + \Delta x) - f(x) = 5\Delta x$

$\dfrac{f(x + \Delta x) - f(x)}{\Delta x} = 5$

$\displaystyle\lim_{\Delta x \to 0} \dfrac{f(x + \Delta x) - f(x)}{\Delta x} = 5$

$f'(x) = 5$

$f'(-2) = 5$

2. $f(x) = \sqrt{x + 3}$

$f(x + \Delta x) = \sqrt{x + \Delta x + 3}$

$f(x + \Delta x) - f(x) = \sqrt{x + \Delta x + 3} - \sqrt{x + 3}$

$\dfrac{f(x + \Delta x) - f(x)}{\Delta x} = \dfrac{1}{\sqrt{x + \Delta x + 3} + \sqrt{x + 3}}$

$\displaystyle\lim_{\Delta x \to 0} \dfrac{f(x + \Delta x) - f(x)}{\Delta x} = \dfrac{1}{2\sqrt{x + 3}}$

$f'(x) = \dfrac{1}{2\sqrt{x + 3}}$

$f'(1) = \dfrac{1}{4}$

3. $f(x) = 3x - x^2$

$f(x + \Delta x) = 3x + 3\Delta x - x^2 - 2x\Delta x - (\Delta x)^2$

$f(x + \Delta x) - f(x) = 3\Delta x - 2x\Delta x - (\Delta x)^2$

$\dfrac{f(x + \Delta x) - f(x)}{\Delta x} = 3 - 2x - \Delta x$

$\displaystyle\lim_{\Delta x \to 0} \dfrac{f(x + \Delta x) - f(x)}{\Delta x} = 3 - 2x$

$f'(x) = 3 - 2x$

$f'(4) = -5$

4. $f'(x) = 0$ **5.** $f'(x) = 19$ **6.** $f'(x) = 4x^3 - 9x^2 - 10x$

7. $f'(x) = \dfrac{3}{x^{3/4}}$ **8.** $f'(x) = -\dfrac{8}{x^3}$ **9.** $f'(x) = -\dfrac{2}{x^{6/5}} - \dfrac{3}{x^4}$

10. $f'(x) = -\dfrac{5}{(3x + 2)^2}$ **11.** $f'(x) = -6x^2 + 8x - 2$

12. $f'(x) = 15x^2 + 26x + 14$ **13.** $f'(x) = \dfrac{-4(x^2 - 3)}{(x^2 + 3)^2}$

14. Average rate: 0

Instantaneous rates: $f'(0) = -3$, $f'(3) = 3$

15. Average rate: 1

Instantaneous rates: $f'(-1) = 3$, $f'(1) = 7$

16. Average rate: $-\dfrac{1}{30}$

Instantaneous rates: $f'(-5) = -\dfrac{1}{75}$, $f'(-2) = -\dfrac{1}{12}$

17. Average rate: $\dfrac{1}{19}$

Instantaneous rates: $f'(8) = \dfrac{1}{12}$, $f'(27) = \dfrac{1}{27}$

18. (a) \$11.63   (b) \$11.61

(c) The results are approximately equal.

19. $y = -4x - 6$

20. $y = -\dfrac{3}{8}x + \dfrac{5}{2}$

21. $y = 10x - 8$

22. $y = 22x - 31$

23. (a) $\dfrac{dS}{dt} = -0.05166t^2 + 1.4666t - 7.657$

(b) 2008: \$0.77/yr
   2011: \$2.22/yr
   2012: \$2.50/yr

Section 2.5 *(page 135)*

Skills Warm Up *(page 135)*

1. $(1 - 5x)^{2/5}$ **2.** $(2x - 1)^{3/4}$

3. $(4x^2 + 1)^{-1/2}$ **4.** $(2x^3 + 9)^{-1/6}$

5. $x^{1/2}(1 - 2x)^{-1/3}$ **6.** $(2x)^{-1}(3 - 7x)^{3/2}$

7. $(x - 2)(3x^2 + 5)$ **8.** $(x - 1)(5\sqrt{x} - 1)$

9. $(x^2 + 1)^2(4 - x - x^3)$

10. $-(x^2 - 6)(x^3 - 7)$

	$y = f(g(x))$	$u = g(x)$	$y = f(u)$
1.	$y = (6x - 5)^4$	$u = 6x - 5$	$y = u^4$
3.	$y = \sqrt{5x - 2}$	$u = 5x - 2$	$y = \sqrt{u}$
5.	$y = (3x + 1)^{-1}$	$u = 3x + 1$	$y = u^{-1}$

7. $y' = 32x + 56$ **9.** $y' = -\dfrac{x}{\sqrt{3 - x^2}}$

11. $y' = \dfrac{40x^3 - 4}{3\sqrt[3]{5x^4 - 2x}}$ **13.** c   **15.** b   **17.** d

19. $6(2x - 7)^2$ **21.** $6x(6 - x^2)(2 - x^2)$ **23.** $\dfrac{1}{2\sqrt{t + 1}}$

25. $\dfrac{4t + 5}{2\sqrt{2t^2 + 5t + 2}}$ **27.** $\dfrac{54}{(2 - 9x)^4}$ **29.** $-\dfrac{7x}{\sqrt{(x^2 + 11)^9}}$

31. $y = 216x - 378$ **33.** $y = \dfrac{8}{3}x - \dfrac{7}{3}$

35. $y = x - 1$

37. $f'(x) = \dfrac{1 - 3x^2 - 4x^{3/2}}{2\sqrt{x}(x^2 + 1)^2}$

The zero of $f'(x)$ corresponds to the point on the graph of $f(x)$ where the tangent line is horizontal.

39. $f'(x) = -\dfrac{\sqrt{(x + 1)/x}}{2x(x + 1)}$

$f'(x)$ has no zeros.

In Exercises 41–55, the differentiation rule(s) used may vary. A sample answer is provided.

41. $\dfrac{2x}{(4 - x^2)^2}$; Chain Rule

43. $\dfrac{5(t - 8)}{(t + 8)^3}$; Quotient Rule and Chain Rule

45. $-6(3x^2 - x - 3)$; Product Rule

47. $-\dfrac{1}{2(x + 2)^{3/2}}$; Power Rule

49. $27(x - 3)^2(4x - 3)$; Product Rule and Chain Rule

51. $\dfrac{3(x + 1)}{\sqrt{2x + 3}}$; Product Rule and Chain Rule

53. $\dfrac{t(5t - 8)}{2\sqrt{t - 2}}$; Product Rule and Chain Rule

55. $\dfrac{2(6 - 5x)(5x^2 - 12x + 5)}{(x^2 - 1)^3}$; Chain Rule and Quotient Rule

57. $y = 4x - 3$ **59.** $y = \frac{8}{3}t + 4$

61. $y = -6t - 14$ **63.** $y = -2x + 7$

65. $\left(0, \sqrt[3]{4}\right)$ **67.** $(1, 1)$

69. (a) $74.00 per 1%
   (b) $81.59 per 1%
   (c) $89.94 per 1%

71. $V = \dfrac{10{,}000}{\sqrt{t + 1}}$; about $-$1767.77 per year; $-$625.00 per year

73. (a) $r'(t) = \dfrac{1.2068t^3 - 28.971t^2 + 194.7t - 266.8}{2\sqrt{0.3017t^4 - 9.657t^3 + 97.35t^2 - 266.8t - 242}}$;
   Chain Rule

   (b)

   (c) Changing the most: $t = 8$
   Changing the least: $t \approx 9.60$, $t \approx 12.57$

Section 2.6   *(page 142)*

Skills Warm Up   *(page 142)*

1. $t = \pm\dfrac{\sqrt{73}}{2}$ **2.** $t = 0, t = 5.5$ **3.** $t = -2, t = 10$

4. $t = \dfrac{9 \pm 3\sqrt{10{,}249}}{32}$ **5.** $\dfrac{dy}{dx} = 6x^2 + 14x$

6. $\dfrac{dy}{dx} = 8x^3 + 18x^2 - 10x - 15$

7. $\dfrac{dy}{dx} = \dfrac{2x(x + 7)}{(2x + 7)^2}$ **8.** $\dfrac{dy}{dx} = -\dfrac{6x^2 + 10x + 15}{(2x^2 - 5)^2}$

9. Domain: $(-\infty, \infty)$ **10.** Domain: $[7, \infty)$
   Range: $[-4, \infty)$ Range: $[0, \infty)$

1. 0   **3.** 2   **5.** $2t - 8$ **7.** $\dfrac{24}{t^5}$

9. $18(2 - x^2)(5x^2 - 2)$ **11.** $\dfrac{4}{(x - 1)^3}$ **13.** $60x^2 - 72x$

15. $120x + 360$ **17.** $-\dfrac{45}{x^7}$ **19.** 260   **21.** $-\dfrac{1}{648}$

23. 12   **25.** $4x$ **27.** $\dfrac{80}{x^6}$ **29.** $12x^2 + 4$

31. $f''(x) = 6(x - 3) = 0$ at $x = 3$.

33. $f''(x) = \dfrac{x(2x^2 - 3)}{(x^2 - 1)^{3/2}} = 0$ at $x = \pm\dfrac{\sqrt{6}}{2}$.

35. (a) $s(t) = -16t^2 + 144t$ (b) $s(3) = 288$ ft
   $v(t) = -32t + 144$ $v(3) = 48$ ft/sec
   $a(t) = -32$ $a(3) = -32$ ft/sec^2

   (c) 4.5 sec; 324 ft

   (d) $v(9) = -144$ ft/sec; This is the same speed as the initial velocity.

37.

t	0	10	20	30	40	50	60
$\dfrac{dv}{dt}$	0	45	60	67.5	72	75	77.1
$\dfrac{d^2v}{dt^2}$	9	2.25	1	0.56	0.36	0.25	0.18

As time increases, velocity increases and acceleration decreases.

39. (a)

(b) The degree decreases by 1 for each successive derivative.

(c)

The degree decreases by 1 for each successive derivative.

(d) The degree decreases by 1 for each successive derivative.

41. (a) $y(t) = -21.944t^3 + 701.75t^2 - 6969.4t + 27,164$

(b) $y'(t) = -65.832t^2 + 1403.5t - 6969.4$
$y''(t) = -131.664t + 1403.5$

(c) $y'(t) > 0$ on $[8, 13]$

(d) 2010 ($t \approx 10.66$)

43. True

45. Answers will vary.

Section 2.7   *(page 149)*

Skills Warm Up   *(page 149)*

1. $y = x^2 - 2x$ **2.** $y = \dfrac{x - 3}{4}$

3. $y = 1, x \neq -6$ **4.** $y = \dfrac{3x^2 - 7}{4 - x^2}$

5. $y = \pm\sqrt{5 - x^2}$ **6.** $y = \pm\sqrt{6 - x^2}$

7. $\dfrac{8}{3}$ **8.** $-\dfrac{1}{2}$ **9.** $\dfrac{1}{7}$

---

1. $-\dfrac{3y}{x}$ **3.** $-\dfrac{x}{y}$ **5.** $\dfrac{6}{4y^3 - 2y + 7}$ **7.** $-\dfrac{y^2 + 4y}{2xy + 4x}$

9. $-\dfrac{1}{6}$ **11.** $\dfrac{4y^2 + 12}{2 - 8xy}$ **13.** 0   **15.** $-\dfrac{1}{4}$

17. Undefined   **19.** -4 **21.** $-\dfrac{5}{4}$ **23.** $\dfrac{1}{4}$ **25.** $\dfrac{1}{3}$

27. 3   **29.** $-\sqrt{3}$ **31.** $-\dfrac{2}{3}$ **33.** $\dfrac{1}{2y}$; $-\dfrac{1}{2}$

35. At $(8, 6)$: $y = -\dfrac{4}{3}x + \dfrac{50}{3}$

At $(-6, 8)$: $y = \dfrac{3}{4}x + \dfrac{25}{2}$

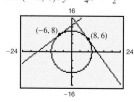

37. At $(1, \sqrt{5})$: $15x - 2\sqrt{5}y - 5 = 0$

At $(1, -\sqrt{5})$: $15x + 2\sqrt{5}y - 5 = 0$

39. At $(0, 2)$: $y = 2$ **41.** At $(2, 2)$: $y = 2x - 2$

At $(2, 0)$: $x = 2$ At $(2, -2)$: $y = -2x + 2$

43. $-\dfrac{2}{p^2(0.00003x^2 + 0.1)}$ **45.** $-\dfrac{4xp}{2p^2 + 1}$

47. (a) -2

(b)

As more labor is used, less capital is available.
As more capital is used, less labor is available.

49. (a)

The numbers of cases of chickenpox decreases from 2008 to 2012.

(b) 2008

(c)

t	8	9	10	11	12
y	30.40	20.51	15.39	14.51	13.40
y'	-11.79	-7.71	-2.54	-0.06	-3.26

2008; The result is the same.

Section 2.8   *(page 156)*

Skills Warm Up   *(page 156)*

1. $A = \pi r^2$ **2.** $V = \dfrac{4}{3}\pi r^3$ **3.** $S = 6s^2$

4. $V = s^3$ **5.** $V = \dfrac{1}{3}\pi r^2h$ **6.** $A = \dfrac{1}{2}bh$

7. $-\dfrac{x}{y}$ **8.** $\dfrac{2x - 3y}{3x}$ **9.** $\dfrac{-3x^2 - 2y + 12}{2x - 4}$

10. $-\dfrac{y^2 - y + 1}{2xy - 2y - x}$

---

1. (a) $\frac{3}{4}$ (b) 20   **3.** (a) $-\frac{5}{8}$ (b) $\frac{3}{2}$

5. (a) 36π in.^{2}/min   (b) 144π in.^{2}/min

7. No; If $\frac{dr}{dt}$ is constant, then $\frac{dA}{dt} = 2\pi r \frac{dr}{dt}$. So, $\frac{dA}{dt}$ is proportional to r.

9. (a) $\frac{5}{2\pi}$ ft/min   (b) $\frac{5}{8\pi}$ ft/min

11. (a) $112.50 per week
   (b) $7500 per week
   (c) $7387.5 per week

13. (a) $13,800 per day
   (b) $6900 per day

15. (a) 72 cm^3/sec
   (b) 1800 cm^3/sec

17. -10.4 ft/sec; As $x \to 0$, $\frac{dx}{dt}$ increases.   **19.** 300 mi/h

21. (a) -750 mi/h   (b) 20 min

23. About 37.7 ft^3/min   **25.** 4 units/wk

Review Exercises for Chapter 2   *(page 162)*

1. -2 **3.** 0

5. Answers will vary. Sample answer:
   $t = 8$: slope ≈ 225; Revenue was increasing by about $225 million per year in 2008.
   $t = 10$: slope ≈ 350; Revenue was increasing by about $350 million per year in 2010.

7. Answers will vary. Sample answer:
   $t = 1$: slope ≈ 65; The number of visitors to the national park is increasing at about 65,000 visitors per month in January.
   $t = 8$: slope ≈ 0; The number of visitors to the national park is neither increasing nor decreasing in August.
   $t = 12$: slope ≈ -1000; The number of visitors to the national park is decreasing at about 1,000,000 visitors per month in December.

9. -3 **11.** 6   **13.** $\frac{1}{4}$ **15.** -1 **17.** 9

19. $-x + 2$ **21.** $\dfrac{1}{2\sqrt{x - 5}}$ **23.** $-\dfrac{5}{x^2}$

25. All values except $x = -1$; The function has a vertical tangent at $x = -1$.

27. All values except $x = 0$; A function is not differentiable at a discontinuity.

29. 0   **31.** $7x^6$ **33.** $8x$ **35.** $\dfrac{15x^2}{4}$ **37.** $8x^3 + 6x$

39. $2x + 6$ **41.** -0.125 **43.** 5

45. (a) $y = 5x - 7$
   (b) and (c)

47. (a) $y = x - 1$
   (b) and (c)

49. (a) 2008: $m \approx 218.8$
        2010: $m \approx 359.0$
   (b) Results should be similar.
   (c) The slope shows the rate at which revenue was increasing or decreasing in a particular year.

51. Average rate of change: 4
   Instantaneous rate of change at $t = -3$: 4
   Instantaneous rate of change at $t = 1$: 4

53. Average rate of change: $\frac{3}{7}$
   Instantaneous rate of change at $x = 1$: $\frac{2}{3}$
   Instantaneous rate of change at $x = 8$: $\frac{1}{3}$

55. (a) -94 ft/sec
   (b) $t = 1$: -62 ft/sec
        $t = 3$: -126 ft/sec
   (c) 5.26 sec
   (d) -198.32 ft/sec

57. $\dfrac{dC}{dx} = 320$ **59.** $\dfrac{dC}{dx} = \dfrac{1.275}{\sqrt{x}}$

61. $\dfrac{dR}{dx} = -1.2x + 150$ **63.** $\dfrac{dR}{dx} = -12x^2 + 4x + 100$

65. $\dfrac{dP}{dx} = -0.0006x^2 + 12x - 1$

67. (a) $10   (b) $9.95
   (c) Parts (a) and (b) differ by only $0.05.

In Exercises 69–89, the differentiation rule(s) used may vary. A sample answer is provided.

69. $15x^2(1 - x^2)$; Power Rule

71. $16x^3 - 33x^2 + 12x$; Product Rule

73. $\dfrac{3}{(x + 3)^2}$; Quotient Rule

75. $\dfrac{2(3 + 5x - 3x^2)}{(x^2 + 1)^2}$; Quotient Rule

77. $30x(5x^2 + 2)^2$; Chain Rule

79. $-\dfrac{1}{(x + 1)^{3/2}}$; Quotient Rule

81. $\dfrac{2x^2 + 1}{\sqrt{x^2 + 1}}$; Product Rule

83. $80x^4 - 24x^2 + 1$; Product Rule

85. $18x^5(x + 1)(2x + 3)^2$; Chain Rule

87. $\dfrac{2x(x - 7)^{1/5}(8x - 35)}{5}$; Product and Chain Rules

89. $\dfrac{3(9t + 5)}{2\sqrt{3t + 1}(1 - 3t)^3}$; Quotient Rule

91. (a) $t = 1$: -6.63 $t = 3$: -6.5
        $t = 5$: -4.33 $t = 10$: -1.36
   (b)

The rate of decrease is approaching zero.

93. 6   **95.** $\dfrac{360}{x^7}$ **97.** $\dfrac{15}{\sqrt{x}}$ **99.** $2 + \dfrac{6}{x^3}$

101. (a) $s(t) = -16t^2 + 5t + 30$; $\dfrac{ds}{dt} = -32t + 5$; $\dfrac{d^2s}{dt^2} = -32$

(b) About 1.534 sec

(c) About -44.09 ft/sec

(d) About -32 ft/sec^2

103. $-\dfrac{2x + 3y}{3(x + y^2)}$ **105.** $\dfrac{2x - 8}{2y - 9}$ **107.** $y = \dfrac{1}{3}x + \dfrac{1}{3}$

109. $y = \dfrac{4}{3}x + \dfrac{2}{3}$

111. (a) 12π in.^{2}/min

(b) 40π in.^{2}/min

113. $\dfrac{1}{64}$ ft/min

Test Yourself *(page 166)*

1. $f(x) = x^2 + 3$

$f(x + \Delta x) = x^2 + 2x\Delta x + (\Delta x)^2 + 3$

$f(x + \Delta x) - f(x) = 2x\Delta x + (\Delta x)^2$

$\dfrac{f(x + \Delta x) - f(x)}{\Delta x} = 2x + \Delta x$

$\lim\limits_{\Delta x \to 0} \dfrac{f(x + \Delta x) - f(x)}{\Delta x} = 2x$

$f'(x) = 2x$

$f'(3) = 6$

2. $f(x) = \sqrt{x} - 2$

$f(x + \Delta x) = \sqrt{x + \Delta x} - 2$

$f(x + \Delta x) - f(x) = \sqrt{x + \Delta x} - \sqrt{x}$

$\dfrac{f(x + \Delta x) - f(x)}{\Delta x} = \dfrac{1}{\sqrt{x + \Delta x} + \sqrt{x}}$

$\lim\limits_{\Delta x \to 0} \dfrac{f(x + \Delta x) - f(x)}{\Delta x} = \dfrac{1}{2\sqrt{x}}$

$f'(x) = \dfrac{1}{2\sqrt{x}}$

$f'(4) = \dfrac{1}{4}$

3. $f'(t) = 3t^2 + 2$ **4.** $f'(x) = 8x - 8$ **5.** $\dfrac{3\sqrt{x}}{2} + \dfrac{3}{\sqrt{x}}$

6. $10x + \dfrac{9}{x^4}$ **7.** $f'(x) = 3x^2 + 10x + 6$

8. $f'(x) = \dfrac{5 + x}{2\sqrt{x}} + \sqrt{x}$ **9.** $f'(x) = 36x^3 + 48x$

10. $f'(x) = -\dfrac{1}{\sqrt{1 - 2x}}$

11. $f'(x) = \dfrac{(10x + 1)(5x - 1)^2}{x^2} = 250x - 75 + \dfrac{1}{x^2}$

12. $y = 2x - 2$

13. (a) \$13.37 billion/yr

(b) 2010: \$6.68 billion/yr

2012: \$11.63 billion/yr

(c) The annual sales of CVS Health increased from 2010 to 2012 by an average of about \$13.37 billion per year. The instantaneous rates of change for 2010 and 2012 are \$6.68 billion per year and \$11.63 billion per year, respectively.

14. (a) $P = -0.016x^2 + 1460x - 715,000$ (b) \$1437.60

15. 0 **16.** $-\dfrac{3}{8(3 - x)^{5/2}}$ **17.** $-\dfrac{96}{(2x - 1)^4}$

18. $s(t) = -16t^2 + 30t + 75$ $\quad s(2) = 71$ ft

$v(t) = -32t + 30$ $\quad v(2) = -34$ ft/sec

$a(t) = -32$ $\quad a(2) = -32$ ft/sec^2

19. $\dfrac{dy}{dx} = -\dfrac{1 + y}{x}$ **20.** $\dfrac{dy}{dx} = -\dfrac{1}{y - 1}$

21. $\dfrac{dy}{dx} = -\dfrac{8x + 3x^2y}{x^3 - 6y}$

22. (a) 3.75π cm^3/min (b) 15π cm^3/min

Chapter 3

Section 3.1 *(page 175)*

Skills Warm Up *(page 175)*

1. $x = 0$, $x = 8$ **2.** $x = 0$, $x = \dfrac{27}{5}$ **3.** $x = \pm 5$

4. $x = 0$ **5.** $(-\infty, 3) \cup (3, \infty)$ **6.** $\left(-\infty, \dfrac{5}{2}\right)$

7. $(-\infty, -2) \cup (-2, 5) \cup (5, \infty)$ **8.** $\left(-\sqrt{3}, \sqrt{3}\right)$

9. $x = -2$: -6 **10.** $x = -2$: 60

$x = 0$: 2 $\qquad\qquad$ $x = 0$: -4

$x = 2$: -6 $\qquad\qquad$ $x = 2$: 60

11. $x = -2$: $-\dfrac{1}{3}$ **12.** $x = -2$: $\dfrac{1}{18}$

$x = 0$: 1 $\qquad\qquad$ $x = 0$: $-\dfrac{1}{8}$

$x = 2$: 5 $\qquad\qquad$ $x = 2$: $-\dfrac{3}{2}$

---

1. Increasing on $(-\infty, -1)$

Decreasing on $(-1, \infty)$

3. Increasing on $(-1, 0)$ and $(1, \infty)$

Decreasing on $(-\infty, -1)$ and $(0, 1)$

5. $y' > 0$ on $(-\infty, 0)$ **7.** $y' < 0$ on $(-\infty, 2)$

$y' < 0$ on $(0, \infty)$ $\qquad$ $y' > 0$ on $(2, \infty)$

9. $x = \dfrac{3}{4}$ **11.** $x = 0$, $x = -3$ **13.** $x = \pm 5$

15. No critical numbers **17.** Critical number: $x = 3$

Increasing on $(-\infty, \infty)$ $\quad$ Decreasing on $(-\infty, 3)$

$\qquad\qquad\qquad\qquad\qquad$ Increasing on $(3, \infty)$

19. Critical number: $x = 1$

Increasing on $(-\infty, 1)$

Decreasing on $(1, \infty)$

21. Critical numbers: $x = -1$, $x = -\dfrac{5}{3}$

Increasing on $\left(-\infty, -\dfrac{5}{3}\right)$ and $(-1, \infty)$

Decreasing on $\left(-\dfrac{5}{3}, -1\right)$

23. Critical numbers: $x = 0$, $x = \frac{3}{2}$
Decreasing on $\left(-\infty, \frac{3}{2}\right)$
Increasing on $\left(\frac{3}{2}, \infty\right)$

25. Critical number: $x = -2$
Decreasing on $(-\infty, -2)$
Increasing on $(-2, \infty)$

27. Critical number: $x = \frac{1}{5}$
Increasing on $\left(-\infty, \frac{1}{5}\right)$ and $\left(\frac{1}{5}, \infty\right)$

29. Critical number: $x = 0$
Increasing on $(-\infty, 0)$ and $(0, \infty)$

31. Critical numbers: $x = 0$, $x = \pm 1$
Decreasing on $(-\infty, 0)$
Increasing on $(0, \infty)$

33. Critical number: $x = -2$
Increasing on $(-\infty, -2)$ and $(-2, \infty)$

35. Critical numbers: $x = -1$, $x = -\frac{2}{3}$
Decreasing on $\left(-1, -\frac{2}{3}\right)$
Increasing on $\left(-\frac{2}{3}, \infty\right)$

37. Critical numbers: $x = -3$, $x = 3$
Decreasing on $(-\infty, -3)$ and $(3, \infty)$
Increasing on $(-3, 3)$

39. No critical numbers
Discontinuity: $x = 5$
Decreasing on $(-\infty, 5)$ and $(5, \infty)$

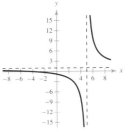

41. Critical number: $x = 1$
Discontinuity: $x = 0$
Decreasing on $(-\infty, 0)$ and $(0, 1)$
Increasing on $(1, \infty)$

43. No critical numbers
Discontinuities: $x = \pm 4$
Increasing on $(-\infty, -4)$, $(-4, 4)$, and $(4, \infty)$

45. Critical number: $x = 0$
Discontinuity: $x = 0$
Increasing on $(-\infty, 0)$
Decreasing on $(0, \infty)$

47. Critical numbers: $x = -1$, $x = 0$
Increasing on $(-\infty, -1)$ and $(0, \infty)$
Decreasing on $(-1, 0)$

49. No critical numbers
$S'(t) = -0.386t + 5.39$
$S'(t) > 0$ on $(9, 13)$ so $S(t)$ is increasing on $(9, 13)$.

51. (a)

Decreasing from 1998 to 2001
Increasing from 2001 to 2012
(b) $y' = -0.0759t^2 + 2.862t - 22.63$
Critical number: $t = 11.3$
Therefore, the model is decreasing from 1998 to 2001 and increasing from 2001 to 2012.

53. (a) $P = -0.00005x^2 + 2.65x - 7500$
(b) Critical number: $x = 26,500$
Increasing on $(0, 26,500)$
Decreasing on $(26,500, 50,000)$
(c) 26,500 hamburgers; Because the function changes from increasing to decreasing at $x = 26,500$, the maximum profit occurs at this value.

Section 3.2   *(page 184)*

Skills Warm Up   *(page 184)*

1. $0, \pm 3$ **2.** $-2, 5$ **3.** 1 **4.** $0, 125$
5. $-4 \pm \sqrt{17}$ **6.** No solution
7. Negative   **8.** Positive   **9.** Positive
10. Negative   **11.** Increasing   **12.** Decreasing

1. Relative maximum: $(1, 5)$ **3.** Relative minimum: $(3, -9)$
5. Relative minimum: $(9, -2187)$ **7.** No relative extrema

9. Relative maximum: $(0, 3)$

Relative minimum: $\left(\frac{4}{3}, -\frac{79}{27}\right)$

11. Relative maximum: $(-1, 2)$; Relative minimum: $(0, 0)$

13.

Relative maximum: $(0, 0)$
Relative minimum: $(8, -8)$

15.

Relative maximum:
$\left(-1, -\frac{3}{2}\right)$

17.

No relative extrema

19. Minimum: $(-1, 6)$

Maximum: $(-4, 15)$

21. Maximum: $(0, 5)$

Minimum: $(-3, -13)$

23. Minimum: $(4, -32)$

Maxima: $(0, 0), (6, 0)$

25. Maximum: $(2, 1)$

Minimum: $\left(0, \frac{1}{3}\right)$

27. Maxima: $(-2, -1), (1, -1)$

No minima

29. Maximum: $(-7, 4)$

Minimum: $(1, 0)$

31. 2, absolute maximum (and relative maximum)

33. 1, absolute maximum (and relative maximum);

2, absolute minimum (and relative minimum);

3, absolute maximum (and relative maximum)

35. Maximum: $(5, 7)$

Minimum: $(2.69, -5.55)$

37. Maximum: $(2, 2.\overline{6})$

Minima: $(0, 0), (3, 0)$

39. No relative minimum

Maximum: $(0, 8)$

41. Maximum: $\left(2, \frac{1}{2}\right)$

Minimum: $(0, 0)$

43. Answers will vary. Sample answer:

45. (a) Population tends to increase each year, so the minimum population occurred in 2000 and the maximum population occurred in 2013.

(b) Maximum population: 316.07 million

Minimum population: 281.8 million

(c) The minimum population was about 281.8 million in 2000 and the maximum population was about 316.07 million in 2013.

47. 82 units   **49.** 3500 bottles, $2.25

Section 3.3 *(page 193)*

Skills Warm Up *(page 193)*

1. $f''(x) = 48x^2 - 54x$ **2.** $g''(s) = 12s^2 - 18s + 2$

3. $g''(x) = 56x^6 + 120x^4 + 72x^2 + 8$

4. $f''(x) = \dfrac{4}{(3x - 7)^{2/3}}$ **5.** $h''(x) = \dfrac{190}{(5x - 1)^3}$

6. $f''(x) = -\dfrac{42}{(3x + 2)^3}$ **7.** $x = \pm\dfrac{\sqrt{6}}{6}$

8. $x = 0, x = 3$ **9.** $t = \pm 4$ **10.** $x = 0, x = \pm 5$

1. Sign of $f'(x)$ on $(0, 2)$ is positive.

Sign of $f''(x)$ on $(0, 2)$ is positive.

3. Sign of $f'(x)$ on $(0, 2)$ is negative.

Sign of $f''(x)$ on $(0, 2)$ is negative.

5. Concave downward on $(-\infty, \infty)$

7. Concave upward on $(-\infty, 1)$

Concave downward on $(1, \infty)$

9. Concave upward on $(-\infty, -1)$ and $(1, \infty)$

Concave downward on $(-1, 1)$

11. Concave upward on $(-\infty, -2)$ and $(2, \infty)$

Concave downward on $(-2, 2)$

13. Concave downward on $(-\infty, 3)$

Concave upward on $(3, \infty)$

Point of inflection: $(3, 0)$

15. Concave downward on $\left(-\infty, \frac{1}{2}\right)$

Concave upward on $\left(\frac{1}{2}, \infty\right)$

Point of inflection: $\left(\frac{1}{2}, -\frac{3}{2}\right)$

17. Concave downward on $(-\infty, \infty)$

No inflection points

19. Concave upward on $(-\infty, 2)$ and $(4, \infty)$

Concave downward on $(2, 4)$

Points of inflection: $(2, -16), (4, 0)$

21. Relative maximum: $(3, 9)$

23. Relative minimum: $(3, -25)$

25. Relative minimum: $(0, -3)$

27. Relative minimum: $(0, 1)$

29. Relative maximum: $(0, 3)$

31. Relative maximum: $(0, 4)$

33. No relative extrema

35. Relative maximum: $(2, 9)$

Relative minimum: $(0, 5)$

37. Relative maximum: $(0, 0)$

Relative minima: $(-0.5, -0.052), (1, -0.\overline{3})$

39. Relative maximum: $(-2, 16)$

Relative minimum: $(2, -16)$

Point of inflection: $(0, 0)$

41. No relative maximum
Relative minimum: $(4, 4)$
Point of inflection:
$$\left(12, \frac{8\sqrt{3}}{3}\right)$$

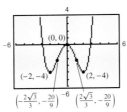

43. Relative maximum: $(0, 0)$
Relative minima: $(\pm 2, -4)$
Points of inflection:
$$\left(\pm\frac{2\sqrt{3}}{3}, -\frac{20}{9}\right)$$

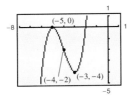

45. Relative maximum: $(-5, 0)$
Relative minimum: $(-3, -4)$
Point of inflection: $(-4, -2)$

47. Relative minimum: $(-2, -2)$
No inflection points

49. Relative maximum: $(0, 4)$
Points of inflection:
$$\left(\pm\frac{\sqrt{3}}{3}, 3\right)$$

51. Answers will vary. Sample answer:

53. Answers will vary. Sample answer:

55.

(a) f': Positive on $(-\infty, 0)$
f: Increasing on $(-\infty, 0)$
(b) f': Negative on $(0, \infty)$
f: Decreasing on $(0, \infty)$
(c) f': Not increasing
f: Not concave upward
(d) f': Decreasing on $(-\infty, \infty)$
f: Concave downward on $(-\infty, \infty)$

57. (a) f': Increasing on $(-\infty, \infty)$
(b) f: Concave upward on $(-\infty, \infty)$
(c) Relative minimum: $x = -2.5$
No inflection points

59. (a) f': Increasing on $(-\infty, 1)$
Decreasing on $(1, \infty)$
(b) f: Concave upward on $(-\infty, 1)$
Concave downward on $(1, \infty)$
(c) No relative extrema
Point of inflection: $x = 1$

61. $(200, 320)$ **63.** 8:30 P.M.

65.

Relative minimum: $(0, -5)$
Relative maximum: $(3, 8.5)$
Point of inflection:
$$\left(\tfrac{2}{3}, -3.2963\right)$$

When f' is positive, f is increasing. When f' is negative, f is decreasing. When f'' is positive, f is concave upward. When f'' is negative, f is concave downward.

67.

Relative maximum: $(0, 2)$
Points of inflection:
$(0.58, 1.5), (-0.58, 1.5)$

When f' is positive, f is increasing. When f' is negative, f is decreasing. When f'' is positive, f is concave upward. When f'' is negative, f is concave downward.

69. 120 units

71. (a)

(b) 2010
(c) 2013
(d) Greatest: 2013
Least: 2008

CHAPTER 3

73. (a)

(b) Concave upward on $(5, 8.0903)$
Concave downward on $(8.0903, 13)$

(c) Point of inflection: $(8.0903, 40.7015)$

(d) The point of inflection is where the change in the population living below poverty starts to decrease.

75. Answers will vary. Sample answer:

$f(x) = x^4$
$f''(x) = 12x^2$
So $f''(0) = 0$, but $(0, 0)$ is not a point of inflection.

Section 3.4   *(page 201)*

Skills Warm Up   *(page 201)*

1. $x + \frac{1}{2}y = 12$ **2.** $3xy = 36$ **3.** $xy = 24$
4. $\sqrt{(x_2 - x_1)^2 + (y_2 - y_1)^2} = 10$
5. $x = -2$ **6.** $x = -\frac{2}{3}, x = 1$ **7.** $x = \pm 5$
8. $x = 4$ **9.** $x = \pm 1$ **10.** $x = \pm 3$

---

1. $l = w = 20$ m   **3.** $l = w = 7$ ft
5. $x = 20$ ft, $y = \frac{80}{3}$ ft
7. (a) Proof
(b) $V_1 = 99$ in.^{3}
$V_2 = 125$ in.^{3}
$V_3 = 117$ in.^{3}
(c) 5 in. × 5 in. × 5 in.
9. (a) $l = w = h = 20$ in.
(b) 2400 in.^{2}
11. Width of rectangle: $\dfrac{100}{\pi} \approx 31.8$ m
Length of rectangle: 50 m
13. $l = w = 2\sqrt[3]{5} \approx 3.42$
$h = 4\sqrt[3]{5} \approx 6.84$
15. $V = 16$ in.^{3} **17.** 9 in. by 9 in.
19. Length: 3 units
Width: 1.5 units
21. Length: $5\sqrt{2}$ units
Width: $\dfrac{5\sqrt{2}}{2}$ units
23. Radius: about 1.32 in.
Height: about 2.64 in.

25. $(1, 1)$ **27.** $\left(3.5, \dfrac{\sqrt{14}}{2}\right)$ **29.** 18 in. × 18 in. × 36 in.

31. Radius: $\sqrt[3]{\dfrac{750}{\pi}} \approx 6.204$ ft
Height: about 24.808 ft

33. Radius of circle: $\dfrac{8}{\pi + 4}$
Side of square: $\dfrac{16}{\pi + 4}$

35. (a) $A(x) = \left(1 + \dfrac{4}{\pi}\right)x^2 - \dfrac{8}{\pi}x + \dfrac{4}{\pi}$
(b) $0 \le x \le 1$
(c)

(d) The total area is minimum when 2.24 feet is used for the square and 1.76 feet is used for the circle.
The total area is maximum when all 4 feet is used for the circle.

37. 4.75 weeks; 135 bushels; $3645

Quiz Yourself   *(page 204)*

1. Critical number: $x = 4$
Increasing on $(4, \infty)$
Decreasing on $(-\infty, 4)$
2. Critical numbers: $x = -4, x = 0$
Increasing on $(-\infty, -4)$ and $(0, \infty)$
Decreasing on $(-4, 0)$
3. Critical numbers: $x = 5, x = -5$
Increasing on $(-5, 5)$
Decreasing on $(-\infty, -5)$ and $(5, \infty)$
4. Relative minimum: $(0, -5)$
Relative maximum: $(-2, -1)$
5. Relative minima: $(2, -13), (-2, -13)$
Relative maximum: $(0, 3)$
6. Relative minimum: $(0, 0)$
7. Minimum: $(-2, -16)$
Maximum: $(-5, -7)$
8. Minimum: $(3, -54)$
Maximum: $(-3, 54)$
9. Minimum: $(0, 0)$
Maximum: $(1, 0.5)$
10. (a) Decreasing from 2000 to 2001 and from 2012 to 2014
Increasing from 2001 to 2012
(b) Greatest: 2012
Least: 2001
11. Point of inflection: $(2, -2)$
Concave downward on $(-\infty, 2)$
Concave upward on $(2, \infty)$
12. Points of inflection: $\left(-3, -\frac{135}{4}\right), \left(3, -\frac{135}{4}\right)$
Concave upward on $(-\infty, -3)$ and $(3, \infty)$
Concave downward on $(-3, 3)$

13. Relative minimum: $(1, 9)$
   Relative maximum: $(-2, 36)$
14. Relative minimum: $(3, 12)$
   Relative maximum: $(-3, -12)$
15. \$120,000 $(x = 120)$ **16.** 50 ft by 100 ft
17. 8 in. by 8 in.

Section 3.5 (page 212)

Skills Warm Up (page 212)

1. 1   **2.** $\frac{6}{5}$ **3.** 2   **4.** $\frac{1}{2}$

5. $\frac{dC}{dx} = 1.2 + 0.006x$ **6.** $\frac{dP}{dx} = 0.02x + 11$

7. $\frac{dP}{dx} = -0.9x^2 + 16x - 4$ **8.** $\frac{dC}{dx} = 1 + 0.09x^2$

9. $\frac{dR}{dx} = 14 - \frac{x}{1000}$ **10.** $\frac{dR}{dx} = 3.4 - \frac{x}{750}$

1. 2000 units   **3.** 150 units   **5.** 200 units
7. 50 units   **9.** \$60   **11.** \$40
13. 3 units

$\overline{C}(3) = 17; \frac{dC}{dx} = 4x + 5;$ When $x = 3, \frac{dC}{dx} = 17.$

15. (a) \$55   (b) \$30.32
17. The maximum profit occurs when $s = 10$ (or \$10,000).
   The point of diminishing returns occurs at $s = \frac{35}{6}$ (or \$5833.33).
19. 350 players   **21.** \$50
23. C = cost under water + cost on land
   $= 25(5280)\sqrt{x^2 + 0.25} + 18(5280)(6 - x)$
   $= 132,000\sqrt{x^2 + 0.25} + 570,240 - 95,040x$

The line should run from the power station to a point across the river approximately 0.52 mile downstream.

$\left(\text{Exact: } x = \frac{9\sqrt{301}}{301} \text{ mi}\right)$

25. $v \approx 63.5$ mi/h
27. -1, unit elastic

Elastic: $(0, 60)$
Inelastic: $(60, 120)$

29. $-\frac{2}{3}$, inelastic

Elastic: $\left(0, 83\frac{1}{3}\right)$
Inelastic: $\left(83\frac{1}{3}, 166\frac{2}{3}\right)$

31. $-\frac{25}{23}$, elastic

Elastic: $(0, \infty)$

33. (a) Elastic: $[0, 500)$
      Unit elasticity: $x = 500$
      Inelastic: $(500, 1000]$
   (b) The revenue function increases on the interval $[0, 500)$, then is flat at 500, and decreases on the interval $(500, 1000]$.
35. 500 units $(x = 5)$
37. No; When $p = 8, x = 540$ and $\eta = -\frac{2}{3}$.
   Because $|\eta| = \frac{2}{3} < 1$, demand is inelastic.
39. (a) Maximum sales: 2011
      Minimum sales: 2004
   (b) 2006
   (c) 2013
   (d)

41. Proof   **43.** Answers will vary.

Section 3.6 (page 223)

Skills Warm Up (page 223)

1. 3   **2.** 1   **3.** -11 **4.** 15   **5.** $-\frac{1}{2}$
6. -2 **7.** 0   **8.** 1

9. $\overline{C} = \frac{150}{x} + 3$ **10.** $\overline{C} = \frac{1900}{x} + 1.7 + 0.002x$

$\frac{dC}{dx} = 3$ $\frac{dC}{dx} = 1.7 + 0.004x$

11. $\overline{C} = 0.005x + 0.5 + \frac{1375}{x}$ **12.** $\overline{C} = \frac{760}{x} + 0.05$

$\frac{dC}{dx} = 0.01x + 0.5$ $\frac{dC}{dx} = 0.05$

1. Vertical asymptote: $x = 0$
   Horizontal asymptote: $y = 1$
3. Vertical asymptotes: $x = -1, x = 2$
   Horizontal asymptote: $y = 1$
5. Vertical asymptote: none
   Horizontal asymptote: $y = \frac{3}{2}$

CHAPTER 3

7. ∞ **9.** $-\infty$ **11.** ∞ **13.** $x = 0, x = -3$

15. $x = -3$ **17.** $x = 4$ **19.** ∞ **21.** $-\infty$

23. ∞ **25.** 1   **27.** -8 **29.** $y = 2$ **31.** $y = 0$

33. No horizontal asymptote   **35.** $y = 5$

37. d   **38.** b   **39.** a   **40.** c

41. (a) ∞ (b) 5   (c) 0

43. (a) 0   (b) 1   (c) ∞

45.

x	10^0	10^1	10^2	10^3
$f(x)$	0.646	11.718	800.003	29,622.777

x	10^4	10^5	10^6
$f(x)$	980,000	31,422,776.6	998,000,000

$$\lim_{x \to \infty} \sqrt{x^3 + 6} = 2x = \infty$$

47.

x	10^0	10^1	10^2	10^3
$f(x)$	2.000	0.348	0.101	0.032

x	10^4	10^5	10^6
$f(x)$	0.010	0.003	0.001

$$\lim_{x \to \infty} \frac{x + 1}{x\sqrt{x}} = 0$$

49.

51.

53.

55.

57.

59.

61.

63. (a) $\overline{C} = 1.15 + \dfrac{6000}{x}$

(b) $\overline{C}(600) = 11.15$;
$\overline{C}(6000) = 2.15$

(c) \$1.15; The cost approaches \$1.15 as the number of units produced increases.

65. (a) $\overline{P} = 35.4 - \dfrac{15,000}{x}$

(b) $\overline{P}(1000) = \$20.40$; $\overline{P}(10,000) = \$33.90$;
$\overline{P}(100,000) = \$35.25$

(c) \$35.40; Explanations will vary.

67. (a) 25%: \$176 million; 50%: \$528 million;
75%: \$1584 million

(b) ∞; The limit does not exist, which means the cost increases without bound as the government approaches 100% seizure of the illegal drug entering the country.

69. (a)

n	1	2	3	4	5
P	0.5	0.74	0.82	0.86	0.89

n	6	7	8	9	10
P	0.91	0.92	0.93	0.94	0.95

(b) 1

(c)

The percent of correct responses approaches 100% as the number of times the task is performed increases.

Section 3.7 *(page 233)*

Skills Warm Up *(page 233)*

1. Vertical asymptote: $x = 0$
Horizontal asymptote: $y = 0$

2. Vertical asymptote: $x = 4$
Horizontal asymptote: $y = 0$

3. Vertical asymptote: $x = -3$
Horizontal asymptote: $y = 40$

4. Vertical asymptotes: $x = 1, x = 3$
Horizontal asymptote: $y = 1$

5. Decreasing on $(-\infty, -2)$
Increasing on $(-2, \infty)$

6. Increasing on $(-\infty, -4)$
Decreasing on $(-4, \infty)$

7. Increasing on $(-\infty, -1)$ and $(1, \infty)$
Decreasing on $(-1, 1)$

8. Decreasing on $(-\infty, 0)$ and $(\sqrt[3]{2}, \infty)$
Increasing on $(0, \sqrt[3]{2})$

9. Decreasing on $\left(-\infty, \frac{5}{2}\right)$ and $\left(\frac{5}{2}, \infty\right)$

10. Decreasing on $(-\infty, -3)$ and $\left(\frac{1}{3}, \infty\right)$
Increasing on $\left(-3, \frac{1}{3}\right)$

1.

3.

5.

7.

9.

11.

13.

15.

17.

19.

21.

23.

25.

27.

29.

31.

33.

35.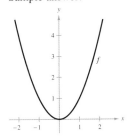

37. Answers will vary. Sample answer:

39. Answers will vary. Sample answer:

41. Answers will vary. Sample answer:

43. Answers will vary. Sample answer: $y = \dfrac{1}{x-3}$

45. (a) The model fits the data well.
(b) $1470.32
(c) Yes, because the benefits continue to increase at a steady rate for the next several hundred years.

CHAPTER 3

47.

Absolute maximum: $(7, 80.69)$
Absolute minimum: $(1, 35.10)$
The maximum temperature of 80.69°F occurs in July.
The minimum temperature of 35.10°F occurs in January.

49.

The rational function has the common factor $3 - x$ in the numerator and denominator. At $x = 3$, there is a hole in the graph, not a vertical asymptote.

51. (a) $f(x) = \dfrac{x^2 - 2x + 4}{x - 2} = \dfrac{x^2 - 2x}{x - 2} + \dfrac{4}{x - 2}$
$= \dfrac{x(x - 2)}{x - 2} + \dfrac{4}{x - 2} = x + \dfrac{4}{x - 2}$

(b)

As x approaches negative infinity or positive infinity, the graph of f approaches the line $y = x$.

(c) A slant asymptote is a line that is neither horizontal nor vertical that the graph of a function approaches as x approaches negative infinity or positive infinity.

Section 3.8   (page 240)

Skills Warm Up   (page 240)

1. $\dfrac{dC}{dx} = 0.18x$ **2.** $\dfrac{dC}{dx} = 0.15$

3. $\dfrac{dR}{dx} = 1.25 + 0.03\sqrt{x}$ **4.** $\dfrac{dR}{dx} = 15.5 - 3.1x$

5. $\dfrac{dP}{dx} = -\dfrac{0.01}{\sqrt[3]{x^2}} + 1.4$ **6.** $\dfrac{dP}{dx} = -0.12x^2 + 950$

7. $\dfrac{dA}{dx} = \dfrac{\sqrt{3}}{2}x$ **8.** $\dfrac{dA}{dx} = \dfrac{4}{9}x$ **9.** $\dfrac{dC}{dr} = 2\pi$

10. $\dfrac{dP}{dw} = 4$ **11.** $\dfrac{dS}{dr} = 8\pi r$ **12.** $\dfrac{dP}{dx} = 2 + \sqrt{2}$

13. $A = \pi r^2$ **14.** $A = x^2$

15. $V = x^3$ **16.** $V = \frac{4}{3}\pi r^3$

---

1. $dy = 0.15$ **3.** $dy = 0.32$ **5.** $dy = 0.075$
$\Delta y = 0.1655$ $\Delta y \approx 0.3224$ $\Delta y \approx 0.0745$

7.

$dx = \Delta x$	dy	Δy	$\Delta y - dy$	$\dfrac{dy}{\Delta y}$
1.000	12.0000	19.0000	7.0000	0.6316
0.500	6.0000	7.6250	1.6250	0.7869
0.100	1.2000	1.2610	0.0610	0.9516
0.010	0.1200	0.1206	0.0006	0.9950
0.001	0.0120	0.0120	0.0000	1.0000

9.

$dx = \Delta x$	dy	Δy	$\Delta y - dy$	$\dfrac{dy}{\Delta y}$
1.000	-0.1875	-0.0880	0.0995	2.1307
0.500	-0.0938	-0.0610	0.0328	1.5377
0.100	-0.0188	-0.0170	0.0018	1.1059
0.010	-0.0019	-0.0019	0.0000	1.0000
0.001	-0.0002	-0.0002	0.0000	1.0000

11.

$dx = \Delta x$	dy	Δy	$\Delta y - dy$	$\dfrac{dy}{\Delta y}$
1.000	0.14865	0.12687	-0.02178	1.17167
0.500	0.07433	0.06823	-0.00610	1.08940
0.100	0.01487	0.01459	-0.00028	1.01919
0.010	0.00149	0.00148	-0.00001	1.00676
0.001	0.00015	0.00015	0.00000	1.00000

13. $5.20; The actual change in cost is $5.25, which is a difference of $0.05.

15. $7.50; The actual change in revenue is $7.35, which is a difference of $0.15.

17. $-$1250; The actual change in profit is $-$1325.50, which is a difference of $75.50.

19. $dy = 24x^3\, dx$

21. $dy = (15x^2 - 6)\, dx$ **23.** $dy = -\dfrac{3}{(2x - 1)^2}\, dx$

25. $dy = \dfrac{-x}{\sqrt{9 - x^2}}\, dx$ **27.** $dy = (3x - 7)^2(12x - 7)\, dx$

29. $y = 28x + 37$
For $\Delta x = -0.01$, $f(x + \Delta x) = -19.281302$ and $y(x + \Delta x) = -19.28$
For $\Delta x = 0.01$, $f(x + \Delta x) = -18.721298$ and $y(x + \Delta x) = -18.72$

31. $y = x$
For $\Delta x = -0.01$, $f(x + \Delta x) = -0.009999$ and $y(x + \Delta x) = -0.01$
For $\Delta x = 0.01$, $f(x + \Delta x) = 0.009999$ and $y(x + \Delta x) = 0.01$

33. (a) $dP = 1160 (b) Actual: $1122.50

35. (a) $65.20; The actual change in revenue is $65.00, which is a difference of $0.20.

(b) $40.00; The actual change in revenue is $39.80, which is a difference of $0.20.

37. Approximately 19 deer

39. $R = -\frac{1}{3}x^2 + 100x$; $6   **41.** $\pm\frac{3}{4}$ in.^{2} **43.** True

2.08%

Review Exercises for Chapter 3   *(page 246)*

1. $x = -\frac{5}{2}$ **3.** $x = -3, x = 3$ **5.** $x = 1, x = \frac{7}{3}$

7. Critical number: $x = -\frac{1}{2}$

Increasing on $\left(-\frac{1}{2}, \infty\right)$

Decreasing on $\left(-\infty, -\frac{1}{2}\right)$

9. Critical numbers: $x = 0, x = 4$

Increasing on $(0, 4)$

Decreasing on $(-\infty, 0)$ and $(4, \infty)$

11. Critical number: $x = 6$

Increasing on $(6, \infty)$

Decreasing on $(-\infty, 6)$

13. Critical numbers: $x = 0, x = \pm 5$

Increasing on $(-5, 0)$

Decreasing on $(0, 5)$

15. Critical numbers: $x = \pm 7$

Increasing on $(-7, 7)$

Decreasing on $(-\infty, -7)$ and $(7, \infty)$

17. The only critical number is $t \approx -12.05$. Any $t > -12.05$ produces a positive dS/dt, so the sales were increasing from 2009 to 2013.

19. Relative maximum: $(0, -2)$

Relative minimum: $(1, -4)$

21. Relative maximum: $(5, 29)$

23. Relative maxima: $(-1, 1), (1, 1)$

Relative minimum: $(0, 0)$

25. Relative maximum: $(0, 6)$

27. Relative maximum: $(-1, 2)$

Relative minimum: $(1, -2)$

29. Maximum: $(0, 6)$

Minimum: $\left(-\frac{5}{2}, -\frac{1}{4}\right)$

31. Maxima: $(-2, 17), (4, 17)$

Minima: $(-4, -15), (2, -15)$

33. Maximum: $(1, 1)$

Minimum: $(9, -3)$

35. Maximum: $(1, 1)$

Minimum: $(-1, -1)$

37. $r \approx 1.58$ in.

39. Concave upward on $(2, \infty)$

Concave downward on $(-\infty, 2)$

41. Concave upward on $\left(-\frac{2\sqrt{3}}{3}, \frac{2\sqrt{3}}{3}\right)$

Concave downward on $\left(-\infty, -\frac{2\sqrt{3}}{3}\right)$ and $\left(\frac{2\sqrt{3}}{3}, \infty\right)$

43. Concave upward on $(-\infty, 0)$ and $(4, \infty)$

Concave downward on $(0, 4)$

Points of inflection: $(0, 0), (4, -128)$

45. Concave upward on $(1, \infty)$

Concave downward on $(-\infty, 1)$

Point of inflection: $(1, 1)$

47. No relative extrema

49. Relative maximum: $\left(-\sqrt{3}, 6\sqrt{3}\right)$

Relative minimum: $\left(\sqrt{3}, -6\sqrt{3}\right)$

51. Relative maxima: $\left(-\frac{\sqrt{2}}{2}, \frac{1}{2}\right), \left(\frac{\sqrt{2}}{2}, \frac{1}{2}\right)$

Relative minimum: $(0, 0)$

53. $\left(50, 166\frac{2}{3}\right)$ **55.** $l = w = 15$ m   **57.** 144 in.^{3}

59. $x = 900$ **61.** $x = 150$ **63.** (a) $24   (b) $8

65. (a) For $0 < x < 750$, $|\eta| > 1$ and the demand is elastic.

For $750 < x < 1500$, $|\eta| < 1$ and the demand is inelastic.

For $x = 750$, the demand has unit elasticity.

(b) From 0 to 750 units, revenue is increasing.

From 750 to 1500 units, revenue does not increase.

67. $x = -9, x = 0$ **69.** $x = -\frac{1}{2}$ **71.** $-\infty$

73. ∞ **75.** $y = -\frac{3}{4}$ **77.** $y = 0$

79. (a) $\overline{C} = 0.75 + \dfrac{4000}{x}$

(b) $\overline{C}(100) = 40.75$

$\overline{C}(1000) = 4.75$

(c) The limit is 0.75. As more and more units are produced, the average cost per unit will approach $0.75.

81. (a) 20%: $62.5 million

50%: $250 million

90%: $2250 million

(b) The limit is ∞, meaning that as the percent gets very close to 100, the cost grows without bound.

83.

85.

87.

89.

91.   **93.**

95. (a)

The model fits the data well.
(b) About 2434 bacteria
(c) Answers will vary.

97. $dy = 0.08$, $\Delta y = 0.0802$

99. $dy = -2.1$, $\Delta y = -2.191$

101. $800; The actual change in cost is $840, which is a difference of $40.

103. $15.25; The actual change in revenue is $15.26, which is a difference of $0.01.

105. $4.52; The actual change in profit is $4.52, which is the same.

107. $dy = 2.4x^2 \, dx$

109. $dy = 18x(3x^2 - 2)^2 \, dx$

111. $dy = -\dfrac{7}{(x + 5)^2} \, dx$

113. (a) $164
(b) The actual change in profit is $163.20, which is a difference of $0.80.

115. $B = 0.1\sqrt{5w}$

$\dfrac{dB}{dw} = \dfrac{0.05\sqrt{5}}{\sqrt{w}}$

$\Delta B \approx dB = \dfrac{0.05\sqrt{5}}{\sqrt{w}} \, dw$

$= \dfrac{0.05\sqrt{5}}{\sqrt{90}} (5)$

$\approx 0.059 \text{ m}^2$

Test Yourself   *(page 250)*

1. Critical number: $x = 0$
Increasing on $(0, \infty)$
Decreasing on $(-\infty, 0)$

2. Critical numbers: $x = -3$, $x = 3$
Increasing on $(-3, 3)$
Decreasing on $(-\infty, -3)$ and $(3, \infty)$

3. Critical number: $x = 6$
Increasing on $(6, \infty)$
Decreasing on $(-\infty, 6)$

4. Relative minimum: $(3, -14)$
Relative maximum: $(-3, 22)$

5. Relative minima: $(-1, -7)$ and $(1, -7)$
Relative maximum: $(0, -5)$

6. Relative maximum: $\left(0, \frac{8}{3}\right)$

7. Minimum: $(-3, -1)$
Maximum: $(0, 8)$

8. Minimum: $(0, 0)$
Maximum: $(2.25, 9)$

9. Minimum: $\left(2\sqrt{3}, 2\sqrt{3}\right)$
Maximum: $(1, 6.5)$

10. Concave upward on $(2, \infty)$
Concave downward on $(-\infty, 2)$
Point of inflection: $(2, -288)$

11. Concave upward on $\left(-\infty, -\dfrac{2\sqrt{2}}{3}\right)$ and $\left(\dfrac{2\sqrt{2}}{3}, \infty\right)$

Concave downward on $\left(-\dfrac{2\sqrt{2}}{3}, \dfrac{2\sqrt{2}}{3}\right)$

Points of inflection: $\left(-\dfrac{2\sqrt{2}}{3}, \dfrac{15}{8}\right)$ and $\left(\dfrac{2\sqrt{2}}{3}, \dfrac{15}{8}\right)$

12. No point of inflection
The graph is concave upward on its entire domain.

13. Concave upward on $(-\infty, -3)$ and $(3, \infty)$
Concave downward on $(-3, 3)$
Points of inflection: $(-3, -175)$ and $(3, -175)$

14. Relative minimum: $(6, -166)$
Relative maximum: $(-2, 90)$

15. Relative minimum: $(3, -97.2)$
Relative maximum: $(-3, 97.2)$

16. Vertical asymptote: $x = 5$
Horizontal asymptote: $y = 3$

17. Horizontal asymptote: $y = 0$

18. Vertical asymptote: $x = 1$

19.   **20.**

21.

22. $dy = 0.3$
$\Delta y = 0.3005$

23. (a) $250
(b) $\overline{C}(50) = \$82.40$; $\overline{C}(650) = \$80.18$
(c) Elastic: $[0, 350)$
Inelastic: $(350, 700]$
Unit elasticity: $x = 350$
(d) $1120; The actual change in revenue is $1080, which is a difference of $40.

Chapter 4

Section 4.1   *(page 256)*

1. (a) 46,656   (b) $\frac{1}{9}$ (c) 390,625   (d) $\frac{1}{125}$
3. (a) $\frac{1}{7}$ (b) 27   (c) 5   (d) 64
5. 2 g
7.
9.
11.
13.
15.
17.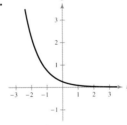

19. (a) About 28.62 million
    (b) About 31.38 million
21. (a) $80,634.95   (b) $161,269.89
23. $36.93

25. $V(t) = 30{,}000\left(\frac{4}{5}\right)^t$

$9830.40

27. (a)

Year	2008	2009	2010
Actual	104,260	106,630	109,380
Model	104,395	106,827	109,316

Year	2011	2012	2013
Actual	112,160	114,950	116,500
Model	111,863	114,470	117,137

The model fits the data well. Explanations will vary.

(b)   (c) 2018

Section 4.2   *(page 264)*

1. (a) e^7 (b) e^{12} (c) $\dfrac{1}{e^6}$ (d) 1
3. (a) e^5 (b) $e^{5/2}$ (c) e^6 (d) e^7
5. f   **6.** e   **7.** d   **8.** b   **9.** c   **10.** a
11.   13.

15.

17.

Horizontal asymptote: $N = 0$
Continuous on the entire
real number line

19.

Horizontal asymptote: $g = 0$
Continuous on about
$(-\infty, 6.13)$

21.

No horizontal asymptotes
Continuous on the entire
real number line

23.

Horizontal asymptote: $y = 1$
Discontinuous at $x = 0$

25. (a)

The graph of $g(x) = e^{x-2}$ is
shifted horizontally two units
to the right.

(b)

The graph of $h(x) = -\frac{1}{2}e^x$
decreases at a rate slower
than the rate at which the
graph of $f(x) = e^x$ increases.

(c)

The graph of $q(x) = e^x + 3$
is shifted vertically three
units upward.

27.

n	1	2	4	12
A	\$4031.75	\$4040.57	\$4045.05	\$4048.06

n	365	Continuous compounding
A	\$4049.53	\$4049.58

29.

n	1	2	4	12
A	\$11,235.95	\$11,400.40	\$11,485.38	\$11,543.09

n	365	Continuous compounding
A	\$11,571.31	\$11,572.28

31.

t	1	10	20
P	96,078.94	67,032.00	44,932.90

t	30	40	50
P	30,119.42	20,189.65	13,533.53

33.

t	1	10	20
P	95,132.82	60,716.10	36,864.45

t	30	40	50
P	22,382.66	13,589.88	8251.24

35. \$165,347.65
37. (a) 9%   (b) 9.20%   (c) 9.31%   (d) 9.38%
39. \$6450.04
41. (a) \$849.53   (b) \$421.12   (c) $\lim\limits_{x\to\infty} p = 0$
43. (a) 0.1535   (b) 0.4866   (c) 0.8111
45. (a) The model fits the data well.
(b) $R = 0.618t - 3.61$; The linear model fits the data well,
but the exponential model fits the data better.
(c) Exponential model: 2020
Linear model: 2039
47. Amounts earned:
(a) \$5267.71   (b) \$5255.81   (c) \$5243.23
You should choose the certificate of deposit in part (a) because
it earns more money than the others.
49. (a) 0.536   (b) 0.666
(c)

11 trials
(d) Yes, $\lim\limits_{n\to\infty} \dfrac{0.83}{1 + e^{-0.2n}} = 0.83$.
51. (a)

4 days

(b) Yes, $\lim\limits_{t\to\infty} \dfrac{925}{1 + e^{-0.3t}} = 925.$

(c) $\lim\limits_{t\to\infty} \dfrac{1000}{1 + e^{-0.3t}} = 1000$

Models similar to this logistic growth model, where $y = \dfrac{a}{1 + be^{-ct}},$ have a limit of a as $t\to\infty.$

Section 4.3   *(page 273)*

Skills Warm Up   *(page 273)*

1. $\dfrac{1}{2} e^x(2x^2 - 1)$ 2. $\dfrac{e^x(x + 1)}{x}$ 3. $e^x(x - e^x)$

4. $e^{-x}(8e^{2x} - x^3)$ 5. $-\dfrac{6}{7x^3}$ 6. $6x - \dfrac{1}{6}$

7. $6(2x^2 - x + 6)$ 8. $\dfrac{t + 12}{2(t + 5)^{3/2}}$

9. Relative maximum: $\left(-\dfrac{4\sqrt{3}}{3}, \dfrac{16\sqrt{3}}{9}\right)$

   Relative minimum: $\left(\dfrac{4\sqrt{3}}{3}, -\dfrac{16\sqrt{3}}{9}\right)$

10. Relative maximum: $(0, 5)$
    Relative minima: $(-1, 4), (1, 4)$

---

1. 0 3. $5e^{5x}$ 5. $-e^{3-x}$ 7. $\dfrac{14}{x^3} e^{-1/x^2}$

9. $4x^2 e^{-x}(3 - x)$ 11. $-\dfrac{6(e^x - e^{-x})}{(e^x + e^{-x})^4}$

13. $-\dfrac{2e^x}{(e^x - 1)^2}$ 15. $xe^x + e^x + 4e^{-x}$ 17. 4 19. $-\dfrac{1}{2}$

21. $y = -3x + 10$ 23. $y = \dfrac{4}{e^2}$ 25. $y = 24x + 8$

27. $\dfrac{dy}{dx} = \dfrac{10 - e^y}{xe^y + 3}$ 29. $\dfrac{dy}{dx} = \dfrac{e^{-x}(x^2 - 2x) + y}{4y - x}$

31. $6(3e^{3x} + 2e^{-2x})$ 33. $32e^{4x}(x + 1)$

35.

No relative extrema
No points of inflection
Horizontal asymptote to the right: $y = \dfrac{1}{2}$
Horizontal asymptote to the left: $y = 0$
Vertical asymptote: $x \approx -0.693$

37.
Relative minimum: $(0, 0)$
Relative maximum: $\left(2, \dfrac{4}{e^2}\right)$
Points of inflection:
$\left(2 - \sqrt{2}, 0.191\right),$
$\left(2 + \sqrt{2}, 0.384\right)$
Horizontal asymptote to the right: $y = 0$

39. $x = -\dfrac{1}{3}$ 41. $x = 9$

---

43. (a) The critical number $x = 0$ yields a relative minimum of the function on the interval $[-25, 25].$

(b)

The lowest point on the wire occurs at $(0, 50),$ which is midway between the two towers.

(c) About 6.4 ft

45. (a)

(b) $-\$5028.84$ per year
(c) $-\$217.01$ per year

47. (a) $\$433.31$ per year
(b) $\$890.22$ per year
(c) $\$21,839.26$ per year

49. (a) $f(x) = \dfrac{1}{120\sqrt{2\pi}} e^{-(x-513)^2/28,800}$

(b)

(c) $f'(x) = -\dfrac{1}{1,728,000\sqrt{2\pi}}(x - 513)e^{-(x-513)^2/28,800}$

(d) Answers will vary.

51.

As σ increases, the graph becomes flatter.

53. Proof; maximum: $\left(0, \dfrac{1}{\sigma\sqrt{2\pi}}\right)$

Quiz Yourself   *(page 275)*

1. $65,536$ 2. 216 3. 27 4. $\sqrt{15}$
5. e^7 6. $e^{11/4}$ 7. e^6 8. e^3

CHAPTER 4

9.   **10.**

11.   **12.**

13.   **14.**

15. 22.69 g   **16.** \$31.06

17. (a) \$2465.10   (b) \$2466.60   (c) \$2467.36

18. \$10,379.21   **19.** $7e^{7x}$ **20.** e^{x-4} **21.** $5e^{x+2}$

22. $e^x(2 - x)$ **23.** $y = -2x + 1$

24.

Relative maximum: $(4, 8e^{-2})$
Relative minimum: $(0, 0)$
Points of inflection:
$$\left(4 - 2\sqrt{2}, 0.382\right),$$
$$\left(4 + 2\sqrt{2}, 0.767\right)$$
Horizontal asymptote to the right: $y = 0$

25. (a) \$19.87   (b) \$993,274

Section 4.4   *(page 282)*

Skills Warm Up   *(page 282)*

1. $f^{-1}(x) = \frac{1}{5}x$ **2.** $f^{-1}(x) = x + 8$

3. $f^{-1}(x) = \dfrac{x - 2}{3}$ **4.** $f^{-1}(x) = \frac{4}{3}(x + 9)$

5. $x > -4$ **6.** Any real number x

7. $x < -1$ or $x > 1$ **8.** $x > \frac{5}{2}$

9. \$3462.03   **10.** \$3374.65

---

1. $e^{1.6094\cdots} = 5$ **3.** $e^{-1.6094\cdots} = 0.2$

5. $\ln(20.0855\ldots) = 3$ **7.** $\ln(0.0498\ldots) = -3$

9. c   **10.** d   **11.** b   **12.** a

13.   **15.**

17.   **19.**

21. x^2 **23.** $5x + 2$ **25.** $2x - 1$

27. $\ln 2 - \ln 3$ **29.** $\ln x + \ln y - \ln z$

31. $\frac{1}{3}\ln(2x + 7)$ **33.** $\ln z + 2\ln(z - 1)$

35. $\ln 3 + \ln x + \ln(x + 1) - 2\ln(2x + 1)$

37. (a) 1.7917   (b) -0.4055 (c) 4.3944   (d) 0.5493

39. $\ln\dfrac{x - 2}{x + 2}$ **41.** $\ln\dfrac{x^3 y^2}{z^4}$ **43.** $\ln\left[(x - 6)^5\sqrt{4x + 1}\right]$

45. $\ln\left[\dfrac{x(x + 5)}{(x^2 + 4)^4}\right]^{2/3}$

47. $x = 4$ **49.** $x = \ln 8 - 2 \approx 0.0794$

51. $t = \dfrac{\ln 7 - \ln 3}{-0.2} \approx -4.2365$

53. $x = \frac{1}{2}\left(1 + \ln\frac{3}{2}\right) \approx 0.7027$ **55.** $x = 1$

57. $x = \dfrac{e^{2.4}}{2} \approx 5.5116$ **59.** $x = e^3 \approx 20.0855$

61. $x = \dfrac{6e^3}{e^3 - 1} \approx 6.314$ **63.** $x = \dfrac{\ln 15}{2\ln 5} \approx 0.8413$

65. $t = \dfrac{\ln 2}{\ln 1.07} \approx 10.2448$

67. $t = \dfrac{\ln 3}{12\ln\left[1 + (0.07/12)\right]} \approx 15.7402$

69. Answers will vary.   **71.** Answers will vary.

73. (a) 8.15 yr   (b) 12.92 yr

75. (a) 14.21 yr   (b) 13.89 yr   (c) 13.86 yr   (d) 13.86 yr

77. (a) About 254.9 thousand   (b) 2023

79. 9395 yr   **81.** 12,484 yr

83. (a) 80   (b) 57.5   (c) 10 mo

85. (a) About 991 units   (b) About 319 units

87.

x	y	$\dfrac{\ln x}{\ln y}$	$\ln\dfrac{x}{y}$	$\ln x - \ln y$
1	2	0	-0.6931	-0.6931
3	4	0.7925	-0.2877	-0.2877
10	5	1.4307	0.6931	0.6931
4	0.5	-2	2.0794	2.0794

89.

91. False. $f(x) = \ln x$ is undefined for $x \le 0$.

93. False. $f\left(\dfrac{x}{2}\right) = f(x) - f(2)$ **95.** False. $u = v^2$

97. Options (b) and (c) will give you the same amount, but it makes more sense to double the rate, not the time. So option (b) is better than option (c). If you are looking for a long-term investment, choose option (a).

99.   Answers will vary.

Section 4.5 *(page 291)*

Skills Warm Up   *(page 291)*

1. $2\ln(x + 1)$ **2.** $\ln x + \ln(x + 1)$

3. $\ln x - \ln(x + 1)$ **4.** $3[\ln x - \ln(x - 3)]$

5. $\ln 4 + \ln x + \ln(x - 7) - 2\ln x$

6. $3\ln x + \frac{1}{2}\ln(x + 1)$

7. $-\dfrac{y}{x + 2y}$ **8.** $\dfrac{3 - 2xy + y^2}{x(x - 2y)}$

9. $-12x + 2$ **10.** $-\dfrac{1}{x^4}$

1. $\dfrac{8}{x}$ **3.** $\dfrac{2x}{x^2 + 3}$ **5.** $\dfrac{3}{4(3x - 5)}$ **7.** $\dfrac{4}{x}(\ln x)^3$

9. $2\ln x + 2$ **11.** $\dfrac{4x^6 - 2}{x(x^6 - 2)}$ **13.** $\dfrac{1}{x(x + 1)}$

15. $\dfrac{2}{3(x^2 - 1)}$ **17.** $-\dfrac{4}{x(4 + x^2)}$ **19.** $e^{-x}\left(\dfrac{1}{x} - \ln x\right)$

21. $\dfrac{e^x - e^{-x}}{e^x + e^{-x}}$ **23.** 2   **25.** -3 **27.** 3   **29.** 1.404

31. 5.585   **33.** -0.631 **35.** $(\ln 3)3^x$ **37.** $\dfrac{1}{x\ln 2}$

39. $(2\ln 4)4^{2x-3}$ **41.** $\dfrac{2x + 6}{(x^2 + 6x)\ln 10}$ **43.** $2^x(1 + x\ln 2)$

45. $y = 3x - 3 = 3(x - 1)$ **47.** $y = 4e^2 x - 3e^3$

49. $y = -\frac{8}{5}x - 4$ **51.** $y = \dfrac{1}{27\ln 3}x - \dfrac{1}{\ln 3} + 3$

53. $\dfrac{2xy}{3 - 2y^2}$ **55.** $\dfrac{y(1 - 6x^2)}{1 + y}$ **57.** $x(6\ln x + 5)$

59. $\dfrac{6\ln x - 5}{3x^4}$ **61.** $(\ln 5)^2 5^x$

63. $\dfrac{d\beta}{dI} = \dfrac{10}{(\ln 10)I}$, so for $I = 10^{-4}$, the rate of change is about 43,429.4 dB/W/cm^2.

65. Relative maximum: $(1, -1)$

67. Relative maximum: $(-2, \ln 5)$

69.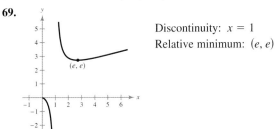

Discontinuity: $x = 1$
Relative minimum: (e, e)

71.

Relative minimum: $\left(4e^{-1/2}, \dfrac{-8}{e}\right)$

Point of inflection: $\left(4e^{-3/2}, \dfrac{-24}{e^3}\right)$

73. $-\dfrac{1}{p}, -\dfrac{1}{10}$

75. $p = 1000e^{-x}$

$\dfrac{dp}{dx} = -1000e^{-x}$

At $p = 10$, rate of change $= -10$.

$\dfrac{dp}{dx}$ and $\dfrac{dx}{dp}$ are reciprocals of each other.

77. (a) $\overline{C} = \dfrac{500 + 300x - 300\ln x}{x}$

(b) Minimum of 279.15 at $e^{8/3}$

79. (a)

(b) About 29.39 thousand per year

81. (a)

(b) $t \approx 30$; $503,434.80   (c) $t \approx 20$; $386,685.60

(d) About -0.081; About -0.029

(e) For a higher monthly payment, the term is shorter and the total amount paid is smaller.

83. (a) $s(t) = 84.66 - 11.00 \ln t$

(b)

The model fits the data well.

(c) $t = 2: -5.5$ $t = 5: -2.2$

(d) After 2 months, the average score is decreasing at a rate of 5.5 points per month. After 5 months, the average score is decreasing at a rate of 2.2 points per month.

Section 4.6   *(page 299)*

Skills Warm Up   *(page 299)*

1. $-\dfrac{1}{4} \ln 2$ **2.** $\dfrac{1}{5} \ln 2$ **3.** $-\dfrac{\ln(25/16)}{0.01}$

4. $-\dfrac{\ln(11/16)}{0.02}$ **5.** $7.36e^{0.23t}$ **6.** $1.296e^{0.072t}$

7. $-33.6e^{-1.4t}$ **8.** $-0.025e^{-0.001t}$ **9.** 4

10. 4 **11.** $2x + 1$ **12.** $x^2 + 1$

---

1. $y = 2e^{0.1014t}$ **3.** $y = 4e^{-0.4159t}$

5. $y = 4\sqrt[3]{2}e^{[(\ln 0.5)/3]t}$ **7.** $y = 10e^{2t}$, exponential growth

9. $y = 30e^{-4t}$, exponential decay

11. 6.48 g; 0.13 g   **13.** 2.16 g; 1.62 g   **15.** 8.66 g; 7.48 g

17. About 68%   **19.** 15,642 yr

21. $k_1 = \dfrac{\ln 4}{12} \approx 0.1155$, so $y_1 = 5e^{0.1155t}$.

$k_2 = \dfrac{1}{6}$, so $y_2 = 5(2)^{t/6}$.

Explanations will vary.

23. (a) 1350   (b) $\dfrac{5 \ln 2}{\ln 3} \approx 3.15$ h

(c) No. Answers will vary.

25. 5.78 yr; $3320.12; $20,085.54

27. 8.66%; $1783.04; $6535.95

29. 9.50%; 7.30 yr; $5375.51

31. $4869.79; 15.40 yr; $7637.35   **33.** $49,787.07

35. (a) Answers will vary.   (b) Answers will vary.

37. Answers will vary.

39. (a) $y = 33,000 - 4048x$ (b) $y = 33,000e^{-0.153x}$

(c) Linear Model: Exponential Model:
    after 1 year: $28,952   after 1 year: $28,318.28
    after 4 years: $16,808   after 4 years: $17,894.75

(d)

The exponential model depreciates slightly faster.

(e) After the third year, a buyer would gain an advantage by using the linear model, because it yields a lower value for the vehicle. A seller would want to use the exponential model, because it yields a higher value for the vehicle.

41. (a) $C = 30$

$k = \ln\left(\tfrac{1}{6}\right) \approx -1.7918$

(b) $30e^{-0.35836} = 20.9646$ thousand, or 20,965 units

(c)

43. About 36 days

45. (a) $C \approx 81.090$, $k = \dfrac{\ln(45/40)}{-200} \approx -0.0005889$

(b) $x = 1/0.0005889 \approx 1698$ units, $p = 29.83

47. Answers will vary.

Review Exercises for Chapter 4   *(page 306)*

1. (a) 16,384   (b) 40,353,607   (c) 0.0625   (d) 81

3.   **5.**

7.

9. (a) About 8530.75 thousand

(b) About 9388.31 thousand

11. (a) $69,295.66   (b) $233,081.88

13. (a) e^{21} (b) $\dfrac{1}{e^2}$ (c) $e^{14/3}$ (d) $\dfrac{1}{e^8}$

15.   **17.**

19.

n	1	2	4	12
A	$1216.65	$1218.99	$1220.19	$1221.00

n	365	Continuous compounding
A	$1221.39	$1221.40

21.

n	1	2	4	12
A	4231.80	4244.33	4250.73	4255.03

n	365	Continuous compounding
A	4257.13	4257.20

23. b
25. (a) 4%   (b) 4.04%   (c) 4.06%   (d) 4.07%
27. $9744.69
29. (a) $8276.81   (b) $7697.12   (c) $7500
31. (a) The model fits the data well.
   (b) $y = 162.93t - 474.7$
      The linear model fits the data moderately well.
      The exponential model is a better fit.
   (c) Exponential: $5033.45 million
      Linear: $2458.04 million
33. (a) $P \approx 1049$ fish
   (b)

   13 months

   (c) Yes, P approaches 10,000 fish at t approaches ∞.

35. $18x^5 e^{x^6}$ **37.** $\dfrac{1 - 2x}{e^{2x}}$ **39.** $-\dfrac{10e^{2x}}{(1 + e^{2x})^2}$

41. $y = 3 - x$ **43.** $y = \dfrac{27}{e^3}$

45.

Relative minimum: $(-3, -1.344)$
Inflection points: $(0, 0), (-3 + \sqrt{3}, -0.574)$,
   and $(-3 - \sqrt{3}, -0.933)$
Horizontal asymptote: $y = 0$

47.

Relative maximum: $(-1, -2.718)$
Horizontal asymptote: $y = 0$
Vertical asymptote: $x = 0$

49. $e^{2.4849} \approx 12$ **51.** $\ln 4.4816 \approx 1.5$
53.

55.

57. $\ln x + \frac{1}{2}\ln(x - 1)$ **59.** $2 \ln x - 3 \ln(x + 1)$
61. $3[\ln(1 - x) - \ln 5 - \ln x - \ln z]$

63. $\ln(2x^2 - x - 15)$ **65.** $4 \ln\left(\dfrac{x^5 - x^2}{x - 5}\right)$

67. 12   **69.** $e^3 \approx 20.09$ **71.** $-\dfrac{\ln(0.25)}{1.386} \approx 1.0002$

73. $\frac{1}{2}(1 + \ln 8) \approx 1.5397$ **75.** $-\dfrac{e^5}{2 - 3e^5} \approx 0.3348$

77. $\dfrac{3 + \sqrt{13}}{2} \approx 3.3028$ **79.** $\dfrac{\ln 1.1}{\ln 1.21} = 0.5$

81. (a) About 28.07 yr   (b) About 27.75 yr
   (c) About 27.73 yr   (d) About 27.73 yr
83. (a) 75   (b) 65.34   (c) About 11 months

85. $\dfrac{2}{x}$ **87.** $\dfrac{4x}{3(x^2 - 2)}$ **89.** $\dfrac{3x^2}{x^3 - 5}$

91. $\dfrac{1 - 4 \ln x}{x^5}$ **93.** $\dfrac{x + 2}{x(x + 1)}$

95. $\dfrac{2}{x} + \dfrac{1}{2(x + 1)}$ **97.** $\dfrac{1}{1 + e^x}$

99. 2   **101.** 0
103. 1.594   **105.** 1.500   **107.** $(2 \ln 5)5^{2x + 1}$

109. $\dfrac{2}{(2x - 1) \ln 3}$ **111.** $\dfrac{-1}{\ln 10} \cdot \dfrac{1}{x} = -\dfrac{1}{x \ln 10}$

113. Relative maximum: $(3, \ln 27 - 3)$
115. Relative minimum: $(-1, \ln 6)$
117. 2010: About 11.7 million
   2012: About 9.7 million
119. $y = 3e^{-0.27465t}$
121. 5.19 g; 0.10 g
123. 20.18 g; 17.88 g
125. 2.47 g; 1.85 g
127. 8.66 yr; $1335.32; $4433.43
129. 2%; 34.66 yr; $24,730.82
131. (a) $D = 500e^{-0.38376t}$
   (b) 107.72 mg/mL
   (c)

Test Yourself   *(page 310)*

1. 1   **2.** 4   **3.** $e^{9/2}$ **4.** e^{20}

5.

6.

7.

8.

9.

10.

11. $\ln 3 - \ln 2$ **12.** $\ln z + \frac{1}{2}\ln(x + y)$

13. $3[\ln(x + 1) - \ln y - \ln z]$ **14.** $\ln[y(x + 1)]$

15. $\ln\dfrac{x^3}{(x - 1)^2}$ **16.** $\ln\dfrac{xy^4}{\sqrt{z + 4}}$

17. $x \approx 11.038$ **18.** $x \approx 1.750$ **19.** $x \approx 58.371$

20. (a) 17.67 yr   (b) 17.36 yr   (c) 17.33 yr   (d) 17.33 yr

21. $-3e^{-3x}$ **22.** $7e^{x+2} + 2$ **23.** $3(\ln 6)(6^{3x+1})$

24. $-\dfrac{3x^2}{7 - x^3}$ **25.** $\dfrac{2}{x(x + 2)}$ **26.** $2e^{x^4}\left[\dfrac{1 + 2x^4\ln(4x^2)}{x}\right]$

27. (a) About \$26.88 billion
    (b) About \$2.19 billion/yr

28. About 65%   **29.** About 39.61 yr

Chapter 5

Section 5.1   *(page 319)*

Skills Warm Up   *(page 319)*

1. $x^{-1/2}$ **2.** $(2x)^{4/3}$ **3.** $5^{1/2}x^{3/2} + x^{5/2}$

4. $x^{-1/2} + x^{-2/3}$ **5.** $(x + 1)^{5/2}$ **6.** $x^{7/6}$

7. -12 **8.** -10 **9.** 14   **10.** 0

1–5. Answers will vary.   **7.** $u + C$ **9.** $6x + C$

11. $\frac{7}{2}x^2 + C$ **13.** $-3t^3 + C$

15. $-\dfrac{5}{2x^2} + C$ **17.** $\frac{2}{5}y^{5/2} + C$

	Rewrite	Integrate	Simplify
19.	$\displaystyle\int x^{2/3}\,dx$	$\dfrac{x^{5/3}}{5/3} + C$	$\dfrac{3}{5}x^{5/3} + C$
21.	$\displaystyle\int x^{-3/2}\,dx$	$\dfrac{x^{-1/2}}{-1/2} + C$	$-\dfrac{2}{\sqrt{x}} + C$
23.	$\dfrac{1}{2}\displaystyle\int x^{-3}\,dx$	$\dfrac{1}{2}\left(\dfrac{x^{-2}}{-2}\right) + C$	$-\dfrac{1}{4x^2} + C$

25. $\dfrac{x^2}{2} + 3x + C$ **27.** $\frac{1}{6}x^6 - 8x$

29. $\frac{1}{3}x^3 + \frac{5}{2}x^2 + x + C$ **31.** $\frac{3}{4}x^4 - 2x^3 + 2x + C$

33. $2x + \dfrac{1}{2x^2} + C$ **35.** $3x^{2/3}(x + 2) + C$

37. $x^3 + \frac{1}{2}x^2 - 2x + C$

39.

41.

43. $f(x) = 3x^2 + 8$ **45.** $f(x) = x^3 + 4x - 1$

47. $f(x) = 5x^4 - 2x^2$ **49.** $f(x) = -\dfrac{1}{x^2} + \dfrac{1}{x} + \dfrac{1}{2}$

51. $f(x) = x^2 + x + 4$ **53.** $f(x) = \frac{9}{4}x^{4/3}$

55. $C = 85x + 5500$ **57.** $C = \frac{1}{10}\sqrt{x} + 4x + 750$

59. $R = 225x - \frac{3}{2}x^2, p = 225 - \frac{3}{2}x$

61. $P = -9x^2 + 1650x$ **63.** $P = -12x^2 + 805x + 68$

65. $s(t) = -16t^2 + 6000$; about 19.36 sec

67. $v_0 = 8\sqrt{555} \approx 188.47$ ft/sec

69. (a) $C = x^2 - 12x + 125$

$$\overline{C} = x - 12 + \dfrac{125}{x}$$

    (b) \$2025
    (c) \$125 is fixed. \$1900 is variable. Examples will vary.

71. (a) $P(t) = 72.76t^2 + 2002.3t + 69{,}017.86$
    (b) 360,384; Yes, this seems reasonable. Explanations will vary.

73. (a) $M(t) = 0.0306t^3 - 0.994t^2 + 10.28t - 1.3624$
    (b) 45.009 million; Yes, this seems reasonable. Explanations will vary.

75. (a)

$$\dfrac{dB}{dx} = -19.9x + 351$$
$$\dfrac{dC}{dx} = 5.38x^2 - 40.6x + 182$$

    (b) $B(x) = -9.95x^2 + 351x$;
    $C(x) = 1.79x^3 - 20.3x^2 + 182x + 425$

    (c) $(2.32, 12.00)$; The company should produce from 3 to 11 units.

Section 5.2 *(page 329)*

Skills Warm Up *(page 329)*

1. $\frac{1}{2}x^4 + x + C$ **2.** $\frac{3}{2}x^2 + \frac{2}{3}x^{3/2} - 4x + C$

3. $-\dfrac{1}{4x^4} + C$ **4.** $-\dfrac{1}{8x} + C$

5. $\frac{4}{7}t^{7/2} + \frac{2}{5}t^{5/2} + C$ **6.** $\frac{4}{5}x^{5/2} - \frac{2}{3}x^{3/2} + C$

7. $\dfrac{5x^3 - 4}{2x} + C$ **8.** $\dfrac{-6x^2 + 5}{3x^3} + C$

9. $\frac{2}{5}\sqrt{x}(8x^2 + 15) + C$

$\displaystyle\int u^n \frac{du}{dx}\,dx$	u	$\dfrac{du}{dx}$
1. $\displaystyle\int (5x^2 + 1)^2(10x)\,dx$	$5x^2 + 1$	$10x$
3. $\displaystyle\int \sqrt{1 - x^2}\,(-2x)\,dx$	$1 - x^2$	$-2x$
5. $\displaystyle\int \left(4 + \frac{1}{x^2}\right)^5\left(\frac{-2}{x^3}\right) dx$	$4 + \dfrac{1}{x^2}$	$-\dfrac{2}{x^3}$
7. $\displaystyle\int \left(1 + \sqrt{x}\right)^3\left(\frac{1}{2\sqrt{x}}\right) dx$	$1 + \sqrt{x}$	$\dfrac{1}{2\sqrt{x}}$

9. $\frac{1}{5}(x - 1)^5 + C$ **11.** $\frac{1}{3}(7 - 2x)^3 + C$

13. $\frac{1}{2}(x^2 + 3x)^2 + C$ **15.** $\frac{2}{3}(4x^2 - 5)^{3/2} + C$

17. $\dfrac{1}{3(5 - 3x^2)^3} + C$ **19.** $\dfrac{(2x^3 - 1)^5}{30} + C$

21. $\frac{1}{3}(t^2 + 6)^{3/2} + C$ **23.** $\dfrac{1}{20(3 - 2x^5)^2} + C$

25. $\frac{1}{10}(x^2 - 6x)^5 + C$ **27.** $-\dfrac{1}{3(x^3 + 9x - 4)} + C$

29. $-\dfrac{15}{8}(1 - x^2)^{4/3} + C$ **31.** $-\dfrac{3}{2(1 + x^2)} + C$

33. $-3\sqrt{2t + 3} + C$ **35.** $\frac{3}{16}(4x + 3)^{4/3} + C$

37. $\frac{1}{48}(6x^2 - 7)^4 + C$ **39.** $\frac{4}{7}\sqrt{7x - 1} + C$

41. $\frac{2}{3}\sqrt{x^3 + 3x + 4} + C$

43. (a) $\frac{1}{3}x^3 - x^2 + x + C_1 = \frac{1}{3}(x - 1)^3 + C_2$
 (b) Answers differ by a constant: $C_1 = C_2 - \frac{1}{3}$
 (c) Answers will vary.

45. (a) $\frac{1}{6}x^6 + x^4 + 2x^2 + C_1 = \frac{1}{6}(x^2 + 2)^3 + C_2$
 (b) Answers differ by a constant: $C_1 = C_2 - \frac{1}{6}$
 (c) Answers will vary.

47. $f(x) = \frac{1}{5}[(2 - x^3)^5 + 3]$

49. $f(x) = \frac{1}{12}(4x^2 - 10)^3 - 8$

51. (a) $C = 8\sqrt{x + 1} + 18$ (b) \$75.13

53. $x = \frac{1}{3}(p^2 - 25)^{3/2} + 24$ **55.** $x = \dfrac{6000}{\sqrt{p^2 - 16}} + 3000$

57. (a) $h = \sqrt{17.6t^2 + 1} + 5$ (b) 26 in.

59. (a) $Q = (x - 24{,}999)^{0.95} + 24{,}999$

(b)

x	25,000	50,000	100,000	150,000
Q	25,000	40,067.14	67,786.18	94,512.29
$x - Q$	0	9932.86	32,213.82	55,487.71

(c)

61. $-\frac{2}{3}x^{3/2} + \frac{2}{3}(x + 1)^{3/2} + C$

Section 5.3 *(page 336)*

Skills Warm Up *(page 336)*

1. $x + 2 - \dfrac{2}{x + 2}$ **2.** $x - 5 + \dfrac{1}{x - 3}$

3. $x + 8 + \dfrac{2x - 4}{x^2 - 4x}$ **4.** $x^2 - x - 4 + \dfrac{20x + 22}{x^2 + 5}$

5. $\frac{1}{4}x^4 - \dfrac{1}{x} + C$ **6.** $\frac{1}{2}x^2 + 2x + C$

7. $\frac{1}{2}x^2 - \dfrac{4}{x} + C$ **8.** $-\dfrac{2}{x^2} + \dfrac{5}{3x^3} + C$

---

1. $e^{7x} + C$ **3.** $\frac{1}{3}x^3 - 4e^x + C$ **5.** $\frac{1}{5}e^{5x-3} + C$

7. $-\frac{9}{4}e^{-2x^2} + C$ **9.** $\frac{5}{3}e^{x^3} + C$ **11.** $e^{x^2+x} + C$

13. $4\ln|x| + C$ **15.** $\ln|5x + 2| + C$

17. $-\frac{1}{2}\ln|3 - 2x| + C$ **19.** $\frac{2}{3}\ln|3x + 5| + C$

21. $\ln\sqrt{x^2 + 1} + C$ **23.** $\frac{1}{3}\ln|x^3 + 1| + C$

25. $\frac{1}{2}\ln|x^2 + 6x + 7| + C$ **27.** $\ln|\ln x| + C$

29. $\ln|1 - e^{-x}| + C$

31. $\frac{1}{4}x^2 - 4\ln|x| + C$; General Power Rule and Log Rule

33. $8x + 3\ln|x| - \dfrac{3}{x^2} + C$; General Power Rule and Log Rule

35. $e^x + 2x - e^{-x} + C$;
 Exponential Rule and General Power Rule

37. $-\frac{2}{3}(1 - e^x)^{3/2} + C$; Exponential Rule

39. $\ln|e^x + x| + C$; Log Rule

41. $\frac{1}{7}\ln(7e^{5x} + 1) + C$; Log Rule

43. $\frac{1}{2}x^2 + 3x + 8\ln|x - 1| + C$;
 General Power Rule and Log Rule

45. $x - 6\ln|x + 3| + C$;
 General Power Rule, General Log Rule

47. $f(x) = \dfrac{-e^{2/x}}{2} + \dfrac{e^{1/2}}{2} + 6$

49. $f(x) = \frac{1}{2}x^2 + 5x + 8\ln|x - 1| - 8$

51. (a) $P(t) = 1000[1 + \ln(1 + 0.25t)^{12}]$
 (b) $P(3) \approx 7715$ bacteria (c) $t \approx 6$ days

53. (a) $p = -50e^{-x/500} + 45.06$

(b)

The price increases as the demand increases.

(c) 387

55. (a) $R(t) = 0.293e^{0.2387t} + 0.276$

(b) About $5.4 billion

57. False. $\ln x^{1/2} = \frac{1}{2}\ln x$

Quiz Yourself   *(page 338)*

1. $3x + C$ **2.** $5x^2 + C$ **3.** $-\dfrac{1}{5x^5} + C$

4. $\dfrac{x^3}{3} - x^2 + 15x + C$ **5.** $\dfrac{(6x + 1)^4}{4} + C$

6. $\dfrac{1}{50}(5x^2 - 2)^5 + C$ **7.** $\dfrac{(x^2 - 5x)^2}{2} + C$

8. $-\dfrac{2}{3(x^3 + 3)^2} + C$ **9.** $\dfrac{4}{21}(7x + 4)^{3/2}$

10. $f(x) = 8x^2 + 1$ **11.** $f(x) = 2x^3 + x + 6$

12. (a) $C = -0.03x^2 + 16x + 9.03$ (b) $9.03   (c) $509.03

13. $f(x) = 2(x^2 + 16)^{5/4} - 63$

14. (a) 1000 bolts   (b) About 8612 bolts

15. $e^{10x+7} + C$ **16.** $3e^{x^3} + C$

17. $\frac{1}{2}e^{(x^2-6x)} + C$ **18.** $\ln|9x - 4| + C$

19. $-\frac{1}{8}\ln|3 - 8x| + C$ **20.** $\frac{1}{6}\ln(3x^2 + 4) + C$

21. (a) $R(t) = 17.69t^2 - 612.804\ln(t) + 771.69$

(b) About $1129.66 million

Section 5.4   *(page 348)*

Skills Warm Up   *(page 348)*

1. $\frac{3}{2}x^2 + 7x + C$ **2.** $\frac{2}{5}x^{5/2} + \frac{4}{3}x^{3/2} + C$

3. $\dfrac{1}{5}\ln|x| + C$ **4.** $-\dfrac{1}{4e^{4x}} + C$

5. $C = 0.008x^{5/2} + 29{,}500x + C$

6. $R = x^2 + 9000x + C$

7. $P = 0.01x^3 + 4600x + C$

1.
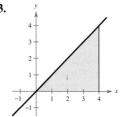

Area = 10

3.

Area = 8

5.

Area $= \dfrac{9\pi}{2}$

7. (a) 8   (b) 4   (c) -24 (d) 0

9. $\dfrac{1}{6}$ **11.** $\dfrac{1}{2}$ **13.** $6\left(1 - \dfrac{1}{e^2}\right)$ **15.** $8\ln 2 + \dfrac{15}{2}$

17. 10   **19.** $-\frac{5}{2}$ **21.** 38   **23.** $-\frac{15}{4}$ **25.** -4

27. $-\frac{27}{20}$ **29.** $6\ln 2 \approx 4.16$ **31.** 2

33. $-e^{-1} + 1 \approx 0.63$ **35.** $\frac{1}{3}\left[(e^2 + 1)^{3/2} - 2\sqrt{2}\right] \approx 7.157$

37. $\frac{1}{8}\ln 17 \approx 0.354$ **39.** 10   **41.** 39   **43.** 10

45. $4\ln 3 \approx 4.394$ **47.** $6.75   **49.** $22.50   **51.** $3.97

53. Average $= 9$

$x = \dfrac{3}{2}$

55. Average $= \dfrac{8}{3}$

$x = \pm\dfrac{2\sqrt{3}}{3} \approx \pm 1.155$

57. Average $= e - e^{-1} \approx 2.3504$

$x = \ln\left(\dfrac{e - e^{-1}}{2}\right) \approx 0.1614$

59. Average $= \frac{3}{4}\ln\frac{7}{3} \approx 0.6355$ **61.** $\frac{6}{5}$ **63.** 0

$x = \dfrac{4}{\ln(7/3)} - 2 \approx 2.721$

65. (a) $\frac{1}{3}$ (b) $\frac{2}{3}$ (c) $-\frac{1}{3}$

Explanations will vary.

67. $1925.23   **69.** $16,605.21

71. $2500   **73.** $4565.65

75. (a) $137,000   (b) $214,720.93   (c) $338,393.53

77. $2623.94

79. (a) $M(t) = -1.079t + 1.7033t^{1.5} - 0.37714t^2$
$\qquad + 0.0000061e^t + 5.1$

(b) About $8.98 trillion

Section 5.5   *(page 357)*

Skills Warm Up   *(page 357)*

1. $-x^2 + 3x + 2$ **2.** $-2x^2 + 4x + 4$

3. $-x^3 + 2x^2 + 4x - 5$ **4.** $x^3 - 6x - 1$

5. $(0, 4), (4, 4)$ **6.** $(1, -3), (2, -12)$

7. $(-3, 9), (2, 4)$ **8.** $(-2, -4), (0, 0), (2, 4)$

---

1. 36   **3.** 9   **5.** $e - 2$ **7.** $\frac{3}{2}$

9.

11.

13. d

15.

Area $= 2\frac{1}{6}$

17.

Area $= \frac{64}{3}$

19.

Area $= \frac{4}{5}$

21.

Area $= \frac{8}{5}$

23.

Area $= 21\frac{1}{12}$

25.

Area $= -\frac{1}{2}e^{-1} + \frac{3}{2}$

27.

Area $= (2e + \ln 2) - 2e^{1/2}$

29.

Area $= \frac{7}{3} + 8 \ln 2$

31.

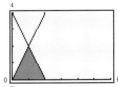

Area $= \displaystyle\int_0^1 2x\, dx + \int_1^2 (4 - 2x)\, dx$

33.

Area $\displaystyle\int_1^2 \left(\frac{4}{x} - x\right) dx + \int_2^4 \left(x - \frac{4}{x}\right) dx$

35.

Area $= \frac{32}{3}$

37.

Area $= \frac{1}{6}$

39.

Area $= \frac{128}{15}$

41. 8

43. Consumer surplus $= 1600$
Producer surplus $= 400$

45. Consumer surplus $= 500$
Producer surplus $= 2000$

47. Consumer surplus $= 640.00$
Producer surplus ≈ 426.67

49. R_1; \$14.625 billion

51. \$609 million; Explanations will vary.

53. (a)

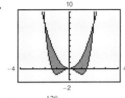

(b) 19.894 fewer pounds

55. \$333.33 million

57. $CS = \$700{,}000$
$PS = \$1{,}375{,}000$

59. (a) \$2148.95

(b)

Quintile	Lowest	2nd	3rd	4th	Highest
Percent	2.58	6.64	14.14	26.60	45.54

Section 5.6   *(page 364)*

Skills Warm Up   *(page 364)*

1. $\frac{1}{6}$ **2.** $\frac{3}{20}$ **3.** $\frac{7}{40}$ **4.** $\frac{13}{12}$ **5.** $-\frac{59}{30}$ **6.** $\frac{53}{18}$

7. $\frac{2}{3}$ **8.** $\frac{4}{7}$ **9.** 0   **10.** 5

1. Approximation: 2
Exact area: 2

3. Approximation: 0.6730
Exact area: $\frac{2}{3} \approx 0.6667$

5. Approximation: 1.245
Exact area: $\frac{5}{4} = 1.25$

7. 71.25

9. 1.079

11. 20.37   **13.** 24.28

15. 9.875   **17.** 24.56

19. 1.39

21.

n	10	20	30
Approximation	15.4543	15.4628	15.4644

n	40	50
Approximation	15.4650	15.4652

23.

n	10	20	30
Approximation	5.8520	5.8526	5.8528

n	40	50
Approximation	5.8528	5.8528

25. Area $\approx$ 54.6667,   **27.** Area $\approx$ 0.9163,
$n = 33$ $n = 3$

29. 9920 ft^2 **31.** 381.6 mi^2

33. Midpoint Rule: 3.1468
Graphing utility: 3.141593

Review Exercises for Chapter 5 *(page 370)*

1. $16x + C$ **3.** $\frac{3}{10}x^2 + C$ **5.** $x^3 + C$

7. $x^{2/3} + C$ **9.** $\frac{2}{3}x^3 + \frac{5}{2}x^2 + C$ **11.** $\frac{3}{7}x^{7/3} + \frac{3}{2}x^2 + C$

13. $\frac{4}{9}x^{9/2} - 2\sqrt{x} + C$ **15.** $6x^2 - 3$ **17.** $\frac{2}{3}x^{3/2} - 14$

19. $s(t) = -16t^2 + 80t$; 5 sec

21. $\frac{1}{4}(x + 4)^4 + C$ **23.** $\frac{1}{5}(5x + 1)^5 + C$

25. $\frac{1}{6}x^6 + 4x^5 + 32x^4 + \frac{256}{3}x^3 + C$ or $\frac{1}{6}(x^2 + 8x)^3 + C_1$

27. $\frac{4}{27}(3x^3 + 1)^3 + C$ **29.** $\frac{-1}{12(2x^3 - 5)^2} + C$

31. $\frac{4}{7}\sqrt{7x - 1} + C$

33. (a) 30.54 board-feet   (b) 125.2 board-feet

35. $e^{4x} + C$ **37.** $-\frac{1}{5}e^{-5x} + C$

39. $\frac{7e^{3x^2}}{6} + C$ **41.** $\ln|x - 6| + C$

43. $\frac{2}{3}\ln|6x - 1| + C$ **45.** $-\frac{1}{3}\ln|1 - x^3| + C$

47.   **49.**

Area $= 8$ Area $= 8$

51. (a) 13   (b) 7   (c) 11   (d) 50

53. $\frac{32}{3}$ **55.** 8   **57.** 2 ln 2   **59.** 16   **61.** 0

63. 34   **65.** 5 ln 3 $\approx$ 5.49   **67.** $6 - 2\sqrt{5} \approx 1.53$

69. 1.899   **71.** Average value $= 3$; $x = 1$

73. Average value $= \frac{2}{3}(1 - e^3) \approx -12.724$;
$x = \ln\left[-\frac{1}{3}(1 - e^3)\right] \approx 1.850$

75. Average value $= \frac{2}{5}$; $x = \frac{25}{4}$ **77.** 0   **79.** 115.2

81. \$17,492.94   **83.** Increases by \$700.25   **85.** \$565.10

87.   **89.**

          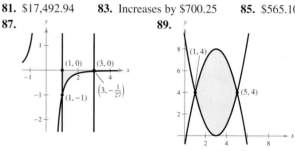

Area $= \frac{4}{9}$ Area $= \frac{64}{3}$

91.   **93.**

          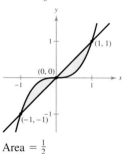

Area $= 20$ Area $= \frac{1}{2}$

95. Consumer surplus $= 1417.5$
Producer surplus $= 202.5$

97. Consumer surplus $= 1250$
Producer surplus $= 1250$

99. R_2; \$112.5 million

101. \$300 million

103. Approximation: 1.5
Actual area: 1.5

105. 3.875

107. 1.738

109. 0.987

111. 36.5

113. 7.6615

115. 9840 ft²

18.

Area = $\frac{5}{12}$

19.

Area = 8

20. Consumer surplus = 20 million
Producer surplus = 8 million

21. Midpoint Rule: $\frac{63}{64} \approx 0.9844$
Exact area: 1

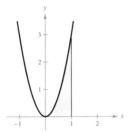

22. Midpoint Rule: $\frac{37}{8} = 4.625$
Exact area: $\frac{14}{3} \approx 4.667$

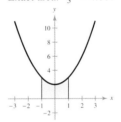

Test Yourself   *(page 374)*

1. $3x^3 - 2x^2 + 13x + C$ **2.** $\dfrac{(x + 1)^3}{3} + C$

3. $\dfrac{(x^4 - 7)^{3/2}}{6} + C$ **4.** $\dfrac{10x^{3/2}}{3} - 12x^{1/2} + C$

5. $5e^{3x} + C$ **6.** $\frac{3}{4} \ln|4x - 1| + C$

7. $f(x) = \frac{5}{2}x^2 + 2x - 18$ **8.** $f(x) = e^x + x$

9. 8   **10.** 300   **11.** $\frac{2}{3}$

12. $2\sqrt{5} - 2\sqrt{2} \approx 1.644$ **13.** $8(e^{1/2} - e^{1/4}) \approx 2.918$

14. ln 6 ≈ 1.792

15. (a) $S(t) = 3.0156e^{0.096t} + 0.00002185,\ 4 \le t \le 13$
  (b) \$7.03 billion

16.

Area = $\frac{343}{6}$

17.

Area = $\frac{16}{3}$

Chapter 6

Section 6.1   *(page 383)*

Skills Warm Up   *(page 383)*

1. $\dfrac{1}{x + 1}$ **2.** $\dfrac{2x}{x^2 - 1}$ **3.** $3x^2 e^{x^3}$

4. $-2xe^{-x^2}$ **5.** $e^x(x^4 + 4x^3)$ **6.** $e^{-4x}(1 - 4x)$

7. $\frac{64}{3}$ **8.** $\frac{4}{3}$ **9.** 36   **10.** 8

1. $u = x;\ dv = e^{9x}\,dx$ **3.** $u = \ln 2x;\ dv = x\,dx$

5. $\frac{1}{3}xe^{3x} - \frac{1}{9}e^{3x} + C$ **7.** $\dfrac{x^4}{16}(4 \ln x - 1) + C$

9. $x \ln 5x - x + C$ **11.** $-x^2 e^{-x} - 2xe^{-x} - 2e^{-x} + C$

13. $\frac{2}{3}x^2(x + 4)^{3/2} - \frac{8}{15}x(x + 4)^{5/2} + \frac{16}{105}(x + 4)^{7/2} + C$

15. $\frac{1}{3}x^3(\ln x)^3 - \frac{1}{3}x^3(\ln x)^2 + \frac{2}{9}x^3 \ln x - \frac{2}{27}x^3 + C$

17. $\frac{1}{4}e^{4x} + C$ **19.** $\frac{1}{4}xe^{4x} - \frac{1}{16}e^{4x} + C$

21. $-4e^{-x/4}(x^2 + 8x + 32) + C$

23. $\frac{1}{2}t^2 \ln(t + 1) - \frac{1}{2}\ln(t + 1) - \frac{1}{4}t^2 + \frac{1}{2}t + C$

25. $-e^{1/t} + C$ **27.** $\frac{1}{2}x^2(\ln x)^2 - \frac{1}{2}x^2 \ln x + \frac{1}{4}x^2 + C$

29. $\frac{1}{3}(\ln x)^3 + C$ **31.** $-\frac{1}{x}(\ln x + 1) + C$

33. $\frac{2}{3}x(x - 3)^{3/2} - \frac{4}{15}(x - 3)^{5/2} + C$

35. $2x\sqrt{x - 1} - \frac{4}{3}(x - 1)^{3/2} + C$ **37.** $\frac{e^{2x}}{4(2x + 1)} + C$

39. $\frac{5}{36}e^6 + \frac{1}{36} \approx 56.060$ **41.** $\frac{3}{2} \ln 3 - 1 \approx 0.648$

43. $\frac{1192}{15} \approx 79.467$ **45.** $\frac{1}{4}e^2(5e^2 - 1) \approx 66.400$

47. Area $= 4e - e^{-2} \approx 10.738$

49. Area $= \dfrac{e - 2}{e} \approx 0.264$

51. Area $= \frac{1}{25}(4e^5 + 1) \approx 23.786$

53. Proof   **55.** $\dfrac{e^{5x}}{125}(25x^2 - 10x + 2) + C$

57. $-\dfrac{1}{9x^3}(1 + 3 \ln x) + C$ **59.** $\dfrac{3}{128} - \dfrac{379}{128}e^{-8} \approx 0.022$

61. $\dfrac{1,171,875}{256}\pi \approx 14,381.070$

63.

(a) Increase
(b) 113,212 units
(c) 11,321 units/yr

65. (a) $3.2 \ln 2 - 0.2 \approx 2.018$
(b) $12.8 \ln 4 - 7.2 \ln 3 - 1.8 \approx 8.035$

67. \$18,482.03   **69.** \$931,265.10   **71.** \$4955.34

73. \$1,141,592.33   **75.** (a) \$1,200,000   (b) \$1,094,142.27

77. (a) \$18,000,000   (b) \$16,133,084   **79.** \$45,957.78

81. (a) \$17,378.62   (b) \$3681.26   **83.** About 4.254

Section 6.2   *(page 391)*

> **Skills Warm Up**   *(page 391)*
>
> **1.** $x^2 + 8x + 16$ **2.** $x^2 - 2x + 1$
> **3.** $x^2 + x + \frac{1}{4}$ **4.** $x^2 - \frac{2}{3}x + \frac{1}{9}$
> **5.** $2e^x(x - 1) + C$ **6.** $x^3 \ln x - \dfrac{x^3}{3} + C$

1. $\dfrac{1}{9}\left(\dfrac{2}{2 + 3x} + \ln|2 + 3x|\right) + C$

3. $x(2x^2 + 16)\sqrt{x^2 + 16} - 256 \ln\left|x + \sqrt{x^2 + 16}\right| + C$

5. $\ln\left(x^2 + \sqrt{x^4 - 9}\right) + C$ **7.** $\frac{1}{2}(x^2 - 1)e^{x^2} + C$

9. $\ln\left|\dfrac{x}{1 + x}\right| + C$ **11.** $-\dfrac{1}{7}\ln\left|\dfrac{7 + \sqrt{x^2 + 49}}{x}\right| + C$

13. $-\dfrac{1}{2}\ln\left|\dfrac{2 + \sqrt{4 - x^2}}{x}\right| + C$ **15.** $\frac{3}{4}x^2(-1 + 2 \ln 3x) + C$

17. $3x^2 - \ln(1 + e^{3x^2}) + C$ **19.** $\dfrac{1}{24}\ln\left|\dfrac{x^2 - 6}{x^2 + 6}\right| + C$

21. $\frac{2}{35}(x + 3)^{3/2}(5x^2 - 12x + 24) + C$

23. $\frac{1}{2}\left[x\sqrt{3 + x^2} + 3 \ln\left|x + \sqrt{3 + x^2}\right|\right] + C$

25. $-\dfrac{1}{2}x(2 - x) + \ln|x + 1| + C$ **27.** $-\dfrac{\sqrt{1 - x^2}}{x} + C$

29. $\dfrac{1}{2(9 - 2x)^2} - \dfrac{6}{(9 - 2x)^3} + \dfrac{81}{4(9 - 2x)^4} + C$

31. $\frac{4}{9}x^3(3 \ln 2x - 1) + C$

33. $\dfrac{1}{27}\left(3x - \dfrac{25}{3x - 5} + 10 \ln|3x - 5|\right) + C$

35. $\frac{1}{9}(3 \ln x - 4 \ln|4 + 3 \ln x|) + C$

37. $\dfrac{-2\sqrt{2} + 4}{3} \approx 0.3905$ **39.** $\dfrac{3}{4} + \dfrac{7}{16}\ln\dfrac{17}{5} \approx 1.2854$

41. $12\left(2 + \ln\left|\dfrac{2}{1 + e^2}\right|\right) \approx 6.7946$

43. $-\dfrac{1}{5}\left(\dfrac{\sqrt{14}}{3} - \sqrt{6}\right) \approx 0.2405$

45. Area $= \dfrac{1}{8\sqrt{3}} \approx 0.0722$ **47.** Area $= \dfrac{3 \ln \frac{4}{5} + 1}{36} \approx 0.0092$

49. Area $= \frac{1}{4}[21\sqrt{5} - 8 \ln(\sqrt{5} + 3) + 8 \ln 2] \approx 9.8145$

51. $x\left(\ln\dfrac{x}{3} - 1\right) + C$ **53.** $\dfrac{2}{147}(7x + 6)\sqrt{7x - 3} + C$

55. (a) 0.483   (b) 0.283

57.

[graph: 6000 on vertical axis, 10 on horizontal axis]

Average value: 401.40

59. \$1138.43

61. Consumer surplus: About 17.92
Producer surplus: 24

Quiz Yourself   *(page 393)*

1. $\frac{1}{5}xe^{5x} - \frac{1}{25}e^{5x} + C$ **2.** $3x \ln x - 3x + C$

3. $\frac{1}{2}x^2 \ln x + x \ln x - \frac{1}{4}x^2 - x + C$

4. $\frac{3}{14}(x + 6)^{4/3}(2x - 9) + C$

5. $\dfrac{x^2}{4}\ln x - \dfrac{x^2}{8} + C$ **6.** $-\dfrac{1}{2}e^{-2x}\left(x^2 + x + \dfrac{1}{2}\right) + C$

7. (a) 282,016 units   (b) 56,403 units

8. (a) \$784,000   (b) \$673,108.31

9. $\frac{1}{4}(2x - \ln|1 + 2x|) + C$ **10.** $10 \ln\left|\dfrac{x}{0.1 + 0.2x}\right| + C$

11. $\ln\left|x + \sqrt{x^2 - 81}\right| - \dfrac{\sqrt{x^2 - 81}}{x} + C$

12. $\dfrac{1}{2}\ln\left|\dfrac{\sqrt{4 + 9x} - 2}{\sqrt{4 + 9x} + 2}\right| + C$

13. $\frac{1}{4}\left[4x^2 - \ln\left(1 + e^{4x^2}\right)\right] + C$ **14.** $x^2e^{x^2 + 1} + C$

15. (a) \$84,281,126.52   (b) \$257,392,429.72

16. $\dfrac{8}{e} - 4 \approx -1.0570$ **17.** $10 \ln 2 - \dfrac{15}{4} \approx 3.1815$

18. $\frac{64}{3}\left(\sqrt{2} - 1\right) \approx 8.8366$ **19.** About 9.8182

20. $\dfrac{\sqrt{5}}{18} \approx 0.1242$ **21.** $\dfrac{1}{4}\left(\ln\dfrac{17}{19} - \ln\dfrac{7}{9}\right) \approx 0.0350$

Section 6.3   *(page 400)*

> ### Skills Warm Up   *(page 400)*
>
> **1.** $\dfrac{6}{x^4}$ **2.** $-\dfrac{96}{(2x+1)^4}$ **3.** $-\dfrac{12}{x^4}$ **4.** $6x-4$
>
> **5.** $81e^{3x}$ **6.** $e^{x^2}(4x^2+2)$ **7.** $(3,18)$ **8.** $(1,8)$
>
> **9.** $n > 5\sqrt{10}$ or $n > 15.81$ **10.** $n > 5$

	Trapezoidal Rule	*Simpson's Rule*	*Exact Value*
1.	2.7500	2.6667	2.6667
3.	-0.7500	-0.6667	-0.6667
5.	5.2650	5.3046	5.3333
7.	0.6970	0.6933	0.6931
9.	3.8643	3.3022	3.1809

11. (a) 0.783   (b) 0.785   **13.** (a) 5.090   (b) 5.177

15. (a) 1.470   (b) 1.463   **17.** (a) 2.540   (b) 2.541

19. (a) 1.879   (b) 1.888   **21.** 19.5215   **23.** 3.6558

25. \$678.36   **27.** $0.3413 = 34.13\%$

29. $0.4998 = 49.98\%$ **31.** $89{,}500$ ft^2

33. (a) $|E| \le \frac{1}{12} \approx 0.0833$ (b) $|E| \le 0$

35. (a) $|E| \le \dfrac{5e}{64} \approx 0.212$ (b) $|E| \le \dfrac{13e}{1024} \approx 0.035$

37. (a) $n = 566$ (b) $n = 16$

39. (a) $n = 3280$ (b) $n = 60$

41. \$21,831.20; \$21,836.98

43. (a) 36.9 yr   (b) 36.9 yr   (c) The results are the same.

45. 58.912 mg   **47.** 1878 subscribers

49. Answers will vary.

Section 6.4   *(page 410)*

> ### Skills Warm Up   *(page 410)*
>
> **1.** 9   **2.** 0   **3.** $-\frac{1}{8}$ **4.** Limit does not exist.
>
> **5.** Limit does not exist.   **6.** -4
>
> **7.** (a) $\frac{32}{3}b^3 - 16b^2 + 8b - \frac{4}{3}$ (b) $-\frac{4}{3}$
>
> **8.** (a) $\dfrac{b^2 - b - 11}{(b-2)^2(b-5)}$ (b) $\dfrac{11}{20}$
>
> **9.** (a) $\ln\left(\dfrac{5 - 3b^2}{b+1}\right)$ (b) $\ln 5 \approx 1.609$
>
> **10.** (a) $e^{-3b^2}(e^{6b^2} + 1)$ (b) 2

1. Improper; The integrand has an infinite discontinuity when $x = \frac{2}{3}$, and $0 \le \frac{2}{3} \le 1$.

3. Not improper; continuous on $[0, 1]$

5. Improper; The upper limit of integration is infinite.

7. Converges; $\frac{1}{3}$ **9.** Diverges   **11.** Diverges

13. Diverges   **15.** Diverges   **17.** Converges; $\dfrac{1}{2(\ln 4)^2}$

19. Converges; 0

21. 1   **23.** $\dfrac{7}{8}$ **25.** ∞

27. (a) 0.9050   (b) 0.0912   (c) 0.00383

29. (a) $\displaystyle\int_{35.5}^{\infty} f(x)\,dx \approx 0.9938$ (b) $\displaystyle\int_{35.9}^{\infty} f(x)\,dx \approx 0.6915$

31. \$200,000   **33.** \$66,666.67

35. Yes, \$360,000 < \$400,000.

37. (a) \$4,637,228   (b) \$5,555,556

39. (a) \$748,367.34   (b) \$808,030.14   (c) \$900,000.00

41. (a) \$453,901.30   (b) \$807,922.43   (c) \$4,466,666.67

Review Exercises for Chapter 6   *(page 416)*

1. $xe^x + C$ **3.** $2\sqrt{x}\ln x - 4\sqrt{x} + C$

5. $\frac{2}{15}(x-5)^{3/2}(3x+10) + C$ **7.** $x^2 e^{2x} - xe^{2x} + \frac{1}{2}e^{2x} + C$

9. $3e^2 - \dfrac{3(e^2-1)}{2} \approx 12.584$ **11.** $16 - 20e^{-1/4} \approx 0.4240$

13. \$90,634.62   **15.** \$865,958.50

17. (a) \$1,200,000   (b) \$1,052,649.52

19. $\frac{1}{54}(9x^2 - 12x + 8\ln|3x+2|) + C$

21. $-\dfrac{1}{8}\ln\left|\dfrac{8 + \sqrt{x^2+64}}{x}\right| + C$

23. $\dfrac{1}{2}\left(\dfrac{1}{2+3x} + \dfrac{1}{2}\ln\left|\dfrac{x}{2+3x}\right|\right) + C$

25. $\sqrt{x^2+25} - 5\ln\left|\dfrac{5 + \sqrt{x^2+25}}{x}\right| + C$

27. $\dfrac{1}{28}\ln\left|\dfrac{2x-7}{2x+7}\right| + C$

29. $2\sqrt{1+x} + \ln\left|\dfrac{\sqrt{1+x}-1}{\sqrt{1+x}+1}\right| + C$ **31.** 0.0841

33. (a) 0.675   (b) 0.290

35. Trapezoidal: 0.7050   **37.** Trapezoidal: 4.6875
Simpson's: 0.6715   Simpson's: 4.6667
Exact: $\frac{2}{3} \approx 0.6667$ Exact: $\frac{14}{3} \approx 4.6667$

39. Trapezoidal: 1.7652   **41.** (a) 0.741   (b) 0.737
Simpson's: 1.7299
Exact: $2 - 2e^{-2} \approx 1.7293$

43. (a) 0.305   (b) 0.289   **45.** (a) 2.961   (b) 2.936

47. (a) $|E| \le \dfrac{3e^3}{64} \approx 0.9415$ (b) $|E| \le \dfrac{9e^3}{5120} \approx 0.0353$

49. (a) $n = 3486$ (b) $n = 48$

51. Converges; $-\frac{1}{4}$ **53.** Diverges   **55.** Diverges

57. $A \approx 4$ **59.** $A = \infty$ **61.** \$266,666.67   **63.** Yes

Test Yourself   *(page 418)*

1. $xe^{x+1} - e^{x+1} + C$ **2.** $\frac{1}{9}x^3(3\ln 9x - 1) + C$

3. $-3x^2 e^{-x/3} - 18xe^{-x/3} - 54e^{-x/3} + C$

4. (a) About \$69.81 billion   (b) About \$13.96 billion

5. $\frac{1}{4}\left(\dfrac{7}{7+2x} + \ln|7+2x|\right) + C$ **6.** $x^3 - \ln(1 + e^{x^3}) + C$

7. $\frac{1}{16}\left[x^2(2x^4 - 9)\sqrt{x^4 - 9} - 81\ln\left|x^2 + \sqrt{x^4-9}\right|\right] + C$

8. $-1 + \frac{3}{2}\ln 3 \approx 0.6479$

9. $8\frac{2}{3}$ **10.** -7.9691

11. Trapezoidal: 18.28
Exact: 18.0

12. Simpson's: 41.3606; Exact: 41.1711

13. (a) $n = 24$ (b) $n = 6$

14. Converges; $\frac{1}{3}$ **15.** Diverges

16. Diverges   **17.** Converges; $\frac{1}{2}$

18. (a) \$7,585,446.71   (b) \$12,000,000

Chapter 7

Section 7.1   *(page 425)*

1.   **3.**

5. $(-3, 4, 5)$ **7.** $(10, 0, 0)$ **9.** 0 **11.** $3\sqrt{2}$
13. 12 **15.** $(6, 4, 7)$ **17.** $\left(2, -\frac{3}{2}, \frac{5}{2}\right)$ **19.** $(6, -3, 5)$
21. $(1, 2, 1)$ **23.** $3, 3\sqrt{5}, 6$; right triangle
25. $\sqrt{41}, \sqrt{14}, \sqrt{41}$; isosceles triangle
27. $(0, 0, -5), (2, 2, -4), (2, -4, -1)$
29. $x^2 + (y - 2)^2 + (z - 2)^2 = 4$
31. $\left(x - \frac{3}{2}\right)^2 + (y - 2)^2 + (z - 1)^2 = \frac{21}{4}$
33. $(x - 3)^2 + (y + 2)^2 + (z + 3)^2 = 16$
35. $(x + 1)^2 + \left(y - \frac{1}{2}\right)^2 + \left(z - \frac{3}{2}\right)^2 = \frac{69}{2}$
37. $(x + 4)^2 + (y - 3)^2 + (z - 6)^2 = 36$
39. Center: $(6, 0, 0)$ **41.** Center: $(-2, 1, -4)$
   Radius: 6 Radius: 5
43. Center: $(1, 3, 2)$
   Radius: $\dfrac{5\sqrt{2}}{2}$
45. $(x - 1)^2 + (y + 3)^2 = 25$ **47.** $(x - 3)^2 + (y - 5)^2 = 4$

49. $(y - 4)^2 + z^2 = 16$ **51.** $(y - 2)^2 + (z - 3)^2 = 25$

53. (a)   (b)

55. (a)   (b)

57. $x^2 + y^2 + z^2 = 6806.25$

Section 7.2   *(page 434)*

1. $(3, 0, 0), (0, 3, 0), (0, 0, 3)$ **3.** $(3, 0, 0), (0, 6, 0), (0, 0, 2)$

5. $(2, 0, 0), (0, -4, 0), \left(0, 0, \frac{4}{3}\right)$ **7.** $(0, 0, 8)$

9. $(6, 0, 0), (0, 6, 0)$ **11.** $(3, 0, 0), (0, 0, -1)$

13. Perpendicular   **15.** Parallel
17. Neither parallel nor perpendicular
19. Parallel   **21.** Perpendicular
23. c; Ellipsoid   **24.** e; Hyperboloid of two sheets
25. f; Hyperboloid of one sheet   **26.** b; Elliptic cone

27. d; Elliptic paraboloid   **28.** a; Hyperbolic paraboloid

29. (a) $x = \pm y$; Lines

(b) $z = 9 - y^2$; Parabola

(c) $z = x^2$; Parabola

Hyperbolic paraboloid

31. (a) $\dfrac{x^2}{4} + y^2 = 1$; Ellipse   (b) $\dfrac{x^2}{4} + z^2 = 1$; Ellipse

(c) $y^2 + z^2 = 1$; Circle

Ellipsoid

33. (a) $z^2 - \dfrac{x^2}{3} = 1$; Hyperbola

(b) $\dfrac{3}{7}z^2 - \dfrac{3}{56}y^2 = 1$; Hyperbola

(c) $\dfrac{x^2}{45} + \dfrac{y^2}{120} = 1$; Ellipse

35. Ellipsoid   **37.** Elliptic paraboloid

39. Hyperboloid of two sheets   **41.** Hyperbolic paraboloid

43. Hyperboloid of two sheets   **45.** Elliptic cone

47. Hyperboloid of one sheet

49. $\dfrac{x^2}{3963^2} + \dfrac{y^2}{3963^2} + \dfrac{z^2}{3950^2} = 1$

51. (a)

Year	2008	2009	2010
x	38.3	37.2	38.8
y	24.3	25.2	26.3
z (actual)	19.4	18.8	19.2
z (approximated)	19.3	18.8	19.4

Year	2011	2012	2013
x	41.3	44.4	46.5
y	28.3	28.4	28.7
z (actual)	20.4	21.2	22.3
z (approximated)	20.2	21.4	22.2

The approximated values of z are very close to the actual values.

(b) According to the model, increases in expenditures of recreation types y and z will correspond to an increase in expenditures of recreation type x.

Section 7.3   *(page 441)*

Skills Warm Up   *(page 441)*

1. 11   **2.** -16 **3.** 7   **4.** 4   **5.** $(-\infty, \infty)$

6. $(-\infty, -3) \cup (-3, 0) \cup (0, \infty)$

7. $[5, \infty)$ **8.** $(-\infty, -1] \cup [1, \infty)$

9. 55.0104   **10.** 6.9165

1. (a) 1   (b) 1   (c) -17 (d) $9 - y$

(e) $2x - 1$ (f) $13 - t$

3. (a) 5   (b) $3e^2$ (c) $2e^{-1}$ (d) $4e^y$ (e) $6x$ (f) te^t

5. (a) $\frac{2}{3}$ (b) 0   **7.** (a) 90π (b) 50π

9. (a) \$20,655.20   (b) \$1,397,672.67   **11.** (a) 0   (b) 6

13. (a) $x^2 + 2x\,\Delta x + (\Delta x)^2 - 2y$ (b) -2, $\Delta y \neq 0$

15. Domain: all points (x, y) inside and on the circle

$x^2 + y^2 = 16$

Range: $[0, 4]$

17. Domain: all points (x, y)

Range: $[0, \infty)$

19. Domain: all points (x, y) such that $y \neq 0$

Range: $(0, \infty)$

21. Domain: the half-plane below the line $y = -x + 5$

Range: $(-\infty, \infty)$

23. Domain: all points (x, y) inside and on the ellipse

$3x^2 + y^2 = 9$

Range: $[0, 3]$

25. Domain: all points (x, y) such that $x \neq 0$

Range: $(-\infty, \infty)$

27. Domain: all points (x, y) such that $x \neq 0$ and $y \neq 0$

Range: $(-\infty, 0)$ and $(0, \infty)$

29. Domain: all points (x, y) such that $y \geq 0$

Range: $(-\infty, \infty)$

31. b   **32.** d   **33.** a   **34.** c

35. The level curves are parallel lines.

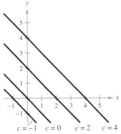

37. The level curves are circles.

39. The level curves are hyperbolas.

41. The level curves are circles.

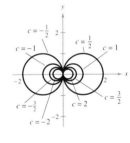

43. 114,870 units   **45.** Proof

47. (a) \$15,250   (b) \$18,425   (c) \$30,025

49.

R \ I	0	0.03	0.05
0	\$5187.48	\$3859.98	\$3184.67
0.28	\$4008.46	\$2982.67	\$2460.85
0.35	\$3754.27	\$2793.53	\$2304.80

51. (a) C (b) A (c) B

53. (a) \$531.5 million

(b) y; Explanations will vary. Sample answer: The y-term has a greater influence on shareholders' equity because the absolute value of its coefficient is greater than that of the coefficient of the x-term.

55. Option (a): $1003.73, $240,895.20;
    Option (b): $798.36, $287,409.60;
    Option (c): $1078.59, $194,146.20;
    Answers will vary.

Section 7.4   *(page 452)*

Skills Warm Up   *(page 452)*

1. $12x - 9$ **2.** $-6x(3 - x^2)^2$ **3.** $e^{t^3}(3t^3 + 1)$

4. $\dfrac{e^{2x}(2 - 3e^{2x})}{\sqrt{1 - e^{2x}}}$ **5.** $-\dfrac{2}{3 - 2x}$ **6.** $\dfrac{3(t^2 - 2)}{2t(t^2 - 6)}$

7. $-\dfrac{10x}{(4x - 1)^3}$ **8.** $-\dfrac{(x + 2)^2(x^2 + 8x + 27)}{(x^2 - 9)^3}$

9. 8   **10.** $\dfrac{7}{2}$

---

1. $\dfrac{\partial z}{\partial x} = 3;\ \dfrac{\partial z}{\partial y} = 5$ **3.** $\dfrac{\partial z}{\partial x} = 6 - 2xy;\ \dfrac{\partial z}{\partial y} = -x^2 + 16y$

5. $h_x(x, y) = -2xe^{-(x^2 + y^2)};\ h_y(x, y) = -2ye^{-(x^2 + y^2)}$

7. $f_x(x, y) = \dfrac{x}{\sqrt{x^2 + y^2}};\ f_y(x, y) = \dfrac{y}{\sqrt{x^2 + y^2}}$

9. $\dfrac{\partial z}{\partial x} = 2y^3 e^{2xy};\ \dfrac{\partial z}{\partial y} = 2xy^2 e^{2xy} + 2ye^{2xy}$

11. $g_x(x, y) = \dfrac{2x}{x^2 + y^2};\ g_y(x, y) = \dfrac{2y}{x^2 + y^2}$

13. $f_x(x, y) = \dfrac{4x^3 y}{3y + 2};\ f_y(x, y) = \dfrac{2x^4}{(3y + 2)^2}$

15. $\dfrac{\partial z}{\partial x} = 3;\ \dfrac{\partial z}{\partial y} = 2$

17. $f_x(x, y) = 6x + y, 13;\ f_y(x, y) = x - 2y, 0$

19. $f_x(x, y) = e^x y^2, 12;\ f_y(x, y) = 2e^x y, 12$

21. $f_x(x, y) = \dfrac{3}{3x + 5y}, 1;\ f_y(x, y) = \dfrac{5}{3x + 5y}, \dfrac{5}{3}$

23. $f_x(x, y) = -\dfrac{y^2}{(x - y)^2}, -\dfrac{1}{4};\ f_y(x, y) = \dfrac{x^2}{(x - y)^2}, \dfrac{1}{4}$

25. (a) 2   (b) 1   **27.** (a) -2 (b) -2

29. $w_x = y^2 z^4 + 9y$
    $w_y = 2xyz^4 + 9x$
    $w_z = 4xy^2 z^3 - 1$

31. $w_x = -\dfrac{2z}{(x + y)^2}$
    $w_y = -\dfrac{2z}{(x + y)^2}$
    $w_z = \dfrac{2}{x + y}$

33. $w_x = 2z^2 + 3yz, 2$
    $w_y = 3xz - 12yz, 30$
    $w_z = 4xz + 3xy - 6y^2, -1$

35. $w_x = \dfrac{x}{\sqrt{x^2 + y^2 + z^2}}, \dfrac{2}{3}$
    $w_y = \dfrac{y}{\sqrt{x^2 + y^2 + z^2}}, -\dfrac{1}{3}$
    $w_z = \dfrac{z}{\sqrt{x^2 + y^2 + z^2}}, \dfrac{2}{3}$

37. $w_x = 4xy^3 z^2 e^{2x^2}, -8\sqrt{e}$
    $w_y = 3y^2 z^2 e^{2x^2}, 12\sqrt{e}$
    $w_z = 2y^3 z e^{2x^2}, -4\sqrt{e}$

39. $w_x = \dfrac{5}{5x + 2y^3 - 3z}, \dfrac{1}{5}$
    $w_y = \dfrac{6y^2}{5x + 2y^3 - 3z}, \dfrac{6}{25}$
    $w_z = -\dfrac{3}{5x + 2y^3 - 3z}, -\dfrac{3}{25}$

41. $(-6, 4)$ **43.** $(1, 1)$

45. $\dfrac{\partial^2 z}{\partial x^2} = 6x$ **47.** $\dfrac{\partial^2 z}{\partial x^2} = 12x^2$

$\dfrac{\partial^2 z}{\partial y^2} = -8$ $\dfrac{\partial^2 z}{\partial y^2} = 18y$

$\dfrac{\partial^2 z}{\partial y \partial x} = \dfrac{\partial^2 z}{\partial x \partial y} = 0$ $\dfrac{\partial^2 z}{\partial y \partial x} = \dfrac{\partial^2 z}{\partial x \partial y} = -2$

49. $\dfrac{\partial^2 z}{\partial x^2} = 108x^2(3x^4 - 2y^3)(11x^4 - 2y^3)$

$\dfrac{\partial^2 z}{\partial y^2} = 36y(2y^3 - 3x^4)(3x^4 - 8y^3)$

$\dfrac{\partial^2 z}{\partial x \partial y} = \dfrac{\partial^2 z}{\partial y \partial x} = 432x^3 y^2(2y^3 - 3x^4)$

51. $\dfrac{\partial^2 z}{\partial x^2} = -\dfrac{y}{x^3}$

$\dfrac{\partial^2 z}{\partial x \partial y} = -\dfrac{x^2 - y^2}{2x^2 y^2}$

$\dfrac{\partial^2 z}{\partial y \partial x} = -\dfrac{x^2 - y^2}{2x^2 y^2}$

$\dfrac{\partial^2 z}{\partial y^2} = \dfrac{x}{y^3}$

53. $f_{xx}(x, y) = 12x^2 - 6y^2, 12$
    $f_{xy}(x, y) = -12xy, 0$
    $f_{yy}(x, y) = -6x^2 + 2, -4$
    $f_{yx}(x, y) = -12xy, 0$

55. $f_{xx}(x, y) = e^{x^2}(4x^2 y^3 + 2y^3), -6e$
    $f_{xy}(x, y) = f_{yx}(x, y) = 6xy^2 e^{x^2}, 6e$
    $f_{yy}(x, y) = 6ye^{x^2}, -6e$

57. $f_{xx}(x, y, z) = 2$
    $f_{xy}(x, y, z) = f_{yx}(x, y, z) = -3$
    $f_{xz}(x, y, z) = f_{yy}(x, y, z) = f_{zx}(x, y, z) = 0$
    $f_{yz}(x, y, z) = f_{zy}(x, y, z) = 4$
    $f_{zz}(x, y, z) = 6z$

59. $f_{xx}(x, y, z) = -\dfrac{8yz}{(x + y)^3}$
    $f_{xy}(x, y, z) = \dfrac{4z(x - y)}{(x + y)^3}$
    $f_{xz}(x, y, z) = \dfrac{4y}{(x + y)^2}$
    $f_{yy}(x, y, z) = \dfrac{8xz}{(x + y)^3}$
    $f_{yx}(x, y, z) = \dfrac{4z(x - y)}{(x + y)^3}$
    $f_{yz}(x, y, z) = -\dfrac{4x}{(x + y)^2}$
    $f_{zz}(x, y, z) = 0$
    $f_{zx}(x, y, z) = \dfrac{4y}{(x + y)^2}$
    $f_{zy}(x, y, z) = -\dfrac{4x}{(x + y)^2}$

61. (a) At $(120, 160)$, $\dfrac{\partial C}{\partial x} \approx 154.77$; At $(120, 160)$, $\dfrac{\partial C}{\partial y} \approx 193.33$

    (b) Racing bikes; Explanations will vary. Sample answer: The absolute value of dC/dy is greater than the absolute value of dC/dx at $(120, 160)$.

63. (a) About 113.72   (b) About 97.47

65. Complementary

67. (a) $\dfrac{\partial z}{\partial x} = 0.391$; $\dfrac{\partial z}{\partial y} = -0.069$

    (b) For every increase of 1 billion dollars in expenditures on amusement parks and campgrounds, the expenditures for spectator sports will increase by 0.391 billion dollars. For every increase of 1 billion dollars in expenditures on live entertainment (excluding sports), the expenditures for spectator sports will decrease by 0.069 billion dollars.

69. $IQ_M(M, C) = \dfrac{100}{C}$, $IQ_M(12, 10) = 10$; For a child who has a current mental age of 12 years and a chronological age of 10 years, the IQ is increasing at a rate of 10 IQ points for every increase of 1 year in the child's mental age.

$IQ_C(M, C) = \dfrac{-100M}{C^2}$, $IQ_C(12, 10) = -12$; For a child who has a current mental age of 12 years and a chronological age of 10 years, the IQ is decreasing at a rate of 12 IQ points for every increase of 1 year in the child's chronological age.

71. $V_I(0.03, 0.28) \approx -14{,}478.99$

$V_R(0.03, 0.28) \approx -1391.17$

The rate of inflation has the greater negative influence on the growth of the investment because $|-14{,}478.99| > |-1391.17|$.

73. (a) $U_x = -10x + y$ (b) $U_y = x - 6y$

    (c) When $x = 2$ and $y = 3$, $U_x = -17$ and $U_y = -16$. The person should consume one more unit of product y, because the rate of decrease of satisfaction is less for y.

    (d)

The slope of U in the x-direction is 0 when $y = 10x$ and negative when $y < 10x$. The slope of U in the y-direction is 0 when $x = 6y$ and negative when $x < 6y$.

Section 7.5 *(page 461)*

Skills Warm Up *(page 461)*

1. $(3, 2)$ **2.** $(11, 6)$ **3.** $(1, 4)$ **4.** $(2, 6)$

5. $(2, -1)$ **6.** $(3, -2)$ **7.** $(0, 0), (-1, 0)$

8. $(-2, 0), (2, -2)$

9. $\dfrac{\partial z}{\partial x} = 12x^2$ $\dfrac{\partial^2 z}{\partial y^2} = -6$

$\dfrac{\partial z}{\partial y} = -6y$ $\dfrac{\partial^2 z}{\partial x \partial y} = 0$

$\dfrac{\partial^2 z}{\partial x^2} = 24x$ $\dfrac{\partial^2 z}{\partial y \partial x} = 0$

10. $\dfrac{\partial z}{\partial x} = 10x^4$ $\dfrac{\partial^2 z}{\partial y^2} = -6y$

$\dfrac{\partial z}{\partial y} = -3y^2$ $\dfrac{\partial^2 z}{\partial x \partial y} = 0$

$\dfrac{\partial^2 z}{\partial x^2} = 40x^3$ $\dfrac{\partial^2 z}{\partial y \partial x} = 0$

11. $\dfrac{\partial z}{\partial x} = 4x^3 - \dfrac{\sqrt{xy}}{2x}$ $\dfrac{\partial^2 z}{\partial y^2} = \dfrac{\sqrt{xy}}{4y^2}$

$\dfrac{\partial z}{\partial y} = -\dfrac{\sqrt{xy}}{2y} + 2$ $\dfrac{\partial^2 z}{\partial x \partial y} = -\dfrac{\sqrt{xy}}{4xy}$

$\dfrac{\partial^2 z}{\partial x^2} = 12x^2 + \dfrac{\sqrt{xy}}{4x^2}$ $\dfrac{\partial^2 z}{\partial y \partial x} = -\dfrac{\sqrt{xy}}{4xy}$

12. $\dfrac{\partial z}{\partial x} = 4x - 3y$ $\dfrac{\partial^2 z}{\partial y^2} = 2$

$\dfrac{\partial z}{\partial y} = 2y - 3x$ $\dfrac{\partial^2 z}{\partial x \partial y} = -3$

$\dfrac{\partial^2 z}{\partial x^2} = 4$ $\dfrac{\partial^2 z}{\partial y \partial x} = -3$

13. $\dfrac{\partial z}{\partial x} = y^3 e^{xy^2}$ $\dfrac{\partial^2 z}{\partial y^2} = 4x^2y^3 e^{xy^2} + 6xy e^{xy^2}$

$\dfrac{\partial z}{\partial y} = 2xy^2 e^{xy^2} + e^{xy^2}$ $\dfrac{\partial^2 z}{\partial x \partial y} = 2xy^4 e^{xy^2} + 3y^2 e^{xy^2}$

$\dfrac{\partial^2 z}{\partial x^2} = y^5 e^{xy^2}$ $\dfrac{\partial^2 z}{\partial y \partial x} = 2xy^4 e^{xy^2} + 3y^2 e^{xy^2}$

14. $\dfrac{\partial z}{\partial x} = e^{xy}(xy + 1)$ $\dfrac{\partial^2 z}{\partial y^2} = x^3 e^{xy}$

$\dfrac{\partial z}{\partial y} = x^2 e^{xy}$ $\dfrac{\partial^2 z}{\partial x \partial y} = xe^{xy}(xy + 2)$

$\dfrac{\partial^2 z}{\partial x^2} = ye^{xy}(xy + 2)$ $\dfrac{\partial^2 z}{\partial y \partial x} = xe^{xy}(xy + 2)$

1. Relative minimum: $(-4, 6, -55)$

3. Relative maximum: $\left(-1, -\frac{1}{4}, \frac{49}{4}\right)$

5. Relative minimum: $(0, 0, 1)$

7. Relative minimum: $(0, 0, 2)$

9. Relative minimum: $(1, 3, 0)$

11. Relative maximum: $(8, 16, 74)$

13. Saddle point: $(0, 0, 1)$

    Relative maximum: $\left(\frac{4}{3}, \frac{4}{3}, \frac{59}{27}\right)$

15. Saddle point: $(0, 0, 0)$

17. Relative maximum: $\left(\frac{1}{2}, \frac{1}{2}, e^{1/2}\right)$

    Relative minimum: $\left(-\frac{1}{2}, -\frac{1}{2}, -e^{1/2}\right)$

19. Insufficient information   **21.** $f(x_0, y_0)$ is a saddle point.

23. $f(x_0, y_0)$ is a relative minimum.

25. Saddle point: $(0, 0, 0)$

    Second-Partials Test fails at $(0, 0)$.

27. Relative minima: $(a, 0, 0), (0, b, 0)$

    Second-Partials Test fails at $(a, 0)$ and $(0, b)$.

29. Relative minimum: $(0, 0, 0)$

    Second-Partials Test fails at $(0, 0)$.

31. Relative minimum: $(1, -3, 0)$

33. 15, 15, 15   **35.** 20, 20, 20   **37.** $x_1 = 3$, $x_2 = 6$

39. $p_1 = 2500$, $p_2 = 3000$ **41** $x_1 \approx 94$, $x_2 = 157$

43. $\frac{100}{27}$ cubic units   **45.** 32 in. $\times$ 16 in. $\times$ 16 in.

47. Base dimensions: 3 ft $\times$ 3 ft

    Height: 2 ft; Minimum cost: $5.40

CHAPTER 7

49. $x = 1.25$, $y = 2.5$; \$4.625 million

51. 500 smallmouth bass; 200 largemouth bass

53. Proof   **55.** True

Quiz Yourself   *(page 464)*

1. (a)   (b) 3   (c) $\left(0, \frac{5}{2}, 1\right)$

2. (a)   (b) $2\sqrt{35}$ (c) $(1, -2, -1)$

3. (a)   (b) $\sqrt{51}$ (c) $\left(\frac{3}{2}, -\frac{1}{2}, \frac{5}{2}\right)$

4. $(x + 1)^2 + (y - 7)^2 + (z - 4)^2 = 36$

5. $(x - 1)^2 + (y - 4)^2 + (z + 2)^2 = 11$

6. Center: $(4, 1, 3)$; radius: 7

7. $(3, 0, 0)$, $(0, 2, 0)$, $(0, 0, 6)$ **8.** $(4, 0, 0)$, $(0, 0, -2)$

9. $(0, 3, 0)$ **10.** Ellipsoid

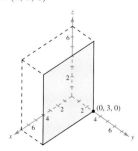

11. Hyperboloid of two sheets   **12.** Elliptic paraboloid

13. $f(1, 0) = 1$ **14.** $f(1, 0) = 2$
$f(4, -1) = -5$ $f(4, -1) = 3\sqrt{7}$

15. $f(1, 0) = 0$
$f(4, -1) = \ln 6 \approx 1.79$

16. Domain: all points (x, y) inside and on the circle $x^2 + y^2 = 81$
Range: $[0, 9]$

17. Domain: all points (x, y)
Range: $(0, \infty)$

18. The level curves are parabolas.

19. \$473.80; \$34,113.60

20. $f_x = 2x - 3$; $f_x(-2, 3) = -7$
$f_y = 4y - 1$; $f_y(-2, 3) = 11$

21. $f_x = \dfrac{y(3 + y)}{(x + y)^2}$; $f_x(-2, 3) = 18$
$f_y = \dfrac{-2xy - y^2 - 3x}{(x + y)^2}$; $f_y(-2, 3) = 9$

22. $f_x = 3x^2 e^{2y}$; $f_x(-2, 3) = 12e^6 \approx 4841.15$
$f_y = 2x^3 e^{2y}$; $f_y(-2, 3) = -16e^6 \approx -6454.86$

23. $f_x = \dfrac{2}{2x + 7y}$; $f_x(-2, 3) = \dfrac{2}{17} \approx 0.118$
$f_y = \dfrac{7}{2x + 7y}$; $f_y(-2, 3) = \dfrac{7}{17} \approx 0.412$

24. Relative minimum: **25.** Relative maximum:
$(1, 0, -3)$ $\left(\frac{4}{3}, \frac{4}{3}, \frac{59}{27}\right)$
        Saddle point: $(0, 0, 1)$

26. $x = 80$, $y = 20$; \$20,000

Section 7.6   *(page 470)*

Skills Warm Up   *(page 470)*

1. $\left(\frac{7}{8}, \frac{1}{12}\right)$ **2.** $\left(-\frac{1}{24}, -\frac{7}{8}\right)$ **3.** $\left(\frac{55}{12}, -\frac{25}{12}\right)$

4. $\left(\frac{22}{23}, -\frac{3}{23}\right)$ **5.** $\left(\frac{5}{3}, \frac{1}{3}, 0\right)$ **6.** $\left(-\frac{57}{53}, \frac{40}{53}, \frac{117}{53}\right)$

7. $f_x = 2xy + y^2$ **8.** $f_x = 50y^2(x + y)$
$f_y = x^2 + 2xy$ $f_y = 50y(x + y)(x + 2y)$

9. $f_x = 3x^2 + 3y^2 - z^2$ **10.** $f_x = yz + z^2$
$f_y = 6xy$ $f_y = xz + z^2$
$f_z = -2xz$ $f_z = xy + 2xz + 2yz$

1. $f(7, 7) = 49$ **3.** $f(3, 1) = 10$ **5.** $f(2, 4) = -12$

7. $f(25, 50) = 2600$ **9.** $f(1, 1) = 2$

11. $f(2, 2) = e^4$ **13.** $f(2, 2, 2) = 8$

15. $f\left(\frac{1}{3}, \frac{1}{3}, \frac{1}{3}\right) = \frac{1}{3}$ **17.** $f\left(\frac{\sqrt{3}}{3}, \frac{\sqrt{3}}{3}, \frac{\sqrt{3}}{3}\right) = \sqrt{3}$

19. 30, 30, 30   **21.** 40, 40, 40   **23.** $3\sqrt{2}$ **25.** $\sqrt{5}$

27. $\sqrt{3}$

29. 15 units $\times$ 10 units $\times$ 6 units   **31.** 12 ft $\times$ 12 ft $\times$ 8 ft

33. $x_1 = 145$ units, $x_2 = 855$ units

35. (a) $f\left(\frac{3125}{6}, \frac{6250}{3}\right) \approx 147{,}314$ (b) 1.473

   (c) 184,142 units   (d) 515,599 units

37. (a) $x \approx 317$ units, $y \approx 68$ units   (b) Answers will vary.

39. (a) 50 ft × 120 ft   (b) $2400

41. $x = \sqrt[3]{0.065} \approx 0.402$ L

   $y = \frac{1}{2}\sqrt[3]{0.065} \approx 0.201$ L

   $z = \frac{1}{3}\sqrt[3]{0.065} \approx 0.134$ L

43. About 190.7 g

45. (a) Cable television: $3600

      Newspaper: $1800

      Radio: $2700

   (b) About 3718 responses

Section 7.7   *(page 478)*

Skills Warm Up   *(page 478)*

1. 5.0225   **2.** 0.0189

3. $y = 2x + 1$ **4.** $y = -x + 4$

5. $y = \frac{1}{2}x - 2$ **6.** $y = -\frac{1}{4}x - 3$

7. $S_a = 2a - 4 - 4b$ **8.** $S_a = 8a - 6 - 2b$

   $S_b = 12b - 8 - 4a$ $S_b = 18b - 4 - 2a$

1. $S = 1.6$; $S = 0.8259$; quadratic model

3. $S = 1.44$; $S = 1.9368$; linear model

5. $y = x + \frac{2}{3}$ **7.** $y = -2.3x - 0.9$

9. $y = 0.8x + 2$ **11.** $y = -1.1824x + 6.385$

13. $y = 1.0129x - 2.417$

15. (a) $y = 8.52x - 5.4$ (b) About $139.4 billion   (c) 2022

17. (a) $y = 0.138x + 22.1$ (b) 44.18 bushels/acre

19.

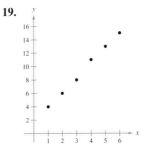

Positive correlation, $r \approx 0.9981$

21.

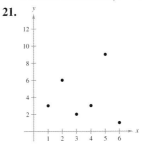

No correlation, $r = 0$

23.

Negative correlation, $r \approx -0.9907$

25. False; The data modeled by $y = 3.29x - 4.17$ have a positive correlation.

27. True   **29.** True   **31.** Answers will vary.

Section 7.8   *(page 486)*

Skills Warm Up   *(page 486)*

1. 1   **2.** 6   **3.** 130   **4.** $\frac{1}{2}$ **5.** $\frac{19}{4}$

6. $\frac{16}{3}$ **7.** $\frac{16}{45}$ **8.** 4   **9.** $\ln 5$ **10.** $\ln(e - 1)$

11. $\frac{e}{2}(e^4 - 1)$ **12.** $\frac{1}{2}\left(1 - \frac{1}{e^2}\right)$

13.   **14.**

15.   **16.**

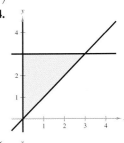

1. $\dfrac{3x^2}{2}$ **3.** $\dfrac{x}{2}(x^2 - 1)$ **5.** $y^4 + \dfrac{1}{2}y^2 - 8y - 2$

7. $x^2\sqrt{x} + x^{3/2} - x^5 - x^9$ **9.** $\dfrac{y^3}{2}$ **11.** 3   **13.** -15

15. 64   **17.** 5   **19.** $\dfrac{21}{2}$ **21.** $\dfrac{1}{3}$ **23.** 4   **25.** 24

27. 8   **29.** $\dfrac{16}{3}$ **31.** 36   **33.** 5   **35.** 2

37.

$$\int_0^1 \int_0^2 dy\,dx = \int_0^2 \int_0^1 dx\,dy = 2$$

39.

$$\int_0^1 \int_{2y}^2 dx\,dy = \int_0^2 \int_0^{x/2} dy\,dx = 1$$

41.

$$\int_0^2 \int_{x/2}^1 dy\,dx = \int_0^1 \int_0^{2y} dx\,dy = 1$$

43.

$$\int_0^1 \int_{y^2}^{\sqrt[3]{y}} dx\,dy = \int_0^1 \int_{x^3}^{\sqrt{x}} dy\,dx = \frac{5}{12}$$

45. $\dfrac{125}{6}$ square units   **47.** $\dfrac{25}{6}$ square units

49. (a) Answers will vary; $\frac{1}{2}(e^9 - 1) \approx 4051.042$

(b) Answers will vary; $\frac{1}{2}(1 - e^{-4}) \approx 0.491$

51. 0.6588   **53.** 8.1747   **55.** 0.4521

57. 1.1190   **59.** True

Section 7.9   *(page 494)*

Skills Warm Up   *(page 494)*

1.

2.

3.

4.

5. 1   **6.** 6   **7.** $\dfrac{1}{3}$ **8.** $\dfrac{40}{3}$ **9.** $\dfrac{28}{3}$ **10.** $\dfrac{7}{6}$

1. $\dfrac{125}{6}$ cubic units   **3.** $\dfrac{16}{3}$ cubic units   **5.** $\dfrac{1}{27}$ cubic unit

7. $\displaystyle\int_0^3 \int_0^5 xy\,dy\,dx = \int_0^5 \int_0^3 xy\,dx\,dy = \dfrac{225}{4}$

9. $\displaystyle\int_0^2 \int_x^{2x} \dfrac{y}{x^2 + y^2}\,dy\,dx = \int_0^2 \int_{y/2}^y \dfrac{y}{x^2 + y^2}\,dx\,dy$

$\qquad + \displaystyle\int_2^4 \int_{y/2}^2 \dfrac{y}{x^2 + y^2}\,dx\,dy = \ln\dfrac{5}{2}$

11. $\displaystyle\int_0^1 \int_0^{3x} \dfrac{1}{\sqrt{x^2 + 4}}\,dy\,dx = \int_0^3 \int_{y/3}^1 \dfrac{1}{\sqrt{x^2 + 4}}\,dx\,dy = 3\left(\sqrt{5} - 2\right)$

13. 4   **15.** 12   **17.** 4   **19.** $\dfrac{40}{3}$ **21.** $\dfrac{64}{3}$

23. $\dfrac{134}{3}$ **25.** 10,000; no   **27.** $\dfrac{3}{2}$ **29.** $\dfrac{8}{3}$

31. \$13,400   **33.** \$75,125   **35.** 25,645.24

Review Exercises for Chapter 7   *(page 500)*

1.

3. $\sqrt{22}$ **5.** $(-1, 4, 6)$

7. $x^2 + (y - 1)^2 + (z - 9)^2 = 49$

9. $(x - 2)^2 + (y + 2)^2 + (z + 3)^2 = 9$

11. Center: $(-4, -1, 7)$; radius: 9

13. $(x + 2)^2 + (y - 1)^2 = 16$ **15.** $(6, 0, 0), (0, 3, 0), (0, 0, 2)$

47. $f_x = ye^x + e^y$ **49.** $w_x = yz^2$
$f_y = xe^y + e^x$ $w_y = xz^2$
                                                       $w_z = 2xyz$

51. (a) -9 (b) -6 **53.** (a) -2 (b) -2

55. $f_{xx} = 6$
$f_{yy} = 12y$
$f_{xy} = f_{yx} = -1$

57. $f_{xx} = f_{yy} = f_{xy} = f_{yx} = \dfrac{-1}{4(1 + x + y)^{3/2}}$

59. $f_{xx} = 10yz^3$ $f_{yx} = 1 + 10xz^3$ $f_{zx} = 30xyz^2$
$f_{xy} = 1 + 10xz^3$ $f_{yy} = -18yz$ $f_{zy} = 15x^2z^2 - 9y^2$
$f_{xz} = 30xyz^2$ $f_{yz} = 15x^2z^2 - 9y^2$ $f_{zz} = 30x^2yz$

17. $(0, 2, 0), (0, 0, 4)$

61. (a) $C_x(500, 250) = 99.50$
$C_y(500, 250) = 140$

(b) Downhill skis; This is determined by comparing the marginal costs for the two models of skis at the production level (500, 250).

19. Ellipsoid   **21.** Hyperboloid of two sheets
23. Hyperboloid of one sheet   **25.** Top half of a circular cone
27. (a) 18   (b) 0   (c) -245 (d) -32
29. Domain: all points (x, y) inside or on the circle $x^2 + y^2 = 1$
Range: $[0, 1]$
31. Domain: all points (x, y) above or below the y-axis
Range: $(-\infty, \infty)$
33. The level curves are lines of slope $-\frac{2}{5}$.

63. Relative minimum: $(0, 0, 0)$
65. Relative minimum: $(0, 0, 1)$
67. Relative minimum: $(0, 0, 1)$
69. Saddle point: $(0, 0, 0)$
Relative minimum: $\left(\frac{3}{2}, \frac{9}{4}, -\frac{27}{16}\right)$
71. $x_1 = 2, \ x_2 = 4$
73. At $(3, 6)$, the relative maximum is 36.
75. At $(2, 2)$, the relative minimum is 8.
77. At $\left(2, \frac{2}{3}, 2\right)$, the relative maximum is $\frac{40}{3}$.
79. $x_1 = 378$ benches; $x_2 = 623$ benches

81. $y = -\frac{3}{2}x + \frac{3}{4}$ **83.** (a) $y = -2.6x + 630$
(b) 123 speakers
(c) \$180.77

35. The level curves are hyperbolas.

85. 1   **87.** $\frac{7}{4}$ **89.** $\frac{32}{3}$ **91.** $\frac{9}{2}$ **93.** 20   **95.** 8
97. $\frac{4096}{9}$ **99.** 8   **101.** \$5700   **103.** 16,282.28

Test Yourself   *(page 504)*

37. (a) 104,199   (b) 140,931
39. (a) \$1.66

(b) x; Explanations will vary. Sample answer: The x-term has a greater influence on the earnings per share because the absolute value of its coefficient is greater than that of the coefficient of the y-term.

41. $f_x = 2xy + 3y + 2$
$f_y = x^2 + 3x - 5$

43. $\dfrac{\partial z}{\partial x} = 2(xy + 2x + 4y)(y + 2)$ **45.** $f_x = \dfrac{5}{5x + 4y}$

$\dfrac{\partial z}{\partial y} = 2(xy + 2x + 4y)(x + 4)$ $f_y = \dfrac{4}{5x + 4y}$

1. (a)   **2.** (a)

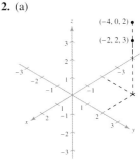

(b) $2\sqrt{2}$ (b) 3
(c) $(2, -2, 0)$ (c) $(-3, 1, 2.5)$

CHAPTER 7

3. (a)

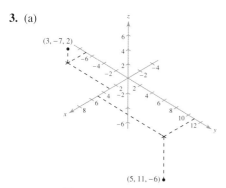

(b) $14\sqrt{2}$

(c) $(4, 2, -2)$

4. Center: $(10, -5, 5)$; radius: 5

5. Hyperboloid of one sheet

6. Elliptic cone   **7.** Hyperbolic paraboloid

8. $f(1, 4) = 6$ **9.** $f(1, 4) = -9$

$f(-6, -2) = 49$ $f(-6, -2) = \frac{5}{8}$

10. $f_x = 6x + 9y^2$; $f_x(10, -1) = 69$

$f_y = 18xy$; $f_y(10, -1) = -180$

11. $f_x = (x + y)^{1/2} + \dfrac{x}{2(x + y)^{1/2}}$; $f_y(10, -1) = \dfrac{14}{3}$

$f_y = \dfrac{x}{2(x + y)^{1/2}}$; $f_x(10, -1) = \dfrac{5}{3}$

12. $\dfrac{\partial^2 z}{\partial x^2} = 12x$

$\dfrac{\partial^2 z}{\partial y^2} = 10$

$\dfrac{\partial^2 z}{\partial y \partial x} = \dfrac{\partial^2 z}{\partial x \partial y} = 0$

13. $\dfrac{\partial^2 w}{\partial x^2} = 2y$

$\dfrac{\partial^2 w}{\partial y^2} = -18y$

$\dfrac{\partial^2 w}{\partial z^2} = -4y$

$\dfrac{\partial^2 w}{\partial x \partial y} = 2x$

$\dfrac{\partial^2 w}{\partial x \partial z} = 0$

$\dfrac{\partial^2 w}{\partial y \partial x} = 2x$

$\dfrac{\partial^2 w}{\partial y \partial z} = -4z$

$\dfrac{\partial^2 w}{\partial z \partial x} = 0$

$\dfrac{\partial^2 w}{\partial z \partial y} = -4z$

14. Relative minimum: $(1, -2, -23)$

15. Saddle point: $(0, 0, 0)$

Relative maxima: $(1, 1, 2), (-1, -1, 2)$

16. About 128,613 units

17. $y = 0.52x + 1.4$

18. $\frac{3}{2}$

19. 1

20. $\frac{4}{3}$ units^2

21. 48

22. $\frac{11}{6}$

Appendix A

Section A.1   *(page A7)*

1. Rational   **3.** Rational   **5.** Irrational   **7.** Rational

9. (a) Yes   (b) No   (c) Yes

11. (a) Yes   (b) Yes   (c) No

13. $x \geq 12$

15. $x \leq -\frac{1}{2}$

17. $x > 1$

19. $-\frac{1}{2} < x < \frac{7}{2}$

21. $-\frac{3}{4} < x < -\frac{1}{4}$

23. $x \leq -1$ or $x \geq 4$

25. $-\frac{3}{2} \leq x \leq 2$

27. $4.1 \leq E \leq 4.25$

29. $p \leq 0.51$

31. $120 \leq r \leq 180$

33. $x \geq 36$

35. (a) False   (b) True   (c) True   (d) False

Section A.2   *(page A12)*

1. (a) 11   (b) 7   **3.** (a) 8   (b) $\frac{35}{3}$

5. (a) 51   (b) -51 (c) 51

7. (a) 14.99   (b) -14.99 (c) 14.99

9. (a) $\frac{128}{75}$ (b) $-\frac{128}{75}$ (c) $\frac{128}{75}$

11. $|x - 5| < 3$ **13.** $|y - a| \geq 2$

15. $-4 < x < 4$

17. $-12 \leq x \leq 2$

19. $x < 4$ or $x > 5$

21. $x \leq -7$ or $x \geq 13$

23. $a - b \leq x \leq a + b$

25. $x < \frac{2}{5}(a - b)$ or $x > \frac{2}{5}(a + b)$

27. 16   **29.** 1.25   **31.** $\frac{1}{8}$ **33.** $|p - 33.15| \leq 2$

35. 20.75 oz and 19.25 oz

37. (a) $|4750 - E| \leq 500$,

$|4750 - E| \leq 237.50$

(b) Within

39. (a) $|20{,}000 - E| \leq 500$,

$|20{,}000 - E| \leq 1000$

(b) Within

41. $\$11{,}759.40 \leq C \leq \$17{,}639.10$

Section A.3   *(page A18)*

1. -108 **3.** $\frac{5}{8}$ **5.** 4 **7.** 44 **9.** 5 **11.** 256

13. $\frac{1}{2}$ **15.** $\frac{1}{4}$ **17.** 908.3483 **19.** -3.7798

21. $10x^4$ **23.** $\dfrac{3}{4y^{14}}$ **25.** $7x^5$ **27.** $\frac{5}{2}(x+y)^5, x \neq -y$

29. $3x, x > 0$ **31.** $2\sqrt{2}$ **33.** $3x\sqrt[3]{2x^2}$

35. $\dfrac{2x^3z}{y}\sqrt[3]{\dfrac{18z^2}{y}}$ **37.** $3x(x+2)(x-2)$ **39.** $\dfrac{7x^4+1}{x^{1/2}}$

41. $(x-2)^{1/3}(x+1)^{1/2}(13-5x)$ **43.** $\dfrac{8x^5+x^2}{7}$

45. $x \geq 4$ **47.** $(-\infty, \infty)$ **49.** $(8, \infty)$

51. $x \neq 1, x \geq -2$ **53.** $\$19{,}121.84$ **55.** $\$11{,}345.46$

57. $\dfrac{\sqrt{2}}{2}\pi$ sec or about 2.22 sec

Section A.4   *(page A24)*

1. $\frac{1}{6}, 1$ **3.** $\frac{3}{2}$ **5.** $-2 \pm \sqrt{3}$ **7.** No real solution

9. $(x-2)^2$ **11.** $(x+3)(x-6)$ **13.** $(2x+1)^2$

15. $(3x-1)(x-1)$ **17.** $(x-2y)^2$

19. $(3+y)(3-y)(9+y^2)$ **21.** $(x-1)(x^2+x+1)$

23. $(y+3)(y^2-3y+9)$ **25.** $(x-y)(x^2+xy+y^2)$

27. $(x-4)(x-1)(x+1)$ **29.** $(2x-3)(x^2+2)$

31. $(x-2)(2x^2-1)$ **33.** $(x+4)(x-4)(x^2+1)$

35. $0, 5$ **37.** ± 9 **39.** $\pm\sqrt{3}$ **41.** $0, 6$ **43.** $-2, 1$

45. $-1, 6$ **47.** $-1, -\frac{2}{3}$ **49.** -7 **51.** ± 1

53. $1, \pm 2$ **55.** $(-\infty, -7] \cup [7, \infty)$

57. $(-\infty, 3] \cup [4, \infty)$ **59.** $(-\infty, -1] \cup \left[-\frac{1}{5}, \infty\right)$

61. $(x+1)(x^2-4x-2)$ **63.** $(x-3)(2x^2-3x+1)$

65. $-4, -2, 6$ **67.** $1, 2, 3$ **69.** 4

71. $-2, -1, \frac{1}{4}$ **73.** $-\frac{2}{3}, -\frac{1}{2}, 3$

75. Two solutions; The solutions of the equation are ± 2000, but the minimum average cost occurs at the positive value, 2000; 2000 units

Section A.5   *(page A31)*

1. $\dfrac{x-4}{x+6}, x \neq 3$ **3.** $\dfrac{x+5}{2x+3}, x \neq 2$ **5.** $\dfrac{x+3}{x-2}$

7. $\dfrac{x^5-4}{x^3}$ **9.** $\dfrac{5x^2-9x+8}{(x-3)(3x+4)}$ **11.** $\dfrac{2-x}{x^2-16}$

13. $\dfrac{1}{(x+2)(x-1)}$ **15.** $-\dfrac{x-2}{x^2+1}$ **17.** $\dfrac{x+2}{(x+1)^{3/2}}$

19. $-\dfrac{3t}{2\sqrt{1+t}}$ **21.** $\dfrac{x(x^2+2)}{(x^2+1)^{3/2}}$ **23.** $\dfrac{2}{x^2\sqrt{x^2+2}}$

25. $\dfrac{1}{2\sqrt{x}(x+1)^{3/2}}$ **27.** $\dfrac{3x(x+2)}{(2x+3)^{3/2}}$ **29.** $\dfrac{\sqrt{10}}{5}$

31. $\dfrac{49\sqrt{x^2-9}}{x+3}$ **33.** $\dfrac{\sqrt{14}+2}{2}$ **35.** $\sqrt{6}-\sqrt{5}$

37. $\sqrt{x}-\sqrt{x-2}$ **39.** $\dfrac{1}{\sqrt{x+2}+\sqrt{2}}$ **41.** $\$200.38$

Appendix B   *(page A39)*

1. 17.5 square units

3. Left Riemann sum: 0.518
   Right Riemann sum: 0.768

5. Left Riemann sum: 0.746
   Right Riemann sum: 0.646

7. Left Riemann sum: 0.859
   Right Riemann sum: 0.659

9. (a)

(b) Answers will vary.
(c) Answers will vary.
(d) Answers will vary.
(e)

n	5	10	50	100
Left sum, S_L	1.6	1.8	1.96	1.98
Right sum, S_R	2.4	2.2	2.04	2.02

(f) Answers will vary.

11. $\displaystyle\int_0^5 3\, dx$

13. $\displaystyle\int_{-4}^4 (4-|x|)\, dx = \int_{-4}^0 (4+x)\, dx + \int_0^4 (4-x)\, dx$

15. $\displaystyle\int_{-2}^2 (4-x^2)\, dx$ **17.** $\displaystyle\int_0^2 \sqrt{x+1}\, dx$

19.

Rectangle

$A = 15$

21.

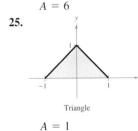

Triangle

$A = 6$

23.

Trapezoid

$A = 14$

25.

Triangle

$A = 1$

27.

Semicircle

$A = 9\pi/2$

29. Answers will vary.

31. $>$; The left endpoint approximation will be greater than the actual area.

Answers to Checkpoints

Chapter 1

Checkpoints for Section 1.1

1   **2**

3 5   **4** $d_1 = \sqrt{20}, d_2 = \sqrt{45}, d_3 = \sqrt{65}$ **5** 25 yd
$$d_1^2 + d_2^2 = 20 + 45 = 65 = d_3^2$$

6 $(-2, 5)$ **7** $14 billion

8 $(-1, -4), (1, -2), (1, 2), (-1, 0)$

Checkpoints for Section 1.2

1   **2**

3 x-intercepts: $(3, 0), (-1, 0)$, y-intercept: $(0, -3)$

4 $(x + 2)^2 + (y - 1)^2 = 25$

5 12,500 units   **6** 4 million units at $99/unit

7

Year	2006	2007	2008	2009
Sales	3.97	4.24	4.64	5.23
Model	3.87	4.28	4.75	5.28

Year	2010	2011	2012	2013
Sales	5.88	6.63	7.39	7.84
Model	5.87	6.52	7.22	7.99

2014 sales: $8.81 billion

Checkpoints for Section 1.3

1 (a)   (b)

(c)

2 Yes, $\frac{27}{312} \approx 0.08654 > \frac{1}{12} = 0.08\overline{3}$.

3 The y-intercept $(0, 1500)$ tells you that the original value of the copier is $1500. The slope of $m = -300$ tells you that the value decreases by $300/yr.

4 (a) 2   (b) $-\frac{1}{2}$ (c) 0   **5** $y = 2x + 4$

6 $y = 28.75t - 210.6$; $163.15

7 (a) $y = \frac{1}{2}x$ (b) $y = -2x + 5$

8 $V = -1375t + 12{,}000$

Checkpoints for Section 1.4

1 (a) Yes, $y = x - 1$.   (b) No, $y = \pm\sqrt{4 - x^2}$.
(c) No, $y = \pm\sqrt{2 - x}$.   (d) Yes, $y = x^2$.

2 (a) Domain: $[-1, \infty)$; Range: $[0, \infty)$
(b) Domain: $(-\infty, \infty)$; Range: $[0, \infty)$

3 $f(0) = 1, f(1) = -3, f(4) = -3$
No, f is not one-to-one.

4 (a) $x^2 + 2x\,\Delta x + (\Delta x)^2 + 3$ (b) $2x + \Delta x, \Delta x \neq 0$

5 (a) $2x^2 + 5$ (b) $4x^2 + 4x + 3$

6 (a) $f^{-1}(x) = 5x$ (b) $f^{-1}(x) = x + 6$

7 $f^{-1}(x) = \sqrt{x - 2}$

8
$$f(x) = x^2 + 4$$
$$y = x^2 + 4$$
$$x = y^2 + 4$$
$$x - 4 = y^2$$
$$\pm\sqrt{x - 4} = y$$

Checkpoints for Section 1.5

1 6   **2** (a) 4   (b) Does not exist   (c) 4

3 (a) 5   (b) 6   (c) 25   (d) -2 **4** 5   **5** 12

6 7   **7** $\frac{1}{4}$ **8** (a) -1 (b) 1   **9** 1

10 $\lim\limits_{x \to 3^-} f(x) = 6$ and $\lim\limits_{x \to 3^+} f(x) = 7$

$$\lim\limits_{x \to 3^-} \neq \lim\limits_{x \to 3^+}$$

11 Does not exist

Checkpoints for Section 1.6

1 (a) f is continuous on the entire real line.
  (b) f is continuous on the entire real line.
  (c) f is continuous on the entire real line.

2 (a) f is continuous on $(-\infty, 1)$ and $(1, \infty)$.
  (b) f is continuous on $(-\infty, 2)$ and $(2, \infty)$.
  (c) f is continuous on the entire real line.

3 f is continuous on $[2, \infty)$.   **4** f is continuous on $[-1, 5]$.

5

6 $A = 10,000(1 + 0.0075)^{\llbracket 4t \rrbracket}$

Chapter 2

Checkpoints for Section 2.1

1 3

2 For the months to the left of July on the graph, the tangent lines have positive slopes. For the months to the right of July, the tangent lines have negative slopes. The average daily temperature is increasing prior to July and decreasing after July.

3 4   **4** 2

5 $m = 8x$
  At $(0, 1)$, $m = 0$.
  At $(1, 5)$, $m = 8$.

6 $2x - 5$ **7** $-\dfrac{4}{t^2}$

Checkpoints for Section 2.2

1 (a) 0   (b) 0   (c) 0   (d) 0

2 (a) $4x^3$ (b) $-\dfrac{3}{x^4}$ (c) $2w$

3 $f'(x) = 3x^2$
  $m = f'(-1) = 3$;
  $m = f'(0) = 0$;
  $m = f'(1) = 3$

4 (a) $8x$ (b) $\dfrac{8}{\sqrt{x}}$ **5** (a) $\dfrac{1}{4}$ (b) $-\dfrac{2}{5}$

6 (a) $-\dfrac{9}{2x^3}$ (b) $-\dfrac{9}{8x^3}$ **7** (a) $\dfrac{\sqrt{5}}{2\sqrt{x}}$ (b) $\dfrac{1}{4x^{3/4}}$

8 (a) $4x + 5$ (b) $4x^3 - 2$ (c) $\dfrac{2x^3 + 1}{x^2}$ **9** -1

10 $y = -x + 2$ **11** $9.44/\text{yr}$

Checkpoints for Section 2.3

1 (a) $0.5\overline{6}$ mg/mL/min   (b) 0 mg/mL/min
  (c) -1.5 mg/mL/min

2 (a) -16 ft/sec   (b) -48 ft/sec   (c) -80 ft/sec

3 (a) When $t = 1.75$, $s'(1.75) = -56$ ft/sec.
  (b) When $t = 2$, $s'(2) = -64$ ft/sec.

4 (a) $\dfrac{3}{2}$ sec   (b) -32 ft/sec

5 When $x = 100$, $\dfrac{dP}{dx} = \$16/\text{unit}$.
  Actual gain $= \$16.06$

6 $p = -1.6x + 529$; $R = -1.6x^2 + 529x$;
  The results are similar.

7 Revenue: $R = 2000x - 4x^2$
  Marginal revenue: $\dfrac{dR}{dx} = 2000 - 8x$; $\$0/\text{unit}$

8 $\dfrac{dP}{dx} = \$1.44/\text{unit}$
  Actual increase in profit $\approx \$1.44$

Checkpoints for Section 2.4

1 $-27x^2 + 12x + 24$ **2** $\dfrac{2x^2 - 1}{x^2}$

3 (a) $18x^2 + 30x$ (b) $12x + 15$ **4** $-\dfrac{22}{(5x - 2)^2}$

5 $y = \dfrac{8}{25}x - \dfrac{4}{5}$

6 $\dfrac{-3x^2 + 4x + 8}{x^2(x + 4)^2}$ **7** (a) $\dfrac{2}{5}x + \dfrac{4}{5}$ (b) $3x^3$

8

t	0	1	2	3	4	5	6	7
$\dfrac{dP}{dt}$	0	-50	-16	-6	-2.77	-1.48	-0.88	-0.56

As t increases, the rate at which the blood pressure drops decreases.

Checkpoints for Section 2.5

1 (a) $u = g(x) = x + 1$, $y = f(u) = \dfrac{1}{\sqrt{u}}$
  (b) $u = g(x) = x^2 + 2x + 5$, $y = f(u) = u^3$

2 $6x^2(x^3 + 1)$ **3** $4(2x + 3)(x^2 + 3x)^3$ **4** $y = \dfrac{1}{3}x + \dfrac{8}{3}$

5 $-\dfrac{8}{(2x + 1)^2}$ **6** $\dfrac{x(3x^2 + 2)}{\sqrt{x^2 + 1}}$ **7** $-\dfrac{12(x + 1)}{(x - 5)^3}$

8 About $\$3.72/\text{yr}$

Checkpoints for Section 2.6

1 $f'(x) = 18x^2 - 4x$, $f''(x) = 36x - 4$, $f'''(x) = 36$, $f^{(4)}(x) = 0$

2 18   **3** $\dfrac{120}{x^6}$

4 Height $= 144$ ft
  Velocity $= 0$ ft/sec
  Acceleration $= -32$ ft/sec^2

5 -9.8 m/sec^2

6

Acceleration approaches zero.

Checkpoints for Section 2.7

1 $-\dfrac{2}{x^3}$

2 (a) $12x^2$ (b) $6y\dfrac{dy}{dx}$ (c) $1 + 5\dfrac{dy}{dx}$ (d) $y^3 + 3xy^2\dfrac{dy}{dx}$

3 $\dfrac{dy}{dx} = -\dfrac{x-2}{y-1}$ **4** $\dfrac{3}{4}$ **5** $\dfrac{5}{9}$

6 $\dfrac{dx}{dp} = -\dfrac{2}{p^2(0.002x + 1)}$

Checkpoints for Section 2.8

1 9   **2** $12\pi \approx 37.7 \text{ ft}^2/\text{sec}$
3 \$1500/day   **4** \$28,400/wk

Chapter 3

Checkpoints for Section 3.1

1 $f'(x) = 4x^3$
$f'(x) < 0$ if $x < 0$, so f is decreasing on $(-\infty, 0)$.
$f'(x) > 0$ if $x > 0$, so f is increasing on $(0, \infty)$.

2 $\dfrac{dB}{dt} = -0.0358t + 0.485 > 0$ when $7 \le t \le 12$, which implies that the consumption of blueberries was increasing from 2007 to 2012.

3 $x = \frac{1}{2}$

4 Increasing on $(-\infty, -2)$ and $(2, \infty)$
Decreasing on $(-2, 2)$

5 Increasing on $(0, \infty)$
Decreasing on $(-\infty, 0)$

6 Increasing on $(-\infty, -1)$ and $(1, \infty)$
Decreasing on $(-1, 0)$ and $(0, 1)$

7 Because $f'(x) = -3x^2 = 0$ when $x = 0$ and because f is decreasing on $(-\infty, 0) \cup (0, \infty)$, f is decreasing on $(-\infty, \infty)$.

8 $(0, 3000)$

Checkpoints for Section 3.2

1 Relative maximum at $(-1, 5)$
Relative minimum at $(1, -3)$

2 Relative minimum at $(3, -27)$

3 Relative maximum at $(1, 1)$
Relative minimum at $(0, 0)$

4 Absolute maximum at $(0, 10)$
Absolute minimum at $(4, -6)$

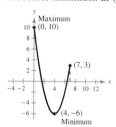

5

x	24,000	24,200	24,300	24,400
P	\$24,760	\$24,766	\$24,767.50	\$24,768

x	24,500	24,600	24,800
P	\$24,767.50	\$24,766	\$24,760

Checkpoints for Section 3.3

1 (a) $f'' = -4$; Because $f''(x) < 0$ for all x, f is concave downward for all x.

(b) $f''(x) = \dfrac{1}{2x^{3/2}}$; Because $f''(x) > 0$ for all $x > 0$, f is concave upward for all $x > 0$.

2 Because $f''(x) > 0$ for $x < -\dfrac{2\sqrt{3}}{3}$ and $x > \dfrac{2\sqrt{3}}{3}$, f is concave upward on $\left(-\infty, -\dfrac{2\sqrt{3}}{3}\right)$ and $\left(\dfrac{2\sqrt{3}}{3}, \infty\right)$. Because $f''(x) < 0$ for $-\dfrac{2\sqrt{3}}{3} < x < \dfrac{2\sqrt{3}}{3}$, f is concave downward on $\left(-\dfrac{2\sqrt{3}}{3}, \dfrac{2\sqrt{3}}{3}\right)$.

3 f is concave upward on $(-\infty, 0)$.
f is concave downward on $(0, \infty)$.
Point of inflection: $(0, 0)$

4 f is concave upward on $(-\infty, 0)$ and $(1, \infty)$.
f is concave downward on $(0, 1)$.
Points of inflection: $(0, 1), (1, 0)$

5 Relative minimum: $(3, -26)$

6 Point of diminishing returns: $x = \$150$ thousand

Checkpoints for Section 3.4

1

Maximum volume $= 108 \text{ in.}^3$

2 $\left(\sqrt{\frac{1}{2}}, \frac{7}{2}\right)$ and $\left(-\sqrt{\frac{1}{2}}, \frac{7}{2}\right)$ **3** 8 in. by 12 in.

Checkpoints for Section 3.5

1 125 units yield a maximum revenue of \$1,562,500.
2 400 units   **3** \$6.25/unit   **4** \$4.00
5 Demand is elastic when $0 < x < 144$.
Demand is inelastic when $144 < x < 324$.
Demand is of unit elasticity when $x = 144$.

Checkpoints for Section 3.6

1 (a) $\displaystyle\lim_{x \to 2^-} \dfrac{1}{x-2} = -\infty$ (b) $\displaystyle\lim_{x \to 2^+} \dfrac{1}{x-2} = \infty$

(c) $\displaystyle\lim_{x \to -3^-} \dfrac{-1}{x+3} = -\infty$ (d) $\displaystyle\lim_{x \to -3^+} \dfrac{-1}{x+3} = \infty$

2 $x = 0, x = 4$ **3** $x = 3$

4 $\displaystyle\lim_{x \to 2^-} \dfrac{x^2 - 4x}{x - 2} = \infty$; $\displaystyle\lim_{x \to 2^+} \dfrac{x^2 - 4x}{x - 2} = -\infty$

5 2

6 (a) $y = 0$ (b) $y = \frac{1}{2}$ (c) No horizontal asymptote

7 $C = 0.75x + 25,000$

$\overline{C} = 0.75 + \dfrac{25,000}{x}$

$\displaystyle\lim_{x \to \infty} \overline{C} = \$0.75/\text{unit}$

8 No, the cost function is not defined at $p = 100$, which implies that it is not possible to remove 100% of the pollutants.

Checkpoints for Section 3.7

1

	$f(x)$	$f'(x)$	$f''(x)$	Shape of graph
x in $(-\infty, -1)$		−	+	Decreasing, concave upward
$x = -1$	−32	0	+	Relative minimum
x in $(-1, 1)$		+	+	Increasing, concave upward
$x = 1$	−16	+	0	Point of inflection
x in $(1, 3)$		+	−	Increasing, concave downward
$x = 3$	0	0	−	Relative maximum
x in $(3, \infty)$		−	−	Decreasing, concave downward

2

	$f(x)$	$f'(x)$	$f''(x)$	Shape of graph
x in $(-\infty, 0)$		−	+	Decreasing, concave upward
$x = 0$	5	0	0	Point of inflection
x in $(0, 2)$		−	−	Decreasing, concave downward
$x = 2$	−11	−	0	Point of inflection
x in $(2, 3)$		−	+	Decreasing, concave upward
$x = 3$	−22	0	+	Relative minimum
x in $(3, \infty)$		+	+	Increasing, concave upward

3

	$f(x)$	$f'(x)$	$f''(x)$	Shape of graph
x in $(-\infty, 0)$		+	−	Increasing, concave downward
$x = 0$	0	0	−	Relative maximum
x in $(0, 1)$		−	−	Decreasing, concave downward
$x = 1$	Undef.	Undef.	Undef.	Vertical asymptote
x in $(1, 2)$		−	+	Decreasing, concave upward
$x = 2$	4	0	+	Relative minimum
x in $(2, \infty)$		+	+	Increasing, concave upward

4

	$f(x)$	$f'(x)$	$f''(x)$	Shape of graph
x in $(-\infty, -1)$		+	+	Increasing, concave upward
$x = -1$	Undef.	Undef.	Undef.	Vertical asymptote
x in $(-1, 0)$		+	−	Increasing, concave downward
$x = 0$	−1	0	− .	Relative maximum
x in $(0, 1)$		−	−	Decreasing, concave downward
$x = 1$	Undef.	Undef.	Undef.	Vertical asymptote
x in $(1, \infty)$		−	+	Decreasing, concave upward

5

	$f(x)$	$f'(x)$	$f''(x)$	Shape of graph
x in $(0, 1)$		$-$	$+$	Decreasing, concave upward
$x = 1$	-4	0	$+$	Relative minimum
x in $(1, \infty)$		$+$	$+$	Increasing, concave upward

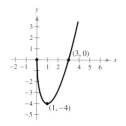

Checkpoints for Section 3.8

1 $dy = 0.32$; $\Delta y = 0.32240801$ **2** $dR = \$22$; $\Delta R = \$21$
3 $dP = \$10.96$; $\Delta P = \$10.98$
4 (a) $dy = 12x^2\,dx$ (b) $dy = \frac{2}{3}\,dx$

(c) $dy = (6x - 2)\,dx$ (d) $dy = -\dfrac{2}{x^3}\,dx$

Chapter 4

Checkpoints for Section 4.1

1 (a) 243   (b) 3   (c) 64
2 (a) 8   (b) $\frac{1}{2}$ (c) $\sqrt{10}$
3 (a) 5.453×10^{-13} (b) 1.621×10^{-13}
(c) 2.629×10^{-14}

4   **5**

Checkpoints for Section 4.2

1

x	-2	-1	0	1	2
$g(x)$	$e^2 \approx 7.389$	$e \approx 2.718$	1	$\frac{1}{e} \approx 0.368$	$\frac{1}{e^2} \approx 0.135$

2 After 0 h, $y = 1.25$ g.
After 1 h, $y \approx 1.338$ g.
After 10 h, $y \approx 1.498$ g.

$\displaystyle\lim_{t \to \infty} \frac{1.50}{1 + 0.2e^{-0.5t}} = 1.50$ g

3 (a) \$4870.38   (b) \$4902.71
(c) \$4918.66   (d) \$4919.21
All else being equal, the more often interest is compounded, the greater the balance.
4 (a) 7%   (b) 7.12%   (c) 7.19%   (d) 7.23%
5 \$16,712.90

Checkpoints for Section 4.3

1 At $(0, 2)$, the slope is 2. At $(1, 2e)$, the slope is $2e$.
2 (a) $3e^{3x}$ (b) $-\dfrac{6x^2}{e^{2x^3}}$ (c) $8xe^{x^2}$ (d) $-\dfrac{2}{e^{2x}}$
3 (a) 0   (b) $3e^{3x+1}$ (c) $xe^x(x + 2)$ (d) $\frac{1}{2}(e^x - e^{-x})$
(e) $\dfrac{e^x(x - 2)}{x^3}$ (f) $e^x(x^2 + 2x - 1)$

4

5 \$18.39/unit (80,000 units); \$1,471,517.77

6

Points of inflection: $(-4, 0.060)$, $(4, 0.060)$

Checkpoints for Section 4.4

1

x	-1.5	-1	-0.5	0	0.5	1
$f(x)$	-0.693	0	0.405	0.693	0.916	1.099

2 (a) 3   (b) $x + 1$
3 (a) $\ln 2 - \ln 5$ (b) $\frac{1}{3}\ln(x + 2)$
(c) $\ln x - \ln 5 - \ln y$ (d) $\ln x + 2\ln(x + 1)$
4 (a) $\ln x^4 y^3$ (b) $\ln \dfrac{x + 1}{(x + 3)^2}$ **5** (a) $\ln 6$ (b) $5\ln 5$
6 (a) e^4 (b) e^3 **7** 7.9 yr

Checkpoints for Section 4.5

1 $\dfrac{1}{x}$ **2** (a) $\dfrac{2x}{x^2 - 4}$ (b) $x(1 + 2\ln x)$ (c) $\dfrac{2\ln x - 1}{x^3}$

3 $\dfrac{1}{3(x + 1)}$ **4** $\dfrac{2}{x} + \dfrac{x}{x^2 + 1}$ **5** $y = 4x - 4$
6 Relative minimum: $(2, 2 - 2\ln 2) \approx (2, 0.6137)$

7 $\dfrac{dp}{dt} = -1.3\%/\text{mo}$

The average score would decrease at a greater rate than the model in Example 7.

8 (a) 4 (b) -2 (c) -5 (d) 3

9 (a) 2.322 (b) 2.631 (c) 3.161 (d) -0.5

10   As time increases, the derivative approaches 0. The rate of change of the amount of carbon isotopes is proportional to the amount present.

Checkpoints for Section 4.6

1 About 2113.7 yr **2** $y = 25e^{0.6931t}$

3 $r = \frac{1}{8}\ln 2 \approx 0.0866$ or 8.66% **4** About 12.42 mo

Chapter 5

Checkpoints for Section 5.1

1 (a) $5x + C$ (b) $-r + C$ (c) $2t + C$ **2** $\frac{5}{2}x^2 + C$

3 (a) $-\dfrac{1}{x} + C$ (b) $\dfrac{3}{4}x^{4/3} + C$

4 (a) $\frac{1}{2}x^2 + 4x + C$ (b) $x^4 - \frac{5}{2}x^2 + 2x + C$

5 $\frac{2}{3}x^{3/2} + 4x^{1/2} + C$

6 General solution: $F(x) = 2x^2 + 2x + C$
Particular solution: $F(x) = 2x^2 + 2x + 4$

7 $s(t) = -16t^2 + 32t + 48$. The ball hits the ground 3 seconds after it is thrown, with a velocity of -64 feet per second.

8 $C = -0.01x^2 + 28x + 12.01$
$C(200) = \$5212.01$

Checkpoints for Section 5.2

1 (a) $\dfrac{(x^3 + 6x)^3}{3} + C$ (b) $\dfrac{2}{3}(x^2 - 2)^{3/2} + C$

2 $\frac{1}{36}(3x^4 + 1)^3 + C$ **3** $\frac{1}{9}(x^3 - 3x)^3 + C$

4 $2x^9 + \frac{12}{5}x^5 + 2x + C$ **5** $\frac{5}{3}(x^2 - 1)^{3/2} + C$

6 $-\frac{1}{3}(1 - 2x)^{3/2} + C$ **7** $\frac{1}{3}(x^2 + 4)^{3/2} + C$

8 About \$33,633

Checkpoints for Section 5.3

1 (a) $3e^x + C$ (b) $e^{5x} + C$ (c) $e^x - \dfrac{x^2}{2} + C$

2 $\frac{1}{2}e^{2x+3} + C$ **3** $2e^{x^2} + C$

4 (a) $2\ln|x| + C$ (b) $\ln|x^3| + C$ (c) $\ln|2x + 1| + C$

5 $\frac{1}{4}\ln|4x + 1| + C$ **6** $\frac{3}{2}\ln(x^2 + 4) + C$

7 (a) $4x - 3\ln|x| - \dfrac{2}{x} + C$ (b) $2\ln(1 + e^x) + C\ dx$

(c) $\dfrac{x^2}{2} + x + 3\ln|x + 1| + C$

Checkpoints for Section 5.4

1 $\frac{1}{2}(3)(12) = 18$

2 $\frac{22}{3}$ units^2 **3** 68

4 (a) $\frac{1}{4}(e^4 - 1) \approx 13.3995$ (b) $-\ln 5 + \ln 2 \approx -0.9163$

5 $\frac{13}{2}$ **6** (a) About \$14.18 (b) \$141.79

7 About 3.3 billion; The answer is the same.

8 (a) $\frac{2}{5}$ (b) 0 **9** About \$12,295.62

Checkpoints for Section 5.5

1 $\frac{8}{3}$ units^2

2 $\frac{32}{3}$ units^2 **3** $\frac{9}{2}$ units^2

4 $\frac{253}{12}$ units^2

5 Consumer surplus: 40
Producer surplus: 20

6 The company can save \$63.84 million.

Checkpoints for Section 5.6

1 $\frac{37}{8}$ units^2 **2** 0.436 unit^2 **3** 5.642 units^2

4 About 1.463

Chapter 6

Checkpoints for Section 6.1

1 $\dfrac{1}{2}xe^{2x} - \dfrac{1}{4}e^{2x} + C$ **2** $\dfrac{x^2}{2}\ln x - \dfrac{1}{4}x^2 + C$

3 $x\ln 2x - x + C$ **4** $e^x(x^3 - 3x^2 + 6x - 6) + C$

5 $e - 2$ **6** \$538,145 **7** \$721,632.08

Checkpoints for Section 6.2

1 $\frac{2}{3}(x - 4)\sqrt{2 + x} + C$ (Formula 19)

2 $\sqrt{x^2 + 16} - 4\ln\left|\dfrac{4 + \sqrt{x^2 + 16}}{x}\right| + C$ (Formula 25)

3 $\dfrac{1}{4}\ln\left|\dfrac{x - 2}{x + 2}\right| + C$ (Formula 21)

4 $\frac{1}{3}[1 - \ln(1 + e) + \ln 2] \approx 0.12663$ (Formula 39)

5 $x(\ln x)^3 - 3x(\ln x)^2 + 6x \ln x - 6x + C$ (Formulas 45 and 44)

6 About 18.2%

Checkpoints for Section 6.3

1 3.2608   **2** 3.1956   **3** 1.154

Checkpoints for Section 6.4

1 (a) Converges to $\frac{1}{2}$ (b) Diverges
2 1   **3** $\frac{1}{2}$ **4** 0.0013 or 0.13%
5 No, you do not have enough money to start the scholarship fund because you need \$125,000.   (\$125,000 > \$120,000)

Chapter 7

Checkpoints for Section 7.1

1 (a)–(c)

2 $2\sqrt{6}$ **3** $\left(-\frac{5}{2}, 2, -2\right)$
4 $(x - 4)^2 + (y - 3)^2 + (z - 2)^2 = 25$
5 $(x - 1)^2 + (y - 3)^2 + (z - 2)^2 = 38$
6 Center: $(-3, 4, -1)$; radius: 6
7 $(x + 1)^2 + (y - 2)^2 = 16$

Checkpoints for Section 7.2

1 x-intercept: $(4, 0, 0)$
y-intercept: $(0, 2, 0)$
z-intercept: $(0, 0, 8)$

2 xy-trace: circle, $x^2 + y^2 = 1$
yz-trace: hyperbola, $y^2 - z^2 = 1$
xz-trace: hyperbola, $x^2 - z^2 = 1$
$z = 3$ trace: circle, $x^2 + y^2 = 10$
Hyperboloid of one sheet
3 (a) Elliptic paraboloid   (b) Elliptic cone

Checkpoints for Section 7.3

1 (a) 0   (b) $\frac{9}{4}$ **2** (a) $x^2 + y^2 \le 9$ (b) $0 \le z \le 3$
3 For each value of c, the equation $f(x, y) = c$ is a circle (or point) in the xy-plane.

4 $f(1500, 1000) \approx 127{,}542$ units
$f(1000, 1500) \approx 117{,}608$ units
x, person-hours, has a greater effect on production.
5 $M = \$421.60/\text{mo}$
Total paid $= (30 \times 12) \times 421.60 = \$151{,}776$

Checkpoints for Section 7.4

1 $\dfrac{\partial z}{\partial x} = 4x - 8xy^3$

$\dfrac{\partial z}{\partial y} = -12x^2y^2 + 4y^3$

2 $f_x(x, y) = 2xy^3; f_x(1, 2) = 16$
$f_y(x, y) = 3x^2y^2; f_y(1, 2) = 12$

3 (a) $f_x(1, -1, 49) = 8$ (b) $f_y(1, -1, 49) = -18$
4 Substitutes

5 $\dfrac{\partial w}{\partial x} = xy + 2xy \ln(xz)$

$\dfrac{\partial w}{\partial y} = x^2 \ln(xz)$

$\dfrac{\partial w}{\partial z} = \dfrac{x^2 y}{z}$

6 $f_{xx} = 8y^2$
$f_{yy} = 8x^2 + 8$
$f_{xy} = 16xy$
$f_{yx} = 16xy$

7 $f_{xx} = 0$ $f_{xy} = e^y$ $f_{xz} = 2$
$f_{yx} = e^y$ $f_{yy} = xe^y + 2$ $f_{yz} = 0$
$f_{zx} = 2$ $f_{zy} = 0$ $f_{zz} = 0$

Checkpoints for Section 7.5

1 $f(-8, 2) = -64$: relative minimum
2 $f(0, 0) = 1$: relative maximum
3 $f(0, 0) = 0$: saddle point
4 $P(3.11, 3.81) = \$744.81$ maximum profit
5 $V\left(\frac{4}{3}, \frac{2}{3}, \frac{8}{3}\right) = \frac{64}{27}$ units^3

Checkpoints for Section 7.6

1 $V\left(\frac{4}{3}, \frac{2}{3}, \frac{8}{3}\right) = \frac{64}{27}$ units^3 **2** $f(187.5, 50) \approx 13{,}474$ units
3 About 26,740 units
4 $P(3.35, 4.26) = \$758.08$ maximum profit

Checkpoints for Section 7.7

1 For $f(x)$, $S = 10$. For $g(x)$, $S = 0.76$.
The quadratic model is a better fit.
2 $f(x) = \frac{6}{5}x + \frac{23}{10}$ **3** $y = 16.76t + 128.4$
About 413,320,000 subscribers

Checkpoints for Section 7.8

1 (a) $\frac{1}{4}x^4 + 2x^3 - 2x - \frac{1}{4}$ (b) $\ln|y^2 + y| - \ln|2y|$

2 $\frac{25}{2}$ **3** $\displaystyle\int_2^4 \int_1^5 dx\,dy = 8$ **4** $\frac{4}{3}$

5 (a)

R: $0 \le y \le 2$
$2y \le x \le 4$

(b) $\displaystyle\int_0^4 \int_0^{x/2} dy\,dx$

(c) $\displaystyle\int_0^2 \int_{2y}^4 dx\,dy = 4 = \int_0^4 \int_0^{x/2} dy\,dx$

6 $\displaystyle\int_{-1}^3 \int_{x^2}^{2x+3} dy\,dx = \frac{32}{3}$

Checkpoints for Section 7.9

1 $\frac{16}{3}$ **2** $e - 1$ **3** $\frac{176}{15}$ **4** Integration by parts   **5** 3

Answers to Tech Tutors

Tech Tutor

Section 1.3 *(page 29)*

The lines appear perpendicular in the setting $-9 \leq x \leq 9$ and $-6 \leq y \leq 6$.

Section 1.6 *(page 61)*

Most calculators set in *connected* mode will join the two branches of the graph with a nearly vertical line near $x = 2$. This line is not part of the graph.

Section 4.5 *(page 287)*

Answers will vary.

Section 6.3 *(page 395)*

1.46265

Index

Basic Differentiation Rules

1. $\dfrac{d}{dx}[cu] = cu'$

2. $\dfrac{d}{dx}[u \pm v] = u' \pm v'$

3. $\dfrac{d}{dx}[uv] = uv' + vu'$

4. $\dfrac{d}{dx}\left[\dfrac{u}{v}\right] = \dfrac{vu' - uv'}{v^2}$

5. $\dfrac{d}{dx}[c] = 0$

6. $\dfrac{d}{dx}[u^n] = nu^{n-1}u'$

7. $\dfrac{d}{dx}[x] = 1$

8. $\dfrac{d}{dx}[\ln u] = \dfrac{u'}{u}$

9. $\dfrac{d}{dx}[e^u] = e^u u'$

10. $\dfrac{d}{dx}[\log_a u] = \dfrac{u'}{(\ln a)u}$

11. $\dfrac{d}{dx}[a^u] = (\ln a)a^u u'$

12. $\dfrac{d}{dx}[\sin u] = (\cos u)u'$

13. $\dfrac{d}{dx}[\cos u] = -(\sin u)u'$

14. $\dfrac{d}{dx}[\tan u] = (\sec^2 u)u'$

15. $\dfrac{d}{dx}[\cot u] = -(\csc^2 u)u'$

16. $\dfrac{d}{dx}[\sec u] = (\sec u \tan u)u'$

17. $\dfrac{d}{dx}[\csc u] = -(\csc u \cot u)u'$

Basic Integration Formulas

1. $\displaystyle\int kf(u)\, du = k\int f(u)\, du$

2. $\displaystyle\int [f(u) \pm g(u)]\, du = \int f(u)\, du \pm \int g(u)\, du$

3. $\displaystyle\int du = u + C$

4. $\displaystyle\int a^u\, du = \left(\dfrac{1}{\ln a}\right)a^u + C$

5. $\displaystyle\int e^u\, du = e^u + C$

6. $\displaystyle\int \ln u\, du = u(-1 + \ln u) + C$

7. $\displaystyle\int \sin u\, du = -\cos u + C$

8. $\displaystyle\int \cos u\, du = \sin u + C$

9. $\displaystyle\int \tan u\, du = -\ln|\cos u| + C$

10. $\displaystyle\int \cot u\, du = \ln|\sin u| + C$

11. $\displaystyle\int \sec u\, du = \ln|\sec u + \tan u| + C$

12. $\displaystyle\int \csc u\, du = -\ln|\csc u + \cot u| + C$

13. $\displaystyle\int \sec^2 u\, du = \tan u + C$

14. $\displaystyle\int \csc^2 u\, du = -\cot u + C$

Trigonometric Identities

Pythagorean Identities

$\sin^2\theta + \cos^2\theta = 1$

$\tan^2\theta + 1 = \sec^2\theta$

$\cot^2\theta + 1 = \csc^2\theta$

Sum or Difference of Two Angles

$\sin(\theta \pm \phi) = \sin\theta\cos\phi \pm \cos\theta\sin\phi$

$\cos(\theta \pm \phi) = \cos\theta\cos\phi \mp \sin\theta\sin\phi$

$\tan(\theta \pm \phi) = \dfrac{\tan\theta \pm \tan\phi}{1 \mp \tan\theta\tan\phi}$

Double Angle

$\sin 2\theta = 2\sin\theta\cos\theta$

$\cos 2\theta = 2\cos^2\theta - 1 = 1 - 2\sin^2\theta$

Reduction Formulas

$\sin(-\theta) = -\sin\theta$

$\cos(-\theta) = \cos\theta$

$\tan(-\theta) = -\tan\theta$

$\sin\theta = -\sin(\theta - \pi)$

$\cos\theta = -\cos(\theta - \pi)$

$\tan\theta = \tan(\theta - \pi)$

Half Angle

$\sin^2\theta = \tfrac{1}{2}(1 - \cos 2\theta)$

$\cos^2\theta = \tfrac{1}{2}(1 + \cos 2\theta)$